Ethics Boxes

ABC Video Case Studies

Environmental Science

John A. Gibbs
Room 412
Medger Evers College

Environmental Science ⊕

The Way the World Works

FOURTH EDITION

Bernard J. Nebel
Catonsville Community College

Richard T. Wright
Gordon College

PRENTICE HALL, *Englewood Cliffs, New Jersey* 07632

Library of Congress Cataloging-in-Publication Data

Nebel, Bernard J.
    Environmental science : the way the world works / Bernard J.
    Nebel, Richard T. Wright. — 4th ed.
        p.   cm.
    Includes bibliographical references and index.
    ISBN 0-13-285446-5
    1. Environmental sciences.   I. Wright, Richard T.   II. Title.
GE105.N42   1993
363.7—dc20   92-37954
                                                    CIP

Acquisitions Editor: David Kendric Brake
Development Editor: Irene Nunes
Editorial/production supervision: Barbara DeVries
Marketing Manager: Kelly Albert
Design Director: Florence Dara Silverman
Interior Design and Layout: Andy Zutis
Cover Designer: Richard Stalzer Associates, LTD.
Prepress Buyer: Paula Massenaro
Manufacturing Buyer: Lori Bulwin
Photo Editor: Lorinda Morris-Nantz
Photo Research: Tobi Zausner
Illustrations: Renaissance Studios, Inc.
Copy Editor: Margo Quinto
Editorial Assistants: Mary DeLuca and Nancy Bauer

Credits: Page 177, Table 8-3 photographs; (a) Susan Kulkin/Photo Researchers; (b) John Spragens, Jr./Photo Researchers; (c) European Space Agency/Science Photo Library/Photo Researchers. Page 212, box, National Research Council, Alternative Agriculture. Washington, DC: National Academy Press, 1989. Page 390, Figure 17-4, clockwise from top left: David Stoecklein/The Stock Market; Melissa Hayes English/Science Source/Photo Researchers; Lawrence Migdale/Photo Researchers; John Dominis/The Stock Market. Page 399 (top) Marc Anderson; (bottom) Art Stein/Photo Researchers. Page 550, clockwise from top left: Sven-Olaf Lindblad; Margaret Miller; Jim Goodwin Photo/Researchers; Con Edison. Photo Part I, Michael Melford/The Image Bank; Part II, Larry Doell/The Stock Market; Part III, Pete Turner/The Image Bank; Part IV, Paul McCormick/The Image Bank.

Printed in the United States of America
10 9 8 7 6 5 4 3 2

ISBN 0-13-285446-5

Prentice-Hall International (UK) Limited, *London*
Prentice-Hall of Australia Pty. Limited, *Sydney*
Prentice-Hall Canada Inc., *Toronto*
Prentice-Hall Hispanoamericana, S.A., *Mexico*
Prentice-Hall of India Private Limited, *New Delhi*
Prentice-Hall of Japan, Inc., *Tokyo*
Simon & Schuster Asia Pte. Ltd., *Singapore*
Editora Prentice-Hall do Brasil, Ltda., *Rio de Janeiro*

*To our wives, Janet Nebel and Ann Wright,
whose love and support have made this work
possible, and to our children and students
who will be shaping a sustainable future.*

About the Authors:

BERNARD J. NEBEL is currently a professor of Biology at Catonsville Community College in Maryland. He earned his Bachelor of Arts from Earlham College and his Ph.D. from Duke University. Nebel has taught environmental science for 21 continuous years and is a member of the American Association for the Advancement of Science, the Institute of Biological Sciences, and the National Association of Science Teachers. He served as the 1990 Earth Day Coordinator for Maryland Colleges, and he strives to make a difference on the environment in his personal life as well. He walks or bikes to work, he is experimenting with the organic production of fruits, vegetables, and fish in his backyard, and he volunteers his efforts for Habitat for Humanity, and supports the Natural Resources Defense Fund and the World Wildlife Fund, and other environmental organizations.

RICHARD T. WRIGHT is the chairman of the Division of Natural Sciences, Mathematics, and Computer Science at Gordon College in Massachusetts, where he has taught environmental science for the past 22 years. He earned a Bachelor of Arts from Rutgers and a Master of Arts and Ph.D. from Harvard. Wright has received research grants from the National Science Foundation (NFS) for his work in aquatic microbiology and, in 1981, was a founding faculty member of Au Sable Institute of Environmental Studies of which he is now Academic Chairman. He is a member of the American Association for the Advancement of Science, the American Society for Microbiology, and the American Society for Limnology and Oceanography. In his personal life, Wright strives to have as light an impact on our environment as possible by biking to work, by planting trees and a vegetable garden in his yard, and by working with such groups as Bread for the World, Essex County Greenbelt, and Earthwatch.

About the Cover:

"The Pond" is an original watercolor created especially for this book by Janet Nebel, wife of author Bernard Nebel. The actual pond that inspired this painting is located on the Nebel's property in Catonsville, Maryland. Bernie Nebel, in his own quest to make a difference, took great pleasure in digging this pond, laying the stonework, and introducing the appropriate plants and fishes into this backyard ecosystem. Bernie and Janet often take time each day to look out at their pond to appreciate the beauty, the colors, and the life that now abounds in what was once merely a space in the backyard waiting to be fulfilled.

Brief Contents

Contents

Preface

Environmental science is about much more than science. Environmental science is also about ethics and values, sociology and politics, law and business, motivation and responsibility. Most of all, environmental science is about life and how to sustain it on Planet Earth.

The ecological and environmental crises we face are such a part of the daily news that they need not be enumerated here. Put briefly, humankind is engaged in numerous courses of action that are resulting in nonsustainable environmental impacts. Simple extrapolation shows that they can lead only to breakdown of the basic systems, such as the protective ozone shield, that support life on earth. A sustainable future depends very directly on our developing in the years ahead ways of producing food and manufactured products, acquiring energy, disposing of wastes, and so on that are sustainable. Does this sound like the talk of extremists?

Both of us (the authors) are Ph.D. scientists who have been trained, conducted research, and published papers in various areas of biology. We were drawn into environmental issues in the late 1960s as our understanding of biological and chemical systems and scientific analysis convinced us that the environmental problems coming to the public's attention were indeed real and serious. And our humanitarian concerns and a deep caring for the survival of other forms of life on earth led us to do what we could.

Already established at colleges, we created and started teaching courses in environmental science in the early 1970s—the field did not exist before that time—and we continue as active environmental science professors. For Nebel, teaching expanded into writing the first to third editions of *Environmental Science: The Way the World Works*. With this fourth edition Nebel joins forces with Wright to create a team that effectively doubles the expertise and teaching experience behind this edition. All our own training and experience and the analysis and input of others have only led us to believe that environmental problems are more critical than ever, and our concern has deepened.

This fourth edition of *Environmental Science: The Way the World Works* has involved a thorough rewriting of the entire text. Readers will find an increased global perspective, increased attention to aquatic and marine systems, the introduction of Ethics Boxes, and a new chapter format that provides learning objectives, review questions, and critical thinking questions as well as to the chapter outline. In addition, benefiting from the painstaking work of an editorial developer, the writing style and the logical flow of ideas, which have been the hallmarks of previous editions, are even more "reader-friendly." Also much of the artwork has been redone, all in full color, for increased clarity and visual impact.

THEME OF THIS TEXT

Our theme for the text is sustainability: what humankind needs to do to attain a sustainable future. After this theme has been presented in the introduction as a basic contrast between Cornucopian and environmental world views, it is highlighted and carried throughout the text. How current practices are resulting in nonsustainable environmental impacts and the changes that must be made to attain a sustainable course are pointed out in each area. (Readers will see immediately the interplay between environmental science and ethics and values.)

ORGANIZATION OF THIS EDITION

Because natural ecosystems have been sustaining themselves for millions of years, they are models for the sustainability of living systems. Therefore, in Part I, **What Ecosystems Are and How They Work** (Chapters 2–5), we examine the structure and function of natural ecosystems and how they evolve and adapt to change. The addition of a geology section giving a brief description of plate tectonics shows the inevitability of geological change and provides the basis for appreciating the fact that preserving biodiversity, which is critical for adaptation, is crucial for sustainability.

Through this scientific investigation of natural ecosystems, we bring to light four basic principles for sustainability (see Table 4–1, page 85.) (This elucidation and presentation of basic principles is considerably more focused and clearer than in the third edition.) These four principles become the basic guide posts for evaluating the potential sustainability or nonsustainability of courses of action or choices seen in each subject area throughout the rest of the text. (This portion of the text may be seen as mainly emphasizing scientific investigation and process.)

Part II of the text is now titled **Finding a Balance between Population, Soil, Water, and Agriculture.** Readers of the third edition will recognize that this involves an integration of chapters on maintaining soil and water resources and on pest control with the chapters focusing on the population itself. In addition, a new chapter, *Production and Distribution of Food,* has been added here. This reorganization and development, as well as updating, will enable a better comprehension of the entire population-food-agriculture issue and a better understanding of the directions that policies pertaining to population, development, and agriculture must take to be sustainable. (Throughout this Part it will be evident that environmental science cannot be separated from sociology and political policy.)

In Part III, **Pollution,** we maintain the format of devoting a separate chapter to each of five major categories of pollution. This enables us to present a complete picture of sources, harmful effects, and remedial measures in each case. Past readers will appreciate the clarity and understanding that this presentation engenders. (When several aspects of pollution are combined in a single chapter, confusion often results because sources, effects, and remedial measures differ in each case.) The chapter *Risks and Economics of Pollution* has been expanded to include a discussion of the problems of assessing risks and the conflicts between scientific assessment and public perception of risks. (In this Part, readers will not escape the fact that environmental science includes law and business.)

Major improvements in Part IV, **Resources: Biota, Refuse, Energy, and Land,** include the separation and expansion of the old chapter, *Biological Resources,* into two chapters, *Wild Species: Diversity and Protection* (Chapter 18) and *Ecosystems As Resources* (Chapter 19). This treatment facilitates a more complete discussion and student understanding in these very important areas. The energy chapters have been extensively updated and revised to reflect the most recent developments in the energy picture.

The title of the last chapter has been changed to *Lifestyle and Sustainability,* and it has been extensively revised. While still covering the environmental price of continuing urban sprawl and the related social price of urban blight, the theme of the text is brought to focus on the fact that sustainability will ultimately depend on the lifestyles of all of us. Avenues toward making existing communities and cities more sustainable are offered. We present a model of a fictitious sustainable community, *Sustainsville.* Discussion of this model is a means of summarizing all the sus-

tainable solutions that have been offered throughout the course. It is also a way of saying that a lifestyle that is both sustainable and high-quality is possible. All we need to do is work in that direction. (Here, it is made clear that environmental science is also motivation and responsibility.)

INDIVIDUAL TEXT ELEMENTS

Opening Chapter Outline: Each chapter begins with an outline of the material in that chapter. The outline reflects the section headings within the chapter. Simple as it is, the outline allows students to quickly overview the material in the chapter and to see important connections and relationships.

Learning Objectives: Following the outline is a list of learning objectives that have been carefully written to reflect the main points and concepts to be gleaned and assimilated. Like the outline, the learning objectives provide an excellent platform from which to preview and review the material in each chapter.

Boxed Material and Video Case Study Applications: All chapters include one or more of the boxes or video case study applications described below.

 In Perspective Boxes: These boxes offer interesting sidelights, expansions, or alternative viewpoints to the text material. The goal of these boxes is to put an important concept in an alternative perspective for the student. Just as real-world debates and discussions must accommodate multiple perspectives, so should an environmental science textbook.

 Ethics Boxes: Each chapter features one or more **Ethics Boxes** that serve to focus one's attention on the fact that choices concerning environmental issues often involve conflicting values and difficult moral and ethical decisions. Although our biases may be evident, we leave the answers to these ethical questions open for discussion. We hope that these ethics boxes will establish in each student's mind the need to examine basic values and ethical assumptions pervading the entire subject of environmental science.

 Video Case Studies: Selected from the archives of ABC news, these timely and highly relevant video segments offer the student an overview of a particular environmental issue or controversy. Case study material is found at the end of the text, following the epilogue, but has direct application to particular chapters in the text. The authors have provided a brief synopsis of each video and have included a list of discussion questions in the hopes of fomenting healthy classroom debate and discussion of these topics. A list of the video segments is found in the Video Case Study section as well as on the inside front cover of the book.

Review Questions: Found at the end of each chapter, these questions allow the student to review the basic concepts and core content of the chapter. The focus of these questions is on comprehension rather than application.

Thinking Environmentally: These critical-thinking-oriented questions take the student a step beyond comprehension to the application of concepts learned in that chapter in an effort to solve or assess real-life situations.

Making a Difference: The authors believe that no amount of text-based learning about the environment truly becomes useful until students challenge themselves and those around them to begin "making a difference." With this in mind, each of the four parts of the text concludes with a two-page spread that suggests and depicts courses of action that each student can take to bring about the changes needed for sustainability.

Fold-out Biomes Poster: Each text includes a six-panel fold-out poster depicting both terrestrial and aquatic biomes. This beautifully illustrated poster includes key information about each biome as well as a highlight of environmental issues and challenges associated with each biome. Chapter references make this an ideal study tool, and the overall design makes it appropriate as a dorm-room decoration with a statement.

SUPPLEMENTS

For the Instructor

Instructor's Manual: Developed by John Peck of St. Cloud State University, this manual features a unique annotated lecture outline with presentation suggestions and topics for classroom discussion and debate.

Test Item File: Written by Dennis Woodland of Andrews University. Over 1800 questions including multiple choice, short answer, and essay, all with text page references and coded by level of difficulty.

Computerized Test Item File: Available in three versions: IBM 3.5", IBM 5.25", and Macintosh.

Transparency Pack: A selection of 125 acetates with images from the text as well as 50 masters.

Laboratory Experiment Masters: Approximately 28 experiments to enhance your environmental science laboratory. These can also be given to students as suggested field-oriented projects.

The ABC News/Prentice Hall Video Library: Twenty-five 3–10 minute video segments focusing on recent environmental issues in the news. Selected from the archives of ABC news, each video includes a written summary that ties the video segment to particular sections of the text. Thus, it will be easier than ever to enhance your classroom presentation with timely and relevant video segments.

For the Student

The New York Times/Prentice Hall Contemporary View Program:

THE NEW YORK TIMES and PRENTICE HALL are sponsoring A Contemporary View, a program designed to enhance student access to current information of relevance in the classroom.

Through this program, the core subject matter provided in the text is supplemented by a collection of time-sensitive articles from one of the world's most distinguished newspapers, THE NEW YORK TIMES. These articles demonstrate the vital, ongoing connection between what is learned in the classroom and what is happening in the world around us.

PRENTICE HALL and THE NEW YORK TIMES are proud to co-sponsor A Contemporary View. We hope it will make the reading of both text-books and newspapers a more dynamic, involving process.

Study Guide: Written by Clark Adams, Texas A&M University. An excellent review tool offering both concept and content review exercises.

ACKNOWLEDGMENTS

The authors acknowledge the many people who contributed ideas, information, hard work, and other support that helped to bring this volume to fruition. In particular we acknowledge the following people who painstakingly reviewed the third edition and offered many, comments, suggestions, and constructive criticisms that contributed to the improvements found in this edition.

○ Clark Adams, Texas A&M University

○ Melvin A. Benarde, Temple University

○ Luckett Davis, Winthrop College

○ David Fluharty, University of Washington, Seattle

○ Robert D. Holmes, North Dakota State University, Fargo

○ David I. Johnson, Michigan State University

○ David J. O'Neill, Dundalk Community College

○ John Peck, St. Cloud State University

○ R. H. Pemble, Moorhead State University

○ L. Harold Stevenson, McNeese State University

○ Dennis Woodland, Andrews University

We also thank all the dedicated people at or working for Prentice Hall who had a hand in the production of this volume. In particular: our editor David Brake for his constant enthusiasm and inspiration and many ideas that have been incorporated into this volume as well as for the overall management of the project; our production editor Barbara DeVries, always good natured, who managed untold details of putting the text together; our developmental editor, Irene Nunes, who edited the entire manuscript and has made it an exceptionally "reader-friendly" text; artists Lisa Nocks and Brian Regal of Renaissance Studios who created all the new or rerendered artwork.

In addition Bernard Nebel thanks Richard Wright for providing new perspective and information and for sharing much of the burden of the rewriting and updating of this edition; his help and support were invaluable; Scott Barr, Prentice Hall sales representative, whose ideas and enthusiasm helped launch this fourth edition; Deborah Keene, Beth Beck, and Angie Biederman for untold hours of typing and other office work and for their real caring and support of the project.

All my colleagues at Catonsville Community College and many other close friends for their interest and for providing support in many ways over the years; countless people who have been most generous in providing information, photographs, and reports; and, again, all those who helped on previous editions. I apologize for not listing them individually.

All my students for continually providing incentive and a testing ground for this work.

Last but most important, my wife Janet, who is a constant source of love, inspiration, and support, and whose painting appears on the cover of this volume.

Richard Wright acknowledges the privilege it has been to collaborate with Bernard Nebel in the writing of this edition of his text. I also thank my daughter, Susan W. Mulley, for her fine work on the critical thinking questions and other elements of the book, my colleagues in the biology department at Gordon College for their encouragement, and especially my wife, Ann, who has been with me since the beginning of my career in biology and has provided the emotional base and companionship without which I would be far less of a person and a biologist.

You Can Make a Difference

Simply by purchasing this text you can have a positive impact on the environment. The authors of this text and Prentice Hall are pleased to contribute a portion of this text's purchase price to one of the environmental organizations listed below. We are proud that together we can make a difference.

Directions: Please complete the information below and then check the box next to the organization to which you would like us to make a contribution. Carefully cut this page from your text and mail it to: Biology Editor, Prentice Hall, 113 Sylvan Avenue, Englewood Cliffs, NJ 07632.

Name: _____

School: _____

Year in School: FR __   SO __   JR __   SR __   Graduate Student __   Other __

Your Instructor's Name: _____

Please send my contribution to the organization I have indicated:

○ **Bread for the World**

○ **Habitat for Humanity International**

○ **Natural Resources Defense Council**

○ **Nature Conservancy**

○ **World Wildlife Fund**

1

The Aim of Environmental Science

OUTLINE

LEARNING OBJECTIVES

When you have finished studying this chapter, you should be able to:

1. Describe the basic assumptions underlying the world views of cornucopianism and environmentalism.

2. Describe how these two world views are in conflict regarding environmental issues.

3. Describe how the predominant world view may be changing.

4. Define environmental science and its objective.

5. Define science.

6. Name and describe the steps of the scientific method.

7. Describe what is meant by a controlled experiment and why controlled experiments are necessary.

8. Distinguish between a hypothesis, a theory, and a natural law.

9. Give the roles instruments play in science.

10. Discuss the connection between science and value judgments.

11. Discuss how both science and value judgments are involved in the application of environmental science.

12. List four things to look for when evaluating information.

> *Some men see things as they are and ask: Why? Some men*
> *see things as they never were and ask: Why not?*
> Robert F. Kennedy about John F. Kennedy

Almost everywhere we turn, we see controversy regarding environmental issues. The energy policy proposed by the Bush administration includes opening up the Arctic National Wildlife Refuge in Alaska for oil exploration and exploitation; environmental interests are adamantly opposed. Environmentalists want stronger regulations regarding air and water pollution; industrial interests claim that such regulations are too costly and are undercutting the competitiveness of U.S. business. All over the country various proposals for land development are being opposed by land preservationists. Nuclear power, which some feel is essential to our energy security, is vehemently opposed by others (Fig. 1–1).

Two World Views

In all these disputes over environmental issues, the opposing sides represent two *world views*. A **world view** may be defined as a set of assumptions that a person holds regarding the nature of the world and how it works. The two world views in opposition here are *cornucopianism* and *environmentalism*.

Cornucopianism is the dominant world view that has been held by Western civilization throughout most of history. It embodies the assumption that all parts of the environment (air, water, soil, minerals, and all plant and animal species) are **natural resources** to be exploited for the advantage of humans, either individually or as a society. In addition, this view assumes that these natural resources are essentially infinite. If one is exhausted, another will be found to replace it. Consequently, the history of development of Western civilization, especially the development of the countries of North and South America, is almost synonymous with the stripping of forests, slaughter of wild animals, mining of minerals, and discarding of wastes with little thought to pol-

lution or regard for the long-term impact on the earth or future generations. Cornucopianism is still the dominant world view and has been adopted by most peoples of the world, as witnessed by the fact that these exploitive activities continue in all countries.

In the last three decades or so, however, the second world view, environmentalism, has been gradually gaining ground. **Environmentalism** embodies the assumptions that what we generally view as natural resources are *products of the natural environment*. It follows, then, that resources will be limited by the regenerative capacities of the natural environment. Furthermore, even the limited resources will be provided only insofar as the natural environment is protected and maintained. Thus, our survival literally depends on suitable protection and stewardship of the natural environment. (Actually environmentalism is not a new world view. Native Americans and a number of other cultures hold a similar world view, but these cultures have long since been dominated if not exterminated by Western European culture.)

The cornucopian world view is not without virtue. People who hold it are fond of pointing out that

FIGURE 1–1
Antinuclear demonstration at the Seabrook nuclear power plant in New Hampshire. This is but one example of a confrontation between people who see the best course to a prosperous and safe future in very different ways. (Jim Bourg/Gamma-Liaison.)

Chapter 1 The Aim of Environmental Science

Cornucopians frequently accuse environmentalists of being doomsayers running around like Chicken Little crying, "The sky is falling." Environmentalists take issue with this accusation. It is not doomsaying, they argue, to foresee a problem and want to take measures to avoid it, and there is plenty of evidence that the problems are real. What would be really doomsaying would be to

deny or ignore the problems until we are too deeply into them to extract ourselves. Cornucopians counter that, as the problems get more severe, existing economic forces will bring about the necessary adjustments and adaptations with no planning or efforts on our part. Therefore, they say, the efforts called for by environmentalists are really just a waste of time and money.

Who is right? Time will tell. Yes, but this is not like a sporting event that you can watch and then go home to life as usual. If we place our bets with the Cornucopians and lose, we lose the world. If we place our bets with the environmentalists and achieve a sustainable society, the most we can lose is to never know whether the Cornucopians were right or wrong.

without the exploitation of natural resources, we would still be living in caves and chasing wild animals with spears and clubs. The "good old days" of primitively living in harmony with nature exist only in the imagination. Primitive tribes suffer a high incidence of disease, discomfort, pain, suffering, and infant mortality and a short life expectancy. Thus, Cornucopians emphasize that the exploitation that some environmentalists decry has enabled most of us to enjoy a high standard of living that few of us would give up willingly. Cornucopians assume that continuing exploitation is the only way to achieve further human progress.

We (your authors are environmentalists) do not wish to argue the point that exploitation has been a necessary part of civilization. In any case, the past is past. Where environmentalists take issue with Cornucopians is what road we should take to achieve a bright future and continued human progress.

Environmentalists are firmly convinced that continuing the trend of increasing exploitation is *unsustainable*. **Sustainability** refers to whether or not a process can be continued indefinitely. To say that a process is unsustainable means that it will inevitably reach a dead end. It is like a profligate heir squandering her or his inheritance: Environmentalists see exploitive trends as squandering our inheritance. That is, exploitation of various components of the environment is undercutting the various systems that provide natural resources. As in the case of the profligate heir, this exploitation can lead humanity only to poverty and destitution.

Innumerable points may be listed to back up the argument that humanity is on a collision course with the environment's ability to provide what cornucopians see as resources. Among the most prominent

points, with which you are probably already familiar, are:

○ Carbon dioxide emissions from burning fossil fuels and other pollutants are likely to cause global warming (if it is not already taking place).
○ The protective ozone shield is being depleted by CFC (chlorofluorocarbons) emissions.
○ Thousands of plant and animal species are being lost each year as a result of the cutting of tropical forests and destruction of other natural areas.
○ Traces of toxic chemicals are found throughout the Great Lakes and many other waterways and even the oceans.
○ Acid precipitation caused by sulfur dioxide emissions from coal-burning power plants is falling over very large regions of the world.
○ Water resources are being exhausted by overuse in many regions of the world.
○ Natural waterways everywhere are being polluted and degraded by sewage and agricultural runoff.
○ Everywhere there are mounting problems with the disposal of chemical wastes and refuse.

Because environmentalists make statements to the effect that continuing current trends of exploitation can lead only to impoverishment, they are frequently attacked by Cornucopians as being doomsayers and against progress. Neither of those accusations is true.

First, environmentalists make no pretense of predicting the future. What environmentalists do is make use of the old saying: If you don't like where you are going—change directions. By pointing out where continued exploitation of natural resources may lead, environmentalists hope to get society to change directions. Similarly, there will be no progress except into poverty and deprivation if humanity continues to destroy the natural environment. The only way to maintain human progress in a positive direction is to learn to protect and enhance the natural environment, which supports all life.

Environmentalists are not just antidevelopment, anti-exploitation, or antiwhatever. Instead, they offer and promote *sustainable* alternatives. The principle is analogous to the profligate heir changing living habits in order to live on the interest of the inheritance rather than squandering the capital. The productive capacity of the natural environment is such that, if properly protected and managed, it can yield an abundant interest, and—in theory at least—humanity can live very comfortably on that interest.

The problem is making the changeover. Surveys in the United States show that a large majority of people profess to be environmentalists. Yet, we are all part of a society that has developed and continues to function according to Cornucopian assumptions. For example, although we profess to be environmentalists, we continue to drive cars, which burn fuels that contribute to the global warming effect, because we are part of a society that has created virtually no practical alternatives for everyday transportation.

The changeover to sustainable alternatives will involve much more than simple "Band-aid" mea-

FIGURE 1–2
Golden lion tamarin. The golden lion tamarin, which originally lived in the Atlantic rain forests of eastern Brazil, was reduced to a population of no more than 300 animals because of habitat destruction and poaching. Habitat protection and captive breeding programs spearheaded by the World Wildlife Fund are now saving it from extinction. (Tom McHugh/Photo Researchers.)

sures; it will involve basic changes throughout society in the ways we transport ourselves and our products, in the ways we heat and cool our homes and workplaces, in the ways we produce crops, in the way we manufacture products, handle wastes, and so on.

In the words of William Ruckelshaus, first head of the U.S. Environmental Protection Agency:

[The changeover to a sustainable society] will be a modification of society comparable in scale to only two other changes: the agricultural revolution [of about 10,000 years ago] and the Industrial Revolution of the past two centuries. Those revolutions were gradual, spontaneous and largely unconscious. This one will have to be a fully conscious operation, guided by the best foresight that science can provide—foresight pushed to its limit. If we actually do it, the undertaking will be absolutely unique in humanity's stay on earth.

In other words, assuming that we do make a changeover to become a sustainable society, this current period will be looked back on in history as a major revolution—the environmental revolution—and you all will have been players in this "absolutely unique [time] in humanity's stay on earth." You all will have been players, yes, but some as part of the solution and success of the revolution and others as part of the problem very nearly causing the revolution to fail. The thought is scary, but the challenge also provides an opportunity that few (perhaps no other) generations have ever had or will have.

In accepting the challenge to work toward a sustainable society, you will not be alone. You will be joining millions of people around the world in all walks of life who have already made a similar commitment. Individually and through thousands of organizations, they are mounting actions to protect the natural environment. Actions range from the radical to the highly professional and scientific. For example, members of Earth First have chained themselves to trees to prevent their cutting. Members and leaders of the Natural Resources Defense Council, Environmental Defense Fund, and other groups focus their efforts on promoting legislation for environmental protection; much of U.S. pollution control legislation is in no small part a result of their efforts. Organizations such as World Wildlife Fund and Conservation International are working in cooperation with governments in many countries around the world to set up and manage biological reserves, areas of natural habitat expressly for the purpose of saving particular species from extinction (Fig. 1–2).

Even labor unions, which traditionally have been anti-environment because of fear that environmentalism would lead to the loss of jobs, are beginning to change. For example, a 1990 report of the United Steelworkers of America included the statement: "In the long run, the real choice is not jobs or environment. It's both or neither. What kind of jobs will be possible in a world of depleted resources, poisoned water and foul air, a world where ozone depletion and greenhouse warming make it difficult even to survive?"

In June 1992 the United Nations held a Conference in Rio de Janeiro on the Environment and Development, at which leaders of all member nations pledged to work toward making environmental protection an integral part of furthering development.

While the fact that world leaders and organizations recognize the seriousness of environmental issues is an encouraging sign, leaders and organizations cannot solve the problem by themselves. As noted above, the changes must ultimately involve all of us and the way we conduct our daily lives. Furthermore, using the quote from Ruckelshaus again, "This [change] will have to be a fully conscious operation, guided by the best foresight that science can provide."

This is where *environmental science* enters the picture. **Environmental science** is the science of looking at the cause-and-effect relationships underlying environmental issues. In short, it is the science of understanding how the world works on the level of the natural environment—how the natural environment regenerates natural resources and how this generative capacity is being affected by human activities. Armed with this understanding, the objective of environmental science is to provide the foresight required to make changes toward a sustainable society.

The Scientific Method and Evaluating Information

In our society there is a considerable lack of scientific understanding as well as a considerable mistrust of scientists and scientific information. This is particularly true in environmental science, where there are many controversies. Therefore, we shall take some time describing the methods of science so that you will have not only a better understanding of science but also the ability to evaluate information for yourself.

WHAT IS SCIENCE?

Science refers both to a particular way of gaining knowledge and to the knowledge that is gained by this method. As students in a classroom, we generally gain all kinds of information in essentially the same way. Teachers and textbooks provide the information, and we study and learn it as well as we can. But where and how was that information originally obtained?

In the case of the humanities, we find the origin in some person, someone who, through inspiration and talent, gave us the teachings or composition that so affected us emotionally or spiritually that we have been treasuring and passing on the gifts ever since. While we may try to emulate the famous teacher, artist, writer, or composer, how they conceived their work remains a mystery.

In the case of science, our search also leads back to particular people: the scientists who made the discoveries. Here, however, we can understand how they made their discoveries. In science, discovery is based on observations made by scientists and logical deductions based on those observations. Effectively, anyone could make the same observations and, through logical deduction, reach the same conclusions. Thus, in science, what becomes even more significant than the scientist (though we still respect his or her contributions) is the observations and the recognition that they are open to being checked out and verified. In fact, it is absolutely counter to science to take anything on faith. The key tenet in science is: *Check it out!*

THE SCIENTIFIC METHOD

Over the years the technique of compiling scientific information has been refined into a series of steps that is called the **scientific method.** It begins with observations of things and events and progresses through the formulation and testing of theories.

Observations and Facts

All scientific information is based on observation and is subject to objective verification ("checking it out"). In science, observations are restricted to only impressions gained through one or more of the five basic senses in their normal state: seeing, hearing, touching, smelling, and tasting. This restriction is made because only such observations lend themselves to measurement, testing, and verification. Verification is a most critical aspect because, even with the best of intents, our senses may be deceived. Magic shows capitalize on such illusions or deceptions.

Therefore, before an observation is accepted as factual, it must be confirmed. That is, other investigators must be able to repeat the observation, perhaps using different techniques and tests, and confirm independently that the original observer was accurate. Observations that do not stand this test of verification are not accepted as scientific fact. For example, some people claim to have seen spaceships carrying alien creatures, but this information is not generally accepted as factual because such observations have not been verifiable.

Observations that do stand the test of verification become accepted as scientific fact, but even they remain open to further checking at any time. It is this restriction to verifiable observations and confirmation by others that gives scientific information its reputation of being "exact," "factual," or "objective."

It is interesting to note that science is not factual or objective because of any particular power of science or scientists; it is because science restricts itself to considering only things and events that can be observed in an objective way. Emotional feelings such as love, beauty, and spirituality are just as real and important to us, but since such things do not lend themselves to objective observation and measurement, they remain outside the realm of science.

The rule concerning verification applies not only to observations made but to all steps of the scientific method. It is the most important tool you or anyone else can use to discriminate between accurate and inaccurate information. A good demand to make before accepting any information is: *Show me the evidence!*

Hypotheses and Their Testing

A great deal of scientific information can be accumulated through observations alone without resorting to experiments. An almost automatic response, however, is to seek explanations for our observations. In some cases an answer may be found by simply making further observations, but often the cause of some observed effect is not evident from direct observations. When this is the case, the investigator must use tests or experiments, which really are ways of making indirect observations, to get at the answer.

Let's use as an example: The patient is sick; what is causing the illness? The observation "patient is sick" and the question "what is causing the illness" are the first and second steps in the scientific method. The next step is to make educated guesses regarding the cause and then eliminate the wrong guesses through suitable tests or experiments. By weeding out wrong answers, a scientist is sooner or later left with the right answer(s). Each educated "guess" is called a **hypothesis.**

A specific thought process used in testing the validity of a hypothesis is the *"if . . . then . . ."* phrase. One reasons that *if* the hypothesis is true, *then* such and such should logically follow. If the *"then"* is not found, the hypothesis is discarded and the procedure is repeated on the next hypothesis. For example, a doctor makes an educated guess—a hypothesis—that the patient's illness is due to a particular bacterium. Experiments or tests are then conducted to determine if the bacterium is present in the patient. If the bacterium is found, it lends support to the doctor's hypothesis. If it is not found, the hypothesis is discarded and the doctor moves on to hypothesizing and testing for other possible causes.

The process is a little like constructing a multiple-choice question (the choices being all the possible hypotheses the scientist can think of) and then gradually testing and eliminating the wrong ones. Sooner or later a hypothesis that stands up to all the possible tests is found, and this hypothesis then becomes the accepted answer.

Controlled Experiments

An experiment to test a hypothesis must be carefully designed to stand up to the question: How can you be sure the observed results are due to the factor you hypothesize rather than to other unrecognized factors? The key is that experiments to test hypotheses must be *controlled.* By **controlled** we mean that the experiment must consist of two groups, a test group and a control group. The experiment is designed so that these two groups are treated exactly the same in every respect except for the single factor being tested. A difference in response between the groups can then be attributed to the single factor being tested (Fig. 1–3). Without a control group or with several differences between groups, the experimenter will be unable to interpret the results because she or he will have no way of knowing which factor or combination of factors was responsible for the results.

FIGURE 1–3
A controlled experiment involves two groups that are treated the same in every respect except for one factor being tested. Without the control group for comparison, there is no way to identify which of the many factors involved is responsible for the effect.

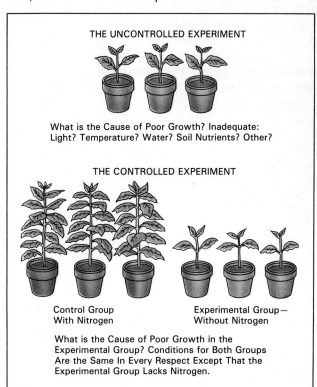

THE UNCONTROLLED EXPERIMENT

What is the Cause of Poor Growth? Inadequate: Light? Temperature? Water? Soil Nutrients? Other?

THE CONTROLLED EXPERIMENT

Control Group With Nitrogen

Experimental Group— Without Nitrogen

What is the Cause of Poor Growth in the Experimental Group? Conditions for Both Groups Are the Same In Every Respect Except That the Experimental Group Lacks Nitrogen.

Two additional points concerning controlled experiments deserve emphasis. First, note that the experiment involves *groups* as opposed to *individuals*. This is because there is always some variability (difference) from one individual to another. Therefore, if the "group" is only one, two, or a few individuals, the experimenter is in danger of confusing natural individual differences with the effects of the treatment. The larger the group, the better. Second, the results from a single experiment, especially when they are based on relatively small groups, should not be taken as conclusive evidence. They also must stand the test of verification through repetition and confirmation.

In some cases, performing controlled experiments would be prohibitively time-consuming, costly, or impossible. Humans, for example, don't readily lend themselves to controlled experiments with regard to the effects of harmful or potentially harmful substances. However, there are so many humans living under such an array of different conditions and engaging in different habits that investigators can generally identify persons in the normal population who are effectively self-selected experimental and control groups. For example, one can find enough smokers and nonsmokers to test hypotheses regarding the effects of smoking on various aspects of health without resorting to experiments in the laboratory. Investigations of this nature are called **epidemiological** studies.

Thus we have another tool for evaluating information: In your demand, "show me the evidence," look for supporting information involving controlled experiments, epidemiological studies, and tests refuting alternative hypotheses, as well as for direct verifiable observations. This process will distinguish sound information from what may be basically speculation.

Formulation of Theories

Individual observations and experiments answer only extremely specific questions. From small pieces of information, however, a larger picture—a theory—emerges. A **theory** is a concept that provides a logical explanation for a certain body of facts. Constructing a theory is a process similar to a detective finding clues (observations) and then fitting them together into a picture (a theory) of "who dunnit." An important point to note is that a theory is not a fact because it does not lend itself to direct observation. Nevertheless, theories may be tested and, pending the results, confirmed or rejected.

Theories are tested in much the same way that hypotheses are tested. The theory will suggest further

aspects that should be observable directly or through experimentation—the *"if . . . then . . ."* reasoning. If observations are made that are contrary to what should be expected on the basis of the theory, the theory is either modified to incorporate the new findings or rejected in favor of an alternative theory that provides a more rational explanation. In either case, a theory is developed that is consistent with *all* observations and experiments and that can be used to reliably predict outcomes in the sense of *"if . . . then."* When theories reach this state, we have every reason to believe that they represent a correct interpretation of reality. For example, we have never seen atoms as such, but innumerable observations and experiments are coherently explainable by the concept that all gases, liquids, and solids consist of various combinations of only slightly more than a hundred kinds of atoms. Hence, we fully accept the *atomic theory of matter*.

Sometimes people will argue that because a theory is not proven fact, one theory is as good as another. This notion is *false!* One theory may have overwhelming supporting evidence, while much evidence may contradict another theory. In evaluating a theory, then, ask: What is the supporting evidence? That is, how much evidence is there to support "who dunnit." Is there more, or less, evidence supporting an alternative theory?

Principles and Natural Laws

In the course of observing natural events and the results of various experiments, we find that matter and energy behave not randomly but, without exception, according to certain principles. We refer to such principles that seem to govern the behavior of matter and energy as **natural laws.** For example, free-falling objects always accelerate at exactly the same rate (exclusive of air resistance). Hence we speak of all bodies being subject to the *Law of Gravity*. Another example comes from the observation that, in all chemical reactions, atoms are only rearranged; they are never created, destroyed, or changed. This principle is referred to as the **Law of Conservation of Matter.** *Laws of Thermodynamics* refer to principles regarding energy changes.

Since matter and energy behave according to the principles we call natural laws without exception, the predictive power of natural laws is tremendous. For example, we are able to carry out successful space flights by making precise predictions of gravitational fields throughout the solar system based on our understanding of the Law of Gravity. We then design rockets with exactly the necessary thrust to overcome the predicted gravitational forces. The fact that our

Putting the Scientific Method to Use

The scientific method may seem arduous, but we all probably use it to a greater or lesser extent in our everyday lives whether we realize it or not. Consider this simple situation: You flip a light switch and the light fails to come on. You make the observation "no light" and immediately ask the question, Why not? Almost as fast comes the first hypothesis: The bulb is burned out. Before you go for a new bulb, however, other hypotheses may occur to you, and you might check them out first. Perhaps the plug got pulled out. Check. No, it is plugged in. Perhaps a fuse is blown. Check. Are other lights functioning? Yes. Then it is not the fuse. At this point you may

go and get a light bulb, put it in, and . . . What? It still doesn't work? At this point you may have to do some real head-scratching, but what are other possibilities? Faulty switch? The wire in the plug has pulled loose?

Of course, it is most likely that you found and solved your problem early on, but do you see how each idea is an educated guess, or hypothesis? Each check is a test of that hypothesis, and gradually, by the process of elimination, you arrive at the solution.

This analogy might be carried even further. Suppose you had the time and skills to trace and check the wires. They would eventually

take you all the way back to the power plant and the "discovery" of the generators, at which point you might publish your "theory of the production and distribution of electricity." Of course, for this situation or anything else human-made, all your work would show only what is well known. It would be much more straightforward to read about it in a book. Still, the process is similar to that of science, which starts from very simple, even trivial observations and traces the connections until there is enough information to present a theory regarding the whole picture.

space flights go as planned (except when the rockets fail) attests to the universality of the Law of Gravity and our understanding of it, as well as to our ability to build rockets.

By the same token, one can predict that any action contrary to a natural law will end in failure, although the failure may not be instantaneous; there is generally a certain amount of "borrowed time" between the violation and the ultimate result. For example, a man leaping from a tall building may have several seconds of borrowed time to enjoy his "flight" and proclaim that he has overcome gravity.

Here is another lesson for evaluating information. When trying to judge whether a course of action is likely to succeed or not, look for consistency with natural laws. This is so obvious in the case of the man jumping off the building that it is laughable. However, even people who should know better persist in ignoring natural laws. The space shuttle *Challenger* was sent off in defiance of the knowledge that low temperatures the night before might cause certain fuel seals to fail. There were 90 seconds of borrowed time before all the astronauts, including one who was to have been the first civilian in space, disappeared in an explosion. Another example, in the mid-1980s, was a person promoting his "backyard" invention of a new generator that produced more power than it consumed, a violation of the First Law of Thermodynamics. Nevertheless, by playing the role of a can-

tankerous genius at odds with an envious "establishment," he gained considerable support in the popular press and raised several million dollars for development of his device while at the same time avoiding any critical testing of it. Finally, even his backers insisted on a demonstration, at which point his device proved to be a hoax. He has since dropped out of sight—along with the several million dollars. The moral of the story is: *Stupidity, ignorance, fraud, and deceit are much more likely than exceptions to natural laws.*

The steps of the scientific method are summarized in Figure 1–4, page 10.

THE ROLE OF INSTRUMENTS IN SCIENCE

Complex instruments often give science an aura of mystery. Yet regardless of complexity, all scientific instruments perform one of three basic functions. First, they may extend our powers of observation—telescopes, microscopes, X-ray machines, and CAT scans, for example. Second, instruments are used to quantify observations; that is, they enable us to measure exact quantities. For example, we may feel cold but a thermometer enables us to measure and quantify exactly how cold it is. Comparisons, communication, and verification of different observations and events would be impossible if it were not for such

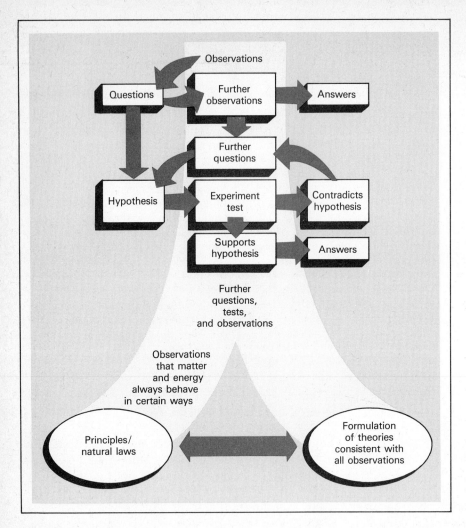

FIGURE 1–4
Steps of the scientific method. Each step involves more observations, which generate more questions. Thus scientific investigation is a never-ending process.

measurement and quantification. Third, instruments such as growth chambers and robots help us achieve conditions and perform manipulations required to make certain observations or to perform experiments.

All instruments used in science are themselves subjected to testing and verification to be sure that they are giving an accurate representation of what is really there as opposed to creating illusions.

ETHICS
The Long and the Short of It

A major and complex ethical problem of modern times concerns the value we place on present satisfaction versus future well-being. A common description of our society is that we seek instant gratification with relatively little concern for future consequences. We see this on all levels, from individual young people taking drugs to the U.S. government spending far beyond its

means and running up the national debt on the order of $300 billion a year. Many if not all environmental problems are a result of the same short-sighted psychology. Today's pollution and degraded land are the price of yesterday's shortcut to gratification and lack of concern for the future.

It should be clear that if we continue on this pathway, the price can

only increase. Reaching a sustainable society will not be possible unless we place more value on future well-being, enough value that we are willing to forego some instant gratification and invest more in conservation, protection, and stewardship of the natural environment.

A particularly important moral dilemma that pervades our society involves the balance between looking out for "number one" and helping or at least not hurting others. It is basically a dilemma between our value that a person has the right to do as he or she chooses on one side and a value that people should be helped and not harmed on the other.

Modern society seems to be fostering undue emphasis on helping number one, with relatively little concern for others. The savings-and-loan debacle, where bank executives, for their own gain, squandered and lost the savings of thousands of customers, is an extreme but significant example of this. It goes on down the scale to cheating on income tax and failing to return library books. Number one gains, but society pays a price. Too much of this and everyone will suffer because we are all part of society. Conversely, we all stand to gain as society gains. This recognition that we stand to gain personally in self-satisfaction and esteem, if not financially, by helping others is referred to as enlightened self-interest.

This is not to say that we should sacrifice ourselves to the point of misery for the sake of others. There is certain truth in the statement that we need to be good to ourselves so that we can be good to others. Thus, the dilemma is finding the optimal balance between helping yourself and helping others.

This same dilemma exists in terms of our interactions with the environment. Just substitute the natural environment for society and others.

SCIENTIFIC CONTROVERSIES

With the scientific method so capable of coming to objective conclusions, why does so much controversy still prevail? There are four main reasons. First, we are continuously confronted by new observations—the hole in the ozone shield, for instance, or the dieback of certain forests. There is considerable time before all the hypotheses regarding the cause can be adequately tested. During this period of time, controversy may be rife as different people tend to favor one hypothesis (suspect one cause) versus another. Such controversies are gradually settled by further observations and testing, but this leads into the second reason for continuing controversy: Phenomena such as the hole in the ozone shield or the dieback of forests do not lend themselves to carefully controlled experiments. Therefore, it is extremely difficult and time-consuming to prove the causative role of one factor or rule out the involvement of another. Gradually, different lines of evidence will generally support one hypothesis over another, however, and enable the issue to be resolved.

This accumulation of additional evidence is a gradual process. When is there enough evidence to say unequivocally that one hypothesis is right and another wrong? Deciding that there is enough evidence to be convincing involves subjective judgment. The biases or vested interests of a person may affect the amount of information required to be convincing. The Tobacco Institute, a lobbying association for the tobacco industry, provides a prime example. It continually makes the point that the connection between smoking and ill health effects is not proven and that more studies are necessary. By harping on the lack of absolute proof and simply ignoring the overwhelming amount of evidence supporting the connection between smoking and ill health effects, the tobacco lobby has succeeded in keeping the issue controversial and thereby has delayed restrictions on smoking. Thus, the third reason for controversy is that there are many vested interests who wish to maintain and promote disagreement because they stand to profit by doing so. The need to keep a watch for this kind of behavior in evaluating the two sides of a controversy is self-evident.

The fourth reason for controversy, which may be seen as a generalization of the third reason, is that subjective value judgments may be involved as well as the scientific issue. This is particularly true in environmental science because it deals with the human response to environmental issues. For example, there is virtually no controversy regarding nuclear power as long as it is considered at the purely scientific level of physics. However, when it comes to the environmental level of deciding whether to promote or shun the further use of nuclear energy to generate electrical power, the controversy is obvious and it stems from different people having vastly different subjective feelings regarding the relative risks and benefits involved.

It is neither possible nor desirable to eliminate

values from decision-making processes. The important thing is to look, as far as possible, at the factual information supporting the two sides and arrive at your own conclusion.

SCIENCE AND VALUE JUDGMENTS

The relationship between science and value judgments, which come together in environmental science, deserves some further attention. Note that the scientific method provides information on the mechanistic, or cause-effect, level—*this* happens because of *that*. Understanding at the cause-effect level gives us tremendous capability to guide our future in the sense of: If we do this, it will have that result. For example, if we build a rocket according to determined principles, it will get us to the moon and back. The same information, though, can also be used to build ballistic missiles.

How much of our money and resources we should spend building rockets for space exploration versus ballistic missiles is a value judgment, which is based not on science but rather on the moral, religious, ethical, and emotional side of our being. Science can help only insofar as it may help us to see the long-term consequences of these two pathways of spending our money and resources. We must be willing to alter our values sufficiently to pursue what our reasoning tells us is the more desirable course. That is, it seems obvious that humanity will gain more in the long run by putting its money and resources into space exploration. Yet, our fears and distrusts (values of another sort) have led us to put much more into ballistic missiles.

Environmental Science and its Application

Let's now relate this discussion of science and value judgments to environmental science and its application. In this text, we (your authors) do our best to convey an understanding of how the natural world works. This understanding comes mainly from the science of **ecology,** the study of the relationships among organisms and between organisms and their environment. The scientists engaged in this area are **ecologists.** From the study of ecology, we can learn some basic principles regarding sustainability—after all, natural groupings of plants and animals have been sustaining themselves for hundreds of millions of years. Furthermore, we shall point out means by which these principles of sustainability may be incorporated into human society and thus support its sustainability.

However, whether or not this information is ultimately used to create sustainable human societies will depend on your values—how you use the information in the choices you make as an individual, consumer, professional, and voter. Our biases will no doubt be evident; nevertheless the choices will be yours.

 Review Questions

1. How do cornucopianism and environmentalism differ in terms of world view and consideration of natural resources?
2. What is the evidence supporting the two world views named in question 1?
3. How does sustainability relate to the two world views named in question 1?
4. Cornucopians accuse environmentalists of being doomsayers and against progress. How do environmentalists respond?
5. Give evidence showing how the balance between cornucopianism and environmentalism is changing throughout the world.
6. Explain why the change to a sustainable society will not be simple. Give a comparison that illustrates the magnitude of the change.
7. What role will environmental science play in making the change to a sustainable society?

8. What is science?
9. What are the steps of the scientific method?
10. What is the difference between a fact, a hypothesis, a theory, and a natural law?
11. What are the requirements for a controlled experiment?
12. What three roles do instruments play in science?
13. Give four reasons that controversies may exist and persist in science.
14. Give four ways in which you might use your understanding of science and the scientific method to evaluate information.
15. Describe the respective roles of scientific understanding and values in making decisions.
16. Distinguish between the different roles that environmental science and values will play in bringing about a sustainable society.

Thinking Environmentally

1. Are you a Cornucopian or an environmentalist? Write a short essay describing your own world view and your justifications for it.

2. Select an issue where you have questions in your own mind as to whether information you have heard or read is correct or not. In light of what you've learned about science in this chapter, how might you evaluate that information?

3. Select an environmental issue that is controversial, and analyze the basis for the controversy. To what degree does the controversy involve lack of infor-mation, conflict of values, and vested or personal interests?

4. Scientific information has the reputation of being true and correct. Are scientists particularly ethical people who never lie, cheat, or make mistakes? Discuss aspects of the scientific process that prevent or eliminate such misinformation.

5. Describe how you might use the scientific method to solve some ordinary, everyday problem you may be confronted with.

6. How might taking this course in environmental science change what you do in your life?

Part One

What Ecosystems Are and How They Work

Tropical rainforests—humid, warm, dense, full of unusual plants and animals. These are the impressions that our senses bring to us when we take the first few steps into a rainforest. However, our senses will not tell us that the tropical rainforests are home to a greater diversity of living things than anywhere else on earth, that they are the result of millions of years of adaptive evolution, or that they are the site of storage of more carbon than the entire atmosphere. This information comes from the work of many scientists who have been asking the basic questions of how such natural systems work, how they came to be, and how they relate to the rest of the natural world.

How important is the information that comes from the work of such scientists? We hope in this first Part to convince you that information about how the natural world works is absolutely crucial to the human enterprise of living on planet earth. Ecosystems—which is the name we give to natural units like the tropical rainforest—are not just the backdrop for human activities, they are the basic context of life on earth, including human life. These are systems that are self-sustaining, and if we accept the notion that the human enterprise should also be self-sustaining (and currently is not), then perhaps we can learn how to construct a sustainable society by studying ecosystems. Environmental science begins by understanding how the natural world works.

2

Ecosystems: What They Are

LEARNING OBJECTIVES

When you have finished studying this chapter, you should be able to:

1. Define and contrast biosphere, biome, ecosystem, ecotone, plant community, and species.

2. Name seven major biomes, and give prime biotic and abiotic characteristics of each.

3. For the biome of your region, name the categories of organisms that make up its biotic structure and describe the role that each plays. Give examples of organisms in each category.

4. Name and describe various feeding relationships, and define and contrast food chains, food webs, and trophic levels.

5. Name and give examples of nonfeeding relationships that exist among organisms.

6. Analyze how competition between different species in the same ecosystem is largely avoided.

7. List abiotic factors that organisms face. For any factor, define the optimum, zones of stress, limits of tolerance, and range of tolerance in terms of the effect on a given species.

8. Analyze the natural distribution of species in terms of limiting factors. Describe an ecotone in terms of limiting factors for species of the two adjacent ecosystems.

9. Define climate, and use the principle of limiting factors to relate different biomes with different climates in various regions.

10. Analyze the diversity of plant species in an ecosystem in terms of microclimates and limiting factors.

11. Describe the origins of the human system and contrast it with natural ecosystems in terms of limiting factors and sustainability.

In 1968 the *Apollo* astronauts returned from their lunar journey with photographs of the earth taken from the moon. These photographs made it clear as never before that the Earth is a sphere in the void of space like a spaceship on an everlasting journey. Indeed, "Spaceship Earth" became a popular term and concept.

Spaceship Earth is unique among all the planets we know in that, in addition to its rock base, it has an oxygen-rich atmosphere, an abundance of liquid water, and, most conspicuous, millions of kinds of living things of which we, *Homo sapiens*, are but one. The layer around the Earth where air (atmosphere), water (hydrosphere), and minerals (lithosphere) interact with living things is called the **biosphere** (Fig. 2–1).

No living organism exists or can exist as an entity unto itself. Each organism is but one member of a particular *species*. The word **species** refers to the total population of a specific *kind* of plant, animal, or microbe (microscopic organisms, mainly bacteria and fungi). All the members of a given species can interbreed to reproduce their kind, and members of different species, by definition, do not usually interbreed. Thus, the life span of individuals is limited, but through continual reproduction and replacement, a species can potentially persist indefinitely.

But no species is an entity unto itself either. Every animal species depends on various plant species for food and for various habitat conditions, such as trees for shelter and nesting sites. Plant species depend on various animal and microbial species to decompose biological waste and release nutrients so that they can be recycled. All species depend on air, water, and nutritive elements from the earth's minerals. But plants, animals, and microbes also play important roles in removing pollutants from air and replenishing its oxygen and carbon dioxide, in purifying water, and in recycling mineral elements. Therefore, we can say that plants support animals; animal activities support plants; and all organisms help to support and maintain air, water, and soil quality.

Thus, the biosphere consists of a virtually infinite number of mutually dependent and mutually supportive interrelationships among all living organisms, air, water, and minerals. Humans are no less a part of and no less dependent on these interrelationships than any other species, although we frequently delude ourselves into thinking that we are. Our very term *natural resources* implies that all these things—air, water, soil, minerals, grasslands,

FIGURE 2–1
Planet earth is unique in the solar system in having a biosphere, a layer around the planet in which living things interact with air, water, and minerals. This is a GOES satellite image of North America. Deserts in the southwest, the Great Lakes, and forests of the eastern United States may be seen. (Earth Satellite Corporation/ Science Photo Library/Photo Researchers.)

Planet Earth is the only planet in our solar system that has conditions suitable for supporting life, but what about elsewhere in our galaxy? No other planetary systems similar to the solar system have yet been discovered, but this is not to say that they do not exist. Astronomers agree that even if the closest star had planets, these planets could not be detected with current technology. Therefore, the existence of other planets similar to earth is open to speculation.

Current theory is that stars arise from the collapse and condensation of gigantic dust clouds, and most astronomers agree that planets should form in the process. Therefore, astronomers speculate that many if not most stars do possess a system of planets. To develop and sustain life, another set of conditions is required. Development of a suitable atmosphere requires a specific gravitational mass in order for undesirable gases to escape while desirable gases and water vapor are held. Then, in order to have liquid water, the planet must be a certain distance from the star so that the planet has the proper temperature. However, there are billions of billions of stars. Even if planets formed in only one case in 100 and even if out of these a planet of the proper size and distance occurred in one case out of a million, these frequencies would still lead to the conclusion that there are literally billions of planets within our galaxy that may support life.

Of course, science fiction writers have been working on this theme for years, even before astronomers held out the plausibility. Some writers even claim that we have been visited by beings from other worlds, but such claims remain unsubstantiated. Serious efforts to determine the existence of extraterrestrial intelligence are beginning, however. In 1992, the U.S. National Aeronautics and Space Administration (NASA) completed a $100 million radio telescope devoted full-time to searching the heavens for tell-tale signals indicating the presence of alien civilizations.

Even if extraterrestrial intelligence is detected, however, will such a finding change our situation here on earth? Travel time would still preclude any meaningful exchange of resources. Even information signals traveling at the speed of light, which is a physical maximum, would take 9 years to reach and return from the *nearest* star. Then, information they send us might be just what we already know.

In short, your future will depend on managing yourselves in ways that create sustainable relationships with your own biosphere.

forests, wildlife—are there simply for our taking with no prices to be paid for their use, misuse, or pollution, and no need to put anything back in return. However, as we noted in Chapter 1, there *are* prices to be paid, and some of the prices, such as potential global warming, depletion of the ozone shield, and loss of tropical forests, may be steeper than we can afford.

It should be clear that sustainability of human society as well as sustainability of other species will depend on maintaining the integrity of the biosphere. To do this, we need to know how the biosphere works. This is an enormous task. However, enough information has been pieced together into a coherent picture so that certain patterns and principles involved in the interrelationships are clear. Our objective here in Part 1 is to gain an overview of how the biosphere functions to support all life. Our study begins with an examination of *ecosystems*.

Definition and Examples of Ecosystems

An **ecosystem** may be defined as a grouping of various species of plants, animals, and microbes interacting with each other and with their environment. (The environment includes temperature, precipitation, amount of moisture, and all other chemical and physical factors to which organisms are exposed.) Furthermore, the interrelationships are such that the entire grouping may perpetuate itself, perhaps indefinitely. This definition is a very condensed description of what is observed in nature and can be best understood by considering some examples of ecosystems with which you are probably already familiar.

A quick tour across the United States shows deciduous (leaf-shedding) forests in the East, which turn brilliant colors in the fall before the leaves drop; prairies or grasslands in the Central States; deserts with distinctive cacti in the Southwest; and coniferous (evergreen) forests in the northern and western

Mountain States. Across northern Canada and on the upper slopes of mountains, there are treeless expanses called tundra, and in equatorial regions we find tropical rainforests. You are probably also familiar with the fact that these different ecosystems are the result of different climates, that is, different environmental conditions of temperature and rainfall.

Looking at these examples more closely, we note that each ecosystem is characterized by a distinctive **plant community**, which is defined as a grouping of particular plants. For example, various species of deciduous trees and associated plants make up the deciduous forest, whereas various kinds of grasses and associated plants dominate the prairies or grasslands. Each plant community supports a particular array of animals. We tend to focus on large animals—deer in eastern deciduous forests, bison on the prairies (before they were killed off by early settlers), moose in conifer forests, and caribou on the arctic tundra, for example. However, smaller animals, such as mice, birds, various species of insects, earthworms, and other such organisms, are far more abundant than the larger animals both in terms of numbers and in terms of total combined weight. Finally, but even less conspicuous, an array of microbes will be found in each ecosystem, feeding on dead plant and animal material.

Each of these examples represents a distinctive grouping of plants, animals, and microbes interacting with one another and with the environment. Thus, each grouping in association with the environment that supports it is an *ecosystem*. Furthermore, we know that these groupings existed long before humans came on the scene and, if not disturbed by humans, would continue to exist for a very long time and perhaps even indefinitely. That is, ecosystems perpetuate themselves.

Major terrestrial ecosystems, such as forest or prairie, are not entirely uniform; each consists of a number of more or less distinct but related ecosystems. Thus, each is also called a **biome**, a term that refers to a number of closely related ecosystems. Six of the major biomes of the earth are described more fully in the foldout attached to the back endpapers of this book. In addition, most ecologists recognize a number of minor biomes.

Within a biome, an ecosystem may include any more-or-less distinctive grouping of organisms interacting with each other in a particular environment. For instance, a patch of woods, an open field, a pond, a stream, each may be considered and studied as an ecosystem.

While it is convenient to divide the biosphere into biomes and ecosystems for study and discussion, it is important to recognize that they seldom have distinct boundaries, and they are definitely not independent of one another. Rather, one ecosystem tends to blend into the next through a transitional region called an **ecotone**, which is a region that contains many of the species and characteristics of the two adjacent systems (Fig. 2–2). The ecotone between adjacent systems may include unique environments that support distinctive plants and animals as well as those that are common to the adjoining ecosystems.

FIGURE 2–2
Ecosystems are not isolated from one another. One ecosystem blends into the next through a transitional region, an ecotone, that contains many species common to the two adjacent systems.

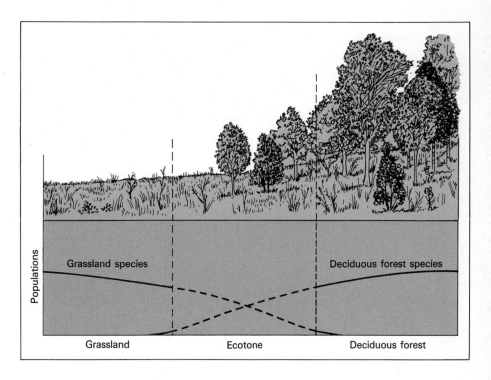

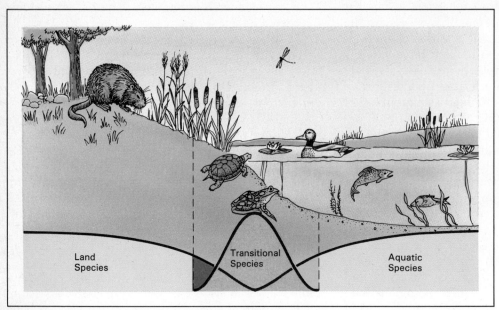

FIGURE 2–3
An ecotone may create a unique habitat that harbors specialized species not found in either ecosystem I or ecosystem II.

Consider a marshy area, which often occurs between the open water of a lake and dry land (Fig. 2–3). Whether they contain distinctive species or not, ecotones may be studied as ecosystems in their own right.

Furthermore, some animal species—migratory birds, for instance—may inhabit different ecosystems at different times of the year, and what happens in one ecosystem used by the migrators will definitely affect any other ecosystems used during other seasons. For example, losses and fragmentation of forests have disrupted migration lanes and resulted in drastic declines in the populations of certain North American song birds. How the loss of all these birds will affect various ecosystems is a question we just cannot answer at this time.

Oceans include a variety of environments depending on temperature, water depth, nature of the bottom, and concentrations of nutrients and sediments. Each of these environments, called **marine environments**, supports a more-or-less distinctive array of seaweeds, plankton (mostly microscopic plant and animal forms that drift with the currents), fish, shellfish, and other marine organisms. Thus different areas of oceans—reefs, continental shelves, deep oceans—may be studied as separate ecosystems, even though their interconnections and interdependencies are obvious. There are conspicuous ecotones between ocean and freshwater systems in the form of estuaries, and between oceans and the land in the form of beaches, wetlands, and rocky coast lines (see the aquatic ecosystems in the end paper foldout).

All ecosystems and biomes are interconnected through the movements of air (wind) and water (see the water cycle, p. 239). Thus, the entire biosphere is really one mammoth ecosystem. It is only for the convenience of study and understanding that we divide it into biomes and smaller ecosystems. Figure 2–4 shows the hierarchy, from biosphere to species, ecologists use when studying ecosystems.

The Structure of Ecosystems

As we are interested in preserving the integrity of the biosphere, we wish to discover how it works. Our study begins with an overview of the *structure* of ecosystems. Structure refers to parts and the way they fit together to make the whole. There are two "sides" in every ecosystem, the organisms on one hand and the environmental factors on the other. All the organisms—plants, animals, and microbes—in the ecosystem are referred to as the **biota** (*bio*, "life"). The way the categories of organisms fit together is referred to as the **biotic structure**. The nonliving chemical and physical factors of the environment (climate, soil quality, and so forth) are referred to as **abiotic** (*a*, "non") factors.

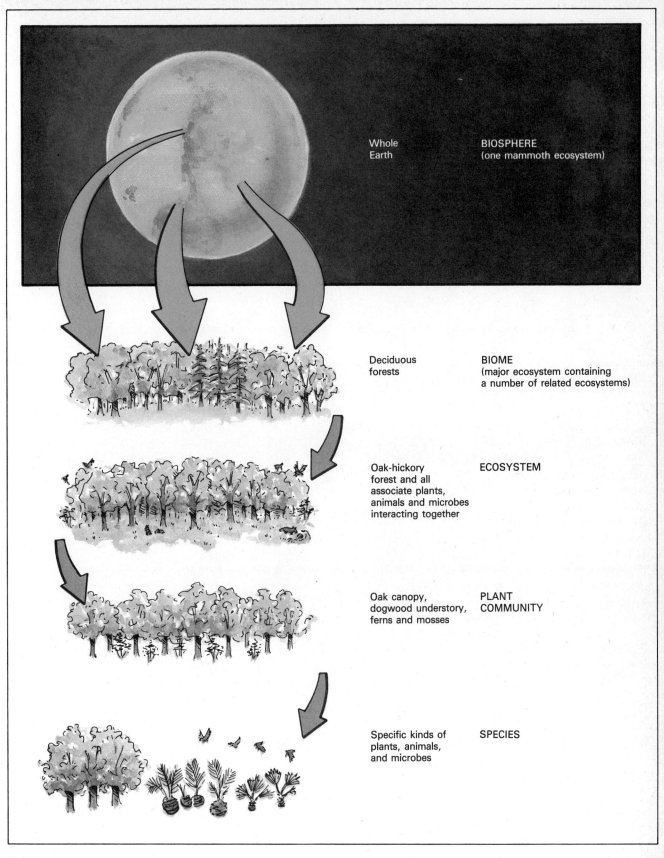

Whole Earth — BIOSPHERE (one mammoth ecosystem)

Deciduous forests — BIOME (major ecosystem containing a number of related ecosystems)

Oak-hickory forest and all associate plants, animals and microbes interacting together — ECOSYSTEM

Oak canopy, dogwood understory, ferns and mosses — PLANT COMMUNITY

Specific kinds of plants, animals, and microbes — SPECIES

FIGURE 2–4
Division of the biosphere into biomes, ecosystems, plant communities, and species.
This hierarchy is used for study only. In actuality, all the levels are interrelated.

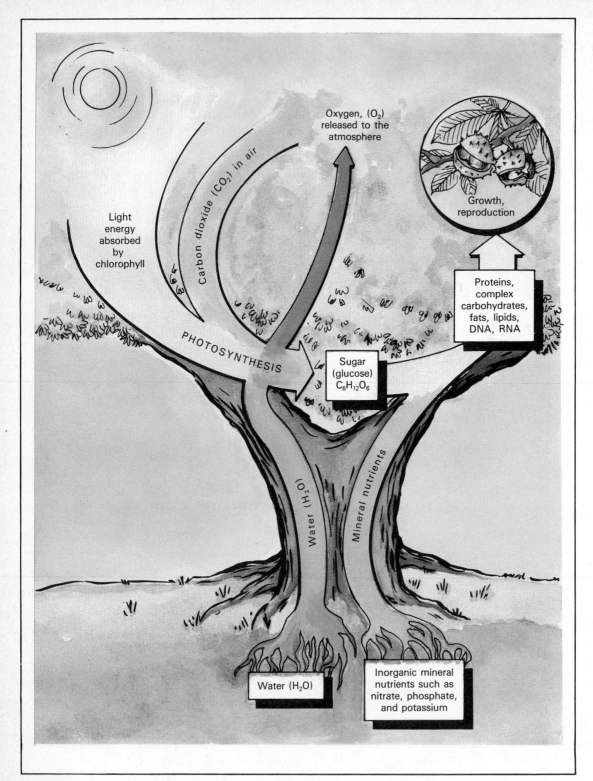

FIGURE 2–5

Producers. The producers of all major ecosystems are green plants, because they contain the green pigment chlorophyll. Chlorophyll absorbs light energy, which is then used to produce the six-carbon sugar glucose from carbon dioxide and water, releasing oxygen as a byproduct. The glucose, along with a few additional mineral nutrients from the soil, is used in the production of all plant tissues leading to growth.

BIOTIC STRUCTURE

Categories of Organisms

Despite the diversity of ecosystems, all have a similar biotic structure. That is, all ecosystems have the same three basic categories of organisms that interact together in the same ways. The major categories of organisms are (1) *producers*, (2) *consumers*, and (3) *detritus feeders and decomposers*.

Producers **Producers** are mainly green plants, which use *light energy* from the sun to convert carbon dioxide (absorbed from air or water) and water to a sugar called glucose and release oxygen as a byproduct. This chemical conversion driven by light energy is called **photosynthesis**. Plants are able to manufacture all the additional complex molecules that make up their bodies from the glucose produced in photosynthesis plus a few additional *mineral nutrients* such as nitrogen, phosphorus, potassium, and sulfur, which they absorb from the soil or from water (Fig. 2–5).

The molecule that plants use to capture light energy for photosynthesis is **chlorophyll**, which is a green pigment. Hence plants that carry on photosynthesis are easily identified by their green color. (In some cases the green may be overshadowed by additional red or brown pigments. Thus, red algae and brown algae also carry on photosynthesis.) They range in diversity from microscopic single-celled algae through medium-sized plants such as grass and cacti to gigantic trees. Thus, every major ecosystem, both aquatic and terrestrial, has its particular producers carrying on photosynthesis.

The protein, fat, and carbohydrate molecules that make up the tissues of plants and animals are referred to as **organic molecules**. The key feature of such molecules is that they are in large part constructed from the carbon atoms captured from carbon dioxide in photosynthesis. In contrast, the simple chemicals that make up air, water, and the minerals of rock and soil are referred to as **inorganic molecules** (Fig. 2–6). Thus, plants that carry on photosynthesis use light as the initial energy source to produce all the complex organic molecules their bodies need from the simple inorganic chemicals (carbon dioxide, water, mineral nutrients) present in the environment.

FIGURE 2–6
Organic and inorganic. Water and the simple molecules found in air and in rocks and soils are *inorganic*. The complex molecules which make up plant and animal tissues are *organic*. Producers, using energy derived from light, convert inorganic substances to organic substances. Organic materials then break down to inorganic materials again through burning or digestion, releasing energy. Chemically, organic compounds involve carbon-carbon and carbon-hydrogen bonds not found in inorganic materials.

As this conversion from inorganic to organic occurs, some of the energy from light is stored in the organic compounds.

Plants that carry on photosynthesis—in other words, producers—are significant to an ecosystem because they produce the organic matter, or "food," that supports all the rest of the organisms in the ecosystem.

Not all plants are producers. A few higher plants, such as Indian pipe (Fig. 2–7) and all the plant-like organisms known as **fungi** (mushrooms, molds, and other such organisms) do not have chlorophyll and do not carry on photosynthesis. Like animals, they gain their energy and nutrients by feeding on other organic matter. Indeed, all organisms in the biosphere can be divided into two categories, *autotrophs* and *heterotrophs*, on the basis of whether they do or do not produce all the organic compounds they need to survive. Those organisms that produce their own organic material from inorganic raw materials plus an energy source are **autotrophs** (*auto*, "self"; *troph*, "feeding"). As stated above, the most important and common autotrophs by far are green plants, which use chlorophyll for photosynthesis. However, a few

FIGURE 2–7
Indian pipe, a flowering plant that is not a producer. It does not carry on photosynthesis but derives its energy from other organic matter as do animals. (Photograph by BJN.)

bacteria use a purple pigment for photosynthesis, and some other bacteria acquire their energy from certain high-energy inorganic chemicals. All other organisms, however, which must *consume* organic material made by autotrophs in order to obtain energy and nutrients, are **heterotrophs** (*hetero*, "other"). Heterotrophs may be divided into numerous subcategories, the two major ones being **consumers** and **detritus feeders** and **decomposers**.

Consumers All organisms that are not producers must feed on complex organic material in order to obtain energy and nutrients, and so are called **consumers**. The term *consumers* embraces a tremendous variety of organisms ranging in size from microscopic bacteria to blue whales and includes such diverse groups as protozoans, worms, fish and shellfish, insects, reptiles, birds, and mammals including humans.

For the purpose of understanding ecosystem structure, consumers are divided into various subgroups according to their food source. Animals that feed directly on producers, whether as large as elephants or as small as mites, are called **primary consumers**. Animals that feed on primary consumers are called **secondary consumers**. For example, a rabbit, which feeds on plants, is a primary consumer; a fox that feeds on rabbits is a secondary consumer. There may also be third, fourth, or even higher levels of consumers, and certain animals may occupy more than one position on the consumer scale. For instance, humans are primary consumers when they eat vegetables, secondary consumers when they eat beef, and third-level consumers when they eat fish that feed on smaller fish that feed on algae.

Primary consumers, those animals that eat only plant material, are also called **herbivores** (*herb*, "grass"). Secondary and higher-order consumers are called **carnivores** (*carni*, "meat"). Consumers that feed on both plants and animals are called **omnivores** (*omni*, "all").

In a relationship in which one animal attacks, kills, and feeds on another, the animal that attacks and kills is called the **predator**; the animal that is killed is called the **prey**. Together, the two animals are said to have a **predator-prey** relationship.

Parasites are another important category of consumers. **Parasites** are organisms—they can be either plants or animals—that become intimately associated with their "prey" and feed on it over an extended period of time, typically without killing it (at least not immediately) but usually doing it harm. The plant or animal that is fed upon is called the **host**; thus we speak of a **host-parasite** relationship.

A tremendous variety of organisms may be parasitic. Various worms are the most well-known examples, but some plants as well as the bacteria and

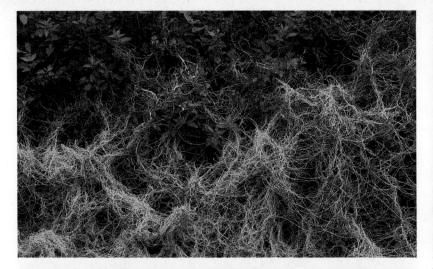

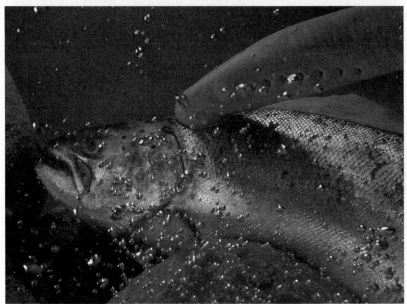

FIGURE 2–8
Diversity of parasites. Nearly every major biological group of organisms has at least some members that are parasitic on others. Shown here are (a) dodder, a plant parasite. The orange "strings" are the dodder stems which suck sap from the host plant; dodder has no leaves and no chlorophyll. (b) Nematode worms (Ascaris lumbricoides), the largest of the human parasites, reach a length of 14 inches (35 cm). (c) Lamprey attached to a salmon. Lampreys are parasitic on fish. ((a) Jeff Lepore/Photo Researchers; (b) Sinclair Stammers/Science Photo Library/Photo Researchers; (c) Runk/Schoenberger/Grant Heilman.)

other microorganisms that cause disease in plants or animals are also parasites. Many serious plant diseases and some animal diseases (such as athlete's foot) are caused by parasitic fungi. Virtually every major group of organisms, including mammals (vampire bats) and flowering plants (Fig. 2–8a), has at least some members that are parasitic. Parasites may live inside or outside their host as the examples shown in Figure 2–8 illustrate.

Representative examples of producers and consumers and feeding relationships among them are shown in Figure 2–9.

Detritus Feeders and Decomposers Dead plant material from sources such as leaf fall in forests, the dieback of grasses in unfavorable seasons, the fecal wastes of animals, and occasional dead animals, is called **detritus** (pronounced di-TRI-tus). Many organisms are specialized to feed on detritus, and we refer

to such consumers as **detritus feeders**. Examples include vultures, earthworms, millipedes, crayfish, termites, ants, and wood beetles. As with regular consumers, one can identify *primary* **detritus feeders** (those that feed directly on detritus), *secondary* **detritus feeders** (those that feed on primary detritus feeders), and so on.

An extremely important group of primary detritus feeders is the **decomposers**, namely fungi and bacteria. Much of the detritus in an ecosystem, particularly dead leaves and wood, does not appear to be eaten as such, but rots instead. Rotting is the result of metabolic activity of fungi and bacteria. These organisms secrete digestive enzymes that break down wood, for example, into simple sugars that the fungi or bacteria then absorb. Thus, organic material rots as a result of being consumed by fungi and bacteria. Even though fungi and bacteria are called decomposers because of their unique behavior, they are pri-

FIGURE 2–9
Common feeding relationships among producers and consumers.

Producers | Consumers
Primary Consumers | Secondary Consumers | Third-Order Consumer
Herbivores | Carnivores
Prey
Prey
Predator
Prey
Predator-Prey Relationships
Host
Parasite
Host–Parasite Relationship
Omnivore

FIGURE 2–10
The feeding relationships among primary detritus feeders, secondary detritus feeders, and decomposers. Fungi and bacteria, which feed on detritus, support many other organisms living in the soil, and these, in turn, may be fed upon by larger consumers.

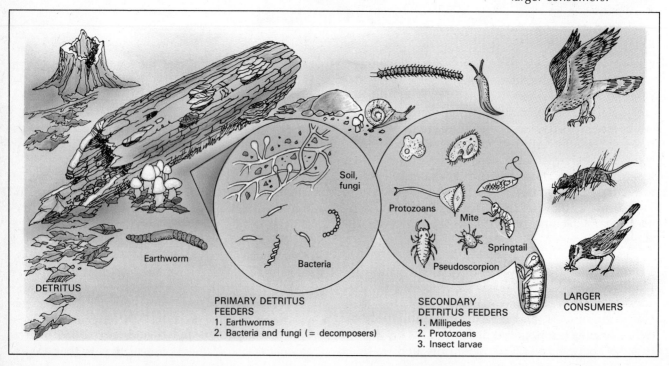

DETRITUS
Earthworm
Soil, fungi
Bacteria
Protozoans
Mite
Springtail
Pseudoscorpion
LARGER CONSUMERS

PRIMARY DETRITUS FEEDERS
1. Earthworms
2. Bacteria and fungi (= decomposers)

SECONDARY DETRITUS FEEDERS
1. Millipedes
2. Protozoans
3. Insect larvae

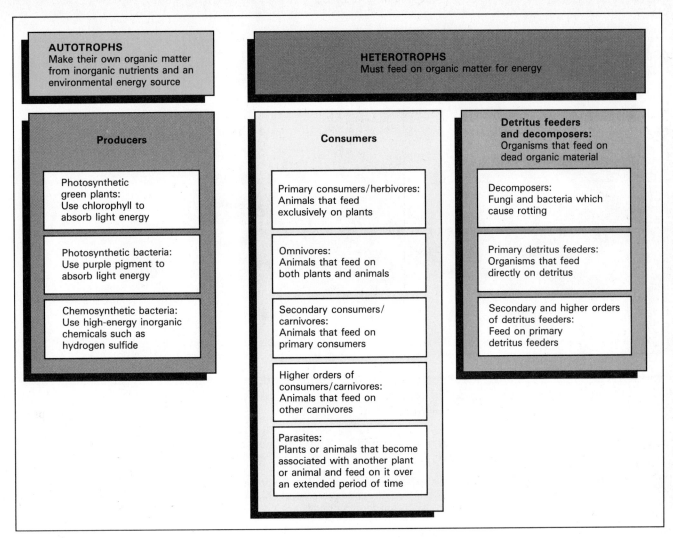

| AUTOTROPHS
Make their own organic matter from inorganic nutrients and an environmental energy source | HETEROTROPHS
Must feed on organic matter for energy | |

Producers

Photosynthetic green plants:
Use chlorophyll to absorb light energy

Photosynthetic bacteria:
Use purple pigment to absorb light energy

Chemosynthetic bacteria:
Use high-energy inorganic chemicals such as hydrogen sulfide

Consumers

Primary consumers/herbivores:
Animals that feed exclusively on plants

Omnivores:
Animals that feed on both plants and animals

Secondary consumers/carnivores:
Animals that feed on primary consumers

Higher orders of consumers/carnivores:
Animals that feed on other carnivores

Parasites:
Plants or animals that become associated with another plant or animal and feed on it over an extended period of time

Detritus feeders and decomposers:
Organisms that feed on dead organic material

Decomposers:
Fungi and bacteria which cause rotting

Primary detritus feeders:
Organisms that feed directly on detritus

Secondary and higher orders of detritus feeders:
Feed on primary detritus feeders

FIGURE 2–11
Producers and consumers.

mary detritus feeders. They are fed upon by such secondary detritus feeders as protozoans, mites, insects, and worms (Fig. 2–10). When a fungus or other decomposer dies, its body becomes part of the detritus and the source of energy and nutrients for yet other detritus feeders and decomposers.

In summary, despite the apparent diversity of ecosystems, they all have a similar *biotic structure*. They can all be described in terms of autotrophs, or producers, which produce organic matter that becomes the source of energy and nutrients for heterotrophs, which are various categories of consumers and detritus feeders and decomposers (Fig. 2–11).

Feeding Relationships: Food Chains, Food Webs, and Trophic Levels

In describing the biotic structure of ecosystems it becomes evident that major interactions among organisms involve feeding relationships. We can identify innumerable pathways where one organism is eaten by a second, which is eaten by a third, and so on. Each such pathway is called a **food chain**. Three simple examples are illustrated in Figure 2–12.

While it is interesting to trace such pathways, it is important to recognize that food chains seldom exist as isolated entities. More commonly, an herbivore population feeds on several kinds of plants and is preyed upon by several kinds of carnivores. Consequently, virtually all food chains are interconnected and form a complex *web* of feeding relationships. Indeed, the term **food web** is used to denote the complex pattern of interconnected food chains.

Despite the number of theoretical food chains and the complexity of food webs, there is a simple overall pattern. All food chains basically lead through a series of steps, or "levels"—namely, from producers to primary consumers (or primary detritus feeders) to secondary consumers (or secondary detritus feeders) and so on. These *feeding levels* are called

FIGURE 2–12
Simple food chains. Food (nutrients and energy) is transferred from one organism to another along pathways known as *food chains*.

trophic levels (*tropic*, "feeding"). All producers belong to the first trophic level. All primary consumers (in other words, all herbivores), whether feeding on living or dead producers, belong to the second trophic level; organisms feeding on these herbivores belong to the third, and so on. We can visualize all feeding relationships as a flow of nutrients and energy through a series of trophic levels. A diagrammatic comparison of a food chain, a food web, and trophic levels is shown in Figure 2–13.

FIGURE 2–13
Three ways of representing the transfer of nutrients and energy. Each single pathway from the bottom to the top is a *food chain*. All the interconnected food chains shown are collectively called the *food web*. The basic feeding levels are *trophic levels*.

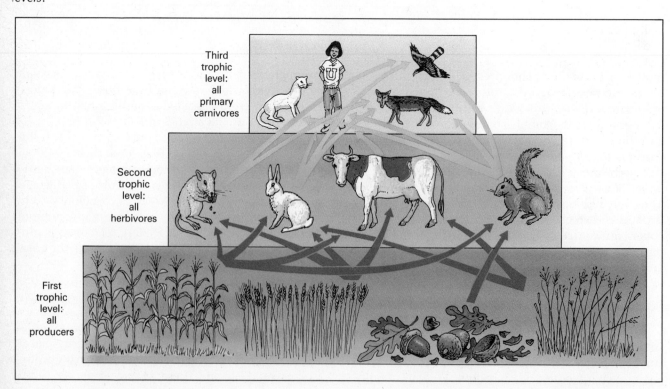

Third trophic level: all primary carnivores

Second trophic level: all herbivores

First trophic level: all producers

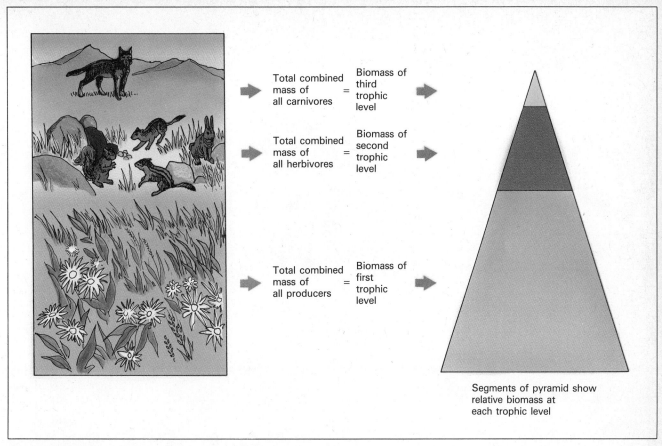

FIGURE 2–14
Biomass pyramid. A graphical representation of the biomass (the total combined mass of organisms) at successive trophic levels has the form of a pyramid.

How many trophic levels are there? There are usually not more than three or four discernible trophic levels in any ecosystem. This answer comes from straightforward observations. The **biomass**, or total combined weight, of all the organisms at each trophic level can be measured by collecting (or trapping) and weighing suitable samples. In terrestrial ecosystems, the biomass is about 90–99 percent less at each higher trophic level. If the biomass of producers in a grassland is 10 tons (1 ton = 2000 lb) per acre, the biomass of herbivores will be no more than 200 pounds and that of carnivores no more than 20 pounds. As you can see, you can't go through very many trophic levels before the biomass approaches zero. Depicting this graphically gives rise to what is commonly called a **biomass pyramid** (Fig. 2–14).

The biomass decreases so much at each trophic level largely because much of the food consumed is not converted to the body of the consumer but is broken down (digested) so that its stored energy can be released and used by the consumer. As this breakdown occurs, the chemical elements are released back to the environment in the inorganic state. Thus, there is a continuous cycle of nutrients from the environment through organisms back to the environment (Fig. 2–15). This cycle will be discussed in greater detail in Chapter 3. For now the main point is: *all food chains must start with producers.* Without producers to constantly replenish the supply of organic matter, consumers, detritus feeders, and decomposers would soon eat themselves into nonexistence.

Nonfeeding Relationships

Mutually Supportive Relationships The overall structure of ecosystems is dominated by feeding relationships, as we have just seen. In any feeding relationships involving two species, we generally think of one species benefiting and the other being harmed to a greater or lesser extent. However, there are many relationships that provide a mutual benefit to both species. This phenomenon is called **mutualism**. A common example is the relationship between flowers and insects. The insects benefit by obtaining nectar from the flowers; the plants benefit by being pollinated in the process. Another example is observed in tropical seas. Clownfish are immune to the predatory tentacles of sea anemones. Thus, these fish are able to feed

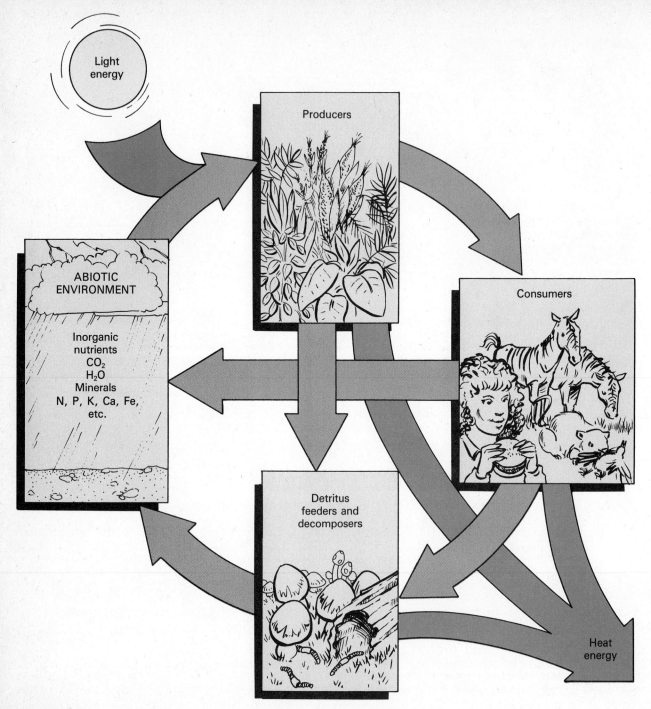

FIGURE 2–15
The movement of nutrients and energy through the ecosystem. Nutrients follow a cycle being used over and over. Light energy absorbed by producers is released and lost as heat as it is "spent."

on detritus around the anemones while at the same time receiving protection from would-be predators that are *not* immune to the anemones. The anemones benefit by being cleaned (Fig. 2–16).

In some cases the mutualistic relationship has become so close that the species involved are no longer capable of living alone. This extreme form of

mutualism is called **symbiosis** (*sym*, "together"; *bio*, "living"). A classic example of symbiosis is seen in the group of plants known as lichens (Fig. 2–17). Lichens actually compromise two organisms, a fungus and an alga. In a beautiful symbiotic pact, the fungus provides protection for the alga, enabling it to survive in dry habitats where it could not live by itself, and

FIGURE 2–16
A mutualistic relationship—one in which both species benefit— is seen between clownfish and sea anemones. The fish, protected by the anemones, can forage without risk of predation; the anemones have detritus removed. (Greg Dimijan/ Photo Researchers.)

the alga, which is a producer, provides food for the fungus, which is a heterotroph.

While not as close as the supportive relationship just discussed, many other such relationships exist in every ecosystem. For example, plant detritus provides most of the food for decomposers and soil-dwelling detritus feeders such as earthworms. Thus, these organisms benefit from plants, but the plants also benefit because the activity of the organisms is instrumental in releasing nutrients from the detritus and in returning them to the soil where they can be reused by the plants. In another example, many birds benefit from vegetation by finding nesting materials and places among trees, while the plant community also benefits because the birds feed on and reduce the populations of many herbivorous insects. Even in predator-prey relationships some mutual advantage

exists. The killing of individual prey that are weak or diseased may benefit the population at large by keeping it healthy. Predators may also prevent prey populations from becoming so abundant that they overgraze the environment.

Competitive Relationships Given the concept of food webs, it might seem that many different species of animals would be in great "free-for-all" competition with each other. In fact, such fierce competition rarely occurs because each species tends to be specialized and adapted to a somewhat different *habitat* and/or *niche*.

Habitat refers to the kind of place—defined by the plant community and the physical environment—where a species lives and thrives. For example, a deciduous forest, a swamp, and an open field are dif-

FIGURE 2–17
Lichens. The crusty-appearing "plants" commonly seen growing on rocks or the bark of trees are actually composed of a fungus and an alga growing in a symbiotic relationship. (Photograph by BJN.)

ferent habitats. Different kinds of forests provide markedly different habitats and support different species of wildlife. When different species occupy the same habitat, competition may still be slight or non-existent because different species, for the most part, have different *niches*. An animal's **niche** refers to what it feeds on, where it feeds, when it feeds, where it finds shelter, nesting sites, and so on. Because niches exist, seeming competitors can coexist peaceably in the same habitat. For example, woodpeckers, which feed on insects in dead wood, are not in competition with birds that feed on seeds. Any species that feeds in tree tops is not in competition with a species that feeds on the ground, even if the food is similar. Bats and swallows both feed on flying insects, but they are not in competition because bats feed at night and swallows feed during the day.

There is often interspecies competition where different habitats or niches overlap to some extent. Where two species do compete directly in every respect, as sometimes occurs when a species is introduced from another continent, one of the two generally perishes in the competition.

All green plants require water, nutrients, and light, and where they are growing in the same location, one species may eliminate others in competition. (Hence, maintaining flowers and vegetables against the advance of weeds is a constant struggle.) However, different plant species are also adapted and specialized to different conditions. Thus each species is able to hold its own against competition where conditions are well suited to it.

ABIOTIC FACTORS

We now turn to the *environmental* side of the ecosystem. As we noted before, the environment involves the interplay of many physical and chemical, or **abiotic factors**, the major ones being rainfall (amount and distribution over the year), temperature (extremes of heat and cold as well as average), light, water, wind, chemical nutrients, pH (acidity), salinity (saltiness), and fire. The degree to which each is present or absent, high or low, profoundly affects the ability of organisms to survive. However, different species may be affected differently by each factor. We shall find that this difference in response to environmental factors determines which species may or may not occupy a given region. In turn, which organisms do or don't survive determines the nature of a given ecosystem.

Optimum, Zones of Stress, Limits of Tolerance

In any study of the abiotic side of ecology, the key observation is that *different species thrive under different conditions*. This principle applies to all living things, both plants and animals. Some like it very wet; others like it relatively dry. Some like it very warm; others do best in cooler situations. Some tolerate freezing, others don't. Some require bright sun; others do best in shade.

Laboratory experiments clearly bear this fact out. Plants may be grown in a series of chambers in which all the abiotic factors are controlled. Thus, a single factor—temperature, say—can be varied in a systematic way, while all other factors are kept constant. (Note that this kind of *controlled experiment* is necessary in order to distinguish the effect of temperature from the effect of other factors.) Experiments show that, as temperature is raised from a low point that fails to support growth, plants grow increasingly well until they reach some maximum. Then, as temperature is raised still further, the plants become increasingly stressed; they do less well, suffer injury, and die. This is pictured graphically in Figure 2–18.

The point that supports the maximum growth is called the **optimum**. Actually, since maximum growth usually occurs over a range of several degrees, we speak of an **optimal range**. The entire span that allows any growth at all is called the **range of tolerance**. The points at the high and low ends of the range of tolerance are called the **limits of tolerance**. Between the optimal range and the high or low limit of tolerance, there are **zones of stress**. That is, as temperature is raised or lowered from the optimal range, the plants experience increasing stress until, at either limit of tolerance, they cannot survive.

Similar experiments have been run to test other factors, and the results invariably follow the same general pattern. Of course, not every species has been tested for every factor; however, the consistency of such observations leads us to conclude that this is a fundamental biological principle: *Every species (both plant and animal) has an optimum range, zones of stress, and limits of tolerance with respect to every abiotic factor.*

This line of experimentation, which applies to animals as well as plants, also demonstrates that different species differ markedly with respect to the values at which the optimum and limits of tolerance occur. For instance, what may be an optimal amount of water for one species may stress a second and result in the death of a third. Some plants cannot tolerate any freezing temperatures, others can tolerate slight but not intense freezing, and some actually require several weeks of freezing temperatures in order to complete their life cycles. Also, some species have a very broad range of tolerance; others have a much narrower range. While optimums and limits of tolerance may differ from one species to another, however, there may be great overlap in the ranges of tolerance for various species. Thus many plants may

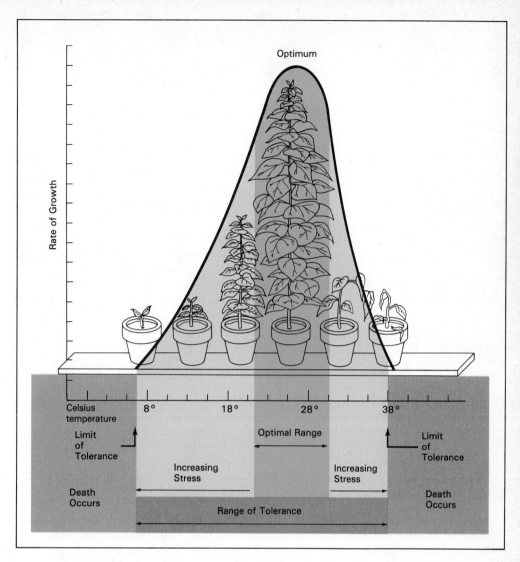

FIGURE 2–18
For every factor influencing growth, reproduction, and survival, there is an optimum level. Above and below the optimum, there is increasing stress until survival is precluded at the limits of tolerance. The total range between the high and low limits is the range of tolerance. Levels at which the optimum, zones of stress, and limits of tolerance occur are different for different species.

grow under the same conditions, although these conditions may not be optimal for all of the plants.

Law of Limiting Factors

In the experiment just described, we focused on varying the temperature while maintaining all other abiotic factors within the optimal range. Since the results just described apply to any and all abiotic factors, we observe what is known as the **Law of Limiting Factors**: *Any one factor being outside its optimal range at any given time will cause stress and limit the growth of an organism.* The factor that is limiting the growth is called the **limiting factor**. It may be any factor that affects the organism.

Keep in mind that the Law of Limiting Factors includes the problem of "too much" as well as the problem of "too little." For example, plants may be stressed or killed by overwatering or overfertilizing as well as by underwatering or underfertilizing, a common pitfall for amateur gardeners. Also note that

the factor that is limiting may change from one time to another. For example, in a single growing season, temperature may be limiting in the early spring, nutrients may be limiting later, and then water may be limiting if a drought occurs. Also, if one limiting factor is corrected, growth will increase only until another factor comes into play. Of course, the organism's genetic potential is an ultimate limiting factor. A daisy will never grow to be the size of a tree regardless of how optimal the environmental factors are.

The Law of Limiting Factors was first presented by Justus von Liebig in 1840 in connection with his observations regarding the effects of chemical nutrients on plant growth. He observed that restricting any one of the many different nutrients at any given time had the same effect: It limited growth. Thus, this law is also called Liebig's Law of Minimums. Observations since Liebig's time, however, show that his law has much broader application. Beyond its application to all abiotic factors, it may be applied to biotic factors as well. Thus the limiting factor for one species

Environmental factors, particularly temperature and rainfall, are always changing. What does it mean to say that rainfall, for example, is a limiting factor when weather is almost always changing? Consideration of a limiting factor should always include the concept of *dose*. **Dose** is defined as the level of exposure multiplied by the length of time over which the exposure occurs. You have probably experienced this kind of thing for yourself. For example, you can enter a walk-in freezer or expose yourself to the intense heat from an oven for a few seconds without particular discomfort, whereas prolonged exposure to such heat or cold would be extremely damaging, if not fatal. You might not be hurt by breathing noxious fumes for a short time, but longer exposures might well kill you.

Similarly, a limiting factor that causes the dieback of a species involves both the intensity of the factor and the duration of exposure. For example, it is not a matter of whether plants can tolerate drought; most can to some degree. It is more a matter of how long they can tolerate it. Cacti and other desert plants can tolerate much longer periods of drought than nondesert species. Likewise, it is not a question of whether or not plants tolerate flooding. Rather it is a question of how long can they tolerate being flooded. Marsh plants, of course, thrive on perpetually flooded land, whereas more than a day or so of flooding will kill many other terrestrial species.

Thus, the balance between ecosystems is not a perfect steady-state balance with a difference in average rainfall as a sharp dividing line. What actually occurs, where forest meets grassland for example, is that trees encroach into grasslands during years of normal rainfall, but then a severe drought occurs, and trees in the grassland are killed back, and the cycle starts again. Thus there is a seesaw effect between adjoining ecosystems, an effect that depends on the occurrence of limiting doses of one or another factor. This seesaw effect is a particularly important consideration in the face of potential global warming. There is no question that most species can probably tolerate the average temperature being a few degrees warmer if this were the only factor involved. The problem is, how will this slightly warmer average temperature affect the occurrence of heat waves and the duration of droughts and floods? Such warming could create limiting doses that have far-reaching albeit unpredictable effects.

may be competition or predation from another. This is certainly the case with our agricultural crops, where it is a constant struggle to keep them from being limited or even eliminated by weeds and "pests."

Finally, while one factor may be determined to be limiting at a given time, several factors being outside the optimum may combine to cause additional stress or even death. Particularly, pollutants may act in a way that causes organisms to become more vulnerable to disease or drought, for example. Such cases are examples of **synergistic effects**, or **synergisms**, which are defined as two or more factors interacting in a way that causes an effect much greater than one would anticipate from the effects of the two acting separately.

Why Do Different Regions Support Different Ecosystems?

Armed with the understanding of limiting factors, we may now address such questions as, Why do different regions support different ecosystems? And what prevents one ecosystem or species from invading and taking over another? The answers to these questions involve the facts (1) that different regions have different climates and (2) that each species is adapted to a more-or-less specific climate. A map showing the global distribution of the terrestrial biomes is given in Figure 2–19.

CLIMATE

The **climate** for a given region is the temperature (average and also high and low extremes) and precipitation that may be expected on each day through the year. A record of climate is compiled by averaging what has occurred each day for the past 30 years or more.

Climates in different parts of the world vary widely. In general, the climate of equatorial regions is continuously warm, with high rainfall and no discernible seasons. Above and below the equator, the climate becomes increasingly seasonal (warm/hot summers and cool/cold winters); the farther we go, the longer and colder the winters become until at the poles it is perpetually winter. Likewise, colder tem-

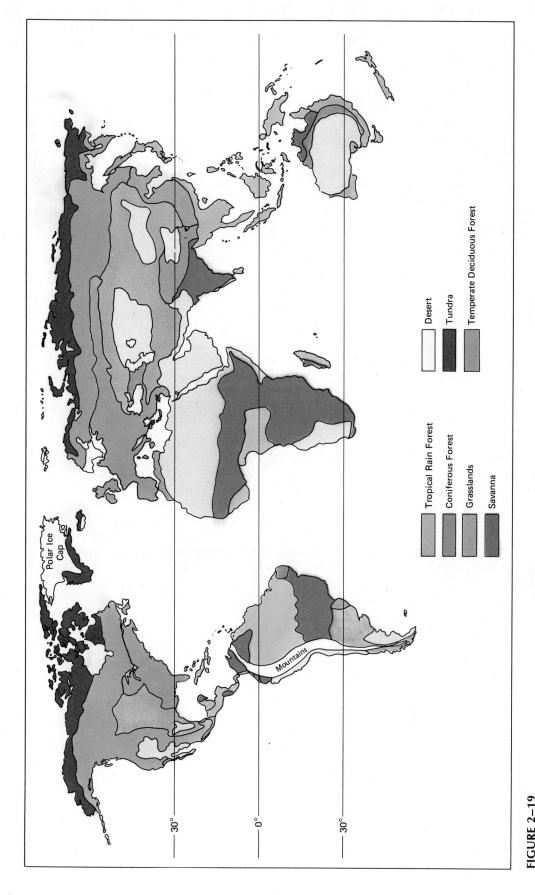

FIGURE 2–19
World distribution of the major terrestrial biomes. (Reproduced by permission from *The Botanical World* by D. K. Northington and J. R. Goodin, Times Mirror/Mosby, 1984.)

Legend:
- Tropical Rain Forest
- Coniferous Forest
- Grasslands
- Savanna
- Desert
- Tundra
- Temperate Deciduous Forest

Polar Ice Cap

Mountains

30°
0°
30°

peratures are found at higher elevations, so that there are a number of snowcapped mountains on or near the equator. Annual precipitation in any area may vary from virtually zero to over 100 inches (250 cm) per year. It may be evenly distributed throughout the year or may all occur in certain months, dividing the year into two seasons, wet and dry. Different temperature and rainfall conditions may occur in almost any combination to give a wide variety of climates.

Any climate will support only those species that find the temperature and precipitation level of the climate optimal or at least within the species range of tolerance. Species for which any of the conditions are outside the limit of tolerance will be precluded from living in that climate. Indeed, the population density of a species (the number of individuals per unit area) is usually greatest where conditions are optimal and falls off as conditions become less favorable (Fig. 2–20).

For example, annual precipitation over most of the eastern United States is 40–60 inches (100–150 cm). It tapers off to 20–30 inches (50–75 cm) per year over the Plains States and in the Southwest declines still more, with large regions receiving less than an

average of 10 inches (25 cm) per year. For most temperate tree species, 46–56 inches (115–140 cm) per year is optimal. Below about 40 inches per year, many trees begin to be stressed, and most reach their limit of tolerance at about 30 inches per year. Grasses, however, have a much lower limit of tolerance, about 10 inches per year, and many species of cacti and other specialized desert plants do well with as little as 2–4 inches (5–10 cm) per year. Thus, the eastern United States supports forests. In the Plains States, forests give way to the grasses, typical of prairies, and in the Southwest grasses give way to deserts with a sparse coverage of cacti, sagebrush, and other species renowned for their drought tolerance.

It is significant to note that these biomes—forests, grasslands, and deserts—are mainly determined by a single limiting factor: amount of rainfall. The effect of temperature, the other parameter of climate, is largely superimposed on that of rainfall. That is, 36 inches (90 cm) or more of rainfall per year will support a forest, but temperature will determine the *kind* of forest. For example, broadleaf evergreen species, which are extremely vigorous and fast-growing but cannot tolerate freezing temperatures, predominate in the tropics. By dropping their leaves and going into dormancy each autumn, deciduous trees are well adapted to freezing temperatures. Therefore, wherever rainfall is sufficient, deciduous forests predominate in temperature latitudes. Most deciduous trees cannot take the extremely hash winters and short summers that occur at higher latitudes and higher elevations, though, and so northern regions and high elevations are dominated by spruce-fir coniferous forests, which are better adapted to these conditions. Temperature by itself limits forests only when it becomes low enough to cause **permafrost** (permanently frozen subsoil). Permafrost prevents the growth of trees because roots cannot penetrate deeply enough to provide adequate support. However, a number of grasses, clovers, and other small flowering plants can grow over permafrost. Consequently, where permafrost sets in, coniferous forests give way to tundra. Of course, at still colder temperatures the tundra gives way to polar ice caps.

The same relationship of rainfall effects being primary and temperature effects secondary applies in deserts. Any region receiving less than about 10 inches of rain per year will be a desert, but the plant and animal species found in hot deserts are different from those found in cold deserts.

Temperature also exerts considerable influence on an ecosystem by its effect on how fast water evaporates. Higher temperatures effectively reduce the amount of available water because more is lost through evaporation. Consequently, the transitions from deserts to grasslands and from grasslands to for-

FIGURE 2–20
Individuals from a population are most abundant where conditions are optimal and less numerous where conditions are less favorable. No individuals are found beyond the limits of tolerance.

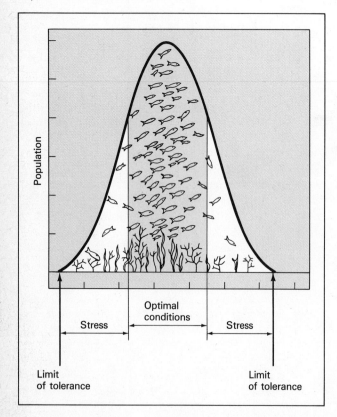

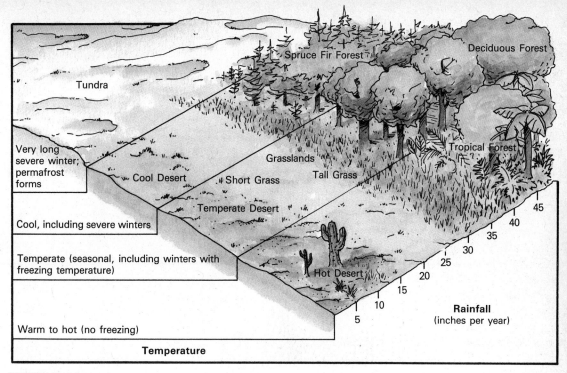

FIGURE 2–21
Climate and major biomes. Moisture is generally the overriding factor determining the type of biome that may be supported in a region. Given adequate moisture, an area will generally support a forest. Temperature, however, determines which *kind* of forest. The situation is similar for grasslands and deserts. At cooler temperatures there is a shift toward less precipitation because lower temperatures reduce evaporative water loss. Temperature becomes the overriding factor only when it is low enough to sustain permafrost.

ests are found at higher precipitation levels in hot regions than in cold regions.

These interactions between temperature and rainfall that determine which biome is to be found where are illustrated in Figure 2–21.

OTHER ABIOTIC FACTORS AND MICROCLIMATE

A specific location may have temperature and moisture conditions that are significantly different from the overall climate for the region, which is necessarily an average. For example, a south-facing slope, which receives more direct sunlight, will be relatively warmer and hence also drier than a north-facing slope. The temperature extremes in a sheltered ravine will be less than those in a more exposed location, and so on. The conditions found in a specific localized area are referred to as the **microclimate** for that location. Different microclimates lead to diversity within a biome.

Soil type or topography or both may also con-

tribute to the diversity found in a biome because these two factors affect the availability of moisture. For example, in the eastern United States, oaks and hickories generally predominate on rocky, sandy soils and hilltops, which retain little moisture, whereas beeches and maples are found on richer soils, which hold more moisture, and red maples and cedars inhabit low, swampy areas. In the transitional region between desert and grassland (10–20 inches of rainfall per year), a soil with good water-holding capacity will support grass, whereas a sandy soil with little ability to hold water will support only desert species (Fig. 2–22).

In certain cases, an abiotic factor other than rainfall or temperature may be the primary limiting factor. For example, the strip of land adjacent to a coast frequently receives a salty spray from the ocean, a factor that relatively few plants can tolerate. Consequently this strip is frequently occupied by a community of salt-tolerant plants (Fig. 2–23). Relative acidity or alkalinity (pH) may also have an overriding effect on a plant or animal community. This fact is particularly significant in view of acid rain.

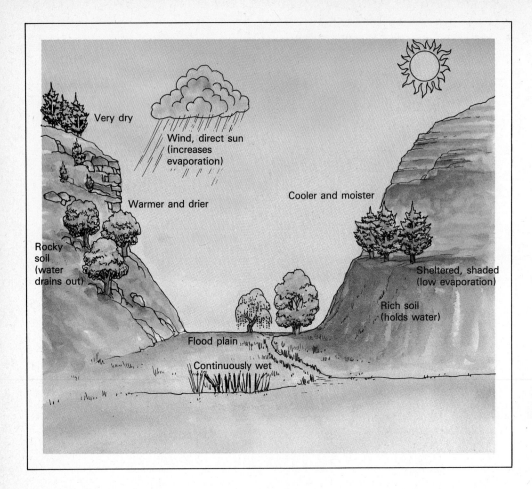

FIGURE 2–22
Abiotic factors such as terrain, wind, and soil type create different microclimates by influencing temperature and moisture in localized areas.

BIOTIC FACTORS

Limiting factors may also be biotic, that is, other species. Grasses thrive when rainfall is more than 30 inches (75 cm). However, when the rainfall is great enough to support trees, grasses get shaded out. Thus, the factor that limits grasses from taking over high-rainfall regions is biotic: overwhelming competition from taller species. Distribution of plants may also be limited by the presence of certain herbivores, particularly insect species and parasitic fungi.

The concept of limiting factors also applies to animals. As with plants, the limiting factor may be abiotic—cold or lack of open water, for example—but it is more frequently biotic in the form of a lack of plant species to provide suitable food or habitat or both.

PHYSICAL BARRIERS

A final factor that may limit species to a particular region is the existence of a physical barrier, such as an ocean, desert, or mountain range, that the species are unable to cross. Thus species making up the eco-

systems on separate continents or remote islands are quite different despite similarity of climates. When such barriers are overcome—for example, by humans transporting a species from one continent to another—the introduced species may make a successful "invasion." However, a successful invasion by a foreign species may cause an ecological disaster because the success of the invader often leads to the displacement of existing species through competition. Examples of such problems with imported species will be explored in Chapter 4.

In summary, the biosphere consists of a great variety of environments, both aquatic and terrestrial. In each environment we find plant, animal, and microbial species that are adapted to all the abiotic factors and also to each other in various feeding and nonfeeding relationships. Each environment supports a more-or-less unique grouping of organisms interacting with each other and with the environment in a way that perpetuates or sustains the entire group. That is, each environment with the species it supports is an ecosystem. Every ecosystem is interconnected with others through ecotones and through some species that cross from one system to another. At the same time, each species and, as a result, each eco-

FIGURE 2–23
Specific abiotic factors may determine the type of vegetation found in a given location. Along the Pacific Coast in central California, usual vegetation is precluded by salt spray, but certain succulent species that are salt-tolerant thrive. (Photographs by BJN.)

Recognizing that everything in the biosphere is interconnected leads to the conclusion that nothing can be changed without affecting everything else to a greater or lesser degree. Obviously, humans have changed and are continuing to change things on a very large scale. What will these changes bring? Many environmentalists, including a number of scientists, promote the idea that we may be on the verge of a sudden and catastrophic "collapse of the biosphere" in which most if not all life may perish.

On the other hand, Cornucopians, also including a number of scientists, point out that there is no solid evidence to support such a "doomsday scenario," much less that it is imminent. Indeed, they argue that all our experience to date should lead to the quite opposite conclusion: Humans have already caused the extinction of thousands of species and made manifold changes over most of the earth, but we and the biosphere are still doing very well. Therefore, they maintain that continuing development as we have been does not pose an environmental threat.

Who is right? You can debate this question, but in the final analysis, does it really matter? Even if collapse of the biosphere is not imminent, does this provide moral justification to continue exploitation of species and destruction of ecosystems? Is imminent death the only justification for changing behavior? Are there not moral and ethical grounds at this stage in human history for working toward creating a society in which humans and all other parts of the natural world can prosper?

---

system, is kept within certain bounds by limiting factors. That is, the spread of each species is at some point limited by its not being able to tolerate particular conditions, compete with other species, or cross some physical barrier. Significantly, nowhere in nature do we find a species restraining its own spread and influence in and of its own volition; restraint is always due to one or more limiting factors.

Do you see how altering any factor, abiotic or biotic, may upset these limits and have far-reaching consequences through a ripple effect? This is a major concern of many environmentalists and is explored in the Ethics Box above and in the text below.

Implications for Humans

How do humans fit into this global picture we have just sketched, and what does our understanding of ecosystems, biomes, and the biosphere tell us about sustaining our society?

Evidence gained through archaeology and anthropology suggests that human ancestry goes back about 3 million years. From that time until about 10 000 years ago (a span of 2,990,000 years), primitive humans survived in small tribes as hunter-gatherers, which means that they "lived off the land," catching wildlife and gathering seeds, nuts, roots, berries, and other plant foods (Fig. 2–24). Settlements were never large and were of relatively short duration because, as one area was "picked over," the tribe was forced to move on. As hunter-gatherers, humans were much like other omnivorous consumers in natural ecosystems. Populations could not expand beyond what natural food sources supported, and deaths from predators, disease, and famine were no doubt common.

About 10 000 years ago, however, a very significant change occurred: Humans began to develop agriculture. Animal husbandry and agriculture are processes of taking particular animal and plant species out of the wild, clearing space, and providing other conditions to grow them preferentially. Plants are protected from competition (weeds) and other would-be consumers, and additional nutrients (fertilizer) and water may be provided. Animals are protected from predators and given food for optimal growth. Over the years, nearly all agricultural plants and animals have been modified greatly through selective breeding so that now they are quite different from their wild ancestors, but the basic process remains unchanged.

The development of agriculture provided a more abundant and reliable food supply, but it created a turning point in human history for other reasons as well. Conducting agriculture not only allows but *requires* permanent (or at least long-term) settlements and the specialization of labor. Some members of the settlement specialize in tending crops and producing food, freeing others to specialize in other endeavors. With this specialization of labor in permanent settlements, there is more incentive and potential for technological development: better tools, better dwellings, better means of transporting water and other materials. Trade with other settlements begins, and thus commerce is born. Also, living in settlements enables

FIGURE 2-24
Before the advent of agriculture about 10 000 years ago, humans subsisted by gathering seeds, nuts, berries, and edible roots from wild plants and on whatever wildlife they could catch. (The Bettmann Archive.)

better care and protection for everyone, and thus the number of deaths is reduced. This reduced mortality rate, coupled with more reliable food production, supports population growth, which in turn, is supported by expanding agriculture. In short, civilization had its origins in the advent of agriculture about 10 000 years ago.

To be sure, this is a grossly simplified picture. Nevertheless, starting with the advent of agriculture and continuing today, we can see a trend of a growing population creating bigger and bigger settlements (cities) supported by ever-expanding agricultural production, with nature being pushed aside with little regard for the consequences. The pace of this expansion has picked up in the last 200 years, since the birth of the Industrial Revolution in the 1700s, with its development of machines and use of fossil fuel energy to perform labor. And, in the last few decades, the pace of growth has accelerated still more. The human population is now growing at a rate of over 90 million people per year, and some 27 million acres (11 million hectares) per year of forests, wetlands, and other natural areas are being converted to agriculture and other kinds of development or otherwise exploited.

Some historians have suggested that the predominant world view changed with the advent of agriculture. Before agriculture and animal husbandry, nature was seen as the provider of all things: food, skins for clothing, materials for building, and so on. With the advent of agriculture, nature became the enemy. First, nature must be cleared away to make room for crops and pastures. Then crops and animals must be protected from wild animals, weeds, and "pests." In essence, this view states that "nature must be overcome for the advance of civilization."

Whether or not a change in world view was involved, the fact remains that the development of agriculture and technology allowed humans to break away from the limiting factors that keep other species and ecosystems within certain bounds. The multitude of ways we humans have devised to overcome these limits is self-evident.

However, does our seeming ingenuity mean that we have permanently overcome all limits and that our "human system" is sustainable? (We shall use the term **human system** to refer to our total system including animal husbandry, agriculture, technology, and supporting activities.) We have seen in this chapter that all life is sustainable only within the context of an ecosystem. Have we created an artificial ecosystem for ourselves that is independent of natural ecosystems and the biosphere?

In some respects, our human system is similar to an ecosystem. It has producers (our crop plants) and herbivores (cattle and poultry), and we ourselves are omnivores. However, it is conspicuously lacking in the detritus feeder/decomposer category: Our system is notably deficient in degrading wastes and recycling the component "nutrients." Consequently various wastes are accumulating and causing numerous pollution problems (Fig. 2-25). Our system also currently depends on the exploitation of fossil fuels and other materials that are finite in quantity (Fig. 2-26).

On these two criteria alone—accumulating wastes and declining fuel supplies—you can see that the sustainability of our system is in doubt unless we make significant changes. However, thinking that we are independent of natural ecosystems and the biosphere is even more erroneous. First, our use of and dependence on natural fisheries, forests, and grass-

FIGURE 2–25
Littered beaches are but one sign of the shortcomings of the human system as an ecosystem. Our human system fails to degrade and recycle wastes like natural systems do. (Tony Suarez/JB Pictures.)

FIGURE 2–26
An open pit copper mine near Bisbee, Arizona, now abandoned because most of the copper ore has been mined out. Our human system depends on continual exploitation of energy and mineral resources. When those are exhausted, what will be left? (Frederick Myers/Tony Stone Worldwide.)

ETHICS

Can Ecosystems Be Restored?

The human capacity for destroying ecosystems is self-evident. To some degree, however, we also have the capacity to restore them. In many cases, restoration involves simply stopping the abuse. For example, it has been found that water quality improves, and fish and shellfish gradually return to previously polluted lakes, rivers, and bays, after the input of pollutants is curtailed. Similarly, forests may gradually return to areas that have been cleared. The limiting factor for such recovery is the availability of surviving populations of the various plants and animals involved.

If at least patches of the original ecosystem containing viable populations remain, these populations may gradually disseminate into the dis-turbed area and return it more-or-less to the original state. Humans can speed up this process by seeding, planting seedling trees, and reintroducing populations of animals that have been eliminated from the area or body of water. For example, wolves are being introduced into Minnesota. A number of organizations such as the World Wildlife Fund, Conservation International, and the Nature Conservancy as well as certain governmental agencies, are engaged in such efforts.

Of course, even with the help of humans the limiting factor is still the availability of surviving populations. If these populations become extinct, there is no way they can ever be brought back. This is one of the great concerns of our times. Ecosys-tem destruction is so rampant that many species are being lost, thus precluding even the possibility of recovery. Hence there is a great cry for conservation and protection to at least preserve the potential for ecological recovery. Still, the forces for protection seem meager in the face of those of exploitation.

What are our values as we seem content to let the forces of extinction go forward? Can such values be analyzed in terms of allowing the self-interests of a few to supersede the interests of society as a whole? As members of society, do we agree that this is right, or do we have a moral right, even an obligation, to stand up for other values?

lands is conspicuous. Second, even as we get increasing proportions of our food and materials from agri- and other forms of controlled culture, such culture remains dependent on the periodic infusion of genes from wild species to maintain its vigor. Many scientists and agricultural experts doubt that we can maintain a viable agricultural system without the backup from nature. (We shall explore the reasons for this more fully in Chapter 5.)

Third, and finally, we now know that all ecosystems are integrated to make the biosphere and that the biosphere, by maintaining predictable climates, atmospheric composition, and so on, supports all ecosystems. The human system cannot divorce itself from these interactions. As we begin to affect the natural world to the extent that basic parameters of the biosphere are altered—for instance, global warming and depletion of the ozone shield to name just two—we will not affect just one ecosystem in a limited area. Instead our meddling will inevitably alter the balances between all species and ecosystems on earth. We cannot predict what all these changes may be much less the final outcome as they ramify through the entire biosphere.

Cornucopians argue that, with our technological capability, we shall be able to adapt to any changes that occur: We'll find new energy sources, develop substitutes for natural products, and whatever else is necessary to maintain the human system. Environmentalists are admittedly more dubious. However, debating whether or not we can find or produce adequate substitutes to survive the loss of natural ecosystems and whether or not we can adapt to the changes in the biosphere completely misses the most important point of all: What kind of world do we want to create for ourselves?

Is there any real advantage in pursuing a trend toward more and more people and development, and less and less nature? Is a purely artificial world, assuming one is possible, what we want to create for ourselves and future generations? Would such a world offer better health, more happiness, more recreational or job opportunities, more conveniences, less stress, or better interpersonal relationships? Or could these ideals be found to an equally high or higher degree in a world with a stable (nongrowing) population that has learned to live in balance with nature rather than destroying it?

In Chapter 3, we shall explore how ecosystems work on the more basic levels of chemistry and energy. This background will provide greater insight into the underlying principles that enable natural ecosystems to be sustainable, and these are the same principles that must be incorporated into our human system if we are to achieve sustainability.

 Review Questions _____

1. Define *ecosystem*.
2. In what ways are the biosphere, biomes, ecosystems, ecotones, and plant communities the same, and how are they different?
3. Identify the biotic components of the biome of your region.
4. Identify and describe the abiotic components of the biome of your region.
5. Name the three main categories of organisms that make the biotic structure of every ecosystem, and describe the role that each plays.
6. Give five categories of consumers present in ecosystems and the role that each plays.
7. Give similarities and differences between detritus feeders and decomposers in terms of what they do, how they do it, and the kinds of organisms involved in each category.
8. All organisms can be divided into two major categories. Name and describe the attributes of these two categories.
9. What is the distinction between food chains, food webs, and trophic levels?
10. Describe three nonfeeding relationships that exist among organisms.

11. How is competition among different species of consumers in an ecosystem reduced?
12. Name five abiotic factors that organisms face. What is the effect that any of these abiotic factors has at the optimum, in the zone of stress, at the limit of tolerance, and outside the limit of tolerance?
13. How are the distribution of species and the formation of ecotones determined by limiting factors?
14. What is climate, and how does it affect the distribution of biomes?
15. What is a microclimate, and how does it affect the species diversity within an ecosystem?
16. What things in addition to abiotic factors may act as limiting factors?
17. How did the development of agriculture cause a major turning point in human history?
18. How have humans overcome the limits that tend to keep natural ecosystems within certain bounds?
19. How is the human system similar to and different from natural ecosystems?

1. Select a woods, field, pond, or other natural area near where you live and discuss how it functions as an ecosystem. Organize your discussion by assigning each species present to a biotic category and describing the role the species plays in the system.

2. From local, national, and international news, compile a list of the many ways humans are altering abiotic and biotic factors on local, regional, and global scales. Analyze the ways in which local changes may affect larger levels and the ways in which global changes may affect local levels.

3. Write a scenario of what would happen to an ecosystem or to the human system in the event of one of the following: (a) all producers were killed by increasing ultraviolet radiation resulting from loss of the protective ozone shield; (b) all parasites were eliminated; (c) decomposers/detritus feeders were eliminated. Support all statements with reasons drawn from your understanding of the way ecosystems function.

4. Consider the various kinds of relationships humans have with other species, both natural and domestic. Give examples of relationships that benefit humans but harm the other species; that benefit both humans and the other species; that benefit the other species but harm humans. Give examples in which the relationship may be changing, for instance from exploitation to protection. Discuss the ethical issues involved in changing relationships.

5. Can the human system be modified into a sustainable ecosystem in balance with (preserving) other natural ecosystems while at the same time maintaining the benefits of modern civilization?

Ecosystems:
How They Work

LEARNING OBJECTIVES

When you have finished studying this chapter, you should be able to:

1. List the key elements of living organisms and identify where these elements occur in the environment through describing air, water, and minerals in terms of atoms and molecules.

2. Give examples of various forms of energy; define and contrast matter and energy by giving examples of the interplay between them.

3. Apply the Law of Conservation of Matter and the Laws of Thermodynamics in presenting an overview of the cycle of life in terms of a movement of atoms and energy between the environment and living organisms.

4. Contrast the function of producers and consumers in terms of matter and energy changes occurring in photosynthesis and cell respiration.

5. Contrast the energy and nutritive roles of food and describe the consequences of over- or undersupplies of either.

6. Distinguish between consumers on the one hand and detritus feeders and decomposers on the other in terms of the ability of the two groups to digest different organic products.

7. Diagram the pathways of carbon, phosphorus, and nitrogen through an ecosystem, naming key compounds in each case and showing how the cycles differ.

8. Diagram the movement of energy through an ecosystem and relate this movement to decreasing biomass at higher trophic levels.

9. Describe two factors in addition to cell respiration that contribute to there being decreasing amounts of biomass at higher trophic levels.

10. State three principles of ecosystem sustainability.

11. Describe and evaluate the human system in terms of how we are adhering or not adhering to the three principles named in objective 10.

In this chapter we explore how ecosystems work at the fundamental level of chemicals and energy. Our look at this basic level will reveal underlying principles that enable natural ecosystems to be sustainable, and it will provide insight into the pathways we must take to make our human system sustainable. Also, understanding at this level will provide a background for understanding agricultural problems, pollution, global warming, and other issues covered in this text.

Elements, Life, Organization, and Energy

The basic building blocks for all **matter** (all gases, liquids, and solids in both living and nonliving systems) are **atoms**. Only 96 different kinds of atoms occur in nature, and these are known as the 96 naturally occurring **elements** (carbon, hydrogen, oxygen, iron, and so on). How can the innumerable different things that make up our world, including ourselves, be made of just 96 elements? The concept is analogous to Lego® blocks: from a small number of basic kinds of blocks, we can build innumerable different things.

Also like "Lego" blocks, nature's materials can be taken apart into the separate atoms, and the atoms can then be reassembled into different materials. All chemicals reactions, whether they occur in a test tube, in nature, or in living things and whether they occur very slowly or very fast (explosions), involve rearrangement of atoms to form different kinds of matter. Like Lego blocks, atoms do not change in the process of disassembling and reassembling different materials. A carbon atom, for instance, will always remain a carbon atom. Furthermore, atoms are not created or destroyed in any chemical reactions. (Recall from Chapter 1 that this is one of the underlying laws of nature, The Law of Conservation of Matter, revealed from innumerable observations and tests.)

On the chemical level, then, the cycle of growth, reproduction, death, and decay of organisms can be seen as a continuous process of taking various atoms from the environment, assembling them into living organisms (growth) and then disassembling them (decay) and repeating the process. This cycle can be visualized in the same way you might visualize Lego blocks being assembled and disassembled to create a succession of forms in a perpetual cycle. Of course, in nature there is no one visible doing the assembling and disassembling. It occurs according to the chemical nature of the atoms themselves and to energy.

The simplicity of the concept need not diminish the wonder of it.

What atoms are involved in living organisms? Where are they found in the environment? How do they get into and make up living organisms? These are the questions we shall now proceed to answer.

ORGANIZATION OF ELEMENTS IN LIVING AND NONLIVING SYSTEMS

A more detailed discussion of the nature of atoms—how they differ from one another, how they bond to form various gases, liquids, and solids, and how we use chemical formulas to describe different chemicals—is given in Appendix C. (p. 589) Studying that appendix first may give you a better comprehension of the material we are about to cover. At the very least, two definitions are essential: molecule and compound. If you are unfamiliar with these two words, look them up before proceeding.

The key elements in living systems and their chemical symbols are: carbon (C), hydrogen (H), oxygen (O), nitrogen (N), phosphorus (P), and sulfur (S). You can remember them by the acronym N. CHOPS. These six elements are the building blocks of all the organic molecules that make up the tissues of plants, animals, and microbes. We shall focus on just these six for now; they as well as other elements found in living systems are listed in Table 3–1. By looking at the chemical nature of air, water, and minerals, we shall see where our six key elements as well as others occur in the environment.

Air of the lower atmosphere is a mixture of molecules of three important gases—oxygen (O_2), nitrogen (N_2), and carbon dioxide (CO_2)—along with trace amounts of several other gases that have no immediate biological importance. Also generally present in air are variable amounts of polluting materials and water vapor. The three main gases found in air are shown in Figure 3–1. Note three of our key elements among these molecules. Thus, air is a source of carbon, oxygen, and nitrogen for all organisms.

TABLE 3–1

Elements Found in Living Organisms and Their Biologically Important Locations in the Environment

Element (Kind of Atom)	Symbol	Biologically Important Molecule or Ion in Which the Element Occurs[a] Name	Formula	Location in the Environment Air	Dissolved in Water	Some Rock and Soil Minerals
Carbon	C	Carbon dioxide	CO_2	X	X	
Hydrogen	H	Water	H_2O		(Water itself)	
Atomic oxygen (required in respiration)	O	Oxygen gas	O_2	X	X	
Molecular oxygen (released in photosynthesis)	O_2	Water	H_2O		(Water itself)	
Nitrogen	N	Nitrogen gas	N_2	X	X	
		Ammonium ion	NH_4^+		X	X
		Nitrate ion	NO_3^-		X	X
Sulfur	S	Sulfate ion	SO_4^{2-}		X	X
Phosphorus	P	Phosphate ion	PO_4^{3-}		X	X
Potassium	K	Potassium ion	K^+		X	X
Calcium	Ca	Calcium ion	Ca^{2+}		X	X
Magnesium	Mg	Magnesium ion	Mg^{2+}		X	X
Trace Elements[b]						
Iron	Fe	Iron ion	Fe^{2+}, Fe^{3+}		X	X
Manganese	Mn	Manganese ion	Mn^{2+}		X	X
Boron	Bo	Boron ion	Bo^{2+}		X	X
Zinc	Zn	Zinc ion	Zn^{2+}		X	X
Copper	Cu	Copper ion	Cu^{2+}		X	X
Molybdenum	Mo	Molybdenum ion	Mo^{2+}		X	X
Chlorine	Cl	Chloride ion	Cl^-		X	X

NOTE: These elements are found in *all* living organisms, plants, animals, and microbes. Some organisms require certain elements in addition to these. For example humans additionally require sodium and iodine.

[a] A molecule is a chemical unit of two or more atoms bonded together. An ion is a single atom or group of bonded atoms that has acquired a positive or negative charge as indicated.

[b] Elements of which only small or trace amounts are required.

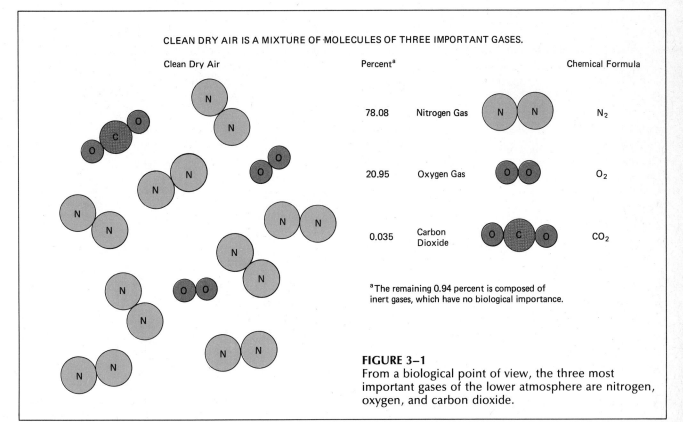

CLEAN DRY AIR IS A MIXTURE OF MOLECULES OF THREE IMPORTANT GASES.

Clean Dry Air

Percent[a]		Chemical Formula
78.08	Nitrogen Gas	N_2
20.95	Oxygen Gas	O_2
0.035	Carbon Dioxide	CO_2

[a] The remaining 0.94 percent is composed of inert gases, which have no biological importance.

FIGURE 3–1
From a biological point of view, the three most important gases of the lower atmosphere are nitrogen, oxygen, and carbon dioxide.

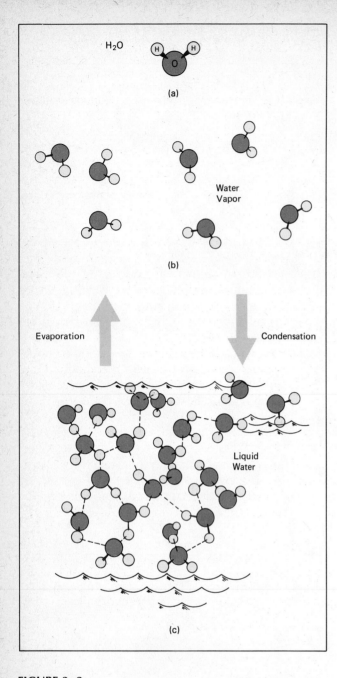

H₂O

(a)

Water
Vapor

(b)

Evaporation   Condensation

Liquid
Water

(c)

The source of the key element hydrogen is water. Each molecule of water consists of two hydrogen atoms bonded to an oxygen atom, as indicated by its formula, H_2O. There is a weak attraction between water molecules known as *hydrogen bonding*. This results in water being a solid (ice) at low temperatures, where hydrogen bonds hold the molecules together, a liquid at moderate temperatures, and a vapor at higher temperatures, where increased energy breaks the hydrogen bonds. Regardless of state, however, water molecules remain H_2O (Fig. 3–2).

All the other elements required by living organisms, as well as the 76 or so elements that are not required, are found in greater or lesser amounts in various rock and soil minerals. A **mineral** refers to any hard, crystalline material of a given chemical composition. Most rocks are made up of relatively small crystals of two or more minerals, and soil generally consists of particles of many different minerals. Each mineral is made up of dense clusters of two or more kinds of atoms bonded together by an attraction between positive and negative charges on the atoms (Fig. 3–3).

There are simple but significant interactions between air, water, and minerals. Gases from the air and ions (charged atoms) from minerals may dissolve in water. Therefore, natural water is usually a **solution** containing variable amounts of dissolved gases and minerals. This solution is constantly subject to change as any dissolved substances may be removed from water solution by various processes or additional materials may dissolve. Molecules of water enter the air by evaporation and come out through condensation (see the water cycle, p. 240). Thus the amount of moisture in air is constantly fluctuating. Wind may carry a certain amount of dust or mineral particles, and this amount is also changing constantly, since the particles gradually settle out from

FIGURE 3–2
(a) Water consists of molecules, each of which is formed by two hydrogen atoms bonded to an oxygen atom, H_2O. (b) In water vapor, molecules are separate and independent. (c) In liquid water, a weak attraction between water molecules, known as hydrogen bonding indicated by dotted lines, gives the liquid property. At freezing temperatures, hydrogen bonding holds molecules firmly giving the solid state—ice.

Saying that air is a **mixture** means that there is no chemical bonding between the molecules involved. Indeed, it is this lack of connection between molecules that results in air being gaseous. Attractions, or bonding, between molecules results in liquid or solid states.

FIGURE 3–3
Minerals, hard crystalline compounds, are composed of dense clusters of atoms of two or more elements. The atoms of most elements gain or lose one or more electrons to become negative (−) or positive (+) ions. The charged atoms, or ions, are held together by the attraction between positive and negative charges.

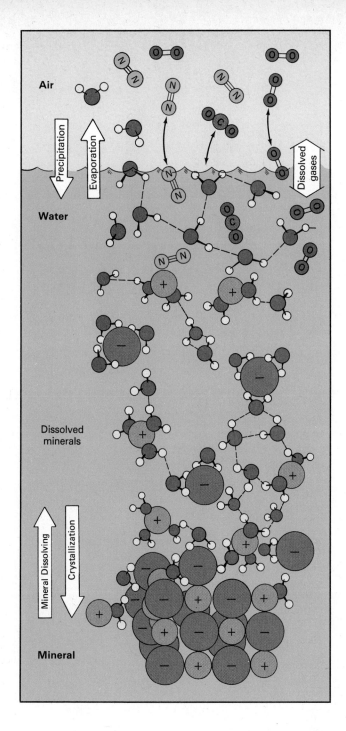

Air

Precipitation

Evaporation

Water

Dissolved gases

Dissolved minerals

Mineral Dissolving

Crystallization

Mineral

FIGURE 3–4
Interrelationships between air, water, and minerals. Minerals and gases dissolve in water, forming solutions. Water evaporates into air, causing humidity. These processes are all reversible: as minerals in solution recrystallize, and water vapor in the air condenses to re-form liquid water.

living things are constructed in large part from carbon atoms bonded together into chains with hydrogen atoms attached. Oxygen, nitrogen, phosphorus, and sulfur are involved also, but the key common denominator is carbon-carbon and/or carbon-hydrogen bonds. Recall (p. 23) that material making up the tissues of living organisms is referred to as *organic*. Hence, these *carbon-based molecules, which make up the tissues of living organisms*, are called **organic molecules**. (Don't miss the similarity between the words *organic* and *organism*.) **Inorganic**, then, refers to molecules or compounds with neither carbon-carbon nor carbon-hydrogen bonds.

Causing some confusion is the fact that all plastics and countless other human-made compounds are based on carbon-carbon bonding and are, chemically speaking, organic compounds—although they have nothing to do with living systems. In places where there is doubt, we resolve this confusion by referring to the compounds of living organisms as **natural organic compounds** and the human-made ones as **synthetic organic compounds**.

In conclusion, we can see that the elements essential to life (C, H, O, and so on) are present in the

FIGURE 3–5
The organic molecules making up living organisms are larger and more complex than the inorganic molecules found in the environment. Glucose and cystine show this relative complexity. (Do *not* memorize these formulas; they are here just to give you a sense of the complexity we are describing.)

$$\begin{array}{ccccccccccccc} & O & & OH & & H & & OH & & OH & & H & \\ & \parallel & & | & & | & & | & & | & & | & \\ H - & C & - & C & - & C & - & C & - & C & - & C & - H \\ & | & & | & & | & & | & & | & & | & \\ & H & & OH & & H & & H & & OH & \end{array}$$

Glucose, a sugar

$$\begin{array}{ccccccc} & H & & H & & O & \\ & | & & | & & \parallel & \\ HS - & C & - & C & - & C & - OH \\ & | & & | & & & \\ & H & & NH_2 & & & \end{array}$$

Cystine, an amino acid used to make proteins

the air. These interactions are summarized in Figure 3–4.

By contrast to the relatively simple molecules that occur in the environment (for example, CO_2, H_2O, N_2), in living organisms we find the key atoms (C, H, O, N, P, S) bonded into very large, complex molecules known as proteins, carbohydrates, lipids, and nucleic acids (Fig. 3–5). Some of these molecules may contain millions of atoms, and their potential diversity is infinite. Indeed, the diversity of living things is a function of the diversity of such molecules.

All the molecules that make up the tissues of

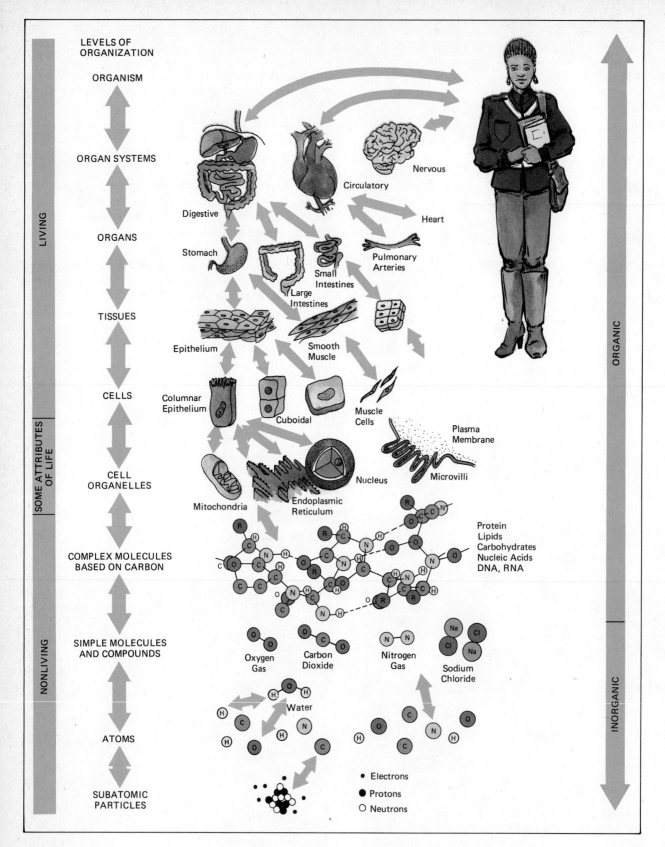

FIGURE 3-6
Life can be seen as a hierarchy of organization of matter. In the inorganic sphere, elements are in very simple arrangements of molecules of the air, water, and minerals. In living organisms, they are arranged in very complex organic molecules, which, in turn, make up cells that make up the whole organism.

environment in relatively simple molecules in air, water, or minerals. In living *organisms*, on the other hand, they are *organized* into very complex *organic* molecules. In turn, these organic compounds make up the various parts of cells, which make up the tissues and organs of the body (Fig. 3–6). Growth, then, may be seen as taking the atoms from the simple molecules that occur in the environment, and using them to construct the complex organic molecules of the organism. Decomposition and decay may be seen as the reverse. We shall look at these processes in more detail later in this chapter. First, however, we must consider another factor, *energy*.

CONSIDERATIONS OF ENERGY

In addition to the rearrangement of atoms, chemical reactions also involve the absorption or release of energy. To grasp this concept, we must first be clear concerning the distinction between matter and energy.

Matter and Energy

The universe is made up of just two things: *matter* and *energy*. A definition of **matter** that is more technical than the one we learned earlier in this chapter is *anything that occupies space and has mass*, that is, can be weighed when gravity is present. This definition obviously covers all solids, liquids, and gases and living as well as nonliving things.

Atoms are made up of protons, neutrons, and electrons, and even these are made of yet smaller particles. Thus, there is debate among physicists as to what the most basic unit of matter is. However, since atoms are the basic units of all elements and since atoms do not change in chemical reactions, it is practical to consider them as the basic units of matter.

Common forms of energy, on the other hand, include *light* and other forms of *radiation, heat, movement,* and *electric power*. Note that, in contrast to matter, forms of energy do not occupy space, nor do they have mass. If you wish to argue this point, just put some light in a container, take it into a dark room, and weigh it. But light, heat, and other forms of energy are real. They *do* things; they *affect* matter, causing changes in either its *position* or its *state*. A release of energy in an explosion causes things to go flying, a change in position. Heating water causes it to boil and change to steam, a change in state. Thus, *energy* is defined in terms of what it does. Physicists define the "doing" as *work*. Thus, **energy** is defined as the *ability to do work*.

FIGURE 3–7
Energy is distinct from matter in that energy neither has mass nor occupies space. It has the ability to act on matter, changing the position of the matter and/or its state. Kinetic energy is energy in one of its active forms. Potential energy refers to systems or materials that have the potential to release kinetic energy.

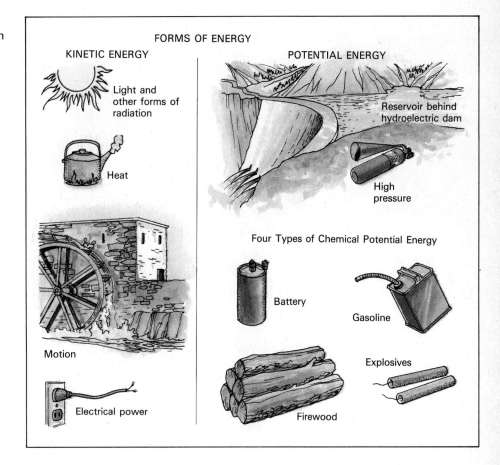

FORMS OF ENERGY

KINETIC ENERGY

Light and other forms of radiation

Heat

Motion

Electrical power

POTENTIAL ENERGY

Reservoir behind hydroelectric dam

High pressure

Four Types of Chemical Potential Energy

Battery

Gasoline

Explosives

Firewood

Energy is commonly divided into two categories: *kinetic* and *potential* (Fig. 3–7). **Kinetic energy** is *energy in action or motion*. Light, heat, motion, and electrical current are all forms of kinetic energy. **Potential energy** is energy in storage. A substance or system with potential energy has the capacity, or *potential*, to release one or more forms of kinetic energy. A stretched rubberband, for example, has potential energy; it can send a paper clip flying. Numerous chemicals, such as gasoline and other fuels, release kinetic energy—heat, light, and movement—when ignited. The potential energy contained in such chemicals and fuels is called **chemical energy**.

There are innumerable ways in which energy may be changed from one form to another. How many examples can you think of in addition to those shown in Figure 3–8? In addition to seeing that potential energy may be converted to kinetic energy, it is especially important to recognize that kinetic energy may be converted to potential energy, as in charging a battery, or pumping water into a high-elevation reservoir. We shall see shortly that photosynthesis is another such process.

Since energy does not have mass or occupy space, it cannot be measured in units of weight or volume, but it can be measured in other kinds of units. One of the most common is the **calorie**, which is defined as the *amount of heat required to raise the temperature of 1 gram of water 1 degree Celsius*. But, since this is a very small unit, it is frequently more convenient to speak in terms of kilocalories (1 kilocalorie = 1000 calories), the amount of heat required to raise one liter of water 1 degree Celsius. Food calories (often capitalized as Calories to distinguish them from ''small'' calories) are actually kilocalories. Any form of energy can be measured in calories by converting it to heat and measuring that heat in terms of a rise in water temperature.

We have said that energy affects matter by changing its position or state. Conversely, no change in position or state of matter can or will occur without absorption or release of energy. Every change in matter—from a few atoms joining or coming apart in a chemical reaction to a major earthquake—involves energy. Changes in matter cannot be separated from respective changes in energy.

Energy Laws: Laws of Thermodynamics

Given that energy can be converted from one form to another, numerous inventors over the years have tried to build machines or devices that would produce

FIGURE 3–8
Any form of energy can be converted to any other form, except heat can only be transferred to something cooler. Therefore, the capacity to convert heat is lost as the temperature drops.

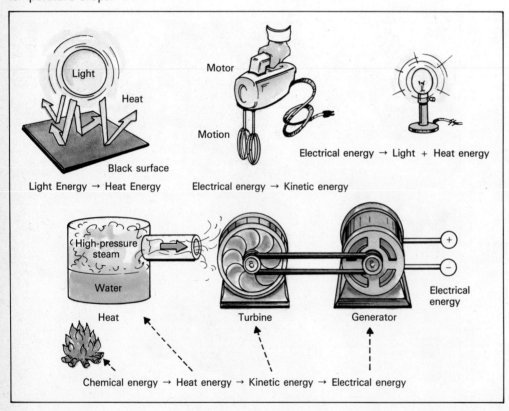

Light Energy → Heat Energy

Electrical energy → Kinetic energy

Electrical energy → Light + Heat energy

Chemical energy → Heat energy → Kinetic energy → Electrical energy

more energy than they consumed. All these devices had one feature in common—they didn't work. What is found when all the inputs and outputs of energy are carefully measured is that they are always *equal*. There is no net gain or loss in total energy. This is now accepted as a fundamental natural law, the **Law of Conservation of Energy**, which is also called the **First Law of Thermodynamics:** *Energy is neither created nor destroyed but it may be converted from one form to another*. It is also commonly stated as: "You can't get something for nothing."

Finding that energy is not destroyed gives the impression that a machine might be built that would run indefinitely by recycling energy, a concept that is commonly referred to as a perpetual motion machine. But countless attempts to build perpetual motion machines have the same results as attempts to build "energy generators"; they don't work. They fail because of two facts that are always observed. First, in every energy conversion, a portion of the energy is converted to heat, and second, heat always flows toward cooler surroundings (Fig. 3–9). A hot object always cools off, losing its heat to the cooler surroundings, never the reverse. Therefore, there is no way of trapping and recycling heat energy since it can flow only "downhill" toward a cooler place. Consequently, in the absence of energy inputs, any and every system will sooner or later come to a stop as

its energy is converted to heat and lost. This is now accepted as another natural law, the *Second Law of Thermodynamics*. Basically it says that in any energy conversion, you will end up with *less* usable energy than you started with. So not only can you not get something for nothing (the first law), you can't even break even.

There is a principle that underlies the loss of heat. It is the principle of increasing *entropy*. **Entropy** refers to the degree of *disorder*: *increasing entropy means increasing disorder*. The principle is that, without energy inputs, everything goes in one direction only and that direction is toward increasing entropy. This principle of ever-increasing entropy is most readily apparent in that all human-made things lacking maintenance tend to deteriorate. We never observe the reverse—a rundown building renovating itself, for example. Students often like to speak of the increasing disorder of their dormitory rooms as the semester wears on as an example of entropy.

The conversion of energy to heat and the loss of heat noted above are both aspects of increasing entropy. Heat is the result of random vibrational motion of atoms and molecules. Thus, it is the lowest (most disordered) form of energy, and its flow to cooler surroundings is a way for that disorder to spread out through the system. Therefore, the **Second Law of Thermodynamics** is nowadays more gen-

FIGURE 3–9
The Second Law of Thermodynamics. Energy cannot be recycled because every conversion involves a loss of heat that cannot be recovered because heat flows only toward cooler surroundings. Consequently, the power output of a generator, for example, will always be less than the power input because of the loss of energy as heat.

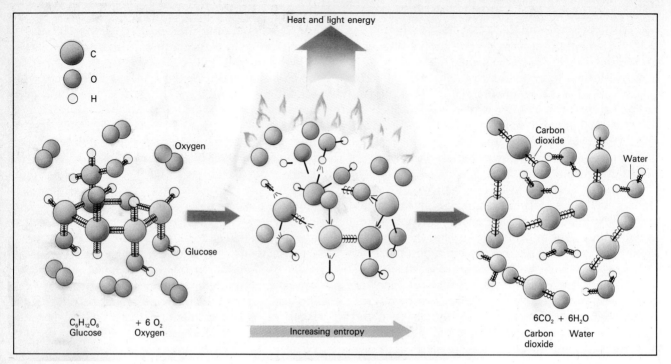

FIGURE 3–10
Systems go spontaneously only in the direction of increasing entropy. When glucose, the building-block molecule of wood, is burned, heat is released and the atoms become more and more disordered, both aspects of increasing entropy. The fact that wood will burn but not form spontaneously is an example of the Second Law of Thermodynamics.

erally stated as: *systems will go spontaneously in one direction only and that direction is toward increasing entropy.* This is also saying that systems will go spontaneously only toward lower potential energy, a direction that releases heat from the system. In practical terms, water always flows downhill, and wood will burn but not form spontaneously (Fig. 3–10).

Very important in the statement of the second law is the word *spontaneously.* It is obviously possible to pump water uphill, charge a battery, or to increase the potential energy of some item in any number of other ways. The point is that going in a direction that increases the potential energy of one item always requires an input of energy from something else. Effectively, energy taken from the "something else" is showing up as the gain in potential energy of the item. Careful measurements always show that the amount of energy lost from the "something else" is always greater than the amount that shows up in the item gaining energy. The "missing" energy is that energy lost as heat, or increasing entropy in accord with the second law.

In conclusion, the practical knowledge you should learn from these energy laws is simply the understanding that: Yes, energy can be converted from one form to another, but wherever the conversion involves an item gaining potential energy, energy must be fed into the item; first law: you can't get something for nothing. Further, the amount of energy "spent" will always be greater than the potential energy gained; second law: you can't even break even.

Let us now relate these concepts of matter and energy to organic molecules, organisms, ecosystems, and the biosphere.

MATTER AND ENERGY CHANGES IN ORGANISMS AND ECOSYSTEMS

Energy and Organic Matter

All organic molecules, which make up the tissues of living organisms, contain *high potential energy.* This is evident by the simple fact that they burn. The heat and light of the flame are their potential energy being released as kinetic energy. On the other hand, try as you might, you will not be able to get energy by burning of inorganic molecules, such as carbon dioxide, water, and minerals. Indeed these materials are used as fire extinguishers. This extreme nonflammability is evidence that such materials have very *low potential energy.* Thus, the production of organic material from inorganic material involves a *gain* in potential energy. We now know from the laws of thermodynamics that this potential energy gain demands a greater-than-equal input of kinetic energy.

IN PERSPECTIVE
To Stop "Wasting" Energy

The Laws of Thermodynamics show that we can't create energy, nor can we convert it from one form to another without losing some. But how much do we have to lose? Certainly not as much as we often do. Many of our machines, motors, lighting fixtures, heating devices, and so on are very inefficient, often delivering only 5–10 percent of the energy

input as useful output. The rest is lost as heat. A standard light bulb, for example, converts to light only about 5 percent of the electrical energy it receives. The other 95 percent simply goes into the heat that will burn your fingers if you touch the bulb.

The major thrust of energy conservation is not to "freeze in the

dark"; it is to improve efficiency. If it could be doubled, we would get twice the work for the same amount of energy. As 100 percent efficiency is approached, limits will be imposed by the Laws of Thermodynamics. However, there is ample opportunity for improvements before that occurs.

In this simple example of burnable organic molecules and nonburnable inorganic molecules, you may see an overview of the energy dynamics of an

ecosystem. Through producers (green plants), light energy from the sun is used to "pump" elements (namely C, H, O, N, P, and S) from their low potential

FIGURE 3–11
Storage and release of potential energy. (a) A simple physical example of the storage and release of potential energy. With suitable energy input, water can be pumped to a higher elevation, thus capturing a portion of the energy input. A portion of the potential energy can then be harnessed to do useful work by letting the water flow back to low potential energy over a turbine. (b) The same principle applies to ecosystems. Through photosynthesis, light energy "pumps" elements from a low-potential-energy state in inorganic materials to a high-potential-energy state in organic materials. The breakdown of these molecules releases the energy, which then drives all the active functions of organisms.

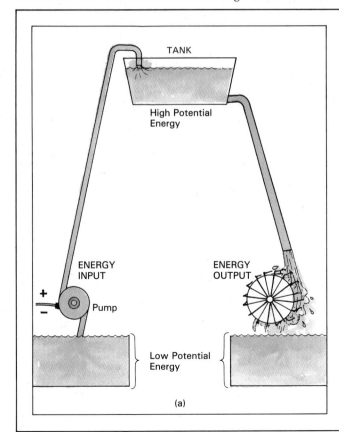
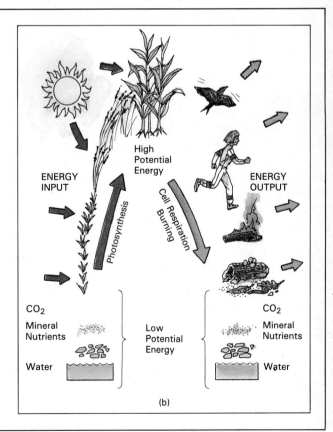

energy states in the environment to a high potential energy state in the organic molecules of the plant. All consumers, detritus feeders, and decomposers, by feeding on organic matter, gain their energy requirements from the reverse process: breaking down the organic molecules (Fig. 3–11). Let's now look at this energy flow in somewhat more detail for each category of organisms.

Producers

Recall from Chapter 2 that producers are green plants, which use light energy in the process of photosynthesis to make sugar (glucose) from carbon dioxide and water and release oxygen gas as a byproduct. The process is expressed by the overall formula:

PHOTOSYNTHESIS

$$6CO_2 + 6H_2O \xrightarrow{\text{light energy}} C_6H_{12}O_6 + 6O_2$$

carbon dioxide (gas)   water   glucose   oxygen (gas)

The kinetic energy of light is absorbed by chlorophyll in the cells of the plant and used to remove the hydrogen atoms from water (H_2O) molecules. The hydrogen atoms are transferred to carbon atoms from carbon dioxide as the carbons are joined in a chain to begin forming a glucose molecule. The oxygen atoms remaining after the removal of hydrogen combine with each other to form oxygen gas, which is released into the air.

Each molecule of glucose is constructed from 6 carbon atoms, 12 hydrogen atoms, and 6 oxygen atoms, hence its formula, $C_6H_{12}O_6$. Thus, construction of one molecule of glucose requires 6 molecules of carbon dioxide to provide the 6 carbon atoms and 6 molecules of water to provide the 12 hydrogen atoms. Among these molecules of carbon dioxide and water, there are 18 oxygen atoms, whereas only 6 are needed. The extra oxygen atoms are given off as molecules of oxygen gas (O_2), 6 for every molecule of glucose formed. This accounting is based on careful quantitative measurements and it supports the Law of Conservation of Matter. Note that oxygen gas, which is essential for the respiration of animals, is a *waste product* of photosynthesis.

The key energy steps in photosynthesis are in removing the hydrogen from water molecules and in joining carbon atoms together to form the high-potential-energy carbon-carbon and carbon-hydrogen bonds of glucose in place of the low-potential-energy bonds in water and carbon dioxide molecules. But the Laws of Thermodynamics are not violated or even strained in this process. Careful measurements show that the rate of photosynthesis (the amount of glucose

formed) is proportional to light intensity and only 2–5 calories worth of sugar is formed for each 100 calories worth of light energy falling on the plant. Thus plants are not particularly efficient "machines" in performing this conversion of light energy to chemical energy.

The glucose produced in photosynthesis plays three roles in the plant. First, either by itself or along with nitrogen, phosphorus, sulfur, and other mineral nutrients absorbed from the soil or water solution, glucose is the "raw material" used in the synthesis of all the other organic molecules (proteins, carbohydrates, and so on) that make up the tissues of the plant body. Second, the synthesis of all these organic molecules requires additional energy, as does absorption of nutrients from the soil and certain other functions. This energy is provided as the plant breaks down a portion of the glucose to release its stored energy in a process called cell respiration, which will be discussed shortly. Finally, a portion of the glucose may be stored for future use in the above two processes. For storage, the glucose is generally converted to starch, as in potatoes, or to oils, as in seeds. These conversions are summarized in Figure 3–12.

Consumers

Energy Role of Food That consumers require energy to move about and to perform such body functions as pumping blood should be self-evident. In addition, consumers also need energy in order to synthesize all the molecules needed in growth, maintenance, and repair of the body. Where does this energy come from? It comes from the breakdown of food molecules and the release of the potential energy locked up in that food. Indeed, in the range of 60–90 percent of food that we or any other consumers eats and digests plays the role of "fuel" to provide energy.

First, the starches, fats, and proteins that you eat are digested in the stomach and/or intestine, which means they are "chopped" into simpler molecules—starches into sugar (glucose), for example. These simpler molecules are then absorbed from the intestine into the bloodstream and transported to individual cells of the body. In each cell, organic molecules may be broken down through a process called *cell respiration* to release the energy required for the "work" done by that cell. Most commonly, **cell respiration** involves the breakdown of glucose, and the overall chemistry is the reverse of that for photosynthesis:

CELL RESPIRATION

$$C_6H_{12}O_6 + 6O_2 \xrightarrow{\Downarrow \atop energy} 6CO_2 + 6H_2O$$

glucose   oxygen   carbon dioxide   water

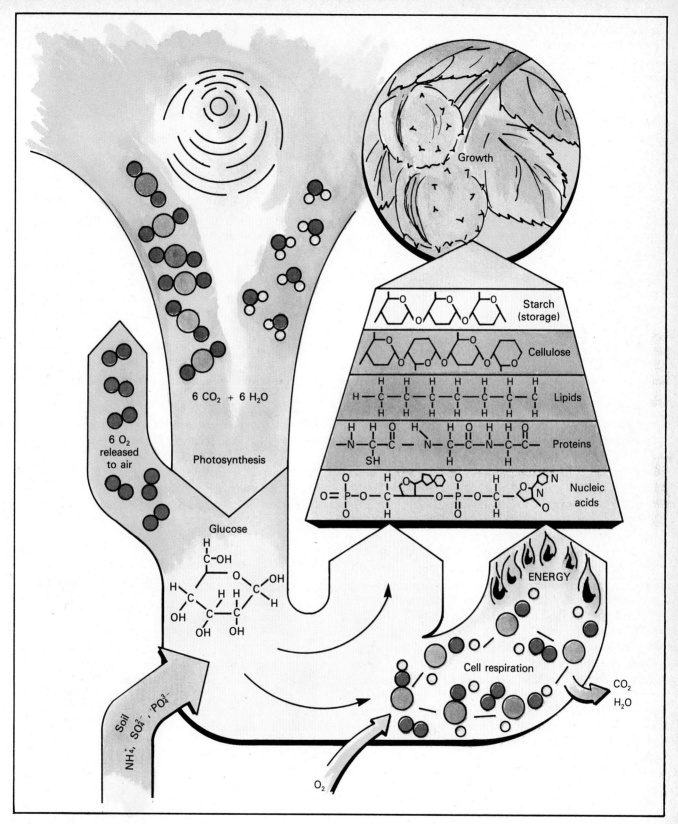

FIGURE 3–12

Producers. Producers are remarkable chemical factories. Using light energy from the sun, they make glucose from carbon dioxide and water, releasing oxygen as a byproduct. Breaking down some of the glucose to provide additional chemical energy, they combine the remaining glucose with certain additional nutrients from the soil to form other complex organic molecules that the plant then uses for growth.

Again, the key point of cell respiration is the release of the potential energy contained in glucose or other organic molecules to perform the activities of the organism. However, other aspects of the chemistry are also significant. Note that whereas oxygen was released in photosynthesis, here it is *required* to complete the breakdown of glucose to carbon dioxide and water. Oxygen is absorbed through the lungs every time we inhale (or gills in the case of fish) and is transported to all cells via the circulatory system. Carbon dioxide, which is formed as a waste product, moves from the cells into the circulatory system and is eliminated through the lungs (or gills) every time we exhale. The other byproduct, water, serves any of the body's needs for water and reduces the need to drink water. A number of desert animals, which are adapted to be very water-conserving, don't need to drink any water because that produced as a result of cell respiration is sufficient. However, the bodies of most animals, including ourselves, are less conserving of water; therefore drinking additional water is necessary.

Again in keeping with the Laws of Thermodynamics, the energy conversions involved in using the potential energy from glucose to do body work are not 100 percent efficient. Considerable waste heat is produced, and this is the source of our body heat. As a result of the waste heat let off during respiration, heat output can even be measured in "cold-blooded" animals and in plants. The heat given off by these two types of organisms is less noticeable than in warm-blooded animals only because the former do not regulate their body temperatures.

The basis of weight gain or loss should be evident here. Only as much organic matter is broken down in cell respiration as is required to meet the energy demands of the body, not one molecule more. Of course this demand changes with your level of exercise and activity. If you consume more food calories than your body needs, the excess is converted to fat and stored, and the result is weight gain. Conversely, the principle of dieting is to eat less and exercise more to create an energy demand that exceeds the amount of energy contained in the food being consumed. This imbalance forces the body to break down its own tissues to make up the difference, and the result is a weight loss. Of course carried to an extreme this imbalance leads to *starvation* and death as the body runs out of anything to break down for its energy needs.

The overall reaction for cell respiration is the same as that for simply burning glucose. Thus it is not uncommon to speak of "burning" our food for energy. Such breakdown of molecules is also called **oxidation**. The distinction between burning and cell respiration is that in cell respiration the oxidation takes place in about 20 small steps so that the energy is released in small "packets" that are suitable for driving the functions of each cell. If all the energy from glucose molecules were released in a single "bang," as occurs in burning, it would be like heating and lighting a room with large firecrackers. Energy, yes, but hardly useful.

We have learned that many organic molecules contain nitrogen, phosphorus, sulfur, and perhaps other elements in addition to the carbon and hydrogen found in all organic molecules. When such molecules are broken down in cell respiration, waste byproducts include compounds of nitrogen, phosphorus, and any other elements present in addition to the carbon dioxide and water. This waste nitrogen, phosphorus, and so on is excreted in the urine (or analogous waste in other kinds of animals). Thus returned to the environment, these elements may be reabsorbed by plants (Fig. 3–13). Here you can begin to see the movement of elements in a cycle between the environment and living organisms. We shall expand on these cycles shortly.

Also, you may begin to visualize a flow of energy that enters as light and exits as heat. Finally, recall the *biomass pyramid* (Fig. 2–14). The amount of biomass inevitably decreases by the amount that is oxidized to provide energy.

Nutritive Role of Food While 60–90 percent of the food that consumers eat, digest, and absorb is oxidized for energy, the other 40–10 percent, which is

FIGURE 3–13
Animal wastes are plant fertilizer. When consumers burn food to obtain energy the waste products are the inorganic nutrients needed by plants. Here dog urine has been deposited on a lawn. The ring of dark green grass is where the urine has been diluted to optimal concentration; the grass in the center has been killed by overfertilization by concentrated urine. (Photograph by BJN.)

converted to the body tissues of the consumer, is equally important. This fraction that is converted to tissue, of course, is the portion that enables growth, maintenance, and repair of the body.

Sugars, carbohydrates (starches), and fats can be oxidized easily by the body to provide energy. Body growth, maintenance, and repair, however, require particular nutrients—namely, the various vitamins, minerals, and protein—that are not present in sugar, carbohydrates, or fats. If any one or more of the specific nutrients is absent from the diet, various "diseases" of **malnutrition** will develop. Herein perhaps you can see the problem of overconsumption of highly processed "junk foods" such as potato chips, sodas, candies, baked goods such as cake and doughnuts, and alcohol. Rich in fat or sugar or both, these items are very high in calories but they contain little or none of the necessary nutrients. Consequently, a diet high in such items may easily oversupply calories causing weight gain while, at the same time, leading to serious disorders of malnutrition.

Material Consumed but Not Digested A portion of what is ingested by consumers is not digested but simply passes through the digestive system and out as fecal wastes. For consumers that eat plants, such waste is largely the material of plant cell walls, **cellulose**. We refer to it as *fiber, bulk,* or *roughage.* Some fiber is a necessary part of the diet in order for the intestines to have something to push through to keep them clean and open.

In summary, organic material (food) ingested by any consumer follows one of three pathways: (1) more than 60 percent of what is digested and absorbed is oxidized to provide energy, and waste by-products are released back to the environment; (2) the remainder of what is digested and absorbed goes into body growth, maintenance, and repair; and (3) the portion that is not digested or absorbed passes out as fecal waste (Fig. 3–14). Of course, in an ecosystem, food that becomes body tissue of a consumer effectively becomes food for the next organism in the food chain, and category 3 becomes food for detritus feeders and/or decomposers.

Detritus Feeders and Decomposers

Recall that detritus is mostly dead leaves, woody plant parts, and animal fecal wastes. As such, it is largely cellulose, which is unusable by most consumers because they can't digest it. Nevertheless, it is still organic and high in potential energy for those organisms that can digest it—namely the decomposers we learned about in Chapter 2: various species of fungi and bacteria and a few other microbes. Beyond

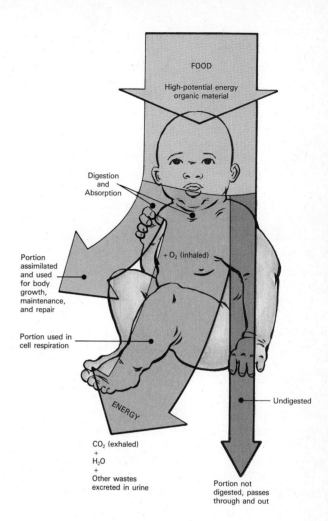

FIGURE 3–14
Consumers. Only a small portion of the food ingested by a consumer is assimilated into body growth, maintenance, and repair. A larger amount is used in cell respiration to provide energy for assimilation, movements, and other functions; waste products are carbon dioxide, water, and various mineral nutrients. A third portion is not digested but instead passes through, becoming fecal waste.

this ability to digest cellulose that is to us indigestible, decomposers act as any other consumer, using the cellulose as a source of both energy and nutrients. Termites and some other detritus feeders can digest woody material by virtue of maintaining decomposing microorganisms in their guts. It is a symbiotic relationship. The termite (detritus feeder), for example, provides a cozy home for the microbes (decomposers) and "shovels" in the cellulose. The microbes digest it for both their own and the termites' benefit.

Most decomposers use the process of cell respiration described above. Thus the detritus is broken down to carbon dioxide, water, and mineral nutrients. Likewise there is a release of waste heat. You may observe a manure pile "steaming" on a cold day.

Some decomposers (certain bacteria and yeasts) can meet their energy needs through the partial oxidation of glucose that can occur in the absence of oxygen. This modified form of cell respiration, which occurs in the absence of oxygen, is called **fermentation** and it results in such end products as alcohol (C_2H_6O), methane gas (CH_4), and vinegar (acetic acid, $C_2H_4O_2$). You may recognize that the commercial production of these compounds is achieved by growing the particular organism on suitable organic matter in a vessel excluding oxygen. In nature **anaerobic**, or *oxygen-free*, environments commonly exist at the bottom of marshes or swamps or buried in the earth where oxygen does not penetrate readily. Methane gas is commonly produced in such locations. Also a number of large grazing animals including cattle maintain fermenting bacteria in their digestive systems and are methane producers as a result. It has been suggested that this may have implications in global warming in that methane is a "greenhouse" gas (see Chapter 16).

For simplicity our orientation in this discussion has been toward terrestrial ecosystems. It is significant to realize that exactly the same processes occur in aquatic ecosystems. Carbon dioxide and mineral nutrients are drawn from water solution by aquatic plants, and the oxygen they produce in photosynthesis dissolves in the water. Aquatic consumers draw on the dissolved oxygen for cell respiration and release carbon dioxide and other nutrients into the water.

Principles of Ecosystem Function

NUTRIENT CYCLING

When you look at the various inputs and outputs of producers, consumers, detritus feeders and decomposers, how they fit together should be conspicuous. The waste, or byproducts, of each is the food or essential nutrients for the other. Specifically, the organic material and oxygen produced by green plants are the food and oxygen required by consumers and other heterotrophs. In turn, the carbon dioxide and other wastes generated when heterotrophs break down their food are exactly the nutrients needed by green plants. Herein is the **first basic principle of ecosystem sustainability**:

For sustainability, ecosystems dispose of wastes and replenish nutrients by recycling all elements.

Do you see how this principle is in harmony with the Law of Conservation of Matter? Since atoms are neither created nor destroyed, nor converted one into another, they can be reused indefinitely. This is exactly what natural ecosystems do; they recycle the same atoms over and over again. We can see this even more clearly by focusing on the pathways of three key elements: carbon, phosphorus, and nitrogen. Since these pathways do lead in a circle, they are known as the *carbon cycle*, the *phosphorus cycle*, and the *nitrogen cycle*.

The Carbon Cycle

For descriptive purposes, it is convenient to start the carbon cycle (Fig. 3–15) with the "reservoir" of carbon dioxide molecules present in the air and dissolved in water. Through photosynthesis and further metabolism, carbon atoms from carbon dioxide become the carbon atoms of all the organic molecules making up the plant's body. Through food chains, the carbon atoms then move into and become part of the tissues of all the other organisms in the ecosystem. However, it is unlikely that a particular carbon atom will be passed through many organisms on any one cycle because at each step there is a considerable chance that the consumer will break down the organic molecule in cell respiration. As this occurs, the carbon atoms are released back to the environment in molecules of carbon dioxide, thus completing one cycle, but of course ready to start another. Likewise, burning organic material returns the carbon atoms locked up in the material to the air in carbon dioxide molecules.

No two successive cycles of a particular carbon atom are likely to be the same. Nor are the two cycles likely to be within the same ecosystem because in the atmosphere wind will carry them around the globe. Some important sidelights on the carbon cycle are given in the In Perspective Boxes, pp. 63 and 65.

The Phosphorus Cycle

The phosphorus cycle is illustrated in Figure 3–16. Phosphorus exists in various rock and soil minerals as the inorganic *phosphate* ion (PO_4^{3-}). As rock gradually breaks down, phosphate and other nutrient ions are released. Phosphate dissolves in water but does not enter air. Plants absorb phosphate from the soil or water solution, and as it is bonded into organic compounds by the plant it is frequently referred to as **organic phosphate**. Through food chains, organic phosphate is transferred from producers to the rest of the ecosystem. As with carbon, at each step there is a high likelihood that the organic phosphate will be broken down in cell respiration, releasing inor-

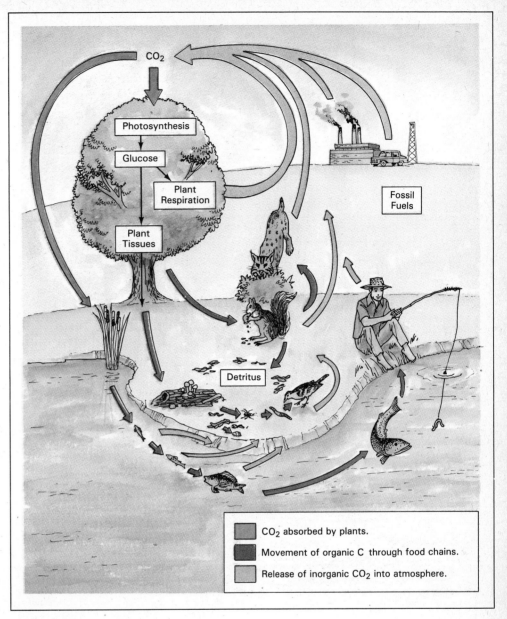

FIGURE 3–15
The carbon cycle. See text for explanation.

ganic phosphate in urine or other waste. The phosphate may then be reabsorbed by plants to start another cycle.

There is an important difference between the carbon cycle and the phosphorus cycle. No matter where carbon dioxide is released, it will mix into and maintain the concentration of carbon dioxide in the air. Mineral nutrients, however, which do not have a gas phase, are recycled only insofar as the wastes that contain them are deposited on the soil *from which the nutrients originally came*. This is basically what happens in a natural ecosystem. However, humans are prone to upset this cycle. A very serious case of humans interfering with the phosphorus cycle is seen in the cutting of tropical rainforests. This type of eco-

system is supported by a virtually 100 percent efficient recycling of nutrients. In other words, there are little or no reserves in the soil. When the forest is cut and burned, the nutrients that were locked up in the trees are readily washed away, and the land is rendered unproductive. Also, in the human system, phosphate from agricultural croplands makes its way, in large part, into waterways—either directly by way of runoff from croplands or indirectly by way of discharge of sewage effluents. Since there is essentially no return of phosphate from water, this addition results in overfertilization of bodies of water. Meanwhile, phosphorus is replaced on croplands by mining phosphate rock—a process that will ultimately result in depletion.

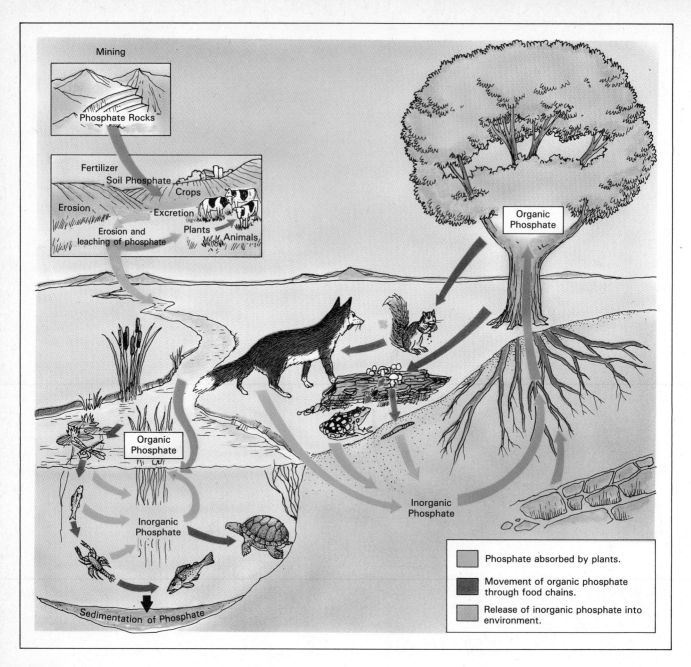

FIGURE 3–16
The phosphorus cycle. See text for explanation.

The Nitrogen Cycle

The nitrogen cycle (Fig. 3–17) is more complex than the carbon and phosphorus cycles because it has both a gas phase and a mineral phase. The main reservoir of nitrogen is the air, which is about 78 percent nitrogen gas (N_2). Plants cannot utilize nitrogen gas directly from the air, however; instead the nitrogen must be in a mineral form, such as ammonium ion (NH_4^+) or nitrate ion (NO_3^-). A number of bacteria and also certain blue-green algae, which are actually bacteria (cyanobacteria), can convert nitrogen gas to the ammonium form, a process called biological **nitrogen fixation**. Most important among these nitrogen-fixing organisms is a bacterium called *Rhizobium*, which lives in nodules on roots of legumes, members of the pea-bean family of plants (Fig. 3–18). This is another example of symbiosis. The legume provides the bacteria with a place to live and with food (sugar) and gains a source of nitrogen in return. Fixed organic nitrogen is passed from the legumes to other organisms in the ecosystem through food chains.

The cycling of carbon and other elements from the environment through organisms and back to the environment is more than just theory. Atoms can be made radioactive (see Appendix C) so that they behave chemically exactly like normal atoms except that they give off radiation, which can be detected with suitable monitors. Introducing radioactive compounds into a cycle is analogous to planting a small transmitter on a vehicle to monitor where the vehicle goes. By this technique, scientists have verified that carbon atoms move from carbon dioxide in the atmosphere into glucose, then into various macromolecules making up the plant tissues. Similarly, the progress of these labeled atoms down food chains and their return to the atmosphere via cell respiration can be observed.

Through these studies, a striking fact becomes evident: All tissues of animal bodies, including our own, are "turned over" quite rapidly. That

is, although we maintain our general appearance from year to year, every tissue in our body is constantly being broken down, oxidized, and replaced with newly made molecules derived from the food we eat. Atom for atom, molecule for molecule, our bodies are entirely replaced on the average of every 4 years. Thus we and every other form of life are constantly participating in the cycles of all the atoms.

Imagine having a magnifying glass powerful enough to see the individual carbon atoms making up the protein of the skin on your hand. Focus on a single carbon atom. Where did it come from? From the food you ate a few weeks ago. Before that? From carbon dioxide in the atmosphere that was incorporated into a plant by photosynthesis. Where will it go? In a few weeks, it will in all likelihood be back in the atmosphere as CO_2 as the top layer of your skin is sloughed off and oxidized by microorganisms. Will this be the end of the carbon atom's

travels? No, it will no doubt go on to additional cycles.

If this or any other atom in your body could tell you its "life history," it might go something like this:

I have existed since the formation of the Earth. In my countless cycles from the air through living things and back, I have participated in the bodies of virtually every species that has ever existed anywhere on Earth, including trees and animals of the forests, seaweeds, fishes and other creatures of the oceans, and the dinosaurs that roamed the land 100 million years ago. In more recent times, my travels along food chains have led through quite a few humans as well as other plants and animals that share your environment. But, so long now; I must continue my travels, for this is my fate till the end of time.

In a very real and verifiable way, all life is interconnected through sharing and recycling a common pool of atoms. Generations pass, but atoms remain the same.

As animals break down proteins and other organic compounds containing nitrogen for energy in cell respiration, the nitrogen is excreted, generally in the ammonium ion form. Bacteria in the soil may convert the ammonium ion to the nitrate form, but either form may be reabsorbed by any plants, thus creating an ongoing cycle. However, another kind of bacterium in the soil gradually changes the nitrate ion back to nitrogen gas (indicated by the black arrow in Fig. 3–17). Consequently, nitrogen will not accumulate in the soil. Some nitrogen gas is also converted to the ammonium form by discharges of lightning in the process known as *atmospheric nitrogen fixation* and comes down with rainfall, but this is estimated to be only about 10 percent of the amount of biological nitrogen fixation.

All natural ecosystems, then, depend on nitrogen-fixing organisms; legumes with their symbiotic bacteria are, by far, the most important. The legume family includes a huge diversity of plants, ranging from clovers (common in grasslands) through desert

shrubs to many trees. Every major terrestrial ecosystem, from tropical rainforest to desert and tundra, has its representative legume species, and legumes are generally the first plants to recolonize a burned-over area. Without them, all production would be sharply impaired because of lack of available nitrogen.

The nitrogen cycle in aquatic ecosystems is similar, but there blue-green algae are the most significant nitrogen fixers.

Only humans have been able to bypass the necessity for legumes when nonlegume crops such as corn, wheat, and other grains are being grown. We do this by fixing nitrogen in chemical factories (industrial nitrogen fixing). Synthetically produced ammonium and nitrate compounds are major constituents of fertilizer. However, the high cost of industrially fixed nitrogen and other soil factors are causing many farmers to readopt the natural process of enriching the soil by alternating legumes with nonlegume crops, that is, by **crop rotation**. This topic will be discussed further in Chapter 9.

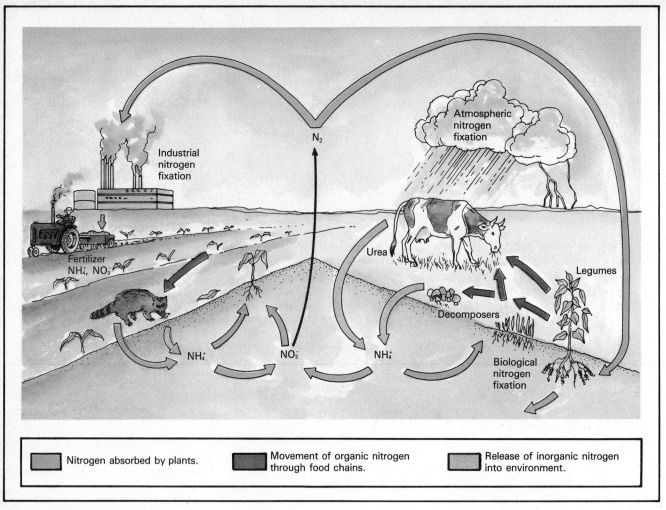

FIGURE 3–17
The nitrogen cycle. See text for explanation.

Nitrogen absorbed by plants.

Movement of organic nitrogen through food chains.

Release of inorganic nitrogen into environment.

RUN ON SOLAR ENERGY

Ecosystems obey the Laws of Thermodynamics. We have seen that ample sources of energy drive both producers and consumers. The initial source of energy for every major ecosystem is light from the sun, light that is captured through photosynthesis. (The only exceptions are ecosystems in deep ocean trenches, where the producers are bacteria that derive energy from the oxidation of hydrogen sulfide that exists in those locations.) As energy is utilized by organisms throughout the food web, the spent energy exits as heat.

FIGURE 3–18
Root nodules. Nitrogen fixation, the conversion of nitrogen gas in the atmosphere to forms that can be used by plants, is carried out by bacteria that live in the nodules of roots of plants of the legume family. (USDA photograph.)

Not all atoms are rapidly recycled as indicated in In Perspective Box, p. 63. Some get "sidetracked" into much longer geological cycles. For example, clams and oysters use carbon in making their shells, which are calcium carbonate (Ca_2CO_3). Deposits of this material left after the organisms die become limestone. Limestone formations—including the trapped carbon atoms—may remain relatively unchanged for hundreds of millions of years. In time, slightly acidic groundwater may gradually dissolve the limestone, returning the carbon to the atmosphere as carbon dioxide. Underground caverns and other rock formations are the result of such removal of limestone.

Another interesting and important aspect of the carbon cycle is that hundreds of millions of years ago, much of the organic matter produced in photosynthesis was neither consumed nor decomposed; instead it accumulated and was gradually buried under sediments. As a result of millions of years under heat and pressure in the earth, this detritus has been converted to crude oil, natural gas, or coal—i.e., our fossil fuels. In burning these materials, we are, in one sense, completing the carbon cycle by freeing carbon atoms that have been sidetracked for eons. However, we are also unbalancing the cycle in that the rate at which we are adding carbon dioxide to the atmosphere far exceeds the reabsorption rate. Consequently the carbon dioxide concentration in the atmosphere is increasing, a situation that has serious implications regarding global warming (see "Greenhouse Effect," Chapter 16).

Realizing that ecosystems run on solar energy gives us the **second basic principle of ecosystem sustainability**:

For sustainability, ecosystems use sunlight as their source of energy.

Solar energy is basic to sustainability because of two attributes: it is *nonpolluting* and it is *nondepletable*.

Nonpolluting Light from the sun is a form of pure energy; it contains no elements or molecules that can or will pollute the environment. All the matter and pollution involved in its production are conveniently left behind on the sun some 93 million miles (150 million kilometers) away in space.

Nondepletable The sun's energy output is essentially constant. How much or how little of this energy is used on earth will not influence, much less deplete, the sun's output. For all practical purposes, the sun is an everlasting source of energy. It is true that astronomers do tell us that the sun will burn out in another 3–5 billion years, but we need to put this figure in perspective. One thousand is only 0.0001 percent of a billion. Thus even the passing of millennia is insignificant on this time scale.

PREVENTION OF OVERGRAZING

We saw on page 56 that each consumer, in order to obtain the energy it needs, breaks down a considerable portion of what it eats. This breakdown obviously contributes to the fact that biomass declines at higher trophic levels, but two additional factors are involved as well. First, a significant portion of what is consumed is not digested but simply passes through and out as fecal waste. This process obviously does not lead to increasing biomass of the consumer, although the waste matter may become the food source for certain decomposers and/or detritus feeders.

The second factor leading to biomass decrease involves the rate at which consumers consume. In a grazing situation, for example, it is readily apparent that if the animals eat the grass faster than the grass regrows, all the animals will starve sooner or later. The situation is known as **overgrazing**. Therefore, in a sustainable situation, no more than what is *produced each year* can be consumed. The rest must remain so that it can regenerate the next year's production. This remaining amount is referred to as the **standing biomass**. In other words, in a sustainable situation consumers eat no more than a small portion of the total. A considerable standing biomass remains and must remain to regenerate. This concept also applies to carnivores; for sustainability a considerable portion, a *standing biomass*, of their prey must remain alive.

Summarizing, we find that for each trophic level, usually only a small portion of the total food available is consumed; the rest remains as standing biomass. Of what is consumed, some simply passes through the consumer and another portion is broken down to release energy. It is only a tiny portion that gets converted to body tissue of the consumer. Therefore, the biomass of the consumer population is inevitably but a small fraction (usually no more than 10

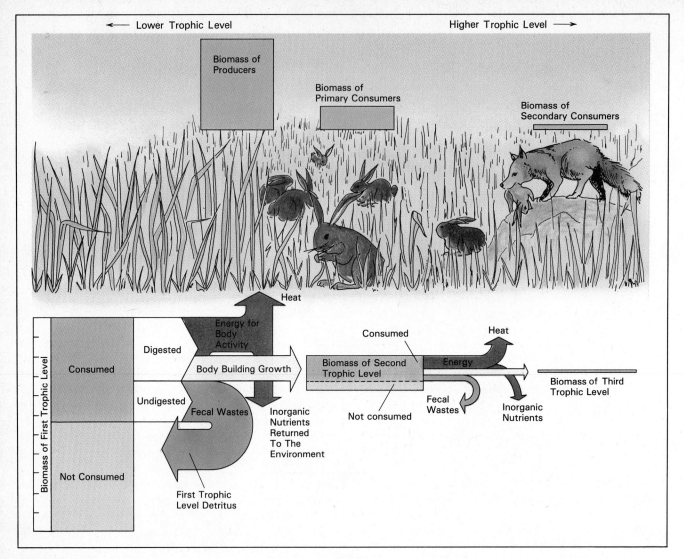

FIGURE 3–19
Decreasing biomass at higher trophic levels. The decrease results from three facts:
(1) much of the preceding trophic level is standing biomass and so is not available
for consumption, (2) much of what is consumed is broken down in order to release
energy, and (3) some of what is consumed passes through.

percent) of the biomass of what it is feeding on. This
sharply declining biomass at each higher trophic level
is symbolized by the "food pyramid." This process is
summarized in Figure 3–19.

It should be conspicuous that if a consumer population gets too large relative to what it is feeding on,
overgrazing will be inevitable and sustainability of the
population will be impossible. This observation
brings us to the **third principle of ecosystem sustainability**:

For sustainability, the size of consumer populations is
maintained such that overgrazing does not occur.

We shall investigate how population sizes are regulated at such levels in Chapter 4. The three basic prin-

ciples of ecosystem sustainability are summarized in
Figure 3–20 on the facing page.

Implications for Humans

We have said that the environmental problems we
face and the questionable long-term viability of our
current human system are caused by our failure to
adhere to basic ecological principles of sustainability.
These principles may show us the direction we need
to take. Let's look at our human system from the point
of view of each of the above principles of ecosystem
sustainability.

An Upside-down Food Pyramid

Marine ecosystems based on phyto-plankton, which are microscopic, free-floating forms of algae, seem to violate the rule of decreasing bio-mass at higher trophic levels. The biomass of phytoplankton, which are the producers, is sometimes considerably less than that of the or-ganisms feeding upon it. This situa-tion is sustainable because the re-generation of the phytoplankton is exceptionally rapid: Under optimal conditions it can double its biomass every 24 hours. Thus, multiplying very rapidly and being eaten at ap-proximately the same rate as it is produced, the relatively small bio-mass of phytoplankton can support a larger biomass of consumers. If the production rate, as well as the standing biomass, of phytoplankton is taken into account, the total bio-mass of phytoplankton is larger than that of the organisms feeding on it. Therefore, basic principles are not violated.

FIGURE 3–20
Nutrient cycling in and energy flow through an ecosystem. Arranging organisms by feeding relationships and depicting energy and nutrient inputs and outputs of each show a continuous recycling of nutrients (gray) in the ecosystem, a continuous flow of energy through it (red), and a decrease of biomass.

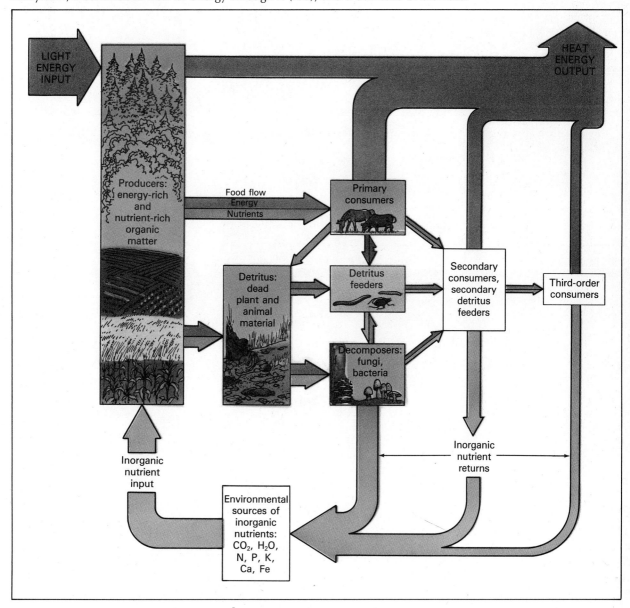

First Principle of Sustainability *For sustainability, ecosystems dispose of wastes and replenish nutrients by recycling all elements.* In contrast to this principle, we have based our human system in large part on a *one-directional flow* of elements. We mine elements in one location and dispose of them in another. For example, phosphate withdrawn from soils by agricultural crops comes to us with our food supplies, but then effluents of our wastes containing the phosphate are discharged into various waterways (rivers, lakes, bays, and estuaries) rather than back into the soil. To make up for the removal of phosphate from soil, phosphate rock is mined at various locations and added to soil as a constituent of fertilizer. Thus, there is basically a one-way flow of phosphate from mine to waterways (Fig. 3–21). The same can be said for such metals as aluminum, mercury, lead, and cadmium, which are the "nutrients" of our industry. We have created a flow of these elements from natural deposits through our systems to dumps and landfills.

This one-way flow leads to two problems: depletion of the resource at one end and pollution at the other. Pollution has proved to be, by far, the more severe problem. Countless waterways around the world and even sections of the ocean are suffering severe ecological disturbances from being oversupplied with nutrients such as phosphate. (This problem, known as eutrophication, will be discussed in Chapter 12.) Likewise, many rivers and other bodies of water are contaminated with toxic elements from various discharges. For example, thousands of kilometers of tributaries of the Amazon are badly contaminated with mercury, a waste product of gold mining. Putting such waste materials into dumps is problematic on two counts. Finding space for dumps and landfills is reaching crisis proportions in many re-

FIGURE 3–21
In contrast to applying the ecological principle of nutrient recycling, human society has developed a pattern of one-directional nutrient flow. There are increasing problems at both ends. This drawing illustrates one-way flow for phosphorus, but the scheme also applies to all other elements we use in our daily lives.

gions. Then, even when such toxic materials are put into dumps, they tend to leak out causing pollution of both ground and surface water.

Aggravating the problem is the fact that we produce and use thousands of products, such as plastics, that are synthetic organic compounds that are **nonbiodegradable**. That is, detritus feeders and decomposers are unable to attack and break them down. Thus enormous amounts of nonbiodegradable products compound the problem of finding dump sites. Also, many such synthetic organic products are toxic and cause pollution in the same way lead and other elements do.

The rapid development of recycling programs in the last few years is an encouraging sign that we are beginning to recognize and implement the first principle of sustainability.

Second Principle of Sustainability For sustainability *ecosystems use sunlight as their source of energy.* In contrast to this principle, our fantastic technological and material progress of the past 200 years has been in large part a story of developing machinery, engines, and heating plants that run on fossil fuels—coal, natural gas, and crude oil. Just consider that virtually all cars, trucks, aircraft, and other vehicles run on fuels refined from crude oil; 70 percent of the electricity in

our country comes from coal-fired power plants, and most homes, buildings, and hot water are heated with natural gas. Even food production, which is basically derived from solar energy (photosynthesis of crop plants), is heavily supported by fossil fuels used in farm machinery, production of fertilizer and pesticides, transportation, processing and canning, refrigeration, and finally cooking. In all, more than 10 calories of fossil fuels are consumed for every calorie of food that is served in the United States.

From meager beginnings in the late 1800s, oil consumption now tops 50 million barrels per day worldwide. (One barrel equals 42 gallons.) The by-products of burning fossil fuels enter the atmosphere and are directly responsible for our most severe air pollution problems—urban smog, acid rain, and, most recently, the potential of global warming. (These problems are all discussed further in later chapters.) Also, we are facing increasingly severe crises because of depletion of present oil reserves and environmental destruction in the effort to find more. Nuclear power is being promoted as an alternative, but this source also seems dubious because of the hazards of its radioactive waste products.

Thus, the danger in continuing to ignore the ecosystem principle regarding solar energy seems clear. In addition to being nonpolluting and nonde-

ETHICS BOX
Are There Limits?

Cornucopians argue that human progress has been a story of overcoming limits. Given human ingenuity and technical capability, these people maintain, there are no limits that cannot be overcome; to impose limits is simply to hold back progress needlessly. This reasoning contains both truth and fallacy. The truth is that, unquestionably, we have overcome many restraints. The fallacy is in the implication that we have nullified or changed natural laws in the process. We have not.

For example, for most of human history, gravity limited us to within a few feet off the ground. Now it is common to fly at 30 000 feet (10 000 meter) altitudes, and astronauts are venturing infinitely higher. Therefore, one can say that we have overcome the limits of gravity. Suppose,

however, that a bus driver declares that he is going to drive you and other passengers off a cliff because we have learned to overcome the limits of gravity. This story makes the point that we definitely have not learned to change gravity or its consequences in any way. In fact, quite the opposite. We have come to understand and appreciate gravity as exactly the force that it is, and then we have developed wings and rockets that produce a force suitable to counter that of gravity.

The same can be said for the principles brought out in this chapter. We can never change the natural Laws of Conservation and Thermodynamics and the consequences they produce any more than the bus driver is likely to change gravity and the consequence of driving off a

cliff. Yet, we can achieve sustainability, just as we have achieved flight, by understanding the principles and adapting our human system accordingly. Plunging ahead blindly, arguing that these principles do not apply to us or that we can nullify them, should be seen as committing the same folly as the bus driver. In this case, however, Planet Earth is the bus and we are all passengers. An important question every one of us needs to ask is, Do I have a moral and ethical obligation to change such behavior if I can? If your answer is yes, then you have a responsibility to understand and appreciate these principles and then pursue development in a manner that is consistent with them.

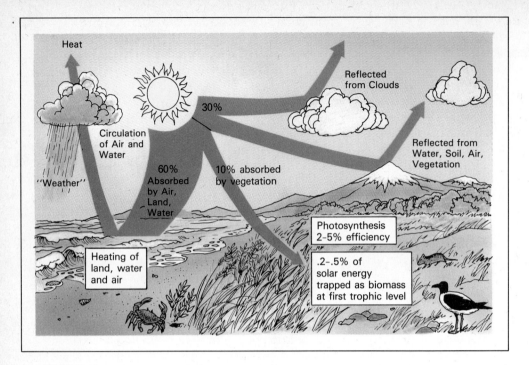

FIGURE 3–22

Of the total solar energy reaching the earth, it is estimated that 30 percent is reflected from clouds, water, and land surfaces. Sixty percent is absorbed by, and goes into heating, land, water, and air. This heating causes evaporation of water and the circulation of air and water resulting in all the factors of weather. Only 10 percent of the solar energy is absorbed by vegetation. (Of course these percentages will vary greatly with the kind of ecosystem, the season of the year, and the location on the earth.) Given the low efficiency of photosynthesis, only 2–5 percent of the 10 percent absorbed by vegetation (0.2–0.5 percent of the total solar energy) is trapped as biomass at the first trophic level, but it is this latter amount of energy that supports all the rest of the ecosystem. It is estimated that just 0.1 percent of the solar energy being harnessed to perform useful work as it is absorbed and converted to heat would supply all human needs and would not affect the dynamics of the biosphere.

pletable, solar energy is also extremely abundant. Green plants, including agricultural crops, utilize a very small fraction of the solar energy that hits the earth. Most of the rest is converted directly to heat as it is absorbed by water or land. In turn, this heated water and land heats the air and causes the evaporation of water. Thus, solar energy is the major driving force behind ocean currents, wind, and rain—i.e., weather (Fig. 3–22). There is ample opportunity to harness some of this energy and put it to work. In our discussion of the Laws of Thermodynamics and energy flow in ecosystems, we observed that the final heat at the end of the line is the same whether the energy is harnessed to perform useful work along the way or not. Therefore, even using solar energy on a vast scale would not change the overall dynamics of the biosphere.

Third Principle of Sustainability *For sustainability the size of consumer populations is maintained such that overgrazing does not occur.* In contrast to this principle, the human population has increased more than five-fold in the past 200 years. It has nearly tripled in just the last 60 years and is continuing to increase at a rate of over 90 *million people* per year (Chapter 6). It can be argued that this ever-accelerating growth rate is irrelevant because humans are supported by a technological agricultural system, not a natural ecosystem. On the other hand, signs of overgrazing are becoming all too evident. First, there is literal overgrazing. Over the world millions of acres of productive grasslands have been badly degraded or even turned into desert because of overgrazing cattle (Chapter 9).

Then, there are any number of examples of overgrazing in a figurative sense. Consider the destruction of tropical and other forests; depletion of groundwater supplies; farming practices that are leading to deterioration of soil and hence loss of productivity; poor people in a number of less-developed countries picking hillsides bare in their quest for firewood, which is their only fuel; depletion of fishing areas, and so on. Perhaps the most serious form of figurative overgrazing, however, may be the ever-expanding human development and exploitation that displace and degrade natural ecosystems and consequently cause the extinction of countless species. The effects that this extinction may have will be discussed further in Chapter 5 and also in Chapter 19. The principle of maintaining a stable (nongrowing) population is a principle that cannot be ignored.

 Review Questions _____

1. What are the six key elements of living organisms and where does each occur in the environment? In four cases identify the specific molecule that the element comes from.

2. Give a simple description of what is happening to the six key elements in the course of growth and decay. What is the "common denominator" that distinguishes organic and inorganic molecules?

3. Give the definitions for matter and energy. Name three main categories of matter and give four forms of kinetic energy.

4. Give five examples that demonstrate different conversions among forms of kinetic energy. Give another five examples that demonstrate conversions between kinetic and potential energy.

5. Name the two energy laws and apply each, describing the relative amounts and forms of energy going into and coming out of the conversions listed in question 4.

6. What is chemical energy? What energy changes are involved in the formation and breakdown of organic molecules?

7. Describe the process of photosynthesis in terms of what is happening to specific atoms (matter) and where energy is coming from and going to. Do the same for cell respiration.

8. Food ingested by a consumer follows three different pathways. Describe each pathway in terms of what happens to the food involved and what products and byproducts are produced in each case.

9. Define and contrast starvation and malnutrition, giving the causes and results of each.

10. Compare and contrast the decomposers with other consumers in terms of matter and energy changes that they perform.

11. Where do carbon, phosphorus, and nitrogen exist in the environment, and how do they move into and through organisms and back to the environment?

12. In what forms does energy enter, move through, and leave ecosystems?

13. How does movement of energy through an ecosystem relate to decreasing biomass at higher trophic levels—i.e., the food pyramid?

14. In addition to the energy movement discussed in question 13, what other factors are involved in causing the decreasing biomass at higher trophic levels?

15. What are the three principles of ecosystem sustainability?

16. How would you rate our human system in terms of adhering or not adhering to each of the principles of sustainability? Cite the evidence for each of your answers.

 Thinking Environmentally _____

1. Describe the consumption of fuel by a car in terms of the laws of conservation of matter and energy. That is, what are the inputs and outputs of matter and energy? (Note: Gasoline is an organic carbon-hydrogen compound.)

2. Relate your level of exercise, breathing hard, and "working up an appetite" to cell respiration in your body. What materials are being consumed, and what products and byproducts are being produced?

3. Using your knowledge of photosynthesis and cell respiration, create an illustration showing the hydrogen cycle and the oxygen cycle.

4. Write a short essay supporting the statement, "Waste" is a human concept and invention; it does not exist in natural ecosystems!

5. Tundra and desert ecosystems support a much smaller biomass of animals than do tropical rainforests. Give two reasons for this fact.

6. Evaluate the sustainability of parts of the human system, such as transportation, manufacturing, agriculture, and waste disposal, by relating them to specific principles of sustainability.

Ecosystems: What Keeps Them the Same? What Makes Them Change?

LEARNING OBJECTIVES

When you have finished studying this chapter, you should be able to:

1. Define and give examples of the factors involved in biotic potential and environmental resistance.

2. Distinguish between reproduction and recruitment, and describe what occurs in all populations if conditions are ideal.

3. Define density dependence, and describe how it relates to population balance between biotic potential and environmental resistance.

4. Give examples of natural enemies and how they serve to maintain herbivore populations in nature.

5. Define territoriality, and describe how it controls certain populations in nature.

6. Analyze principle three of ecosystem sustainability—populations are maintained such that overgrazing does not occur—in terms of mechanisms of population balance.

7. Describe three ways that enable different plant species to coexist in the same region.

8. Give examples of ecological upsets caused by the introduction of foreign species.

9. Give examples of three different kinds of ecological succession, and describe the role of fire and how it may be used in the management of certain ecosystems.

10. Name, draw a graph of, and describe the causes and consequences of two fundamental population curves, and relate these curves to the human system and human impacts on natural ecosystems.

11. State the fourth principle of ecosystem sustainability, and give examples demonstrating the consequences of not maintaining biodiversity.

12. Contrast the ways in which populations are controlled in nature with the methods that are available to humans.

In Chapter 3 we learned that a basic principle of ecosystem sustainability is that consumer populations are maintained such that overgrazing does not occur. Our objective in this chapter is to gain an understanding of how this balance between eaters and eaten is achieved in natural ecosystems. Such understanding will provide insights regarding development of a sustainable human system.

The Key Is Balance

First, the most important point to recognize is that no forces or rigid structures exist that prevent ecosystems from changing. In fact, ecosystems can and do change, even drastically, as conditions are altered. The one thing that enables ecosystems to sustain a given composition of species over long periods of time is that *all the relationships in the system are in a dynamic balance.*

ECOSYSTEM BALANCE IS POPULATION BALANCE

Each species in an ecosystem exists as a **population**—that is to say, an interbreeding, reproducing group. An ecosystem's remaining stable (sustaining itself) over a long period of time implies that the population of each species in the ecosystem remains more or less constant in size and geographic distribution. Any continuing increase or decrease in population would be observed as a change in the ecosystem. In turn, a population's remaining constant over a long time means that reproductive rate is equaled by death rate. Thus, the problem of ecosystem balance boils down to a problem of how birth rate and death rate are balanced for each species in the ecosystem. How is this balance achieved?

Biotic Potential versus Environmental Resistance

Maintaining or increasing a population depends on more than **reproductive rate** (number of live births,

eggs laid, or seeds or spores set in plants) by itself. **Recruitment,** which is defined as making it through the early growth stages to become part of the breeding, reproducing population, is equally important. For example, many fish lay thousands, even millions of eggs, and plants typically set thousands of seed. Yet population increase may be nil because recruitment is so low; in other words, most of the young fish and plants perish in the early stages of growth (Fig. 4–1a). (Note that "low recruitment" is a polite way of saying high mortality of the young.) Conversely, even a relatively low reproductive rate may result in a substantial population increase when recruitment is high. Primates are an outstanding example of this latter strategy (Fig. 4–1b).

Additional factors that influence population growth and geographic distribution are the ability of animals to migrate or of seeds to disperse to similar habitats in other regions, the ability to adapt to and invade new habitats in addition to the one originally occupied, defense mechanisms, and resistance to adverse conditions and disease. All these factors taken together are referred to as the **biotic potential** of the species.

Despite different strategies regarding biotic potential, there is one point in common: Every species has sufficient reproductive capacity to rapidly increase its population *if factors are favorable for a high recruitment.* Indeed, each new generation will be multiplied by the number of females produced. For example, rabbits producing 20 offspring, 10 of which are female, may grow by a factor of 10 each generation: 10, 100, 1000, 10 000. . . . Such a multiplying series is called an **exponential increase**. In populations it is commonly called a **population explosion** (Fig. 4–1c).

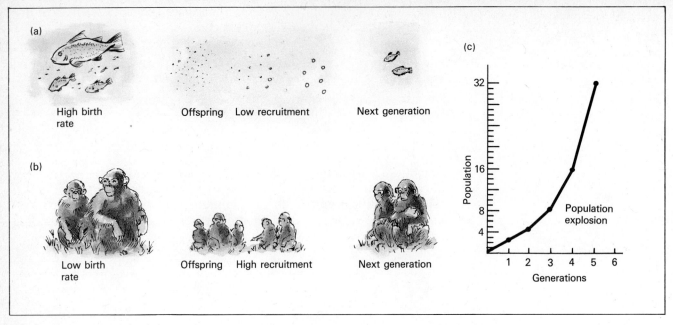

FIGURE 4–1
Two reproductive strategies. (a) High birth rate-low recruitment and (b) low birth rate-high recruitment. Both accomplish the same survival. (c) If recruitment is high, even a relatively low birth rate will lead to rapid population growth. Shown here, a birth rate of just four offspring per couple will double the population each generation if all offspring survive and reproduce. Such exponential growth is called a population explosion.

Populations in natural ecosystems generally do not explode because all conditions are seldom favorable for any extended period of time. One or more abiotic factors, such as unfavorable temperature, amount of available water, pH, or salinity, and/or one or more biotic factors, such as predators, parasites, disease organisms, or lack of sufficient food, become *limiting*. The combination of all these abiotic and biotic factors that may limit population increase is referred to as **environmental resistance**.

You may already foresee the result of the interplay between biotic potential and environmental resistance. Sooner or later, any population increase will be curtailed by one or more factors of environmental resistance. It is important to observe how this curtailment works, however. In general, the reproductive rate for a species remains fairly constant, because that rate is part of the genetic endowment of the species. *What varies tremendously is recruitment.* It is in the early stages of growth that individuals (plants or animals) are most vulnerable to predation, disease, lack of food (or nutrients) or water, and other adverse conditions. Consequently, environmental resistance effectively reduces recruitment. Of course, some adults also perish, particularly the old or weak. If recruitment is at the **replacement level**, just enough to replace these adults, then the population size will remain constant. If recruitment is not sufficient to replace losses in the breeding population, of course, the population size will decline.

In certain situations, environmental resistance may affect reproduction as well as causing mortality (death) directly. For example, loss of suitable habitat often prevents animals from breeding. Also, certain pollutants adversely affect reproduction. However, we can still view these situations as environmental resistance either blocking a population's growth or causing its decline.

In conclusion, whether a population grows, remains stable, or decreases is the result of a *dynamic balance between its biotic potential and environmental resistance* (Fig. 4–2). In general, biotic potential remains constant; it is shifts in environmental resistance that allow populations to increase or cause them to decrease. For example, a number of favorable years (low environmental resistance) will allow a population to increase; then a drought may cause it to die back, and the cycle may be repeated.

We should note that balance is a relative phenomenon. Some balances fluctuate very little, others fluctuate widely, but as long as decreased populations restore their numbers, the system may be said to be balanced. Still, the questions remain: What maintains the balance within a certain range? What

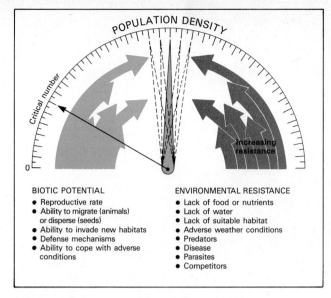

POPULATION DENSITY

Critical number

0

Increasing resistance

BIOTIC POTENTIAL
• Reproductive rate
• Ability to migrate (animals) or disperse (seeds)
• Ability to invade new habitats
• Defense mechanisms
• Ability to cope with adverse conditions

ENVIRONMENTAL RESISTANCE
• Lack of food or nutrients
• Lack of water
• Lack of suitable habitat
• Adverse weather conditions
• Predators
• Disease
• Parasites
• Competitors

FIGURE 4–2
A stable population in nature is the result of a balance between factors tending to increase population (biotic potential) and factors tending to decrease population (environmental resistance). A balance results because many factors of environmental resistance become more intense with increasing population.

prevents a population from going into an explosion or into extinction? Indeed, in nature, neither possibility is ruled out.

Density Dependence and Critical Numbers

In general, populations remain within a certain size range because most factors of environmental resistance are *density-dependent*. That is, as **population density** (the number of individuals per unit area) increases, environmental resistance becomes more intense and causes an increase in mortality such that population growth ceases or declines. Conversely, as population density decreases, environmental resistance is generally mitigated, allowing the population to recover. This balancing act will become clearer as we discuss specific mechanisms of population balance in the next section.

Human impacts, on the other hand, readily result in extinction because they are not density-dependent. Impacts such as ecosystem destruction, habitat alteration, pollution, and exploitation can be just as intense at low population densities as at high. Furthermore, the biotic potential of many species depends on a minimum population base—a herd of deer, a pack of wolves, a flock of birds, or a school of fish, for example. If a population is pushed below a certain **critical number** necessary to support a breeding population, biotic potential fails, and ex-

tinction is virtually assured. Species whose populations are declining rapidly because of human impacts are defined as **threatened**. If the population is approaching or is at what scientists believe to be its critical number, the species may be defined as **endangered**. (See In Perspective Box, p. 418.)

MECHANISMS OF POPULATION BALANCE

Let us turn our attention to some specific mechanisms that provide population balance in nature. For the sake of study, it is necessary to focus on one mechanism at a time, but keep in mind that in natural ecosystems all of the mechanisms are working in concert to create the overall balance. Knowledge of these mechanisms will make us aware of how ecosystems may be upset and the consequences that may result.

Predator-Prey and Host-Parasite Balances

A classic example of population balance is that between the lynx, a member of the cat family, and hares, a member of the rabbit family, as observed in Canada from 1850 to 1930. When the hare population is low, each hare can find abundant food and plenty of places to hide and raise offspring. In other words, the hares' environmental resistance is relatively low, and their population increases despite the presence of the lynx predator. As the hare population increases, however, each hare has relatively less food and fewer hiding places. More hares provide easier hunting for the lynx so that, with plenty of hares to feed lynx young, the lynx population begins to increase. In short, the increasing population density of hares runs into increasing environmental resistance in the form of limited food and shelter and increased predation. As a result, the hare population begins to fall. As the hare population falls, the food and shelter available to each hare again increase. Also, surviving hares are those that are healthiest and best able to escape from the lynx. Hunting becomes harder for the lynx; many of them starve, and their population begins to fall. These factors sum up to lower environmental resistance for the hares, and their population increases again, repeating the cycle. These events explain the fluctuating but continuing balance found between the hare and lynx populations (Fig. 4–3).

While large predators such as the lynx draw our attention, there are relatively few situations where they are the primary controlling factor. Much more abundant and ecologically important in population control are a huge diversity of *parasitic organisms*. Recall from Chapter 2 that these organisms range from tapeworms, which may be a foot or more in length

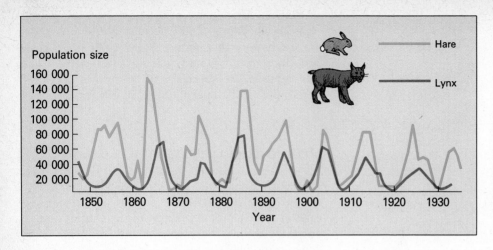

FIGURE 4–3
The predator-prey relationship creates a balance between predator and prey populations. Data based on pelts received by Hudson's Bay Company. (From D. A. MacLulich, University of Toronto Studies, Biological Series No. 43, 1937.)

(Fig. 2–8) to microscopic disease-causing protozoans, fungi, bacteria, and viruses. All species of plants, animals, and even microbes may be infected by parasites.

In terms of population balance, parasitic organisms act in the same way as large predators. As the population density of the host organism increases, parasites and their *vectors* (agents that carry the parasites from one host to another), such as disease-carrying insects, have little trouble finding new hosts, and infection rates increase, causing dieback. Conversely, when the population density of the host is low, transfer of infection is impeded, and there is a great reduction in levels of infection, a condition that allows the population to recover.

Parasites may not kill their host, but they generally weaken it and make it more vulnerable to adverse conditions and to attack by larger predators. It is commonly observed that the animals killed by large predators are infected with parasites, whereas animals killed by hunters are generally healthy.

In a food web, a population of any given organism is affected by a number of predators and parasites simultaneously. Consequently, the balance can be thought of broadly as a balance between the population of an organism and its **natural enemies**. The

---

IN PERSPECTIVE

Pets or Extinction

Buying and owning "pet" specimens of rare plants, animals, birds, or fish is essentially the same as going into the wild and killing them. A specimen removed from the breeding population is the same as dead in terms of being able to provide any benefit toward survival of the population. Buying such pets and hence supporting the pet trade may be even worse than direct hunting because many individuals of the collected species die in the process of trapping and transport. Thus a caged specimen may represent but one of perhaps ten that have died along the way, and of course, it is only awaiting its own death.

This is not a small problem. The black market trade in rare and en-

dangered species is second only to the drug trade. It is threatening the extinction of many species of both plants and animals as people offering to pay tens of thousands of dollars for specimens provide the incentive for poachers to go into the wild and trap them. Often the poachers have little knowledge of how to trap or care for the animals humanely, hence, the species suffer high mortality along the way.

Illegal work by poachers should not be confused with the efforts being undertaken by many modern zoos and other bona fide institutions to save endangered species. Here specimens are being taken from the wild and bred under conditions that maximize the reproduc-

tion of adults and recruitment of the young so that breeding populations can be built back up to sustainable numbers. Then, when natural habitats are suitably reestablished or protected, breeding populations are being returned to the wild. Such programs offer the only hope of avoiding extinction of the species in question. Breaking into such institutions to "free" such animals, as some animal rights activists have done, is certainly counterproductive to all, especially the animals. They cannot survive long outside their natural habitat without the care that the institution provides. Thus, "freeing" such animals may be the final act resulting in extinction of the species.

---

wide swings in populations noted in the hare-lynx case are generally typical of very simple ecosystems involving relatively few species. Balances between an organism and several natural enemies are generally much more stable and less prone to wide fluctuations because different natural enemies come into play at different population densities. Also, when the preferred prey is at a low density, the population of the natural enemy may be supported by its feeding on something else. Thus, the lag time between increase of the prey population and that of the natural enemy is diminished. These factors have a great damping effect on the rise and fall of the prey population.

In all such balances, however, whether simple or complex, it is extremely important to recognize that a high degree of adaptation is involved on the part of both the prey or host and the natural enemy. This adaptation is such that a given natural enemy is incapable of completely eliminating its prey or host but yet is capable of limiting the prey or host population to a certain density. Putting any predator and prey or host and parasite together does not lead to an automatic balance. Such a lack of balance is shown all too clearly by what may occur when a species from one region is introduced into another. Such introductions may lead to what are commonly called ecological disasters as balances fail. Here are some classic examples.

In 1859 rabbits were introduced from England to Australia for sport shooting. The Australian environment proved favorable, but its ecosystem did not contain a carnivore or other natural enemies capable of controlling the rabbits. The result was that the rabbit population exploded and devastated vast areas of rangeland by overgrazing (Fig. 4–4). This devastation was extremely damaging to both native wildlife and sheep ranching. It was finally brought under control by the introduction of a rabbit disease virus that provided a typical host-parasite balance.

Prior to 1900 the dominant tree in the eastern deciduous forests of the United States was the American chestnut, which was highly valued for both its high-quality wood and its prolific production of chestnuts, which were eaten by both wildlife and people. From 1904 to about 1950, however, a fungal parasite, called the chestnut blight, which was accidentally introduced with some Chinese chestnut trees planted in New York, spread through the forests, killing essentially every tree. While oaks filled in where the chestnuts died, the ecological and commercial loss is incalculable. A lot of research has been done to breed a resistant tree or otherwise control the blight to enable reestablishment, but success is still elusive.

Balance problems arise because the introduced species do not have effective natural enemies in the ecosystem into which they are introduced (called the *receiving ecosystem*). For example, many of the most severe insect pests of agricultural crops and forests—the Japanese beetle, fire ants, and the gypsy moth, for example—are species introduced from other ecosystems. Domestic cats introduced into island ecosystems have often proved to be overly effective predators and have exterminated many species of wildlife unique to the islands. Goats introduced onto islands have been devastating because of their overgrazing. A more recent example is the introduction of the zebra mussel to the Great Lakes (see In Perspective Box, p. 414).

From these examples, it should be clear that the introduced species may belong to any category—herbivore, carnivore, or parasite—and it may be large or tiny. The result is the same: ecological upset and loss. Why should introduced species lead to such problems?

FIGURE 4–4
Results of rabbits overgrazing the Australian ecosystem. On one side of a rabbit-proof fence there is lush pasture; on the other side, it is barren. Rabbit-proof fences were built over thousands of miles but proved unsuccessful in stopping movement. (Australian Information Service photograph.)

The explanation lies in the fact that ecosystems on different continents or remote islands have been isolated from each other for millions of years because of physical barriers. Consequently, the species within each system have developed adaptations that provide balance with other species *within their own ecosystem.* While each system has developed balances between predators and prey, hosts and parasites, and so on, the specific balances in one system do not match those in another. Thus, species native to the receiving ecosystem may be poorly adapted to cope with the biotic potential of an introduced species and consequently may be overrun, as the above examples demonstrate.

The seemingly obvious solution to such problems is to introduce a natural enemy of the "pest." Indeed, this approach has been used in a number of cases; rabbit control in Australia is one. Others are discussed in connection with biological control of pests in Chapter 10. However, such control is more easily said than done. Recall that balance is achieved as a result of interplay among *all* the factors of environmental resistance, which often include several natural enemies, as well as *all* the abiotic factors. Thus, a single natural enemy that will control a pest simply may not exist. Also, there is no guarantee that the natural enemy, when introduced into the new ecosystem, will focus its attention on the target pest. For example, control of the rabbit population in Australia was initially attempted by introducing foxes. However, the foxes soon learned that they could catch other Australian wildlife more easily than rabbits and thus went their own way. In short, a great amount of research needs to be done before introducing a natural enemy to prevent doing more harm than good.

Territoriality

In discussing predator-prey balances, we said that in lean times the excess carnivore population—the lynx, for instance—simply starved. Actually, another factor is often involved in the control of carnivore and some herbivore populations: **territoriality**, which refers to individuals or groups claiming a territory and defending it against others of the same species. For example, the males of many species of songbirds stake out a territory at the time of nesting. Their song has the function of warning other males to keep away. Male wolves and other carnivores, including dogs, stake out a territory by spotting it with urine, the smell of which warns other males to stay away. The territory defended is large enough to assure the "owners" of being able to gather enough food to successfully rear a brood. The size of the territory defended varies with resources available. In lean times territories are larger; in good times they are smaller.

The obvious advantage of territoriality is that individuals that are able to successfully claim and defend a territory will have enough resources to rear a well-fed, healthy next generation. Those individuals unable to claim a territory generally meet an unhappy end. Continually chased out of one territory after another, they fall victim to any of the factors of environmental resistance, or at the very least they are unable to breed and raise young.

Territoriality does not change the basic principle of population being a dynamic balance between biotic potential and environmental resistance. In the face of limited resources, however, territoriality creates a mechanism of selecting the strongest and fittest to survive and breed, while eliminating the genes of the weaker individuals.

ETHICS BOX
Humans and Territoriality

Territoriality is an instinctive behavioral trait in many species. By keeping populations in check, territoriality helps to maintain the balance of the ecosystem and thus to ensure survival of the species.

Many people have observed that humans are also a territorial animal, even to an extent that greatly exceeds any other species. Almost all of us aspire to owning a piece of land that we can put a fence around and call ours, and the bigger the piece of land, the better. Mad rulers like Napoleon and Hitler set out to take the world. Throughout history, virtually all wars between nations have involved territorial disputes, and nations continue to arm themselves to the teeth with the most sophisticated weaponry available— always, they say, to defend themselves against the threat of territorial encroachment by their neighbors.

Unlike all other animal species in nature, humans often lose sight of the resource value, and pursue territory as an end in itself. Land and its resources are laid waste while an aggressor is seeking additional land. Spending on armaments diverts funds from preserving or enhancing resources that are already available. Can we learn to aspire toward protecting and enhancing the productivity of the territory we have and help our neighbors do the same, rather than seeking more?

Hunting versus Animal Rights

Hunting and animal rights are increasingly the topic of moral and ethical debates, with hunting enthusiasts and animal rights activists at the extremes of the two sides. Certainly each individual has a right to choose for himself or herself whether or not to engage in legal hunting, eating meat, wearing leather, or otherwise being a party to killing other animals. But which point of view will best serve the long-term interests of society and the biosphere?

Proponents on both sides of the question could profit from and find middle ground in greater ecological understanding and awareness. If maintaining the biodiversity and ecological balances in the biosphere is taken as the primary aim, then there is a rationale for both sides. If we are looking at an endangered species, condoning its further hunting for any reason seems unconscionable because such hunting can only hasten its extinction, causing a permanent loss for both society and the biosphere. This reasoning can be extended beyond hunting to include habitat destruction for development or other purposes, a process that is also causing the extinction of species. Where animal rights projects lend to the support of saving endangered species and their habitats, the efforts of the people involved are commendable.

On the other hand, if we are looking at a species that is overpopulating and overgrazing because its natural enemies have been removed (e.g., deer) or because it is an introduced species without natural enemies, then there is a different answer. Continuing to allow such animals to overpopulate and overgraze can lead only to widespread damage to the ecosystem, damage that will include the death and even extinction of other animals dependent on the same vegetation for food and habitat. It will lead also to the death of the animals that are overpopulating because they will eventually deplete their food supply and die of starvation or disease. Is keeping their population in check by hunting less cruel? Does allowing animals to overgraze in any way serve the value of preserving the integrity of the biosphere? Does one particular animal have rights that exceed those of others or of the ecosystem as a whole?

In conclusion, it makes little sense to argue, much less act on, the ethics of animals rights or hunting as issues in and of themselves. Meaningful resolution can be reached, however, when the debate or action is put in terms of particular species and the ecosystems in which they exist and when preservation of that ecosystem is made the determining value. Preserving the ecosystem is the only way to preserve biodiversity.

Plant-Herbivore Balance and Carrying Capacity

The objective stated at the beginning of this chapter was to understand how *consumer populations in natural ecosystems are maintained at levels such that overgrazing does not occur.* Now that we have studied predator-prey and host-parasite balances, we have almost all the information we need to achieve our objective. To make the picture complete, we need to look at a third major balancing act: that between herbivores and the plants on which they feed. We just saw that predator-prey balances are such that herbivore populations are generally maintained below levels that cause overgrazing, as in the lynx-hare case. Where an herbivore is introduced into an ecosystem in which the herbivore has no natural enemies—goats on islands is one example we learned about—the plant-herbivore balance is again upset, and overgrazing is the inevitable result.

Another way of upsetting the plant-herbivore balance in an ecosystem is to kill off all the natural enemies of the herbivore. For example, in much of the United States, deer populations were originally controlled by wolves, mountain lions, and black bear, all of which have been essentially exterminated because they were felt to be a threat to cattle and even humans. Now, were it not for human hunting in place of these natural predators, deer populations would increase to the point of overgrazing. Indeed, drastic population increases do occur where hunting is prevented. Similarly, prairie dogs and other small rodents are becoming an increasing problem in the western United States as a result of reduction of predators such as coyotes, which ranchers felt were threatening their sheep. In this case, hunting is obviously not practical because who wants to go hunting for mice?

A second factor influencing plant-herbivore balance is that large herbivores—bison in the American West or elephants in Africa, for example—were originally able to roam vast regions. As forage was reduced in one area, a herd would simply move on before overgrazing. As humans have fenced such regions for agriculture and cattle ranching, however, wild herbivores are increasingly confined to areas such as parks and reserves. When the grazers are forced to remain in one area, overgrazing frequently becomes a serious problem.

As humans have upset natural balances, they have increasingly moved in to provide balance in whatever ways possible. This is the essence of **wildlife management**, which may involve means as diverse as heroic efforts to save an endangered species and hunting to prevent overgrazing. Both extremes are fraught with controversy (see Ethics Box, p. 79). In any case, a question that always arises is: What is the *carrying capacity* of the area for a given animal? The **carrying capacity** is defined as the maximum population of an animal that a given habitat will support without the habitat being degraded over the long term. Unfortunately, this population level is not always easy to determine and may also involve controversy. (See In Perspective Box below.)

A third aspect of plant-herbivore balance is seen in situations of *monoculture*. **Monoculture** refers to the growing of a single species over a wide area, a practice commonly followed in agriculture and forestry for economic efficiency. Monocultures are particularly prone to being essentially wiped out by insects, fungal diseases, or other pests. Such organisms have a fantastic biotic potential; they have a generation time of only a few days or weeks, and they often produce thousands of offspring—even millions in the case of fungal spores—in each generation. A monoculture provides a continuous, lush food supply, a situation very conducive to supporting a population explosion. In a monoculture the pest population may explode so fast that natural enemies, if present, cannot keep up with it, and the explosion will not end until the crop is essentially gone. At that point, the pest population will die back in a precipitous crash caused by starvation. Essentially, this scenario is an example of overgrazing—a population explosion causing destruction of the supporting resource and a precipitous dieback as a result. In a sense, this calamity takes care of the consumer population, but the crop or trees that are being grown are also lost in the process. It is for this reason that many farmers and forest managers feel obliged to use chemical pest controls that have undesirable side effects (see Chapter 10).

Insects, fungal diseases, and other parasites are **host-specific**. That is, they are unable to attack species other than their specific host. Consequently, when plants of different species are mixed together, a pest species has trouble getting to its next host. With this limitation, most of the pest offspring perish, and

IN PERSPECTIVE
Maximum versus Optimum Population

We have noted that carrying capacity is defined as the *maximum* population a habitat will support without being degraded over the long term. This population level, however, is not a simple fixed number that can be arrived at easily, and it will vary from year to year in any given ecosystem. Carrying capacity may be considerably higher in wet years than in dry years, which inhibit the growth of vegetation. In addition, the line between degrading and not degrading the habitat over the long term is anything but clear. Degradation may take place slowly and be hardly perceptible from year to year, yet nevertheless occur.

Therefore, managers nowadays focus more on what is the *optimal* rather than the maximum population. The optimal population is large enough to ensure a healthy breeding stock, yet considerably less than the theoretical maximum. It allows flexibility between good years and bad years without upsetting the ecosystem. Indeed, natural balances seem to operate such that populations are held at an optimum rather than a maximum.

The optimum population is not simple or fixed either. It too can be influenced by habitat management. For example, planting certain species, thinning a forest, and starting new trees will increase the amount of browse available and enhance the carrying capacity for deer. Creation of ponds and marshes will attract and support waterfowl.

Another factor becomes evident here, however. What is the minimum size of habitat that is required to support a breeding population? Recall that if a population falls below a critical number, it will perish regardless of conditions. Natural areas isolated from surrounding similar areas proceed to lose species. The smaller the isolated area, the greater the number of species lost. Thus fragmentation of forests by intervening development is leading to many extinctions because the fragments are insufficient to support critical numbers even though the ecosystems are otherwise preserved. It is being found that the loss of species may be reduced to some extent by keeping the remaining fragments connected by "green corridors."

In short, protecting biodiversity requires considerable ecological understanding.

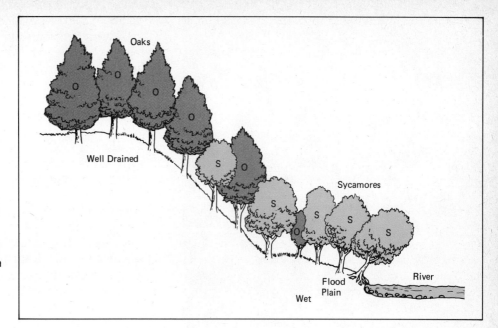

FIGURE 4–5
Competing species coexist in the same area by occupying different habitats. Oaks are better adapted to well-drained soils. Sycamores are better adapted to wet soils and can withstand flooding. Therefore, in deciduous forests of eastern U.S., sycamores are typically found along river banks while oaks are found on the tops of ridges.

the surviving population is held in check by its natural enemies. The fact that monocultures tend to be wiped out by pests while diversified plant communities are not may be one of the reasons that we find diversified ecosystems in nature.

Balances among Competing Plants

A natural ecosystem may contain hundreds or even thousands of species of green plants, all competing for nutrients, water, and light. It is important to understand the balances that exist between such competing plant species.

First, we observed in Chapter 2 that, because of differences in topography, soil type, and so on, the

environment comprises numerous different microclimates. The adaptation of a species to a specific set of conditions enables it to thrive and overcome its competitors in its particular microclimate. An example is the distribution of tree species from riverbank to ridge commonly seen in the eastern United States (Fig. 4–5).

A second factor affecting balance among competing plant species is that a single species generally cannot utilize all of the resources in a given habitat. Therefore, remaining resources may be gathered by other species having different adaptations. For example, grasslands contain both grasses, which have a fibrous root system, and plants that have tap roots (Fig. 4–6). These different root systems enable the

FIGURE 4–6
Plants having fibrous roots may coexist with plants having tap roots because each is drawing water and nutrients from a different part of the soil. (Photograph by BJN.)

FIGURE 4–7
Epiphytes. The plants growing on the tree trunk are
epiphytes. They gain access to light by "perching" on
the branches of trees. They are not parasitic. (Francois
Gohier/Photo Researchers.)

FIGURE 4–8
Balanced herbivory. A plant species may experience a
population explosion as it invades an open area. If it is
killed back and held down by a predator, space is
opened up for a second invader, which may experience
the same fate. Repeating this process for additional
species results in a plant community of many species,
each held in check by its specific herbivores.

plants to coexist because they are getting their water
and nutrients from different layers of the soil. Also,
trees in a forest, while competing with each other for
light in the canopy, leave lots of space near the
ground, and this space may be occupied by plants
(ferns, for example) that can tolerate the reduced light
intensity. Another adaptation is the plethora of
spring wild flowers that inhabit temperate deciduous
forests. Sprouting from perennial roots or bulbs in the
early spring, these plants take advantage of the light
that is abundant before the trees leaf out. In warm
humid climates, tree branches are often loaded with
epiphytes, or air plants (Fig. 4–7). Such plants are not
parasitic; they are simply "perching" on the
branches, where they can get adequate light. They
absorb moisture from the air or rain through their
leaves and survive on the minimal nutrients that seep
from tree leaves or come with rainfall.

A third and very important factor in multiple-
plant balance is called *balanced herbivory*. We observed
above that a monoculture is prone to being wiped out
by an outbreak of a specific insect or fungal disease.
Visualize a monoculture in nature. Its being largely
wiped out by an outbreak of a pest would leave space
that might be invaded by another plant species,
which in turn might be largely wiped out by an outbreak of its pest, again leaving space that might be
occupied by a third, and so on. The end result of this
process would be a diversified plant community with
each species held to a low density by its specific her-
bivore(s) and the herbivores held in check by their
natural enemies (Fig. 4–8). **Balanced herbivory** may
be defined as a diversified plant community held in
balance by various herbivores specific to each plant
species.

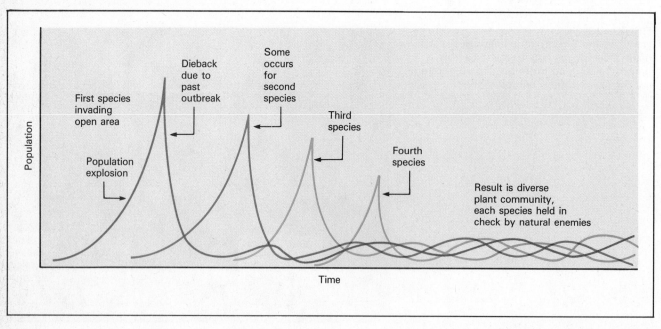

82 _____ *Part One* What Ecosystems Are and How They Work

The best example of balanced herbivory is seen in the tropical rainforests of the Amazon River basin in Brazil. A single acre may contain a hundred or more tree species, but often no more than a single individual of each. The next individual of the same species may be as much as a half mile away. Evidence that this diversity is maintained by balanced herbivory is seen in that attempts to create plantations of single species—rubber plantations, for example—met with failure because outbreaks of various pests proved uncontrollable in the monoculture situation. (However, rubber plantations proved successful in Malaysia, where climatic and soil conditions are different enough to limit pests while still supporting the rubber trees.)

None of these balances is "automatic"; all are the result of adaptations gained over the course of many thousands or millions of years. We see evidence of the specificity of natural balances in numerous ecological upsets that have occurred as a result of introducing foreign plant species. Introduced plants may prove to be more vigorous competitors than native species, and they may lack natural enemies as well. Thus, they "go wild" and may displace any number of native plant species. As this replacement occurs, the entire food web supported by the native plants suffers. Examples abound.

In 1884, the water hyacinth, a plant originally from South and Central America, was introduced into Florida as an ornamental flower. It soon escaped into waterways, where it had little competition and few natural enemies. It has proliferated to the extent of making navigation difficult or impossible (Fig. 4–9a). Millions of dollars have been spent attempting to get rid of this weed, but with little success.

Kudzu, a vigorous vine introduced from Japan in 1876, was widely planted on farms throughout the southeastern United States, with the idea of using it for cattle fodder and also for erosion control. From wherever it is planted, however, kudzu invades and climbs over adjacent forests, smothering everything (Fig. 4–9b). Considerable efforts are now being exerted by the forest service to eradicate it.

Spotted knapweed (Fig. 4–9c), unwittingly introduced into this country probably with alfalfa seed imported from Europe in the 1920s, is taking over vast

FIGURE 4–9
Introduced species do not have balanced relationships with their new partners, so they may overgrow and crowd out natural species. (a) Water hyacinth overgrowing waterways. (U.S. Department of the Army Corps of Engineers photograph.) (b) Kudzu overgrowing forests. (USDA-Soil Conservation Service photograph.) (c) Spotted knapweed, unwittingly introduced from Europe, is taking over rangeland in northwestern United States and southwestern Canada displacing native plants and diminishing grazing of both cattle and wildlife. (Robert Bornemann/Photo Researchers.)

(a)

(b)

(c)

FIGURE 4–10
The range on the left side of the fence has been lightly grazed; the range on the right side has been heavily grazed. Grazing tips the balance between grasses and other plant species such that plants resistant to grazing gain dominance. (USDA Soil Conservation Service photograph.)

areas of rangelands in the northwestern United States and southwestern Canada. Displacing grass and native plants and totally inedible itself, knapweed is endangering wildlife such as elk and reindeer and

rendering the lands worthless for domestic cattle grazing. Control of such weeds by the introduction of plant-eating insects is being investigated. However, finding an insect (or other organism) that will control the target weed but not attack desired species is not easy (see Chapter 10).

The balance between competing plant species may also be upset by the introduction of an herbivore that attacks one plant species but not the other. For example, millions of hectares of pasture and rangelands have become dominated by inedible weeds and scrubs as cattle have eaten back the grass and allowed the competing weeds to flourish (Fig. 4–10).

TWO KINDS OF POPULATION GROWTH CURVES AND THE FOURTH PRINCIPLE OF ECOSYSTEM SUSTAINABILITY

From the above discussion, we may observe that, when population is plotted over time, there are two basic kinds of population curves: S curves and J curves (Fig. 4–11). For example, suppose some abnormally severe years have reduced the population to a low level, but then conditions return to normal. Once normal conditions return, the population may increase exponentially for a time, but then either of two things may occur. One is that natural enemies come into play and cause the population to level off and continue in a dynamic balance at that level. This increase and then leveling off as a dynamic balance

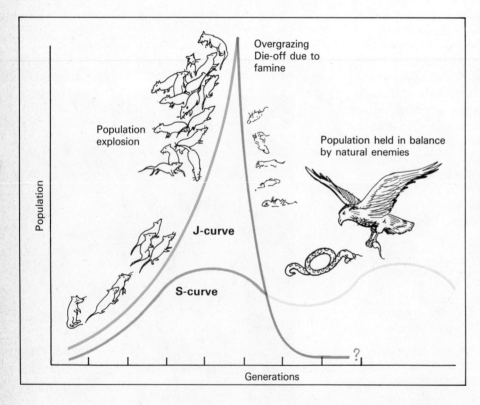

FIGURE 4–11
Two types of growth curves, the S curve (green and yellow) and the J curve (blue and purple). Given optimal conditions, the biotic potential of any species will result in a rapid exponential increase. This initial increase is typical of both curves, but then the curves diverge. In S-curve growth, with effective natural enemies, the population is brought into balance well within the carrying capacity of the system. The dynamic population balance may continue indefinitely (yellow). In J-curve growth, in the absence of natural enemies, the exponential growth continues until overgrazing results in a precipitous dieoff of the population due to famine and malnutrition (purple).

is reached, is known as the S curve (the green and yellow curve in Fig. 4–11). In the absence of natural enemies, however, the second phenomenon will occur. The population will keep growing exponentially until it exhausts essential resources—usually food—and then there will be a precipitous dieoff caused by famine and perhaps diseases related to malnutrition. This exponential growth to a high level followed by a precipitous crash because of overgrazing is known as a J curve (the blue and purple curve in Fig. 4–11).

An important characteristic of J-curve growth is the rapidity with which a population can go from modest levels to the peak and then crash. Consider, for example, that an insect population is doubling (hence doubling the amount it eats) *each week.* Suppose it has taken a given insect population 8 weeks to devour one-half of a crop. How long will it take to devour the second half? The answer is *1 week!* In any doubling sequence, the last doubling necessarily includes one-half of the total. You may test this for yourself by taking a sheet of graph paper and blackening first one square, then two, then four, then eight, and so on, doubling the number each time. You will find that, regardless of the size of the paper or the size of the squares, the last "turn" will involve blackening half the paper. If you double the size of the paper, how many more turns will that provide?

What follows the *J*? There may be any one of three scenarios. First, if the population of herbivores has been reduced to a very low level and the ecosystem has not been too seriously damaged, the producers may recover, and the J phenomenon may be repeated. This scenario is seen in periodic outbreaks of certain pest insects even in natural ecosystems. Second, after the initial J, natural enemies may come into the picture as the ecosystem recovers and bring the pest into an S balance. There is evidence that such a balance is being established in the United States with the introduced gypsy moth. Stands of oak trees that were devastated by the initial invasion of gypsy moths a few years ago are now recovering, and the gypsy moth is remaining at low levels. Finally, damage to the ecosystem may be so severe that recovery does not occur but small surviving populations eke out an existence in a badly degraded environment.

The outstanding feature of natural ecosystems—ones that are more or less undisturbed by human activities—is that they are comprised of populations that are in the dynamic balances of S curves. J curves come about when there are unusual disturbances such as the introduction of a foreign species, the elimination of a predator, or the sudden alteration of a habitat. Nowadays such disturbances are increasingly caused by humans. In short, humans are progressively upsetting the natural population balances that existed in ecosystems and are causing J curves instead.

Again we should visualize the dynamics behind the third principle of ecosystem sustainability: Populations are maintained such that overgrazing does not occur. The dynamic balances at the end of S curves are the sustainable balances between biotic potential and environmental resistance. On the other hand, a J curve is a picture of nonsustainability. The continuing exponential increase necessarily implies the displacement of other species in the ecosystem. Even then, the exponential increase cannot be sustained beyond a relatively few doublings; in animal populations, the exponential increase invariably is followed by precipitous dieback as resources are exhausted.

The fourth principle of ecosystem sustainability becomes evident through this discussion. The dynamic balances at the end of S curves, which provide overall stability and sustainability for the system, depend on the interactions of a great many species. Conversely, we noted that monoculture systems are prone to J curves, which signify lack of sustainability. The term denoting many different species is **biodiversity.** Thus, **the fourth principle of ecosystem sustainability** may be stated:

For sustainability, biodiversity is maintained.

For review, our four principles of ecosystem sustainability are given in Table 4–1.

An ecosystem's biodiversity also underlies its ability to adjust and adapt to various changes. As an example, the diversity of tree species in the deciduous forest ecosystem of the eastern United States enabled the system to adapt to the loss of the American chestnut. What would be left if the American chestnut had been the only tree species? If a system cannot adapt, its sustainability is obviously in jeopardy. Therefore, the concept of adaptability is implied in the fourth principle. The connection between sustainability, adaptability, and biodiversity will become clearer as we study the phenomenon of *ecological succession.*

TABLE 4–1

Principles of Ecosystem Sustainability

For sustainability

- Ecosystems dispose of wastes and replenish nutrients by recycling all elements.
- Ecosystems use sunlight as their source of energy.
- The size of consumer populations is maintained such that overgrazing does not occur.
- Biodiversity is maintained.

ECOLOGICAL SUCCESSION

Given the prevalence of human impacts on the environment, we may tend to think of all ecological change as being caused only by humans. However, the physical environment may also be modified by the growth of organisms themselves. As modification occurs, certain species that were excluded before are able to invade the area. In turn, the growth of the second group of species makes conditions less favorable to the first group; consequently populations of the first group dwindle and may be eliminated altogether. As the plant community changes, the animal community changes also, as different foods and habitats become available. Thus, we observe a *gradual and orderly progression* from one community to another; this process is called **ecological** or **natural succession**.

Succession will proceed until a balanced state among all species and their environment is reached. The final state is referred to as a **climax ecosystem**. The following are three classic examples of natural succession.

Primary Succession

If an area has not previously been occupied, the process of initial invasion and then progression from one ecosystem to another is referred to as **primary succession**. An example is the gradual invasion of a bare rock surface by what eventually becomes a climax forest ecosystem. Bare rock is an inhospitable environment. There are few places for seeds to lodge and germinate, and if they do, the seedlings are killed by lack of water or exposure to wind and sun on the rock surface. However, moss is uniquely adapted to this environment. Its tiny spores, specialized cells that function reproductively, can lodge and germinate in minute cracks, and moss can withstand severe drying simply by becoming dormant. With each bit of moisture, it grows and gradually forms a mat that acts as a sieve, catching and holding soil particles as they are broken from the rock or as they blow or wash by. Thus, a layer of soil, held in place by the moss, gradually accumulates (Fig. 4–12). The mat of moss and soil provides a suitable place for seeds of larger plants to lodge, and the greater amount of water held by the mat permits their germination and growth. The larger plants in turn collect and build additional soil, and eventually there is enough soil to support shrubs and trees. In the process, the fallen leaves and other litter from the larger plants smother and eliminate the moss and most of the smaller plants that initiated the process. Thus, there is a gradual succession from moss through small plants and finally to trees that form a climax forest ecosystem.

The nature of climax ecosystems, of course, differs according to the prevailing abiotic factors of the region. The biomes described in Chapter 2 and in the endpaper foldout are the climax ecosystems typical of each region.

Erosion, earthquakes, landslides, and volcanic eruptions expose new rock surfaces so that primary succession is always occurring somewhere.

Secondary Succession

When an area is cleared, as for agriculture, and then abandoned, the dominant ecosystem of the area will generally return through a series of well-defined stages. Since this process is the reestablishment of an ecosystem that was originally present, the process is termed **secondary succession**. A classic example is the progression from abandoned agricultural fields back to broadleaf trees that occurs in deciduous forest regions of the eastern United States (Fig. 4–13).

On an abandoned agricultural field, crabgrass is predominant among the initial invaders. Crabgrass is particularly well adapted to invading bare soil. Its seeds germinate in the spring, and it grows and spreads rapidly by means of runners; moreover, it is exceptionally resistant to drought. In spite of its vigor on bare soil, however, crabgrass is easily shaded out by taller plants. Consequently, taller weeds and grasses, which take a year or more to develop, eventually take over from the crabgrass. Next, young pine trees, which are well adapted to thrive in the direct

FIGURE 4–12
Primary succession on bare rock. Moss invades bare rock and acts as a sieve, collecting a layer of soil sufficient for additional plants to become established. (Photograph by BJN.)

Part One What Ecosystems Are and How They Work

FIGURE 4-13
Secondary succession. (a) Reinvasion of an agricultural field by a forest ecosystem occurs in stages as shown. (b) Hardwoods (oak species) growing up underneath, and displacing, pines in eastern Maryland. (Photograph by BJN.)

	Year	
Crabgrass	0-1	
Tall Grass/ Herbaceous Plants	1-3	
Pines Come In	3-10	
Pine Forest	10-30	
Hardwoods Come In	30-70	
Hardwood Forest Climax	70+	

FIGURE 4–14
Ravages of erosion in Nepal. Erosion washes away topsoil with seeds and seedlings before they can gain a foothold and leaves an inhospitable subsoil surface which fails to support any new growth. Therefore, severe erosion as seen here prevents secondary succession from getting started and may continue indefinitely. (George Turner/ Photo Researchers.)

FIGURE 4–15
Aquatic succession. Ponds and lakes are gradually filled and invaded by the surrounding land ecosystem.

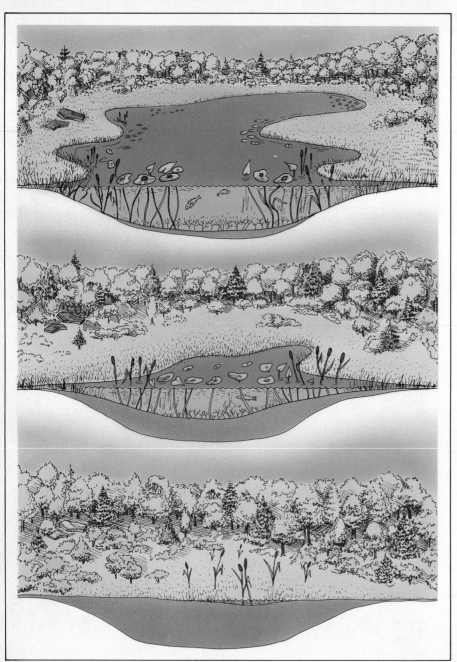

sunlight and heat of open fields, gradually develop and shade out the smaller, sun-loving weeds and grasses, eventually forming a pine forest. But pine trees also shade out their own seedlings. The seedlings of deciduous trees, not pines, develop in the cool shade beneath the pine forest (Fig. 4–13b). Consequently, as the pines die off (their life span is 40 to 100 years), they are replaced by oaks, hickories, beeches, maples, and others that characterize eastern deciduous forests. The seedlings of the latter continue to flourish beneath the cover of their parents, providing a stable balance, the climax ecosystem.

Secondary succession requires that a suitable soil base remain. If this base has been destroyed by erosion, even the initial plants cannot get a start, and, consequently, the bare subsoil may continue to erode and degrade, precluding even the possibility of reforestation (Fig. 4–14).

Aquatic Succession

Another example of natural succession is the gradual conversion of lakes (or ponds) to forest ecosystems. The rivers and streams feeding any lake carry certain amounts of soil washed from the land. Thus, the lake bottom is gradually built up by sediments (as well as by the detritus from the lakeside vegetation), until it gradually becomes more-or-less dry land. As this buildup occurs, terrestrial species are able to advance, and aquatic species move farther out into the lake. In short, the shoreline gradually encroaches toward the center of the lake until finally the lake disappears altogether (Fig. 4–15).

Succession and Biodiversity

In order for natural succession to occur, the spores and seeds of the various invading plants and breeding populations of the various invading animals must be present in the vicinity, so that they can invade the area as conditions become suitable. If this is not the case, natural succession will be blocked or modified. For example, beginning with the early colonization by Norsemen in the 11th and 12th centuries, the forests of Iceland were cut for fuel, a process that accelerated with European colonization in the 18th and 19th centuries. By 1850, not a tree was left standing, and Iceland remained a barren tundra-like habitat because there was no remaining seed source to foster natural regeneration (Fig. 4–16). Tree seedlings are now being imported and planted in Iceland.

In short, natural succession depends on maintaining the biodiversity of the surrounding area.

Fire and Succession

Fire is an abiotic factor that has particular relevance to succession. About 75 years ago, forest and range managers interpreted the potential destructiveness of fire to mean that all fire is bad and embarked on fire prevention programs that eliminated fires from many areas. Unexpectedly, however, fire prevention did not preserve all ecosystems in their existing state. In pine forests of the southeastern United States, for instance, economically worthless scrub oaks and other broadleaf species began to displace the more valuable pines. Grasslands in many situations were gradually taken over by scrubby, woody species that hindered

FIGURE 4–16
Iceland. Forests that originally covered much of Iceland were totally stripped for fuel in the 18th and 19th centuries. Forests could not regenerate through succession because no seed source remained. Therefore, Iceland has remained barren and tundra-like as seen here. Efforts toward reforestation are now beginning. (Ben Simmons/The Stock Market.)

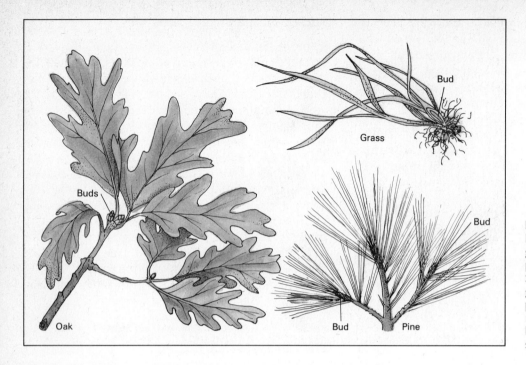

FIGURE 4–17
Position of growing bud gives different degrees of protection against fire. Species such as pines and grasses, which have protected buds, are fire-resistant; oaks do not have protected buds and so are not fire-resistant.

grazing. Pine forests of the western United States that were once clear and open became cluttered with trunks and branches of trees that had died in the normal aging process. This detritus became the breeding ground for wood-boring insects that proceeded to attack live trees. In California, the regeneration of redwood seedlings began to be blocked by the proliferation of broadleaf species.

It is now recognized that fire is a natural abiotic factor and that different species have various degrees of adaptation or resistance to it. In particular, grasses and pines have their growing buds located deep among the leaves or needles, where they are pro-

tected from fire, whereas broadleaf species, such as oaks, have their buds exposed, where they are sensitive to fire damage (Fig. 4–17). Consequently, where these species exist in competition, periodic fires are instrumental in maintaining a balance in favor of pines, grasses, or redwood trees.

In relatively dry ecosystems, where natural decomposition is slow, fire may also play a role in releasing nutrients from dead organic matter. Some plant species actually depend on fire. The cones of lodgepole pine, for example, will not release their seeds until they have been scorched by fire.

Ecosystems that depend on the recurrence of

FIGURE 4–18
Ground fire. Far from being harmful, periodic ground fires are necessary to preserve the balance of pine forests. Such fires remove excessive fuel and kill competing species. Note the absence of dead wood and competing species. (USDA-Soil Conservation Service photograph.)

fire to maintain the existing balance are now referred to as **fire climax ecosystems**. This category includes various grasslands, savannas, and pine forests. Thus, fire is now being used as a tool in the management of such ecosystems. In pine forests, if ground fires occur every few years, there is relatively little accumulation of dead wood. With only small amounts of fuel, there is little danger that a ground fire will ignite a crown fire, and harm to wildlife is minimal (Fig. 4–18). In forests where fire has been prevented for more than 60 years, so much dead wood has accumulated that, if a fire does break out it will almost certainly become a crown fire. This long-term lack of fire was a major factor in the fires in Yellowstone National Park in the summer of 1988. Even those fires, however, served to clear the dead wood and sickly trees that provide a breeding ground for pests, to release nutrients, and to provide for a fresh ecological start. Burned areas soon become productive meadows as secondary succession starts anew. Thus, periodic fires, by creating a patchwork of meadows and forests at different stages of succession, lead to a more varied, healthy habitat that supports a greater diversity of wildlife than the uniform, aging conifer forest (Fig. 4–19).

The devastating fire that spread from brushland into Oakland, California, and destroyed some 1000 homes in the fall of 1991, is another example of lack of fire management. The grasslands on the hills east of San Francisco are a fire climax ecosystem, and preventing fire in this area for many years allowed a superabundance of woody shrubs to grow and dead wood to accumulate. The result was finally a fire that could not be controlled (Fig. 4–20). This tragedy could have been prevented by managed periodic burning. Of course, such burning should never be attempted without the supervision of an experienced manager to be sure that conditions are suitable and backup is available to prevent a fire from getting out of hand.

DEGREE OF IMBALANCE AND RATE OF CHANGE: SUCCESSION, UPSET, OR COLLAPSE

It is important to emphasize that even a climax system is in no way static or rigid; it is simply a system in which all the species involved have reached a balance with one another and with their environment. Any change, such as warming of the climate, increasing acidity due to acid rain, or introduction of foreign species, may set into motion manifold changes as various balances are shifted.

Whether an ecosystem changes rapidly, slowly, or apparently not at all, the *rate* of change is proportional to the degree of imbalance. Natural succession is a gradual, orderly progression from one community

FIGURE 4–19
Fire is a natural abiotic factor and many species are well adapted to it. This photograph was taken one year after the fire in Yellowstone National Park. Note the prolific regeneration of healthy vegetation. Other species will come back in the process of ecological succession. Periodic burning of sections of pine forests actually results in a more diverse and healthy ecosystem. (Renee Lynn/Photo Researchers.)

to another caused by the organisms themselves modifying the environment. Usually, the degree of imbalance at any given time is not great, and hence the rate of change is gradual.

FIGURE 4–20
Oakland, California, fire in 1991 destroyed about 1000 homes. This tragedy could have been prevented with proper fire management of surrounding natural areas. (Michael Jones/Sygma.)

Situations in which one species undergoes a population explosion at the expense of other species, as was seen in cases of introduced species, are referred to as **ecological upsets**. Cases where virtually everything in the ecosystem dies—as may occur with pollution, for example—are referred to as **ecological collapse**. Ecological upsets and collapses occur as a result of sudden or relatively drastic changes.

Implications for Humans

Humans so dominate today's world that there is no ecosystem on earth that has not been influenced to a greater or lesser degree by human activities. And, as long as humans remain on earth, such influence will not end. Sustainability will depend on learning to channel our efforts so that we play a positive supporting role as opposed to a destructive role toward ecosystems. From the understanding gained in this chapter, we should see the potential ecological problems in:

○ introducing species from one ecosystem to another
○ eliminating natural predators
○ losing biodiversity (by clearing land and establishing monocultures)
○ altering abiotic factors
○ misunderstanding the role of fire

A significant fact is that human disturbances of ecosystems by any of the above means do not just gradually modify the dynamic balances. Too often, they are widescale upsets of the ecosystem causing population explosions in some species, perhaps, but losses in many others. In short, ecosystems are easy to upset; restoring balances is much more difficult. It makes more sense to try to preserve the existing balances.

Can you now understand the purpose and value of various governmental and nongovernmental programs aimed at

○ preventing the entry of foreign species
○ eliminating, or at least controlling, the spread of foreign species that have already entered
○ reestablishing natural predators
○ preserving biodiversity
○ preventing alteration of abiotic factors (global warming is a case at point)
○ effective use of fire in management

There are also tremendous implications for us in contrasting the sustainable balances that exist in natural ecosystems with what is occurring in the human system.

First consider the concept of population as a balance between biotic potential and environmental resistance. Until roughly 150 years ago, human populations were largely controlled by such a balance, with periodic epidemics of smallpox, diphtheria, scarlet fever, plague, and other diseases being the major environmental resistance. Then, beginning in the mid-1800s, techniques for the prevention and cure of such diseases began to be discovered and implemented, most significantly vaccinations and improved sanitation and hygiene. The result was that recruitment (survival of babies to adulthood) went from less than 50 percent to over 95 percent today, at least in most countries. The human population explosion, which is currently adding some 93 million people to the population each year, ensued. It is similar to what would occur in any other population released from its natural enemies.

The second and more important point is this: Where is this population explosion headed? Ecologists and environmentalists see a striking similarity between human population growth and the exponential part of a J curve, and they are genuinely concerned that humans may be headed for the consequences that describe all J curves in nature—depletion of essential resources and precipitous dieback as a result. As pointed out at the end of Chapter 3, there is ample evidence that the human species is overgrazing both literally and figuratively. People who believe it is are strong advocates of the necessity to slow population growth and bring it to a steady (no-growth) state. Humans fortunately have an option that is not available to other species—namely, the exercise of self-restraint and the use of contraceptives. Those in the Cornucopian camp, however, are much less, if at all, concerned. They feel that, given our technological and agricultural capabilities, we can support the growing population until such time as it levels out by itself for other reasons that will be discussed in Chapter 6. Extrapolating current trends, this leveling out seems likely to occur toward the end of the next century at between 10 and 14 billion people, two to three times the current population. Can we wait until then?

We shall discuss this issue in much greater depth in Part 2. For now, we wish to examine more closely the question of how species in nature adapt over time to changing biotic and abiotic factors, which is the heart of Chapter 5. Understanding this process will give us additional insight into the extreme importance of preserving biodiversity.

Review Questions _____

1. What are the factors involved in biotic potential and environmental resistance?
2. What is the distinction between reproduction and recruitment, and what occurs in all populations if conditions are ideal?
3. How does population density relate to a balance between biotic potential and environmental resistance?
4. What are the various kinds of natural enemies, and how do they maintain herbivore populations in nature?
5. What is meant by territoriality, and how does it control certain populations in nature?
6. How is the third principle of sustainability—populations are maintained such that overgrazing does not occur—accomplished in natural ecosystems?
7. What are the three main ways that enable different plant species to coexist in the same region?
8. What may occur when plants, herbivores, carnivores, or parasites from foreign regions are introduced into an ecosystem?
9. What is ecological succession?
10. What role may fire play in ecological succession, and how may fire be used in the management of certain ecosystems?
11. What are the two fundamental kinds of population growth curves? What are the causes and consequences of each?
12. What relationship do the two types of growth curves have to the human system and to human impacts on natural ecosystems?
13. What is the fourth principle of ecosystem sustainability?
14. What are potential consequences of not maintaining biodiversity?
15. Do humans need to control their population? If so, what methods are available that are not available to all other animals?

 Thinking Environmentally _____

1. Describe, in terms of biotic potential and environmental resistance, how the human population is affecting natural ecosystems.
2. Analyze the population-balancing mechanisms that are operating among various plants, animals, and other organisms present in a natural area near you.
3. Choose one species (plant or animal) in a natural area near you, and predict what will happen to it if two or three other species native to the area are removed. Then predict what will happen to it if two or three foreign species are introduced into the area.
4. Evaluate such practices as legal hunting, controlling pests with chemical sprays, use or prevention of fire, and poaching of endangered species in terms of supporting sustainable balances.
5. Make an argument, pro or con, regarding sustainability of the human system in terms of the concepts you have learned in this chapter. What new directions do humans need to take, if any, to achieve sustainability?

5

Ecosystems: Adapting to Change—or Not

LEARNING OBJECTIVES

When you have finished studying this chapter you should be able to:

1. Contrast traits and genes, and describe the chemical structure of genes and how they are translated into traits.

2. Relate variations among individuals to genes and alleles and to the gene pool of a species.

3. Define differential reproduction and describe how it may cause changes in the gene pool and traits observed in a population in subsequent generations.

4. Give examples of selective pressures and relate them to differential reproduction.

5. Describe and contrast selective breeding conducted by humans with natural selection in terms of the process and the results obtained.

6. Define mutations and discuss why they may be beneficial, harmful, or neutral.

7. Describe what happens to the three kinds of mutations in natural populations.

8. Define speciation and analyze how it may result from natural selection.

9. Relate natural selection and speciation to the development of a balanced ecosystem.

10. Describe the factors that determine whether a species can adapt or will be forced into extinction by a change.

11. Evaluate which species are most likely and which are least likely to survive changes and why.

12. Discuss the implications of diminishing populations and diminishing biodiversity for the survival of wild species and for agriculture.

In Chapter 4 we observed that ecosystem stability and hence sustainability depend to a high degree on the interplay of many species—in other words, on *biodiversity.* Likewise, both the recovery of damaged ecosystems and succession require biodiversity. We also observed that the balances that exist between species are not automatic but instead are a result of species having gradually adapted over many thousands or even millions of years, and we noted that this ability of species to adapt is a product of biodiversity. These observations gave us the fourth principle of ecosystem sustainability: *For sustainability, biodiversity is maintained.*

Our objective in this chapter is to understand the genetic basis of biodiversity, how it develops, and how it enables adaptation to changing conditions (if the changes are not too great). Likewise, we shall gain a better understanding of how loss of biodiversity may be undercutting the sustainability of our world today.

Gene Pools and Their Change

TRAITS AND GENES

Any organism can be viewed as a combination of *traits.* The term **trait** refers to any particular characteristic of physical appearance (both outward appearance and internal anatomy), metabolism, aptitude, or behavior. Body height, shape of the nose and eye color, bone density, lung capacity and size of the appendix, are examples of physical traits. Metabolic traits include such things as allergic reactions, digestive capacity, tolerance to heat or cold, and disease resistance. Traits pertaining to aptitude refer to any natural talent such as ability in athletics, math, or music. Behavioral traits are the instinctive ways organisms act, such as a spider spinning a web, a bird flying south for the winter, and gentle versus aggressive behavior.

While it is true that many traits may be modified or developed more fully by learning or training, it is also true that the underlying basis of traits is *hereditary, or genetic,* which is to say that individuals are born with certain traits that they inherited from their parents.

Research since about the 1940s has made it abundantly evident that the basis of all genetic traits, regardless of species, resides in molecules of the chemical known as DNA (deoxyribonucleic acid), which resides in the cells of every organism. DNA is a long, chainlike molecule, and the sequence of four subunits along its length provides a "code" analogous to the way the dots and dashes of Morse code provide a code for various letters. The genetic code on the DNA is translated through the cell's metabolism into the production of proteins. In turn, some of the proteins produced, called structural proteins, determine the physical structure of the body—not just the overall structure determining whether the body is that of a mouse, an elephant, a human, or a turnip but down to such details as length of whiskers, nose shape, and hair color. Another set of proteins, called enzymatic proteins, or simply enzymes, specify and control all the chemical reactions that go on in the body. Other proteins either act as hormones or control the production of hormones that in turn control growth and development.

In short, DNA specifies proteins, and the proteins determine everything else (Fig. 5–1). One **gene** can be thought of as a segment of DNA that codes for one particular protein. The physical structure and functioning of every organism are the result of the coordinated interaction of many thousands of genes. In sexual reproduction, one set of genes is provided by the male parent by way of a sperm cell and another set of genes is provided by the female parent by way of an egg cell. The genetic makeup of the new individual is determined as the sperm enters the egg and the two sets of genes combine in the process of fertilization. In each cell division leading to growth of the body, all the genes are replicated (copied) and each daughter cell receives a complete copy of both sets of genes.

SPECIES AND GENE POOLS

It is easy to see that there is *variation* among individuals of the same species. The human species is a good example. We come in a wide range of sizes, shapes, and colors; our bodies have different chemical abilities, such as the ability to digest certain foods; we have different tolerances to various conditions; and we have widely different abilities quite apart from any training or teaching we may receive. Similar variation exists among the individuals of all other species: dogs, elephants, oak trees, mushrooms, and on and

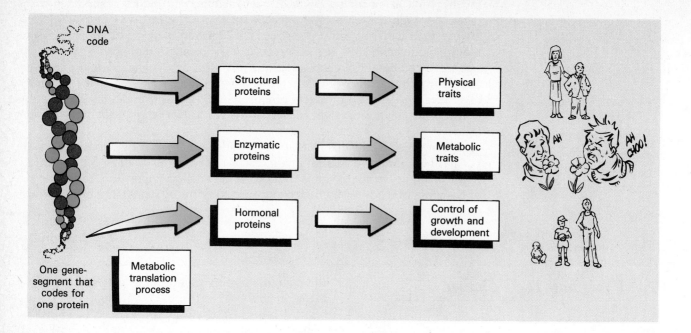

FIGURE 5–1

DNA, genes, and traits. The molecule DNA is the hereditary, or genetic material in all organisms. Each small section of the DNA chain that codes for one protein is a gene. The structure of the DNA molecule is such that the molecule can be replicated and copies passed on through cell division. The sequence of subunits along the DNA chain provides a code that, through cell metabolism, is translated into the synthesis of specific proteins. The proteins of various sorts determine the traits observed in the organism.

on. The only reason we tend to think of all house flies, for example, as identical is that we do not examine them closely. When we do, it is not hard to find slight differences in size, color, or any other trait we choose to focus on.

Variations of a single trait—hair color, for example—result from modifications in a particular gene. These modified genes for a particular trait are called **alleles**. For example, we speak of the gene for hair color; various modifications, or alleles, of this gene may produce blond hair, red hair, or black hair. Another gene controls eye color, and various alleles of this gene result in brown eyes or blue eyes, and so on.

Since each individual has two sets of genes— one from the father and one from the mother—each individual carries two alleles for each gene. The two alleles for any one gene may be different from each other or they may be identical. How they interact determines the trait that the individual shows. Sexual reproduction involves a segregation and recombination of the alleles such that each offspring receives one allele for each gene from its mother and the other allele for that gene from its father. (See In Perspective Box, p. 97.) Thus each offspring has certain similarities to each parent as well as some unique qualities provided by the new combinations of alleles.

Suppose you could take a copy of the genes from every living individual in a population and put all these samples together. What you are visualizing is the *gene pool* for the population. The **gene pool** for a species is the total of all the different alleles of each gene that exist in the entire population of the species. Some alleles in the pool are very common, in which case almost every individual has this particular trait. Other alleles, however, may be relatively rare in the pool, in which case only a few individuals have this particular trait.

In conclusion, we should visualize the gene pool for a species as a pool of alleles, perhaps several for each trait, that is being constantly propagated and mixed through sexual reproduction.

CHANGING GENE POOLS AND TRAITS THROUGH SELECTIVE PRESSURE

If breeding were totally random and if each member of a species produced the same number of offspring, each having the same chance of survival, the gene pool for the species would remain constant in its makeup. However, such constancy is not the case. Some individuals are prolific; others fail to reproduce at all. As a result of this **differential reproduction**, alleles carried by the prolific individuals are repro-

     Part One   What Ecosystems Are and How They Work

Sexual reproduction is at the heart of promoting and maintaining variation among individuals in a population, as well as simply multiplying numbers. Each individual, and this is true whether we are talking about humans, rabbits, or turnips, carries two complete sets of genes. Thus, each individual has *two* alleles for each trait in each of the cells of its body. However, the formation of sex cells (eggs or sperm) is such that each egg or sperm cell receives just *one* set of genes—that is, *one* allele for each trait. In the process of fertilization, the sperm enters the egg to form a cell called the zygote, and the sperm's set of genes joins with the egg's set. The zygote undergoes cell divisions in which all the genes (both sets) are replicated with each division, and the zygote eventually grows into the individual offspring. Thus, each offspring again has two sets of genes. But one allele for each trait came from the father via the sperm cell, and the other allele came from the mother via the egg cell. Each offspring is therefore a "mixture" of traits from its mother and father.

Several different alleles of any given gene may exist in the popula-tion. A well-known example is the three alleles for blood type: A, B, and O. Every individual has two alleles for blood type, which may be two copies of the same allele or any two different alleles—in this case there are six possibilities: AA, BB, OO, AB, AO, BO. Alleles interact in various ways to produce the observed trait. In this case, both AA and AO individuals show A-type blood, and both BB and BO individuals show B-type blood. In other words, A and B alleles are *dominant* over O, and O is said to be a *recessive* allele. To have O-type blood, an individual must be OO.

When sex cells are formed, which of the two alleles goes into any given egg or sperm cell is a random event. Let's look at the possibilities from an AO father and a BO mother. Half of the father's sperm will carry the A allele, and the other half will carry the O allele. Half of the mother's eggs will carry the B allele, and the other half will carry the O allele. Which sperm will fertilize which egg is also a random event. Therefore, our possibilities are:

A sperm + B egg → **AB** offspring
A sperm + O egg → **AO** offspring

O sperm + B egg → **BO** offspring
O sperm + O egg → **OO** offspring

You can see that, for a given trait, any particular offspring may be like the father, like the mother, a combination of both (the AB type), or unlike either (the O type). As these offspring go on to interbreed and produce another generation, every possible combination may occur. Differences in hair color, eye color, body height, nose shape, and so on are similarly based on individuals' having different alleles for the genes involved. In any given mating, there is the simultaneous recombination of alleles for the many thousands of genes constituting each individual's genetic makeup. You can see why no two individuals are exactly alike.

Identical twins are apparent exceptions. They are genetically identical. However they are not really exceptions because they are *not* products of separate fertilizations. They result from the separation of the two cells that result from the first division of the zygote. Each of these cells then develops into a separate embryo. Because the two embryos are derived from the same zygote, they have exactly the same genes.

---

duced and become more abundant in the population, while alleles of individuals that fail to reproduce are eliminated from the population.

What causes some individuals to reproduce more than others? For species other than humans, differential reproduction involves what we call **selective pressures**. (For humans, it involves a host of additional social, economic, cultural, and even political factors. Therefore, for the moment, we must leave humans out of the picture.) For domestic and agricultural species, *selective pressure* is applied by humans; in the natural world, it is applied by nature. In both cases, the result is the same: Some individuals reproduce more than others, and as a result the gene pool, and hence the traits and the species itself, are altered. Let us look at this process in more detail.

Change through Selective Breeding

All agricultural and domestic breeds of both plants and animals have been derived from wild populations through the process of *selective breeding*. The most basic technique behind all such breeding is the following. Breeders first envision the traits they would like to achieve in a given species: a dog with short, squat legs that is able to wiggle into animal burrows to aid hunters, for example. The breeders then examine the existing population of dogs and *select* those individuals that show the sought-after trait (short, squat legs in our example) a little more than other members of the population. The selected individuals are then bred. The offspring tend to be like the parents, but some offspring express the particular trait

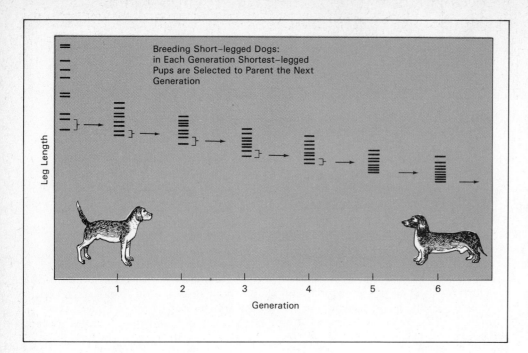

Leg Length

Breeding Short-legged Dogs:
in Each Generation Shortest-legged
Pups are Selected to Parent the Next
Generation

Generation

FIGURE 5–2
Selective breeding is a crucial technique in the development of all agricultural and domestic plants and animals. It entails selecting from the existing population those individuals showing the desired features to the highest degree, breeding them, and preventing others from breeding. As this process is repeated over many generations, the desired features are gradually developed, as shown here for the development of the dachshund.

more than the parents did, others express it less. Those offspring that show the trait most are *selected* as the parents for the next generation, while the other offspring are prevented from breeding, and this process of selection and breeding is repeated over and over until gradually the desired trait is achieved. This process is illustrated in Fig. 5–2, and the final result in this case is the breed you recognize as dachshunds, which was originally developed for the burrowing purpose mentioned earlier.

Breeders do their selective breeding on the basis of visible traits. However, since traits are the manifestation of certain alleles, what is actually occurring is that the breeders are selecting and propagating those alleles which give short, squat leg structure. At the same time, alleles that give taller leg structure are eliminated from the gene pool because the dogs with longer legs are not allowed to reproduce. That the DNA in the gene pool actually changes in the process can be verified by techniques that enable scientists to compare the DNA of the starting population with that of the developed breed.

An important point to remember is that the same starting gene pool can be manipulated through selective breeding for any number of different traits. Thus all the different breeds of dogs, from Great Danes to Chihuahuas, are derived from the same wild dog gene pool, and the same is true for multiple breeds of all other species.

Change through Natural Selection

Populations in nature are also under continuous selective pressure, but here the contest is simply for *survival and reproduction*. Recall from Chapter 4 that the recruitment rate of most species is very low; that is, most offspring do not live to become reproducing adults but instead succumb to various factors of environmental resistance.

Basically, each biotic and abiotic factor that affects a species acts as a selective pressure. Any trait that enables an organism to cope with a given factor more effectively than organisms without the trait benefits survival and thus reproduction. Thus, the alleles responsible for such traits are selected for; they are passed on to the next generation. Conversely, alleles responsible for any trait that handicaps an organism in any way tend to be eliminated from the population because individuals carrying them fail to survive or reproduce.

In the same way that the leg length of successive generations of a dog population diminished under selective pressure applied by humans, species in nature become increasingly well adapted to the various biotic and abiotic factors affecting them. Since the process occurs naturally, it is known as **natural selection**.

What we have described here is exactly the concept presented by Charles Darwin in 1859 in his famous book *The Origin of Species by Natural Selection*. The phrase "survival of the fittest" means the survival of those individuals having traits that best enable the individuals to cope with the biotic and abiotic factors of their environment. Darwin deserves great credit for constructing a picture of what was occurring purely from empirical evidence without any knowledge of genes or genetics, information that wasn't discovered until some years later. Our modern understanding of DNA, mutations, and genetics fully supports Darwin's Theory of Evolution.

The fact that the gene pool of a species changes only through differential reproduction has tremendous practical, ethical, and moral implications for us, the human species. First, we must dispel a thoroughly discredited but still commonly held notion called the *inheritance of acquired characteristics*. This is the *false* notion that characteristics acquired by parents through training, learning, or accident will somehow be inherited by the offspring. Tremendous amounts of experimentation and common experience show that such acquired characteristics are not inherited because they absolutely do *not* affect the DNA of the parent. The offspring will inherit the same genes and alleles no matter what kind of skills the parents have developed.

A special aspect of this false notion concerns genetic disorders. There are many congenital or genetic diseases, defined as conditions that are caused by certain mutant genes, that can be held in check with medicines or treatments. Certain forms of diabetes, sickle cell anemia, and hemophilia are examples. Medicines do not correct the defective gene, however. Thus any offspring may inherit the disorder and thus require the same treatment.

You may have heard (or even expressed) such statements as, "The human race is evolving toward bigger brains because we are required to use them," or "The human race is evolving toward smaller legs because we don't use them." You should recognize that such ideas are basically statements of the discredited notion of inheritance of acquired characteristics. The only way such events would come to pass is if, for example, little-legged people consistently had more children than regular-legged people. The way to arrive at an answer to any question regarding the future evolution of the human race is to reflect upon the following question: Are people with this characteristic consistently reproducing more (or less) than average? If the answer is no, then there is no basis for projecting such change.

In short, the only way the genetics of a population will change from one generation to the next is through differential reproduction. The population will become increasingly dominated by the characteristics of those who reproduce the most, and a smaller and smaller proportion will bear the genetic traits of those who reproduce the least.

Speculate about what might be done with the human species if selective breeding of it were feasible, however. The study of potentially improving the genetic future of humans is known as eugenics and has a cadre of serious advocates, but implementing any eugenics program collides with moral and ethical issues. Consider the ethics of one group dictating how many children another group should or should not produce. Who is to make the decisions, and how are they to be implemented? People who have tried this in the past—Hitler was one—are remembered as "monsters."

Yet, through suitable education, information, and understanding, some people may choose to practice a form of eugenics on a personal level. Prospective parents may obtain genetic counseling regarding the genetic basis of any defect they may have and its probability of being passed to an offspring. They can then choose whether or not to bear their own children. Also, it is now possible to perform DNA analy-ses on a fetus and determine the presence or absence of genetic defects. Parents can then make the decision as to whether or not to have the fetus aborted. Of course, this assumes that having an abortion is a legal and acceptable option, a moral dilemma in itself.

A practice already widely used in animal breeding is to collect sperm samples from certain males, a prize race horse or bull, for example. The sperm is kept frozen and then used to artificially inseminate females whenever and wherever desired. Thus, the genes of the prize male are passed on to a much larger portion of the next generation than would be possible in natural breeding. How do you feel about using this practice in human breeding? Artificial insemination of women with sperm from an unknown man is already commonly practiced. At least one company in California maintains a sperm bank of samples provided by Nobel prize laureates and other talented people. A woman can choose to make one of these famous men the biological father of her child by choosing to be artificially inseminated with his sperm.

Through advances in genetic engineering, it may be possible in the relatively near future to manipulate and "correct" genes in the developing embryo. To what degree should such manipulation be conducted? Who should make the decision, and who will pay the costs?

In short, technologies for identifying, manipulating, and propagating or not propagating particular genes is becoming increasingly available. More and more we will have to face the ethical dilemmas involved in choices of using or not using these technologies.

Indeed, virtually all traits of all species can be looked at in terms of adaptations that support the survival of the species. Adaptations can be grouped accordingly:

1. Adaptations for coping with climatic and other abiotic factors.
2. Adaptations for obtaining food and water, in the case of animals, or nutrients, energy, and water in the case of plants.
3. Adaptations for escaping from or protection against predation and for resistance to disease-causing or parasitic organisms.
4. Adaptations for reproduction—for finding or attracting mates in animal populations or for pollination and setting seed in plant populations.
5. Adaptations for animal migration or dispersal of plant seed.

The fundamental question for any trait is: Does it support survival and reproduction of the organism? If the answer is yes, the trait will be maintained through natural selection. Consequently, various organisms have evolved different traits to accomplish the same function. For example, the ability to run fast, to fly, or to burrow and features such as quills, thorns, and obnoxious smell or taste all support the function of reducing predation and are seen in various organisms. Likewise, when we look at other functions, we find that different organisms have evolved a wide variety of traits which accomplish that function (Fig. 5–3).

You may have observed that in this discussion we have spoken of "survival and reproduction" together. This is because in the process of breeding the two can't be separated. It is obvious that an organism won't reproduce if it doesn't survive. However, it should be equally clear that survival by itself is not enough. An individual that lives to a "ripe old age" without ever reproducing is the same as one that died in infancy in terms of passing its genes on to future generations. Any effect on the genetics of future generations is accomplished only through reproduction. Furthermore, the genetic effect of an individual on the next generation is directly proportional to the number of offspring that it produces that survive and reproduce.

Mutations—The Source of New Alleles

If selection were the only process at work, change would first level off and then cease as the most effective combination of *existing* alleles was reached. Experience shows, however, that this does not occur.

Regardless of how far the development of a given trait is pursued, breeders always observe that there is still variation to provide the basis for further selection and development. Thus, chickens having tail feathers 15 feet (5 m) long have been bred, and still the breeding for longer feathers continues. Breeders have found they can produce a dachshund with legs so short that it can't walk. Thus, there is a practical limit but not a genetic limit.

Modern knowledge of DNA allows us to understand this lack of genetic limit. In the course of cell division and reproduction, the DNA molecules are generally copied exactly, but occasionally *mutations* occur. A **mutation** is a change in the DNA molecule. A small change causes the protein coded by the changed segment to be altered. In turn, the trait determined by that protein will be modified in one way or another. In other words, mutations introduce into the population new alleles and hence new modifications of traits. (If the mutation involves a large change in the DNA, one or more proteins are so altered that the organism fails to function and dies as a result. Mutations that result in death are termed **lethal mutations**.)

Very important, however, is the fact that *mutations are random events*. There is no way, so far as is known, either in nature or through human technology, of causing a specific mutation to cope with a particular selective pressure. Indeed, like randomly turning screws in an engine, most mutations result in modifications that are harmful; only rarely is one beneficial. How does nature cope with the problem of sorting out the good mutations from the bad? Again it is the processes of natural selection and differential reproduction. Any individual having a harmful mutation will usually perish before it can reproduce. Thus, the harmful allele is eliminated from the gene pool. The individual with the rare mutation (allele) that enhances survival, however, is likely to produce offspring. As these offspring, which inherit the new trait, reproduce in their turn, the population will comprise more and more individuals that carry the new allele. Mutations that are neutral, resulting in neither benefit nor harm, may remain in the population and simply lend to the variation among individuals.

DEVELOPMENT OF SPECIES AND ECOSYSTEMS

Now let us relate this concept of change through natural selection to the development and particularly the sustainability of ecosystems. Again, recognize that all the abiotic and biotic factors present in an ecosystem

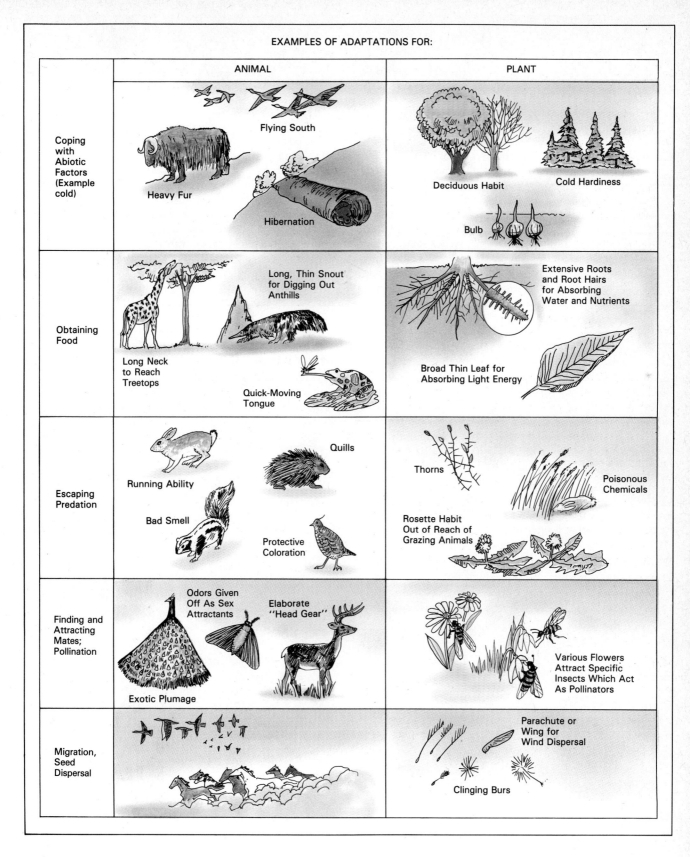

FIGURE 5–3

Examples of adaptations. Every species can be viewed as a complex of adaptations that increase the chances of survival. In each case a multitude of features will accomplish the same function.

We have seen that mutations are the source of new alleles, which ultimately enable evolutionary adaptation. Yet, factors that cause mutations, *such as high-energy radiation from nuclear wastes and various drugs and chemicals,* are considered to be exceedingly dangerous. Why? Because mutations involve *random* changes in DNA and therefore are most likely to result in changes that are harmful rather than beneficial. This fact causes a particular problem for humans. Through natural selection, wild species can "throw away" any number of harmful mutations and capitalize on the occasional beneficial one. Indeed, plant breeders frequently subject batches of seed to radiation specifically to cause mutations. They then grow out the seed, pick the occasional desirable variant, and relegate the rest to the compost heap.

Obviously, application to the human species of anything approaching this practice is ethically repugnant. We are ethically bound to provide the best care possible to each human regardless of deformity. Consequently, each mutation, insofar as it is likely to be harmful, tends to burden society with additional health-care costs and suffering of the individuals afflicted. Our best recourse then is to try to avoid anything that will cause additional mutations.

act as selective pressures on each species. Consequently, through natural selection over generations, populations are either modified to cope with existing biotic and abiotic factors or they become extinct—the latter are the species we know only from the fossil record.

One of the most severe selective pressures in nature is competition between members of the same species. For example, if a population takes to browsing on leaves overhead as its main source of food, it will be the individuals with the longest necks that get the most food and leave the most offspring; shorter

FIGURE 5–4

Selective pressure. A selective pressure is any factor that enables individuals with a particular variation to survive and reproduce more than individuals without that variation. In the evolution of the giraffe, we can readily visualize that once the behavior of browsing on trees started, there was a strong selective pressure for longer necks in that longer-necked individuals got to eat and shorter-necked individuals didn't. When a longer neck is of no additional benefit, selection pressure and further modification cease.

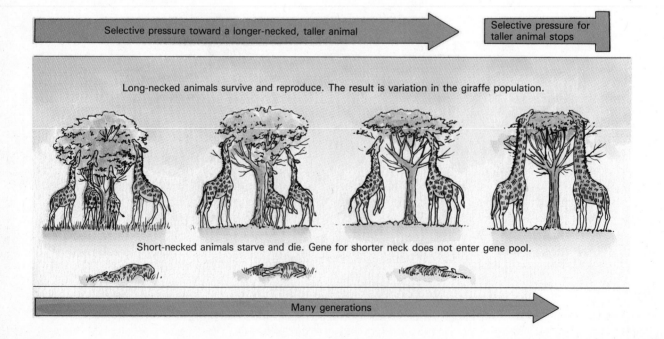

Selective pressure toward a longer-necked, taller animal

Selective pressure for taller animal stops

Long-necked animals survive and reproduce. The result is variation in the giraffe population.

Short-necked animals starve and die. Gene for shorter neck does not enter gene pool.

Many generations

FIGURE 5–5
Giant anteater. The major herbivores (in terms of biomass consumption) on the savannas of Brazil are termites, which build large earthen mounds as nests. The giant anteater, actually a termite eater, has evolved huge forelimbs and claws that it uses to break open the mounds and a head–mouth structure adapted to feeding on the termites. (Photo Researchers.)

individuals will perish. As this weeding out occurs over many generations, the end result, as you may guess in this case, will be the giraffe (Fig. 5–4). Similarly, a population that takes to feeding on ants and termites will, over generations, be modified to be more and more proficient in this regard, and we will have the giant anteater of Brazil (Fig. 5–5). The important point here is that, in addition to adapting species to cope with all the biotic and abiotic factors of the ecosystem, natural selection also generally results in species becoming highly specialized to specific habitats and niches, a feature we observed in Chapter 4.

In this process of adaptation, the final "product"—giraffe, anteater, whatever—may be so different from the population that started the process that it is considered a different species. This is one aspect of the process of **speciation**.

Speciation may also result in two or more species developing from one. This may occur when populations of the same species become physically isolated from one another, and then different biotic or abiotic factors or both act on the two populations to cause their modification in different directions. Isolation of the two populations is necessary because if the two populations continue to interbreed, there will be enough mixing of alleles to ensure that all the members remain as one species. Take foxes, for example. It is assumed that long ago an ancestral population broke into two subpopulations, one migrating into the Arctic and the other remaining in the south. In the Arctic, selective pressures favor individuals that have heavier fur, shorter tail, legs, ears, and nose (all of which help conserve body heat), and white color (which helps animals hide in snow). In the southern regions, selective pressures regarding adaptation to temperature and background color are the reverse. Adaptations for the Arctic would actually be

harmful in southern animals because in warmer climates animals need to dissipate excessive body heat, and a white coat would make animals more conspicuous. The result of this selection over many generations is the two species known today as the arctic fox and the gray fox (Fig. 5–6).

Plant height offers another example of speciation. In alpine regions (high elevations above the treeline), virtually all plants have a low, dense, mosslike vegetative structure. The flowers on these plants, however, are very similar if not identical to those of tall plants found at lower elevations, indicating a close genetic relationship between the tall and short plants. One can visualize that, as the plants dispersed up the mountain, the abrasive force of ice crystals driven by strong winds sheered off and killed taller plants. Thus there was a strong selective pressure for the low, dense, mosslike habit. In some cases, populations showing all the intermediate stages continue to coexist (Fig. 5–7).

This model for how new species form explains why in nature we generally find groups of closely related species rather than single species that have no close relatives. Perhaps the most studied example is a group of some 14 species of finches living on the Galapagos Islands, located in the Pacific Ocean about 650 miles (1000 km) west of Ecuador (South America) and described by Darwin in *The Origin of Species.* That all of these species belong to the finch family is easy to see by observing overall body characteristics. The different species of finches occupy various niches differentiated by feeding habit. One species feeds on small seeds, another on large seeds, another on cactus vegetation, and so forth, all the way up to one species that uses a cactus thorn to pry insects from crevices. Apparently some 10,000 years ago a population of migrating finches from the South American mainland

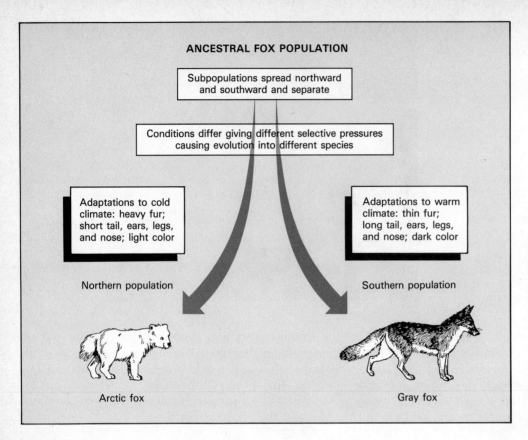

ANCESTRAL FOX POPULATION

Subpopulations spread northward and southward and separate

Conditions differ giving different selective pressures causing evolution into different species

Adaptations to cold climate: heavy fur; short tail, ears, legs, and nose; light color

Northern population

Arctic fox

Adaptations to warm climate: thin fur; long tail, ears, legs, and nose; dark color

Southern population

Gray fox

FIGURE 5–6
A population spread over a broad area may face different selective pressures in different regions. If the population splits so that interbreeding among the subpopulations does not occur, the different selective pressures may result in the subpopulations' evolving into different species, as shown here for the arctic fox and the gray fox.

were blown westward by a freak storm and thus were the first terrestrial birds to arrive on the relatively new volcanic islands. As the initial population grew and members faced increased competition from each other, groups dispersed to nearby islands, where they were separated from the main population. These

FIGURE 5–7
Speciation in progress. All these plants are still one species, as demonstrated by the fact that they are all capable of interbreeding naturally. Environmental factors clearly favor different variations in different areas, however. Given time (1000+ generations), the high-altitude and low-altitude plants may diverge enough to become separate species. (Redrawn from J. Clausen, D. Keck, and W. Hiesey, Carnegie Institute of Washington Publication No. 581, Washington, D.C., 1948.)

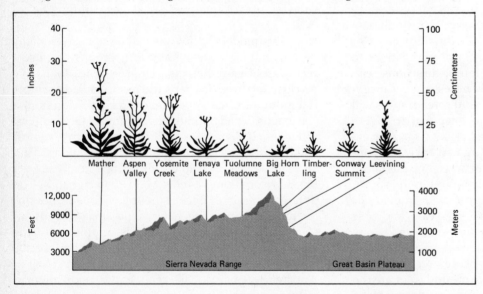

Mather · Aspen Valley · Yosemite Creek · Tenaya Lake · Tuolumne Meadows · Big Horn Lake · Timberling · Conway Summit · Leevining

Sierra Nevada Range · Great Basin Plateau

A species was originally, and to a considerable extent still is, defined simply as a distinct kind of organism. The problem is: What constitutes a "distinct kind"? For example, two populations of tree frogs, one in New England and one in Georgia, appear to be distinct enough to be classified as different species. In the intervening states, however, we find tree frog populations that, when all put together, create a smooth transition of variations from the New England to the Georgia populations. Should all these frogs be classified as one large species with much variation, as two species, or as many very similar species? This problem exists with numerous kinds of plants and animals, and there is no clear-cut answer.

Taxonomists, those who make a profession of classifying organisms, fall into two philosophical camps, "splitters" and "groupers." Splitters recommend dividing groups, such as the frogs noted above, into multiple species; groupers recommend putting them all into one species.

In an attempt to remedy the situation, the definition of species was amended to include the aspect of interbreeding. If interbreeding occurs among individuals, they are one species regardless of how different they may appear. By the same token, if interbreeding does not occur between two groups, they should be considered as separate species. While this definition may help in an intellectual sense, it often does not help in a practical sense, mainly because interbreeding is often impractical or impossible to test or observe. Also, natural interbreeding may not occur between two populations of similar animals, defining them as separate species. When members of the two populations are placed together *in captivity*, however, they may interbreed readily. This situation is even more problematic in plants that generally do not cross-pollinate in nature but can be readily crossed by artificial pollination. The offspring from populations that generally do not interbreed in nature but can be induced to do so are termed **hybrids**.

So, what is a species, then? We can conclude only that a species, as a particular kind of organism distinct from all others, is a hypothetical construct of the human mind, which likes to categorize things and put them in discrete pigeonholes. However, if species are undergoing a process of separation and change as we have described, then examples such as those described above are what we should expect to observe. For the sake of discussion, we still need to use the term *species* to refer to kinds of organisms. However, the problem in defining precisely what a species is should be taken as reinforcement of the concept that species are always in the process of changing and creating new species.

---

subpopulations encountered different selective pressures and became specialized for feeding on different things. In time, when these changed populations dispersed back to their original island, they were different enough from the parent species to be distinguishable as new species and different enough that interbreeding among them did not occur (Fig. 5–8).

For study purposes, it is necessary to focus on the adaptation and change of one species at a time. However, one species in an ecosystem cannot change without altering its relationships with other species in the ecosystem. Thus all the species and hence the ecosystem itself undergo simultaneous adaptation and change through natural selection.

Selective pressures at the ecosystem level lead toward the various kinds of balanced relationships discussed in Chapter 4, because only balanced relationships are sustainable. If there is a lasting imbalance between two species, the disadvantaged species will obviously be forced into extinction. So will the better-adapted species unless it succeeds in establishing balanced relationships with remaining species. As unbalanced relationships are thus weeded out, the ecosystem will inevitably evolve toward one of balanced relationships among the species involved. Thus, the ecosystems we studied in earlier chapters, and all the balanced relationships among species that they encompass, should be recognized as the product of selective pressures acting over many thousands or even millions of years.

In addition to selective pressures, another feature profoundly affects the way ecosystems develop: the species present at any given stage. Contrary to science fiction, neither nature nor humans have any way of creating from scratch an entirely new gene pool or an entirely new species. Selective pressures and mutations can only modify what already exists. *What exists can be modified; what does not exist cannot be created.*

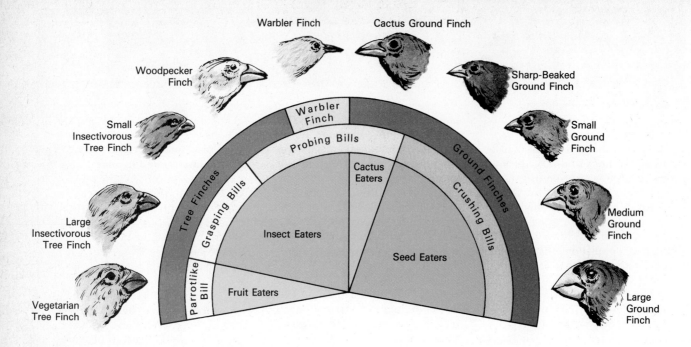

FIGURE 5–8
Darwin's finches. The similarities among these birds attest to their common ancestor. Selective pressures adapting subpopulations to feed on different foods has caused their modification and speciation. (Adapted from Raven, P. H. and G. Johnson. 1988. Biology, ed. 2. St. Louis, Mosby-Year Book, Inc. Original illustration by George Venable.)

We have just seen how a single population of finches on the Galapagos Islands diversified through natural selection to fill the available niches—seed eating, vegetation eating, woodpecker habit—which on the South American mainland are occupied by different species of birds. In other words, natural selection worked to modify what existed—finches—to fill the ecological niches. Consider what the situation today might be if a different species of terrestrial bird had been the first to get to the Galapagos Islands.

The Galapagos Islands are perhaps even better known for their gigantic land tortoises, which are the primary large herbivore of the islands (Fig. 5–9). This is another example of ecosystem development. The volcanic origin of the islands and the invasion by plant species—certain seeds can be carried incredible distances by wind or migrating birds—provided a grazing niche. Typical mammal species had no way of getting to the Galapagos, but sea turtles did. Thus, it was turtles that evolved into the grazing niche. (In more recent times, mammals have been introduced. Goats and dogs in particular have all but caused the extinction of the tortoises. The goats are much more

FIGURE 5–9
On the Galapagos Islands, mammals were not present but turtles were. Therefore tortoises evolved as the prime grazing animal. (Miguel Castro/Photo Researchers.)

Part One What Ecosystems Are and How They Work

efficient grazers than the tortoises and thus consume all the food the tortoises would ordinarily eat, and dogs dig up and eat the turtle eggs.)

As noted, the Galapagos Islands had a relatively recent (only 10 000 years ago) volcanic origin. On a time scale of millions of years, even the major continental land masses are in a state of continuous flux, sometimes splitting apart and sometimes coming together. Land bridges have come and gone as sea level has varied. These changes will be described further in the geology section at the end of this chapter. For our purposes at the moment, it is sufficient to recognize that different land masses, large and small, have become available or splintered apart in different times. Which species were either present at the time of separation or else gained access later profoundly influenced the subsequent course of speciation and ecosystem development. Indeed, it is the influence of the original species that result in the species makeup of an ecosystem on one continent or remote island being very different from that on another continent or island, despite similar climates in the two locations. A familiar example is seen in comparing Australia with North America. The climates on the two continents are similar, but the large grazing mammals—various species of kangaroos in Australia; bison, deer, and antelope in North America—are vastly different (Fig. 5–10).

Despite all the different species found in different ecosystems, every ecosystem develops the same basic biotic structure of producers, consumers, and detritus feeders and decomposers, and each functions in the same way as we observed in earlier chapters. Therefore the four principles of ecosystem sustainability are universal; they apply to all ecosystems regardless of the species involved. Furthermore, different ecosystems coming into balance under particular climatic conditions develop overall similarities despite different species being involved. For example, the grasslands in Australia, South America, and North America have similar outward appearance and are all categorized as grassland biomes, although in each case the particular species of grasses, other plants, grazing animals, and carnivores are different.

In conclusion, we see that selective pressures acting on the gene pools of all species gradually lead to the development of a balanced ecosystem. Once a balanced status is reached, selective pressures tend to preserve the status quo—*as long as conditions remain constant.*

But conditions do not remain constant. The earth is not a static body. Through various geological processes, climate, sea level, and other abiotic factors are always changing slowly (generally over tens of thousands to millions of years). As one or more factors change, selective pressures are shifted, and thus each species is pushed again toward either further speciation or extinction. While the fossil record is rife with extinct species, the natural world of today is proof that ecosystems have managed to cope with geological changes.

A major reason that species and hence ecosystems are able to cope with geological changes is that geological changes occur slowly over the course of

FIGURE 5–10
In Australia the marsupials present (mainly kangaroos) evolved as the prime grazing animal. (Australian Information Center photograph.)

FIGURE 5–11
Because of isolation from other ecosystems, the ecosystem on the Hawaiian Islands developed thousands of species that are found nowhere else. More than 500 of these unique species are in danger of extinction as a result of the introduction of more-vigorous competing species. Shown here is the silversword. (Photo Researchers/Stan Goldblatt, 1980.)

thousands and millions of years. Gene pools may be modified by selective pressures rapidly enough to keep abreast of such slow changes. A major concern of environmentalists is that now humans are causing changes in both biotic and abiotic factors that are not slow; instead, these changes are on orders of magnitude (one order of magnitude is a factor of 10) faster than natural processes. For example, if projections for global warming come to pass (Chapter 16), the climatic changes we cause in the next 100 years will be greater than all the climatic changes that have occurred over the past 10 000 years, that is, 100 times faster. Another comparison: Before human transport, the Hawaiian ecosystem was invaded by a new species from the mainland perhaps once every 10 000 years. Now several hundred species of various plants and animals have been introduced in the past 200 years. Is it any wonder that the natural Hawaiian ecosystem is being overwhelmed and exterminated by the new invaders (Fig. 5–11)?

The critical question, then, is: *How fast can species adapt?*

LIMITS OF CHANGE—ADAPTATION, MIGRATION, OR EXTINCTION

As any biotic or abiotic factor in a given ecosystem is changed (for instance, a species from another ecosystem is introduced, or the climate changes), each species that is ill-adapted to the new situation faces three possibilities:

IN PERSPECTIVE
Punctuated Evolution

We have seen that, through natural selection, species gradually adapt to the biotic and abiotic factors of their ecosystem (if they don't adapt, they become extinct), and the ecosystem itself evolves balances that are sustainable. As long as factors remain constant—that is, within the normal range of climatic fluctuations—a climax ecosystem may sustain itself with little change for many thousands of years. This is because, as species become adapted to particular habitats and niches and as balances are established, selective pressures work in large part to preserve the status quo. For example, selective pressure maintains the gi-

raffe population at a height that can graze to the tops of available trees. Shorter necks are eliminated by natural selection for obvious reasons, but so are longer necks because the longer the neck, the more difficult it is for the animal to get down to drink water and perform other maneuvers. This kind of argument can be made for all species in a balanced ecosystem.

Darwin conceived of evolution as a process that was always going on, slowly but steadily. We now recognize that evolution may in fact go in "steps." As ecosystems reach a balanced state, there is little change. Then one or another factor is

changed. This shift alters selective pressures and sets into motion fairly rapid changes in almost all, if not all, species in the ecosystem until a new balance is reached. This start-and-stop model is referred to as **punctuated evolution**.

Even the relatively "rapid" changes seen in punctuated evolution occur over the course of tens of thousands of years. In other words, this view of evolution in no way changes the basic mechanism: Regardless of whether we picture a smooth, continuous process or one of fits and starts, natural selection is still the driving force.

---

1. **Readaptation** and perhaps speciation through natural selection to adjust to the new conditions

2. **Migration** and finding another area where conditions are the same as they were previously

3. **Extinction.** Failing possibilities 1 and 2, extinction is inevitable.

What factors determine whether a species can or cannot adapt to new conditions? Our understanding of natural selection enables us to answer this question. Recall from our earlier discussion that natural selection works only on existing variations. In our example of the giraffes, for instance (Fig. 5–4), variation in neck length already existed. Adaptation through natural selection was a process of short-necked animals starving and longer-necked animals surviving and reproducing. This process was repeated over many generations.

An important point to realize is that organisms other than humans can never anticipate future changes. In addition no organism is able to initiate a mutation that will allow it to adapt to a new environmental challenge. Also, a new environmental challenge can never cause a specific mutation that will provide adaptation to the challenge. In short, the only way a species can survive and adapt to a new situation is if some individuals *already have* variations (alleles) that enable them to cope with the new condition(s). Only if those individuals survive and reproduce will the next generation be better adapted through inheriting those alleles. As this selection is repeated over many generations, a population that is better and better adapted will gradually be produced.

A good example of species adapting to new conditions is the development of populations of insect pests that are highly resistant to the insecticides that have been used against them. At first these insecticides were highly effective, killing over 99 percent of the insects. There were some survivors, however, and it has been clearly shown that alleles for resistance to the insecticides, which these survivors carried, existed before the insecticides were used. With their phenomenal reproductive capacity, the insect populations recovered. The new population, which arose from that 1 percent who survived the spraying, was more resistant than the original population.

Obviously, if no members of a species can survive the new condition, the species becomes extinct unless a population of the species can migrate or already exists in another region where the change has not occurred.

From this understanding, we can identify several parameters that determine whether a species will be able to survive a change:

1. **Rate and degree of change.** If a change in climate, for example, occurs very slowly (over thousands of years), it is likely that populations of most species will have many members that can survive the first increments of change; at first, only the most sensitive members will die off. As the more resistant members of the population reproduce, the growing resistance in the population may well keep pace with the change. If the rate of change at any point exceeds the rate at which natural selection leads to adaptation, however, extinction will occur. Even if changes occur slowly, there are biological limits to adaptability, and so degree of change is as important as rate. As temperatures go above 110°F (45°C), for example, proteins begin to break down. Since all organisms need protein in order to function, they will be unable to adapt to temperatures consistently above this point. Likewise, any number of other factors could go beyond the biological limits of adaptability.

2. **Genetic variability in the starting population.** If the starting population is large and has considerable genetic variation, the chances are that at least some individuals will have alleles that enable them to cope with and survive the change even if it is a fairly significant one. Thus, the potential for further reproduction, natural selection, and adaptation exists. Conversely, if the starting population is highly uniform or small, there obviously will be fewer alleles in its gene pool. Thus there is less chance that there will be members who can survive even the initial change. Even if some do survive the initial change, further adaptation is impeded because there are then fewer alleles to work with.

3. **Biotic potential of the species.** If some individuals have the traits that enable them to survive the initial change, the next question is: How fast can they reproduce to bring the population back to original levels? In other words, how many young does each female produce, and how long does it take the young to reach reproductive age? Recall that natural selection works only by "weeding out" sensitive members in each generation. Therefore, the more offspring produced and the sooner they reach maturity, the faster adaptation can occur. For example, a pair of insects commonly produces several hundred offspring each of which completes a life cycle from egg to reproducing adult in 2 weeks. Thus insects have a reproductive capacity that is much, much greater than that of birds, which may raise only two to six fledglings per year. Thus insects may accomplish the same degree of adaptation to new conditions in 1 year that birds do in 1000 years. Is it surprising then that insect

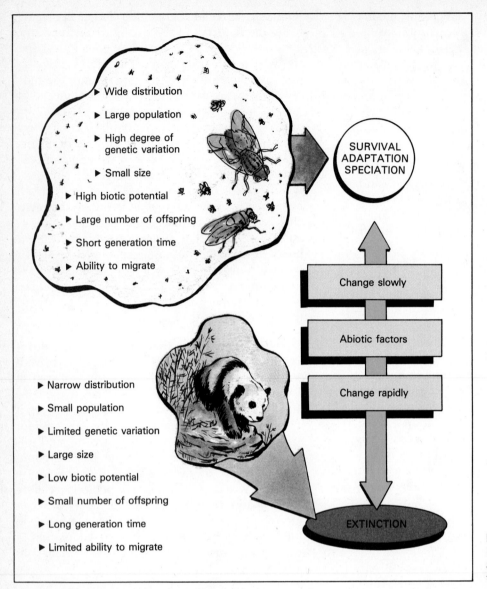

- ▶ Wide distribution
- ▶ Large population
- ▶ High degree of genetic variation
- ▶ Small size
- ▶ High biotic potential
- ▶ Large number of offspring
- ▶ Short generation time
- ▶ Ability to migrate

SURVIVAL
ADAPTATION
SPECIATION

Change slowly

Abiotic factors

Change rapidly

- ▶ Narrow distribution
- ▶ Small population
- ▶ Limited genetic variation
- ▶ Large size
- ▶ Low biotic potential
- ▶ Small number of offspring
- ▶ Long generation time
- ▶ Limited ability to migrate

EXTINCTION

FIGURE 5–12
Factors supporting the survival and adaptation of species versus their extinction are summarized.

populations rapidly adapted to pesticides, while bald eagles almost became extinct and were saved only by a banning of the pesticides (see Chapter 10)?

4. **Size of the organism.** Smaller organisms—flies, say—can find a suitable habitat and suitable conditions for breeding in almost any garbage pail, while panda bears need many square miles of bamboo forest. Thus small organisms have a distinct advantage over large ones. By the same token, populations of small organisms generally are both much larger and reproduce much faster than populations of large organisms. Thus, the small organisms have the additional advantages of factors 2 and 3. Is it surprising that pandas are threatened with extinction, while flies aren't?

5. **Geographic distribution.** A population that has a wide geographic distribution is likely to contain a considerable degree of genetic diversity and vice versa. Also, some areas of a wide geographic range are likely to be more-or-less isolated from the change so that populations of the species may survive in some regions while being eliminated in others.

These factors are summarized in Fig. 5–12.

The Force Behind Climate Changes

Throughout this discussion, we have referred to climatic change as a condition behind the evolution of ecosystems. What causes the climate to change? One major factor is that continents are continually on the move toward different relative positions on the earth and climates change accordingly. Let's look further at this movement of continental land masses.

The interior of the earth is molten rock and minerals kept hot by the radioactive decay of unstable isotopes still remaining from the time when the solar system was formed about 5 billion years ago. The earth's crust, which includes the bottom of oceans as well as the continents, is a relatively thin (no more than 65 miles [100 km] thick) layer that can be visualized as huge slabs of rock floating on the molten core, much like crackers float on a bowl of soup. These slabs of rock are called **tectonic plates**; there are about half a dozen major plates and two dozen minor ones involved in making up the earth's crust.

The tectonic plates are not stationary. Within the molten core, convection currents exist because hotter material rises toward the surface and spreads out at some locations, while cooler material sinks toward the interior at other locations. Riding atop these currents, the plates move slowly but inexorably with respect to one another, much like crackers might move if the soup were gently stirred from below.

The average rate of movement is only 1–2 centimeters (about one-half inch) per year, but over 100 million years this adds up to between one and two thousand kilometers. Also, while the underlying movement is slow and gradual, where the plates contact each other it goes in "jumps," which we experience as earthquakes. You can understand this process by pressing your thumb firmly on a table top and simultaneously sliding it across the surface. You will find it moves in jerks as it sticks, jumps, sticks, jumps. Likewise, the gradual movement of tectonic plates is manifested as sudden dramatic events—namely earthquakes and volcanic eruptions—as pressures from the movement gradually build until a break oc-

FIGURE 5–13
Pangaea. (a) Similarities of rock types, distribution of fossil species and other lines of evidence indicate that 200 million years ago all the present continents were formed into one huge land mass called Pangaea. (b) Slow but steady movement of the tectonic plates over the intervening time caused the breakup of Pangaea and has brought the continents to their present positions.

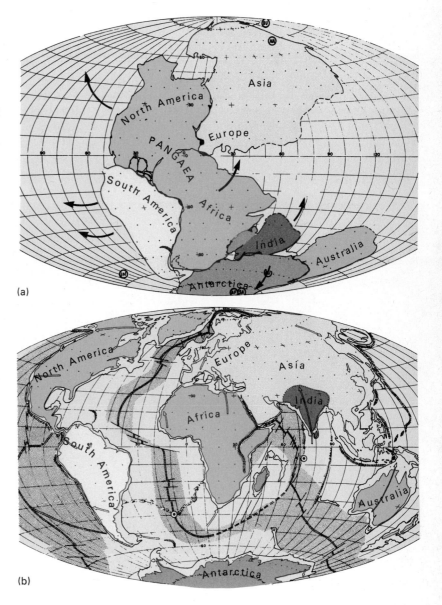

(a)

(b)

curs. The earthquake or volcanic eruption releases the pressures and is thus followed by a quiescent period during which pressures gradually build again to eventually cause another break or eruption.

Adjacent tectonic plates may move with respect to each other in four basic ways. First, where rising convection currents of molten material reach the surface, plates are forced apart and the gap between is filled with solidification of molten material. Such regions are presently seen as mid-ocean ridges, and there is considerable volcanic activity along these ridges. Various lines of evidence indicate that about 200 million years ago all the continents were positioned as one major continent called Pangaea (Fig. 5–13). The spreading process has brought the continents to their present positions and is the basic driving force behind the other interactions between tectonic plates.

The second kind of interaction is that two plates may gradually slide past each other. The best example of this sliding that can be seen in the United States is the San Andreas fault line in California, which marks the boundary between the Pacific plate, which is moving in a northwesterly direction relative to the American plate, which forms the bulk of the United States (Fig. 5–14). Along this fault line there is a major earthquake every 50 to 100 years as the pressures built up by the 1–2 cm per year movement is suddenly released in a jump of 1–2 meters. The last major earthquake on the central California region of the San An-dreas fault was in 1906. Therefore you can understand the anticipation for the "big one" at any time.

In the third kind of interaction, one plate may slide under another. This is taking place in the northwestern United States as part of the Pacific plate is sliding under Oregon and Washington. Again there are periodic earthquakes because the sliding is not smooth, but in this case another dramatic event also occurs. As the edge of the plate that is sliding under is forced down into the molten core, it melts and creates such pressure that molten material periodically erupts to the surface. Thus, there is a line of five major volcanoes through Oregon and Washington as a result. The most recent major eruption on this line was Mount St. Helens in 1980 (Fig. 5–15).

The fourth kind of interaction between tectonic plates is a head-on collision. Again, the manifestation is periodic earthquakes, but the gradual result is crumpling and uplifting of the plates into mountain ranges similar to the hoods of colliding cars. The world's prime example of this is the Himalayas, which result from the plate which bears India, which was originally attached to Africa, moving northward and crashing into the plate which comprises most of Asia.

In addition to the periodic catastrophic destruction that may be caused in localized regions by earthquakes and volcanic eruptions, these geological events may gradually lead to major shifts in climate. First, as continents gradually move to different po-

FIGURE 5–14
The narrow valleylike scar running from the top to the bottom of the picture is the San Andreas fault as it crosses the Carizo Plain about 300 miles south of San Francisco. Every 50 to 100 years mounting pressures of plate movement cause the fault to rupture, and the left side moves away from the viewer relative to the right side a distance of 3 to 6 feet in a major earthquake. Hills on either side of the fault are pressure ridges formed as a result of many fault movements. (David Parker/Science Photo Library/Photo Researchers.)

Part One What Ecosystems Are and How They Work

FIGURE 5–15
Volcanic eruptions such as this one of Mt. St. Helens on May 18, 1980 are a result of tectonic plate movement in which one plate is sliding under another. (John H. Meehan/Science Source/Photo Researchers.)

sitions on the globe, their climates will change accordingly. Second, the movement of continents alters the direction and flow of ocean currents, which in turn have an effect on climate. Third, the uplifting of mountains alters the movement of air currents, which also affects climate (for example, see the rain shadow effect p. 243). Additionally, there appear to be other factors involved in climatic change which are not well understood.

The fact that volcanic eruptions and earthquakes continue to occur is evidence that the tectonic plates are continuing to move today, much as they have over the past several hundred million years. Corresponding climatic changes are bound to gradually occur as well. Thus, change is inevitable. We have seen in this chapter that through adaptation, speciation, and extinction, ecosystems are capable of adapting to, and

evolving with, these gradual geological changes. The big questions are, will they be able to adapt to the much more rapid changes brought about by humans, and how long will we humans be around to record either the fast or the slow changes?

Implications for Humans

Our fourth principle of ecosystem sustainability is: *For sustainability, biodiversity is maintained.* After reading this chapter, you should see clearly that biodiversity is gradually developed through selective pressures acting on populations to gradually modify and adapt them to various habitats and niches. Increasing differences lead to speciation in the process. But, if this diversity is not maintained, can you see the problems that are created?

Let's review the ways in which the resulting biodiversity supports sustainability. First, we saw in Chapter 4 that a more stable population balance is achieved when a species is interacting with several natural enemies. Simple systems are prone to wide population swings, and monoculture systems are prone to being wiped out by a pest outbreak. In short, diversity creates a biotic structure that is relatively stable and hence sustainable.

Second, the diverse ecosystem is best able to recover from untoward impacts. Recall, also from Chapter 4, how oaks filled in as the American chestnut trees died out, thus preserving the overall integrity of the United States' eastern deciduous forest ecosystem and hence most of the species in that ecosystem. Also, reestablishment of ecosystems—in other words, succession—following volcanic eruptions, earthquakes, and landslides depends on biodiversity.

Third, it should be clear from this chapter that every species is a potential "seed" for further diversification and speciation. The greater the diversity present in each species and the greater the number of species, the greater the options for future diversification and speciation. Conversely, the absence of biodiversity sharply limits the potential for future speciation and adaptation to change.

Currently humans are pursuing a course that is greatly diminishing the biodiversity of the earth. Loss of species is occurring at a rate estimated to be on the order of many thousands of species each year because of four main factors. The first factor is direct assaults on ecosystems (cutting forests) and on species (poaching of various endangered species). The second factor is changing conditions, ranging from draining wetlands and creating reservoirs to potential global warming. The third is the myriad forms of pollution.

The fourth factor is introduction of foreign species, which may take over and eliminate many native species, as we learned in Chapter 4. Even without extinction, diminishing the populations of species makes them increasingly vulnerable to extinction because the gene pool of the reduced population will lack many of the alleles that may be necessary for adaptation. This factor causes the critical number (discussed in Chapter 4) to be much larger than might otherwise be supposed.

The loss of biodiversity that is occurring has particular implications for agriculture. A major trend of modern agriculture around the world, particularly in the last 30 years, has been to replace a tremendous diversity of local varieties of crop plants and animal breeds with a single high-yielding variety or breed. This practice has increased production and has been a major factor in the world's being able to support the growing human population (see Green Revolution, Chapter 8). However, as discussed in Chapter 4, such monocultures are extremely vulnerable to outbreaks of pests and diseases. Also, with loss of the gene pools represented by all the local varieties and breeds, future options for adapting agricultural species to changing conditions or to the challenges of diseases or pests will be sharply curtailed. (See In Perspective Box, page 116.)

The argument that the loss of biodiversity does not matter because the biosphere is obviously still functioning despite considerable loss seems dubious at best. One could also argue, more convincingly, that the biosphere has had enough biodiversity to fill in the gaps and handle our impacts *up to this point*. Should we continue on a course that will determine "experimentally" and irreversibly just how much biodiversity is necessary to support and sustain the biosphere?

Which wild species are most likely to withstand human impacts, and which are most likely to perish? Species that have those attributes (listed above) that enable survival and rapid readaptation to changes are cockroaches, mosquitoes, ticks, termites, mice, rats, weeds, and other species that we consider pests. Indeed, they are pests *because* they have all the features that enable them to adapt rapidly to overcome whatever we do to try to dispose of them. On the other hand, panda bears, leopards, rhinoceroses, many species of monkeys, sea turtles, whales, condors, storks, whooping cranes, and thousands of other species are endangered because they are ill-suited to adapting to rapid change. These species may well perish despite our best efforts to save them. Are we creating a world in which we share the planet with only such species as cockroaches, house flies, rats, and weeds and the few domesticated animals we are able to save?

Moreover, when we cause the extinction of a species, we are not causing the extinction of just that species. We are terminating forever that genetic line and all species that might have evolved from it. Remember that the processes of change can cause incremental changes only on species that exist. Consequently, reducing the world to those species with exceptional biotic potential means that those species are all that evolution has to work with. Cockroaches may speciate, but all they will ever evolve into is more species of cockroaches and closely related creatures. Science fiction aside, they will never evolve into another panda.

In looking at the development and extinction of species, it is natural to ask: When did the human species arise, and how long may we persist on earth? The fossil record shows that the origin of humans on earth is a very recent phenomenon. Just how recent can be appreciated by doing the following: Imagine the entire time period of organic evolution (some 4 billion years) condensed into a single year, with each day representing about 11 million years (Fig. 5–16). Most of the year, up to mid-September, is taken up by the evolution of primitive bacteria-like organisms. On about September 1 (1.4 billion years ago), the first complex cells typical of present-day plants and animals were formed, and then the pace quickened. All the major invertebrate groups of marine organisms developed through September and October. The first vertebrates developed during the first part of November (450 million years ago). The vertebrate body structure "rapidly" gave rise to fish, and various species of fish became the dominant animals on earth during the first 2 weeks of November—the 100-million-year Age of Fish.

About mid-November, amphibians evolved, allowing large animals to move from the sea onto the land, which was already occupied by plants and invertebrate animals, including insects. Thus, the Age of Fish gave way to the Age of Amphibians, which persisted another 100 million years, bringing us up to about December 1. During the last few days of November, the vertebrate structure was developing further and giving rise to reptiles. Amphibians remained tied to bodies of water because they must lay their

FIGURE 5–16
Contrasting the geological time scale with a single year gives an appreciation for the relative amount of time taken for various evolutionary stages. Note that two-thirds of the time is taken in the development of cells; then the pace quickens. Humans developed in just the last 8 hours, civilization since the advent of agriculture occurred in the last 2 minutes, and progress since the Industrial Revolution occupies only the last 2 seconds of the year.

eggs in water, and their young must develop there. However, reptiles are independent of water except for drinking, and some reptiles get sufficient water from just their food. Therefore, reptiles were able to displace amphibians in domination of the land. By December 1, amphibians gave way to the giant rep-

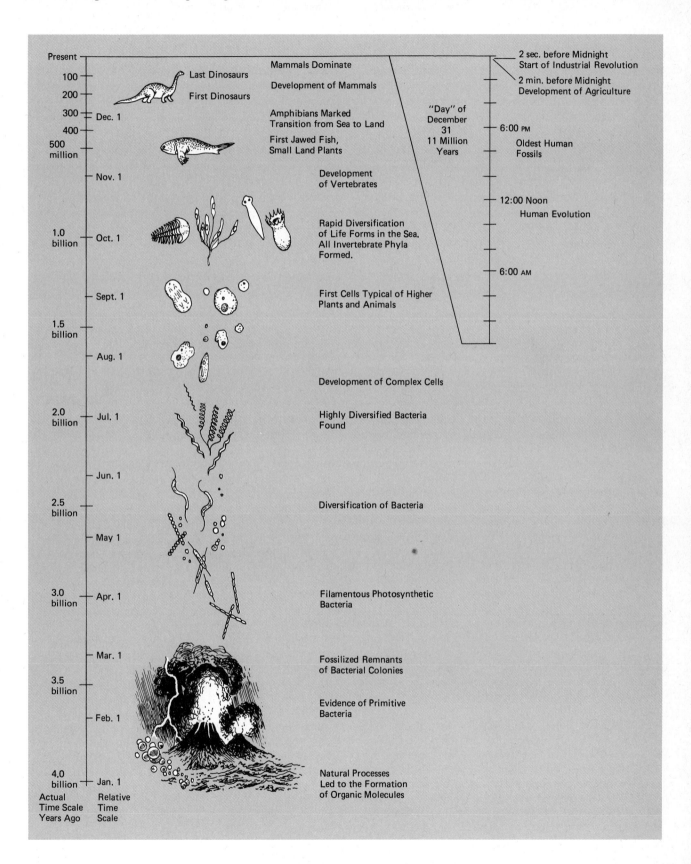

Present
100
200
300 — Dec. 1
400
500 million
— Nov. 1
1.0 billion — Oct. 1
1.5 billion
— Sept. 1
— Aug. 1
2.0 billion — Jul. 1
— Jun. 1
2.5 billion
— May 1
3.0 billion — Apr. 1
— Mar. 1
3.5 billion
— Feb. 1
4.0 billion — Jan. 1

Actual Time Scale Years Ago

Relative Time Scale

Last Dinosaurs
First Dinosaurs

Mammals Dominate

Development of Mammals

Amphibians Marked Transition from Sea to Land

First Jawed Fish, Small Land Plants

Development of Vertebrates

Rapid Diversification of Life Forms in the Sea. All Invertebrate Phyla Formed.

First Cells Typical of Higher Plants and Animals

Development of Complex Cells

Highly Diversified Bacteria Found

Diversification of Bacteria

Filamentous Photosynthetic Bacteria

Fossilized Remnants of Bacterial Colonies

Evidence of Primitive Bacteria

Natural Processes Led to the Formation of Organic Molecules

"Day" of December 31 11 Million Years

2 sec. before Midnight Start of Industrial Revolution
2 min. before Midnight Development of Agriculture

6:00 PM
Oldest Human Fossils

12:00 Noon
Human Evolution

6:00 AM

tiles (dinosaurs) that dominated the earth during the first half of December, the Age of Reptiles. Then, toward the end of the third week of December (65 million years ago), dinosaurs, along with many other species, suddenly became extinct. (Scientists are now in general agreement that this sudden extinction was caused by an enormous meteorite or asteroid striking the earth. This event caused such ecological upsets that many species, particularly large ones, could not adapt, and they became extinct.)

Among surviving animals, however, was another recently developed vertebrate organism, the forerunners of mammals. In the few "days" following the asteroid impact, mammals spread and speciated. Finally, the first humanlike creatures (upright body posture) appeared about 4:00 P.M. on December 31 (3 million years ago). Agriculture was established only in the last 2 minutes of the year (10 000 years ago), and the explosion of technology and knowledge of the last 200 years is represented by the last 2 *seconds*.

Thus, humans are a very recent development on the evolutionary scene. However, in this brief time (geologically speaking), we have so completely dominated the earth that the present era is often referred to as the Age of Humans. The Age of Fish and Age of Amphibians each lasted about 100 million years, and the Age of Dinosaurs lasted about 200 million years. If we consider the age of humans as starting with the advent of agriculture, the Age of Humans has thus far lasted 0.005 percent as long as the Age of Dinosaurs. Do we have only 100 to 200 million years to go?

We should see from what we have learned that species last only as long as a balanced relationship with their ecosystem is maintained. As conditions change—as occurred with the development of new organisms or the impact of an asteroid—extinction is the result for some species; others adapt and evolve. The change that leads to the displacement of a dominant species does not necessarily come from an outside factor, however. We learned in studying succes-

---

IN PERSPECTIVE
Preserving Genes for Agriculture

There are two basic approaches to agriculture: You can select and grow plants suited to the conditions, or you can change the conditions to suit the plants. Having limited ability to control the conditions, premodern cultures selected and cultivated plants that had the ability and tenacity to produce even in unfavorable conditions, without the benefits of irrigation, pesticides, or fertilizer.

Along the Missouri River in North Dakota, Indian families grew numerous species of corn and squash. On the dry, wind-shifting sands of the Colorado Plateau, Hopi farmers tended sunflowers. In the dry lands of the southwest, Indians grew hundreds of species of beans. By sending roots up to six feet into the earth to find moisture, these beans had a high yield despite scorching 115 F heat and almost no rain. In Peru, each valley grew its particular variety of potato or other tuber. Thus, ancient people survived for millennia by growing remarkably adapted crops. These crops are often referred to as "heirloom vege-

tables" by horticulturists, because they were grown only in one particular locality and their seeds were passed down through the generations.

Modern agriculture, however, has taken the approach of modifying conditions to fit the plant. It has focused on using relatively few high-yielding varieties and obtaining maximum yields by intensive use of pesticides, fertilizer, and irrigation. Agriculturalists are well aware of the vulnerability of this monoculture system. In 1980, for example, 20 percent of the U.S. corn crop was lost to a fungus disease because the disease was able to spread so rapidly through the monoculture. Disease, pest resistance, and vigor of modern varieties can be improved through the infusion of genes from the old heirloom varieties but only so long as those varieties continue to survive—new genes cannot be "invented."

But, in the process of shifting to modern agriculture, many of the old varieties and species have been

abandoned and many lost forever. Of all the food plants that were grown by Native Americans at the time of Columbus, roughly 75% no longer exist and many more are at risk. The loss of these species and the alleles they contain puts modern agriculture increasingly at risk. Aware of this problem, a worldwide network of "seed banks," institutions devoted to the collection and preservation of seeds, has been established under the auspices of the World Bank, Consultative Group on International Agricultural Research. Also a number of nongovernmental organizations such Seed Savers Exchange, Decorah, Iowa and Native Seed/SEARCH, Phoenix, Arizona, and the National Gardening Association operate seed exchanges whereby members can obtain, grow, and keep alive heirloom varieties. The viability of future agriculture may depend on the efforts of those who are creating and maintaining these "arks."

---

sion (Chapter 4) that a species dominating an ecosystem may be the cause of changes that result in another species coming in and displacing the formerly dominant species.

The conclusion is that the Age of Humans can last only as long as we maintain balanced relationships with the biosphere—and no longer. Some people, looking at current trends of environmental impacts, speculate that the Age of Humans may be a "flash in the pan," the shortest age the world has ever seen. We, your authors, are more optimistic, however. We believe that humans can establish the balances necessary for a sustainable future. How to achieve such balances is implied in the four principles of ecosystem sustainability (Table 4–1) set forth in these first five chapters. In the following chapters, we shall get down to specifics regarding what we need to do to achieve a sustainable society.

 Review Questions _____

1. What is the chemical structure of genes, and what is their relationship to traits?
2. What are alleles, and how are they related to variations seen among individuals of the same species?
3. What is differential reproduction, and how may it cause changes in the gene pool and the traits observed in a population?
4. What is selective pressure, and how does it relate to differential reproduction?
5. What is the most basic technique used by breeders in the development of new breeds or varieties of plants and animals?
6. What is natural selection, and what are the results of this process acting on a population?
7. What are mutations, and what kinds of change do they cause?
8. What effect do beneficial, harmful, and neutral mutations have on a population?
9. What is speciation, and how may it result from natural selection and mutations?

10. How may natural selection and speciation lead to the development of a balanced ecosystem?
11. What will happen to any unbalanced relationship in an ecosystem?
12. What factors determine whether a species can adapt or will be forced into extinction by a change?
13. Which species are most likely and which are least likely to survive changes, and why?
14. What are the implications of diminishing biodiversity for the future of species and natural ecosystems?
15. What are the implications of diminishing biodiversity for the future of agriculture?
16. Contrast the Ages of Fish, Amphibians, and Reptiles with the Age of Humans.
17. How long is the Age of Humans likely to last? What factors will determine the length of the Age of Humans?

 Thinking Environmentally _____

1. Select a particular species, and describe how its various traits support its survival and reproduction and how these traits may have developed and are maintained through natural selection.
2. How do strains of disease-causing organisms become resistant to medicines used against them?
3. Why is it biologically impossible for a "Ninja turtle" to arise from a single mutation?
4. To what extent is differential reproduction occurring in the human population, and what may be the long-term results?
5. Speculate as to the future course of evolution, including the evolution of humans, if substantial biodiversity is lost.
6. Discuss how we humans will determine the length of the Age of Humans, in terms of our ability to abide by each of the four basic principles for ecosystem sustainability.

"Nobody made a greater mistake than he who did nothing because he could do little."
Edmund Burke

Making a Difference

1 Begin to find out where and how your school, college, community or city gets its water and power, how it disposes of sewage and refuse, and what is the trend of land development and preservation around the area.

2 Find out if there is an environmental organization or club on your campus or in your community. Join if there is, or create one if there isn't.

3 Learn the common names of trees and other plants, birds and mammals found in your area. This knowledge is a prerequisite to understanding and describing an actual ecosystem. Then, pass your knowledge on to children, who are often curious about the natural world.

4 Take courses offered by colleges or local environmental organizations that give you a greater appreciation for the natural world; amaze and influence your friends by becoming a local expert in some taxonomic group, such as birds, wildflowers, butterflies.

5 Consider small animals, birds, butterflies and other wildlife that might be present in your yard, school or campus grounds were it not for human-created limiting factors. Investigate and begin a project to create natural habitats that will attract and support additional wildlife.

6 Create a list of the things you do and the things that you use and throw away. Write a brief evaluation of each in terms of the principles of sustainability. Consider how you might begin to change any of these habits to move toward a more sustainable option. Begin by making one such change.

7 Rather than using chemical pesticides and fertilizers to maintain your lawn, which is an "unnatural" monoculture, introduce clover and other low-growing flowering plants that will create a sustainable balance with only mowing.

8 Read local papers, make contact with environmental organizations and become aware of efforts to protect natural areas locally. Support efforts to protect such areas locally, regionally and globally.

9 Select a current environmental issue related to biodiversity or endangered species, and write your congresspersons to express your concern and ask for their support for legislation that effectively addresses the issue.

10 Support and join efforts and organizations, such as the World Wildlife Fund and Conservation International, that are devoted to protecting endangered and threatened species.

11 Continue your education toward a profession that is important for environmental concerns, and make your life work one that helps rather than hinders the environmental revolution.

12 Use your citizenship to support environmentally sound practices and policies by voting in your local, state and national elections for candidates who are clearly environmentally aware.

Part Two

Finding a Balance Between Population, Soil, Water, and Agriculture

A mountainside on the island of Bali is terraced for rice cultivation; tall-grass prairies of Illinois and Iowa are plowed under to raise corn; tropical rainforests in Brazil are converted to cattle pasture; dry woodlands in Kenya are grazed by goats. Wherever there are people, the environment is brought into service to produce food. More than five and a half billion people now depend on the environment to bring them their daily bread. Soil, water, nutrients, sunlight—the basis for productivity in natural ecosystems—are diverted toward meeting human needs through agricultural enterprises or through direct harvesting from natural systems. Many of these same systems also produce fuel, building materials, fabrics, and much else for human use. The global economy is largely based on extracting food and products from the natural world.

The population continues to increase by more than 90 million per year and, thankfully, soil, water, trees, crops, and animals are all "renewable" resources—if this were not so, we would long since have run out of them. Yet their capacity to renew is limited, and in our efforts to put nature to use, we can easily take faster than these systems can give. A moment's thought is enough to conclude that at some point, the human population must come into a basic balance with resources for food and other needs. In this Part, we begin to make serious application of the principles of sustainability, as we consider how many of us there are and how we have used and misused the natural world, and what we must do to move to a sustainable future.

6

The Population Explosion: Causes and Consequences

OUTLINE _____

The Population Explosion
  Just Numbers
  Different Worlds
    Rich Nations and Poor Nations
    Different Population Problems
  Consequences of Exploding Population
    Population and Poverty
    Population and Affluence

Dynamics of the Population
  Population Profiles
  Population Profiles and Projections
    For Developed Countries
    For Less-developed Countries
  Population Momentum
  Changing Fertility Rates and the
    Demographic Transition

LEARNING OBJECTIVES _____

When you have finished studying this chapter, you should be able to:

1. Draw a graph showing the change in human population from historical times through the present and projected into the future.

2. On a world map, point out developed and less-developed countries and discuss economic disparities between them.

3. Contrast current population growth rates in less-developed countries with those in highly developed countries.

4. Give the three factors and the relationship among them that determine the impact that humans have on the environment. Describe how impact will vary with change in each of the three factors.

5. List and describe the consequences of rapid population growth in less-developed countries.

6. Give specific examples showing how negative environmental impacts are made worse by increasing affluence.

7. Describe what data are shown and how they are displayed in a population profile.

8. Describe how a population profile is used to predict future numbers of deaths and births and other changes in a population.

9. Describe and contrast population profiles, fertility rates, and future population projections for more-developed and less-developed nations.

10. On the basis of objective 9, suggest what future goals of different nations might be regarding population.

11. Define population momentum and explain why it occurs.

12. Describe the four phases of the demographic transition and contrast developed nations with less-developed nations in terms of their progressing through the demographic transition.

122

The third basic principle for ecosystem sustainability is that consumer populations are maintained such that overgrazing does not occur (Chapter 3). We noted in Chapter 3 that the human population, in contrast to this principle, is growing phenomenally, and there are many signs of overgrazing in both a literal and a figurative sense. If humans cannot achieve a sustainable balance between their own numbers and the soil and water resources necessary to produce food, then all other environmental issues are moot.

In this part of the text we focus on these parameters. Chapters 6 and 7 concentrate on human population growth and the potential for achieving a balance. Chapters 8 through 11 focus on various aspects of achieving a sustainable agricultural system. Our objective here in Chapter 6 is to gain a clear understanding of the dynamics of human population growth—past, present, and future—and its impacts on the biosphere.

The Population Explosion

JUST NUMBERS

Over the last 200 years or so, the human population has grown and continues to grow at a phenomenal, explosive rate. The numbers speak for themselves. From the dawn of human history until the beginning of the 1800s, the population increased very slowly. A high reproductive rate, probably 8 to 10 babies per woman, was largely offset by high infant and childhood mortality—in other words, low recruitment—so that population growth was slow at best. Also, occasional setbacks occurred in the form of intermittent famines and outbreaks of such diseases as plague, smallpox, and typhoid fever.

It was roughly 1830 before world population reached the 1 billion mark. During the 1800s, however, a remarkable change in the growth rate occurred, triggered by major advances in sanitation, medical knowledge, agriculture, and industry. (The advent of vaccination, improvements in sanitation, and, more recently, antibiotics brought infectious diseases under control and vastly increased the survival rate of infants and children.) Human population changed from a condition of relative stability to explosive growth. By 1930, just 100 years after reaching the first billion, the population had doubled to 2 billion. Barely 30 years later (1960), it reached 3 billion. And in only 15 more years (1975), it had climbed to 4 billion, thus doubling from 2 billion in just 45 years. Then, 12 years later (1987), it crossed the 5 billion mark! In mid-1992 the world population stood at 5.42 billion and continues to grow at the rate of about 93 million people per year. This rate is equivalent to fitting into the world each year the combined populations of New York, Los Angeles, Chicago, Philadelphia, Detroit, Dallas, Boston, and 10 other metropolitan areas in the United States.

During the last couple of decades, the percentage rate of growth has begun to slow as a result of *decreasing birth rates*. Still, with more and more children growing up to become parents in turn, even the lower birth rate continues to add absolute numbers faster than at any other time in history. On the basis of current trends, the Population Reference Bureau projects that the 6 billion mark will be crossed in 1998, the 7 billion mark in 2009, the 8 billion mark in 2020, the 9 billion mark in 2033, and the 10 billion mark in 2046 before the declining birth rate, assuming it continues to decline, causes population to level off at around 12 billion people by the end of the next century (Fig. 6–1, page 124).

These projections into the future, however, are simply mathematical extensions of current trends. They do not consider any of the looming ecological questions of whether the biosphere can sustain these numbers. Where are all the additional millions, indeed billions, of people going to live, and how are they going to be fed, clothed, housed, educated, and otherwise cared for? Will there be enough energy and material resources for them to fulfill their aspirations? To answer these questions, we must recognize that we live in a world in which there is tremendous economic disparity among nations.

DIFFERENT WORLDS

Rich Nations and Poor Nations

The world is commonly divided into three main economic categories:

1. High-income, highly developed, industrialized countries: mainly, the United States, Canada, Japan, Australia, and the countries of western Europe and Scandinavia.

2. Middle-income, moderately developed countries: mainly the countries of Latin America (Mexico,

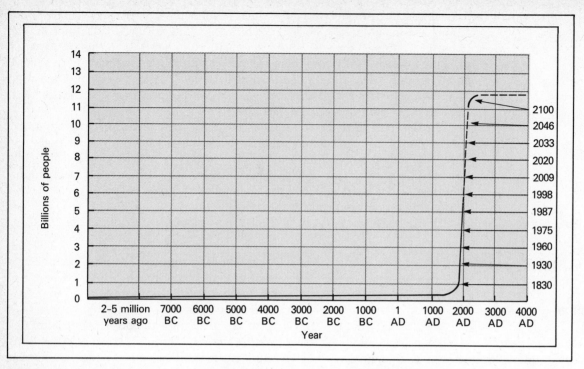

FIGURE 6–1

The world population explosion. For most of human history, the population grew very slowly, but in modern times it has suddenly "exploded." Further, the still high but gradually declining birth rates will carry the population to about 12 billion by the end of the next century. This is simply a mathematical projection based on current population statistics. It does not address the question, Will the biosphere be able to support 12 billion people. (From Joseph A. McFalls, Jr., "Population: A Lively Introduction," *Population Bulletin*, 46, no. 2 [Washington, D.C.: Population Reference Bureau, Inc., Oct. 1991], p. 4.)

Central America, and South America), northern and western Africa, and eastern Asia.

3. Low-income countries: mainly the countries of eastern and central Africa, India, and other countries of central Asia. (The People's Republic of China is still placed in this category, but it may soon move into category 2.)

Comparable data are not available for the Commonwealth of Independent States (the former USSR). However, general living standards suggest that most of these countries are in the middle income category. The world map shown in Figure 6–2 shows these groups of countries with their populations and data regarding average incomes.

IN PERSPECTIVE

How Many More for Dinner?

Two hundred and seventy-five thousand additional people are coming to dinner tonight, and 275000 more tomorrow night, and another 275000 more the night after that, and . . . This huge and every-growing figure is the result of simply dividing the world population growth rate of 93 million per year by 365. These new people are not merely coming to dinner; they are babies and so are staying to be clothed, housed, educated, and trained. Each of them will want her or his full share of energy and material resources. If a large percentage of these people are to spend their lives in anything better than absolute poverty, governments will have to do a much better job of planning and "setting the table" than they have in the past. But all sorts of evidence tells us that the "cupboard" (natural resources) is being stripped bare.

Is it possible not to invite so many to dinner? In other words, is it possible to lower fertility rates in less-developed countries?

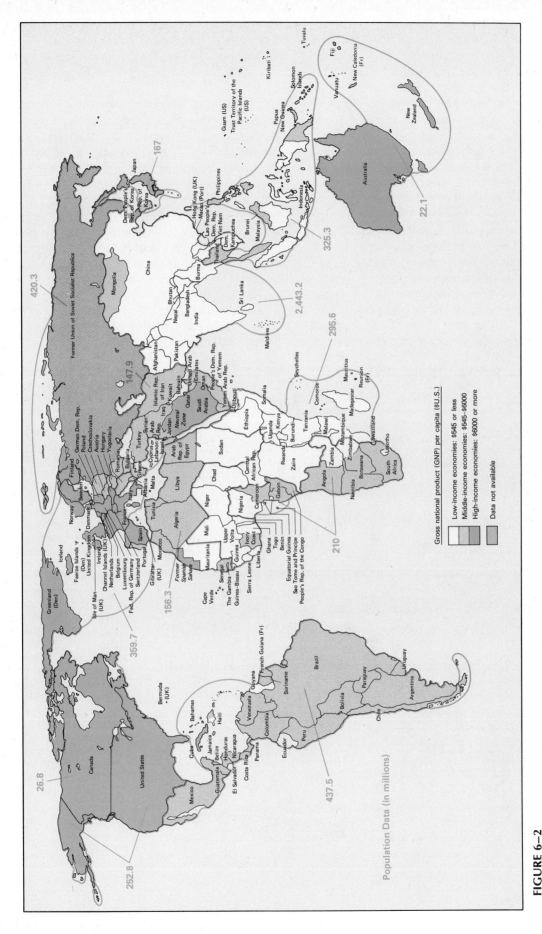

FIGURE 6–2
Nations of the world grouped according to gross national product (GNP) per capita, a general indicator of standard of living. Population of various regions is also shown. (From the *World Development Report 1990.* Copyright © 1990 by the International Bank for Reconstruction and Development/The World Bank. Reprinted by permission of Oxford University Press, Inc., New York. Population data from the Population Reference Bureau Population Data Sheet 1991.)

Gross national product (GNP) per capita ($U.S.)

Low-income economies: $545 or less
Middle-income economies: $545–$6000
High-income economies: $6000 or more
Data not available

Population Data (in millions)

The high-income nations are commonly referred to as developed countries, while middle- and low-income countries are often grouped together and referred to as developing countries. The highly developed countries are sometimes called HDCs, and middle- and low-income countries are grouped as less-developed countries and are called LDCs. Finally, the less-developed nations are also referred to as the **Third World**, or as Third World nations.

The disparity in distribution of wealth among the nations of the world is mind-boggling. Highly developed nations hold just 25 percent of the world's population, yet they control about 80 percent of the world's wealth. Thus, less-developed countries, which have 75 percent of the world's population, have only about 20 percent of the world's wealth. This disparity is illustrated in the following analogy. Imagine the world's economic wealth as a plate of 20 cookies on a table. Twenty persons surround the table. Five of the 20, representing the populations of the highly developed nations, take 16 of the cookies (80 percent). This leaves just 4 cookies for the other 15 people, who represent the poorer 75 percent of the world's population. But these 15 (the developing na-

tions) do not divide the 4 remaining cookies equally. Five people—representing the populations of the moderately developed countries—take 3 cookies, leaving just 1 cookie for the remaining 10 people representing the 2.5 billion people living in poor countries.

Of course, the distribution of wealth within each country, is also disproportionate. Between 10 and 15 percent of the people in highly developed countries are recognized as poor (unable to afford adequate food, shelter, and/or clothing), and about 10 percent of those in less-developed countries are wealthy.

Bear in mind that "rich" and "poor" are relative terms. On a world scale, being "rich" is simply being able to afford a comfortable home or apartment, a car, and whatever one chooses to eat. At the other end of the spectrum, about a billion people—one-fifth of the world's population, predominantly in low- and middle-income nations—live in "absolute poverty," a condition defined by Robert McNamara, president of the World Bank in 1978 as, "A condition of life so limited by malnutrition, illiteracy, disease, squalid surroundings, high infant mortality, and low life expectancy as to be beneath any reasonable definition

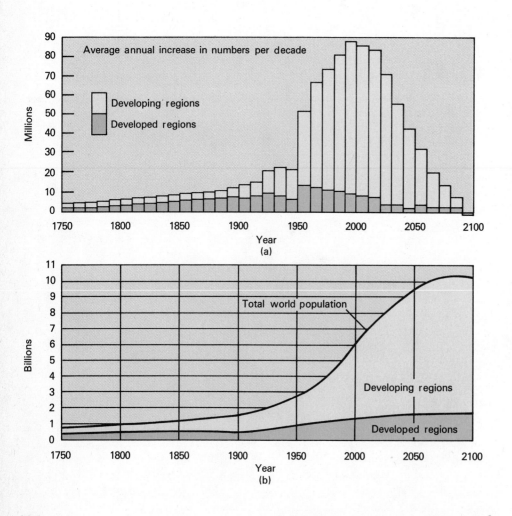

FIGURE 6–3
Population increase in developed and less-developed nations. (a) Because of higher populations and higher birth rates, most of the population growth is occurring and will continue to occur in less-developed nations. (b) Because of the disparity in growth rates, less-developed countries are coming to represent a larger and larger share of the world's population. If current trends continue, developing nations will have 90 percent of the world population by 2075. (From Thomas W. Merrick, "World Population in Transition," *Population Bulletin*, 41, no. 2 [Washington, D.C.: Population Reference Bureau, Inc., Jan. 1988, reprint], p. 4.)

of human decency." For these people, every day is simply a struggle for survival. An estimated billion more live along a "subsistence margin," which precludes much beyond minimal necessities.

Since World War II, developed nations have felt some responsibility toward improving conditions in less-developed nations by providing both humanitarian aid and economic assistance to help developing nations create their own sources of income. During the 1960s and 1970s such development efforts resulted in some mitigation of the world's economic disparity. Since the 1980s, however, various factors, which will be discussed later, combined to reverse the trend. In the last decade, rich nations have been growing richer while many poor nations have been falling further behind, with increasing numbers of the people living in these poor nations joining the ranks of the poor. The World Bank speaks of the 1980s as a "lost decade" and notes that the downward slide continues in many Third World nations.

Different Population Problems

Yet it is in the low-income countries that the population explosion is most intense (Fig. 6–3a). A key factor in population growth is the **total fertility rate**—the average number of children each woman has over her lifetime. In many less-developed countries, total fertility rates are such that, if sustained, they will result in a doubling of the population over the next 20 to 30 years. In contrast, barring immigration, populations in highly developed nations are approaching, and a few are already at, stability (no growth) (Table 6–1). This difference in growth rates is causing less-developed nations to represent a larger and larger fraction of the world's population (Fig. 6–3b). However, this is not to say that only the less-developed countries need to be concerned about population problems.

The major concern for all of us must be the overall impact of humans on the biosphere. Everyone, simply by the nature of human existence, demands certain resources—even if just food and water—and produces certain wastes. Hence every person has a certain amount of negative impact; therefore negative impact increases with population. However, the negative impact of individuals is multiplied many times by the consumptiveness of their lifestyle, because each increment of consumption involves a further demand on resources and greater amounts of wastes being produced. However, these negative impacts may be moderated to large extent by a factor we call **environmental regard**. For example, suitable attention to conservation and recycling may offset, to some extent, the negative impact of a consumptive lifestyle.

TABLE 6–1

Population Data for Selected Countries

Country	Total Fertility	Doubling Time (Years)
Less-developed Nations		
Average (excluding China)	4.4	30
Egypt	4.5	24
Kenya	6.7	18
Madagascar	6.6	22
India	3.9	34
Iraq	6.4	26
Viet Nam	4.0	31
Haiti	6.4	24
Brazil	3.3	36
Mexico	3.8	30
More-developed Nations		
Average	1.9	137
United States	2.1	88
Canada	1.7	96
Japan	1.5	210
Denmark	1.6	—
Former West Germany	1.5	—
Italy	1.3	1155
Spain	1.3	308

Data from: 1991 World Population Data Sheet. Population Reference Bureau, Inc.

The relationship between these factors may be expressed as

$$\text{Negative environmental impact} \propto \frac{\text{Population} \times \text{Consumptiveness of lifestyle}}{\text{Environmental regard}}$$

This "equation" should be read: Negative environmental impact is proportional to the population multiplied by the consumptiveness of the population's lifestyle moderated by the environmental regard of the population. This is not a strictly mathematical equation because there is no way to assign meaningful numbers to any of the factors other than population. Also, you may note that numerous people who are devoting their lives to environmental protection in one way or another are exercising a degree of environmental regard that offsets their negative impacts; these people have a highly positive impact on the environment. Still, the relationship among the factors expressed by the equation is valid.

Looking at developed and less-developed countries in terms of our environmental impact formula, we can see two distinct population problems. In poor nations, the problem is increasing numbers of people straining basic soil, water, and forest resources to the extent that even survival seems increasingly precarious. In rich nations, the problem is the negative impact of fewer people being multiplied by highly consumptive lifestyles. This two-pronged problem will become clearer in the following section as we discuss specific consequences of the exploding human population.

CONSEQUENCES OF EXPLODING POPULATION

Population and Poverty

In less-developed nations, population growth, poverty, and environmental degradation are entwined in a number of ways.

Population Pressures in the Countryside Prior to the advent of the population explosion and the Industrial Revolution in the late 1700s and early 1800s, most of the human population survived through subsistence agriculture. That is, families or tribal groups lived on the land and produced enough food for their own consumption and perhaps enough extra to barter for other essentials. Natural forests provided firewood, structural materials for housing, and wild game for meat. With a stable population, this system is basically sustainable. As an older generation passes away, the land remains adequate to support the next generation, and so on. The need for additional land to cultivate is minimal, and the forests and wildlife regenerate fast enough to sustain the use. Indeed, many cultures once sustained themselves in this way over thousands of years, and today considerable areas of Latin America, Africa, and Asia maintain this tradition.

With the commencement of the population explosion, however, each new generation is larger than the one before. As they become adults, how are the "extra" children to be accommodated? There are three basic alternatives:

1. Farms can be subdivided so that, as the generations pass, each person has less land than his/her elders did.
2. The extra people in each generation can either seek new land to farm or intensify cultivation of the existing farmland.
3. The extra people can move to cities and seek employment.

All of these alternatives are being played out to varying degrees in Third World countries. Unfortunately, they all have severe environmental and social consequences.

Over wide areas of Asia and Africa, plots of land have been divided and redivided to the point that, in

1988, the United Nations Food and Agriculture Organization estimated that over a billion rural people live in households that have too little land to adequately meet even their own needs for food and fuel, much less producing extra for income or barter. Thus, with a growing population, rural people become locked into a cycle of increasing poverty.

Secondly, the overflow population creates pressure to bring new lands into cultivation. Unfortunately, most land the world over that is well suited for agriculture is already in cultivation. The new land being farmed is "marginal," which means it is not well suited to agriculture or is ecologically sensitive or both. It consists of tropical forests, steep slopes, and arid (dry) lands. For example, it is estimated that two-thirds of the tropical deforestation that is occurring in Brazil and Central America is for the purpose of increasing agricultural production (Fig. 6–4a). Much of this deforestation is done by poor young people who are seeking an opportunity to get ahead but are unskilled and untrained in the unique requirements of maintaining tropical soils. Consequently, beyond the loss of biodiversity inherent in destroying the forest, we have the additional problem that between a third and a half of the cleared land becomes unproductive within 3–5 years, again leaving the people in absolute poverty.

Likewise, the attempt to cultivate steep slopes traditionally used for seasonal grazing leads to massive soil erosion; that is, the soil is washed down the slopes and into streams and rivers. The consequences of this are manifold. The loss of topsoil diminishes crop productivity. It also results in more water running off rather than soaking into the ground, a situation that aggravates flooding in the lowlands and diminishes groundwater. The soil washing into streams and rivers destroys fisheries, clogs channels, and aggravates flooding. (These effects are explained further in Chapter 11.)

Intensification of cultivation leads to similar problems. For example, traditional subsistence farming in Africa used to involve rotating cultivation among three plots. This way the soil in each plot, after being cultivated for 1 year, had 2 years to regenerate. With pressures for increasing crop production in order to feed more and more people, plots have been put into continuous production, with no time off. The result has been deterioration of soil, decreased productivity, and erosion. Cultivation of arid lands leads to wind erosion (the blowing away of topsoil) and decreased productivity.

Adding to the problems is the quest for firewood. Some 3 billion people, or 60 percent of the world's population, do not have gas, electricity, or even kerosene stoves; they depend on firewood for daily food preparation. Population growth is causing

(a)   (b)

FIGURE 6–4
(a) Millions of acres of rain forest in Central and South America are being cut down
each year to make room for agriculture as shown in this photograph from Peru.
Unfortunately agricultural benefits are meager, because a very thin topsoil is soon
washed away leaving only a hard, nutrient-poor subsoil that is nearly impossible to
till and yields little produce. (Asa C. Thresen/Photo Researchers.) (b) Like the
Nepalese woman seen here, some 60 percent of the world's population still depends
on gathering firewood for cooking and other fuel needs. Resulting deforestation is
causing both ecological and human tragedy. (Zviki-Eshet/The Stock Market.)

forests to be cut faster than they can regenerate. Much
of East Africa, Nepal, and Tibet as well as many
slopes of the Andes in South America have been deforested as a result (Fig. 6–4b). In 1983 the U.N. Food
and Agriculture Organization estimated that 1.3 billion people are able to meet current demand for firewood only by cutting trees faster than they are
regrowing, and that by the year 2000 as many as 3
billion poor people will face acute shortages of firewood. Already, women in many Third World countries spend a large portion of each day on increasingly
long treks to gather firewood. Worse, as the forests
are removed, massive soil erosion occurs, causing all
the ecological problems described earlier in the discussion on cultivating steep slopes.

In summary, increasing population is driving
many rural peoples of the world into increasing hardship, poverty, and deprivation. Of course, it is more
than just the poor who will suffer ultimately from the
loss of biodiversity and productivity of land and
waterways.

Population Pressures in Cities Faced with the poverty
and hardship of the countryside, many hundreds of
millions of people in less-developed nations continue

to migrate to cities in search of employment and a
better life. The result is that a number of Third World
cities are now among the world's largest and are still
growing rapidly (Fig. 6–5). Unfortunately, expansion
of housing, water systems, sewer systems, schools,
and other elements of an urban infrastructure have
not kept pace. The result is that many Third World
cities share the following situation described for Sao
Paulo, the industrial center of Brazil:

Too many people in the wrong places. The shacks and
shanty towns surround Sao Paulo in concentric circles—
called the rings of misery—of millions of people living
below the poverty level, trying to earn, beg or steal a living
with virtually no hope of aid from a government that feels
it has a long way to go before it can consider social welfare
programs. And most of those people are young; each year
three million Brazilians enter the job market. They come
to the cities because, bad as life is there, it is better than
in the desolate rural area where many were born. It is expected that the population of Sao Paulo—as well as of Rio
de Janeiro with five million people—will double by the
end of the century. Brazil's economy must grow rapidly if
the country is to keep from eating itself alive.[1]

[1] Brian Kelly and Mark London, *Amazon.* (New York: Harcourt
Brace Jovanovich, 1983), p. 19.

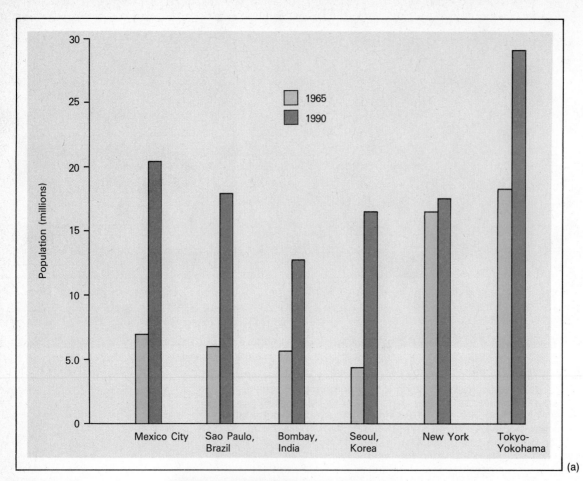

(a)

FIGURE 6–5

Growth of Third World cities. Since 1965 Third World cities have grown phenomenally, and a number of them are now among the world's largest. Unfortunately, in many cases the growth rate has far surpassed the ability to provide adequate housing, sewer and water systems, schools, and so forth, leaving many people to live in despicable shanty towns as seen here on the outskirts of Sao Paulo, Brazil. By contrast, growth of developed world cities has been more modest. (a) From the *Christian Science Monitor* Dec. 11, 1990, p. 12. (b) (Nickelsberg/ Gamma-Liaison.)

(b)

FIGURE 6-6
Many poor in Third World cities, including mothers and children, subsist only by scrounging through refuse for bits of food and items they can resell. (Jerry Cooke/Photo Researchers.)

Or, of Bombay, India:

More than a million people crouch cheek by jaw along a maze of suffocating narrow footpaths. Naked children play in black puddles of stagnant sewage beside tents of rags and corrugated tin huts. A choking stench, unending noise and festering disease are everywhere.[1]

Seven million of Mexico City's 20 million people are not connected to a sewer system. As a result, water supplies are being seriously overdrawn and polluted.

It is not coincidental that an epidemic of cholera broke out in Peru in 1990 and is making its way through Latin America and perhaps around the world. Cholera is caused by a bacterium spread via sewage in unsanitary conditions. It causes extreme vomiting and diarrhea, which result in such loss of body fluids as to be frequently fatal if not properly treated.

Worse, these cities do not provide the jobs people seek. Indeed, the high numbers of rural immigrants in the cities dilute the value of the one thing they have to sell—their labor. A common wage for a day's unskilled work is often equivalent to no more than 1–2 dollars a day, not enough for food, much less housing and clothing and other amenities. Thousands, including many children, make their living by scrounging dumps to find items they can salvage, repair, and sell. Many survive by begging and stealing, and others turn to drug dealing, poaching endangered species, and other illicit but lucrative activities.

Women and Children The hardships and deprivation of poverty fall most heavily on women and children.

[1] Robert Benjamin, *Baltimore Sun*, June 16, 1991.

Men are free to roam and pick up whatever work is available but then frequently keep their wages for themselves. They often take no responsibility at all for the women they make pregnant, much less the children that are produced. Even many married men under the stress of poverty abandon wives and children. Many Third World countries do not have any kind of welfare system that will provide care in such situations. Too often the women cannot cope with the children, or a new man in their lives throws the children out.

What happens to these children? If they survive at all, it is by begging, scrounging through garbage, stealing, and finding shelter in any hole or crevice they can find (Fig. 6–6). The problem is not small. Nearly every sizable Third World city has thousands of these "stray" children, on the order of 20 million in all by some estimates, and their numbers are growing. As these children grow up, you can speculate as to kind of adults they will become.

All these factors tend to lock the poor into a cycle of illiteracy and squalid conditions defining absolute poverty.

Emigration/immigration Facing the stresses and limited opportunities in their own countries, many people of Third World nations see emigration to developed countries as their best hope for a brighter future. Historically, the New World was colonized by the overflow population of Europe, and there is no question that immigrants contributed immensely to the development and economic growth of the United States, Canada, and other countries. However, times have obviously changed.

Is it feasible for the United States or any other developed nation to simply open its doors to all who

Roughly 150 years ago Charles Dickens wrote novels and magazine stories describing the horrors and inhumanity of the child labor sweatshops that developed in the early part of the Industrial Revolution. Young children were forced to work at machines or perform simple repetitive tasks for 12–14 hours a day 7 days a week. Often they ate and slept under their worktables. When they grew old enough to become rebellious, they were simply turned out onto the streets,—with no education and no skills—to face a world they had never seen.

In the early years of the twentieth century, the public's becoming aware of such atrocities led to the child labor laws we now have and the generally improved working conditions we in the developed world have come to expect. Thus we would like to think that the conditions Dickens wrote about have

passed into history. Yet, just in 1989 the United Nations held a "Convention on the Rights of the Child" and unanimously adopted resolutions outlawing child labor and trafficking in children.

Why did the United Nations find such a convention necessary in 1989? Because conditions every bit as horrendous as what Dickens described were (and still are) occurring in many places in the developing and also the developed world. According to the Anti-Slavery Society, there are 200 million "child slaves" around the world in sweatshops. In Tamil Nadu, India, there are 45000 children, some of them only 4–5 years old, at work in the fireworks and match industry. Many of the beautiful, intricate, handmade rugs from Asia, which one can buy at a "very reasonable price" are made by children forced to work. In Bangkok, Thailand, according to the Thai

Center for the Protection of Children's Rights, there are some 800 000 girl prostitutes between the ages of 12 and 15. As recently as 1990, children were being sold near Bangkok's main railway station.[1]

Is there a cause-and-effect relationship between these occurrences, population growth, and poverty? Recall from the text that, in the circles of poverty in less-developed countries, it is not uncommon for a man to take no responsibility for the woman he makes pregnant, much less for the children that are produced. Also, generally there is no social welfare system to provide even minimal benefits. In such situations, how is a woman to cope with children?

[1]Jonathan Power, "Where Death Squads Hunt Stray Children," *Baltimore Sun*, October. 25, 1991, p. 15A.

would flee from their poverty in the Third World? The current population of the United States is about 256 million. In contrast, there are many hundreds of millions who would seek access. The fact is, developed nations, including the United States, are becoming increasingly forceful in restricting immigration and in locating and deporting illegal aliens. As population pressures in Third World countries continue to mount, the problems and moral dilemma of limiting immigration and coping with illegal aliens seem certain to mount (see Ethics Box, p. 133).

Population and Affluence

We now turn to ways in which environmental pressures are intensified by affluence.

Affluence and global pollution Affluence, almost by definition, is synonymous with increased production, increased consumption, and increased use of material and energy resources and inevitably places increasing stresses on the biosphere. For example, carbon dioxide, the primary gas leading to potential global warming, is the inevitable waste product when fossil fuels

(coal, oil, and natural gas) are burned. The amount of energy used in driving, heating and cooling spacious homes and office buildings, using electrical equipment, and so on in the United States is such that, with just 4.6 percent of the world's population, the United States is responsible for about 25 percent of global emissions of carbon dioxide. In other words, the average affluent person living in the United States is responsible for 20–30 times the carbon dioxide emissions of an average person in a Third World nation. Much the same can be said of other global pollution issues. The emission of CFCs, which are causing degradation of the ozone shield, the emission of chemicals causing acid rain, hazardous chemical and nuclear wastes, all are primarily byproducts of affluent lifestyles.

Insofar as Third World countries are gaining affluence, they are contributing more and more to these pollution issues. For example, traffic congestion and pollution resulting from cars are horrendous in many Third World cities. These problems are made even worse because many cars are older models and are not well maintained, and pollution control standards are either very lax or nonexistent.

For people trapped by poverty or lack of opportunity in their homeland, emigration to another country has always held allure as a way to achieve a better life. As the New World opened up, many millions of people recognized this dream by emigrating to the United States and other countries. The United States is largely a country of immigrants and their descendants. Until 1875 there was no such thing as illegal immigration into the United States; all who could manage to arrive could stay and become citizens. This openness was epitomized by the inscription on the Statute of Liberty, which reads in part *"Give me your tired, your poor, your huddled masses yearning to breathe free, the wretched refuse of your teaming shore. Send these, the homeless, tempest-tost to me. . . ."*

Emigration to the New World both relieved population pressures in European countries and helped development of the New World. It is apparent, however, that a totally open policy toward immigration would be untenable today. The United States, with its current population of 256 million, is no longer a vast open land awaiting development; yet hundreds of millions of people would like to come to this "land of opportunity" if they could. The questions and the moral dilemma are: How much immigration should be permitted? and, Should some groups be favored over others? For example, in 1882 the U.S. Congress passed what was called the Chinese Exclusion Act, which barred the immigration of Chinese laborers but not of Chinese teachers, diplomats, students, merchants, or tourists. This act remained in effect until 1943, when China and the United States became allies in World War II. However, the United States continued an immigration policy that made it relatively easy for trained people to gain citizenship and relatively difficult for untrained people to do so. This policy created what is commonly referred to as a "brain drain." Brain power, many pointed

out, is the "export" developing nations can least afford.

Under the Immigration Reform Act of 1991, the United States can now accept up to 830 000 new immigrants per year, a larger number than we have received at any time since the 1920s and a number larger than is accepted by all other countries combined. Further, a lottery system was adopted for selection. Under this act, immigration accounts for about 30 percent of the U.S. population growth, which is presently 2.7 million per year. The remainder of the population growth is due to population momentum because the U.S. fertility rate has been at or slightly below replacement since 1972. If the U.S. fertility rate remains low, immigration will account for a growing portion of population growth.

We are speaking, thus far, of legal immigration; illegal immigration is another matter. There are hundreds of thousands who, unable to gain access through legal channels, try to slip in. The United States maintains an active Border Patrol, especially along the border with Mexico, where each night several thousand people try to get across the border. Most are caught and returned, but an undetermined number don't get caught.

The Haitian boat people who received headlines in 1991 and 1992 presented a somewhat different issue, that of granting asylum. Under terms of a 1951 United Nations convention, all European Community countries and the United States pledged to grant asylum to persons who can show a "well-founded fear of being persecuted for reasons of race, religion, nationality, membership in a particular social group or political opinion." People qualifying for asylum may stay regardless of other restrictions on immigration. It is anticipated that the number seeking asylum will be trivial compared with those seeking a better life; however, often the distinction is not clear.

Haiti, an island republic in the

Caribbean, has the notorious reputation of being the poorest country in the Western Hemisphere. Environmental stresses are severe. Less than 2 percent of the original forest cover remains. Soil erosion has already severely undercut productivity and remains rampant. The crystal blue of the Caribbean near the Haitian shore is often brown with eroding soil, and the mud settling on the coral has destroyed fishing in many areas. Yet, Haiti's population of 6.3 million maintains a fertility rate of 6.4. Haiti's environmental and population disaster stems in no small part from a long line of repressive dictatorial regimes. Hopes brightened with the election of a democratic government in 1991, but then the forceful takeover by yet another military dictatorship drove thousands of Haitians to despair. They crowded into anything that would float and headed for the United States in the hope of being granted asylum. They argued that their plight was political and that they would face further political repression if they were returned.

In February 1992 the U.S. Supreme Court ruled that the Haitian boat people were *economic* refugees, that is, they were only seeking a better life, and they were deported back to Haiti to face a dubious future. While some people in the United States lauded this Court decision, others decried it as utterly lacking in compassion and making a mockery of what we say we stand for on the Statue of Liberty.

As population pressures in the Third World continue, the questions of how many immigrants to accept, from what groups, and where to draw the line regarding asylum seem certain to become more and more pressing. Social, economic, and environmental consequences, both national and global, of various alternatives must be weighed in the decision, in addition to compassion. Where do you stand?

Pressures on the Land Another result of increasing affluence is the increase in meat consumption. The poor of Third World countries subsist primarily on a vegetarian food diet (rice, beans, corn, and potatoes and other root vegetables) simply because it is less expensive than a meat diet. It is found that, as people become more affluent, one of the first ways they indulge themselves is to eat more meat. (Of course, this trend happens only to a point. In developed nations, we have generally become aware that there are certain health risks in a high-meat diet, and consequently we have reversed the trend in recent years.)

The environmental impact of increasing meat consumption goes back to our understanding of the biomass pyramid discussed in Chapter 3. As we consume more meat, we are effectively moving to a higher trophic level and utilizing additional primary plant production and land required to support the growing of that meat. Thus, even without increasing population, increasing affluence demands that more and more land be converted from natural ecosystems to grain-growing or to grazing acreage. Where the natural ecosystem is already grassland and overgrazing is avoided, the impact may be minimal. However, this is too often not the case; overgrazing is occurring in a great many countries. In Brazil, for example, hundreds of thousands of square miles of rainforest have been cut and put into grass for low-intensity grazing. The result is a depressing loss of biodiversity for minimal benefit in terms of supporting people; yet the trend continues (Fig. 6–7). Well-managed cultivation of food, fiber, and forest products could support many more people in terms of jobs, income, and food production.

Pressures on Forests and Endangered Species Finally, increasing affluence, combined with lack of ecological regard, is creating a growing market for exotic pets, including rare fish, birds, reptiles, and plants; skins, furs, and various other artifacts made from parts of rare species; and tropical hardwoods for furnishings. Numerous species of both plants and animals seem headed toward extinction as a result of being collected for trade because the countries where they exist do not have suitable regulations to protect them. Where such regulations do exist, poaching and black market trade are epidemic, as noted in Chapters 5 and 18. Tropical hardwoods are obtained by the wholesale cutting of tropical forest with no pretense of replanting, controls on erosion, or anything else.

Summing Up

In conclusion, none of the trends discussed above bode well for sustainability. Both growing population and increasing affluence, even without population growth, are putting increasing stresses on the biosphere. Reaching a sustainable balance will demand four things:

1. The need to stabilize the human population is probably self-evident. It is difficult to see how any of the problems we just discussed can be lessened by more people.

2. Even to sustain the existing human population at existing levels, overgrazing, deforestation, soil degradation, and erosion must be brought under control. This means that people, both rich and poor, must adopt practices of sustainable agriculture and of forest ecosystem management and rehabilitation. The details of such practices are discussed in Chapters 8 and 9.

3. To sustain and improve general living standards (affluence), we must find ways to provide needs

FIGURE 6–7
Much of eastern Brazil was originally covered by Atlantic Coastal rain forest. This ecosystem, which covered hundreds of thousands of square miles, has been 97 percent cut and converted to low-productivity rangeland as shown here. (Photograph by BJN.)

and desires (comfortable housing, transportation, and so on) in more environmentally benign ways—for example, by substituting solar energy for fossil fuels.

4. For all of these things to occur, there needs to be a much higher level of environmental understanding and regard throughout the world—in a nutshell, environmental education.

In the remainder of this chapter, we examine some of the dynamics of population growth. Once you understand these dynamics, you will be in a better position to understand the potential of bringing the human population into balance, the subject of Chapter 7.

Dynamics of the Population

Whether and how fast a population grows depends on three factors. First, it depends on the **age structure** of the population: how many people are old, middle-aged, young adults, and children. Second, it depends on the total fertility rate. Third, it depends on the degree of infant and childhood mortality. We can understand how all these factors work together by examining the dynamics of a population.

POPULATION PROFILES

Understanding the dynamics of population begins with a look at the **population profile**, which is a bar graph showing the age structure of a population. Figure 6–8a shows the population profile for the United States (1990). As shown in the figure, ages are generally grouped into 5-year increments starting with 0–4-year-olds at the bottom and moving through 5–9, 10–14, and so on until the oldest age in the population is reached. The data are collected through censuses, which for the United States and most other countries are taken every 10 years. Population profiles are used in making projections regarding future population trends.

In order to see how this projection is done, first recognize that, every 5 years, each bar moves up one level as everyone ages, and each bar is shortened by the number of people in that group who die during the interval. Because, in the modern world, relatively few people die before older ages, an approximation of deaths may be achieved by simply removing the uppermost bar (Fig. 6–8b). A new bar is added at the bottom for the number of children born during that 5 years. How many children are born depends on the total fertility rate and the number of women in the reproductive age groups. A woman may bear children from her early teens through about age 45. Professional **demographers**, those who study populations and make such projections professionally, take this long time span into consideration and calculate the probable numbers of births accordingly. However, a reasonable approximation of births can be made by multiplying the number of women at the average childbearing age (20 to 24 in Fig. 6–8) by the total fertility rate.

How do we know what the fertility rate will be for the next 5 years? We know what it is now from current statistics of how many babies women are having. We cannot be certain what it will be in the future, but we can make various projections based on the

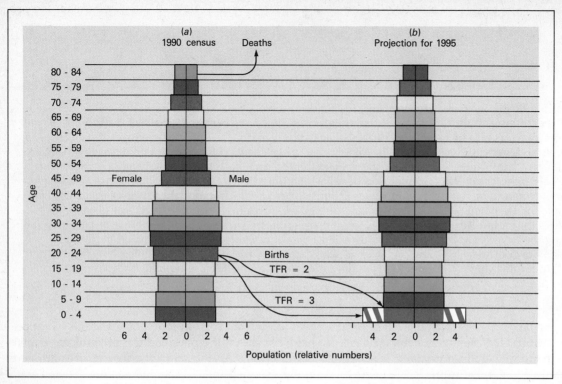

FIGURE 6–8

Population profile. (a) Each bar shows the number of persons in the respective age group, determined by a census, females on the right, males on the left. The repetition of colors indicates the approximate parent-child relationship; for instance, each red bar is mainly the offspring of the red bar above. (b) An approximation of how a population's profile will change in 5 years is obtained by moving each bar up one level, removing the uppermost bar, and adding a bar at the bottom for new births. The number of new babies in the bottom bar will be roughly the number of women in the group of average reproductive age (here we have used the 20–24-year-old group) times the total fertility rate (TFR). Two estimates are shown: one for TFR = 2 (solid) and one for TFR = 3 (hatched extension). The same process may be used to extrapolate the population into the future as far as desired. The profile shown here is for the United States. The bulge in the middle age groups is the "baby boom" generation, which resulted from high birth rates following World War II. (Profile shown in [a] from Joseph A. McFalls, Jr., "Population: A Lively Introduction," *Population Bulletin*, 46, no. 2 [Washington, D.C.: Population Reference Bureau, Inc., Oct. 1991], p. 4.)

assumption that it will either remain constant or follow certain trends of increase or decrease. The bottom bar in Figure 6–8b shows the results of two different assumptions of fertility.

This process may be repeated any number of times, each time moving the bars up one level, removing the uppermost bar (deaths) and adding a bar at the bottom (births), the size of which is calculated by multiplying the total fertility rate by the number

of women at the average childbearing age. (Note that this is a different group of women each time, because a new group of women moves into the 20–24 group each time the bars are moved up. After four times, representing the passage of 20 years, the childbearing group would be mainly the group that started as 0–4-year olds. The number of women in the childbearing age group will change, then, according to the number of children born roughly 20 years before.) Thus, in addition to seeing how a total population will change, you can see how the number of people in each age group will change over the years ahead. Of course a population profile will also be influenced by immigration and emigration, and demographers take this into account as well.

Can you see from this exercise that a total fertility rate of two (two children per woman) will exactly equal and eventually replace the number of parents as they die? This assumes that all children born survive to adulthood, which is obviously not the case. It is not far off, however, because the percentage of children who die—even in less-developed countries—is relatively small. Taking **infant mortality** (the number of babies that die before age 1) into account, **replacement fertility**, the fertility actually required to replace the parents, is 2.03 for developed nations and 2.16 for less-developed nations.

These figures for replacement fertility are based on current levels of **infant mortality**. Particularly in less-developed countries, however, many children who do reach their first birthday die of malnutrition or disease before reaching age 16. It is nonsensical if not immoral to argue that a higher fertility rate is necessary to replace these older children who die after age 1 and therefore are not reflected in infant-mortality statistics. Any number of children above what can be supported and cared for simply condemns that many more to die of starvation and disease. Conversely, eliminating hunger and providing adequate treatment for curable diseases will immediately lead to a higher survival rate and a replacement fertility rate closer to 2.

Observe that fertility rates continuing at the replacement level will gradually generate a column-shaped population profile and a stable population as each generation is just replaced by its offspring. The profile for the United States, which we have been looking at in Figure 6–8, is an example of such a column, albeit not perfect. The bulge in the middle is the "baby boom" generation, which resulted from the high fertility rates that followed World War II. The smaller age groups following the baby-boomers are the result of the U.S. fertility rate dropping to slightly below replacement (1.8); currently in the United States it is again slightly above replacement (2.1).

If fertility rates remained below replacement, the result would be fewer and fewer children. Eventually population would decline because the number of children being born would be less than the number of people reaching old age and dying. Fertility rates continuing *above* replacement will generate a pyramid-shaped profile whose base gets larger and larger as each generation is larger than the one before.

Business people and governmental officials, with the aid of demographers, use population profiles continuously to project future needs for additional schools, housing units, old age facilities, innumerable manufactured products, and, perhaps most of all, food production. If there is a failure to make suitable population projections, gross and perhaps tragic errors may be committed. Let's suppose for a moment that we are demographers advising governments as to such future needs. Let's look at example population profiles for a developed and a less-developed country, make projections, and see what advice we might give.

POPULATION PROFILES AND PROJECTIONS

For Developed Countries

The population profile for Denmark, a highly developed country in northern Europe, is shown in Figure 6–9a. Denmark's total fertility rate is currently 1.6. Assuming that this rate remains constant, and

projecting over the next 20 years, we obtain the profiles shown in Figure 6–9b–e. We see an increase in the numbers of older people but, with one exception, a marked decline in the number of children and young people. The one exception is the slightly larger group of children who will be born between 1990 and 1995 because of the larger group of parents 20–24. What should we advise the Danish government on this basis?

Over the next 20 years and further, Denmark will have a growing need for facilities for older people and for doctors, nurses, and other personnel trained in caring for them. However, there will be a declining need for schools, teachers, toys, and new housing units. (What do these two facts say for people in or contemplating going into those businesses and professions?) Perhaps these demographic changes provide a greater opportunity for improving education, for instance. Rather than laying off teachers, class sizes can be reduced, and more special education classes can be offered with the money that previously went into constructing new buildings. Likewise, construction money, rather than going into building more schools, might be directed into improving existing schools. Social services for young people can also be improved because there are fewer people coming along to be served.

For the next 20 years the total population of Denmark will remain relatively constant because the number of deaths (removal of top bars) will continue to be balanced by the reduced number of children being born. Following 2010, however, we can predict a considerable population decline as the large numbers of people currently in their thirties and forties reach their seventies and eighties and die as fewer and fewer children are being born. Indeed, the number of children being born will decline even more rapidly than the existing profile indicates because there will be fewer and fewer people rising into the childbearing years.

Unless a smaller population is the goal, our advice here might well be to encourage, and provide incentives for, Danish couples to bear more children, certainly not fewer. We might also advocate allowing more immigration; otherwise who will take the jobs to keep the economy going?

The 1990 population profile for the United States is shown in Figure 6–8, and the total fertility rate is 2.1. Can you make projections through the year 2010 and offer advice for the United States?

In summary, developed countries have population profiles with a columnar shape and total fertility rates very close to or below replacement. Therefore, their future, barring immigration, is toward stable or declining populations unless fertility rates are increased.

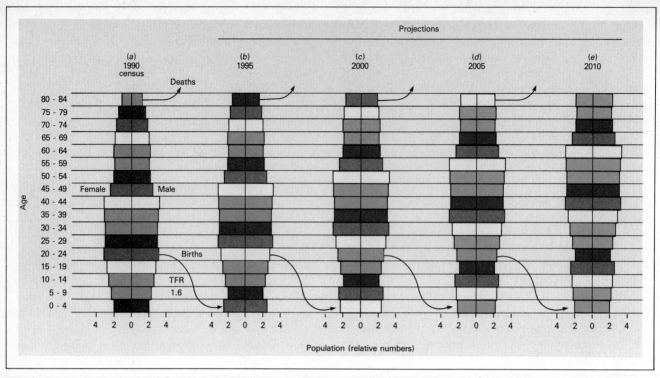

FIGURE 6–9
(a) 1990 population profile for Denmark, a highly developed country. (b–e)
Projections for Denmark's population made as described for Figure 6–8 assuming
TFR remains at its current 1.6. Note how larger numbers of persons are moving into
older age groups, while there is a diminishing number of children. (Profile shown in
[a] from Joseph A. McFalls, Jr., "Population: A Lively Introduction," *Population
Bulletin*, 46, no. 2 [Washington, D.C.: Population Reference Bureau, Inc., Oct.
1991], p. 4.)

For Less-developed Countries

Less-developed countries are in a vastly different situation. Their population profiles have a pyramidlike shape, and their total fertility rates, with few exceptions, are well above 2. For example, the 1990 population profile for the African country of Kenya is shown in Figure 6–10a. This nation's total fertility rate is 6.7, which means that each woman on average has between 6 and 8 children. Assuming this fertility rate remains constant, projecting Kenya's population ahead just 10 years gives us the profile shown in Figure 6–10b. Even with high mortality, each new generation is much larger than the one before. As these children reach reproductive age and the fertility rate remains at about 6.7, this pattern will be maintained so that the pyramid base will get broader and broader. This swelling Kenyan population will double in 18 years, as noted in Table 6–1. In 2000, 50 percent of the population will still be under 20 years of age.

Another factor exacerbating growth in Kenya and most other less-developed countries is a shorter generation time. Many women in less-developed

countries start bearing children in their early teens, whereas many women in developed countries wait until their late twenties, early thirties or even longer.

What should we advise? Just to provide the same level of goods and services per person, Kenya will need to double its numbers of schools, housing units, hospitals, roads, sewage collection and treatment facilities, and telephones; double its output (or import) of everything from soap to television sets; and double its food production in the next 18 years. If the country fails to do this, the average person—both adult and child—can only sink into deeper poverty. Kenya may be an extreme example, but it is important that you realize that the Kenyan profile describes the general situation confronting *all* developing countries, which include 75 percent of the world's population over most of Latin America, Africa, and Asia.

Our advise could be, "Yes, quickly double everything," but this immediately poses two problems. First, where is the money to come from in countries that are already desperately poor? Second, are the natural resources available? It is policies that lead in this direction of ever-increasing goods and services that have already caused the pressures resulting in

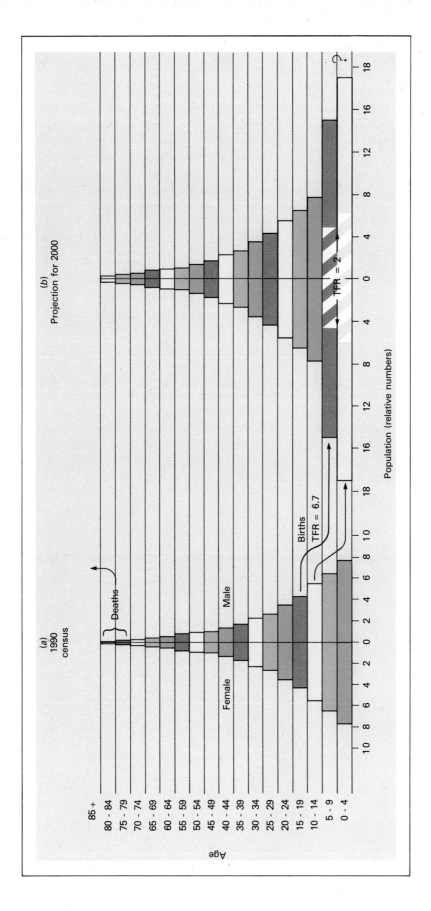

FIGURE 6–10
(a) 1990 population profile for Kenya, a less-developed country. (b) Projection to 2000 based on the assumption that TFR will continue at its current 6.7. Even if TFR immediately dropped to 2, the number of births (hatched inner portion of the bottom two bars) would still greatly exceed the number of deaths because there are so few persons in the upper age groups. (Profile shown in [a] from Joseph A. McFalls, Jr., "Population: A Lively Introduction," *Population Bulletin*, 46, no. 2 [Washington, D.C.: Population Reference Bureau, Inc., Oct. 1991], p. 4.)

overgrazing, overcultivation, deforestation, overdrawing water resources, and so on, discussed above. And even if a less-developed country succeeds in doubling all outputs in the same time its population doubles: has it won? Unfortunately, such doubling means that a country has only managed to stay in the same place relative to poverty—except that now there are twice as many people and children. Then, because of the shape of the population profile and the high fertility rates, the population can only head toward doubling again. Then what?

It should be conspicuous that the situations facing developed and less-developed countries are vastly different and the advice needs to differ accordingly. In our Kenyan example, our advice might be: While making every effort to accommodate growing numbers, reduce fertility rates. If population growth is to be reduced, efforts in this direction cannot be started too soon because of a phenomenon known as *population momentum*.

POPULATION MOMENTUM

A rapidly growing population, such as Kenya's, will continue to grow for 50–60 years even after the total fertility rate is reduced to the replacement level. This phenomenon is called **population momentum**, and it occurs because of the current pyramidlike profile of the population. Because such a small portion of the population is in the upper age groups where most

deaths occur, the number of deaths per year will be modest. There are so many young people coming into their reproductive years, however, that even if they have only two children per woman, it will still result in a number of births that far exceeds the number of deaths.

This imbalance, which is the cause of population momentum, is shown by the projection of Kenya's population (Fig. 6–10b). Note that if Kenya's fertility rate were 2, the numbers of children indicated by the inner hatched portion of the bars would be produced. Even this number would far exceed the number of deaths because of the very small percentage of elderly people in the population. By projecting further, you can see that the growth will continue until the current children reach the upper age brackets—in 50–60 years!

CHANGING FERTILITY RATES AND THE DEMOGRAPHIC TRANSITION

We have seen (Table 6–1) that developed countries already have low fertility rates, perhaps even too low. If we understand the causes for disparity in fertility rates between developed and developing countries, we may be able to understand how fertility rate can be influenced. To gain this understanding, we need to look back at the origin of the population explosion.

Recall that the historical norm for the human population was high fertility rates. The human population was held in check by high infant and child-

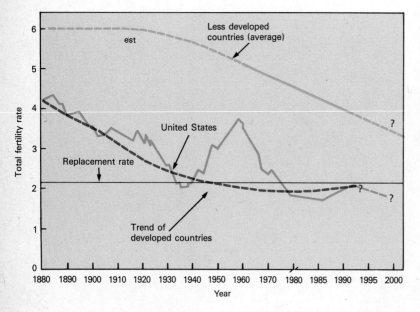

FIGURE 6–11
Change in fertility rates. Historically, high fertility rates were the norm for the human population. In more modern times, there has been a trend toward lower fertility rates. In developed countries the trend started earlier and has progressed, except for the anomalous post-World War II baby boom, to or even slightly below replacement fertility. In less-developed countries the trend toward lower fertility started later and has not progressed as far. (Data from the Population Reference Bureau, Inc., Washington, D.C.)

TABLE 6–2

Crude Birth and Death Rates for Selected Countries

Country	Crude Birth Rate	Crude Death Rate	Annual Rate of Increase (%)	Doubling Time (Years)
Less-developed Nations				
Average (excluding China)	34	10	2.4	30
Egypt	38	9	2.9	24
Kenya	46	7	3.9	18
Madagascar	45	13	3.2	22
India	31	10	2.1	34
Iraq	41	15	2.6	26
Viet Nam	32	9	2.3	31
Haiti	45	16	2.9	24
Brazil	27	8	1.9	36
Mexico	29	6	2.3	30
More-developed Nations				
Average	14	9	0.5	137
United States	17	9	0.8	88
Canada	15	7	0.8	96
Japan	10	7	0.3	210
Denmark	12	12	0.0	—
Former West Germany	11	11	0.0	—
Italy	10	9	0.1	1155
Spain	11	8	0.3	308

Data from: 1991 World Population Data Sheet. Population Reference Bureau, Inc.

hood mortality, as occurs in populations of other species. The human population explosion was initiated by a drop in the infant and childhood mortality, not by any increase in fertility. Then, in more a recent times, another trend has begun: a gradual *decline* in fertility rates. In developed countries, this trend has progressed to the point where fertility rates are very close to or even below replacement. In less-developed countries, the trend has progressed also but not as far (Fig. 6–11).

In human societies, economic development brings about more than just a lower death rate resulting from better health care; it also brings about a decline in fertility rate. With development, human societies move from a primitive population stability, in which high birth rates are offset by high infant and childhood mortality, to a modern population stability, in which low infant and childhood mortality are balanced by low birth rates. This gradual shift from the primitive to the modern condition that occurs with development is known as the **demographic transition**.

To understand the totality of demographic transition, we need to introduce two new terms, **crude birth rate** (**CBR**) and **crude death rate** (**CDR**). These terms are defined as the number of births or deaths per 1000 of population per year. By giving the data in terms of "per 1000," populations of different coun-

tries can be compared regardless of their total size. The term *crude* is used because there is no consideration given to what proportion of the population is old or young, male or female. Subtracting the CDR from the CBR gives the increase or decrease per 1000 per year. Dividing this result by 10 puts it in terms of "per 100," or, as it is more commonly known, percent:

CBR	–	CDR	= Natural increase
Number of		Number of	(or decrease)
births per		*deaths* per	in population
1000 per year		1000 per year	per 1000
			per year

$$\div 10 = percent$$

Of course a stable population is achieved if and only if CBR and CDR are equal. Some statistics are given in Table 6–2.

Doubling time, the number of years it will take a population growing at a constant percent per year to double, is calculated by dividing the percentage rate of growth into 70. The answer is the number of years it will take the population to double. (The 70 has nothing to do with population. It is simply the number that works to give the result.)

The demographic transition has four phases, as shown in Figure 6–12. **Phase I** is the "primitive sta-

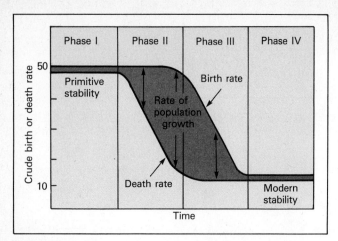

FIGURE 6–12
The human population is undergoing a demographic transition from a primitive, Phase I population stability (high birth rates balanced by high infant and childhood mortality) to a modern, Phase IV population stability (low death rates resulting from very reduced infant and childhood mortality balanced by low birth rates). High rates of population growth—indicated by the difference between death and birth rates—occur between Phases I and IV because reduction in childhood mortality, which leads to lower death rates (Phase II), occurs before the reduction in birth rates (Phase III). Developed countries have already reached Phase IV, while less-developed nations are still in various stages of Phase III. (From Joseph A. McFalls, Jr., "Population: A Lively Introduction," *Population Bulletin*, 46, no. 2 [Washington, D.C.: Population Reference Bureau, Inc., Oct. 1991], p. 33.)

bility" resulting from high CBR being offset by equally high CDR. **Phase II** is marked by declining CDR brought about by the reduction of infant and childhood mortality. Because fertility and hence CBR remain high during Phase II, this is a phase of ac-

celerating population growth. **Phase III** is a phase of declining CBR resulting from declining fertility rate. Finally, **Phase IV** is reached, where "modern stability" is achieved by a continuing low CDR but equally low CBR.

Basically, developed nations have completed the demographic transition. The population momentum from higher birth rates of the past is still causing some growth, but stability will soon be reached as this "bulge" passes. Less-developed nations, on the other hand, are still in Phase III. Death rates (infant and childhood mortality) have declined markedly, and fertility and birth rates are declining but are still considerably above replacement levels. Therefore, populations in less-developed countries are still growing rapidly, as we have noted.

On the basis of the demographic transition, we might predict that the world population problem will take care of itself and that there is nothing to worry about. The fallacy in this notion, as we have seen, is that in many countries the population is growing so rapidly that development—the factors that bring about lower fertility—is being undercut. Further, we have noted that projecting current trends leads to an estimated global population of 12 billion before the world as a whole achieves Phase IV of the demographic transition. It is highly questionable that the biosphere can sustain the negative impacts caused by those numbers.

Therefore, the question must be asked: What (if anything) can we do to hasten the demographic transition (encourage lower fertility rates) in less-developed nations? In Chapter 7, we shall examine the factors that contribute to lowering fertility and consider the potential for promoting lower fertility rates in less-developed countries.

IN PERSPECTIVE
How Life Is Getting Longer

It is commonly said that, with the introduction of modern techniques of disease control, average longevity increased from about 40 to 65 years of age. While mathematically correct, this statement is misleading. It tends to imply that nearly everyone used to die around age 40; now everyone lives to around 65. However, the years when a person is 35 to 45 are generally the healthiest period of human life. With or without

modern medicine, a relatively small portion of the population dies in this age range. The "average longevity of 40" before modern techniques of disease control was a function of a large fraction of the population dying in childhood and so counterbalancing another fraction that lived into their sixties and beyond.

Through disease control, most of those who formerly would have died in childhood now live past 50. Ex-

tending the life span of this group that used to die young shifts the average age of death of the population. But the basic lifespan of the human species has changed little if at all as a result of modern medicine. All modern medicine has done is to increase the proportion of people who get to or near the maximum.

1. Beginning with the appearance of the first humans and projecting into the future, describe the changes that have occurred, are now occurring, and are likely to occur in the human population.

2. What are the main features of nations defined as developed? Of nations defined as less-developed? What are the differences in average lifestyles in the two types of nations?

3. Which group of countries, the HDCs or the LDCs, is growing most rapidly? Which group is growing least rapidly?

4. What three factors and what relationship among them determine human impact on the environment?

5. Describe how changes in the three factors given in question 4 influence the degree of and even the direction of environmental impact.

6. What are the consequences of rapid population growth is less-developed countries?

7. Define total fertility rate.

8. Explain how high fertility rates tend to trap less-developed countries in a vicious cycle of increasing poverty.

9. What information is given by a population profile? How is the information presented?

10. How are numbers of future deaths and future births predicted from a population profile? What simplifying assumptions can be made to enable you to approximate a projection of the future size and structure of a population?

11. How do the population profile, fertility rate, and population projection of a developed country differ from those of a less-developed country?

12. On the basis of question 11, how might future population goals of developed and less-developed nations contrast?

13. What is meant by population momentum, and what is its cause?

14. Define crude birth rate and crude death rate.

15. Describe how CBR and CDR are used to calculate the percentage rate of growth of a population.

16. Describe the demographic transition in terms of relative CBR and CDR during each of the four phases.

17. How do developed and less-developed nations differ regarding their current positions in the demographic transition?

 Thinking Environmentally

1. Make a cause-and-effect "map" showing the many social and environmental consequences linked to unabated population growth. Include crossovers between the developed and less-developed worlds.

2. It has been proposed that excess human populations may be accommodated by building orbiting space stations that would house about 10 000 persons each. Each station would be able to produce its own food. How many space stations would be required to accommodate the world's projected population growth over the next 10 years? What kind of population policy would have to be enforced on the space stations? Are space stations a logical solution to the population problem?

3. Starting with a hypothetical population of 14 000 people and an even age distribution (1000 in each age group from 0–4 to 65–70), assume that this population initially has a total fertility rate (TFR) of 2 and an average longevity of 70 years. Project how this population will change over the next 60 years under each of the following situations:

   a. TFR and longevity remain constant.
   b. TFR changes to 4; longevity remains constant.
   c. TFR changes to 1; longevity remains constant.
   d. TFR remains at 2; longevity increases to 100.
   e. TFR remains at 2; longevity decreases to 50.

4. Contrast the long-term effect on population growth caused by changes in TFR with that caused by changes in longevity.

5. A country such as Denmark, which faces a decreasing population, has the following options:

   a. Accept decreasing population.
   b. Encourage women to have more children.
   c. Accept more immigration.

   Discuss the long-term (50 years) economic and social implications of each option.

6. From the given crude birth and crude death rates, calculate the percentage rate of population growth and the population doubling time for the each of the following countries.

Algeria	CBR = 35	CDR = 8
Egypt	CBR = 38	CDR = 9
India	CBR = 31	CDR = 10
Iran	CBR = 41	CDR = 8
Italy	CBR = 10	CDR = 9
France	CBR = 14	CDR = 9
Australia	CBR = 15	CDR = 7

7

Addressing the Population Problem

OUTLINE

Development and Declining Fertility Rates
Why Fertility Declines with Development
Economic Development
Large-Scale, Centralized Projects
Small-Scale, Decentralized Projects:
Appropriate Technology

Reducing Fertility
Factors Affecting Fertility Directly
The Missing Element
Family Planning Policy
The Abortion Controversy
Family Planning Services and the Abortion Debate
Contraceptive Technology

LEARNING OBJECTIVES

When you have finished studying this chapter, you should be able to:

1. List and describe factors inherent in industrial development that tend to lower fertility rates.

2. Name the two kinds of development and give the general characteristics contrasting the two.

3. Describe two major pitfalls of large-scale development.

4. Relate the concept of appropriate technology to agricultural and lending practices.

5. Discuss drawbacks of appropriate technology.

6. List the three factors that are most specifically related to declines in fertility rates and discuss how they are mutually interdependent.

7. Describe the most important step that might be taken to bring fertility rates down and discuss how women and leaders of most less-developed countries feel about the issue.

8. Outline the policy and activities of family-planning agencies.

9. Outline the positions of the two sides of the abortion debate.

10. Evaluate how government policies regarding the support or curtailment of family-planning clinics and services affect the poor and how they affect the affluent.

11. Discuss the pros and cons of giving more support to development of better contraceptives and better sex education.

144

I cannot believe that the principal objective of humanity is to establish experimentally how many human beings the planet can just barely sustain. But I can imagine a remarkable world in which a limited population can live in abundance, free to explore the full extent of man's imagination and spirit.

Philip Handler, past president of the U.S. National Academy of Sciences

In Chapter 6, we noted that less-developed countries are caught in a cycle of exploding population, poverty, and environmental degradation. It would seem that humanity is indeed embarked on "establishing experimentally how many human beings the planet can just barely sustain." On the other hand, we have seen that fertility rates have fallen in developed countries such that, except for immigration, the population in those countries is leveling off. This seems to say that (assuming we control pollution) a "remarkable world in which a limited population can live in abundance" is within the realm of possibility. The question is, How do we get there?

Some observers argue that population control is a problem for each country to solve by itself. If a country cannot solve its population problem, let it "sink" (see Ethics Box, p. 183). Moral implications aside, this suggestion is analogous to one where rich people in the bow of the ship look at poor people crowded in the stern and inform them that their end of the boat is sinking! The analogy is not far-fetched. We are all in one ship, Spaceship Earth. The environmental degradation and loss of biodiversity that is occurring will ultimately affect all of us.

IN PERSPECTIVE
How Many People Are Enough?

Not everyone shares the idea that population is a problem.

The only constraint upon our capacity to enjoy unlimited raw materials at acceptable prices is knowledge. People generate that knowledge. The more people there are, the better off the world will be. (Julian Simon, Professor of business administration at the University of Maryland, College Park, and leading spokesperson for the Cornucopian school of thought)

How does Simon come to the conclusion expressed above? He backs up his argument by citing statistics from developed countries, especially the United States, showing that over the last century development and technological advances have paralleled population growth. He ignores all evidence that fails to support his point of view, however. For example, if there were a direct relationship between population and increasing technological progress, why are not India and China, which have populations three and five times that of the United States, the world's technological leaders? Why are Third World countries with exploding populations not booming with technological innovations?

To prove his connection between unlimited raw materials and knowledge, Simon cites mineral resources only. Increasing knowledge and technology do expand the base of mineral resources because they allow the exploitation of lower-grade ores from more-remote locations—deeper in the earth and under the sea, for example. Since the earth is effectively one vast ball of minerals, there is no way we will run out. But does this argument hold for tropical forests, elephants, fertile soil, and other biological resources that are being destroyed under the pressures of growing population?

Biologists and ecologists point out that the concept of an optimum presented in Chapter 2 (p. 32) applies to populations as well as to abiotic factors that affect populations. Thus, a population at very low levels will have difficulty surviving, and increasing numbers will be beneficial. There is an optimum population, however, and increases above it lead to overgrazing, which also threatens survival. It is unquestionable that a certain number of people are necessary to support a technological society. Increasing population up to that number may be a benefit. But should we forget about the concepts of an optimum and overgrazing? Most who look at the problem conclude that the human population on Planet Earth is already considerably above the optimum.

Therefore, enlightened self-interest, if not humanitarian reasons, should bring us to address this manifold problem of population, poverty, and environmental degradation. The objective of this chapter is to examine what directions and policies we might promote toward this end.

Development and Declining Fertility Rates

In Chapter 6, we observed that fertility rates decline with development (see Changing Fertility Rate and the Demographic Transition, p. 140). What are the specific factors that cause this decline? Understanding the answer to this question may enable us to promote such factors in less-developed countries.

WHY FERTILITY DECLINES WITH DEVELOPMENT

Fertility, the number of children a couple has, depends on two basic factors: First, the number of children the couple desires; second, assuming the number desired is less than reproductive capacity, the availability and use of effective contraceptive techniques.

A large number of social and economic factors, in addition to love and fulfillment, come to bear on a couple's decision regarding family size. Many of these socioeconomic factors hinge on the differences between an agrarian society and an industrial society.

For a number of reasons, an agrarian society tends to favor large families, while an industrial society tends to favor small families.

1. **Children: An economic asset or liability?** On a family farm, children can be a great economic asset (Fig. 7–1). A child as young as 4 can provide significant help with lighter chores, such as feeding the chickens and collecting eggs, and 12-year-olds often do the same work as an adult. Children are also a tremendous help in taking the produce to market and selling it. In an urban setting, however, opportunities for children to contribute to the economic welfare of the family are extremely limited, and the costs of feeding, clothing, and educating children are conspicuous. Stated simply, agrarian families tend to desire more children because they are an economic asset, and urban families desire fewer children because they are an economic liability.

2. **Old-age security.** Industrialized societies have come to offer, and even require that people belong to, various health-care and pension plans that will provide for their care in times of sickness and in old age. In agrarian societies people are self-employed; no such plans exist. It is traditional, and

FIGURE 7–1
Children working with adults in the fields in Bali, Indonesia. In most Third World countries children perform the same work as adults and thus contribute significantly to the income of the family. Far from being an economic expense, in such situations children are seen as an economic asset. (Bruno J. Zehnder/ Peter Arnold, Inc.)

expected, that grown children will provide whatever care and support their elderly parents may need. Consequently, people in agrarian societies look toward children as their old-age security. In conditions where there is high infant and childhood mortality, there is a particular desire to have many children to ensure that some will survive to provide the desired security.

3. **Educational and career opportunities.** In agrarian societies, there are few career opportunities, especially for women, other than raising families of their own. Therefore, people raised in agrarian societies tend to marry and start having children earlier, leading to larger numbers overall. Conversely, industrial societies and urban settings generate a wide variety of educational and career opportunities. If people, especially women, can take advantage of these offerings, they tend to delay marriage and put off having children. The result is that they forgo the most fertile years of their lives (the late teens and early twenties) and ultimately have fewer children than might otherwise be the case.

4. **Status of women.** In agrarian societies, it is generally felt that the first and foremost role of a woman is to bear and rear children. Girls are frequently not sent to school, and women are frequently ostracized and even legally barred from many careers. The only fully acceptable role for women is to rear a large family. Women, therefore, have little choice but to focus their abilities on doing so. Conversely, the opening up of education and careers to women which has occurred in industrialized countries has been an important factor influencing women to limit child bearing and pursue careers instead.

5. **Religious beliefs.** Some religions promote large families. People in urban settings, perhaps as a result of being more exposed to different cultural values, tend to set aside such dogma, if not the entire religion. On the other hand, people in agrarian settings tend to maintain such traditions.

6. **Infant and childhood mortality.** Contrary to common belief, high infant and childhood mortality lead to higher fertility, not lower. This is because, on experiencing the death of a child, parents generally desire more children, both to replace the loss and as extra "insurance" to meet all of the above goals.

7. **Availability of contraceptives.** Finally, desires for fewer children are unlikely to be met unless safe, acceptable, and effective contraceptives are readily available. In industrialized countries, such contraceptives are readily available, whereas in less-developed countries, especially in rural areas, they often are not.

Prior to the Industrial Revolution, all societies were basically agrarian. They had to be because, without mechanization, the majority of the population had to be farmers to produce enough food for the society. During the last half of the 1800s through the first half of the 1900s, three things occurred more-or-less simultaneously in what are now the highly developed countries. As we have already noted, advances in disease prevention and cure were bringing down death rates (childhood mortality). In the same period, however, farms were being mechanized and cities were becoming more and more industrial, so that labor was moving from the farms to take jobs in the cities. As more people were adopting the urban lifestyle, fertility rates were coming down more or less in parallel with death rates. The result is that a severe population explosion never developed in what are now the industrialized countries (Fig. 7–2a). Finally, abundant opportunities for emigration to the New World, particularly the United States and Canada, took the population pressure off European countries.

Today's less-developed countries, however, are largely the former subjects of eighteenth- and nineteenth-century colonialism. While the countries of Europe and North America were industrializing, these nations were basically kept agrarian to produce cash crops (coffee, sugar, tea, tobacco, cotton, bananas, and so on) for the "mother" country. Most of these countries gained independence only in the mid-1900s. Consequently, the condition of high birth rates and high death rates persisted, in large part, until about 1950. Following World War II, there was a massive effort to improve world health. The result was a sudden and precipitous drop in childhood mortality as children received vaccinations and antibiotics and other measures for the cure and prevention of infectious diseases. This sudden drop in infant and childhood mortality in the absence of all those factors that bring fertility rates down precipitated a sudden population explosion in less-developed countries. To a greater or lesser extent, this situation persists to this day (Fig. 7–2b).

ECONOMIC DEVELOPMENT

The fact that fertility rates come down with development indicates that the answer to both overpopulation and poverty is: *promote development*. In addition to the humanitarian goals, development of Third World countries will create additional markets for both obtaining and selling materials and will thus en-

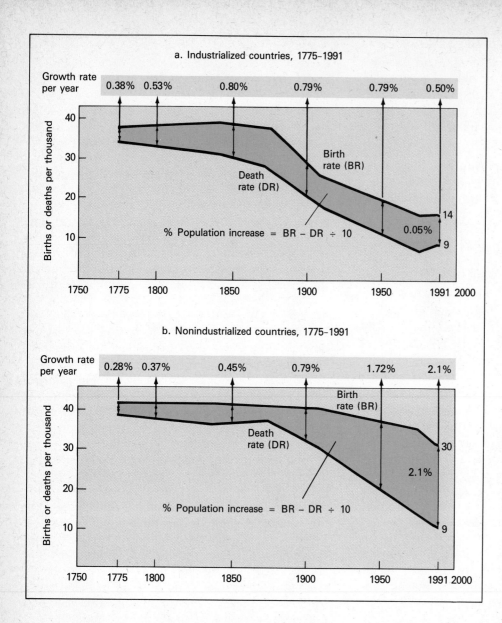

a. Industrialized countries, 1775–1991

Growth rate per year: 0.38% 0.53% 0.80% 0.79% 0.79% 0.50%

Births or deaths per thousand

Birth rate (BR)

Death rate (DR)

% Population increase = BR − DR ÷ 10

0.05%

14

9

1750 1775 1800 1850 1900 1950 1991 2000

b. Nonindustrialized countries, 1775–1991

Growth rate per year: 0.28% 0.37% 0.45% 0.79% 1.72% 2.1%

Births or deaths per thousand

Birth rate (BR)

Death rate (DR)

% Population increase = BR − DR ÷ 10

2.1%

30

9

1750 1775 1800 1850 1900 1950 1991 2000

FIGURE 7–2
The increase in world population has been due to a drop in death rate, not to an increase in birth rate. The increasing rate of population growth is seen in the increasing gap between birth rate and death rate in less-developed countries. (*a*) In developed countries, the decrease in birth rates proceeded soon after and along with the decrease in death rates, so very rapid population growth never appeared. (*b*) In less-developed nations, birth rates remained high longer, causing very rapid population growth. (Redrawn with permission of Population References Bureau, Inc., Washington, D.C.)

hance world trade and prosperity for both developed and less-developed countries.

Indeed, with aid from developed countries, Third World countries have actively pursued development in the past 50 years. Notable success has been achieved. According to United Nations statistics:

○ Child death rates have been reduced by half everywhere but in sub-Saharan Africa, resulting in average life expectancy increasing from 40 to 60 years.

○ The proportion of children attending primary school and thus becoming literate has doubled or more.

○ Over most of the world, food production increased faster than population.

○ The proportion of families in the developing world practicing family planning has increased from 9 to 45 percent.

○ Developing countries increased their share of world industrial output from 5 percent to 20 percent.

○ Immunization of children against common diseases went from near zero to 70 percent, a major factor responsible for lower childhood mortality.

Through the 1980s, however, success has been mixed. Some countries, notably South Korea, Taiwan, and Singapore, have done phenomenally well and now are essentially part of the developed world. Other countries, most notably those of Africa and Latin America, have regressed, and many have come into the 1990s in various states of "economic disaster" with rampant inflation, unemployment, and the poor worse off than ever. It is worth reviewing what went wrong in order that we may get back on track.

Development efforts can be divided into two categories: *large-scale, centralized projects* and *small-scale, decentralized projects* (also commonly known as *appropriate technology*).

Large-Scale, Centralized Projects

Examples of large-scale, centralized projects are hydroelectric dams, power plants, industrial plants, modern hospitals, and high-speed highways. Projects begin with negotiations with a bank and the arrangement of a loan to pay for the project. The bank is generally The World Bank, which acts as a conduit to channel investment moneys from banks in the developed world to projects in the Third World. The advantages of large-scale, centralized projects are:

○ they are relatively easy to administer; lenders need only interact with a few people in the recipient nation;

○ progress on such projects is easy to measure and monitor;

○ they result in very conspicuous end products that both grantors and recipients can point to with pride.

These projects may substantially increase the gross national product (GNP) of the recipient nation so that, on the books, it looks as if the country has advanced in terms of development. Unfortunately, the growth in GNP may be deceiving. Too often the benefits of such projects go mainly to those who are already wealthy. The plight of the poor may even be worsened. For example, The World Bank lent India nearly a billion dollars to create a huge electric generating facility at Singraali, consisting of five coal-burning power plants, and to develop open-pit coal mines to support them. However, the increased power does little for the poor, who cannot afford electrical hookups. The project displaced over 200 000 rural poor, who had farmed the fertile soil of the region for generations, and moved them to a much less fertile area without giving them any say in the matter and little if any compensation. Finally, the project has caused extensive air and water pollution.

Nowhere are the failing and environmental destructiveness of large-scale projects more evident than in agriculture. The World Bank funneled 1.5 billion dollars into Latin America from 1963 to 1985 for clearing millions of acres of tropical forests. Most of the cleared land was put into large cattle operations for producing beef for export and thus increasing the GNP. However, no type of agriculture requires less labor per acre than ranching. Million-plus-acre spreads are run by millionaire cattle barons checking their herds by aircraft. Meanwhile, the poor are pushed into more marginal lands or into cities, as described in Chapter 6. Because the soil of cleared tropical forests is so poor, some of these ranches have already been abandoned, and most of the rest are only marginally profitable. The linkage between agriculture, poverty, and population is examined further in Chapter 8.

Development or Exploitation of the Poor? In many cases of industrial plants built in the Third World, there is much debate as to whether the projects should be classified as development or as exploitation of the poor. Third World countries lack strong labor unions. Consequently, they do not have the wages, fringe benefits, or other protections that unions in developed nations have won for workers over the years. Also, the less-developed countries lack or fail to enforce pollution regulations.

An example of the result may be seen just across the U.S. border in Mexico. Here a number of U.S. firms have set up plants (with the government of Mexico holding a 20 percent share) to assemble items such as television sets, which are mostly returned to the U.S. market for sale. At these plants, people work about 50 hours per week under poor conditions for 60–70 *cents* per hour (1991) and no health insurance or benefits of any kind. The plants have spawned surrounding communities where whole families live in single-room cinder block "huts" with no indoor plumbing or electricity. Water is obtained from a single community pipe. Outdoor privies commonly overflow. Both air and water are often fouled by pollution from the plant (Fig. 7–3, page 150).

Why do the people put up with such deplorable conditions? The answer is the poverty-population problem. Workers are mostly young people desperate for work of any kind. Many are children in their early teens who lie about their age (working is technically illegal till age 16), quit school, and go to work to help their struggling and often fatherless families. According to the Mexican Center for Children's Rights, there are some 12 million such working children in Mexico.

On the U.S. side of the border, it is not hard to see what happens to U.S. workers as production plants are transported to the Third World—and what happens to the U.S. plant owners as they save immensely on wages, costs of pollution control, safety equipment, and so on.

FIGURE 7–3
Raynosa, Mexico, two miles from the U.S. border. U.S. firms have built factories and employ Mexicans for 60-70 cents per hour to make consumer items mostly for the U.S. market. The workers on this wage can only afford to live in the shacks and slum conditions seen here. (Nubar Alexandrian/Woodfin Camp & Associates.)

The Debt Crisis Another consequence of large-scale projects is the debt crisis, which has come to a head in recent years. All these projects were financed by loans that presumably would be paid back through the increased GNP generated by the project. A number of things went wrong, not the least of which were corruption and mismanagement. Other factors, however, were beyond the developing nation's control. The quadrupling of oil prices in the 1970s put severe financial burden on all but the oil-exporting countries. Then, high interest rates combined with recession in the late 1970s and early 1980s. Caught between rising costs and falling export earnings, developing countries were unable to pay interest in full. Developed countries added unpaid interest to the principal,

causing the debt of developing countries to balloon without providing any additional benefit to the Third World.

The less-developed countries now have an outstanding debt to the developed world exceeding 1.3 trillion dollars. In some cases the per capita debt exceeds 10 times the average annual per capita income. (The U.S. national debt per capita, which many people consider alarming, is about one-half the average annual income.) With this kind of debt, it is understandable that lending institutions are reluctant to lend more money. No one expects most Third World countries to be able to repay all they owe. Yet most are making valiant attempts at repayment in order to maintain some standing with the world economic

---

ETHICS
Forgiving Debts

In the text, we pointed out that there is currently a net flow of $50 billion a year going from the Third World to the developed world. Despite these payments, Third World debt is growing because these countries have not been meeting interest payments. Some policy makers argue that the best thing we could do for Third World development, humanity, and the forests and wildlife that are being exploited to pay the interest is to forgive Third World debt. Such a policy makes sense,

they argue, because the original loan amounts have been more than repaid in full.

Since no one seriously believes most Third World countries ever will be able to pay the amount they owe, the losses have already been absorbed and written off in a practical sense. "IOUs" for debt are bought and sold much like stocks on the stock market. The value of the IOU depends on the buyer's evaluation of the debtor's ability or intention to pay. Thus, much of the

Third World debt is currently traded on the "debt market" for 10–20 cents on the dollar. That is, a speculator can buy (or sell) a $1000 IOU from a Third World country for $100–$200. This effectively means that 80–90 percent of the face value has already been written off and absorbed by those who traded the IOU on the way down. What do you think? Should Third World debt be forgiven?

community. Therefore, because the countries are paying interest but little or no new money is coming in, the net flow of capital is currently from less-developed countries to developed countries at the rate of about 50 billion dollars per year. A number of countries are paying more than half of their export earnings back in interest, effectively giving away a large share of their products. Many have already paid back in interest considerably more than they originally borrowed; yet having failed to pay the full interest, their debt has continued to grow rather than diminish.

The debt crisis has been and continues to be an economic, social, and ecological disaster for less-developed countries. In order to keep up even partial interest payments, Third World governments have done one or more of the following:

1. Focused agriculture on large-scale growing of cash crops for export. This has occurred at the expense of peasant farmers growing food. Thus, hunger and malnutrition have increased and so has poverty as peasants have been pushed from the land.

2. Adopted austerity measures. Government expenditures have been drastically reduced so that income can go to pay interest. What is cut? Schools, health clinics, police protection in poor areas, building and maintenance of roads in rural areas, and other goods and services that benefit not only the poor but the country as a whole.

3. Invited the rapid exploitation of natural resources (e.g., logging of forests, extraction of minerals) for quick cash. With the emphasis on quick cash, few if any environmental restrictions are imposed. Thus, the debt crisis has caused disaster for the environment. Ironically, this forced exploitation of material resources has placed such oversupplies of commodities on the market that prices have been severely depressed so that actual earnings are low.

You can see that the brunt of these measures falls on the poor. Thus, many observers point out it is the poor, who gained nothing from the development, who are now being required to pay for it (see Ethics Box, page 150).

To be effective in alleviating poverty and bringing fertility rates down, development must benefit the poor. This is the focus of small-scale, decentralized projects using appropriate technology.

Small-Scale, Decentralized Projects: Appropriate Technology

The term *appropriate technology* does not refer to any specific technologies or techniques. Instead, it is used to describe any project that helps poor people to help themselves. There are several attributes common to all forms of appropriate technology:

○ It is directly relevant to the economic and social needs of the recipients, and implementing it does not upset the existing social structure.

○ It does not require a high degree of training or skill but involves the kinds of techniques that one person can teach others.

○ It utilizes local resources that are plentiful and inexpensive.

○ It does not require expensive, centralized workplaces; existing homes, farms, and small shops suffice.

○ It allows many individuals to use their own talents and minds, express their creativity, and develop self-reliance.

Here are a few examples.

Handloom versus Textile Mill Introduction of an improved handloom allows women to produce more cloth in their homes while caring for children, thus increasing their incomes without changing the social structure. In contrast, putting in a modern textile mill would undercut an important source of income for many women and provide only a few jobs in return.

Handmade Brick Dwellings versus a Large Apartment Complex Inappropriate technology was almost employed in a situation where a government planned to clear a poor slum area and build a huge complex of apartments. The appropriate technology involved teaching people to make bricks from the local mud. The technique spread quickly, and soon nearly every family in the area was making bricks and building a simple but adequate dwelling.

More-Efficient Wood Stoves We have noted that many people in less-developed countries are dependent on firewood for cooking and that women spend much of their day gathering wood, denuding the landscape in the process. Introduction of a clay stove that people could mold themselves and that consumes only half the wood of the open fires they were originally using is an appropriate technology that alleviates the situation.

Appropriate Technology and Agriculture Nowhere can appropriate technology be applied with more advantage than in increasing plant and animal production on small plots of land using techniques that are sustainable.

FIGURE 7–4
Open air market in Latacunga, Ecuador. Numerous small independent farmers bring their produce and crafts to these huge marketplaces for sale and trade. The system fosters independence and a great diversity of food products year round and avoids middleperson costs. (Thomas Ives/The Stock Market.)

Historically, the highest production per acre is achieved on small plots of land that are carefully tended and harvested by hand. Also, wherever land is farmed by numerous small, independent farmers, a great diversity of vegetables and small animals is grown. Further, variations in times of planting and harvest, especially in the tropics, where few seasonal restrictions exist, ensure food availability year-round. Finally, individual farmers sell and trade fresh produce, small animals, and fish in large, open markets (Fig. 7–4). This direct farmer-to-consumer exchange eliminates the need for and costs of brokers and storage facilities.

The earnings of the farmers underpin the economy of the rest of the community because the farmers are able to buy other goods. While these people are not rich by U.S. standards, at least their basic needs are met and they have an active social life. In many

ways this goes back to or maintains the tradition of subsistence agriculture (described in Chapter 8). However, modern understanding of plant nutrition, soil, and water management may allow yields (and profits) to be increased several-fold through application of the appropriate technologies. These ideas are the focus of Chapter 9.

Attempts to mechanize Third World agriculture, a common objective of the developed world, has been anything but appropriate for the poor. Large machinery, not usable on small plots, benefits only owners of large tracts, who are generally cultivating cash crops. It enables them to lay off laborers and acquire more land, thus forcing the poor off the land. You can see how this practice leaves most of the rural people without food or income. All they can do is migrate to the shanty towns of cities and hope to get a job even if it pays only 60 cents per hour or less. Finally, the high inputs of fertilizers, pesticides, and irrigation involved in mechanized agriculture may not be sustainable, as we shall learn in Chapter 8.

Microlending Very often, no outside ideas or technical assistance is required for an appropriate technology project—only a small loan at a reasonable rate of interest to enable a person to make a start. The idea of "microlending" was started by an economics professor, Muhammad Yunus, in Bangladesh in 1976. The bank he started lends only to the poor, and amounts average just U.S. $67. Several hundred thousand persons have taken advantage of these loans to start small businesses, ranging from raising chickens to weaving baskets. Ninety-eight percent repay these loans. In addition, many loan recipients are women. Yunus observes, "When women borrow, the beneficiaries are the children and the household. In the case of a man, too often the beneficiaries are himself and his friends."

Loans from Yunus's bank have had outstanding results when applied to small-scale agriculture. In a rural area of Bangladesh, small loans along with horticultural advice are now enabling peasant farmers to raise tomatoes and other vegetables for sale to the cities. These people have doubled their incomes in three years. A number of other governmental and nongovernmental organizations are now spreading the concept of microlending throughout the Third World.

The Down Side of Appropriate Technology The concept of appropriate technology, however, is not entirely without drawbacks. The first is simply the magnitude of the task. Recall that, worldwide, 2 billion people live in absolute or marginal poverty. The prospect of reaching this number of people with appro-

On November 6, 1980, 150 Honduran peasants first attempted to claim the unused land they now own and work. The landlord forced them off. They tried again and again. On the twenty-second attempt, they succeeded. After holding the land for two years, they won legal title.

Sixth of November Cooperative now boasts 54 small but comfortable cement block houses and 350 acres of farm land. Three-fourths of the co-op's land is divided into plots for individual households. Each family farms three or four sites so that all have some quality land near the river. They grow corn, rice, beans and vegetables. Every family also contributes time to community upkeep and farming of collective land.

Grassroots cooperatives are the driving force of land reform in Honduras. In a country where the best land is controlled by foreign banana companies, an uncharacteristically liberal government began an ambitious land-reform program in 1962. When the following years brought little change, peasant organizations such as the Union Nacional de Campesinos decided to take matters into their own hands.

Land that met the legal requirements for redistribution was identified. Groups of peasants organized themselves into cooperatives, occupied the land and began farming.

The process continues today. In 1987 over 50 000 acres of land was reclaimed in nationally coordinated recoveries.

Sixth of November no longer depends on middlemen to market its products. Having confronted a powerful landowner and having demystified basic mathematics, negotiating prices in Tegucigalpa or San Pedro Sula markets is within its stride.

The co-op has built a granary where it accumulates grain during the harvest for use later in the year when prices are high. On the average, families now earn $3000 per year, 10 times their earnings as landless day workers.

The most encouraging news is that in the last few years not one child has died of malnutrition.

(Reprinted by permission of Pueblo to People, 2105 Silber Rd., Suite 101, Houston, TX 77055.)

---

priate technology is daunting, to say nothing of the fact that their numbers are likely to double over the next 30 years. Another drawback is that appropriate technology may be viewed by people of developing nations as an effort to prevent them from entering the modern industrialized world.

In fact, some mix of appropriate technology and large-scale, or perhaps we should say modest-scale, projects is probably in order. One commonly cited example is the need to provide paved roads and bridges. This does not mean the building of new high-speed highways noted above as an inappropriate large-scale project. It simply means providing a road surface that can be used year-round. Throughout the less-developed world and the Commonwealth of Independent States (formerly the USSR), well over 1 billion people live in cities or communities where the only access to the rest of the world is by way of dirt roads that can turn into deep mud, becoming impassable, and swollen rivers that are impossible to cross during rainy seasons. This often results in produce spoiling because there is no way to get it to market and many activities shutting down because it is impossible to bring in needed supplies such as fuel during these periods.

Even where development efforts are well focused and administered, however, in many cases the ranks of the poor are still growing, because of high birth rates, faster than they are being diminished by development efforts. Therefore, while in no way denying the need to help people improve their lives, we shift our focus to the question of how fertility rates may be more directly affected.

Reducing Fertility

The correlation between development and lower fertility rates is far from perfect, particularly for the low-income groups. Figure 7–5 on the next page shows that countries with equally low average incomes per person have markedly different fertility rates. Why?

FACTORS AFFECTING FERTILITY DIRECTLY

At the beginning of this chapter we observed that development leads to lower fertility only indirectly, by providing factors such as old-age security, edu-

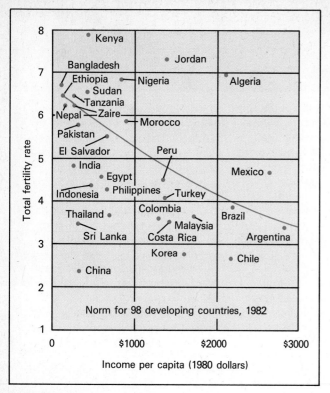

FIGURE 7–5
Fertility in relation to income for selected less-developed countries. There is only a very weak correlation between income and lower total fertility. Factors affecting fertility more directly are health care, education for women, and availability of contraceptive information and services. (From the *World Development Report, 1984.* Copyright 1984 by the International Bank for Reconstruction and Development/The World Bank. Reprinted by permission of Oxford University Press, Inc.)

cational and career opportunities, contraceptives, and so on. Further analysis has shown that low fertility rates are more closely correlated with the following four factors:

1. good health and nutrition, particularly for children and mothers,
2. availability of education, especially for women,
3. rights of women, and
4. availability of contraception.

The first three factors influence people's choice to have fewer children; the fourth enables fulfillment of the choice. The four factors work together as follows. Even very poor parents can readily see that two or three healthy children are more likely to fulfill their needs for economic assistance and old-age security than are a half dozen sickly children. This is particularly so if the children can gain an education and look forward to better jobs. Therefore, when avenues toward better health and education are clear, parents

focus their scant resources on seeing that two children are well nourished and provided with shoes, clothing, paper, pencils, and other school needs, rather than having more children. Then, as young women gain education and have access to jobs and careers, they are much more likely to delay having children and to choose to have fewer. Countries where women's rights are denied typically have the highest fertility rates. The need for availability of contraception is self-evident.

Thus, a "package" can be designed to promote and provide three of these basic elements: health care, education, and contraceptive information and materials. In this context, health care does not refer to geriatrics. It refers to providing basic information and instruction on good nutrition; hygiene steps, such as boiling water to avoid the spread of disease, as water supplies are frequently contaminated; and treatment of common ailments such as diarrhea. (In Third World countries, diarrhea is a major killer of young children but is easily treated by giving suitable liquids, a technique called oral rehydration therapy.) The care emphasizes pre- and postnatal care of the mother as well as that of the children.

Education in this context emphasizes simply learning to read, write, and do simple calculations. Illiteracy rates among the poor women in less-developed countries are commonly between 50 and 70 percent, in part because the education of women is not considered important and in part because the population explosion has overwhelmed school systems and means of getting children to schools. Providing basic literacy will enable people to read "how-to" pamphlets on anything from health care to new techniques for growing chickens and will greatly facilitate the exchange of information.

The emphasis is on educating women because they not only bear the children, they are also the primary providers of nutrition, child care, hygiene, and, potentially, education as they help their children learn. Then, it is the girls growing up who will be most critical in determining the number and welfare of the next generation.

The entire package may be promoted through posters, radio advertisements, and popular songs conveying the message that a two-child family is a happy, healthy, prosperous family. As parents begin to see this becoming a reality in their own or neighboring families, they are even more inclined to focus their limited resources on fewer children, and their desire for more children will diminish. It is important to emphasize that it is good health, not poor health, that brings about a desire for fewer children. This observation brings us to contraception.

Contraception refers to any means of preventing a pregnancy. However, a matter of growing urgency

AIDS is an acronym for acquired immune deficiency syndrome. It is caused by a virus known as HIV (human immunodeficiency virus).

Most viruses cause the body's immune system to produce antibodies. The measles and chickenpox viruses work this way, for instance. Assuming the patient doesn't die from the initial acute infection, the antibodies gradually overcome the virus and make the body immune to further infection. Vaccinations work on the principle of stimulating your body to produce antibodies without making you sick from the disease.

HIV also elicits the production of antibodies; the AIDS test involves testing for the presence of those antibodies. A positive test for them indicates that the person harbors the HIV virus (is HIV-positive). But the course of infection with HIV is very different from that of typical viral diseases. The virus escapes antibodies by retreating into cells of the body's immune system. Consequently, at first it does not make the person sick or even give any warning symptoms. Over a period of time, though, which may vary from a few months to 10 years or longer, the body's immune system is gradually impaired such that is cannot ward off even a simple common cold virus. As this occurs, the person is said to have AIDS as opposed to simply being HIV-positive.

A person with AIDS eventually dies from one or more of these secondary infections, not from HIV itself. Research to develop a vaccination for HIV is under way, but it will probably be at least the late 1990s before one is available.

The fact that HIV infection does not cause any warning symptoms at first makes it particularly insidious. It means that an infected person may act as a carrier and pass HIV on to others for some months or years without knowing it. On the other hand, it is largely a disease of choice. HIV does not survive in the open air outside the body for longer than a minute. Therefore, it is impossible to catch AIDS by simply being around someone who has the disease or is HIV-positive. The only way it can be transmitted is through direct transfer of blood or other body secretions from the HIV-positive person into the blood of the HIV-negative person.

At first, AIDS was thought to be a disease entirely of homosexual men, intravenous drug users (who passed the virus by sharing needles), and a few people who received contaminated blood transfusions. (Blood is now tested to prevent transmission by transfusions.) Then, babies from infected mothers were added to the HIV-positive group. In recent years, it has become increasingly clear that HIV is being transferred through heterosexual contacts, as in the case of Erwin "Magic" Johnson. Except for babies born HIV-positive, each individual can make a choice not to get AIDS. Abstention until you enter into a monogamous relationship with a person who has similarly abstained or who has not had sex for at least 3 months before an HIV-negative test offers no risk. Using condoms minimizes risk but not to zero because there is a certain failure rate. No other contraceptive offers any protection whatsoever.

How serious a problem is AIDS? World health leaders are extremely alarmed. Surveys indicate that as much as 20 percent of the adult population in certain parts of Africa, where the disease originated, may be infected. In Africa there are already an estimated 7 million "AIDS orphans"—children both of whose parents have died of AIDS—and it is expected that this figure may rise to 20 million by the year 2000. How African nations that are already finding it impossible to cope with numbers of children can handle this additional burden is yet to be determined. The disease is beginning to make serious inroads into Latin America and Asia as well. In the United States there are an estimated 2 million HIV-positive persons in addition to those who already have AIDS. Worldwide it is estimated that there are between 10 and 20 million HIV-positive persons. You can see the cause for alarm. Is it responsible or even ethical to be cutting back on sex education, provision of condoms, and other family planning information and materials at this time?

that is intricately involved with contraception is the control and prevention of sexually transmitted disease (STD), especially AIDS (see Ethics Box above). The ideal method of contraception and prevention of STDs is abstention. However, for those who decide to be sexually active, numerous contraceptive techniques or devices have been developed. Those most commonly used are listed in Table 7–1, with their effectiveness and other advantages and disadvantages. With suitable information and education, persons or couples can choose the method that best meets their needs and preferences and can be supplied with materials or treated accordingly. Beyond abstention, only condoms (male and female) are effective in preventing STDs.

None of the three elements in our package (health care, education, and availability of contraception) involves great expense. The health care is mostly

TABLE 7–1

Contraceptive Methods

Method	What It Is	How It Works	Advantages	Disadvantages
Abstention	Total abstinence from sexual intercourse	Absence of contact between sexual organs	Total protection from STD and unwanted pregnancies	Requires exceptional determination/will
Calendar Rhythm Method	Abstinence during natural fertility times in cycle	Assumes infertility at specified time following onset of menstruation	Clinician involvement unnecessary: nonintrusive, acceptable to most religions	Notoriously ineffective because of cycle variations: requires abstinence during fertile times
Natural Family Planning (Symptothermal Method)	Uses body signs to determine fertility and ovulation	Body temperature and cervical mucus changes recorded daily, along with other fertility signs; abstinence during fertile times	Clinician involvement unnecessary; nonintrusive; acceptable to most religions	Mucus can be disguised by other body fluids; body temperature can vary; requires training and diligence; abstinence during fertile times
Fertility Awareness Method (Ovulation Method or Billings Method)	As in NFP, with emphasis on cervical mucus	As in NFP, emphasis on mucus; barrier methods for intercourse during fertile times	Clinician involvement unnecessary; allows intercourse throughout cycle	Mucus can be disguised by body fluids and barriers; requires training and diligence
Condom	Thin, flexible sheath, usually of latex, worn by man during intercouse	Fitted over erect penis; used with spermicide prevents transmission of semen to vagina; used once and discarded	Available at drugstore; protects against STDs	Requires diligence; latex deteriorates with time and heat; can break during ejaculation; perceived interference with mood and sensation
Diaphragm	Shallow rubber cap with flexible spring rim	Used with spermicide and positioned over cervix, prevents sperm from entering uterus	Less intrusive during intercourse than condom	Requires diligence in use and care; must be fitted by clinician every 2 years; must be inserted up to 6 hours before intercourse
Cervical Cap	Small rubber cap that fits snugly over cervix, creating suction seal	Used with spermicide; blocks and kills sperm	Fits more tightly than diaphragm; can remain in place for up to 48 hours	Fitted by clinician; requires diligence in use and care; limited sizes available; may become dislodged during intercourse
Sponge	Absorbent, 2-inch polyurethane sponge containing spermicide	Moistened with water and inserted in back of vagina; blocks or kills sperm for up to 24 hours; used once and discarded	Available at drugstore; can remain in place for 24 hours	Health risks; possible difficulty in placement and removal; poor effectiveness for women who have given birth

Source: Reprinted with permission: East West Natural Health, Box 1200, 17 Station St., Brookline Village, MA 02147. All rights reserved.

Prevent STDs?	Cost	Negative Health Effects	Accidental Pregnancy in 1st Year of Use	
			Method Failure During Ideal Use	Combined Method and User Failure during Typical Use
Total protection	None	None	0	0 for those who manage it, 100% for those who don't.
No	None	None	9%	14%
No	$25–$100 for training	None	2%	13%
No	$25–$100 for training	Possible side effects from barrier methods	3%	10%
Rubber gives almost complete protection; lambskin gives none	About $.60 to $2.60 apiece	Possible side effects from spermicide; possible allergic reaction to rubber	2%	12%
Some protection	Doctor's office visit rate for fitting; $15–$35 for device	Same as condom; risk of TSS, PID (pelvic inflammatory disease), bladder infections	6%	18%
No	$150–$250 for doctor's office visit and device	Same as diaphragm	6%	18%
No	About $1.25 apiece	Possible side effects of spermicide; high risk of TSS, PID, bladder infections	9% among women who have given birth 6% among those who haven't	28% among women who have given birth 18% among those who haven't

TABLE 7–1

(Continued)

Method	What It Is	How It Works	Advantages	Disadvantages
Female Condom	Worn by women, pre-lubricated polyure-thane sheath held be-tween two rings—one over cervix, one ex-tending beyond geni-talia	Inserted like a dia-phragm; prevents transmission of sperm to vagina; used once and discarded	Stronger and more durable than male condom; can be in-serted before initiat-ing intercourse; broad coverage against STDs	Requires diligence; perceived intrusive-ness; cumbersome; outer ring unattractive; non-tampon users may have difficulty inserting
Spermicides	Creams, jellies, foams, suppositories	Kill or immobilize sperm; foams and sup-positories inserted into vagina, creams and jel-lies applied on barriers	Available at drugstore	Requires diligence; must be used with bar-riers
The Pill	Synthetic version of female hormones es-trogen and progester-one	Estrogen inhibits devel-opment of egg; proges-terone creates environ-ment inhospitable to sperm	Requires minimal dili-gence; doesn't inter-fere with intercourse	Requires doctor's pre-scription; not recom-mended for women over 35; risk of life-threatening side effects
Norplant (FDA ap-proval pending)	Synthetic progestin; carried in 6 small sili-cone rod capsules, surgically implanted under skin	Suppresses ovulation, fertilization, and im-plantation; thickens cervical mucus and blocks sperm	Releases hormone di-rectly into blood-stream daily for up to 5 years; may cause fewer side effects than pill	Requires minor surgery for placement and re-moval; dose may be in-adequate for women heavier than 154 lbs
IUD (Intrauterine de-vice)	Small, variably shaped device made of plastic or copper, with a string tail, inserted in uterus	Impairs or suppresses implantation; probably causes low-grade infec-tion that destroys both sperm and fertilized egg	Once fitted, remains in place; easy to use and care for; highly effective until re-moved	Inserted by physician; high health risks; in-creasingly unavailable
Tubal Ligation	In women—surgical severing or blocking of fallopian tubes	Prevents egg from de-scending to meet sperm	Permanent; does not impair hormone pro-duction, sex drive, or intercourse; elim-inates worry about pregnancy	Requires surgery, usu-ally in hospital; more extensive than vasec-tomy; should be con-sidered irreversible
Vasectomy	In men—surgical sev-ering of vas deferens, the tubes that carry sperm from the testi-cles	Prevents sperm from mixing with semen, so egg cannot be fertilized	Same as for tubal liga-tion; also simple and less expensive—often performed in doctor's office	Requires surgery; should be considered irreversible

Prevent STDs?	Cost	Negative Health Effects	Accidental Pregnancy in 1st Year of Use	
			Method Failure During Ideal Use	Combined Method and User Failure during Typical Use
Almost total protection	Estimated $1.50 to $1.75 apiece	None found so far	Studies incomplete, but estimates similar to male condom	Studies incomplete, but estimates similar to male condom
No	$.25 to $.50 per use	May irritate urethra; may cause urinary tract infections and skin irritations of vagina and penis	3%	21% (if used without barrier)
No	Yearly doctor's office visit; rate; prescription; $15–$25 per month	Known side effects: heart attack, stroke, clotting disorders, pulmonary embolism; long-term effects not known	0.1–0.5% depending on type	3%
No	Estimated at equivalent to 2 years of the pill	irregular menstruation bleeding and spotting between periods, headache, depression, weight changes; long-term effects not known	0.04%	0.04%; 9% for women over 154 lbs
No	$250–$300 (doctor's office visit for placement)	Cramps, excessive bleeding during menstruation; risk of PID, atopic pregnancy, sterility, septic abortion; risk of uterine wall perforation	0.8%–2.0% depending on type	0.8%–2.0% depending on type
No	$1,000 or more	Heavy, irregular bleeding and increased menstrual pain—may require hysterectomy to alleviate; risks attendant with any surgical procedure	0.2%	0.4%
No	$250 or more	Risks attendant with any minor surgical procedure; long-term effects unknown	0.1%	0.15%

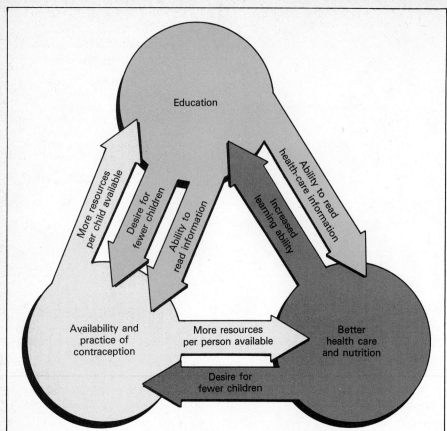

FIGURE 7–6
Education, the availability of contraceptive information and services, and better health care and nutrition create an integrated "package" leading to lower fertility.

FIGURE 7–7
Trends in contraceptive prevalence, 1970–1987, in selected countries. Note that although contraceptive use is increasing, it is still very low in most developing countries. Thus, there is a tremendous opportunity to decrease fertility through promoting family planning and supplying contraceptives. (From the *World Development Report, 1984*. Copyright 1984 by the International Bank for Reconstruction and Development/ The World Bank. Reprinted by permission of Oxford University Press, Inc.; and *Levels and Trends of Contraceptive Use* [as assessed in 1988], United Nations: New York, 1989.)

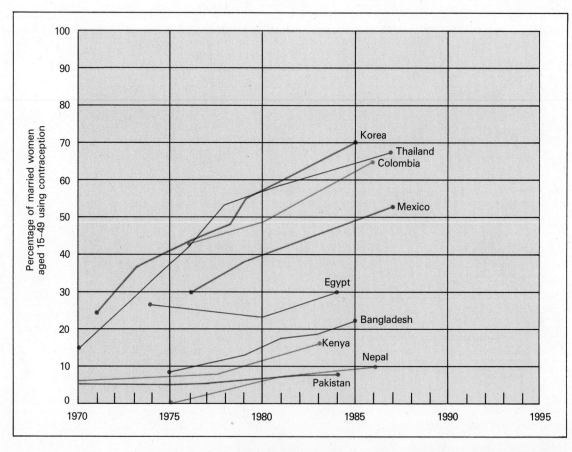

a matter of providing information that can be transmitted by sharing leaflets and by word of mouth. Education can occur in outdoor classes on the ground. Very important, though, the three elements are intimately related to one another, as shown in Figure 7–6. Together, they lead to generally increasing standards of living. If any one element is missing, the effectiveness of the other elements is severely undercut. In situations where one element is absent, great gains can be achieved by providing just that missing element.

THE MISSING ELEMENT

A survey conducted in 61 less-developed countries revealed that most women are having *more* children than they want. In particular, 50 percent of the women interviewed wanted no more children, and another 25 percent wanted to delay their next pregnancy for at least 2 years. But 60 percent of the women who wanted to delay or prevent future pregnancies did not know of any source for contraception information and materials, and the 40 percent who did know often experienced inadequate service. The place for obtaining the information and materials was frequently too far away, too busy to see them, or out of supplies, or the supplies were too costly. Thus contraceptive use in less-developed countries, while increasing in some, is still low (Fig. 7–7). That 60 percent of the women who want family planning information cannot get it tells us that simply filling the existing demand for contraceptive information and materials would go a long way toward bringing fertility rates down and relieving the pressures of growing population.

Third World governments recognize the problem and are asking for assistance in this respect. In 1987, a "Statement on Population Stabilization" was signed by 45 heads of state of developing countries and presented to the U.S. Congress. The statement reads in part:

Degradation of the world environment, income inequality, and the potential for conflict exist today because of over-consumption and over-population. If this unprecedented population growth continues, future generations of children will not have adequate food, housing, medical care, education, earth resources, and employment opportunities. . . .

We believe the time has come now to recognize the worldwide necessity to stop population growth within the near future and for each country to adopt the necessary policies and programs to do so. . . . We call upon donor nations and institutions to be more generous in their support of population programs in those developing nations requesting such assistance.

In their call for donors to be "more generous in their support," developing nations are not looking for a free ride. Already 75 percent of the funding for such programs is provided by the developing countries themselves. Fifty percent of the population of these countries is under 20 years of age, largely poor, and lacking in education. In all, there are over two billion young people who will be coming into reproductive age over the next 20 years. Reaching this population with contraceptive information is imperative and will require help from developed countries.

An example of the success that is possible with full support of family planning is seen in the case of Thailand, an Asian country of 55 million people. The fertility rate in Thailand has dropped from over 7 in 1960 to 2.2, now among the lowest in the less-developed countries (Fig. 7–8). Much of the remarkable success has been achieved through the vigorous promotion and free distribution of contraceptives, although some additional economic incentives were also used, in particular, making small loans contingent on the use of contraceptives. The percentage of Thai people in reproductive age groups using modern, effective methods of contraception is among the highest in the world, about equal to that in the United States and many European countries.

FAMILY PLANNING POLICY

For those who can pay, contraceptive information, materials, and treatments are readily available from private doctors and health-care institutions. The

FIGURE 7–8
Decrease in total fertility rate of Thailand. This remarkable decline over the last 30 years is attributed mainly to vigorous promotion and free distribution of contraceptives. Different lines in the graph represent separate surveys, but all show close agreement. (From *Thailand Demographic and Health Survey 1987*, Population Reference Bureau, Inc., 1991.)

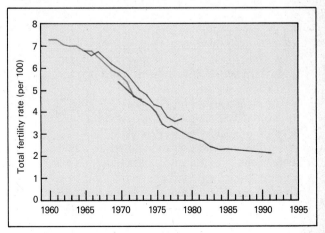

poor, however, must depend on family planning agencies, which are supported by a combination of private donations, government funding, and small amounts the clients may be able to afford. The best known such agency is Planned Parenthood, which has clinics in the United States and throughout the world.

The stated policy of family planning agencies (or private services) is, as the name implies, to enable people to plan family size, that is, to have children only when they want them. In addition to helping people avoid having children against their wishes, planning often involves determining and overcoming

fertility problems for those couples who are having reproductive difficulty. More specifically, family planning services include:

○ Counseling and education of singles, couples, and groups regarding the reproductive process, the hazards of STDs (AIDS in particular), and the benefits and risks of various contraceptive techniques.

○ Counseling and education on achieving the best possible pre- and postnatal health for mother and child. Emphasis is on good nutrition, sanitation, and hygiene.

IN PERSPECTIVE
Economic Incentives for Changing Fertility

The choice of how many children to have is always made within the context of a person's economic, social, and cultural climate. Thus fertility rates change with those factors quite apart from any plan or intent on the part of governments. For example, in the United States there was a conspicuous drop in fertility during the Great Depression of the 1930s and a marked increase in fertility, the baby boom, following World War II (1947–1960). Recognizing this fact, many governments have made and continue to make conscious attempts to influence fertility by providing various economic incentives (or disincentives) toward having more or fewer children. The U.S. income tax deduction for each child may be seen as an economic incentive, albeit small, toward having children. A ban on abortions is typically used by governments that want to increase their populations. Now Third World governments are increasingly turning toward economic incentives and disincentives for limiting population growth. At what point such measures are seen as outright coercion or undue meddling in the private lives of people is a matter of debate and opinion.

China, with its current population

of 1.2 billion (a fifth of the world's population), offers the most comprehensive example of extensive economic incentives and disincentives for reducing population growth. Some years ago, China's leaders recognized that, unless population growth was stemmed, the country would be unable to live within its resource limits. Because of inevitable population momentum, the leaders felt the country could not even afford a fertility of two; they set a goal of a one-child family, and to achieve that goal they instituted an elaborate array of incentives and deterrents. The prime incentives are as follows:

○ Paid leave to women who have fertility-related operations, namely sterilization or abortion procedures
○ A monthly subsidy to one-child families
○ Job priority for only children
○ Additional food rations for only children
○ Housing preferences for single-child families
○ Preferential medical care to parents whose only child is a girl. (There is a strong preference for

sons in China, and parents generally wish to have children until at least one son is born.)

Penalties for an excessive number of children in China include:

○ Repayment to the government of bonuses received for the first child if a second is born.
○ Payment of a tax for a second child.
○ Payment of higher prices for food for a second child.
○ Denial of maternity leave and paid medical expenses after the first child.

Along with improving economic opportunities, these incentives and deterrents have helped China achieve a precipitous drop in its fertility rate, from about 4.5 in the mid-1970s to 2.3 currently. (The one-child policy has not been consistently promoted in many rural areas; therefore fertility in those areas offsets a fertility below 2 in cities.)

We may consider the measures taken in China to be bordering on coercion. If you were faced with China's problems, what would you do?

○ Counseling and education regarding the health advantages of spacing children. Emphasis is on breast feeding, which both provides ideal nutrition and acts as a natural contraceptive, since nursing mothers generally do not ovulate. Nursing, and the contraceptive protection it provides, may be extended till the child is 2–3 years old. Such spacing may cut fertility in half.

○ Provision of materials and treatments after people have been properly instructed about all alternatives.

○ Because all contraceptive techniques have certain drawbacks and none is 100 percent effective even when used properly, some unwanted pregnancies do occur. Women with an unwanted pregnancy may come to family planning agencies for abortions. Further, there are high-risk pregnancies that jeopardize the health or even the life of the mother. Namely: when the mother is under age 16 or over age 40; when a pregnancy occurs less than 1 year after a previous birth; when there have already been four or five previous full-term pregnancies. In such situations, risks are pointed out and, if the client chooses, referrals for abortions may be made. In other words, abortion is used as a last-resort means of family planning.

Governments can do a great deal in terms of providing incentives for people to use family planning services. However, government policy should be seen as distinct from family planning policy per se (see In Perspective Box, page 162).

The topic of abortion has become the center of extreme controversy in the United States in recent years. Since many of you reading this book will have the opportunity to vote either on this issue or for candidates who, if elected, will be making decisions about the availability of abortion, it is worth looking at the two sides.

THE ABORTION CONTROVERSY

On one side are "right-to-life" advocates who believe that, from conception onward, the fetus (developing baby) is a human being who has the right to live and should be granted the same protection under the law accorded to people already born. Hence, abortions at any stage should be outlawed. Supporters of this position point out that the process of development that begins at conception and progresses through birth, growth, aging, and death is a continuum. There is no

obvious point at which you can say: now it is a human baby; before this it is not. Indeed, all the genetic components (DNA) of the embryo are determined at fertilization, and distinctly human characteristics become evident by 7 weeks (see Ethics Box, page 164). Right-to-life advocates believe that, since we are dealing with a human life from conception on, no one—not even the mother—should have the right to terminate that life before birth any more than after birth; therefore, abortions should be illegal.

On the other side, "pro-choice" advocates believe a woman has the right to determine her own future, and this includes when and whether to have a baby. When it comes to an unwanted pregnancy, pro-choice advocates believe that the woman should have the legal right to choose to have an abortion. Broader issues raised by pro-choice advocates are: What kind of care will an unwanted child be likely to receive? What psychological or behavioral problems may develop, and What will be the pressures on and costs to society as a result?

A person can be pro-choice without being pro-abortion. Most choice advocates readily state that abortion is the *least* desirable way to avoid having an unwanted child—abstention or other methods of contraception would be preferable—and many would never choose an abortion for themselves. Nevertheless, unwanted pregnancies do occur and the stand of pro-choice advocates is that each woman should be allowed to make for herself the difficult choice of whether or not to have an abortion. No one else should make the choice for her.

The basic conflict, then, is the rights of the fetus/baby versus the rights of the woman. One demands the sacrifice of the other; it is this choice between two evils—sacrifice the fetus/baby or sacrifice the rights of the woman—that makes the issue a moral dilemma. Ironically, the issue might almost go away if more people practiced abstinence and all sexually active people used contraceptives responsibly, because unwanted pregnancies would be relatively few. The fact that many men and women are engaging in irresponsible sex and then relying on abortion as an after-the-fact contraceptive is indicated by the fact that some 1 600 000 pregnancies (1991 figure) are terminated yearly in the United States, nearly 80 percent of them involving unwed women, and of these 57 percent are under age 25. This is considered shameful by most pro-choice advocates as well as right-to-life advocates.

In Baltimore, Maryland, the two sides of the abortion controversy have recognized that common ground and are working together to promote abstinence and safe sex. At this stage the program includes billboard advertising (Fig. 7–9) and providing free condoms through schools.

FIGURE 7–9
Right-to-life and pro-choice advocates are working together in Baltimore, Maryland, on a campaign to promote greater sexual responsibility. One action is billboard "advertising." (Photograph by BJN.)

FAMILY PLANNING SERVICES AND THE ABORTION DEBATE

Family planning services, both national and international, get caught in the middle of the abortion debate. On the international level, from World War II to 1984, the United States provided world leadership and financial support for family planning programs. Then, in 1984, at the International Conference on Population in Mexico City, just as the rest of the world

ETHICS
Whose Religion Is Right?

The Religious Coalition for Abortion Rights (RCAR), as the name implies, is a coalition of various religious denominations that hold that whether or not to have an abortion should be the right of every woman to decide for herself. RCAR members maintain that abortion is basically a religious issue, and many religions do not make abortion a sin.

Two philosophical/religious viewpoints that lead to an acceptance of abortion, at least at an early stage, while still believing in the sanctity of life are the following.

Eggs and sperm, which carry the genetic components for new individuals, move, metabolize, and respond to stimuli. Biologically speaking, they are as much alive as the fertilized egg that results when they combine. So where does life really begin? The question makes us recognize that life is not just a continuum from conception to death; it is also a continuum from one generation to the next. We might then conclude that all sperm and eggs should be protected, but this is an obvious absurdity. The biological process of reproduction has built into it the inevitable sacrifice of most eggs and sperm. Further, a large percentage of fertilized eggs are spontaneously aborted. Since

which sperm fertilizes which egg is a random process, the argument goes that one particular conception is not more sacred than another that may occur at a later time if the first one is aborted. Thus, an abortion may be seen as making room for another equally sacred conception at another time that might otherwise be denied.

The second argument involves the common belief that there is a soul that exists apart from the body; it enters a new individual at some stage and exits at death. At what stage does it enter or leave? Since medical technology has become capable of keeping the biological functioning of a body going indefinitely, we have come to accept the cessation of brain waves as the point of clinical death and grounds for discontinuing life support. Some people contend that the beginning of brain waves, which start when the embryo is about 7 weeks old, is when the soul enters the body and thus marks the beginning of the individual's life. Others contend that the soul does not enter until later; for instance, when the baby is able to live independently outside the mother's body, about 25 weeks, or at the point of birth itself. The argument is that abortions are not wrong

if performed before the time the soul enters, whatever that time may be.

These arguments are presented here just to point out that religious and philosophical points of view can lead to various opinions regarding the acceptability of abortion. Indeed, to a majority of the world's population, abortion is acceptable. This acceptance is witnessed by the fact that in 37 countries—including Italy, France, the Netherlands, England, Germany, Sweden, Japan, India, and China—abortion is legal and accepted with few if any restrictions and without significant opposition.

The point made by RCAR is that the belief that the biological entity that will develop into a new individual has legal rights from the moment of conception is but one of many religious points of view. What is occurring in the United States with the outlawing of abortion, RCAR maintains, is that the government is imposing the religious view of one group on all others. This is an abridgment of the First Amendment of the U.S. Constitution, which guarantees separation of church and state, and undercuts our basic freedom of religion. What do you think?

leaders were recognizing the extreme urgency of the population problem and asking for family planning assistance, the United States, siding with the growing right-to-life movement, announced an about-face. Worldwide, less than 10 percent of U.S. support for family planning was being used to perform abortions, and these abortions were mostly in high-risk pregnancies. Yet, because some funds were going for abortions, the U.S. government withdrew support from international family planning organizations, including The United Nations Fund for Population Activities, International Planned Parenthood Federation, and Family Planning International Assistance. This withdrawal has severely undercut the activities of these and other such organizations. The Worldwatch Institute estimates that more than 340 million couples in 65 countries are failing to receive contraceptive information and services as a result of the cutbacks in U.S. funding.

Many observers note that the most conspicuous losses caused by the curtailment of family planning activities are education and counseling on reproduction in general, on contraception methods other than abortion, and on STDs, nutrition, and hygiene. It is these activities that account for 90 percent of the budget. Without such education and counseling, these observers say, unwanted pregnancies are more than likely to increase, creating more demand for abortions, not less. For example, Denmark has vigorously supported family planning programs, focusing on prevention of unwanted pregnancies and making contraceptive services freely available to all, including teenagers. The result is that abortion rates among teens in Denmark are about a third of the rate in the United States, where support of family planning has diminished over the last decade.

If abortions are not legal, pro-choice advocates claim, the result will be an increase in illegal abortions and attempts by women to self-induce abortions. Such abortions are extremely risky because they are often performed by untrained practitioners using crude, often unsterile implements in unclean surroundings. Therefore the risk of outright injury and secondary infections is high. Right-to-life proponents respond that simply because some individuals will pursue an action illegally is no reason to make it legal. They use addictive drugs as an example. Further, they minimize the hazards of illegal abortions, saying that the same practitioners would perform them in either case and infections can be controlled with modern antibiotics. However, Worldwatch Institute estimates that 115 to 204 thousand women die each year from infections and injuries received in illegal or self-induced abortions. In some poor African and Asian countries, the chance of death during pregnancy is as high as 3000 for every 100 000 live births, compared

with between 2 and 10 per 100 000 in developed countries. At least half of these deaths in less-developed countries result from illegal or self-induced abortions. In addition, for every woman who dies, many other suffer serious long-term, often permanent health problems including sterility resulting from injury or infection.

Conversely, when abortions are legal and available they are generally performed according to "hospital" standards, and the risk of the mother, while not zero is minimal. Jodi Jacobson of the Worldwatch Institute concluded after an exhaustive study of numbers of abortions and maternal deaths in relation to abortion policy: "It is the number of maternal deaths, not abortions, that is most affected by legal codes."

However, the U.S. government has continued to move in a direction of curtailing the availability of legal abortions. Because family planning organizations were getting around the funding restrictions by referring patients to nongovernment-funded clinics for abortions, the government added what is called the "gag rule." Under this rule, federal funding may be withheld from any family planning organization that even mentions to women, regardless of circumstances, that abortion might be one of their options. The legality of the gag rule was upheld by the U.S. Supreme Court in 1991, which set aside arguments that the rule constitutes an abridgment of free speech. Family planning officials argue that the gag rule places them in an impossible compromise. Doctors are obligated to do all they can to protect the lives and health of their patients. How can they fulfill this obligation if they cannot mention abortion even in the case of a high-risk pregnancy? Rather than living with this restriction, a number of family planning organizations have chosen to forgo U.S. funding and curtail operations as necessary. In 1992 the gag rule was relaxed slightly by allowing that doctors can mention abortion but other family planning personnel cannot.

Many people point out that an irony in attempting to limit abortions by withholding funds from family planning organizations is in who is affected. Information regarding abortions is still openly provided by private doctors to private patients. Consequently, it is only the poor, who depend on publicly funded clinics, who are affected. As one United States senator said in his vote against the gag rule despite his right-to-life stance: "The issue is not about abortion. It is about who has access to the information." Likewise, if abortions are made illegal, the more affluent will still get safe abortions even if they have to travel to other states or countries to do so. Only those who cannot afford such travel will be restricted.

In conclusion, it is possible, indeed quite normal, to hold very mixed feelings and beliefs regarding abortion. What is important to recognize is that the

FIGURE 7–10
Confrontation between pro-life and pro-choice advocates in Buffalo, NY, April 24, 1992. Confrontations such as this are being played out in many cities across the United States. Is the issue as clear-cut as it seems? (Joe Travers/Gamma-Liaison.)

abortion issue is but one facet of a very complex problem involving the future of all people of the world. And there are many levels of the abortion question: your decision as to what is right for you personally; your decision as to what is right for other people (freedom to choose or imposition of a particular viewpoint); what may be right for the United States or other developed countries; and what may be right for less-developed countries. At each level, we need to consider the total ramifications of the decision in terms of trying to determine the lesser of the evils (Fig. 7–10).

CONTRACEPTIVE TECHNOLOGY

The perfect contraceptive—one that is 100 percent effective, does not require remembering, is fully and immediately reversible, has no painful or harmful side effects, prevents STDs, and is freely available to all women and men—would make the abortion issue largely moot. Unfortunately, such a perfect contraceptive does not exist yet, but technology is advancing slowly. (In the United States, research in this field has virtually been abandoned again because of various controversies and the resulting difficulty of getting new contraceptives approved.)

Perhaps the most promising new contraceptive device is Norplant, a matchstick-sized capsule that, when implanted in a woman's arm, releases hormonal chemicals that prevent pregnancy. One implant may provide protection for up to 5 years but the protection may be reversed at any time by removing the implant. Norplant has been approved by the United Nations World Health Organization and is being introduced into Asia, Africa, and Latin Amer-

ica. It is extremely cost-effective since once an area in a developing country has been serviced by a medical team, there is no need to return at frequent intervals. Norplant was approved for use in the United States in 1990. Other similar implants and contraceptive injections are being developed.

A French pharmaceutical company has developed an "after-the-fact" drug known as RU486, which prevents a pregnancy from developing if taken within 6 weeks of the first missed period. The drug has been approved for use in France and China and is undergoing tests in other countries. In 1988, however, the company received vigorous protests and threats of boycotts from U.S. right-to-life groups because the drug effectively causes a spontaneous abortion. This protest movement led the company to announce that it would halt production, but the French government intervened, saying that the company had an obligation to produce the drug to protect the welfare of women who might otherwise seek unsafe methods of abortion. However, with right-to-life opposition, there seems little chance that RU486 will be approved for use in the United States in the foreseeable future.

In closing our discussion on population, we wish to return to our overall theme of sustainability. There is little hope for sustainability of civilization unless humans can achieve a balanced population (our principle 3). However, achieving a balanced population that is not "overgrazing" the supporting systems of the biosphere will involve much more than just reducing birth rates. It also demands breaking the entire cycle of poverty, illiteracy, poor health, and lack of opportunity that is intertwined with high birth rates and environmental degradation. Most of all, this is not something that can simply be handed to, or forced upon, the populations of less-developed countries. They themselves must be the key players; the question is how to best involve them in the "play."

A fitting summary of what we have discussed in this chapter is a quote from the World Bank, which now recognizes the excesses of large-scale development:

Rapid and politically sustainable progress on poverty has been achieved by pursuing a strategy that has two equally important elements. The first element is to promote the productive use of the poor's most abundant asset—labor. It calls for policies that harness market incentives, social and political institutions, infrastructure, and technology [i.e. appropriate technology] to that end. The second is to provide basic social services to the poor. Primary health care, family planning, nutrition, and primary education are especially important. (World Development Report, 1990.)

What would all this cost? The World Bank notes that if a 10 percent reduction in military spending by the North Atlantic Treaty Organization (NATO) were directed to the poor, it would double global aid. Thus, the potential, if we change our priorities, is self-evident.

The most important area in which to bring together the abundant available labor, appropriate technology, and market incentives toward improving health, and so on, is in the development of sustainable agricultural systems. Of course sustainable agricultural systems are just as important for the developed world. The following Chapters of this Part of our text will focus on requisites for sustainable agriculture.

 Review Questions _____

1. Which of the factors inherent in industrial development tend to lower fertility rates?
2. What are the two major types of development and the general characteristics of each?
3. What are two major pitfalls of large-scale development?
4. What are examples of appropriate technology?
5. How may appropriate technology be applied to agricultural and lending practices?
6. What are the drawbacks of appropriate technology?
7. What factors are most specifically related to lower fertility, and how are they interdependent?
8. What is the first and most important step that can bring down fertility rates in Third World countries? How do the women in these countries feel about taking this step? How do the leaders in those countries feel?
9. What are the various activities of family planning agencies?
10. What are the positions of the two sides of the abortion debate?
11. How are decisions regarding the legality of abortions and family planning policy likely to affect poor women? How are they likely to affect affluent women?
12. What are the pros and cons of giving more support to development of better contraceptives and better sex education?
13. In what area could people on both sides of the abortion issue work together?

 Thinking Environmentally _____

1. Is the world population below, at, or above the optimum? Defend your answer by pointing out things that may get better and things that may get worse by increasing population.
2. Suppose you are the head of an agency responsible for Third World development with a budget of $100 million. Describe how you would spend the money (kinds of projects, loans, staff hiring, and so on).
3. List and discuss the benefits and harms of writing off Third World debt.
4. Suppose you are the head of a Third World island nation with a poor, growing population, and the natural resources of the island are being degraded. What kinds of programs would you initiate, and what help would you ask for to try to provide a better, sustainable future for your nation's people?
5. Describe what you think will be long-term results (for people, the society, and the environment) of cutting back family planning activities in Third World nations.
6. To combat the problems of AIDS, other STDs, and unwanted pregnancies, more responsible sexual behavior, including abstention, is called for. If you were in charge, which programs and agencies would you expand? Which would you close down? Give a rationale for your actions.

The Production and Distribution of Food

LEARNING OBJECTIVES

When you have finished studying this chapter, you should be able to:

1. Describe how industrialized agriculture developed, and analyze its environmental costs.

2. Describe the origins and impact of the Green Revolution.

3. List the distinctive features of subsistence agriculture.

4. Explain how animal farming has a detrimental impact on the environment.

5. Evaluate the prospects for increasing food production in the future.

6. Analyze global patterns of food trade, and explore the consequences of those patterns.

7. Explore the dimensions of responsibility for meeting food needs at familial, national, and global levels.

8. Define and describe the extent of hunger, malnutrition, and undernutrition in the world.

9. Understand the relationship between hunger and poverty.

10. Explain the two immediate causes of famines, and discuss the geographical areas affected.

11. Explore the usefulness of food aid.

12. Apply the four principles of ecosystem sustainability to agriculture.

As the world population continues its relentless rise (Fig. 6–1), no resource is more directly affected than food. Can the world's farmers and herders produce food fast enough to keep up with population growth? After all, our planet holds only a finite amount of all the resources needed for food production: suitable land, water for irrigation, energy, and fertilizer. In 1798, when the world population was still less than 1 billion, British economist Thomas Malthus pointed out that population has the capacity to increase geometrically, as shown in Figure 6–1, while agriculture, which is dependent on a finite amount of arable land, is limited. Malthus predicted that in the absence of restraints on human reproduction, the world was headed toward catastrophic famines as the number of people outpaced agricultural capacity. Thomas Malthus was right about human population growth, but he underestimated the potential for raising food.

By many measures, human societies have done very well at putting food on the table since Malthus's time. More people are being fed than ever before, with more nutritional food, and a lively world trade in foodstuffs forms the bulk of economic production for many nations. True, chronic hunger and malnutrition exist, even occasional famines, but these are the exceptions to an otherwise remarkable accomplishment. Many observers are convinced that, if and when the world population levels off (at 12 billion?), there will be enough food produced to sustain a population that large.

As we start scratching beneath the surface, however, this rosy picture begins to peel away, and we see human and environmental dimensions of agriculture and food distribution that some observers call disastrous. The environmental impacts of our food-producing activities are enormous, and, most important, nonsustainable. Inequities in food availability, both within one nation and from one nation to another, are great, leaving a pattern of hunger, disease, and death that is unjustifiable. Given this perspective, the predictions of Malthus begin to look contemporary; how can we feed 12 billion people if we aren't even doing well at fewer than half that number?

This chapter takes a hard look at food production systems, problems surrounding the lack of food, and questions of how to build sustainability as well as justice into the agricultural enterprise.

Crops and Animals: Major Patterns of Food Production

THE DEVELOPMENT OF MODERN INDUSTRIALIZED AGRICULTURE

Until 150 years ago, the majority of people in the United States lived and worked on small farms. Human and animal labor turned former forests and grasslands into systems that produced enough food to supply a robust and growing nation. Farmers used traditional approaches to combat pests and soil erosion: Crops were rotated regularly, many different crops were grown, and animal wastes were returned to the soil. The land was good, and farming was efficient enough to allow a substantial segment of the population to leave the farm and join the growing ranks of merchants and workers living in cities and towns. This migration was important, because in the mid-1800s the Industrial Revolution came to the United States, and it had a major impact on farming.

The Transformation of Traditional Agriculture

The Industrial Revolution contributed to a revolution in agriculture so profound that today less than 3 percent of the U.S. work force produces enough food for all the nation's needs plus a substantial amount for trade on world markets. Indeed, this revolution has achieved such gains in production that today the United States has policies to cope with surpluses of many crops.

The agricultural revolution involved the following developments:

○ shifting from animal labor to machinery powered by fossil fuels

○ bringing additional land into cultivation

○ increasing the use of chemical fertilizers and pesticides

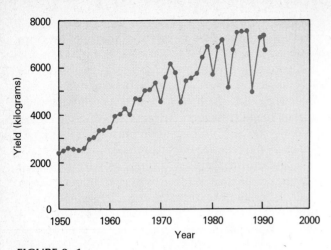

FIGURE 8–1

U.S. corn yield in kilograms per hectare (8000 kg/hectare = 3.5 tons/acre). This graph demonstrates two phenomena: the long-term rise in yields and the effects of droughts in 1970, 1973, 1980, 1983, and 1988. (Worldwatch Institute, Washington, DC.)

○ increasing the use of irrigation

○ substituting new varieties of crops

Virtually every industrialized nation has experienced the agricultural revolution. The pattern of development in U.S. agriculture could just as well describe that of France, Australia, or Japan. The combination of these developments in agricultural practice has raised crop yields and crop production to new heights, doubling or tripling yields per acre (Fig. 8–1). However, each development carries an environmental cost. According to many agriculture experts, expanding production by these methods has reached, or even exceeded, sustainable limits. Let us examine the costs.

Machinery The shift from animal labor to machinery has created a dependency on fossil fuel energy that adds significantly to the energy demands of the industrial societies (Chapter 21). For example, calculations indicate that 4 percent of total energy use in the United States is farm-related. Further, continued use of farm machines for plowing, planting, and harvesting causes soil compaction (Chapter 9).

Land Under Cultivation Much of the increased production in the United States before 1960 came from bringing new land into production. Since then, attempts have been made to increase the land used to raise grain, but these new areas were not well-suited for agriculture and are now being abandoned because erosion or depletion of water resources has rendered them no longer productive. Current farm policy now reimburses farmers for "retiring" erosion-prone land

and planting it to trees or grasses. Essentially all of the good cropland in the United States is now under cultivation. Worldwide, cropland on a per capita basis is on the decline as population continues to rise. Any significant future cropland expansion will come at the expense of forests and wetlands, which are both economically important and ecologically fragile.

Fertilizers and Pesticides When fertilizers were first used, 15 to 20 additional tons of grain were gained for each ton of fertilizer used. Now, however, farmers are applying near-optimal levels of fertilizers, and there is little to be gained from adding more. High levels of fertilizer make plants more vulnerable to attack by insects and other pests, and the washing away of fertilizer leads to water pollution (Chapter 12).

Chemical pesticides have provided significant control over insect and plant pests, but the pests have become resistant to most of the pesticides. Also, there are efforts to reduce the use of pesticides because of side effects to human and environmental health. As we shall see in Chapter 10, progress is being made toward developing natural means of control that will be environmentally safe, but these new methods will be unlikely to increase yields.

Irrigation Worldwide, irrigation acreage increased about 2.6 times from 1950 to 1980. It is still expanding, but at a much slower pace because of limits on water resources. More ominous, much of present irrigation is not sustainable because groundwater resources are being depleted. In addition, production is being adversely affected on as much as one-third of the world's irrigated land because of waterlogging and accumulation of salts in the soil, consequences of irrigating where there is poor drainage (Chapter 9).

High-Yielding Varieties Several decades ago, plant geneticists developed new varieties of wheat, corn, and rice that gave yields of double to triple those of traditional varieties (see below). As these new varieties were introduced throughout the world, production soared. However, most of their potential has been realized, and plant geneticists do not have any more "super-high-yielding" varieties in the wings to repeat the performance. Even breakthroughs in genetic engineering, such as introducing nitrogen-fixing genes into wheat and corn, may enable farmers to reduce their costs for fertilizer, but they will not increase yield. The widespread use of genetically identical crops has given rise to major pest damage, as pests have become adapted to the new varieties and resistant to pesticides.

The same technologies that gave rise to the agricultural revolution in the industrialized countries were eventually introduced to the Third World, where they gave birth to the Green Revolution.

The Green Revolution

In 1944, The Rockefeller Foundation sent four U.S. agricultural scientists, headed by agricultural expert Norman Borlaug, to Mexico with the objective of exporting U.S. agricultural technology to a less developed nation that had serious food problems. Their aim was to improve the traditional crops grown in Mexico. One crop that received special attention was wheat. Mexican wheat was well adapted to the subtropical climate, but it gave low yields and responded to fertilization by growing very tall stalks that were easily blown over. Using wheat from other areas of the world, Borlaug and his co-workers bred a dwarf hybrid with large heads and thick stalks that did well in warm weather when provided with fertilizer and sufficient water (Fig. 8–2). The program was highly successful. By the 1960s, Mexico had closed the gap between food production and food needs, wheat production had tripled, and Mexican wheat appeared on the export market.

Research workers with the Consultative Group on International Agricultural Research (CGIAR) extended the work done in Mexico and introduced both high-yielding wheat and high-yielding rice to other Third World countries. To cite just one success, India imported hybrid Mexican wheat seed in the mid-1960s, and in 6 years, India's wheat production tripled. Within a few years, many of the world's most populous countries turned the corner from being grain importers to achieving stability and in some cases becoming grain exporters. Thus, while the world population was increasing at its highest rate (2 percent per year), the production of rice and wheat underwent a remarkable increase. This achievement, called the **Green Revolution**, has probably done more than any other single advance in preventing hunger and malnutrition. Norman Borlaug was awarded the Nobel Peace Prize in 1970 in recognition of his contribution.

The high-yielding grains are now cultivated throughout the world and have become the basis of food production in China, Latin America, the Middle East, southern Asia, and of course, the industrialized nations. But as remarkable as it was, the Green Rev-

(a)

(b)

FIGURE 8–2
Comparison of an old variety of wheat, shown growing in Rwanda, (a) with a new high-yielding variety of dwarf wheat growing in Mexico (b). The new varieties have short stalks and large heads and are much more responsive to fertilizers. (Dr. Nigel Smith/Earth Scenes.)

olution is not a panacea for all of the world's food-population difficulties, for the following reasons:

1. Most of its potential has been realized; many of the most populous countries are reaching a plateau in their grain production and in acreage planted to high-yielding varieties.

2. Without irrigation, it does not work in drought-prone lands, and it requires constant inputs of fertilizers, pesticides, and mechanized labor, all of which are often in short supply in Third World countries.

3. Because it is patterned after agriculture in the developed world, Green Revolution agriculture tends to benefit larger landholders. Many farm laborers and small landholders become displaced and then migrate to the cities and join the ranks of the unemployed.

4. Unlike corn, wheat, and rice, the most important African food crops (sorghum, millet, and yams) are not commonly used in the developed world, and so they have not benefited from the Green Revolution technology.

The fact is, the Green Revolution has had little impact on the large part of the Third World where another kind of agriculture, subsistence agriculture, is practiced.

SUBSISTENCE AGRICULTURE IN THE THIRD WORLD

In most of the Third World, plants and animals continue to be raised for food by *subsistence farmers*, using traditional agricultural methods. These farmers represent the great majority of the rural populations. **Subsistence farmers** live off small parcels of land that provide them with the food for their households and, it is hoped, a small cash crop. From the point of view of the modern world, such farmers are very poor, although they may not all consider themselves to be so. Like the U.S. agriculture of the past, subsistence farming is labor-intensive and lacks practically all of the inputs of industrialized agriculture. It is often practiced on land that is marginally productive (Fig. 8–3). In some areas, subsistence agriculture involves the clearing of tropical forests in what is called slash-and-burn agriculture, where the cleared land supports a few years of crops until the soil is depleted of nutrients and the farmers move on (see Fig. 6–4a).

Typically, a family will own a small parcel of land for growing food and will maintain a few goats, chickens, or some cattle. Such a family is making the best use of very limited resources, and very often the people are adapted well enough to the prevailing social and environmental conditions to provide a livelihood for a household. An important fact to re-

FIGURE 8–3
Mexican subsistence farmer plowing marginally productive land with a team of oxen. Subsistence farming feeds more than 1.4 billion people in the Third World. (Dick Davis/ Photo Researchers.)

TABLE 8–1

Parallels Between Plant and Animal Farming		
	Plant	**Animal**
Major products	Grains, fruits, and vegetables for food	Meat, dairy products, and eggs for food
Other important products	Oils, fabrics, rubber, specialty crops (spices, nuts, etc.)	Labor, leather, wool, manure, lanolin
Modern practices	Industrialized agriculture on former grasslands and forests	Ranching, dairy farming, and stall-feeding
Traditional practices	Subsistence agriculture on marginally productive lands	Pastoral herding on nonagricultural land
Current global land use	1.5 billion hectares (3.7 billion acres) (11% of land surface)	3.1 billion hectares (7.6 billion acres) (25% of land surface)

member, however, is that subsistence agriculture is practiced in regions experiencing the most rapid population growth, even though this kind of agriculture is best suited for low population densities. An estimated 1.4 billion people in Latin America, Asia, and Africa—over one-third of the people there—are sustained by subsistence agriculture.

Because subsistence agricultural practice varies with the local climate, it is difficult to draw sweeping generalities. It is fair to say, however, that because of a rapidly expanding population and the fact that more and more of the better land is being diverted to industrialized agriculture, pressure on the remaining land to produce more food is increasing. This pressure leads to practices that are at best unsustainable and at worst ecologically suicidal. In region after region in developing countries, woodlands and forests are being cleared for agriculture or removed for firewood and animal fodder, leaving the soil susceptible to erosion and forcing the gatherers to travel farther and farther from their homes. The scarcity of firewood leads the peasants to burn animal dung for cooking and heat, thus diverting nutrients from the land. Erosion-prone land that is suited only to growing grass or trees is planted to annual crops. Good land is forced to produce multiple crops instead of being left fallow for nutrient recovery. Growing populations force continued subdivision of existing land, leading to a diminishing of the land's ability to support a single household. All these factors tend to increase the poverty that is characteristic of populations supported by subsistence agriculture, and, in a relentless circle, the added poverty in turn puts increased pressures on the land to produce food and income.

ANIMAL FARMING AND ITS CONSEQUENCES

Raising livestock—sheep, goats, cattle, buffalo, and poultry—has many parallels to raising crops (Table 8–1), and there are also direct connections between the two. Fully one-fourth of the world's croplands are used to feed animals; in the United States alone, 70 percent of the grain crop goes to animals. Indeed, in many Third World countries farmers are switching from food crops to feed crops, to satisfy the growing market demand for meat. In essence, this practice removes the land's potential to alleviate hunger and malnutrition. In another important connection, manure from animals normally returns vital nutrients to the land, but this cycle is broken when fuels are scarce and dung is collected for fuel, or when the livestock are enclosed.

An estimated 15 billion domestic animals are tended by a human population one-third that number. The care, feeding, and "harvesting" of all those animals constitute one of the most important economic activities on earth. The primary force driving this livestock economy is the growing number of the world's people who enjoy eating meat and dairy products—primarily, the developed world and the affluent people in less-developed nations.

A thriving livestock economy does not mean that all is well, however. Animal farming impacts the environment in a host of ways that are not sustainable. Since so much of the plant crop is fed to animals, all of the problems of industrialized agriculture apply to animal farming. In addition, rangelands are susceptible to overgrazing, either because of mismanagement of prime grazing land or because the land being used for grazing is marginal dry grasslands being used because the better lands have been converted to crop use. For example, overstocking on the rangelands of the western United States has reduced the carrying capacity by an estimated 50 percent. Much of this is public land leased at subsidized fees that easily lead to overgrazing.

In Latin America, over 20 million hectares of tropical rainforests have been converted to cattle pasture. Even though most of this land is best suited for growing rainforest trees, some of it could support a rural population of small farmers producing a diversity of crops. Instead it is held by relatively few ranchers who own huge spreads.

The burning of these rainforests has released an estimated 1.4 billion tons of carbon to the atmosphere, contributing a significant amount of carbon dioxide to the greenhouse effect. Because their digestive process is anaerobic, cows annually belch and eliminate some 80 million tons of methane, another greenhouse gas. Anaerobic decomposition of manure leads to an additional 35 million tons of methane per year. All this methane released by livestock makes up about 3 percent of the gases that are causing global warming (discussed in Chapter 16).

Such significant environmental costs make it clear that animal farming needs to be brought into an ecologically sustainable balance. Essentially, humans are violating the third principle of sustainability: We are maintaining populations of herbivores that are overgrazing the land.

PROSPECTS FOR INCREASING FOOD PRODUCTION

Figure 8–4 gives us some idea of how different parts of the world are coping with the food-population race. The data show that per capita production was rising but in recent years has leveled off in most regions. The one region that stands out as being in serious trouble is Africa. On balance, food production per capita in Africa has been on a downward course since 1970, more than reversing the gains since 1950—a consequence of rapid population growth and poor harvests. Thus with harvests either holding steady or actually headed downward while worldwide population is growing by 93 million per year, Malthusian alarms are being raised again. Malthus was wrong about the imminent prospect of famines in his day, but today his views are seen by many observers as prophetic.

Because the end of Green Revolution yield increases is in sight and essentially all arable land in the world either is now or has recently been in cultivation, we have only two prospects for increasing food production: (1) continue to increase crop yields; (2) begin to grow food crops on land that is now in cash crops and feedstock crops.

Increasing yields further will be difficult for two reasons. First, a significant percentage of cropland, especially in the less developed countries, has been seriously degraded by erosion. Second, weather conditions in years to come may be unfavorable for farming. The climate between 1940 and 1980 was exceptionally stable and ideal for agriculture in most parts

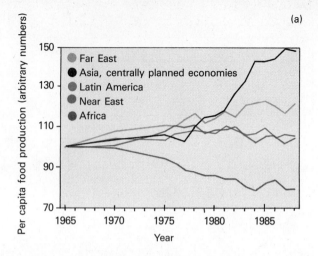

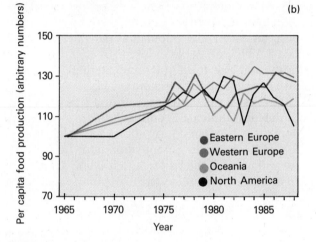

FIGURE 8–4
Changes in per capita food production over time. Numbers are based on 1965 as 100. (a) Developing regions. (b) Developed regions. (World Resources, 1990–1991; Copyright Oxford U. Press.)

of the world. Then three severe droughts occurred, drastically affecting harvests in North America in 1980, 1983, and 1988. Corn harvest in the United States was cut 38 percent by the 1988 drought, an unprecedented loss. Fortunately, the world, and particularly the United States, had ample stocks from previous surpluses. Drawing down these stockpiles averted shortages but left stocks depleted. Another drought hit several regions in 1991, reducing the Midwest corn crop by 10 percent and cutting grain production in Argentina and Australia by as much as 40 percent. Carryover stocks remained low because of these droughts and a shift in U.S. agricultural policy toward keeping more land out of production. With the countries of the former USSR and those of eastern Europe asking for increasing grain aid, the situation will be precarious if the droughts continue.

On the basis of their computer models, some respected climatologists are predicting that droughts will become increasingly commonplace as the climate warms because of the greenhouse effect (Chapter 16). At best, climate cannot do more than return to the "ideal" of the 1940–1980 period. Thus, expecting climatic change to increase yields is out of the question, and the crucial question is, Will there be more droughts in major areas of the world in the years to come?

Converting to food crops land that is now in cash or feedstock crops is a complex undertaking, for it involves such issues as land reform and maintaining a balance of trade. These will be addressed below.

Food Distribution and Trade

For centuries the general rule for basic foodstuffs—grains, vegetables, meat, dairy products—was *self-sufficiency*. Whenever climate or blight (as in the Irish potato famine) or war interrupted the agricultural prowess of a nation or region, the inevitable result was famine and death, sometimes on the scale of millions. Once colonies were established in the New World, timber, furs, tobacco, and fish, and later sugar, coffee, cotton, and other raw materials began to flow back to the Old World. In turn, the Old World exported manufactured goods, which helped to transform the colonies into societies much like the European ones that had given birth to them. With the Industrial Revolution, trade between nations intensified, and soon it became economically feasible to ship basic foodstuffs from one part of the world to another. In time, a lively and important world trade

in foodstuffs arose, and as it did, the need for self-sufficiency in food diminished.

PATTERNS IN FOOD TRADE

Today, agricultural production systems do much more than supply a country's internal food needs. For some nations (such as the United States and Canada), the capacity to produce more basic foodstuffs than the home population needs represents an extremely important entry into the international market. And for many countries (especially those of the Third World), special commodities such as coffee, fruit, sugar, spices, cocoa, and nuts provide the only significant export product (Fig. 8–5). This trade clearly helps the exporter, and it allows importing nations to use foods that they are not able to raise. Given the realities of a market economy, this exchange works well only as long as the importing nation can pay cash for the food, cash earned by exporting raw materials, fuels, manufactured goods, or special commodities. In this way, for example, Japan imports $22 billion worth of food and livestock feed each year, but more than makes up for it by exporting $223 billion worth of manufactured goods (cars, electronic equipment, and so on) annually.

The most important foodstuff on the world market is the grains: wheat, rice, corn, barley, rye, and sorghum. It is instructive to examine the pattern of global grain trade over the past half-century (Table 8–2). In 1935, only western Europe was importing grain; Asia, Africa, and Latin America were self-sufficient. By 1950, new patterns were emerging, and today the trade in grains—as well as other basic foodstuffs—represents a development with enormous

FIGURE 8–5
A coffee harvest in East Java. Coffee is one of many commodity crops that produce important income for nations of the Third World. (Sam Abell/ Woodfin Camp & Associates.)

TABLE 8–2

World Grain Trade Since 1935

Region	Amount Exported or Imported (million metric tons)[a]					
	1935	1950	1960	1970	1980	1990[2]
North America	5	23	39	56	131	105
Latin America	9	1	0	4	−10	−11
Western Europe	−24	−22	−25	−30	−16	22
Eastern Europe and former USSR countries	5	0	0	0	−46	−28
Africa	1	0	−2	−5	−15	−27
Asia	2	−6	−17	−37	−63	−74
Australia and New Zealand	3	3	6	12	19	13

[a] No sign in front of a figure indicates net export; a minus sign in front of a figure indicates net import.

Sources: U.N. Food and Agriculture Organization, *FAO Production Yearbook* (Rome: various years); U.S. Department of Agriculture, Foreign Agricultural Service, *World Rice Reference Tables* (unpublished printout) (Washington, D.C.: June 1988); USDA, FAS, *World Wheat and Coarse Grains Reference Tables* (unpublished printout) (Washington, D.C.: June 1988). [2] From *FAO Food Outlook*, Aug 1991. (Worldwatch Institute, Washington, DC.)

economic and political implications. As Table 8–2 shows, North America has become the major source of exportable grains—the world's "breadbasket" or as one observer put it, the Saudi Arabia of food. Asia, Latin America, and Africa show a disturbing trend of increasing dependence on imported grain over the past 40 years. These three regions have in common 40 years of continued, rapid population growth. For example, Mexico, birthplace of the Green Revolution, now must import 4 million metric tons of grain per year; population growth has eaten up all the gains of the Green Revolution. Although most of the food needs of these regions are met by internal production,

the trend toward greater dependency is an ominous signal.

An interesting point to note is there has been no time in recent history when grain supply has run out. In other words, enough food is produced to satisfy the world market. Why, then, are there people in every nation who are hungry and malnourished? Shouldn't every nation make an all-out effort to provide food for its people? And if it can't, shouldn't the rest of the world assume some of the responsibility for providing food to a hungry nation? Where does the responsibility lie for meeting the need for this most basic resource?

IN PERSPECTIVE
National Food Security

Is there a sense of responsibility for food security in most societies? Such responsibility, if it exists, will be seen in a variety of policies:

1. There will be safety net programs, such as those most commonly seen in the wealthier nations and nations with centrally planned economies (socialistic). Farm policies in these nations will promote soil conservation and other sustainable practices.

2. In nations with limited resources (Third World, in particular), policy needs will be directed toward

encouraging agricultural production, especially in areas populated by rural poor.

3. Additional programs that encourage food security are land reform (seeing that the land is distributed broadly and not concentrated in the hands of a wealthy few), effective family planning and health programs, encouragement of a lively market economy (instead of a controlled, centrally planned economy), and old age security programs.

Unfortunately, few nations come close to achieving these policies, for

a variety of reasons. One of the most important is the direction of massive national resources for military purposes. Another is the tendency for power in the society to be in the hands of a wealthy elite, whose main interest is maintaining their power and wealth. For most of the nations where hunger is chronic, natural resources and wealth are simply scarce, and consequently these nations look to the international level for assistance in meeting food and development needs.

TABLE 8–3

Goals and Strategies for Meeting Food Needs

Family	Nation	Globe

Goal: Personal and family food
  security
Policies:
  —*Employment security*
  —*Adequate land or livestock*
  —*Good health and nutrition*
  —*Adequate housing*
  —*Effective family planning*

Goal: Self-sufficiency in food and
  nutrition
Policies:
  —*Just land distribution*
  —*Support of sustainable agriculture*
  —*Effective family planning*
  —*Promotion of market economy*
  —*Avoidance of militarization*
  —*Effective safety net*

Goal: Sustainable food and nutrition
  for all nations
Policies:
  —*Food aid for famine relief*
  —*Appropriate technology in
    development aid*
  —*Aid for sustainable agricultural
    development*
  —*Debt relief*
  —*Fair trade*
  —*Disarmament*

LEVELS OF RESPONSIBILITY IN SUPPLYING FOOD

It is helpful to begin answering this question of responsibility for meeting food needs by examining Table 8–3. The table displays three major levels of responsibility for meeting food needs: family, nation, and globe. At each level, the players are part of a cash economy as well as a sociopolitical system. In the cash economy, food flows in the direction of economic demand. Need is not taken into consideration. In the event that there are hungry cats and hungry children, the food will go to the cats if the owners of the cats have money and the children's parents don't. In other words, the cash economy, following the rules of the market, provides the *opportunity* to purchase food but not the food itself.

Where the economic status of the player (a destitute breadwinner, a poor country) is very low, there is the possibility that the sociopolitical system will provide the needed purchasing power or the food. In the United States, this help is described as the "safety net," and it is represented by a variety of welfare measures such as the food stamp program, Aid to Families with Dependent Children, and the Supplemental Security Income program.

The most important level of responsibility is at the micro-level—the family. The *goal* at this level is **food security**: the ability to meet the food needs of everyone in the family at a nutritional level that grants freedom from hunger and malnutrition. For an individual, there are three legitimate options for attaining food security: (1) You can purchase the food; (2) you can raise the food yourself or gather it from natural ecosystems; or (3) you can have it given to you (dependency).

Of course, option 3 implies that there is an effective safety net—that at the national (or state) level there exist policies with the objectives of meeting the food security needs of all individuals in the society. Thus an appropriate goal at the nation-state level would be *self-sufficiency in food.* The nation can either produce all the food its people need or buy it on the world market. This goal should include eliminating chronic hunger and malnutrition in the society.

At the global level, responsibility for meeting food needs is most often seen as a matter of humanitarian food aid, and whenever famines occur, this direct approach is of course the best way to attack the problem. According to the Food and Agriculture Organization of the United Nations (FAO), some 11 million metric tons in food aid was extended in 1989–

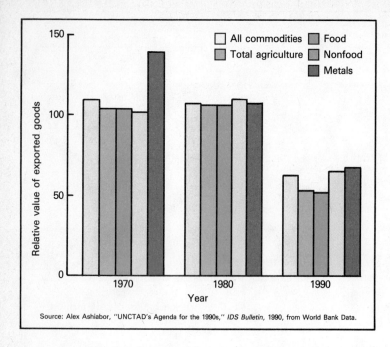

FIGURE 8–6
Declining values of exported goods sold by Third World countries. The figure demonstrates the great decline in the 1980s, which has meant that Third World countries have less money for imports, investment, and debt repayments. (© 1990 Bread for the World Institute on Hunger and Development.)

1990. However, there are some less obvious factors to consider when we are talking about global food needs. One of the most serious is the Third World debt crisis. Interest and principal payments ($50 billion per year) on the $1.2 trillion now owed by Third World nations have imposed an enormous financial burden on those nations. Before 1984, there was still a net flow of funds from rich nations to poorer nations, mostly in the form of loans. Since then, however, the flow has reversed as the loans have come due, and now the poorer nations pay more money than they receive. For example, the Philippines, with a debt burden of $24 billion, must take 44 percent of the national budget for principal and interest payments on its international loans. As a consequence, half of the agricultural land in the Philippines is used to grow export crops, in spite of extensive hunger and malnutrition in that nation. Also, debtor nations are being forced to exploit their natural resources at an unprecedented rate in order to pay foreign creditors. To make matters worse, prices being paid for Third World products are dropping (Fig. 8–6), making it even more difficult for these nations to repay their debts.

Another factor in our discussion of how to meet global nutrition needs is trade barriers in the wealthier countries. The tariffs the European Community (EC) charges against cloth imported from Third World nations is four times the tariffs it charges against cloth imported from wealthy nations. Sugar policies in the EC and the United States guarantee high domestic prices, subsidize exports, and impose import barriers against sugar from the Third World. These discrim-

inatory policies cost an estimated $7 billion per year in lost earnings to the Third World.

In light of the debt crisis and trade barriers, many observers are calling for a restructuring of the economic arrangements between rich and poor countries as the most effective way to address the food needs of the world—and at the same time alleviate much of the environmentally destructive pressure on the poor countries to exploit their natural resources. Thus, an appropriate goal, on the global level, would be for the wealthy nations to adopt policies that promote both food self-sufficiency and sustainable relationships between the poorer nations and their environments.

Hunger, Malnutrition, and Famine

At the United Nations World Food Conference in 1974, delegates from all nations subscribed to the objective "that within a decade no child will go to bed hungry, that no family will fear for its next day's bread, and that no human being's future and capacities will be stunted by malnutrition." Almost two decades have passed since that declaration. What is the extent of hunger, malnutrition, and famine today?

EXTENT AND CONSEQUENCES OF HUNGER

Before we examine the estimates of hunger and malnutrition, it is necessary to give some definitions.

Hunger is the general term referring to a lack of basic food required for energy and meeting nutritional needs, such that the individual is unable to lead a normal, healthy life. **Malnutrition**, as we learned in Chapter 3, is the lack of essential nutrients such as amino acids, vitamins, and minerals, and **undernutrition** is the lack of adequate food energy (usually measured in calories).

Absolutely reliable figures on the worldwide extent of hunger are unavailable, mainly because few governments make any effort to document such figures. A Washington-based organization called Bread for the World Institute on Hunger and Development estimated that more than half a billion people experience *continuous* hunger, while over a billion suffer from various nutritional deficiencies. In their report (see Bibliography) Bread for the World provided a country-by-country breakdown of hunger-related data. The regions most seriously affected were southern Asia (especially Bangladesh), Latin America, and Africa. It is safe to say that at least 20 percent of the world's people suffer from the effects of hunger and malnutrition. Thus we see that, almost 20 years later, the United Nation's 1974 objective is still a dream.

The consequences of malnutrition and undernutrition vary. Those most affected are first children and then women—a reflection of dependency. Hunger can prevent normal growth in children, leaving them thin, stunted, and often mentally and physically impaired (Fig. 8–7). For example, a national survey carried out 10 years ago in the Philippines revealed that 42 percent of the children under 5 were stunted and underweight. In Honduras, an estimated 70 percent of infants are undernourished, and 34 percent are stunted.

Sickness and death are companions of hunger. According to Bread for the World, "Almost 40 000 children under five die each day from malnutrition and infection. . . . The number of deaths is the same as if one hundred jumbo jets, each loaded with 400 infants and young children, crashed to earth each day—one every 14 minutes." Hunger is often a seasonal phenomenon in rural areas that are supported by subsistence agriculture, as people are forced to ration their stored food in order to make it to the beginning of the next harvest. Anyone who travels in the Third World cannot help but notice the fact that few people in the rural areas look well fed; most are thin and spare from a lifetime of limited access to food.

FIGURE 8–7
Malnutrition and hunger affect children most directly. These children show the effects of severe malnutrition brought on by famine in Niger, Africa. (Carl Purcell/Photo Researchers.)

ROOT CAUSE OF HUNGER

By now it should be obvious that *the root cause of hunger is poverty*. Our planet produces enough food for everyone alive today. Hungry and malnourished people lack either the money to buy food or adequate land to raise their own. If by some miracle world food production were to double next year, the status of most of the millions who suffer from extreme poverty and hunger would not change. Remember that any food over and above what the producers need for themselves enters the cash economy and flows in the direction of economic demand not of nutritional need.

Lack of food is only one of many consequences of poverty. Alan Durning of Worldwatch Institute defines **absolute poverty** as "the lack of sufficient income in cash or kind to meet the most basic biological needs for food, clothing, and shelter." On the basis

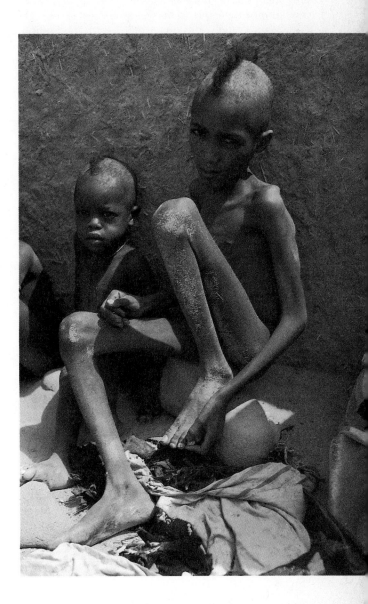

Students confronting the problems of hunger and poverty often want to do something about them. It is difficult to connect directly with needy people in the Third World, but numerous organizations exist with the expressed goal of bringing food and other forms of aid to the hungry. You may be willing to contribute time and money to help those who are in need, but you would like to be assured that your efforts are going to be effective and not eaten up by organizational overhead. Several organizations deserve special mention for their role in *hunger advocacy*—a strategy for hunger relief that takes advantage of the U.S. political system. The advocacy organizations mobilize citizen support of national legislation through phone and letter-writing campaigns to Congress and the Administration and through letters to the editors of local newspapers.

For example, Congress recently increased funding to the Special Supplemental Food Program for Women, Infants and Children (WIC) as a result of a concerted campaign by several advocacy groups: **Interfaith Impact for Justice and Peace, Children's Defense Fund,** and **Bread for the World**. These groups organized grass roots support and endorsements from prominent Protestant, Jewish, and Catholic leaders and so effectively lobbied in favor of this legislation that Congress has increased appropriations in 1991 by $130 million above current levels, and for 1992, President Bush has proposed a further $129 above current services.

The WIC Program targets mothers and children under 5 who are at nutritional risk, and provides federal funding for state and local public health agencies involved in the delivery of health and nutritional information and food aid. The program has enjoyed bipartisan support, as it has been shown to be cost-effective: Every dollar spent saves many more dollars in health care for those who would suffer if the nourishment were not available.

Involvement in support of such organizations makes it possible for you to give something you may not often think of giving: your citizenship. As a participant in the work of these advocacy organizations, you can effectively influence public policy.

of this definition, there are 1.2 billion people in absolute poverty today. Most of these people live in rural villages, are illiterate, spend half or more of their income on food, and represent races, tribes, or religions that suffer discrimination. They are powerless to do anything about their plight, and quite often the society in which they live is content to keep them that way.

Hunger and poverty do not always go from bad to worse. A number of Asian countries, including China, Indonesia, and Thailand, significantly reduced the extent of poverty and hunger during the 1980s. Deliberate public policies and social services have greatly improved the welfare of millions in China, where food security is a matter of high national priority. Indonesia has benefited from oil exports and Green Revolution technology, and it continues to put major emphasis on rural development and social infrastructure. Clearly, it is possible for societies to address the needs of the hungry poor and to make progress in reducing the extent of absolute poverty and hunger. On the other hand, the most severe kind of hunger—famine—is found in societies that are regressing into disorder and chaos, and it is here that international responsibility comes most sharply into focus.

FAMINE

By definition, a **famine** is a severe shortage of food accompanied by a significant increase in the death rate. Famine is a clear signal that a society is either unable or unwilling to distribute food to all segments of its population. Two factors are the immediate causes of famines in recent years—drought and warfare.

Drought is blamed for the famines that occurred in 1968–1974 and again in 1984–1985 in the Sahel region of West Africa (Fig. 8–8). The Sahel is a broad belt south of the Sahara Desert occupied by 50 million people who practice subsistence agriculture or tend cattle, sheep, and goats (such people are called *pastoralists*). The region normally has enough rainfall to support dry grasslands or savannah ecosystems. The rainfall is seasonal and undependable, and it is prone to failure. Making matters worse, population increases in the region have led to unsound agricultural practices and overgrazing by the expanding herds of livestock. Beginning in 1965, the region experienced 20 years of subnormal rainfall, with tragic results. Crops withered, forage for livestock declined, watering places dried up, and livestock died. Both farmers and pastoralists began abandoning their land and mi-

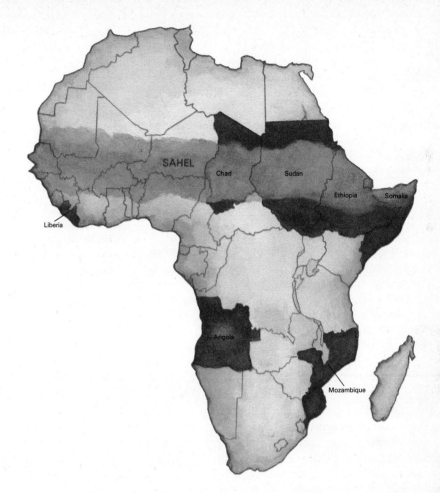

FIGURE 8–8
The geography of famine. Famines have occurred repeatedly in sub-Saharan Africa, especially in the Sahel (a band of dry grasslands that stretches across the continent). The map shows the countries where civil wars and droughts have recently brought on serious famine conditions.

FIGURE 8–9
A refugee camp in Ethiopia. Such camps represent the last resort for refugees from hunger and are often scenes of unthinkable human suffering and death. (David Burnett/Contact Press Images.)

Chapter 8 The Production and Distribution of Food _____ **181**

grating toward urban centers, where they were often herded into refugee camps (Fig. 8–9). Unsanitary conditions in the camps and the already weakened condition of the refugees led to the spread of infectious diseases such as dysentery and cholera, and many thousands died before effective aid could be organized. The latest Sahelian famine is thought to have been responsible for 100 000 deaths; this number would have been in the millions except for the aid extended by Africans and numerous international agencies. The rains have returned to the Sahel, removing the immediate threat of famine, but the region still lacks an environmentally sustainable structure for food security.

Famines continue to threaten several African nations in the early 1990s: Ethiopia, Somalia, Sudan, Mozambique, Angola, Chad, and Liberia (see Fig. 8–8). The common factor in these countries is war. Devastating and prolonged civil warfare has resulted in putting some 20 million Africans at risk of famine. The civil wars disrupt the farmers' normal planting and harvesting and force the displacement of millions of peasants from their homes and food sources. In some areas, the problem is made worse by persistent drought conditions. In Mozambique alone, 900 000 people have died from direct military action or indirect effects of the war there. Governments in power maintain control over food and relief supplies; relief agencies operate under dangerous conditions and frequently experience casualties.

Famines from drought and war are clearly preventable. India, Kenya, and Brazil have coped effectively with droughts in recent years by mobilizing effective relief in the form of food, clothing, and medical assistance. War, however, will undoubtedly continue to create severe hunger and famine conditions. The link between arms and hunger is a significant one. Direct military aid and the international sales of weapons have encouraged the continuation of military conflict in the Third World. Also, military spending competes with spending for human and environmental improvements in both rich and poor nations. The end of the Cold War has the potential to reduce military spending worldwide, and many people see this development as the most promising step at the international level toward reducing hunger. It is likely, however, that international food aid will continue to be necessary.

FOOD AID

What is the proper role of food aid? Clearly, aid is vitally important in saving lives where famines occur. But what about the 20 percent of the world's people who suffer from chronic hunger and malnutrition? Some people ask, Is it right that the more affluent in the world eat meat raised on grain when we know that this is an inefficient way to convert crops to food for humans? Also, should the United States continue to keep agricultural lands out of production (to avoid surpluses) when the food could be given to the hungry? The basic question is, When should food be given to those in need, instead of being distributed according to market economics?

Numerous humanitarian efforts to end world hunger have been mounted in the last 50 years. The United States has been a world leader in giving surpluses or selling them at low prices. As noted above, a number of serious famines have been moderated or averted by these efforts. As virtuous as such efforts seem on the surface, however, routinely supplying food aid in an attempt to alleviate chronic hunger in developing countries has been self-defeating. As S. Wortman and R. Cummings write in their book *To Feed This World*:

The food generosity of industrial countries, whether in their own self-interests (disposing of food surpluses) or under the mantle of alleged distributive justice, has probably done more to sap the vitality of agricultural development in the developing world than any other single factor.

This situation has occurred because people will not pay more than they have to for food. Therefore, free or very cheap foreign food undercuts the local market. In effect, local farmers must compete economically with free or low-cost imported food. When they cannot earn a profit, they stop producing and eventually enter the ranks of the poor. The cycle continues as people who sell goods to the farmer also suffer when the farmer loses buying power. In the long run, the entire local economy deteriorates. Hence, the availability of free food, while well-intended, often aggravates the very conditions that it is meant to alleviate. Meanwhile, population pressures continue to build, and the magnitude of the problem increases (see Ethics Box, p. 183).

Food aid in grains during the late 1980s averaged 11 million metric tons per year. Some aid is strictly humanitarian; Bangladesh received 1.5 million tons per year, almost as much as it purchased on the market. The African continent received 6.6 million tons, which was about 20 percent of the total imported. Africa is now importing 20 percent of its food, and all the signs indicate that this figure will increase. Some food aid is given for political reasons. Egypt received 1.9 million tons per year of grain in food aid during the late 1980s, a result of the Camp David Peace Accord brokered by President Jimmy Carter. Unprecedented amounts of food aid are being sent to eastern Europe and Russia and the other republics of the former USSR in the early 1990s as the Free

The Lifeboat Ethic of Garrett Hardin

Biologist Garrett Hardin has published several provocative essays addressing the worldwide food-population issue. Here we give you an opportunity to respond to Hardin's thinking.

We begin with the concept of carrying capacity—that number of a species that can be supported indefinitely without degrading the environment. For human societies, carrying capacity means the ability to meet food needs over the long term—that is, sustainably. If ecology had a decalogue, Hardin says, the first commandment would be "Thou shalt not transgress the carrying capacity." A look at the world scene reveals that numerous nations are pressing against the limits or have exceeded their carrying capacity. This, says Hardin, is their problem, not ours, and he uses the lifeboat metaphor to show why.

Picture a number of lifeboats after a ship has sunk—some crowded with people and some in which people are riding in relatively uncrowded luxury. Each lifeboat has a limited capacity. The people in the crowded boats are continually falling into the sea, leaving the people on the uncrowded boats with the problem of whether to take them on board or not. Imagine an uncrowded boat with 50 on board and room for 10 more, with 100 people treading in the water and begging to be taken on board. There are several options: (1) Assume that all people have an equal right to survival, and take everyone on board. This, says Hardin, would lead to catastrophe for all. (2) Admit only 10, filling all the space on the boat. Two problems: You lose your safety factor, and, how do you discriminate among all the people in the water? (3) Admit no more to the boat. This option preserves your safety factor and guarantees the long-term survival of the people on your boat; it is the rational solution to the lifeboat problem.

The metaphor, of course, is to be applied to the problem of food aid. Some people would argue that, if less grain went to feeding animals, there would be more available to feed the hungry in the poor nations. Or, since there are often agricultural surpluses in the rich nations, we could use these surpluses to feed the hungry. The real problem, then, is a problem of food distribution, not of the quantity produced. These arguments, says Hardin, are foolish. In giving away food, we would only be encouraging the population escalator—a population growing rapidly reaches the limits of its food capacity and is supplied with food from abroad, encouraging still further growth and necessitating still further food aid, and so on. Our hearts, says Hardin, tell us to send food, but our heads should tell us not to. We only postpone the day of reckoning, and, in the end, the amount of suffering will be greater. Overpopulated, food-poor countries have transgressed the first commandment of ecology. If we want to help, we should direct our aid toward bringing population growth down, according to Hardin. What do you think?

FIGURE 8–10
Food aid from Germany reaches a citizen in St. Petersburg as the free world acts to alleviate hunger in this city of the former Soviet Union. (A. Medevednikov/Leh/Woodfin Camp & Associates.)

World responds to the crumbling of Communist rule, since the transition to a market economy is disrupting food production and distribution (Fig. 8–10). Relief agencies operating in Africa are concerned that aid to the former Communist countries will divert food from Africa's famine regions. Officials from these agencies point out that, although there is need in Russia and other eastern European countries, no trace of famine has been found there.

Food aid will undoubtedly continue to be an international responsibility. It is at best a buffer against famine, and it will probably continue to be awarded to some nations for political reasons. As part of the solution to the chronic hunger and malnutrition among the poor, however, free food is clearly counterproductive. Much more good will be accomplished by a restructuring of the economic arrangements between rich and poor nations and by the extension of loans and aid directed toward fostering self-sufficiency in food and sustainable interactions with the environment.

Building Sustainability into the Food Arena

In ecosystem terms, farmers should be viewed as herbivores who manage their producers, and pastoralists are predators who manage their prey. The principles of ecosystem structure and function presented earlier in the book apply perfectly well to farming and animal husbandry. The major difference between human systems and natural ecosystems is that we do not have to allow nature to take its course—indeed, we cannot do so and expect to harvest crops instead of weeds or hope to have our livestock flourish. Perhaps for this reason we tend to forget that our manipulations of plants and animals are nevertheless subject to ecosystem laws, and if we are not considering those laws, our human systems are likely to behave counter to our best interests.

Food production and distribution and hunger and famine are very much a matter of how human societies interact with their environment as well as with each other. In this last section, we examine the approaches of sustainable agriculture and also look once more at the socioeconomic and political dimensions of hunger.

SUSTAINABLE AGRICULTURE

The goal of sustainable agriculture is to maintain the level of agricultural production needed to feed the human population while not ruining any part of the supporting system or degrading the environment. In *A Green History of the World*, Clive Ponting shows how the downfall of several past civilizations was unsustainable farming and animal grazing. The Sumerians of Mesopotamia raised crops under intense irrigation, and in time salinization led to the collapse of the agricultural base of the society, followed soon by the decline of the Sumerians. Overgrazing and deforestation throughout the Mediterranean basin, beginning as far back as 650 B.C. in Greece, led to soil erosion that ruined agricultural land and greatly lowered the carrying capacity for livestock, and the empires occupying the basin declined accordingly. It is Ponting's view that what happened in the past on a local scale is now occurring in global proportions.

There is a growing consensus that agricultural sustainability must be patterned after natural ecosystems in order to be successful. In Chapters 3 and 4 we presented four principles of sustainability derived from our studies of natural ecosystems. Let us apply these principles to our analysis of agricultural sustainability, differentiating where necessary between *industrialized agriculture* and *subsistence agriculture*.

1. **For sustainability, ecosystems dispose of wastes and replenish nutrients by recycling all elements.** The first principle, that wastes and nutrients are recycled, is augmented by the tendency of the wastes and nutrients to be held in place by plants and soil. Sustainable practices, therefore, emphasize soil health and stability. Chapter 9 discusses soil and the efforts needed to prevent erosion, salinization, and desertification.

   Organic farming (also referred to as low-input farming) involves the regular addition of crop residues and animal manures to build up the organic matter in soil. When crops are harvested, vital mineral elements are removed from the soil, and these are returned to the soil through the application of animal wastes and **green manures** (grasses or legumes that are plowed into the soil at the end of a growing season), instead of through the addition of chemical fertilizers.

   In a society that depends on subsistence agriculture, it is vitally important that the people do not use animal dung for fuel; instead the dung must be returned to the soil. Also, "appropriate technologies" can be adopted that address the problems of soil moisture loss and erosion, such as the use of "rock dams" to hold water and protect the forest cover around croplands. Nutrients can be cheaply added to the soil by mixing legumes with grain or root crops.

2. **For sustainability, ecosystems use sunlight as their source of energy.** Industrialized agriculture will continue to be dependent on mechanization. Subsistence agriculture, however, will do well to continue to use animal energy for working the land because the animals are fed locally and because dependence on costly fossil fuels is avoided. Wind and solar energy can be employed for many farming tasks; look for a return of the windmills once used to pump water and generate small amounts of electricity in farms all over the United States (Fig. 8–11).

   Sustainability is encouraged in the wealthier nations when people use small plots of land to grow their own fruits, vegetables, and small animals. This is a much more desirable use of the land than simply growing grass to mow.

3. **For sustainability, the size of consumer populations is maintained such that overgrazing does not occur.** The most obvious application of principle 3, the prevention of overgrazing, is in livestock management. Mismanagement of herds and overgrazing lead to rangeland deterioration. Therefore sustainable livestock management all over the world must recognize the carrying capacity of the rangeland ecosystems and preserve the soils and plant cover. In some Third World regions forests and woodlands provide fodder for livestock, and in other (tropical forests) the forests are removed to make way for cattle pasture. The sustainable approach is to protect forested areas, recognizing that they are the most stable ecosystems for the site and will benefit human populations more if they are maintained as forests.

Pests are consumers, and pest control is vital to most agriculture. Chapter 10 presents the issues surrounding the use of pesticides. Alternatives to absolute dependence on pesticides are now available, including integrated pest management programs using natural predators. The selection of planting times, crop rotations, and plant residue management can provide the beneficial insects with optimal habitats. The basic strategy is to maintain biological control of pests and diseases.

One obvious application of this principle is for human populations to come to terms with the carrying capacity of the land they occupy. This and the previous two chapters give abundant evidence that some parts of the world already have violated this principle. Clearly, efforts to lower fertility will in themselves promote sustainable agriculture by reducing the pressure put on the land to produce food. Only if these efforts are successful will the world's subsistence farmers have any hope of meeting their family's food needs in the future.

4. **For sustainability, biodiversity is maintained.** Crop rotation is a vital part of sustainable agriculture. For example, the farmer might plant three seasons of alfalfa plowed under (green manure) followed by four crop seasons of wheat, soybeans, wheat, and oats. In this way, weeds and insects are more easily controlled, and plant diseases do not build up in the soil. Crop combinations and mixing crops, trees (agroforestry), and livestock provide a diversity of marketable products that can be an effective buffer against economic and biological risks (Fig. 8–12). In more arid climates, *alley cropping* can be adopted, where rows of shade-producing trees are alternated with rows of food crops.

FIGURE 8–11
A windmill provides significant amounts of electricity for the needs of this farm in Cartagena, Spain. Small windmills are becoming a common sight in many rural and suburban areas. (Inge Morath/Magnum.)

FINAL THOUGHTS ON HUNGER

By now it should be clear that, although food is our most vital resource, we do not treat it as a commons (available to all who need it). Indeed the production and distribution of food is one of the most important economic enterprises on earth. Alleviating hunger, as we have seen, is primarily a matter of addressing the absolute poverty that afflicts one of every five people on earth. To treat food as a commons would be to treat wealth as a commons. Even though this is one of the tenets of socialist systems, it has proven to be a completely unworkable one. We are part of the world economy now, and it is a market economy— for the perfectly good reason that nothing else seems to work, given the realities of human nature.

FIGURE 8–12
A modern farm in Schuylkill County, PA, shows a healthy diversity of crops, woodlands, and hedgerows, important in maintaining the ecological stability of the farm countryside. (W. Eastep/The Stock Market.)

Chapter 8 The Production and Distribution of Food

What we have not done, however, is to bring this market economy under the discipline of sustainability. Short-term profit crowds out long-term sustainable restraint. Self-interest at every level—from the individual to the global community—generates decisions that prevent the sharing of political power, economic goods, and technology, and in the process it guarantees that the environment will continue to be degraded. Why isn't land reform carried out in Third World countries? Why are the rich nations content to maintain the current debt situation, keep tariffs high and Third World commodity prices low? Why do the nations of the world spend $1 trillion a year on military power and arms?

In 1980, a Presidential Commission on World Hunger delivered its report to President Jimmy Carter, with the major recommendations "that the United States make the elimination of hunger the primary focus of its relations with the developing world." The thrust of that report was that if we re-spect human dignity and have a sense of social justice, we must agree that hunger is an affront to both. The right to food must be considered a basic human right. It follows, then, that we as a nation have a moral obligation to respond to world hunger, the report concluded. Today the question is, Has that moral obligation made its way into public policy in the ensuing years?

Our understanding of the situation is quite well developed. No new science or technology is needed in order to alleviate hunger and at the same time promote sustainability as we grow our food. The solutions lie in the realm of political and social action, at all levels of responsibility. Given the current groundswell of concern about the environment and the disappearance of the Cold War between capitalism and communism there may never be a better time to turn things around and take more seriously our responsibilities as stewards of the planet and our brothers' keepers.

 Review Questions

1. What new developments did the agricultural revolution bring to farming?
2. Discuss the environmental cost of each development named in question 1.
3. What was the Green Revolution? What were its limitations?
4. Describe subsistence agriculture.
5. How does animal farming impact on the environment?
6. What remain as our major prospects for increasing food production in the future?
7. Trace the patterns in grain trade between the different world regions over the last 40 years (see Table 8–2).

8. Describe the three levels of responsibility for meeting food needs. At each level, list several ways food security can be improved.
9. Define hunger, malnutrition, and undernutrition. What are their consequences?
10. How are hunger and poverty related?
11. Discuss the immediate cause of famine.
12. Why does food aid often aggravate poverty and hunger?
13. Define sustainable agriculture. How do each of the four principles of sustainability apply to this type of agriculture?

 Thinking Environmentally

1. Although few people would argue against sending food aid to famine victims, what conditions could be attached to the aid in order to foster self-sufficiency and to prevent further dependence?
2. Farmers in the United States currently hold millions of tons of corn and wheat in storage. What should be done with this agricultural surplus?
3. Imagine that you have been sent as a Peace Corps volunteer to a poor African nation that is experiencing widespread hunger. What agricultural technologies would you introduce to increase agricultural productivity?

4. Record your food intake over a 2–3 day period. Analyze the nutritional value of your diet. Which nutrients are lacking? Which are in excess? What changes in your diet would reconcile these differences?
5. Of the methods suggested below for increasing food production, which do you feel are viable options. Why?

clearing jungles
catching more fish in the open sea

increasing the yield per acre of cropland
irrigating arid lands

9

Soil and the Soil Ecosystem

OUTLINE

LEARNING OBJECTIVES

When you have finished studying this chapter, you should be able to:

1. Specify the components of the soil system.

2. List and describe soil aspects that are important for maintaining plant growth.

3. Describe the source of mineral nutrients and how they may be added to, removed from, or lost from the soil.

4. Contrast a natural ecosystem and an agricultural system with respect to the movement of nutrients.

5. Describe two soil aspects that influence the availability of water.

6. Define soil texture and describe how it influences the various aspects of soil.

7. Describe how detritus and soil organisms change the various aspects of a mineral soil.

8. Describe the process of erosion and how it affects specific properties of topsoil. Relate erosion to desertification.

9. Identify three practices that are likely to lead to erosion.

10. Explain what salinization is and discuss its causes and consequences.

11. List the four objectives of sustainable agriculture. Discuss methods involved in achieving each objective.

12. Contrast a diversified farm with a monoculture operation in terms of how the former is suited for sustainable agriculture while the latter is not.

187

No factor is more critical for sustaining civilization than maintaining soil and water resources. As noted in Chapter 8, a number of past civilizations met their demise in large part because they failed to learn this lesson. Yet, modern civilization does not seem to have learned the lesson either. Throughout the world, soils are deteriorating and water resources are being overdrawn, threatening the future productivity of agriculture, forests, and rangelands. This degradation is happening because of poor management, however, not because of any lack of knowledge. Our objective in this chapter is to understand soil dynamics and the management principles necessary to maintain soil productivity.

Plants and Soil

In Chapter 3 we learned that the first principle of sustainability is that ecosystems dispose of wastes and replenish nutrients by recycling all elements. Nutrients are withdrawn from the soil by green plants (producers) and are passed up various food chains. Plant and animal wastes (detritus) are fed upon and broken down by various detritus feeders and decomposers living in the soil, and the nutrients are released to be reabsorbed by plants.

The emphasis here is that soil is far more than just inert dirt. **Soil** is a *dynamic system* involving three components: mineral particles, detritus, and soil organisms feeding on the detritus (Fig. 9–1). This three-component *soil system* is critical for successful completion of nutrient cycles; remove any component and soil quality becomes compromised. In addition to providing nutrients, the soil system also acts as an environment that supports plant growth in other respects. Hence, maintenance of this soil system is critical to the sustainability of the ecosystem as a whole.

To understand the importance of various as-

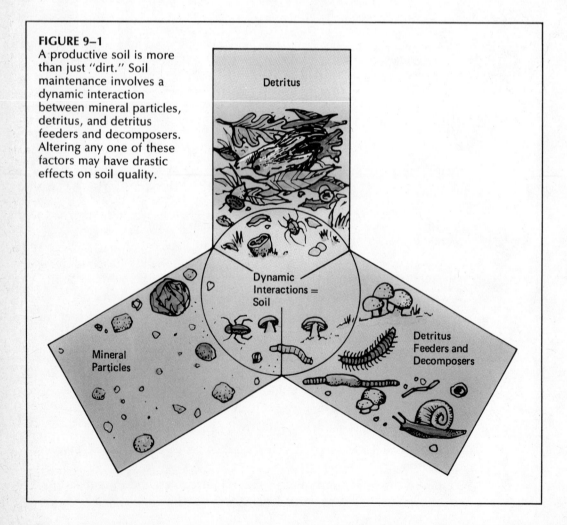

FIGURE 9–1
A productive soil is more than just "dirt." Soil maintenance involves a dynamic interaction between mineral particles, detritus, and detritus feeders and decomposers. Altering any one of these factors may have drastic effects on soil quality.

Detritus

Dynamic Interactions = Soil

Mineral Particles

Detritus Feeders and Decomposers

Hydroponics: Growing Plants without Soil

Plants do not require any secret ingredients from the soil; all they need in order to flourish are nutrients, water, and aeration. This is demonstrated by the practice of **hydroponics**, the culture of plants without soil. In hydroponic systems, roots are constantly wetted with a well-aerated water solution containing optimal amounts of the required mineral nutrients. One common method is to grow plants in a container of pea-sized gravel that is constantly or periodically irrigated with nutrient solution. The roots, supported by the gravel, grow in a film of water around the stones with plenty of air between the stones. The solution is recirculated from the bottom of the container back to drip tubes above.

Another method is to grow the plants in a trough—sections of eave trough work well—covered with black plastic except for holes for the plants. The trough is mounted at a slight incline so that nutrient solution constantly added at one end flows through, drains from the other end, and is recirculated. Plants are supported by strings from above, and their roots simply dangle or lie in the trough.

Because nutrients and water can be kept at optimal levels at all times, the production from hydroponic systems can be quite spectacular. Why worry about preserving topsoil if this is the case? The answer is cost. For all the equipment required, a hydroponic system will cost on the order of $100 000 per acre. When the crop is high-priced fruits and vegetables, the income may be sufficient to justify this cost. However, bread would have to sell for $10 a loaf or more to pay for hydroponically grown wheat. Thus, it seems unlikely that hydroponics will solve the food problem for the poor, but this technique may help in the following way.

Urban populations consume large amounts of salad vegetables, which are often imported from great distances at great expense. At the same time, cities have large numbers of unemployed poor people. Hydroponic production on rooftops or other open areas could make the city largely self-sufficient in salad vegetables and at the same time employ a great many people, as operating hydroponic systems is labor-intensive. In this respect, hydroponic systems are a form of appropriate technology that might well be supported.

Combining a hydroponics operation with fish farming can increase efficiencies still more. The nutrient-rich water resulting from fish excrements can be used for the plants, and the plants effectively filter and clean the water for the fish. Also, waste plant material can be a component of the fish feed. John Reid, founder of Bioshelters in Amherst, Massachusetts, has developed and is marketing such a system.

---

pects of the soil system, we shall begin by considering the needs of the growing plant as they relate to the soil.

CRITICAL FACTORS OF THE SOIL SYSTEM

In order to sustain plants, the soil system must fulfill the plants' need for mineral nutrients, water, and oxygen. The pH (relative acidity) and salinity (salt concentration) of the soil are also critically important. **Soil fertility**, the soil's ability to support plant growth, often refers specifically to the presence of proper amounts of nutrients. However, the soil's ability to fulfill all the other needs of plants is also involved in soil fertility.

Mineral Nutrients and Nutrient-holding Capacity

Mineral nutrients—phosphate (PO_4^{3-}), potassium (K^+), calcium (Ca^{2+}), and other ions—are present in various rocks along with nonnutrient elements. However, plants cannot absorb these nutrient ions as long as they are bonded in the rock structure. The rock must be broken down such that the nutrient ions are released into water solution or into a loosely bound state.

The rock, referred to as **parent material**, is naturally broken down by *weathering*. **Weathering** includes all the physical forces acting on the rock, such as freezing and thawing, heating and cooling, the abrasive action of sand particles, and pressure exerted by roots growing in small cracks in the rock. In addition, various chemical reactions that break down the rock into smaller particles are part of the weathering process.

As nutrient ions are released from rock, they may be absorbed by roots, but they may also be washed away by water percolating through the soil, a process called **leaching**. Leaching not only causes a loss of soil fertility but also contributes to pollution as excess nutrients enter waterways. Consequently the soil's capacity to bind and hold nutrient ions until they are absorbed by roots is just as important as the

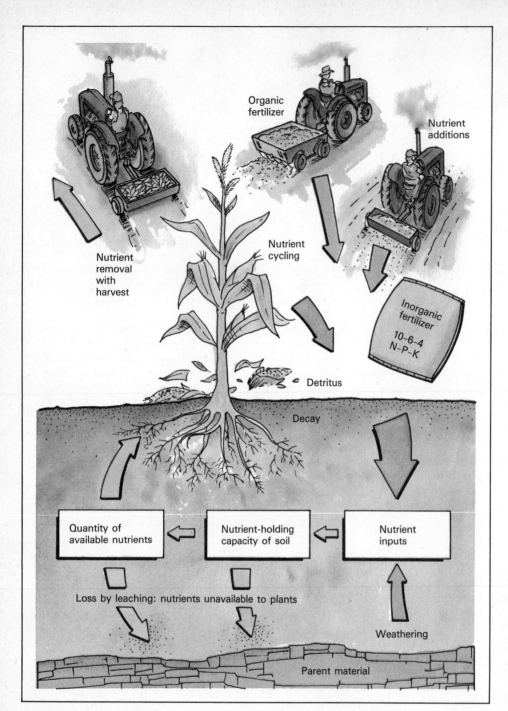

FIGURE 9–2
Plant-soil-nutrient relationships. Soil nutrients may come from decay of detritus, additions of fertilizer, or weathering of parent material. Whether nutrients remain in the soil until they are absorbed by plants or are lost from the system through leaching depends on the nutrient-holding capacity of the soil.

initial supply of those ions. This property is referred to as either the soil's **nutrient-holding capacity** or its **ion-exchange capacity**.

Although weathering is the original source of nutrients, this process is much too slow to support anything approaching normal plant growth. Also, some soils may be essentially devoid of parent materials containing essential nutrients. For natural ecosystems, the major supply of mineral nutrients in the soil comes from the nutrient cycles as described in Figures 3–15, 3–16, and 3–17.

In agricultural systems, there is an unavoidable removal of nutrients from the soil because nutrients are contained in the plant material harvested. Consequently, sustainability of agricultural systems requires that the soil be resupplied with nutrients. This is done through applications of **fertilizer**, material that contains one or more of the necessary nutrients. Fertilizers are generally divided into *organic* and *inorganic*. **Organic fertilizer** includes plant or animal wastes or both; manure is one example. **Inorganic fertilizers** are chemical formulations of required nutrients without any organic matter included. Pros and cons of these fertilizers will be discussed later.

Even with the potential for fertilization, the nutrient-holding capacity of the soil remains vitally important. Without such capacity, added fertilizer will simply leach away, an obvious economic loss. Even worse, leached nutrients go into waterways and cause serious pollution problems as will be described in Chapter 12.

Figure 9–2 summarizes these ideas concerning nutrients.

Water and Water-holding Capacity

Plant leaves have tiny pores that permit the plant to absorb carbon dioxide and release oxygen in photosynthesis. These pores, however, also allow water vapor to escape from the cells inside the leaves. This loss of water vapor from leaves, called **transpiration**, accounts for at least 99 percent of a plant's need for water; less than 1 percent of the water absorbed is used in photosynthesis. If there is inadequate water to replace transpiration loss, plants wilt. The wilted condition conserves water and may hold off total dehydration and death for some time, but it also shuts off absorption of carbon dioxide and growth. Consequently, to keep most plants flourishing requires considerable amounts of water. A field of corn, for example, transpires an equivalent of a layer of water 17 inches (43 cm) deep in a single growing season.

The initial supply of any water in the soil is from natural rainfall or irrigation, but three soil factors are critical in making this water available to plants. First, the soil must be porous so that the water can **infil-**

FIGURE 9–3
Plant-soil-water relationships. Water lost from the plant in transpiration must be replaced from a reservoir of water held in the soil. In addition to the amount and frequency of precipitation, the size of this reservoir depends on the soil's ability to allow water to infiltrate, hold water, and minimize direct evaporation.

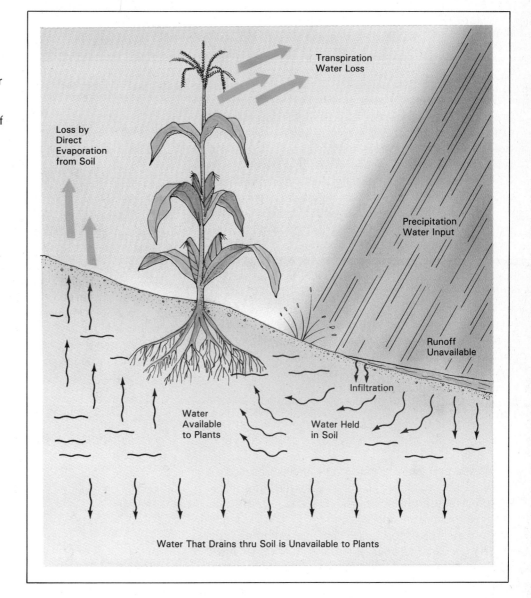

trate, or soak in. If rainfall or irrigation water runs off the surface rather than infiltrating, it won't be useful; worse, it tends to cause erosion, which we shall discuss shortly. Second is **water-holding capacity**, the ability of the soil to soak up and hold the water like a sponge. Since the roots of most plants do not go very deep, water that trickles down through the soil more than a few feet (much less in the case of small plants) becomes unavailable. Between rains, plants depend on the "reservoir" of water that is held in the soil near the surface. With good water-holding capacity, this reservoir may be sufficient to sustain plants over a considerable dry spell. With poor water-holding capacity, plants may suffer during even a short dry spell because the small reservoir is quickly exhausted. The third factor critical in making water available to plants is **evaporation rate** from the soil surface. Factors that reduce evaporative water loss, such as a layer of detritus (mulch) on the surface, make more water available to the plants.

These aspects of the soil-water relationship are summarized in Figure 9–3.

Aeration

While leaves and/or other green parts of plants exposed to light carry on photosynthesis, roots and other nonphotosynthesizing parts derive energy from cell respiration, a process that requires oxygen and produces carbon dioxide as a waste. In general, plants have no means of transporting the oxygen produced in photosynthesis to the roots. Instead, roots depend on oxygen diffusing from the atmosphere through the soil, a property referred to as **soil aeration**. Thus, soil must be loose and porous if there is to be adequate aeration. If soil aeration is reduced through compaction (packing soil down tightly) or flooding, plants will suffer. Indeed, a common pitfall is to kill plants by "drowning," that is, overwatering to the extent that aeration is impeded and roots suffocate and die.

Plants that thrive in the waterlogged soils of wetlands, such as bald cypress, mangrove, Spartina (marsh grass), sedges, and reeds, have adaptations that enable oxygen to diffuse through the stem to the roots.

Relative Acidity (pH)

The term **pH** refers to the units and scale used to measure the acidity or alkalinity (basicity) of any solution. A solution that is neither acidic nor alkaline is said to be neutral and has a pH of 7. The pH scale, which runs from 1 to 14, will be discussed more fully in Chapter 16. For now, it suffices to say that most plants (as well as animals) require a pH near neutral, and most natural environments provide this.

Salt and Water Uptake

To function properly, all living cells must contain a certain amount of water. Plant cells get their water through plant roots (except in the case of certain specialized plants, such as epiphytes, which may absorb water through their leaves). A buildup of salts in the soil makes it impossible for the roots to take in water. Indeed, if salt levels in the soil get high enough, water can be drawn out of the plant, resulting in dehydration and death. This is why highly salted soils are virtual deserts supporting no life at all. We shall see the importance of this problem later when we study how irrigation may lead to the accumulation of salt in soil.

THE SOIL SYSTEM

How does the soil system—that dynamic interaction of minerals, detritus, and soil organisms—provide the attributes discussed above? To answer this question, we shall first consider the attributes of mineral particles by themselves. Then we shall consider how detritus and soil organisms can greatly enhance soil quality.

Soil Texture: Size of Mineral Particles

As a result of weathering, rock is gradually broken down into smaller and smaller particles that are classified according to size, from relatively large to microscopic, as sand, silt, and clay (Table 9–1). The proportion of sand, silt, and clay making up the mineral portion of any soil is defined as **soil texture**. Some soils are virtually pure sand, pure silt, or pure clay. Or, if one of these predominates, we speak of sandy, silty, or clayey soils. A proportion that is commonly found consists of roughly 40 percent sand, 40 percent silt, and 20 percent clay. A soil with such proportions is called a **loam**.

You can determine the texture of a soil by shaking a small amount with water in a large test tube to

TABLE 9–1

USDA Classification of Soil Particles	
Name of Particle	Diameter (mm)
Very Coarse Sand	2.00–1.00
Coarse Sand	1.00–0.50
Medium Sand	0.50–0.25
Fine Sand	0.25–0.10
Very Fine Sand	0.10–0.05
Silt	0.05–0.002
Clay	Below 0.002

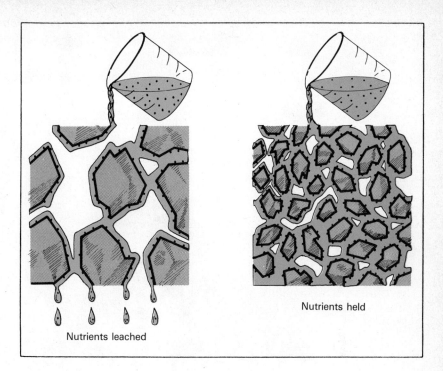

FIGURE 9–4
Soil-holding capacity increases as particle size decreases. Both water and nutrient ions (represented by dots) tend to cling to surfaces, and smaller particles have relatively more surface area.

Nutrients held

Nutrients leached

separate the particles and then allowing them to settle. Since particles settle according to weight, sand particles settle first, silt second, and clay last. The proportion of each can then be seen.

Soil texture has a significant effect on infiltration, aeration, water-holding capacity, and nutrient-holding capacity, for two reasons. First, larger particle size means larger spaces between particles. Consequently, infiltration and aeration are very good with large particle size and worsen as size decreases. Hence, infiltration and aeration are excellent in sandy soils and very poor in clayey soils. Silty soils are intermediate.

Second, holding capacity has the opposite relationship to particle size. It is very poor in sandy soils and improves with diminishing particle size because both water and nutrients are held by virtue of adhering to the particle surface. The more surface available, the greater the holding capacity. Total surface area increases with diminishing particle size. (This can be understood by imagining a rock being broken in half again and again. Every time it is broken, two

new surfaces are exposed, one on each side of the break, but the overall weight and volume of rock do not change.) Thus, in a given volume of soil, silty or clayey soils have relatively greater surface area and therefore greater nutrient- and water-holding capacity than sandy soils (Fig. 9–4).

Soil texture also affects **workability**, the ease with which a soil can be cultivated. This fact has an important bearing on agriculture. Clayey soils are very difficult to work because with even modest changes in moisture content they go from being too sticky and muddy to being too hard to break. Sandy soils are very easy to work because they do not become muddy when wet, nor do they become hard and bricklike when dry.

These relationships between soil texture and various properties are summarized in Table 9–2. Which is the best soil? Recall the principle of limiting factors. The poorest attribute is the limiting factor; the very poor water-holding capacity of sandy soils, for example, may preclude agriculture altogether because they dry out so quickly. The best texture proves

TABLE 9–2

Relationship between Soil Type and Soil Properties

Soil Type	Water Infiltration	Water-Holding Capacity	Nutrient-Holding Capacity	Aeration	Workability
Sand	Good	Poor	Poor	Good	Good
Silt	Medium	Medium	Medium	Medium	Medium
Clay	Poor	Good	Good	Poor	Poor
Loam	Medium	Medium	Medium	Medium	Medium

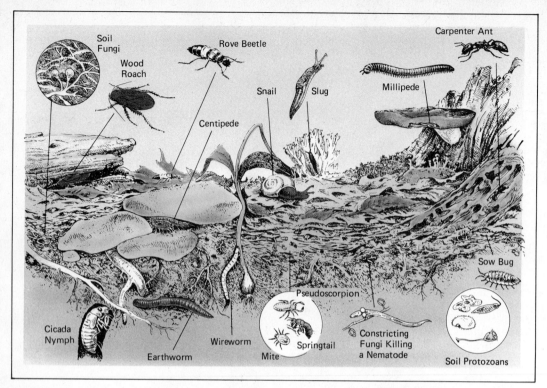

FIGURE 9–5
The major groups of soil organisms. It is the action of these organisms that reduces detritus to humus, intimately mixes the humus with soil, and in the process develops soil structure. (From Robert Leo Smith, *Ecology and Field Biology*, 2nd ed. Copyright 1966, 1974 by Robert Leo Smith. Reprinted by permission of HarperCollins Publishers, Inc.)

to be silt or loam because such limiting factors are moderated in these two types of soil. So are the good qualities, however, and so this "best" is really only "medium." The rest of the soil system—the detritus and soil organisms—is necessary to optimize all attributes.

Detritus, Soil Organisms, Humus, and Topsoil

The accumulation of dead leaves, roots, and other detritus on and in the soil supports a complex food web, including numerous species of bacteria, fungi, protozoans, mites, insects, millipedes, spiders, centipedes, earthworms, snails, slugs, moles, and other burrowing animals (Fig. 9–5). As these organisms feed, the bulk of the detritus is consumed through their cell respiration, and carbon dioxide, water, and mineral nutrients are released as byproducts as described in Chapter 3. However, each organism leaves a certain portion undigested—that is, a certain portion resists breakdown by the organism's digestive enzymes. This residue of organic matter that remains for a time after most of the feeding and digestion have occurred is called **humus**. A familiar example is the black or dark brown, spongy material remaining in a dead log after the center has rotted out (Fig. 9–6).

194

FIGURE 9–6
Formation of humus. Humus is the residue of organic matter, as seen in this rotted log, that remains after the bulk of organic material has been decomposed by fungi and microorganisms. Humus is resistant to the initial digestion of detritus feeders but does eventually decompose to inorganic materials. (Photograph by BJN.)

Composting is being promoted as a great "new" way of converting organic refuse, such as vegetable wastes from the kitchen and raked-up leaves from the yard, into a valuable "soil conditioner." Composting does do this, but it is hardly a new discovery. It is the age-old natural process of letting detritus feeders, mainly fungi and bacteria, decompose organic material to humus.

In the United States, kitchen and yard wastes, on average, make up nearly 20 percent of total refuse and as such constitute a considerable portion of the refuse disposal problem (Chapter 20). You can do your part in alleviating the disposal problem and enhance your soil at the same time by composting such wastes. Any manure from pets can be put in as well and will give a richer compost.

Many companies are marketing composting containers of various sorts. The main purpose of such containers is to make money for the company. None are any more effective and most are less effective than an open pile on the ground, although you may wish to contain the pile with a strip of fencing to prevent material from blowing away and to hide unsightliness.

The conditions for composting (fostering the growth of the fungi and bacteria) are moderate moisture, air, and temperatures above 20°C. A compost pile exposed to the weather will generally have adequate moisture and temperature most of the year. Leaves will usually keep the pile loose enough for adequate aeration. If the pile lacks sufficient air, it will begin to generate the foul odors of anaerobic respiration. If you smell such odors, "turn" the compost with a pitchfork. If leaves are the main constituent, a small amount of fertilizer—manure if it is available, chemical fertilizer if it is not—may facilitate decomposition because dead leaves are nutrient-poor. Lime may also be added to keep the pH up. Humus and a rich topsoil will be generated under the pile as soil organisms do their work. The pile can be lifted to retrieve the humus. Or the pile can be moved every few months, leaving "deposits" of humus-rich topsoil behind.

An increasing number of municipalities are turning to composting the organic portions of refuse as a means of coping with the waste disposal problem. In some cases, paper wastes are being composted with sewage sludges, disposing of both materials and at the same time producing a rich compost. The compost-humus produced is then sold to homeowners, landscapers, and farmers as soil conditioner.

The first principle of sustainability involves recycling nutrients. Composting is one step toward breaking the nonsustainable, one-directional flow of nutrients that we have created and moving toward a sustainable pattern of recycling. It should be accepted and encouraged.

Composting is the process of fostering the decay of organic wastes under more-or-less controlled conditions, and the resulting **compost** is the same as humus (see In Perspective Box above).

The activity of soil organisms integrates humus with mineral particles to create *soil structure*. For example, as earthworms feed on detritus, they ingest inorganic soil particles as well. As much as 15 tons per acre (37 tons per hectare) of soil may pass through earthworms each year. As the mineral particles go through the gut, they become thoroughly mixed and "glued" together with the nondigestible humus compounds. Thus, the sand, silt, and clay particles are bound together with humus into larger clumps and aggregates. The burrowing activity of organisms keeps the clumps loose. This loose, clumpy characteristic is referred to as **soil structure** (Fig. 9–7).

Humus forms and soil structure develops mainly in the upper 4 to 12 inches (10 to 30 cm) of soil, the zone in which soil organisms are active.

Thus, a layer of dark-colored soil with a clumpy, aggregate structure develops on top of the lighter-colored, humus-poor, compacted soil. This layer of humus-rich soil is called **topsoil**; the soil below is **subsoil**. A careful cut through a natural, undisturbed soil reveals this layering, referred to as the **soil profile** (Fig. 9–8).

Humus has phenomenal holding capacity for both water and nutrients, as much as 100-fold greater than clay on the basis of weight. The clumpy aggregate structure of topsoil greatly enhances infiltration, aeration, and workability. Regardless of soil texture, then, attributes are enhanced with humus and the soil structure it imparts. Sandy soils may be given significant water-holding capacity, clayey soils may be given sufficient aeration and infiltration, and loamy and silty soils may be enhanced in all regards.

In addition to humus formation, a number of other interactions between plants and soil biota exist. One is a symbiotic relationship between the roots of

(a)

FIGURE 9–7
Humus and the development of soil structure. (a) On the left is a humus-poor sample of loam. Note that it is a relatively uniform, dense "clod." On the right is a sample of the same loam but rich in humus. Note that it has a very loose structure, composed of numerous aggregates of various sizes. (Photograph by BJN.) (b) A diagrammatic illustration of the difference.

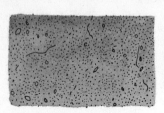

Addition of Organic Matter and Humus-forming Process Involving Numerous Organisms

Lack of Structure Gives Compacted Soil with Poor Aeration and Poor Infiltration

Structured Soil with Excellent Aeration and Excellent Infiltration

(b)

FIGURE 9–8
Soil profile. A cut through soil generally reveals a layer of loose, dark topsoil overlying light-colored, compacted subsoil. The topsoil layer results from the addition of organic matter and the activity of soil organisms. (USDA photograph.)

some plants and certain fungi called **mycorrhizae**. Drawing some nourishment from the roots, mycorrhizae penetrate the detritus, absorb nutrients, and transfer them directly to the plant. Thus, there is no loss of nutrients to leaching. Another important relationship is the role of certain soil bacteria in the nitrogen cycle, as discussed in Chapter 3.

Not all soil organisms are beneficial, however. Nematodes, small worms that feed on living roots, are highly destructive to a number of agricultural crops. In a flourishing soil ecosystem, however, nematode populations may be controlled by other soil organisms, such as a fungus, that forms little snares to catch and feed on nematodes (Fig. 9–9).

In summary, a productive soil must be recognized as the entirety of a dynamic system of mineral particles, detritus, and soil organisms all interacting together in a way that optimizes all the attributes that support plant productivity. An important point to remember is that the system is *dynamic*. Although re-

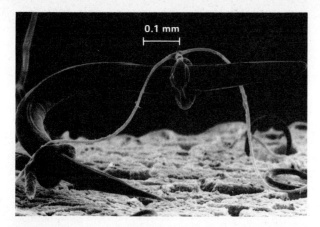

FIGURE 9–9
Soil nematode (roundworm), a root parasite, captured by the constricting rings of the predatory fungus *Arthrobotrys anchonia*. (Courtesy of Nancy Allin and O. L. Barron, University of Guelph)

sistant to digestion, humus does decompose at the rate of about 20 to 50 percent of its volume per year, depending on conditions. Consequently, without additions of sufficient detritus, soil organisms starve,

humus content declines, and there is a loss of soil structure. This loss of humus and the consequent collapse of topsoil is called **mineralization** because what is left is just the gritty mineral content—sand, silt, and clay—devoid of humus. Topsoil is formed and maintained only through continual additions of detritus (Fig. 9–10).

We can readily see how the whole is sustained in a natural ecosystem. Growth of plants, whether grasses or forests, provides a continuous source of detritus which supports soil organisms. In turn, soil organisms support the growth of plants by releasing the nutrients from the detritus and maintaining the other necessary physical/chemical aspects of the soil.

Plants help maintain the soil in other ways, too. A vegetative cover protects the soil from erosion, and a cover of undigested detritus such as dead leaves greatly reduces evaporative water loss while still allowing infiltration.

When land is used for raising either crops or animals, the soil system is at the mercy of our management or mismanagement. The importance of maintaining topsoil and the potential tragedy in its loss should be self-evident. Conversely, restoring a

FIGURE 9–10
Two processes: topsoil formation and topsoil mineralization. Which route a given soil takes depends on whether or not sufficient detritus is added over the years. In order for topsoil to form, a dynamic balance must be maintained between humus formation and humus decomposition.

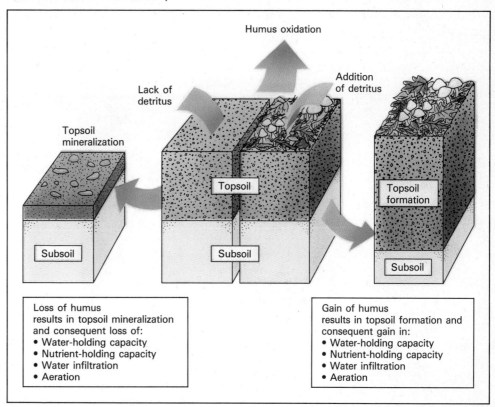

good topsoil from the subsoil base offers the potential of increasing productivity by six- to tenfold. Unfortunately, soil degradation is still the dominant trend in the world today, although there are encouraging signs of change, as we shall see in the following sections.

Losing Ground

The difference between the ability of topsoil and subsoil to support plant growth has been tested by growing plants on adjacent plots, one of which has had the topsoil removed. Results are striking: Yield from plants grown on subsoil are only 10–15 percent of that from plants grown on topsoil. In other words, loss of topsoil results in an 85–90 percent decline in productivity.

Over the years, however, humans have generally focused on maximizing immediate production and have failed to appreciate the importance of maintaining the soil system. Most serious are practices that leave soil exposed to erosion, but overreliance on chemical fertilizers and irrigation are also leading to problems.

BARE SOIL, EROSION, AND DESERTIFICATION

Most destructive to soil is **erosion**, the process of soil particles being picked up and carried away by water

or wind. The removal may be slow and subtle as soil is gradually blown away by wind, or it may be dramatic as gullies are washed out in a single storm (Fig. 9–11).

In natural terrestrial ecosystems other than deserts, a vegetative cover protects against erosion. The energy of falling raindrops is dissipated against the vegetation, and the water infiltrates gently into the loose topsoil without disturbing its structure. With good infiltration, runoff is minimal. Any runoff that does occur is slowed as the water moves through the vegetative or litter mat, and so the water does not have sufficient energy to pick up soil particles. Grass is particularly good for erosion control because when runoff volume and velocity increase, well-anchored grass simply lies down, making a smooth mat over which the water can flow without disturbing the soil underneath. Similarly, vegetation slows the velocity of wind and holds soil particles. (Fig. 9–12a).

When soil is left bare and unprotected, however, it is highly subject to erosion. Water erosion starts as the impact of falling raindrops breaks up the clumpy structure of topsoil. The dislodged particles wash into spaces between other aggregates, clogging the spaces and thereby decreasing infiltration and aeration. The decreased infiltration results in more water running off the surface, causing further stages of erosion.

As further runoff occurs, the water converges into rivulets and streams, which have greater volume, velocity, and energy and hence greater capacity to pick up and remove soil particles. The result is the gullies shown in Figures 9–11 and 9–12b.

FIGURE 9–11
Bare soil is subject to severe erosion as the splash of falling rain seals the surface and resulting runoff readily carries away soil particles. The erosion of the gully in the cultivated field seen here occurred in a single rain. (USDA.)

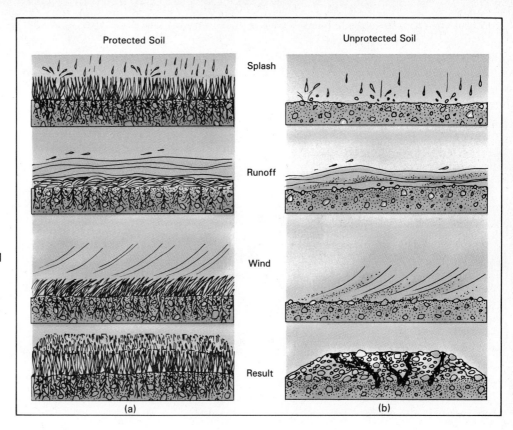

Protected Soil   Unprotected Soil

Splash

Runoff

Wind

Result

(a)   (b)

FIGURE 9–12
Erosion. (a) A vegetative cover protects soil from all forms of erosion. (b) Bare soil is extremely vulnerable to erosion. The splash of falling raindrops breaks up soil aggregates into individual particles. The finer particles of humus, clay, and silt are then readily carried away by runoff or wind, leaving only a layer of coarse sand, stones, and rocks.

IN PERSPECTIVE
Cry the Beloved Country

There is a lovely road that runs from Ixopo into the hills. These hills are grass-covered and rolling, and they are lovely beyond any singing of it. The road climbs seven miles into them, to Carisbrooke; and from there, if there is no mist, you look down on one of the fairest valleys of Africa. About you there is grass and bracken and you may hear the forlorn crying of the titihoya, one of the birds of the veld. Below you is the valley of the Umzimkulu, on its journey from the Drakensberg to the sea; and beyond and behind the river, great hill after great hill; and beyond and behind them, the mountains of Ingeli and East Criqualand.

The grass is rich and matted, you cannot see the soil. It holds the rain and the mist, and they seep into the ground, feeding the streams in every kloof. It is well-tended, and not too many cattle feed upon it; not too many fires burn it, laying bare the soil. Stand unshod upon it, for the ground is holy, being even as it came from the Creator. Keep it, guard it, care for it, for it keeps men, guards men, cares for men. Destroy it and man is destroyed.

Where you stand the grass is rich and matted, you cannot see the soil. But the rich green hills break down. They fall to the valley below, and falling, change their nature. For they grow red and bare; they cannot hold the rain and mist, and the streams are dry in the kloofs. Too many cattle feed upon the grass, and too many fires have burned it. Stand shod upon it, for it is coarse and sharp, and the stones cut under the feet. It is not kept, or guarded, or cared for, it no longer keeps men, guards men, cares for men. The titihoya does not cry here any more.

The great red hills stand desolate, and the earth has torn away like flesh. The lightning flashes over them, the clouds pour down upon them, the dead streams come to life, full of the red blood of the earth. Down in the valleys women scratch the soil that is left, and the maize hardly reaches the height of a man. They are valleys of old men and old women, of mothers and children. The men are away, the young men and the girls are away. The soil cannot keep them any more.

(Alan Paton, 1948)

FIGURE 9–13
Erosion causes loss of soil productivity. The rocky, stony soil to the right of the fence has such poor water-holding capacity that it will no longer support the growth of grass. The condition results from initial destruction of the protective grass cover by overgrazing; then erosion has removed fine material leaving only coarse material. Note the drop of 10–15 cm (4–6 in.) at the fence line. Since the eroded soil now only supports desert species, the process is frequently referred to as *desertification*. (USDA.)

A very important and devastating feature of erosion is that it always involves the *differential* removal of soil particles. This is true of both wind and water erosion. The lighter particles of humus and clay are the first to go, while rocks, stones, and coarse sand remain behind. (People often have the misconception that clay will not erode easily because they visualize clay as hard, bricklike clods. Pour water over a clod of dry clay, however, and you will observe that the water running off is very muddy. This murkiness attests to how easily individual clay particles are separated from the clod and wash away.) Consequently, as erosion removes the finer materials, the remaining soil becomes progressively coarser—sandy, stony, and rocky. Such coarse soils are frequently a reflection of past or ongoing erosion. Did you ever wonder why deserts are full of sand? The sand is what remains; the finer, lighter clay and silt particles have been blown away.

Recall that clay and humus are the most important components for nutrient-holding capacity. As these components are removed, most of the nutrients are removed as well because they are bound to these particles. Then waterways receiving these materials may get an oversupply of nutrients.

Likewise, clay and humus are the chief components providing water-holding capacity. Consequently, water-holding capacity is greatly diminished by erosion. In regions that receive only 10–30 inches (25–75 cm) of rainfall per year, the loss of water-holding capacity is exceedingly serious. Such regions originally supported productive grasslands. With loss of water-holding capacity they are able to support only

drought-resistant, desert species. In other words, grasslands may become deserts as a result of erosion (Fig. 9–13). Indeed, the term **desertification** is used to denote this process.

In summary, soil that is left unprotected erodes as a result of the action of wind and water. As topsoil is destroyed by the differential removal of clay and humus, productivity drops drastically. With diminished productivity, the soil is left unprotected. Further erosion takes place, causing further reduction of productivity, and on and on in a vicious cycle. Thus, desertification, once started, is difficult to slow or reverse. The end point may be nothing more (or less) than a barren desert landscape that supports virtually no growth at all (see In Perspective Box, p. 199).

PRACTICES LEADING TO BARE SOIL AND EROSION

The major practices resulting in bare soil being exposed to erosion are (1) overcultivation, (2) overgrazing, and (3) deforestation.

Overcultivation

Traditionally, the first step in growing crops has been (and to a large extent still is) to plow to control weeds. Without weed control, a grower may get a field of weeds with little if any yield from crop plants. When the top layer is turned upside down, weeds are buried and smothered. The drawback, of course, is that soil is thus exposed to wind and water erosion. Further, it may remain bare for a considerable time before the newly planted crop forms a complete cover; after harvest, the soil again may be left largely exposed to erosion. Runoff and erosion are particularly severe on slopes, but in regions of minimal rainfall wind erosion may extract a heavy toll regardless of topography.

It is ironic that plowing is frequently deemed necessary to "loosen" the soil to improve aeration and infiltration. All too often the effect is the reverse. Splash erosion destroys the soil's aggregate structure and seals the surface so that aeration and infiltration are decreased. The weight of tractors used in plowing may add to the compaction. In addition, plowing accelerates the degradation of humus and evaporative water loss.

The processes of weathering and soil formation vary greatly with climate and composition of the parent material. On average, however, new soil is formed at a rate of about 5 tons per acre (12 tons per hectare) per year, which is equivalent to a layer of soil about 0.5 mm thick. Hence, soils can sustain an erosion rate of up to 5 tons per acre per year and still remain in balance. The hard fact is that much of the globe's cropland—as well as forest and rangeland—is not within this balance.

Where enough measurements have been made to allow realistic estimates, they often show erosion at 2 to 10 times the sustainable rate; in some cases these rates are for entire countries (Table 9–3, Nepal). Nor is erosion resulting from overcultivation a problem only in developing countries. In the early 1980s, U.S. croplands were being eroded at unacceptably high rates. In regions of high erosion, U.S. farmers were sacrificing about 5 tons of topsoil for every ton of grain produced. Over 100 million of some 600 million acres (40 million of some 240 million hectares) of croplands in the countries of the former USSR suffer from extreme erosion. China has lost about 20 percent of its total croplands to erosion, but the erosion continues.

TABLE 9–3

Soil Erosion in Selected Countries

	Extent and Location	Rate of Erosion (metric tons per hectare per year*)
Africa		
Ethiopia	Total cropland (12 million ha)	42
Kenya	Njemps Flats	138
Madagascar	Mostly cropland (45.9 million ha)	25–250
Zimbabwe	304 000 ha	50
North & Central America		
Canada	Cultivated land New Brunswick	40
Dominican Republic	Boa watershed (9330 ha)	346
Jamaica	Total cropland (208 595 ha)	36
United States	Total cropland (170 million ha)	18
South America		
Argentina, Brazil, and Paraguay	La Plata River basin	18.8
Peru	Entire country	15
Asia		
China	Loess Plateau region (60 million ha)	11–251
India	Seriously affected cropland (80 million ha)	75
Nepal	Entire country (13.7 million ha)	35–70
Europe		
Belgium	Central Belgium	10–25
Former USSR	Total cropland (232 million ha)	11

* 1 hectare = 2.5 acres.

Source: World Resources 1988–89. Washington, D.C.: World Resources Institute.

TABLE 9–4

Extent of Desertification

| | Productive Dryland Types | | | |
| | Rangelands | | Rain-fed Croplands | |
	Area (million hectares)*	Percent Desertified	Area (million hectares)*	Percent Desertified
Total	**2556**	**62**	**570**	**60**
Sudano-Sahelian Africa	380	90	90	80
Southern Africa	250	80	52	80
Mediterranean Africa	80	85	20	75
Western Asia	116	85	18	85
Southern Asia	150	85	150	70
Former USSR in Asia	250	60	40	30
China and Mongolia	300	70	5	60
Australia	450	22	39	30
Mediterranean Europe	30	30	40	32
South America and Mexico	250	72	31	77
North America	300	42	85	39

* 1 hectare = 2.5 acres.

Source: World Resources 1989. Washington, D.C.: World Resources Institute.

Overgrazing

Grasslands that do not receive enough rainfall to support cultivated crops have traditionally been used for grazing livestock. Unfortunately, such lands are frequently subjected to overgrazing, and as a result the soil is exposed to erosion. Wind erosion and consequent desertification of such regions are particularly severe.

Overgrazing occurs in many cases because the range lands are "public lands" not owned by the people who own the animals. A herder's income is a function of how many animals he or she raises; thus, the more animals the better. This topic will be discussed further in Chapter 19, but you can see how population pressures lead to more and more animals and overgrazing. An additional factor, which is particularly severe in Africa, is that herders were traditionally nomadic, moving their herds from place to place during the year. As a result of the movement, overgrazing did not occur. As agricultural development has restricted this movement, however, overgrazing of the areas left for the herders has become severe. Between 39 and 90 percent of the productive drylands around the world suffer some degree of desertification (Table 9-4).

Desertification of rangeland is not a linear phenomenon. Ranchers may think their herds are within

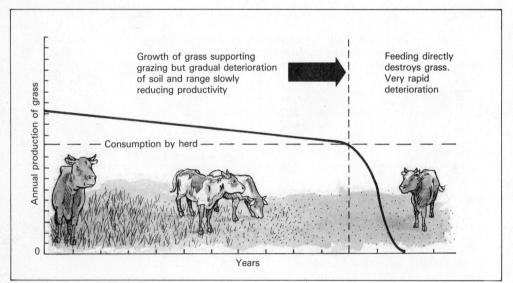

FIGURE 9–14
Overgrazing begins with a slow deterioration of rangeland, a process that gradually reduces the amount of grass produced. The end may come with extreme suddenness, however, as grass production fails to keep up with what is being eaten and cattle graze the land into barrenness.

the carrying capacity, but with intensive grazing there is less detritus to regenerate humus. Thus, there may be gradual mineralization of soil over many years. Then when the land is subjected to stress of some kind—an abnormally dry year, perhaps—grass production drops below the grazing level. As the land is grazed to barrenness, erosion sets in and a vicious cycle of increasing soil degradation and dropping plant productivity accelerates. Thus, apparently productive rangelands may go into desertlike barrenness with surprising suddenness (Fig. 9–14). The huge areas of agricultural and rangelands being already at least partially desertified and the suddenness with which the final stage of desertification can occur do not speak well for the future.

Deforestation

A forest cover is particularly efficient in preventing erosion and holding water because it breaks the fall of raindrops and allows the water to infiltrate into a litter-covered, loose topsoil. Investigators at Hubbard Brook Forest in New Hampshire have found that runoff from a forested slope is as much as 50 percent less than runoff from a comparable grass-covered slope. Forests are particularly efficient at reabsorbing and recycling the nutrients released from decaying detritus. The same investigators found that leaching of nitrogen, for example, increased as much as 45 times after forests were cleared. Thus, not only does clearing forests render the soil subject to erosion, but problems are compounded by nutrient loss.

The tragedy of deforestation is then twofold: first is the loss of biodiversity described previously; second is the resulting erosion and degradation of soils that may preclude the regrowth of forests or anything else. The problem is particularly serious in tropical regions (see Chapter 19). When tropical rainforests are cut, a thin, humus-rich topsoil is easily washed away, exposing a clayey subsoil that is essentially devoid of nutrients, nearly impossible to work, and highly erodible. Most disturbing, perhaps, is the fact that such eroding soil will not support natural regrowth of the forests that were removed, because seeds and seedlings cannot establish themselves on the eroding soil. It just continues to erode indefinitely.

Between clearing for agriculture and obtaining wood for construction, furniture, or fuel, deforestation and the resulting erosion are occurring on an unprecedented scale throughout much of the world despite some notable efforts toward reforestation.

In all, lands suffering from or prone to desertification cover much of the globe (Fig. 9–15). Erosion is a particularly insidious phenomenon because the first 20 to 30 percent of the topsoil may be lost with only marginal declines in productivity, a loss that may be compensated for by additional fertilizer and favorable distribution of rains. However, as loss of topsoil continues, the decrease in productivity and increase in vulnerability to drought become increasingly pronounced.

The Worldwatch Institute estimates that worldwide the loss of soil from crop, range, and deforested lands is about 23 billion tons per year. This is equivalent to all the topsoil on about 23 million acres (9.2 million hectares), an area about the size of Indiana.

THE OTHER END OF THE EROSION PROBLEM

Erosion and soil degradation are just one end of the problem. The other end concerns what happens to the soil and the increased runoff coming from the land. Downstream from an eroded area, lowland areas have a greatly increased probability of being flooded. The **sediments**, as the eroding soil is called, fill reservoirs and clog channels (causing even more flooding) and upset the ecosystems of streams, rivers, bays, and estuaries. Indeed, excess sediments and nutrients resulting from erosion are recognized as the number one pollution problem of surface waters in many regions of the world. This problem will be addressed further in Chapter 12. In addition, water running off rather than soaking in leads to a depletion of groundwater resources, as will be described in Chapter 11.

IRRIGATION, SALINIZATION, AND DESERTIFICATION

Irrigation, supplying water to croplands by artificial means, has dramatically increased crop production in regions that typically receive inadequate rainfall. Irrigated lands in both developed and developing countries total about 325 million acres (130 million hectares), a square 700 miles (1100 km) on a side. Traditionally, water has been diverted from rivers through canals and flooded through furrows in fields, a technique known as **flood irrigation** (Fig. 9–16). In recent years, **center pivot irrigation**, a procedure in which water is pumped from a central well through a gigantic sprinkler that slowly pivots itself around the well, has become much more popular (see Fig. 11–8).

In either case, irrigation may lead to **salinization**, an intolerable increase in salinity (saltiness) of the soil. Salinization occurs because even the freshest irrigation water contains at least 200–500 parts per million of salts dissolved from the earth. Additional salts may be leached from the minerals in the soil. As

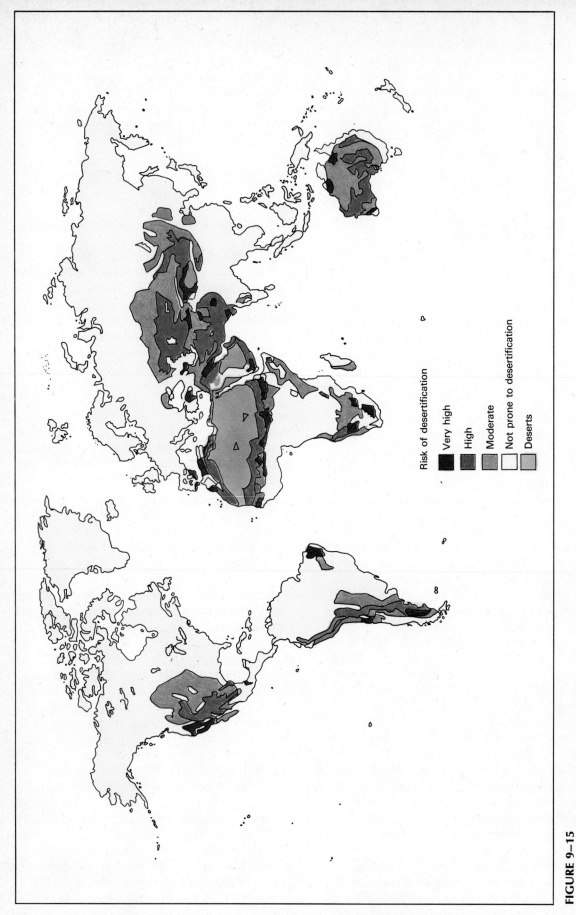

FIGURE 9–15
Deserts and areas subject to desertification. Throughout the world, overgrazing and deforestation are causing vast low-rainfall areas to degenerate into deserts. (Reprinted with permission from "Desertification: Its Causes and Consequences," United Nations Conference on Desertification, Nairobi, 1977. Copyright 1977, Pergamon Books Ltd.)

Risk of desertification

- Very high
- High
- Moderate
- Not prone to desertification
- Deserts

FIGURE 9–16
Flood irrigation. The traditional method of irrigation is to flood furrows between rows with water from an irrigation canal. This method is extremely wasteful since most water either evaporates or percolates beyond the root zone, or the water table rises to waterlog the soil and prevent crop growth. (Betty Derig/Photo Researchers.)

FIGURE 9–17
Millions of acres of irrigated land are now worthless because of the accumulation of salts left behind as the water evaporates, a phenomenon known as *salinization*. (John M. White/Texas Agricultural Extension Service.)

irrigation water leaves the cropland by evaporation or transpiration, salts in solution remain behind and may accumulate in and on the soil to the point of precluding plant growth (Fig. 9–17). Salinization is considered a form of desertification.

Salinization can be avoided, and even reversed, if sufficient water is applied to leach the salts down through the soil. However, unless there is suitable drainage, the soil will become a waterlogged quagmire in addition to being salinized. Artificial drainage may be installed (Fig. 9–18) at great expense, but then attention must be paid to where the salt-laden water

FIGURE 9–18
Drainage of irrigated land. To prevent salinization, excess water with salts must be drained away. This procedure may necessitate a system of tiles as shown here, and additional environmental problems may be caused by the discharge of such high-mineral-content drainage into natural waterways.

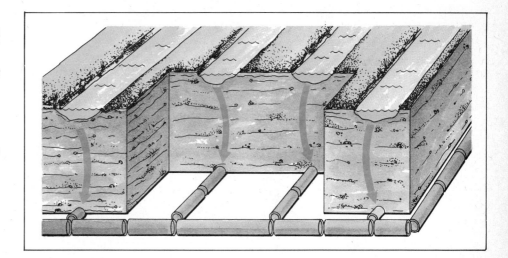

TABLE 9–5

Irrigated Land Desertified by Salinization		
	Area (million hectares)*	Percent Desertified
Total	**131**	**30**
Sudano-Sahelian Africa	3	30
Southern Africa	2	30
Mediterranean Africa	1	40
Western Asia	8	40
Southern Asia	59	35
Former USSR in Asia	8	25
China and Mongolia	10	30
Australia	2	19
Mediterranean Europe	6	25
South America and Mexico	12	33
North America	20	20

* 1 hectare = 2.5 acres.

Source: World Resources 1989. Washington, D.C.: World Resources Institute.

is draining. The wildlife in the Kesterson Wildlife Preserve in California has been all but destroyed by pollution from irrigation drainage.

As Table 9–5 shows, it was estimated in the late 1980s that 30 percent of all the irrigated land in the world has already been salinized, negating the value of both the irrigation project and the land. Further, it is estimated that an additional 1 to 1.5 million hectares is salinized each year. In the United States, the problem is especially acute in the lower Colorado River Basin area and in the San Joaquin Valley of California, areas in which a total of 400 thousand acres (160 thousand hectares) has been rendered nonproductive, representing an economic loss of more than $30 million per year. Adding to the problems, water supplies are fast being depleted by withdrawals for irrigation.

OTHER FACTORS HARMING THE SOIL ECOSYSTEM

In addition to erosion and salinization, there are other factors leading to soil degradation. The most serious of these other factors is overreliance on chemical fertilizers and on chemicals for pest and weed control.

Organic versus Inorganic Fertilizer

There is no question that optimal amounts of required nutrients can be efficiently provided by suitable application of inorganic chemical fertilizer. The disadvantage of chemical fertilizer comes in mistakenly thinking it will substitute for all the benefits of organic material. Without sufficient detritus, soil organisms

starve, humus content declines, and all the desirable attributes of the soil decline as the topsoil mineralizes. Then, with the soil's loss of nutrient-holding capacity, applied inorganic fertilizer is prone to simply leach into waterways.

This is not to say that chemical fertilizers do not have a place in growing plants. Exclusive use of organic material may provide insufficient amounts of one or more nutrients required to support plant growth. What is required is for anyone growing plants to understand the different roles played by organic material and inorganic nutrients and then to use each type as necessary. The federal Agricultural Extension Service, which has an office in every county in the United States, will test soil samples, determine which if any nutrients are missing, and make recommendations accordingly. Following such recommendations, plant growers—both farmers and home gardeners—can judiciously use inorganic fertilizer *along with organic material* and gain the benefits of both.

Pest and Weed Control with Chemicals

In addition to chemical fertilizers, modern agriculture relies heavily on chemicals to control weeds and pests. Results have been less than fully satisfactory, as target pests have become increasingly resistant to pesticides and additional insects have become pests as their natural enemies have been killed. These problems and alternative biological control methods will be explored in Chapter 10.

Agricultural chemicals, both fertilizers and pesticides, are prone to leaching. The U.S. Environmental Protection Agency has identified agriculture as the largest nonpoint source of surface water pollution. (A **nonpoint source** means the pollution does not come from a specific point, such as the outlet of an industrial drain.) Not only do pesticides and nutrients from fertilizers contaminate surface water, they are also detected in the groundwater in many agricultural regions.

CONCLUSIONS

The failing behind all the problems discussed is that, in agriculture as in other areas, humans have seen and attacked each problem as a separate entity. Need to clear the soil? Plow! Need nutrients? Add chemical fertilizer! Need water? Irrigate! Pests a problem? Use pesticides! There has been a general failure to recognize that we are dealing with a complex, integrated system, in other words, an ecosystem. What is done in one area inevitably will have an impact on other areas. That this one-thing-at-a-time approach has had considerable success in meeting the food demands of a growing population should not be underestimated (see Chapter 8). However, that this approach is also proving nonsustainable needs no elaboration.

Making Agriculture Sustainable

It goes without saying that sustainability will depend on controlling erosion, overgrazing, deforestation, and salinization. For the moment, let us focus on erosion from croplands. Policies affecting overgrazing and deforestation will be considered in Chapter 19.

CONTROLLING EROSION FROM CROPLANDS

Erosion is far from a new problem. In the early 1930s in the United States, lack of soil conservation combined with drought conditions to create the infamous Dust Bowl, which extended from Texas to Illinois. As John Steinbeck wrote in *The Grapes of Wrath*, "The dust lifted out of the fields and drove great plumes into the air like sluggish smoke. . . . Dawn came but no day . . . and the wind cried and whimpered over the fallen corn." Farmers went bankrupt, tenant farmers migrated en masse to California, and the dusty earth settled on policy makers as far east as Washington, D.C., providing the impetus to pass the Soil Conservation Act of 1935, which created the Soil Conser-

vation Service (SCS). The SCS assisted farmers in establishing what are now the standard soil conservation techniques: contour strip cropping and establishment of shelterbelts (Fig. 9–19). These techniques represented a major step toward stabilizing the nation's topsoil and restoring the economy of the Great Plains. The invaluable service continues through the nationwide network of Agricultural Extension Service offices mentioned earlier. However, the SCS can only provide assistance; it has no "teeth" to enforce conservation.

During the 1970s, farmers were under economic pressures to increase production. This pressure, coupled with the introduction of huge tractors that increase the efficiency of farming on vast, uninterrupted acreages and dimmed memories of the Dust Bowl era, led many farmers to abandon soil conservation techniques. Some 12 million hectares of windrows, fallow strips, and slopes were sacrificed and put into cultivation. Not only did production and short-term profits increase, however, but so did erosion—by about 50 percent, to rates equaling if not exceeding those of the Dust Bowl period.

Once again there was an effective response to this episode of drastic soil erosion. The problem was recognized and publicized by environmentally focused organizations, notably the American Farmland Trust, Rodale Press, and the Worldwatch Institute. These groups, along with many others, including the Sierra Club, the National Audubon Society, and the Conservation Foundation, lobbied Congress and won passage of the Food Security Act of 1985. Under this

FIGURE 9–19
Shelter belts. Belts of trees around farm fields break the wind and protect the soil from erosion. (Dan McCoy/The Stock Market.)

act, 40 million acres (16 million hectares) of highly erodible cropland were put into a "Conservation Reserve" of forest and grass. Farmers are paid about $50 per acre ($125 per hectare) per year for land placed in the reserve. Farmers with erodible land are required to develop and implement a soil conservation program to remain eligible for price supports and other benefits provided by the government.

While the Food Security Act has reduced erosion in the United States, the problem still persists because not all land is covered by the Act, traditional erosion control techniques are not perfect, and a farmer's compliance is voluntary. Also, the program does not address fertilizer and pesticide overuse, deterioration of soil from lack of sufficient detritus, and salinization. The overall problem is that each remedy is still single-focus in nature. To have a really sustainable agriculture, in the United States or anywhere else, we must look at the total picture.

PRACTICES FOR SUSTAINABLE AGRICULTURE

The objectives of any sustainable agriculture are:

○ Use lower amounts of chemical fertilizers and pesticides.
○ Keep food safe and wholesome.
○ Maintain a productive topsoil.
○ Keep agriculture economically viable.

Our understanding of soil as a system should enable us to appreciate how the following practices meet the objectives for sustainable agriculture.

Crop Rotation and Organic Fertilizer Legume crops, such as clover and alfalfa, are nitrogen fixers (p. 64). Also, their roots penetrate deeply and help bring other nutrients to the surface. Growing a legume crop, then plowing it in or cutting it and leaving it on the surface as a mulch both adds nitrogen and maintains soil structure by nurturing the soil organisms. Alternating a legume crop with two or three nonlegume crops—a practice known as **crop rotation**—maintains a healthy soil. Applications of organic fertilizer, such as animal manure, enhance these benefits.

Protection from Erosion Without protection from erosion, all other measures are wasted. Short of a full vegetative cover, the best protection from erosion is a mulch of dead organic matter such as leaves and grass clippings. In addition to protecting the soil from erosion, a mulch cover suppresses weed growth, reduces evaporative water loss, and preserves soil structure. A number of tillage methods referred to as **conservation tillage**, have been developed that allow cultivation to control weeds but at the same time leave the dead plant material on the ground.

Contour strip cropping is an effective aid in erosion control and also fosters crop rotation (Fig. 9–20). Strips follow the contours of hills so that any runoff from a recently cultivated strip is caught by the strip below, particularly if the lower strip is grass for hay. You can readily see how strip cropping goes hand in hand with crop rotation. The farmer is growing all crops simultaneously; she or he is just rotating strips. Of course, steep slopes, which are highly erodible, should not be cultivated at all but rather should be left in grass or forest. On wide, level areas, shelterbelts, hedgerows of trees and shrubs, are effective in breaking the wind and reducing wind erosion (Fig. 9–19).

Water Management The best water management tactics are (1) to maintain a good topsoil that allows maximum infiltration and holding of natural rainfall and (2) to maintain a mulch cover that reduces evaporative

FIGURE 9–20
Contour farming. Cultivation up and down a slope encourages water to run down furrows and may lead to severe erosion. The problem is reduced by plowing and cultivating along the contours at a right angle to the slope. In this photograph, strip cropping is also being utilized: the light green bands are one type of crop, the dark bands are another. (USDA photograph)

water loss. Also, it helps to grow only crops adapted to rainfall of the region, so that no irrigation is needed. Where irrigation is required, drip irrigation systems, which consist of a network of tubes that drip just the requisite amount of water on each plant may be used (see Fig. 11–17).

Biological Pest Management Instead of chemical pesticides and herbicides, biological methods for controlling pests should be used so far as they are available. This type of pest control is facilitated by growing a suitable mixture of crops and rotating crops. This topic will be explored further in Chapter 10.

IN PERSPECTIVE
Home Gardening

The principles and practices of sustainable agriculture can easily be applied to home lawns and gardens. Homeowners are even more likely to conduct nonsustainable practices than are farmers. On average, homeowners use two to three times more fertilizer and pesticides per unit area than commercial farmers do. And notice the pathway of nutrients: A person buys fertilizer, applies it to a lawn, mows, then bags the clippings and puts them in the trash, which goes to a landfill. Furthermore, excess pesticides are often used because overfertilized and overwatered grass and other plants are more prone to attack by diseases and insects.

Alternatively, growing clover with the grass will make a lawn nitrogen-self-sufficient. And don't collect the clippings; they are the detritus, which will decay naturally and re-

turn the nutrients to the soil. The only time it may be necessary to rake clippings is when they are so heavy that they totally cover the grass and cut off the light. Avoid this by mowing more frequently. If clippings are raked, compost them.

Or why grow grass at all? Many people are finding that it decreases work and increases beauty and satisfaction to convert their lawn (or at least portions of it) to beds of perennial wild flowers that are suited to the area and hence require no fertilizer, pesticides, or watering. For example, increasing numbers of people in the arid Southwest are turning to xeroscaping with various species of cacti, yucca, aloe, and other desert plants. Natural landscaping also attracts birds and butterflies of the area. You can also grow an assortment of dwarf fruit trees and vegetables even in a relatively small yard

that gets eight or more hours of sunlight a day. Again, use all pulled weeds, leaves, and other yard wastes to create compost, or use them as a mulch to protect and regenerate the soil. Rotate locations of planting legumes (beans or peas) with other vegetables. Use biological controls of pests as far as possible or give up on growing pest-susceptible plants.

All kinds of assistance are available from your county extension agent, from organic gardening magazines and books, and from people you will find who are already interested in gardening. Some things may not grow as well as you might like, but other things are likely to surpass your expectations. In all, it can be a wonderfully rewarding experience and an important lesson in learning to care for the earth.

THE DIVERSIFIED FARM

You can see that the above practices can all be a logical part of the operation of a *diversified farm*, that is, one growing several different crops along with animals. Production of animal products—meat, milk, eggs, cheese, leather—may be integrated into the operation. Legumes, hay, and the waste from other crops can all be used as animal feed. The animal manure may be applied to the soil.

The diversified farm also provides a higher degree of economic stability for the farmer in that "all the eggs are not in one basket." A loss of one crop may be offset by better-than-average yield in another. Studies have shown that gross income from a diversified farm may be somewhat less than from a farm growing only one crop because in the former case some of the land is out of production in the rotation. However, profitability is often greater because of lower inputs of fertilizer and pesticides. Also, diversified farming offers an interest and challenge to the manager not found in monoculture farming.

Diversified family farms were the rule until World War II. Since then there has been a continuous trend toward highly specialized monoculture operations. Animal farming has become divorced from crop production. In the developed world, animals are raised largely in feed lots, where they are fed grain produced perhaps hundreds of miles away. The disposal of manure from such feed lots is a tremendous problem because it is uneconomical to truck it back to the croplands (Fig. 9–21). The ironic result is that feed growers depend on inorganic fertilizer, and manure wastes too often go into waterways, causing pollution.

Paradoxically, monoculture is now commonly referred to as **traditional farming**, while the low-chemical-input, diversified methods are referred to as **alternative farming**.

FARM POLICY

After exhaustive study, the National Research Council concludes in a 1989 report titled *Alternative Agriculture*:

As a whole, federal policies work against environmentally benign practices and the adoption of alternative agricultural systems, particularly those involving crop rotations, certain soil conservation practices, reductions in pesticide use, and increased use of biological and cultural means of pest control. These policies have generally made a plentiful food supply a higher priority than protection of the resource base.

Effectively, federal programs that provide acreage allowances and guaranteed prices for certain export crops are an economic incentive locking farmers into a monoculture operation. A farmer stands to make the most by maximizing yield. This reality leads to the generous use of agricultural chemicals, a practice that is supported and promoted by a vast agri-

FIGURE 9–21
Most beef in the United States is fattened in feedlots, such as this one in western Kansas. Disposal of manure from such feedlots is a problem because costs of hauling it back to croplands is not cost effective. Unused, it becomes a pollution problem. (J.P. Jackson/Photo Researchers.)

Who Is Responsible for Maintaining the Soil?

The lives of all of us depend on the abundant production of agriculture. Yet, who is responsible for maintaining the topsoil that underlies that production?

Over the years since the start of the Industrial Revolution, people have become increasingly divorced from soil and the production of food. In every developed country, countless diversified family farms have given way to huge specialized monocrop operations. In the United States, for instance, less than 2 percent of the population is engaged directly in crop production. From the field to storage and then to flour mills, baking companies, and supermarket shelves, grains travel an average of 1000 miles before they are consumed. Less than one penny of the dollar you pay for a loaf of bread goes to the farmer. The same is true for other foods.

With this specialization of production and its separation from consumers, who should be responsible for seeing that soil is properly protected and nurtured as a sustainable resource? Not the 98 percent who are off-farm consumers, surely, but not the 2 percent who are growers

either. Far removed from the final consumers and receiving minimal compensation, most farmers consider production to be simply a business. If it fails, they will be forced to move on to something else.

In other words, we have unwittingly created a food production system in which virtually no one is in a position to see the whole system as an *ecosystem*, much less manage it as such.

Many people point with pride to the U.S. agricultural system, saying, "Look how efficient it is; people in this country generally spend a smaller portion of their incomes on food and yet eat better than in any other country of the world." If the system is not sustainable, however, its efficiency is a moot point.

In their book *For the Common Good*, Herman Daly and John Cobb suggest that each region of the country, a region to be no larger than a small state, should be required to be agriculturally self-sufficient (disaster-relief excluded). Such a policy would force each region to adopt a diversified agriculture and would bring producers and con-

sumers closer together. If such a policy were ever put into effect, the people of each region would take more interest in the management of their soil, because they would see a healthy soil system as crucial to their own survival and well-being.

What would be the ramifications of Daly and Cobb's plan?

○ Would each region adopt a more diversified, sustainable agriculture?

○ Would a more diversified agriculture facilitate recycling of animal wastes, compost, and sewage sludges back to the soil?

○ Would growers develop a greater sense of responsibility toward consumers, their customers?

○ Would consumers develop a greater interest in and responsibility toward how their food was grown?

○ Would people face the consequences of converting more and more farmland to housing developments and office parks?

○ Would prices of food skyrocket?

cultural chemical industry. The government has also built innumerable dam-and-canal projects to supply farmers with irrigation water at far below actual cost. This has encouraged overuse of water and salinization as a consequence. The cost to the U.S. taxpayers for these price supports and subsidies, which run on the order of $30 billion per year, should not go unnoted.

But how to change? The good news is that people on all levels are becoming increasingly aware of the shortfalls of "traditional" farming and are at work developing alternatives. A small but growing number of farmers are experimenting and developing alternative systems on their own (see In Perspective, p. 212). In 1988 the U.S. Department of Agriculture

started the Low Input Sustainable Agriculture program, which provides funding for "alternative" methods. Agricultural colleges and universities across the country are revising curricula to emphasize sustainable agriculture.

However, the trend could be vastly hastened by suitable federal policies. The National Research Council concludes that there is a vital need to either change policies or adopt new ones that will encourage and support farmers in efforts to adopt alternative practices. We noted above the success of environmental groups in obtaining passage of the Food Security Act of 1985. Your involvement and support of such organizations can likewise hasten the movement toward sustainable agriculture.

IN PERSPECTIVE
An Example of a Small Diversified Farm

In the gently rolling area of southwestern Iowa is a small (160 acre) diversified farm run by Clark BreDahl and his wife Linda, a teacher able to devote full time to the farm during the summer. Less than half the size of the average farm in the area, it is among the top 10 percent in profitability and generally provides full support for the BreDahl family, allowing them to save Linda's income. Crops include corn, alfalfa, and turnips to feed animals; soybeans for market; pedigreed oats sold as seed (which bring double the feed price); sheep for wool; and lambs and pigs for market.

Crops are grown in 100-foot-wide contour strips and follow a 5-6 year rotation sequence of corn–soy–beans–corn–oats/turnips–alfalfa. Soybeans and alfalfa, which are nitrogen fixers, and manure provide most fertilizer needs. The strips are fenced, so that animal feeding is made more efficient by letting the animals graze (and deposit manure) in the strips as feed crops are growing, allowing the use of damaged crops and crop residue as feed resources.

A unique part of the system is the use of turnips. Planted after the oats are harvested in July, turnips mature in the fall. The sheep are turned into the turnip strips in September and graze on the turnip tops and the above-ground roots. They then cup-out the tap root below the surface. The holes left in the soil after grazing fill with water, snow, and ice, helping add moisture and preventing runoff. Thus turnips provide highly nutritious fodder for the sheep into the new year, and the soil benefits in the process. Using "organic methods," the soil was found in excellent condition, whereas soils on most other county farms were classified as moderately to severely eroded; and crop yields per acre on the BreDahl farm were consistently above the average for the area.

With intelligent management, a family can make a comfortable living from a small diversified farm, following sustainable practices. (National Research Council.)

Review Questions

1. What are the components of the soil system?
2. Which aspects of soil are important for maintaining plant growth?
3. What is the source of mineral nutrients, and how are they added to, removed from, or lost from the soil? What is leaching?
4. How does a natural ecosystem differ from an agricultural system with respect to the movement of nutrients?
5. Which two aspects of soil influence the availability of water?
6. What is meant by soil texture, and how does it influence the various aspects of soil?
7. How do detritus and soil organisms change the various aspects of a mineral soil? What is humus? What is soil structure?
8. What is erosion, and how does it affect topsoil? How does erosion result in desertification?
9. Which farming practices are most likely to lead to erosion, and what are the consequences of erosion in addition to soil degradation?
10. What is salinization, and what are its causes and consequences?
11. What are the principles of sustainable agriculture?
12. What are the distinctions between a diversified farm and a monoculture operation? Why does one lend itself to sustainability and the other does not?

Thinking Environmentally

1. Suppose you are going into farming. Describe the type of operation you would create, and give a rationale for each measure in terms of sustainability.
2. Explain how degradation of topsoil affects plant growth, specifically relating plant needs to various aspects of the soil.
3. Why do human societies, past and present, seem to place so little value in maintaining topsoil? What might be done to change this?
4. Evaluate the following argument. Erosion is always with us; mountains are formed and then erode; rivers erode canyons. You can't ever eliminate soil erosion, and so it is foolish to ask farmers to do so.
5. Explain why diversified farms are necessary for the cost-effective use of manure as fertilizer. Why can't manure be used cost effectively in our present system of monoculture farms and feed lots? (Fifty pounds of inorganic fertilizer costs a farmer roughly $4.00 and has roughly the same nutrient content as 500 pounds of manure. Estimate and compare the costs—fuel, labor, and so on—of trucking manure 50 miles from a feed lot to a farm with the cost of using an equivalent amount of inorganic fertilizer, assuming the manure is free.)

10

Pests and Pest Control

OUTLINE

LEARNING OBJECTIVES

When you have finished studying this chapter you should be able to:

1. Understand what pests are and why they need to be controlled.
2. Describe the two basic philosophies of pest control.
3. Trace the development of synthetic organic pesticides.
4. Define the problems resulting from use of chemical pesticides.
5. Describe how nonpersistent pesticides differ from other kinds.
6. Describe the life cycle of an insect and the natural control methods that apply to the different stages.
7. Discuss several methods of biological pest control.
8. Evaluate the impact of economic factors on decisions about use of pesticides.
9. Describe integrated pest management.
10. Understand the major laws dealing with pesticides and their shortcomings.

Since earliest times, humans have suffered the frustration and food losses brought on by destructive pests. Genesis, the first book of the Bible's Old Testament, speaks of the "thorns and thistles" thwarting Adam's attempts to eat the plants of the field, and later biblical passages speak of locust invasions that

213

destroyed crops and caused famines. To this day, both farmers and pastoralists must wage a constant battle against the insects, plant pathogens, and weeds that compete with them for the biological use of crops and animals.

If we adopt a dictionary definition of **pest** as "any organism that is noxious, destructive, or troublesome," we see that the term includes a broad variety of organisms that interfere with humans or with their social and economic endeavors. The principal categories of pests are:

1. Organisms that cause disease in humans or domestic plants and animals. These pests include viruses, bacteria, and such parasitic organisms as intestinal worms and flukes.

2. Organisms that are annoying to people and domestic animals and that may transfer disease by biting or stinging. Common examples are flies, ticks, bees, and mosquitoes.

3. Organisms that feed on ornamental plants or agricultural crops, both before and after the crops are harvested. The most notorious of these organisms are various insects, but certain worms, snails, slugs, rats, mice, and birds also fit into this category.

4. Animals that attack and kill domestic animals, such as coyotes, foxes, and weasels.

5. Organisms that cause wood, leather, and other materials to rot and food to spoil. Bacteria and fungi, especially molds, are largely responsible for this spoilage, but in warm, moist climates, termites are the primary culprit in the destruction of wood.

6. Plants that compete with agricultural crops, forests, and forage grasses for light and nutrients. A plant in any of these roles is often referred to as a weed (Fig. 10–1). Some weeds poison cattle; others simply detract from the aesthetics of lawns and gardens.

Although all six types of pests are highly important in human affairs, this chapter will confine its coverage to those pests that interfere with human agricultural crops, grasses, and animals and will examine the ways in which we work to bring such pests under control.

The Need for Pest Control

Part of the credit for the prosperity of humans can be attributed to pest control. We would still live under extremely precarious conditions—our food supply and physical health at the mercy of all the organisms we call pests—if it were not for our ability to control them. Indeed, pests represent a major component of the environmental resistance that—until recently—kept human populations from expanding rapidly.

CROP LOSSES DUE TO PESTS

Insects, plant pathogens, and weeds destroy an estimated 37 percent of potential agricultural production in the United States, at a yearly loss of $64 billion.

Efforts to control these losses involve the use of 500 000 metric tons of **herbicides** (chemicals that kill plants) and **pesticides** (chemicals that kill animals and insects considered to be pests) annually, with a direct cost of $4 billion per year (Fig. 10–2). In vast areas of Africa and Amazonia, insect-borne diseases such as sleeping sickness and malaria have all but precluded human endeavors. Pesticides used to combat these diseases have become important public-health tools, in addition to their agricultural use.

Many of the changes in agricultural technology such as monoculture and widespread use of genetically identical crops that have boosted crop yields have also brought on an increase in the proportion of crops lost to pests—from 31 percent in the 1950s to 37 percent today. During these past 40 years, herbicide and pesticide use has multiplied manyfold, leading to a dependency that has disturbed many ob-

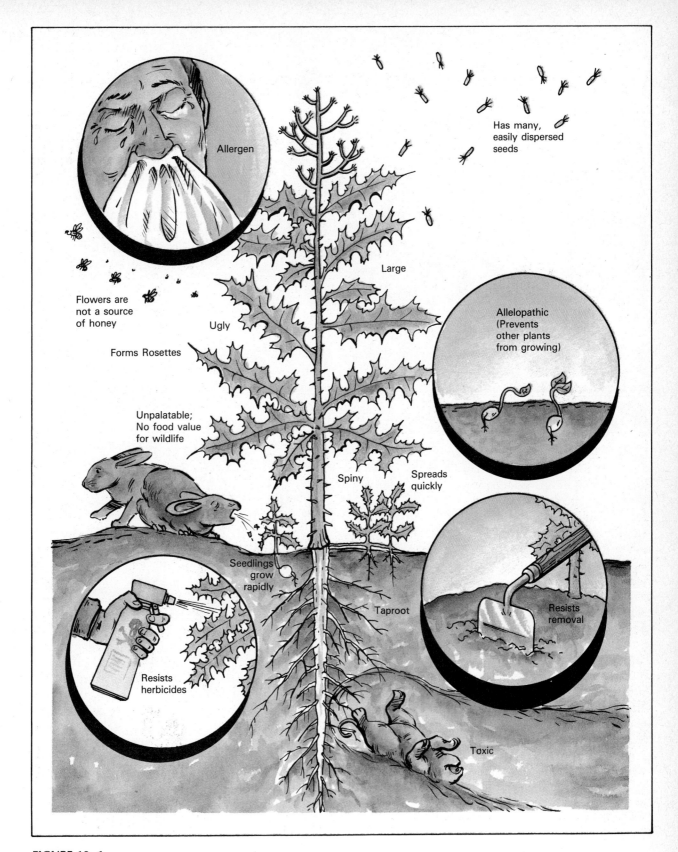

FIGURE 10–1
Weeds are plant pests with features that interfere with human purposes. This drawing shows a hypothetical "ultimate weed," a plant with just about all the undesirable properties of weeds wrapped up into one. (From *Biological Control of Weeds*, by G.A. Strobel. Copyright © 1991 by Scientific American, Inc. All rights reserved.)

FIGURE 10–2
A crop being sprayed with pesticides to keep insects under control. Spraying is the basic technique still used in most control programs, although nowadays the pesticides in use do not persist in the environment for more than a week or two. (EPA photo.)

servers who are concerned with the indirect effects of the chemicals.

DIFFERENT PHILOSOPHIES OF PEST CONTROL

Medical practice employs two basic means of treating infectious diseases. One approach is to give the exposed patient a massive dose of antibiotics, hoping to eliminate the pathogen causing the problem or to stop that pathogen before it can get established. The other approach is to stimulate the patient's immune system with a vaccine to produce long-lasting protection against any future invasion.

The same two basic philosophies are used to control agricultural pests. The first is **chemical technology.** Like the use of antibiotics, chemical technology seeks a "magic bullet" that will eradicate or exterminate the pest organism. Although it has had much success, this approach gives only short-term protection. Furthermore, the chemical often has side effects that are highly damaging to other organisms.

The second philosophy is **ecological pest management.** Like stimulating the body's immune system, this approach seeks to give long-lasting protection by developing control agents based on knowledge of the pest's life cycle and its interrelationships in the ecosystem. Such agents, which may be either other organisms or chemicals, work in one of two ways. Either they are highly specific for the pest species being fought, or they manipulate one or more aspects of the ecosystem. Ecological pest management emphasizes the protection of people and domestic plants and animals from pest damage rather than eradication of the pest organism. Thus, the ben-

efits of pest control can be obtained while the integrity of the ecosystem is maintained.

Promises and Problems of the Chemical Approach

DEVELOPMENT OF CHEMICAL PESTICIDES AND THEIR SUCCESSES

Pesticides are categorized according to the group of organisms they aim to kill. There are insecticides (insect killers), rodenticides (mice and rat killers), fungicides (fungi killers), and so on. None of these chemicals, however, is entirely specific for the organisms it is designed to control; each poses hazards to other organisms, including humans. Therefore, pesticides are sometimes referred to as **biocides,** a name that emphasizes that they may endanger many forms of life.

Finding effective materials to combat pests is an ongoing endeavor. The early substances used included toxic heavy metals such as lead, arsenic, and mercury. These inorganic compounds are frequently referred to as **first-generation pesticides.** We now recognize that these substances may accumulate in soils and inhibit plant growth. Animal and human poisonings are possible also. In addition, these chemicals lost their effectiveness as pests became increasingly resistant to them. For example, in the early 1900s, citrus growers were able to kill 90 percent of injurious **scale insects** (minute insects that suck the juices from plant cells) by placing a tent over the tree and piping in deadly cyanide gas for a short time. By

1930, this same technique killed as few as 3 percent of the pests.

With agriculture expanding to meet the needs of a rapidly increasing population and first-generation (inorganic) pesticides failing, the farmers of the 1930s were begging for new pesticides. These new **second-generation pesticides,** as they came to be called, were found in synthetic organic chemicals. The science of organic chemistry began in the early 1800s. Over the next century, chemists synthesized thousands of organic compounds, but, for the most part, these compounds sat on shelves because uses for them had not been developed.

The DDT Story

In the 1930s a Swiss chemist, Paul Muller, began systematically testing some of these chemicals for their effect on insects. In 1938 he hit upon the chemical dichlorodiphenyltrichloroethane (DDT), a chlorinated hydrocarbon that had first been synthesized some 50 years before.

DDT appeared to be nothing less than the long-sought magic bullet, a chemical that was extremely toxic to insects and yet seemed relatively nontoxic to humans and other mammals. It was very inexpensive to produce. At the height of its use in the early 1960s, it cost no more than about 20 cents a pound. It was broad-spectrum, meaning that it was effective against a multitude of insect pests. It was persistent, meaning it did not break down readily in the environment and hence provided lasting protection. This last attribute provided additional economy by eliminating both the material and labor costs of repeated treatments.

DDT quickly proved successful in controlling important insect disease carriers. During World War II, for example, the military used DDT to control body lice, which spread typhus fever among the men living in dirty battlefield conditions. As a result, World War II was one of the first wars in which fewer men died of typhus than of battle wounds. The World Health Organization of the United Nations used DDT throughout the tropical world to control mosquitoes and thereby greatly reduced the number of deaths caused by malaria. There is little question that DDT saved millions of lives. In fact, the virtues of DDT seemed so outstanding that Muller was awarded the Nobel Prize in 1948 for his discovery.

Postwar uses of DDT expanded dramatically. It was sprayed on forests to control defoliating insects such as the spruce budworm. It was routinely sprayed on salt marshes to deal with nuisance insects. It was sprayed on suburbs to control the beetles that spread Dutch elm disease. And, of course, DDT proved very effective in controlling agricultural insects. It was so effective, at least in the short run, that

many crop yields dramatically increased. Growers could ignore other more painstaking methods of pest control such as crop rotation and destruction of old crop residues. They could grow varieties that were less resistant but more productive. They could grow certain crops in warmer and/or moister regions that had formerly been precluded because of devastating pest damage. In short, DDT gave growers more options for growing the most economically productive crop.

It is hardly surprising that DDT ushered in a great variety of synthetic organic pesticides: chlorinated hydrocarbons such as aldrin, lindane, chlordane, heptachlor, and dieldrin; organic phosphorus compounds such as parathion and malathion; and carbamates such as aldicarb and carbaryl. Not all of these are in use today, and for this we can thank Rachel Carson.

Silent Spring

Rachel Carson was a U.S. government biologist who also happened to be an accomplished chronicler of nature. In the 1950s, she began to read disturbing accounts in the scientific literature of DDT's effects on wildlife. Fish-eating birds were dying from DDT received through the food chain. Robins were dying as a result of eating worms from soil under trees sprayed with DDT. Carson was finally galvanized into action by a letter from a friend distressed over the large number of birds killed when the friend's private bird sanctuary was sprayed for mosquito control. By 1962 Rachel Carson had finished a book-length documentary of the effects of the almost uncontrolled use of insecticides across the United States. The book, *Silent Spring*, became an instant best seller. Its basic message was that if insecticide use continued as usual, there might some day come a spring with no birds—and with ominous impacts on humans as well.

Silent Spring triggered a national debate that continues today. It was immediately attacked by representatives of the agricultural and chemical industries as a hysterical and unscientific account that, if taken seriously, would lead to a halt of human progress and eventually famine and death. On the other side, however, the book was hailed by many distinguished scientists and government officials as an unparalleled breakthrough in environmental understanding. Thirty years later, Rachel Carson is credited with the creation of the Environmental Protection Agency (which eventually banned DDT) and with starting the environmental movement and laying the groundwork for much of pollution science and regulation. *Silent Spring* has become a classic, and the regulation of insecticides and other toxic chemicals is in a sense

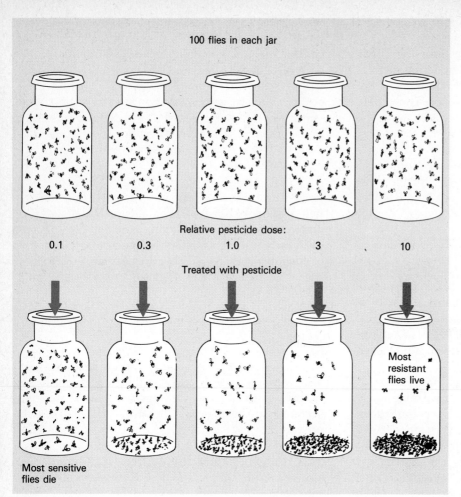

100 flies in each jar

Relative pesticide dose:

0.1   0.3   1.0   3   10

Treated with pesticide

Most resistant flies live

Most sensitive flies die

FIGURE 10–3
Genetic variation in resistance to pesticides exists in all populations. Here, the flies that died with the smallest amount of pesticide (0.1 relative dose) are the ones most sensitive to the pesticide. The flies still alive after the heaviest dosage (10) are the most resistant members of the populations. These few surviving individuals will pass on their resistance to future generations, and therefore future generations will be harder than ever to kill with pesticides.

a monument to Rachel Carson, who died of cancer only 2 years after her book was published.

PROBLEMS STEMMING FROM CHEMICAL PESTICIDE USE

Problems associated with synthetic organic pesticides can be placed in three categories:

○ Development of pest resistance
○ Resurgences and secondary pest outbreaks
○ Adverse environmental and human health effects

Development of Pest Resistance

The most fundamental problem for growers is that chemical pesticides gradually lose their effectiveness. Over the years, larger and larger quantities or new and more potent pesticides or both are required to obtain the same degree of control. Synthetic organic pesticides fared no better than first-generation pesticides in this respect. Numbers illustrate the situation. In 1946, 1 kg of pesticides provided enough pro-

tection to produce about 60 000 bushels of corn. In 1971, 64 kg were used for the same production, and losses due to pests actually *increased* during this period.

Resistance builds up because pesticides destroy the sensitive individuals of a pest population, while the more resistant individuals continue to breed, creating a new population of more-resistant pests (Fig. 10–3). Resistant insect populations develop rapidly because insects have a phenomenal reproductive capacity. A single pair of house flies, for example, can produce several hundred offspring that may mature and themselves reproduce only 2 weeks later. Consequently, repeated pesticide applications result in the unwitting selection and breeding of genetic lines that are highly, if not totally, resistant to the chemicals that were designed to eliminate them. Cases have been recorded in which the resistance of a pest population has increased as much as 25 000-fold.

Over the years of pesticide use, the number of resistant species has climbed steadily (Fig. 10–4). About 25 major pest species are resistant to all of the principal pesticides. Interestingly, as a pest population becomes resistant to one pesticide, it also may

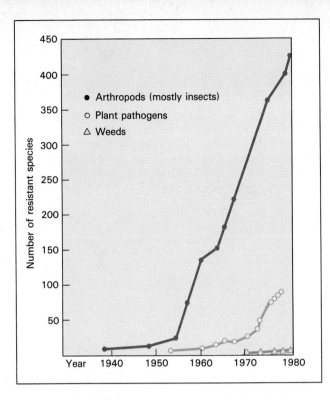

FIGURE 10–4
Number of species resistant to pesticides, 1940–1980. Using pesticides causes selection (survival of the fittest) for those individuals that are resistant. This graph shows how an increasing number of species became resistant during the first four decades of organic pesticide use. As we continue to use pesticides, we are continuing to breed insects and other pests that are increasingly resistant to the pesticides used against them.

Resurgences and Secondary Pest Outbreaks

The second problem with the use of synthetic organic pesticides is that after a pest outbreak has been virtually eliminated with a pesticide, the pest population not only recovers, it explodes to higher and more severe levels. This phenomenon is known as a **resurgence.** To make matters worse, populations of insects that were previously of no concern because of their low numbers suddenly start to explode, creating new problems. This phenomenon is called a **secondary pest outbreak.** For example, with the use of synthetic organic pesticides, mites have become a serious pest problem, and the number of serious pests on cotton has increased from 6 to 16.

At first, pesticide proponents denied that resurgences and secondary pest outbreaks had anything to do with the use of pesticides. However, careful investigations have shown otherwise (see In Per-

gain resistance to other unrelated pesticides even though it has not been exposed to the other chemicals (see In Perspective Box, p. 219).

IN PERSPECTIVE
The Ultimate Pest?

Fortunately, no weed growing anywhere meets all the criteria of the ultimate weed, illustrated in Fig. 10–1. If we were to use this "ultimate" idea to conjure up the ultimate insect pest, we might imagine the following characteristics. The ultimate insect pest would

1. attack a broad variety of plants and fruits,
2. be resistant to all of the usual pesticides,
3. be highly prolific and have a rapid life cycle,
4. lack natural predators and parasites, and
5. become a nuisance in other ways

(it would interfere with breathing, cover car windshields, and so forth).

Such a pest, if unleashed on agriculture, could cause millions of dollars in crop losses and devastate many farmers.

Unfortunately, recent news from California suggests that something very close to the ultimate pest might already be with us. The poinsettia whitefly is a tiny white insect that was probably introduced by accident from the Middle East and has become established in California. It has all the characteristics listed above and has been dubbed the Superbug by farmers who have en-

countered it. It is known to eat at least 500 species of plants—just about everything except asparagus and onions. It thrives on roadside weeds. Total crop losses in 1991 to this insect are above $90 million. The insects swarm all over the plants, sucking them dry and leaving them withered and rotten. Pesticides have proven useless—it is resistant to all usable pesticides in the state. Swarms become so dense that they interfere with breathing and vision. Entomologists are frantically searching for a natural predator or other pesticides to bring this pest under control. Before they do, we all may be eating more onions than ever before.

IN PERSPECTIVE
A Recipe for Red Scale

As the text mentions, pesticides can be more of a hindrance than a help in controlling crop damage. Entomologist Paul DeBach has gone so far as to demonstrate pesticide "recipes" for fostering pests. For example, here is his recipe for growing the California red scale insect on citrus trees.

"Take four pounds of 50 percent wettable DDT and mix into 100 gal-lons of water. Using a three-gallon garden sprayer, spray one or two quarts of this mixture lightly over the tree, repeating at monthly intervals until the tree is defoliated or dying as a result of the increase in red scale. This may require from about six to twelve applications, depending upon the degree of initial infestation; the higher the numbers [of insects] initially, the faster the [population] explosion occurs. Most growers do not like this sort of test and have usually insisted that we stop before trees were killed, but it is a wonderful way to raise red scale."

(Paul DeBach, *Biological Control by Natural Enemies*, p. 2. London: Cambridge University Press, 1974)

Insect Food Chains

FIGURE 10–5
(a) Food chains exist among insects just as they do among higher animals. (b) Aphid lion (on right) impaling and eating a larval aphid. (Courtesy of David Pimentel, Cornell University.)

(a)

(b)

spective, p. 220). Resurgences and secondary pest outbreaks occur because the insect world involves a complex food web. Populations of plant-eating insects are frequently held in check by other insects that are parasitic or predatory on them (Fig. 10–5). Pesticide treatments often have a greater impact on these natural enemies than on the plant-eating insects. Consequently, with natural enemies suppressed, both the population of the original target pest and populations of other plant-eating insects explode.

To illustrate the seriousness of resurgences and secondary pest outbreaks, a recent study in California surveyed a sequence of 25 major pest outbreaks, each of which caused more than $1 million worth of damage. All but one involved resurgences or secondary pest outbreaks. Of course, the species involved in secondary outbreaks quickly became resistant to pesticides, thus compounding the problem.

The chemical approach fails because it is contrary to basic ecological principles. It assumes that the ecosystem is a static entity in which one species, the pest, can simply be eliminated. In reality, the ecosystem is a dynamic system of interactions, and a chemical assault on one species will inevitably upset the system and produce other undesirable effects. The path to sustainability demands that we understand how ecosystems work and adapt our interventions accordingly.

Adverse Environmental and Human Health Effects

Perhaps of greatest concern to most people, however, is the potential for adverse effects to human and environmental health. The now-classic story of DDT, used so widely during the 1940s and 1950s, illustrates the hazards.

In the 1950s and 1960s, ornithologists (people who study birds) observed drastic declines in populations of many species of birds that feed at the top of the food chains. Fish-eating birds such as the bald eagle and osprey (Fig. 10–6) were so affected that extinction was feared. Investigators at the U.S. Fish and Wildlife National Research Center near Baltimore, Maryland, showed that the problem was reproductive failure; eggs were breaking in the nest before hatching. They also showed that the fragile eggs contained high concentrations of DDE, a product of the partial breakdown of DDT by the body. DDE interferes with calcium metabolism, causing birds to lay thin-shelled eggs. Further study revealed that birds were acquiring high levels of DDT and DDE by **biomagnification,** the process of accumulating higher and higher doses through the food chain (see p. 316).

In addition, tissue assays showed that DDT was accumulating in the body fat of humans and virtually all other animals, including Arctic seals and Antarctic

FIGURE 10–6
American osprey. Populations of fish-eating birds like the osprey, brown pelican, and bald eagle were decimated in the 1950s and 1960s by the effects of widespread spraying with DDT. With the banning of DDT, populations have greatly recovered. (Tom Blesoe/Photo Researchers.)

penguins even though those animals were far removed from any point of DDT application. While harmful effects on humans have not been substantiated, experimental testing has shown that compounds related to DDT are carcinogenic, mutagenic (cause mutations), and teratogenic (cause birth defects).

Concerns about environmental and long-term health effects led to the banning of DDT in the United States and most other industrialized countries in the early 1970s. Numerous other related chlorinated hydrocarbon pesticides (chlordane, dieldrin, endrin, lindane, heptachlor) were also banned because of their propensity for bioaccumulation in the environment and potential for causing cancer.

In the years since the banning of DDT, observers have noted a marked recovery in the populations of birds that were adversely affected. However, this does not mean that the situation is under control. Because of increasing resistance, resurgences, and secondary pest outbreaks, the kinds and quantities of pesticides in use continue to grow. Approximately 70 percent of all cropland in the United States receives some pesticide application. The total amount used yearly in the United States has declined somewhat since 1976, not because of decreased use but because newer pesticides are much more effective. Globally, some 4 billion pounds of pesticides are used annually, but this level may decrease as more and more farmers turn to biological controls.

Many pesticides in use are far more toxic to people than DDT and are responsible for poisoning an

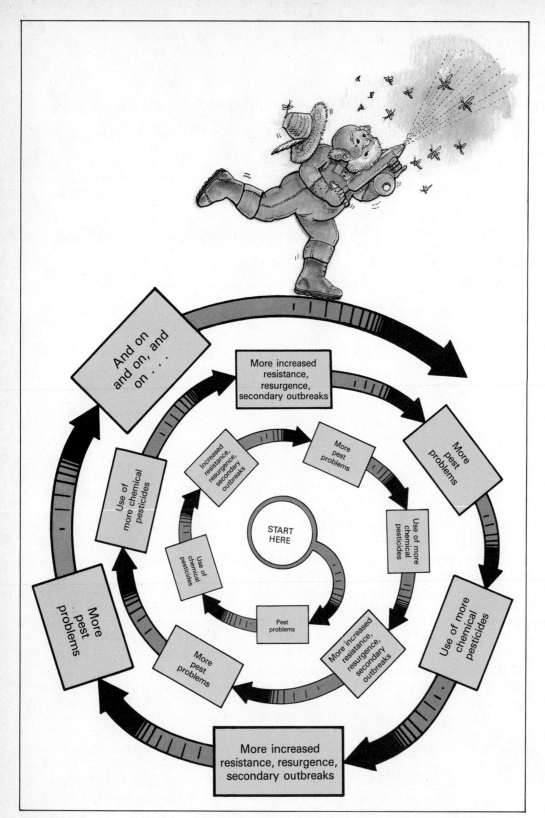

FIGURE 10–7
The pesticide treadmill. Use of chemical pesticides aggravates many pest problems.
Continued use demands ever-increasing dosages of pesticides, which further
aggravate pest problems and produce more contamination of foodstuffs and
ecosystems.

estimated 45 000 persons annually in the United States. Some 3000 of these victims require hospitalization, and approximately 50 of them die each year from pesticide toxicity. The World Health Organization conducted a survey to determine the extent of pesticide poisoning in humans and estimates that, worldwide, 500 000 poisonings occur each year, of which 5000 are fatal. Use by untrained persons is considered to be the major cause of these poisonings.

The Pesticide Treadmill

The late entomologist Robert van den Bosch coined the term **pesticide treadmill** to describe attempts to eradicate pests with synthetic organic chemicals. It is an apt term. The chemicals do not eradicate the pests. They increase resistance and secondary pest outbreaks, which lead to the use of new and larger quantities of chemicals, which in turn lead to more resistance and more secondary outbreaks, and so on. The process is an unending vicious cycle constantly increasing the risks to human and environmental health (Fig. 10–7). It is clearly not sustainable.

NONPERSISTENT PESTICIDES: ARE THEY THE ANSWER?

A key factor underlying the tendency of chlorinated hydrocarbons to contaminate the environment and to biomagnify is their persistence, that is, their slowness to break down. Because these chemicals have such complex structures, soil microbes are unable to metabolize them. DDT, for example, has a half-life on the order of 20 years—half the amount applied is still present and active 20 years later; after 20 more years, half of that amount, or one-quarter of the amount applied, remains, and so on.

Recognizing this persistence factor, the agrochemical industry has in large measure substituted *nonpersistent* pesticides for the banned compounds. For example, the synthetic organic phosphates malathion and parathion, as well as carbamates such as aldicarb and carbaryl, are now used extensively in place of the chlorinated hydrocarbons. These compounds break down into simple nontoxic products within a few days or weeks after application. Thus, there is no danger of their migrating long distances through the environment and affecting wildlife or humans long after being applied. For several reasons, however, nonpersistent pesticides are not as environmentally sound as their proponents claim.

First of all, total environmental impact is a function of persistence along with three other important factors: toxicity, dosage applied, and location where applied. Many of the nonpersistent pesticides are far more toxic than DDT. This higher toxicity, combined with the frequent applications needed to maintain control, presents a significant hazard to agricultural workers and others exposed to these pesticides.

Second, nonpersistent pesticides may still have far-reaching environmental impacts. For example, to control outbreaks of the spruce budworm in New Brunswick, Canada, forests were sprayed with a nonpersistent organophosphate pesticide that was promoted as environmentally safe. After spraying, however, an estimated 12 million birds died. These birds may have died by direct poisoning or by the loss of their food supply since a bird eats nearly its own weight in insects each day. In either case, visitors commented on the eerie silence and the numerous dead warblers littering the ground after spraying.

Third, desirable insects may be just as sensitive as pest insects to these substances. Bees, for example, which play an essential role in pollination, are highly sensitive to nonpersistent pesticides. Thus, use of these compounds creates an economic problem for beekeepers as well as jeopardizing pollination. Regular spraying of neighborhoods with malathion to control mosquitoes leads inevitably to great declines in butterflies and fireflies.

Finally, nonpersistent chemicals are just as likely to cause resurgences and secondary pest outbreaks as are persistent pesticides, and pests become resistant to nonpersistent chemicals just as quickly as they do to persistent pesticides.

Alternative Pest Control Methods

Numerous ecological and biological factors affect the relationship between a pest and its host (the plant or animal attacked). Ecological pest management seeks to manipulate one or more of these natural factors so that crops are protected without upsetting the rest of the ecosystem or jeopardizing environmental and human health. Since ecological pest management involves working with natural factors instead of synthetic chemicals, the techniques are referred to as **natural control** or **biological control** methods. This natural approach, unlike the chemical technology approach, depends on an understanding of the pest and its relationship with its host and with its ecosystem. The more we know about the organisms involved, the greater our opportunities for natural control.

To illustrate, the life cycle of moths and butterflies is shown in Fig. 10–8. Many groups of insects

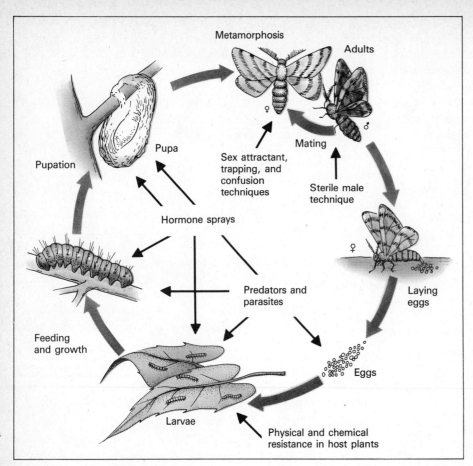

Metamorphosis

Adults

Pupa

Pupation

Mating

Sex attractant,
trapping, and
confusion
techniques

Sterile male
technique

Hormone sprays

Predators and
parasites

Laying
eggs

Feeding
and growth

Eggs

Larvae

Physical and chemical
resistance in host plants

FIGURE 10–8
Life cycle of moths and butterflies. Most insects have a complex life cycle that includes a larval stage and an adult stage. Biological control methods recognize the different stages and attack the insect, using knowledge of its needs and life cycle.

have a similarly complex life cycle. The development of each stage may be influenced by numerous abiotic factors, and at each stage the insect may be vulnerable to attack by a parasite or predator. Proper completion of each stage depends on internal chemical signals provided by hormones. Locating mates, finding food, and other behaviors depend on external chemical signals. All these findings suggest ways in which pest populations may be controlled without resorting to synthetic chemical pesticides.

The four general categories of natural or biological pest control are:

○ Cultural control
○ Control by natural enemies
○ Genetic control
○ Natural chemical control

CULTURAL CONTROL

A **cultural control** is a nonchemical alteration of one or more environmental factors in such a way that the pest finds the environment unsuitable or is unable to gain access to it.

Cultural Control of Pests Affecting Humans

We routinely practice many forms of cultural control against diseases and parasitic organisms. Some of these practices are so familiar and well entrenched in our culture that we no longer recognize them as such. For instance, proper disposal of sewage and the avoidance of drinking water from "unsafe" sources are cultural practices that protect against waterborne disease-causing organisms. Combing and brushing the hair, bathing, and wearing clean clothing are cultural practices that eliminate head and body lice, fleas, and other parasites. Regular changing of bed linens protects against bedbugs. Proper and systematic disposal of garbage and keeping a clean, tight house with good screens are effective methods for keeping down populations of roaches, mice, flies, mosquitoes, and other pests. Sanitation requirements in handling and preparing food are cultural controls designed to prevent the spread of diseases. Refrigeration, freezing, canning, and drying of foods are cultural controls that inhibit the growth of organisms that cause rotting, spoilage, and food poisoning.

If such practices of personal hygiene and sanitation are compromised, as they may be in any major disaster situation, there is very real danger of additional widespread mortality resulting from outbreaks of parasites and diseases. Also, where these practices are not broadly pursued in a society, as in some less-developed countries, sickness and death are the consequences. For example, it is estimated that 80 percent of human illness in less-developed countries results from contaminated drinking water.

Cultural Control of Pests Affecting Lawns, Gardens, and Crops

Homeowners are prone to use excessive amounts of pesticides to maintain a weed-free lawn. Weed problems in lawns are frequently a result of cutting the grass too short. If grass is not cut to less than 3 in high, it will usually maintain a dense enough cover to keep out crabgrass and many other noxious weeds. Thus, monitoring grass height is a form of cultural weed control. Many homeowners allow plant diversity in their lawns, tolerating a variety of plants some people might consider weeds.

Some plants are particularly attractive to certain pests; others are especially repugnant. In each case, the effect may spill over to adjacent plants. A gardener may control many pests by paying careful attention to eliminating plants that act as attractants and growing those that act as repellents. Marigolds and chrysanthemums are justly famous for being insect repellents.

Some parasites require an alternate host. They can be controlled by eliminating the alternate host (Fig. 10–9). Hedgerows, fencerows, and shelterbelts can provide refuges where natural enemies of pests (birds, amphibians, mantises, and so on) can be maintained.

Management of any crop residues not harvested is important. Spores of plant disease organisms and insects may overwinter or complete part of their life cycle in the dead leaves, stems, or other plant residues that remain in the fields after harvest. Plowing under or burning the material may be very effective in keeping pest populations to a minimum. In gardens, a clean mulch of material such as grass clippings or hay will keep down weed growth and protect the soil from drying and erosion.

Growing the same crop on a plot of land year after year keeps the pest's food supply continuously available. Crop rotation, the practice of changing crops from one year to the next, may provide control because pests for the first crop cannot feed on the second crop and vice versa. Crop rotation is especially effective in controlling root nematodes (roundworms that live in the soil and feed on roots) and other pests that do not have the ability to migrate appreciable distances.

For economic efficiency, agriculture has moved

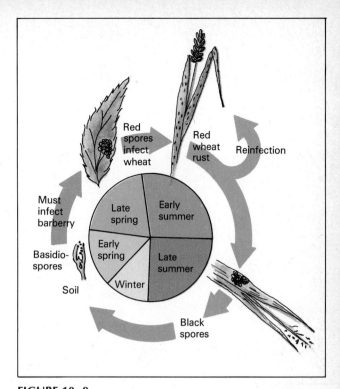

FIGURE 10–9
The life cycle of wheat rust, a parasitic fungus that is a serious pest on wheat. Since part of the life cycle requires that the rust infest barberry, an alternate host, the elimination of barberry in wheat-growing regions has been an important cultural control (along with the development of resistant varieties of wheat).

progressively toward monoculture—the Corn Belt, Cotton Belt, and so on. Recall the fourth principle of sustainability: diversity provides stability. (Conversely, a simple system is ecologically unstable.) When a pest outbreak occurs, monoculture is most conducive to its rapid multiplication and spread. Other natural controls, even if present, may be overwhelmed by the avalanche of spreading pests. On the other hand, spread of a pest outbreak is impeded and other natural controls may be more effective if there is a mixture of species, some of which are not vulnerable to attack. Indeed, much of our prodigious use of pesticides may be seen as a desperate attempt to maintain unstable monocultures, and today agriculturalists are experimenting with various systems of polyculture.

Most of our pests that are hardest to control are species that have been unwittingly imported from other parts of the world, and we realize that many species existing in other regions would be serious pests if introduced here. Therefore emphasis is placed on keeping would-be pests out. This is a major function of the customs office. Biological materials that may carry pest insects or pathogens are either prohibited from entering the country or are subjected to quarantines, fumigation, or other treatments to ensure that they are free of pests. The cost of such procedures is small in comparison with the costs that could be incurred if such pests gained entry and became established.

CONTROL BY NATURAL ENEMIES

The following examples illustrate the range of possibilities for controlling pests with natural enemies:

○ Scale insects, potentially devastating to citrus crops, have been successfully controlled by vedalia (ladybird) beetles, which feed on them.

○ Various caterpillars have been controlled by parasitic wasps (Fig. 10–10).

○ Japanese beetles are controlled in part with the bacterium *Bacillus thuringiensis,* which produces toxic crystals when soil containing the bacterium is ingested by the larvae.

○ Prickly pear cactus and numerous other weeds

FIGURE 10–10
Parasitic wasps used to control caterpillars. (a) The life cycle of the parasitic wasp that uses the gypsy moth as its host. (b) A wasp depositing eggs in a gypsy moth pupa. (c) Another insect parasite, a braconid wasp, lays its eggs on the pest known as the tomato hornworm, which is the larva of the hawk moth. The wasp larvae feed on the caterpillar, and shortly before the caterpillar dies they emerge and form the cocoons seen here. (Scott Camazine/Photo Researchers.)

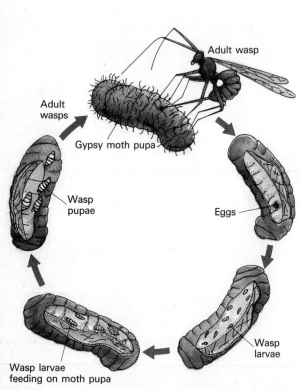

(a)

(b)

(c)

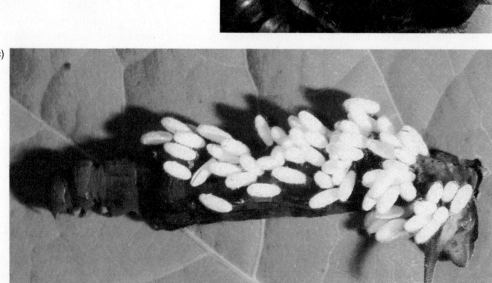

IN PERSPECTIVE
Wasps 1, Mealybugs 0

The cassava (manioc) plant originated in South America but has been cultivated throughout the tropical world. Currently, it is the primary food for more than 200 million people in sub-Saharan Africa, one-third of the human population in that region. It is a high-yielding crop that requires no high technology, one that is raised by subsistence farmers everywhere.

An insect not previously seen in Africa, the mealybug, appeared in Congo in the early 1970s and spread across sub-Saharan Africa, leaving a trail of ruined harvests and hunger in its wake. Zaire and Congo, unable to afford pesticides, turned for help to the International Institute for Tropical Agriculture in Nigeria.

This group formed the Biological Control Program and began to look for natural enemies of the mealybug.

Returning to the land of origin of cassava, a researcher found the mealybug in Paraguay and observed that the insect was kept under control by natural predators and parasites. After extensive testing, researchers identified a parasitic wasp, *Epidinocarsis lopezi*, as the prime candidate for controlling the bugs. The female wasp seeks out mealybugs, paralyzes them with a sting, and deposits eggs that hatch inside the mealybug and eat their way through the insect.

Once this wasp-bug relationship was identified, the wasps were reared by the thousands on captive mealybugs and then spread across the continent over a period of 8 years. The Biological Control Program has trained 400 workers to monitor control, and all indications suggest that the battle has been won. It is estimated that every dollar invested in the control program has yielded $149 in crops saved from destruction. The best news is that the control is permanent and does not require the repeated application of expensive and environmentally damaging pesticides.

(Reported by Peter Weber, *World Watch*, March–April, 1991.)

FIGURE 10–11
Biological control of the weed prickly pear cactus by a cactus-eating moth in Queensland, Australia. (a) The land shown here, once settled, had to be abandoned because of prickly pear infestation. (b) Same land reoccupied following destruction of prickly pear by the moth. ((a) Queensland Department of Primary Industries; (b) Australian Information Service.)

(a)

(b)

have been controlled by plant-eating insects (Fig. 10–11). More than 30 weed species worldwide are now limited by introduced insects.

○ Rabbits in Australia are controlled by an infectious virus.

○ Aquatic weeds in the American Midwest are controlled by a vegetarian fish native to South America.

○ Mealybugs in Africa are controlled by a parasitic wasp (see In Perspective, p. 227).

The problem in using natural enemies is in finding organisms that provide control of the target species without attacking other, desirable, species. Entomologists estimate that, of the 50 000 known species of plant-eating insects that have the potential of being serious pests, only about 1 percent actually are. The populations of the other 99 percent are held in check by one or more natural enemies so that they do not do significant amounts of damage. Therefore, the first step in using natural enemies for control should be *conservation*, protecting the natural enemies that already exist. Conservation means avoiding the use of broad-spectrum chemical pesticides, which may affect natural enemies even more than the target pests. Elimination or considerable restriction of the use of broad-spectrum chemical pesticides will, in many cases, allow natural enemies to reestablish themselves and control secondary pests, those that became serious problems only after the use of pesticides.

However, effective natural enemies are not always readily available. In some cases, the lack of natural enemies is the result of accidentally importing the pest without its natural enemy. Quite often, effective natural enemies have been found by systematically combing the home region of an introduced pest and finding its various predators or parasites. The great advantage of this approach is the specificity of the natural enemy for its target. Of more than 100 insects introduced to new regions for weed management, none have switched their diet.

Yet, the potential for utilizing natural enemies has barely been tapped. Of more than 2000 serious insect pest species, 90 percent remain for which effective natural enemies have not been found mainly because no one has looked for such enemies. Finding suitable natural enemies is costly, and the profit margin for this kind of work is not as great as it is for pesticides. Fortunately, the U.S. Department of Agriculture is increasing its funding of research into biological control agents. After a potential natural enemy is located, it must be propagated and carefully tested before it is released, to be sure that it will not harm other organisms. This testing often takes many years of painstaking research. When effective natural enemies are found and deployed, however, they may provide control indefinitely, saving millions of dollars per year without further inputs.

GENETIC CONTROL

Most plant-eating insects and plant pathogens attack only one species or a few closely related species. This specificity implies a genetic incompatibility between the pest and species that are not attacked. The essence of genetic control is to develop genetic traits in the host species that provide the same incompatibility, that is, resistance to attack. This technique has been extensively utilized in connection with plant "diseases"—fungal, viral, and bacterial parasites. For example, in the years 1845–1847 the potato crop in Ireland was devastated by late blight, a fungal parasite. Nearly a million people starved, and another million people emigrated to escape the same fate. Now protection against such disasters is provided in large part

by growing potato varieties that are resistant to the blight. It is no overstatement to say that the world owes much of its production of potatoes, corn, wheat, and other cereal grains to the painstaking work of plant geneticists who selected and bred disease-resistant varieties.

The same potential exists for breeding plants that are resistant to insect pests. Traits that provide resistance may be categorized in two groups: chemical barriers and physical barriers.

A chemical barrier is some chemical produced by the plant we want to protect that is lethal or at least repulsive to the would-be pest. Once they have identified such a barrier, plant breeders are then able, through selection and cross-breeding, to enhance this trait in the desirable plant. The relationship between wheat and the Hessian fly provides an example. This fly lays its eggs on wheat leaves, and the larvae move down the leaves and into the main stem as they feed. The weakened stem either dies or is broken in the wind. The Hessian fly was introduced into the United States in the straw bedding of Hessian soldiers during the Revolutionary War. The fly eventually spread throughout much of the Midwest, causing widespread devastation until scientists at the University of Kansas developed a variety of wheat that causes the larvae to die as they feed on the leaves.

Increasing resistance through breeding may not provide 100 percent protection, but even partial protection can make the difference between profit and loss for the grower. In addition, any degree of resistance lessens the need for chemical pesticides.

Physical barriers are structural traits that impede the attack of a pest. For example, leafhoppers are significant worldwide pests of cotton, soybeans, alfalfa, clover, beans, and potatoes, but they can damage only plants with relatively smooth leaves. Hooked hairs on the leaf surfaces of some plants tend to trap and hold immature leafhoppers until they die (Fig. 10–12). Similarly, alfalfa weevil larvae are fatally entrapped by glandular hairs that exude a sticky substance. Such traits can be enhanced through selective breeding.

Unfortunately, pests may develop the ability to overcome genetic controls—both chemical and physical—in the same way they develop resistance to pesticides. This means that breeders must continually develop new resistant varieties to substitute for old varieties. This substitution process has occurred seven times in the case of wheat and the Hessian fly. Such substitution often takes place without the public ever knowing that a potential catastrophe is being averted. Note again how important it is to maintain biological diversity to make this substitution possible.

Another genetic control strategy involves flooding a natural population with sterile males that have

sential features of screwworm flies: (1) Their populations are never very large, and (2) the female fly mates just once, lays her eggs, then dies. Knipling reasoned that if the female mated with a sterile male, no offspring would be produced. His hypothesis was correct, and today sterile males are routinely used to control this pest. Huge numbers of screwworm larvae are grown on meat in laboratories. The pupae are then subjected to just enough high-energy radiation to render them sterile. These sterilized pupae are then air-dropped into the infested area. Ideally, 100 sterile males are dropped for every normal female in the natural population, giving a 99 percent probability that wild females will mate with one of the sterile males.

This technique successfully eliminated the screwworm fly from Florida in 1958–1959 and continues to be used to control the problem in the Southwest. The savings to the cattle industry is estimated at more than $300 million a year. The technique has also been used to eradicate infestations of imported pests before they gained a strong foothold. Populations of such insects are maintained in facilities around the world so that sterile males may be called up on very short notice if the need arises. As a recent example, 1.3 billion sterilized males, released by U.N. Food and Agriculture Organization planes into Libya in 1991, successfully eradicated screwworm flies from that region. If the fly had moved south of the Sahara, it could have had devastating effects on native animals and livestock, threatening food supplies in that region.

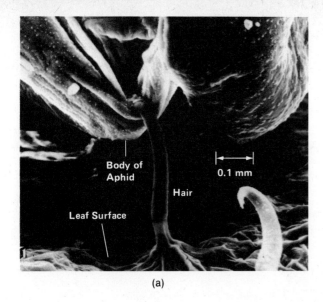

(a)

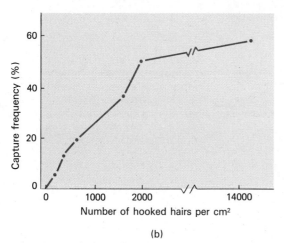

(b)

FIGURE 10–12
Genetic control by development of physical barriers. (a) The leafhopper is controlled by hairs on the leaf that hook immature leafhoppers. These hairs are a genetic trait of the plant. (b) Relationship between the number of hairs and the frequency of capture. ((a) Courtesy of E.A. Pillemar and W.M. Tingey, New York State Agricultural Experiment Station, Ithaca, N.Y. Science 193 (August 6, 1976), 482–84. © 1976 by the American Association for the Advancement of Science.)

been reared in laboratories. Combating the screwworm fly provides a prime illustration. This fly, closely related and similar to the housefly, lays its eggs in open wounds of cattle and other animals. The larvae feed on blood and lymph, keeping the wound open and festering. Secondary infections frequently occur and often lead to the death of the animal.

Before 1940, this problem became so severe that cattle ranching from Texas to Florida and northward was becoming economically impossible. In studying the situation, Edward Knipling, an entomologist with the U.S. Department of Agriculture, observed two es-

NATURAL CHEMICAL CONTROL

As in humans and other animals, each stage in insect development is controlled by **hormones,** chemicals that are produced in the organism and provide "signals" that control developmental processes and metabolic functions. In addition, insects produce many **pheromones,** chemicals secreted by one individual that influence the behavior of another individual of the same species.

The aim of **natural chemical control** is to isolate, identify, synthesize, and then use an insect's own hormones or pheromones to disrupt its life cycle. Advantages of natural chemicals are: (1) They are highly specific to the pest in question (they do not affect natural enemies to any appreciable extent), and (2) they are nontoxic. When eaten by another organism they are simply digested. Ways in which natural chemicals may be used are illustrated by the following.

Scientists have discovered that caterpillar pupation is triggered by a decrease in the level of the chemical called *juvenile hormone*. If this chemical is sprayed on caterpillars, pupation does not occur. The larvae simply continue to feed and grow, become grossly oversized, and eventually die. Hormone sprays are now being used as one means of controlling gypsy moths in some urban areas.

Adult female insects secrete pheromones that attract males for the function of mating. Once identified and synthesized, these **sex-attractant** pheromones may be used in either of two ways: (1) the

trapping technique or (2) the **confusion technique.** In the trapping technique, the pheromone is used to lure males into traps or poisoned bait (Fig. 10–13). In the confusion technique, the pheromone is dispersed over the field in such quantities that males become confused, cannot find the females, and thus fail to mate.

The theoretical potential of natural chemicals for controlling insect pests without causing ecological damage or upsets is enormous and has been recog-

FIGURE 10–13
The trapping technique, a form of natural chemical control. Control of pink bollworms using their sex-attractant pheromone. The pheromone synthesized in the laboratory is packaged in tiny fibers, a microscopic close-up of which is shown in (a). The outside of the fiber is treated with a minute amount of insecticide. A single handful of fibers (b) is enough to treat two acres. The male bollworm moths are attracted to the fibers when pheromones evaporate from the open end (c). In attempting to mate with the fiber, the moth contacts the insecticide and dies. (Courtesy of Jack Jenkins, Scentry Inc., Buckeye, Arizona.)

(a)
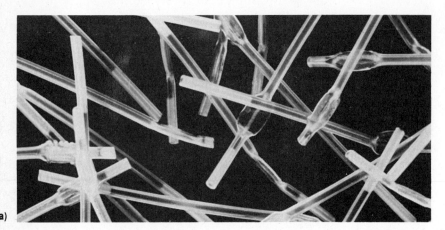

(b)
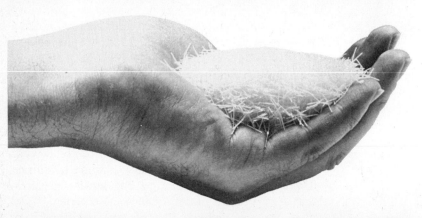

(c)

nized for at least 30 years. However, the background research and necessary testing are just now reaching fruition. Tests are extremely promising, and it appears that the great expectations in this field are about to be realized as over 800 natural chemicals have been identified and more than 250 are being produced commercially.

Socioeconomic Issues in Pest Management

With any method of pest control, it is important to keep in mind that a species becomes a pest only when its population multiplies to the point of causing significant damage. Natural controls are generally aimed at keeping pest populations below damaging levels, not at total eradication. By keeping pest populations down, natural controls avert significant damage while preserving the integrity and balance of the ecosystem.

Therefore, the question to be asked in facing any pest species is: Is it causing significant damage? Damage should be deemed significant only when the cost of the damage considerably outweighs the cost of a pesticide application. This point is called the **economic threshold** (Fig. 10–14). If significant damage is not occurring, natural controls are already operating, and the situation is probably best left as is. Spraying with synthetic chemicals at this stage is more than likely to upset the natural balance and make the situation worse through resurgences and upsets. On the other hand, if significant damage *is* occurring, a pesticide treatment may be in order.

It makes little difference whether the threat of loss as a result of pest infestation is real or imagined, close at hand or remote; what is important is how the grower perceives the threat. Even if there is no evidence of immediate pest damage, a grower who believes his or her plantings are at risk is likely to succumb to **insurance spraying,** the use of pesticides "just to be safe." Few such insurance treatments actually prove necessary, but they all have negative impacts on the environment.

Consumers also put pressure on growers to use pesticides. From customers to supermarket chains to canneries, there is a tendency to select the best-looking fruits and vegetables, leaving the remainder to be sold at lower prices or trashed. Like Snow White, we all tend to reach for the perfect, unblemished apple. Growers know that blemished produce means less profit, and so they indulge in **cosmetic spraying**—the use of pesticides to control pests that harm only the item's outward appearance. Cosmetic spraying accounts for a significant fraction of pesticide use, does

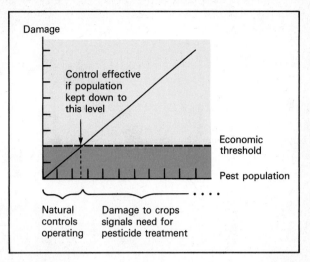

FIGURE 10–14
The economic threshold. The objective of pest control should not be to eradicate a pest totally. All that is needed is to keep population levels below the economic threshold. When the population rises into the yellow part of the graph, damage to crops is significant and pesticides are needed. Any time the population is in the blue area, however, natural controls are succeeding in keeping the pest population down; therefore no pesticides are needed.

nothing to increase yield or nutritional value, and results in an increase in pesticide residues remaining on the produce.

Unfortunately, the chemical companies that market the pesticides exploit these concerns of the growers in order to increase their profits. Through advertising and paid-by-commission field representatives, pesticide producers attempt to convince growers that the threat of pests is much greater than it is and that spraying pesticides is good insurance. Likewise, they emphasize the enhanced cosmetic quality that can be obtained with pesticides. The fact that unnecessary treatments aggravate pest problems only serves to benefit the chemical companies by creating a market for larger quantities and numbers of pesticides.

ORGANICALLY GROWN FOOD

There is strong public feeling against the use of chemical pesticides. Cosmetic or other unnecessary spraying persists in large part because consumers are kept ignorant about the kinds or amounts of pesticides used. Experience shows that when consumers are informed, many of them abandon the Snow White attitude. Food outlets selling **organically grown** produce (produce that is grown without synthetic chemical pesticides or fertilizers) are thriving, although the

FIGURE 10–15
A roadside farm stand in Concord, Massachusetts, specializes in organic produce.
Organically grown foods represent an expanding market in the United States.
(Timothy Lucas/Tony Stone Worldwide.)

produce is not as cosmetically perfect and may cost more than chemically treated produce (Fig. 10–15). As a result, many growers now find it more profitable to rely on natural controls and sell through these specialized markets. Only 2 percent of the U.S. farm production is organic, but the movement is growing rapidly.

For example, Purepak is a company in California that grows, packs, and ships only organically grown vegetables. Once Purepak obtained their state-administered organic certification (an essential procedure now available in most states), their business took a sharp rise. Growers for Purepak use machine and hand cultivation instead of herbicides to keep weeds under control. Or, if an insect infestation appears, they spray soap instead of insecticides and introduce ladybugs into the fields to prey on the insects. Purepak officials admit that organic farming costs more, but as more and more growers join the movement, the costs will likely decline.

INTEGRATED PEST MANAGEMENT

The approach known as **integrated pest management** (IPM) aims to minimize use of synthetic organic pesticides without jeopardizing crop protection through addressing all the sociological, economic, and ecological factors involved. IPM is not a technique in and of itself. Rather, it is an approach that integrates many techniques (Fig. 10–16).

Making the economic benefits of natural controls known to growers is an important aspect of integrated pest management. Because pesticides increased yields and profits when they were first used, many farmers still cling to them, believing that they offer the only way to bring in a profitable crop. In some instances, however, the rising costs of pesticides, along with their tendency to aggravate pest problems, have eliminated their economic advantage.

The Center for the Biology of Natural Systems at Washington University compared 16 "organic" farms with 16 similar "conventional" farms in the Corn Belt. Organic farms relied on crop rotation and other cultural techniques to control pests; conventional farms used pesticides. Crop production per acre was virtually identical in the two cases. The total value of crops produced was slightly higher on conventional farms because pesticides permitted monocropping and hence a large percentage of the land was in high-value corn. However, this gain was offset by the cost of pesticides and fertilizer so that the economic return to the farmers was not significantly different. Productivity of crops on Amish farms, where

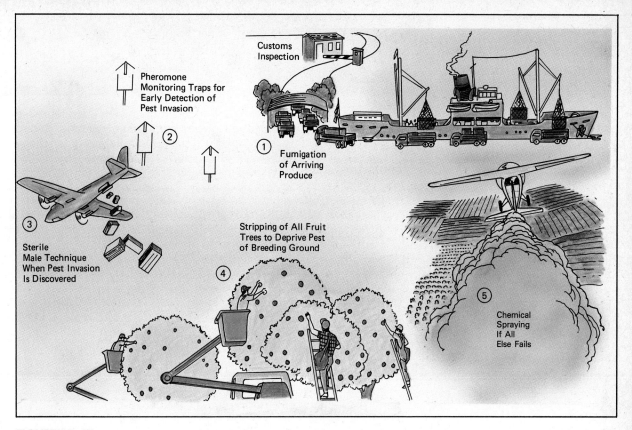

FIGURE 10–16
Integrated pest management involves the integration of a number of techniques, none of which would be completely effective by itself. The illustration depicts the techniques, or "lines of defense," used to keep the Mediterranean fruit fly out of California.

pesticides have never been used because of religious beliefs, is as high as or higher than that of farms where pesticides are used.

Devastating pest infestations on organic farms generally occur only when certain weather and crop conditions coincide, making circumstances particularly favorable for the pests to reproduce. At other times, pest populations usually remain below the economic threshold.

Most growers are not trained entomologists and cannot be expected to recognize all the conditions that make a pest outbreak a real rather than an imagined threat. Thus, they rely on information from other people as to whether a pesticide application is necessary or not. Through the 1970s, at least 90 percent of such information that growers received was from pesticide producers or marketers. You can guess the bias of the information.

This situation is being remedied by *field scouts* employed by local agricultural extension services, farm cooperatives, or acting as independent consultants. **Field scouts** are persons trained in monitoring pest populations (traps baited with pheromones are

used for this purpose) and other factors and in determining whether the pest population is exceeding the economic threshold. Many growers have cut pesticide use dramatically without crop loss by using information from field scouts. In addition, **pest-loss insurance,** which pays the farmer in the event of loss due to pests, has enabled growers to refrain from unnecessary insurance spraying.

While the aim of IPM is to reduce reliance on chemical pesticides, such pesticides still remain an integral part of the whole approach. Indeed, some entomologists feel that IPM is overly reliant on the use of synthetic organic pesticides and is thus still contributing to driving the pesticide treadmill. These entomologists advocate renewed emphasis on the study of pest ecology and the development and implementation of natural controls—in short, a renewed emphasis on the concepts of ecological pest management. This is the logical pathway to a sustainable future, where pesticides are not polluting groundwater, contaminating food, killing pollinating insects, and in the end, creating new and more-resistant pests.

The U.S. program to protect domestic crops against the Mediterranean fruit fly, or "Medfly," exemplifies how IPM can work. It also illustrates an ethical dilemma, where the state is caught between growers who want aerial spraying to control the Medfly and citizens who strongly protest having the whole Los Angeles basin sprayed with pesticides.

Unlike the common fruit fly, which is attracted only to very ripe fruit, the Medfly lays its eggs on unripe fruits and vegetables in the field. The feeding maggots thus cause extensive damage before harvest and during storage and transport. Worldwide, this insect is one of the most destructive pests known. If it became established in the United States, it would present an enormous economic threat.

Medflies probably brought in on imported produce have invaded Florida, Texas, and California on several occasions, and whenever they have been detected, an intensive eradication campaign has been mounted. This diligence seems to have been successful in Florida and Texas, but there are signs that the Medfly may now be established in California, leading to increasingly frequent but controversial sprayings. The 1990 appearance of Medflies in the Los Angeles basin prompted an aerial spraying of the pesticide malathion, a spraying bitterly opposed by residents of the area.

Malathion spraying represents the last resort in a battery of integrated pest management techniques. First, all imported produce is checked and then fumigated if officials detect any trace of the insect. Second, a network of traps baited with a sex attractant is maintained throughout the agricultural area. Monitoring these traps provides an early warning of the presence of Medflies that have slipped through customs. If Medflies are found in the traps, the sterile male technique is called on as the third line of defense. Stocks of sterile Medflies are maintained in South America, and batches of them can be delivered on short notice and dropped over the infected area. In addition, fruit is stripped from the trees in the infected area to prevent the Medfly from reproducing. Officials continue to use pheromone-baited traps to monitor the program's success. The final line of defense is aerial spraying.

Entomologist James Carey has presented convincing evidence that the Medfly is now established in California and argues that strategies other than aerial spraying are needed to keep it from becoming a major pest in the state. In particular, more research is needed to study the basic biology of the Medfly so that effective natural controls might be devised. Although malathion is not very toxic to humans, Los Angeles residents feel that aerial spraying is an unacceptable alternative for controlling the Medfly. They argue that the benefits to the growers are not outweighed by the potential harm to the environment and to humans. As a consequence, the control program has become a very hot political issue in California. How do you feel about it?

Public Policy

IMPORTANT LEGISLATION

The key legislation to control pesticides in the United States is the Federal Insecticide, Fungicide, and Rodenticide Act, commonly known as FIFRA. This law, administered by the EPA, requires manufacturers to register pesticides with the government before marketing them. The registration procedure involves testing to determine toxicity to animals (and, by extrapolation, to humans). From the test results, usage standards are set. For example, highly toxic compounds such as chlordane, which is used to control termites, are not authorized for use on crops. If health hazards appear after a substance has been registered and marketed, the act provides for "deregistration," whereby the pesticide may be banned from one or more uses.

Another law, the Federal Food, Drug and Cosmetic Act, requires the EPA to set standards regarding the "safe" amounts of pesticide residues that may be left on food to be eaten by animals and humans. Many foods have been withdrawn from the market because residues of certain pesticides were above the established standard.

SHORTCOMINGS

Unfortunately FIFRA has many shortcomings. The four main shortcomings are inadequate testing, case-by-case bans, pesticide exports, and lack of public input. Let us look at each one.

Inadequate Testing

The perils of biomagnification and the potential for long-term exposure that may cause cancer, birth defects, mutations, and other physiological disorders were not fully appreciated until the late 1960s, when bitter experience with DDT and other chemicals came to the forefront. This experience made officials and the public aware that many pesticides in use have never been adequately tested. Consequently, Congress amended FIFRA in 1972 to require the EPA to reevaluate and reregister all pesticide products then on the market.

When the amendment was adopted, there were already some 1400 chemicals and 40000 formulations (mixtures of different chemicals) on store shelves. In addition, there was and still is tremendous pressure from the chemical industry to register numerous new pesticides. Needless to say, the EPA's office of pesticide programs has been overwhelmed. Ongoing budget restrictions have not helped the situation.

Since the law allows existing pesticides to remain in use unless proven hazardous, many products on the market still have not been subjected to adequate tests, and existing standards may allow hazardous levels of pesticide residues in food. In 1989, the Natural Resources Defense Council (NRDC) published the results of a 3-year study from which they concluded that many fruits and vegetables routinely contain levels of pesticide residues that, although within prescribed standards, "pose an increased risk of cancer, neurobehavioral damage, and other health problems" to children (Fig. 10-17). The focus is on children both because they typically consume fruits and vegetables at a significantly greater rate than adults and because studies have shown that the young are frequently more susceptible than adults to carcinogens and neurotoxins. The principle culprit was the pesticide Alar, used on apples. While the government disputed the NRDC's assessment, it acknowledged that the risk to children had not been previously quantified in a comprehensive manner. In the end, the EPA moved to bring cancellation proceedings against Alar.

Bans Issued Only on a Case-by-Case Basis when Threats Are Proven

Each pesticide must be shown to pose a threat before it is subject to a ban. As the Alar case indicated, however, assessing the risk of a substance is a difficult and often controversial matter. Sometimes proof of danger is so long in coming that the damage has already been done by the time the ban is put into effect. The story of EDB (ethylene dibromide) illustrates this point.

It has been known since the mid-1970s that EDB causes cancer, birth defects, and other illnesses in laboratory animals. Yet its widespread use as a soil fumigant to control root nematodes (small worms that attack roots) continued. By 1982, 4.5 million pounds

FIGURE 10–17
Because of possible impacts on small children, actress Meryl Streep spearheaded a protest against the use of Alar to spray apples. The publicity forced EPA and applegrowers to stop using Alar. (UPI/Bettmann Newsphotos.)

of EDB were being pumped into the soil each year. In addition, EDB was increasingly used to fumigate and protect grains and other crops in storage. It could have easily been predicted that this practice would eventually result in groundwater contamination. But EDB was not banned until 1984, after traces were found in hundreds of wells in Florida, California, Massachussetts, and other regions, and unacceptable residue levels were found in flour and other food products.

Clearly, waiting until after the consequences to ban such chemicals is hardly a prudent way to protect public health. Yet, this is precisely how the system operates. A few persistent halogenated hydrocarbon pesticides remain in use, and the use of some, notably lindane, is increasing. Perhaps most serious is the widespread and increasing use of herbicides. They now account for more than half of all pesticides used.

Large numbers of people are being exposed to these compounds with unknown effects.

Pesticide Exports

Another ominous loophole in FIFRA is the lack of a requirement for registering pesticides intended for export. The United States currently exports more than 200 000 metric tons of these chemicals to less-developed countries each year. Some 25 percent of this total consists of products banned in the United States. Chemical companies are free to promote their products in other countries; thus, less-developed countries hear only how pesticides will solve their pest problems. These countries are not informed about resurgences or secondary pest outbreaks, much less about hazards to human and environmental health. Ironically, a large portion of these pesticides are used by foreign countries on export crops, many of which are imported by the United States.

Lack of Public Input

FIFRA provides no mechanism for public input. Therefore, the legislation primarily reflects the chemical industry's interests, which are conveyed through intense lobbying efforts as opposed to public forums. The bias is obvious when one considers that the law imposes a $1000 penalty on those who misuse a pesticide and a $10 000 penalty on those who reveal a trade secret about a pesticide's formulation.

NEW POLICY NEEDS

The shortcomings of FIFRA are obvious. The EPA is aware of the problems and is engaged in revising its regulatory rules to address many of them. On EPA's agenda are the following: (1) implementing a strategy to protect groundwater from infiltration by pesticides, (2) improving methods for ensuring food safety, (3) determining the extent of exposure to home and garden pesticides, (4) expanding worker protection standards to ensure the safety of all pesticide handlers, (5) expanding the requirements for testing to include impacts on natural environments.

Some of the new policy needs will require new legislation. A "citizen suit" provision should be provided under FIFRA, so that the public can go to court to force the government to uphold the law. This provision has given EPA added clout where it has been made part of other federal environmental statutes. The amendment giving the Secretary of Agriculture the opportunity for special review before any changes in EPA regulations should be stricken. This amendment now provides the often pro-pesticide agricultural interests with powerful political pressure and can be seen as a serious deterrent to the rigorous protection of humans and the environment by the EPA. Integrated pest management and other elements of ecological pest management strategy need to be given more encouragement in the form of research support and extension support. One group of leading entomologists has called for a 50 percent reduction in U.S. pesticide use, supporting their recommendation with the economic and environmental benefits of such a reduction. This goal could become public policy if included in legislative or regulatory law.

As with so many of the issues we consider in this book, significant progress in pesticide control will very likely depend on grassroots action and pressure from public interest groups and nongovernmental organizations. Such groups are pressing for a continued movement in the direction of ecological pest management, the path to sustainability. The simple monoculture crop system and the chemical approach keep farmers on the pesticide treadmill and guarantee that our food will continue to contain pesticide residues. All of us will be better off if we can kick the pesticide habit!

 Review Questions

1. Into what six basic categories can pests be grouped?
2. Discuss the two basic philosophies of pest control.
3. What were the apparent virtues of the synthetic organic pesticide DDT?
4. Describe four possible problems that might arise when pests are sprayed repeatedly with one pesticide.
5. What adverse environmental and human health effects might occur as a result of pesticide use?
6. Why are nonpersistent pesticides not as environmentally sound as first thought?
7. Describe the life cycle of an insect. What natural control methods can be applied at each stage?
8. Describe the four categories of biological control, and give an example of each.
9. What economic factors could influence a farmer's decision to continue extensive use of pesticides?
10. Describe how integrated pest management works. Give an example of how several natural

controls may be combined in a pest control program of this type.

11. Explain the shortcomings of FIFRA.

12. What amendments to public policy would promote pest control that is environmentally safer than the methods used today?

🌐 *Thinking Environmentally*

1. U.S. companies export pesticides that have been banned or restricted in this country. Should this practice be allowed to continue? Defend your answer.

2. Japanese beetle traps, which contain a chemical attractant, are a common sight in many neighborhoods today. Explain how they might work.

3. Almost one-third of the chemical pesticides bought in the United States are for use in houses and on gardens and lawns. What should be done by manufacturers and users to ensure their limited and prudent use?

4. Should the government give farmers economic incentives to switch from pesticide use to integrated pest management? Why or why not?

11

Water, the Water Cycle, and Water Management

OUTLINE

LEARNING OBJECTIVES

When you have finished studying this chapter, you should be able to:

1. Distinguish between evaporation, condensation, and precipitation in terms of energy changes and bonding of water molecules.

2. Describe the water cycle, including alternative pathways over and through the ground and explain why and how water quality may differ at different stages of the cycle.

3. Define runoff, infiltration, surface water, capillary water, gravitational water, percolation, groundwater, leaching, water table, aquifer, recharge area, spring, seep.

4. Describe why different regions receive different amounts of precipitation.

5. Identify major uses of water and how humans obtain water supplies.

6. List and discuss the civic and ecological consequences of overdrawing both surface and groundwater supplies.

7. Describe and evaluate the options for expanding water supplies.

8. Describe and evaluate various options for reducing the amount of water used.

9. Describe and evaluate the potential for reusing water.

10. Analyze how the quantity and quality of water in various parts of the water cycle change as a result of development.

11. Describe and evaluate various methods of stormwater management.

12. Evaluate the potential for sustainability of various water-use policies.

Water is essential for life. It is a universal solvent that carries all nutrients into cells. Within living cells, all metabolic functions occur in a water medium. Then water carries wastes out of the cells and out of the body.

There is plenty of water on Spaceship Earth; about 70 percent of the earth's surface is covered by oceans and seas, but this water is salty. All major terrestrial ecosystems and humans depend on **fresh water,** water that has a salt content of less than 0.01 percent (100 parts per million). The supply of fresh water on our planet is much more limited: less than 1 percent of the global water supply. Yet, as fresh water constantly drains from the land, down rivers, through lakes, and into the sea, it is replenished through precipitation as the biosphere recycles water.

Precipitation and hence freshwater resources are far from evenly distributed on the earth, however. In Chapter 2, we saw that the natural distribution of forests, grasslands, and deserts is basically a function of rainfall. Likewise, where cities can exist and where crops can be grown depend on the availability of fresh water.

Since water is constantly recycled in the biosphere, it is theoretically a sustainable resource. However, as in other situations, this is true only insofar as usage rates are within the capacity of the natural cycle to replenish supplies. The sobering fact is that, in many situations, humans are overdrawing and depleting water resources. This practice is threatening the sustainability of agriculture in many areas and even human habitation in some. There are also manifold ecological effects as humans divert water from water-rich areas to water-poor areas.

We have three objectives in this chapter. The first is to understand the natural water cycle, its capacities and its limitations. The second is to understand how we are overdrawing certain water resources and the consequences of this action. The third is to understand how water must be managed if we are to achieve sustainable supplies.

The Water Cycle

The earth's **water cycle,** also called the **hydrological cycle,** is represented in Figure 11–1. The cycle basically consists of water entering the atmosphere through either evaporation or transpiration and returning through condensation and precipitation. However, there are additional aspects that bear more consideration.

EVAPORATION, CONDENSATION, AND PURIFICATION

In Chapter 3 we learned that there is a weak attraction known as *hydrogen bonding* between water molecules (H_2O) that tends to hold them together. Below 32°F (0°C), the kinetic energy of the molecules is so low that the hydrogen bonding is enough to hold the molecules in place with respect to one another, and the result is ice. At temperatures above freezing but below boiling, the kinetic energy of the molecules is such that hydrogen bonds keep breaking and re-

forming with different molecules; the result is liquid water. As water molecules absorb energy from sunlight or an artificial source, the kinetic energy they gain may be enough to allow them to break away from other water molecules entirely and enter the atmosphere. This is the process we know as evaporation, and the water molecules are said to be in the gaseous state.

We speak of water molecules in the air as **water vapor,** and the amount of water vapor in the air is commonly spoken of (and felt) as **humidity.** The amount of water vapor a given volume of air can hold increases with rising temperature. Therefore humidity is generally measured as **relative humidity,** the amount of water vapor in a volume of air expressed as a percentage of the total amount it can hold at that temperature. For example, a relative humidity of 60 percent means that the air contains 60 percent of the amount of water vapor it could hold at that temperature.

As the temperature and hence the kinetic energy of water molecules decrease, hydrogen bonding again holds the molecules together as they come in contact with each other. Thus, as temperature drops, water vapor collects in droplets. This is **condensation.** If the

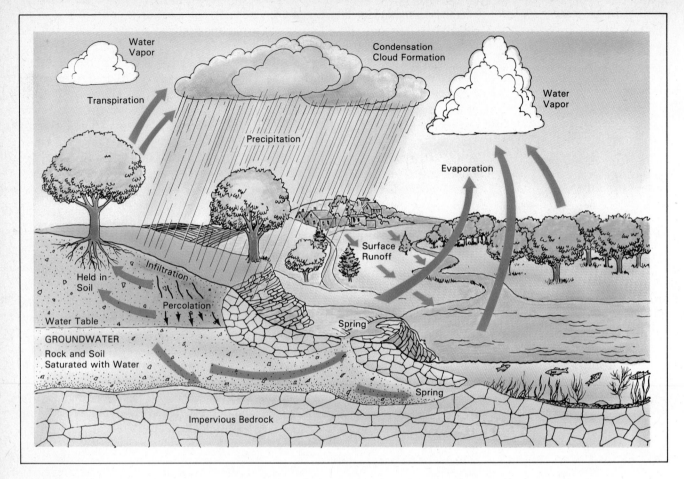

FIGURE 11-1

The water cycle. All the water on earth is continuously recycled through evaporation or transpiration, condensation, and precipitation. There are additional movements of water over and through the ground as shown.

droplets form in the atmosphere, the result is fog or mist. (Masses of fog or mist at a distance are seen as clouds.) If the droplets of condensing vapor form on the cool surfaces of vegetation, the result is dew. If the temperature is below freezing as condensation occurs, the water vapor forms directly into ice crystals and the result is snow or frost.

Water evaporates from oceans, lakes, rivers, soil, and any other moist surfaces, and it transpires from vegetation. Evaporated or transpired water vapor rises in the atmosphere. At increasing altitudes, temperature drops as heat is radiated and lost to outer space. Thus, condensation starts and clouds form. With continuing condensation, water droplets or ice crystals become large enough to fall, resulting in precipitation (Fig. 11–2).

One very important aspect of evaporation and condensation is that these processes result in natural *water purification*. When water evaporates, only water molecules leave the surface; salts and other solids in solution remain behind. (We noted this when we dis-

cussed the problem of salinization in Chapter 9.) The condensed water is thus purified water—except insofar as it picks up pollutants in the air. (The most chemically pure water for use in laboratories is obtained by distillation, a process of boiling water and recondensing the vapor.) Thus, evaporation and condensation of water vapor are the source of all natural fresh water on earth. Fresh water from precipitation falling on the land gradually makes its way through streams, rivers, and lakes to oceans or seas, and salts from the land are constantly flushed into seas as a result. Salts accumulate in seas, making them salty, because there is no way for water to leave except by evaporation.

PRECIPITATION

The distribution of precipitation over the earth, which ranges from near zero in some areas to more than 100 inches (2.5 m) per year in others, basically depends on patterns of rising or falling air currents. As air

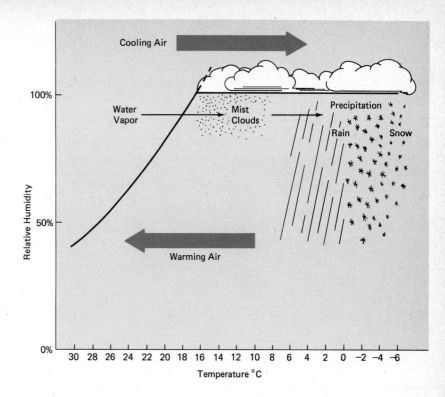

FIGURE 11–2
For a given amount of water in the air, relative humidity (RH) increases as temperature drops. When air cools below the point of saturation (100% RH), condensation and precipitation result. When air warms, there is a drop in relative humidity.

rises, cooling and condensation occur, and precipitation results. As air descends, it tends to become warmer, causing an increase in evaporation and dryness.

There are two situations that cause rising air (hence high precipitation) and descending air (hence low precipitation) more or less continuously. First are global convection currents. Solar heating of the earth is most intense over and near the equator, where sunlight hits the earth most perpendicularly. As the air at the equator is heated by the warm earth, it expands and rises. As it rises, it cools, and condensation and precipitation occur. Thus, equatorial regions have high rainfall, which, along with continuously warm temperatures, supports tropical rainforests.

Rising air over the equator is just half of the convection current, however. The air must come down again. It literally "spills over" to the north and south of the equator and descends over subequatorial regions (25 to 35 degrees north and south of the equator), resulting in subequatorial deserts (Fig. 11–3). The Sahara of Africa is the prime example.

The second situation causing air to rise and fall occurs where trade winds (winds that blow almost continuously from the same direction) hit mountain ranges. As the moisture-laden air in the trade winds encounters a mountain range, the air is deflected upward, causing cooling and high precipitation on the windward side of the range. As the air crosses the range and descends on the other side, it becomes warmer and increases its capacity to pick up moisture. Hence, deserts occur on the leeward side of mountain ranges. The dry region downwind of a mountain is referred to as a **rain shadow** (Fig. 11–4). The severest deserts in the world are caused by the rainshadow effect. For example, the westerly trade winds, full of moisture from the Pacific Ocean, strike the Sierra Nevada mountain range in California. As the winds rise over the mountains, large amounts of water precipitate out, supporting the lush forests on the western slopes. Immediately east of the Sierra Nevada range lies Death Valley, a result of the rain shadow.

WATER OVER AND THROUGH THE GROUND

As precipitation hits the ground, it may follow either of two alternative pathways. It may soak into the ground, **infiltration,** or it may run off the surface, **runoff.** We speak of the amount that soaks in compared with the amount that runs off as the **infiltration-runoff ratio.**

Runoff flows over the ground surface into streams and rivers, which make their way to the ocean or to inland seas. All ponds, lakes, streams, rivers, and other waters on the surface of the earth are referred to as **surface waters.**

For water that infiltrates, there are another two

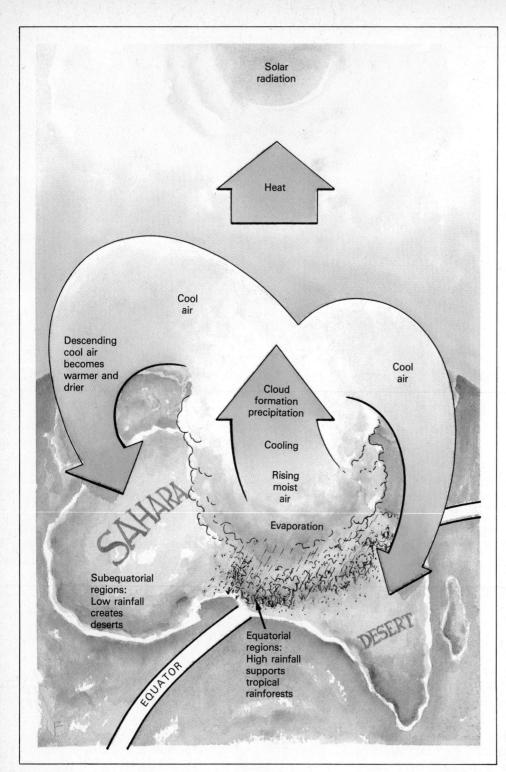

Solar
radiation

Heat

Cool
air

Cool
air

Descending
cool air
becomes
warmer and
drier

Cloud
formation
precipitation

Cooling

Rising
moist
air

Evaporation

SAHARA

Subequatorial
regions:
Low rainfall
creates
deserts

Equatorial
regions:
High rainfall
supports
tropical
rainforests

DESERT

EQUATOR

FIGURE 11–3
Equatorial tropical
rainforests and
subequatorial deserts. Solar
radiation causes maximum
heating in equatorial
regions and produces rising
currents of moist air. As the
moist air cools, there is
heavy precipitation over the
equatorial regions,
supporting tropical
rainforests. As the air
descends, it becomes
warmer and drier, resulting
in subequatorial deserts.

alternatives. The water may be held in the soil, the amount held depending on the water-holding capacity of the soil, as was discussed in Chapter 9. This water, called **capillary water,** returns to the atmosphere either by way of evaporation from the soil or by being taken up by plants and escaping by transpiration. The combination of evaporation and transpiration is referred to as **evapotranspiration.**

Infiltrating water that is not held in the soil is called **gravitational water** because it trickles, or percolates, down through pores or cracks in the earth under the pull of gravity. Sooner or later, however, gravitational water comes to an impervious layer of rock or dense clay. There it accumulates, completely filling all the spaces above the impervious layer. This accumulated water is called **groundwater,** and its

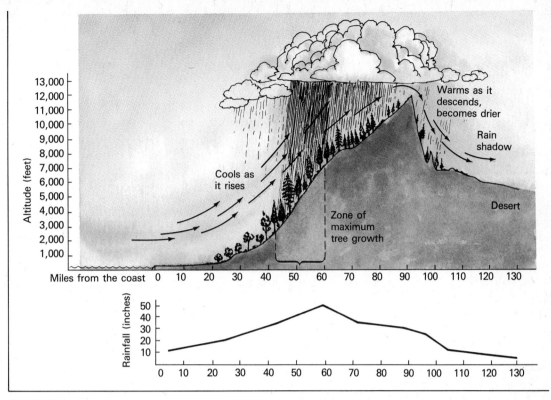

FIGURE 11–4
Rain shadow. Moisture-laden air in a trade wind cools as it rises over a mountain range, resulting in high precipitation on the windward slopes. Desert conditions arise on the leeward side as the descending air warms and tends to evaporate water from the soil. (Redrawn from *Trees: The Yearbook of Agriculture, 1949*. Washington, D.C.: USDA.)

upper surface is the **water table** (Fig. 11–5). Gravitational water becomes groundwater as it hits the water table in the same way that rainwater becomes lake water as it hits the surface of the lake. Wells must be dug to below the water table; then groundwater, which is free to move, seeps into the well and fills it to the level of the water table.

Underground rock layers frequently slope, and hence groundwater seeps laterally. Where a highway has been cut through rock layers, you can frequently observe groundwater seeping out. The layers of porous material through which groundwater moves are

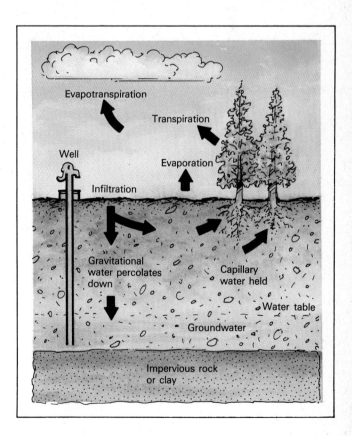

FIGURE 11–5
Pathways of infiltrating water. Water that infiltrates into the soil may be held like water in a sponge. This water may be taken up by plants and returned to the atmosphere by evapotranspiration. When soil reaches its water-holding capacity, additional infiltrating water percolates downward under the pull of gravity and is called *gravitational* water. Water that accumulates above an impervious layer, saturating all the pore spaces in the soil, is called *groundwater*. The upper surface of the groundwater is called the *water table*. Groundwater may be withdrawn by drilling wells to below the water table, or it may percolate horizontally to finally reach a spring or seep.

called **aquifers.** It is often difficult to determine the location of an aquifer. Layers of porous rock are often found between layers of impervious material, and the entire formation may be folded or fractured in various ways. Thus groundwater in aquifers may be found at various depths between layers of impervious rock. Also, the **recharge area,** the area where water enters an aquifer, may be many miles away from where the water leaves the aquifer. If the elevation of the recharge area is higher than the elevation where the water leaves the aquifer, the result is high pressure, which forces water through the aquifer. An aquifer in which the water is under pressure is called an **artesian aquifer.**

Drawn by gravity, groundwater may move through aquifers until it finds some opening to the surface. We observe such natural exits as *springs* or *seeps.* In a **seep,** water seeps out over a relatively wide area; in a **spring,** water exits the ground as a significant flow from a relatively small point. Since springs and seeps feed streams, lakes, and rivers, groundwater joins and becomes part of surface water. A spring will flow, however, only if the water table is higher than the spring. Whenever the water table drops below the level of the spring, the spring will dry up.

SUMMARY OF THE WATER CYCLE

The water cycle consists of evaporation, condensation, and precipitation. There are three principal "loops" in the cycle: (1) the *surface runoff loop,* in which water runs across the ground surface and becomes part of the surface water system; (2) the *evapotranspiration loop,* in which water enters the soil, is held as capillary water and then returns to the atmosphere by way of evapotranspiration; and (3) the *groundwater loop,* in which water enters and moves through the earth, finally exiting through springs, seeps, or wells and rejoins the surface water.

We can visualize all the lands of the earth being continuously bathed with a flow of fresh water coming down as precipitation and gradually moving over and through the earth. Bodies of water—ponds, lakes, and so on—that have an outlet are constantly flushed by fresh water and remain fresh as long as water keeps flowing. Salts and other minerals flushed from the land accumulate only in seas or other locations that have no exit for water other than evaporation. Great Salt Lake in Utah is one example; irrigated croplands that do not have sufficient drainage is another, and of course, all oceans.

IMPLICATIONS FOR WATER QUALITY

Whether precipitation runs off the surface or follows the groundwater pathway has a number of implications regarding relative purity. As precipitation hits the ground, it may pick up impurities such as soil particles, dissolved chemicals, and detritus and the microorganisms feeding on it. Therefore, runoff may be heavily polluted.

In natural ecosystems, there is relatively little runoff, however. Most precipitation infiltrates, and particles of dirt, detritus, and microorganisms are filtered out as the water percolates through soil and porous rock. Dissolved chemicals are not filtered out, however. Instead they are leached into the groundwater. Indeed, as water percolates through the earth, it may dissolve and leach a number of minerals. Underground caverns are a result of millennia of leaching limestone (calcium carbonate). In most natural situations, the minerals that leach into groundwater are not harmful. Indeed, calcium from limestone is con-

TABLE 11–1

Terms Commonly Used to Describe Water	
Term	**Definition**
Water Quantity	The amount of water available to meet desired demands.
Water Quality	The degree to which water is pure enough to fulfill the requirements of various uses.
Fresh Water	Water having a salt concentration below 0.01 percent. As a result of purification by evaporation, all forms of precipitation are fresh water, as are lakes, rivers, groundwater, and other bodies of water that have a throughflow of water from precipitation.
Salt Water	Water, typical of oceans and seas, that contains at least 3 percent salt (30 parts salt per 1000 parts water).
Brackish Water	A mixture of fresh and salt water, typically found where rivers enter the ocean.
Hard Water	Water that contains minerals, especially calcium or magnesium, that cause soap to precipitate, producing a scum, curd, or scale in boilers.
Soft Water	Water that is relatively free of those minerals that cause soap to precipitate causing scale buildup.
Polluted Water	Water that contains one or more impurities that make the water unsuitable for a desired use.
Purified Water	Water that has had pollutants removed or rendered harmless.

sidered beneficial to health. Thus, groundwater is generally high-quality fresh water that is safe for drinking. A few exceptions occur where there is leaching of minerals containing arsenic or other poisonous elements.

As groundwater exits through springs and seeps, it feeds and replenishes surface waters with high-quality fresh water.

Terms commonly used to describe water are given in Table 11–1.

Humans and the Water Cycle

By visualizing the water cycle, you should be able to see that any fresh water we use must come out of the cycle at one point or another. Likewise, all the polluted waste water we put down the drain or throw out goes back into the cycle. Anything we do to the land surface, from development to deforestation, will influence the infiltration-runoff ratio and thus the cycle. Anything we put into the air may end up as a contaminant in precipitation. Finally, any chemicals we put on or bury in the soil are subject to leaching into groundwater.

All these potential problems and impacts can be divided into two categories: *quantitative* and *qualitative*. **Quantitative** refers to such issues as, Is there enough water to meet our needs? and, What are the impacts of diverting water from one point of the cycle to another? **Qualitative** refers to such issues as, Is the water of sufficient purity so as not to harm human or environmental health? In the remainder of this chapter, we shall focus primarily on the quantitative aspects, although you should bear in mind that quality is an ever-present concern. In following chapters, we shall focus on the qualitative aspects; that is, pollution.

SOURCES AND USES OF FRESH WATER

The major categories of freshwater use are given in Table 11–2. Most of the water used in homes and industries is for washing and flushing away undesired materials, and the water used in electrical power production is used for taking away waste heat. Such uses are termed **nonconsumptive** because the water remains available for the same or other uses if its quality is adequate or if it can be treated to remove undesired materials. In contrast, irrigation is called a **consumptive** use because the applied water is lost for further use. It can only percolate into the ground or return to the atmosphere through evapotranspira-

TABLE 11–2

U.S. Demands on Fresh Water	
Use	Gallons (liters) used per person per day
Consumptive	
Irrigation and other agricultural use	700 (2800)
Nonconsumptive	
Electrical power production	600 (2400)
Industrial use	370 (1500)
Residential use	100 (400)

tion. Of course, irrigation water does eventually reenter the water cycle, but this cycling will not bring the water back to the place that is being irrigated.

The primary sources of fresh water for all uses are surface water, namely rivers and lakes, and groundwater. Originally each family dipped water for its own use from a local stream or shallow well. This still occurs to a considerable extent, and many women in Third World countries walk long distances each day to fetch water (Fig. 11–6). Because this water is untreated surface water, it is often contaminated with various pathogens (disease-causing organisms). The Worldwatch Institute estimates that 1.2 billion people in the world lack safe drinking water.

In the developed world, the freshwater sources

FIGURE 11–6
In many villages and cities of the Third World open, untreated wells such as this one in Mali (West Africa) are the only water supply for people. You can see that such sources may often be contaminated with pathogens and other pollutants. (Sam Bryan/Photo Researchers.)

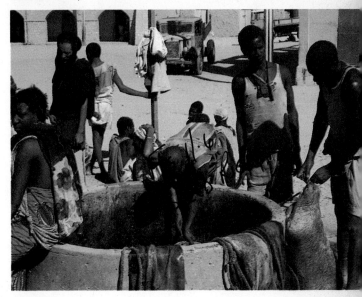

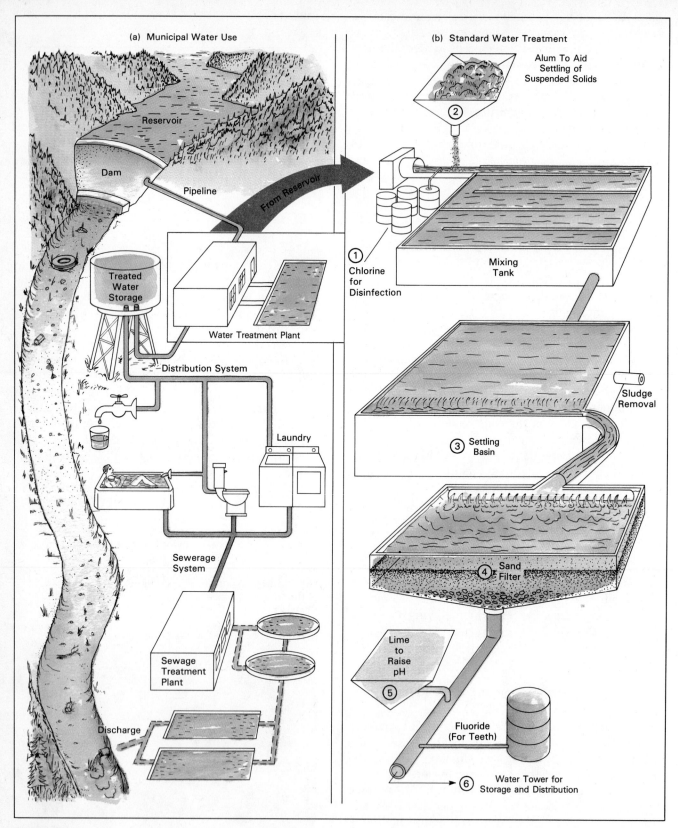

FIGURE 11–7

(a) Municipal water use. Water is generally taken from a river, treated, used, then returned. (b) Water treatment. Water is piped from a reservoir to the treatment plant. At the plant, (1) chlorine is added to kill bacteria, (2) alum (aluminum sulfate) is added to coagulate organic particles, and (3) the water is put into a settling basin for several hours to allow the coagulated particles to settle. It is then (4) filtered through sand filters, (5) treated with lime to adjust pH, and (6) put into a storage water tower or reservoir until distribution to your home.

are still rivers and lakes, but methods for collection, treatment, and distribution are more sophisticated. Dams are built across rivers to create reservoirs, which hold water in times of excess flow and can be drawn down at times of inadequate flow. In addition, dams and reservoirs may provide for power generation, recreation, and flood control. Water for municipal use is piped from the reservoir to a treatment plant, where it is treated to improve quality and kill pathogens, as shown in Figure 11–7 (see also In Perspective below). After treatment, water is distributed through the water system to our homes, schools, and industries. The post-use water, collected by the sewage system, is carried to a sewage treatment plant, where it is treated before being discharged into a nat-

ural waterway. Often it is discharged into the same river from which it was withdrawn.

On major rivers, such as the Mississippi, water is reused many times without good pollution control. Each city along the river takes water, treats it, uses it, and then returns it to the river, often with minimal treatment. Thus, each successive city has a higher load of pollutants to contend with, and ecosystems at the end of the line may be severely affected by the pollution. Pollutants include industrial wastes as well as pollutants from households, since industries utilize the same water system.

Reservoirs created by dams on rivers are also major sources of water for irrigation. In this case no treatment is required. As noted above, the croplands

IN PERSPECTIVE
Water Purification

Polluted water is defined as water that contains one or more materials that make the water unsuitable for a given use. Water purification is any method that will remove one or more such materials. Several methods may be used in combination to obtain water that is sufficiently pure for a given use. Methods that are commonly used in water purification are:

1. Settling. Soil particles and other solid material carried by flowing water may be removed by holding the water more-or-less still and allowing the solids to settle. Clarified water is removed from the top. Settling may be aided by the addition of alum (aluminum sulfate). The +3 charge on aluminum ions pulls clay and other particles, which are negatively charged, into clumps that settle more readily than the individual particles do.

2. Filtration. Filtration is the passage of water through a porous material. Any materials larger than the pores will be filtered out. A bed of sand is often used for this purpose.

3. Adsorption. Certain materials bind and hold other materials on their surface. Passing water

through an adsorbing material will remove certain pollutants. Activated carbon is a material commonly used in this way to remove organic contaminants from water or air.

4. Biological oxidation. Organic material (detritus and organisms) is fed upon by detritus feeders and decomposers, broken down in cell respiration, and thus removed. Passage of water through systems supporting the growth of such organisms accomplishes removal (see Chapter 13).

5. Distillation. Distillation is the evaporation and condensation of water. All materials present in the water before the evaporation step remain behind in the holding tank and are therefore not present when the water vapor is condensed.

6. Disinfection. Treatment with chlorine or other agents that kill disease-causing organisms.

The natural water cycle includes all of these purification methods except disinfection. Sitting in lakes, ponds, or the oceans, water is subject to settling. As it percolates through soil or porous rock, it is filtered. Soil and humus are also good chemical adsorbents. As water flows

down streams and rivers, detritus is removed by biological oxidation. As water evaporates and condenses, it is distilled.

Thus numerous freshwater sources might be safe to drink were it not for human pollution. The most serious threat to human health is contamination with disease-causing organisms and parasites, which come from the excrements of humans and their domestic animals. In human settlements, you can see how these organisms may get into water and be passed onto people before any of the natural purification processes can work.

The World Health Organization estimates that 80 percent of sickness, disease, and deaths of infants and children in less-developed countries is attributable to contaminated water. Conversely, the greatest safeguard to human health, which we tend to take for granted in developed countries, is proper collection and treatment of sewage wastes and suitable protection and treatment of water supplies (See Fig. 11–7). Thus, the greatest step we could take toward improving world health would be to implement these services wherever they are not present.

(a)

(b)

FIGURE 11–8
(a) Center-pivot irrigation. Water is pumped from a central well. A self-powered boom rotates around the well, spraying water as it goes. (b) Aerial photograph shows the extent of center-pivot irrigation in Nebraska. As a result, groundwater is being depleted, which will bring an end to this kind of farming. (Earl Roberge/Photo Researchers)

are the end point for this water except insofar as it reenters the water cycle.

Both to augment surface water supplies and to obtain water of higher quality, there has been an increasing trend in the past few decades of drilling wells and tapping groundwater. Groundwater generally requires no treatment, even for drinking. Since about 1950, hundreds of cities have drilled huge wells for municipal supplies, and millions of wells have been drilled for individual households in suburbs beyond municipal supply systems. Farmers have turned to **center-pivot irrigation,** a system in which water is withdrawn from a central well and applied to the field through a gigantic sprinkler that pivots in a circle around the well (Fig. 11–8). The use of such wells has increased tremendously in the last 20 years, increasing agricultural production but also consuming huge amounts of groundwater. One system may use as much as 10 000 gallons (40 000 liters) per minute.

Since the 1970s, the amount of water used in irrigation in the United States has grown much faster than has the population. This disproportionate growth has occurred because irrigation has been and continues to be a major way of increasing crop production, as noted in Chapter 8.

From our understanding of the water cycle, we can see that each source of water involves a continuous flow; there are always natural inputs as well as outputs. Water from any source represents a sustainable or renewable (self-replenishing) resource. However, while the flow of water through any part of the cycle is continuous, the *volume* of flow during any given time period is limited. If humans attempt to extract volumes that exceed those of natural flow, shortages may be anticipated. Furthermore, our use of water often involves diverting its flow from one location to another, and such diversion is likely to have certain consequences. Using and diverting water in volumes that lead to shortages or other undesirable consequences is spoken of as *overdrawing* or *overdraft* of water resources. Let's turn our attention now to the consequences of overdrawing water resources.

OVERDRAWING WATER RESOURCES

Since the consequences of overdrawing surface waters differ somewhat from those of overdrawing groundwater, we shall consider these two categories separately. However, since all water is tied together in the same overall cycle, you should be aware of similarities and interconnections in the two discussions.

Consequences of Overdrawing Surface Waters

Inevitable Shortages There are wet years and dry years, and surface-water flows vary accordingly. On the average of once every 20 years, surface-water flow may drop to only 30 percent of its annual average. Therefore, the rule of thumb is that no more than 30 percent of a river's average flow can be taken out each year without risking shortfalls on the average of once every 20 years. This rule has not always been heeded, however. In some river systems in the United States, for instance, water demand has grown to 100 percent (and even *more than* 100 percent!) of average flow, making water shortages of increasing length and severity inevitable (Fig. 11–9).

Southern California is a case in point. This region entered its sixth year of drought in the spring

FIGURE 11–9
Droughts occur on an average of every 20 years and may reduce normal water flows by 70 percent. Therefore no more than 30 percent of the average surface-water flow can be counted on to be continuously available. By the year 2000, large areas of the United States will be above the 30 percent level, making severe, recurring water shortages inevitable. (U.S. Water Resources Council)

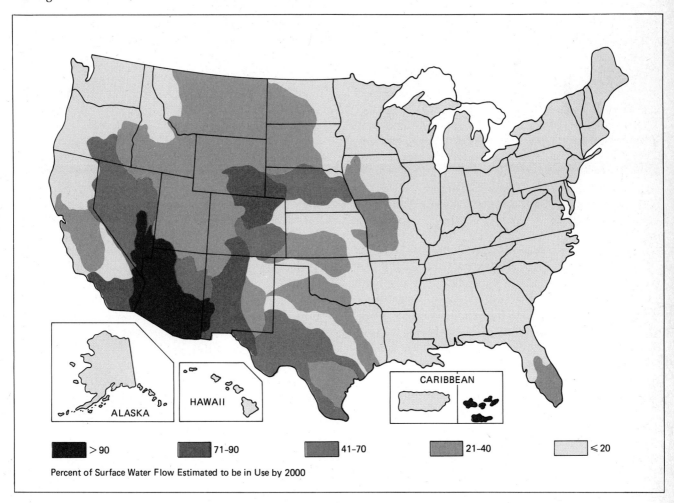

Percent of Surface Water Flow Estimated to be in Use by 2000

Many rivers form borders between or flow through a number of different states or countries, and of course having different governments involved complicates the situation. As water demands shoot up with expanding populations, the issue of who has the right to the water is leading to ever-increasing acrimony between states and between nations.

For example, in a 1922 agreement, the states of the U.S. Southwest divided the flow of the Colorado River among themselves: 7.5 million acre feet to be shared by Colorado, Wyoming, Utah, and New Mexico, all states in which tributaries of the river originate, and 7.5 million acre feet to be shared by Arizona, Nevada, and California, states through which the lower river flows. (An acre foot is the amount of water that will cover one acre to a depth of one foot, about 325,000 gallons.) Another agreement gave Mexico, where the mouth of the Colorado River empties into the Gulf of California, the right to 1.5 million acre feet, and Native Americans have rights to whatever amounts they need. (Their use was almost none at the time.) All went well as long as demands were below the allocated rights. As population has grown, however, each party is demanding its full share. The only problem is that the *total* average flow of the Colorado is only 14.9 million acre feet. Even worse, for the past few years flow has dropped to just 9 million acre feet the past few years.

Another water-allocation problem takes us to the other side of the world. About 90 percent of the tributaries of the Nile River, which flows through Egypt into the Mediterranean Sea, arise in Ethiopia, and Ethiopia is now considering ways to divert more of the water to ease its drought and famine. Egypt, which depends on the Nile water, has blocked a loan for the Ethiopian project.

The conflict between Israel and Jordan over the West Bank, a piece of territory on the western bank of the Jordan River, is as much over access to water as over territory. Many people speculate that the next war(s) in the Mideast may well be over water.

What moral and ethical principles are involved in such disputes? What are the moral dilemmas? How do you think they should be resolved? Who should be the final authorities to enforce agreements?

of 1992, with reservoirs down to 20 to 40 percent of capacity. Of course, people blame the water shortage on the "terrible drought," but is it really the drought? Droughts are not abnormal; they are normal climatic fluctuations and must be taken into account in long-term planning for sustainability.

Demands on the Colorado River are another example, setting the stage for inevitable shortages. (see Ethics Box above).

Ecological Effects When a river is dammed and its flow is diverted to cities or croplands, the river below the dam may become very nearly or totally dry, with obvious impact on fish and other aquatic organisms. The ecological impacts go far beyond the river itself. Wildlife that depends on the water or on food chains involving aquatic organisms—and this includes virtually all wildlife—are also adversely affected. Wetlands along many rivers, no longer nourished by occasional overflows, dry up, resulting in tremendous dieoffs of waterfowl and other wildlife that depended on these habitats (Fig. 11–10). Fish such as salmon, which swim from the ocean up rivers to spawn, are seriously affected by the reduced water level and have a problem getting around the dam, even one equipped with fish ladders. If the fish do get up the river, the hatchlings have similar problems getting back to the sea.

The problems extend to estuaries, which are bays in which fresh water from a river mixes with seawater at the river's mouth. Estuaries are among the most productive ecosystems on earth; they are rich breeding grounds for many species of fish, shellfish, and water fowl. As a river's flow is diverted to other locations, there is less fresh water entering and flushing the estuary. Consequently, the salt concentration increases, profoundly upsetting the ecology. The Colorado River provides a dramatic example. Withdrawals from it are so great that where it used to flow into the Gulf of California is now seldom more than a dry bed. The entire ecology at the upper end of the Gulf of California has been altered as a result of increasing salinity.

Mono Lake, a 63-square-mile (163–km^2) lake in east-central California provides another example—one that also shows the potential for the public to bring rectification. Mono Lake has no outlet, but a substantial inflow of fresh water from snowmelt off the Sierra Nevada Mountains kept its water at moderate salinity. Like other salt lakes around the world, Mono Lake supported numerous species of wildlife including huge flocks of aquatic birds. Beginning in

(a) (b)

the 1940s, however, much of the freshwater inflow was diverted to support water-hungry Los Angeles. As Mono Lake lost more water to evaporation than it was receiving, it shrank rapidly, and salts became more concentrated threatening total ecological collapse (Fig. 11-11). In the later 1970s a group of students formed the Mono Lake Committee, which brought legal action under the Public Trust Doctrine. With public support, the committee won a series of the court battles in the 1980s forcing Los Angeles to release enough stream flow to maintain Mono Lake at about its 1976 level, still some 40 feet (13 m) below its prediversion level.

The problem is not limited to the United States. The southeastern end of the Mediterranean Sea was formerly flushed by water from the Nile River. Because this water is now held back and diverted for irrigation by the Aswan High Dam in Egypt, this part of the Mediterranean is suffering the ecological consequences.

The world's most dramatic example of water mismanagement is the Aral Sea, an inland sea in south-central Russia (see Ethics Box, p. 258).

Consequences of Overdrawing Groundwater

Falling Water Tables and Depletion The total amount of fresh groundwater is estimated to be more than 75 times the amount of fresh surface water. Thus,

FIGURE 11–11
With freshwater input diverted to Los Angeles, Mono Lake was drying up, exposing tufa towers (formations of precipitated calcium carbonate) and threatening ecological collapse. Recent court decisions reduced diversions and require that a balance be struck between ecological needs and the consumptive needs of Los Angeles. (Peter Vorster, Oakland, CA)

groundwater may be visualized as a vast but not un-limited reservoir. As with any other reservoir, the ability of groundwater to sustain demands over a long term depends on the balance between input rates and output rates. When withdrawal rates exceed recharge rates, the water table drops, and eventually the groundwater of the area is depleted.

The lateral flow of groundwater is generally very slow because the water percolates through porous material and may be precluded altogether by various rock formations. Therefore, the groundwater may be depleted in one area but still present in a nearby area. Also, in many arid regions of the world, existing groundwater may be essentially "fossil water"—that is, deposits of water from past times when the climate of the region was wetter. In these arid places, current recharge rates may be nil.

With the increased use of groundwater for both cities and irrigation, withdrawal rates now exceed re-charge rates in many regions in the United States and around the world. In many cases, water tables are dropping a meter or more per year as a result. A prime example is the Great Plains region (Texas, Oklahoma, New Mexico, Colorado, Kansas, and Nebraska), where the Ogallala aquifer is extensively exploited for center-pivot irrigation. It is predicted that 3.5 million acres (1.4 million hectares) in this region will be abandoned or converted to dryland farming (ranching and production of forage crops) over the next 10 years because of water depletion. As discussed in Chapter 8, a reduction in agricultural production seems inevitable.

The recent growth and prosperity of numerous cities around the world are based on overdrawing groundwater and falling water tables. Mexico City is almost entirely dependent on exploiting groundwater, and its water table is falling at a rate approaching 6 feet (2 m) a year; this rate may result in depletion within the next 20 years. What does a city of 22 million people do as it runs out of water?

While eventually running out of water is the obvious conclusion of overdrawing groundwater, falling water tables have a number of other consequences before the water is entirely depleted. Let us now examine some of them.

Diminishing Surface Water Surface waters are also affected by falling water tables. Recall that streams, rivers, and lakes are fed in large part by springs and seeps of outflowing groundwater. As the water table drops, the amount of water exiting from the ground will diminish and finally cease. This is another example of diverting water flows. The same ecological impacts on rivers, wetlands, and estuaries that are caused by diverting surface water may be caused or exacerbated by excessive groundwater withdrawals.

Land Subsidence Over the ages, groundwater has leached cavities in the earth. Because these spaces are filled with water, water plays a role in supporting the overlying rock and soil. As the water table drops, this support is lost, and there may be a gradual settling of the land, a phenomenon known as **land subsidence.** The sinkage rate may be 6–12 inches per year. In some areas of the San Jaoquin Valley in California, land has settled as much as 27 feet because of groundwater removal. Land subsidence causes building foundations, roadways, and water and sewer lines to

FIGURE 11–12
Sinkhole. Removal of groundwater may drain an underground cavern until the roof, no longer supported by water pressure, collapses. The result is the sudden development of a sinkhole, such as this one which consumed a home in Frostproof, Florida, July 12, 1991 (R. Veronica Decker/The Orlando Sentinel/Gamma-Liaison.)

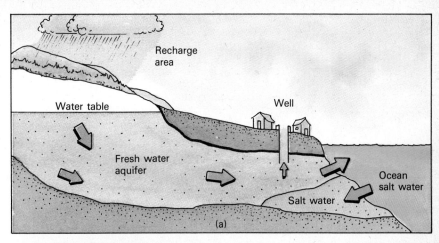

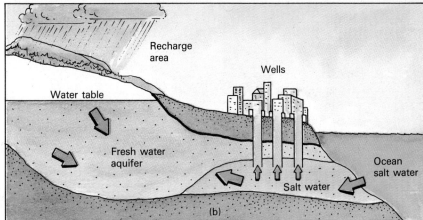

FIGURE 11–13
Saltwater intrusion. (a) Where aquifers open into the ocean, fresh water is maintained in the aquifer by the head of fresh water inland. (b) Excessive removal of water may reduce the pressure, so that salt water moves into the aquifer.

crack. In coastal areas, subsidence causes flooding unless levees are built for protection. For example, a 4000-square-mile (10000-km^2) area in the Houston-Galveston Bay region of Texas is gradually sinking because of groundwater removal, and coastal properties are being abandoned as they are gradually inundated by the sea. Land subsidence is also a serious problem in New Orleans, sections of Arizona, Mexico City, and many other places throughout the world.

Another kind of land subsidence, the occurrence of a **sinkhole,** may be sudden and dramatic (Fig. 11–12). A sinkhole results when an underground cavern, drained of its supporting groundwater, suddenly collapses. Sinkholes may be 300 feet or more across and as much as 150 feet deep. Formation of sinkholes is particularly severe in the southeastern United States, where groundwater has leached numerous passageways and caverns through ancient beds of underlying limestone. An estimated 4000 sinkholes have occurred in Alabama alone, some of which have "consumed" buildings, livestock, and sections of highways.

Saltwater Intrusion Another problem resulting from dropping water tables is **saltwater intrusion.** In coastal regions, springs of outflowing groundwater

may lie under the ocean. As long as the water table on land is higher than the ocean level and pressure is maintained in the aquifer, there is a net flow of fresh water into the ocean. Thus, wells near the ocean yield fresh water (Fig. 11–13a). However, a lower water table or rapid rates of groundwater removal may reduce the pressure in the aquifer, permitting salt water to push back into the aquifer and hence into wells (Fig. 11–13b). Saltwater intrusion is a problem at many locations along U.S. coasts.

Figure 11–14 provides a summary of these effects and where they are occurring most severely in the United States. Of course the problems are by no means unique to the United States.

OBTAINING MORE WATER

Despite the obvious and growing negative impacts of overdrawing water resources, growing populations create an ever-increasing demand for additional water for both irrigation and municipal use. In the United States, very few rivers remain undammed, and many are dammed at several points. Yet governments eager to please both people clamoring for water and businesses having economic interests in dam

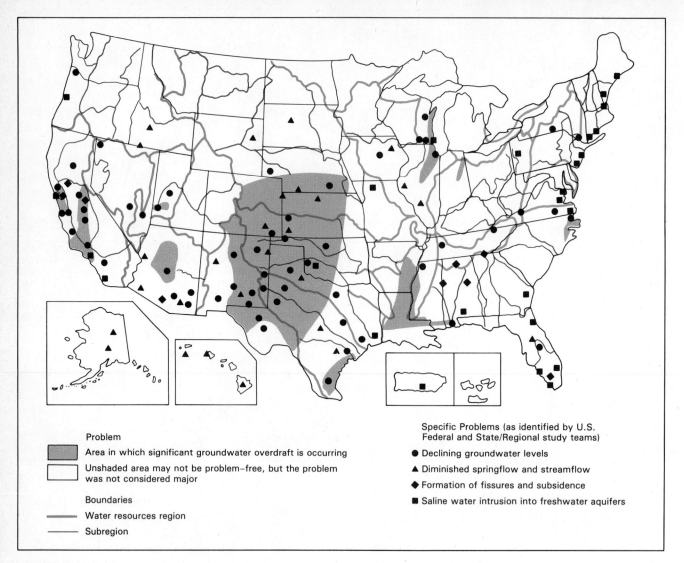

Problem

Area in which significant groundwater overdraft is occurring

Unshaded area may not be problem–free, but the problem was not considered major

Boundaries

Water resources region

Subregion

Specific Problems (as identified by U.S. Federal and State/Regional study teams)

● Declining groundwater levels

▲ Diminished springflow and streamflow

◆ Formation of fissures and subsidence

■ Saline water intrusion into freshwater aquifers

FIGURE 11–14
Declining groundwater levels and related problems in the United States. (From V. J. Pye et al., *Groundwater Contamination in the United States*, Philadelphia: University of Pennsylvania Press, 1983. Used with permission from the Academy of Natural Sciences, Philadelphia, PA.)

construction persist in offering ever more grandiose plans for water projects. One example, which is still being promoted by some, is a plan to dam the Yukon River in Alaska to create a reservoir that would flood nearly a third of the state and bring the water via canal all the way to the American Southwest and northern Mexico.

Increasingly, however, people are recognizing the inevitable tradeoffs that will occur and are considering them to be unacceptable. Intense counter-lobbying by various environmental organizations has, in recent years, led to the rejection of a number of dam proposals, including the Yukon project (at least for the present). In addition, environmentalists won passage of the Wild and Scenic Rivers Act of 1968, which enables certain rivers to be designated as

"wild and scenic" and thus prohibits any future damming of them. Bitter controversies between environmentalists and dam-building interests continue, however.

The situation is different in the Third World. In general, water resources remain to a large extent undeveloped, and, as we have noted, many people still collect water from streams and shallow wells that frequently are polluted. It is easy to make an argument for additional development of water resources in many Third World countries. However, some major water projects in less-developed countries—the Aswan High Dam in Egypt is one—are generally recognized as ecological disasters. In addition, these projects have not really helped the poor. The loss of fisheries and productive wetlands below the dam and the loss of land flooded by the reservoir above have largely canceled out any gains realized from additional irrigation. Such dams are no more than another example of large-scale centralized projects that benefit the already wealthy but displace and further undercut the survival of the poor, as discussed in Chapter 7.

What is necessary is more thinking about the long-term environmental effects of a proposed project and how these effects ultimately impact the population at large. The answer may be to develop water resources, yes, but as a series of multiple small-scale projects that will give local people access to more abundant and cleaner water supplies but will not displace them in the process. Multiple small-scale projects, far from destroying the ecological integrity of the river basin, may increase its productivity and biodiversity.

Another avenue for increasing water supplies in both developed and less-developed countries is **desalinization plants,** plants that distill sea water into high-quality drinking water (see In Perspective Box,

IN PERSPECTIVE
Desalting Sea Water

The majority of the world's population live in coastal cities, and oceans represent a limitless supply of water. Desalination (removal of salt and other impurities) of seawater can provide water of very high quality. Several thousand desalination plants already exist, primarily in Saudi Arabia, Israel, and other countries of the Mideast. The world produces 3.4 billion gallons (13 billion liters) of drinkable water per day through desalination.

Small plants generally use a process called **reverse osmosis,** in which the sea water is forced under great pressure through a membrane that is fine enough to filter out the salt. Large plants, particularly where a source of waste heat is available

(from electrical power plants, for instance), generally use nature's method: evaporation followed by condensation. Additional efficiency is gained by using the heat given off by condensing water to heat the incoming water. Even where waste heat is used, however, the costs of building and maintaining the plant, which is subject to corrosion from sea water, are considerable. Under the best of circumstances, the production of desalinized water costs about 3 dollars per 1000 gallons (4000 liters). This is three to six times what most city dwellers in the United States currently pay, but it is still not a high price to pay for drinking water. The high cost might cause some people to cut back on

watering lawns and implement other conservation measures, but most people in the United States could afford it without undue strain.

Irrigation commonly consumes 500 thousand gallons or more per acre for producing one crop. Farmers currently pay as little as 1 cent per 1000 gallons. Thus, it is cost-effective to irrigate crops that bring in only a few hundred dollars per acre. Three dollars per 1000 gallons and the additional cost of pumping the water from sea level to croplands would be obviously cost-prohibitive. Thus, desalinization is likely to become an increasingly attractive option for water-short coastal cities but not for irrigation.

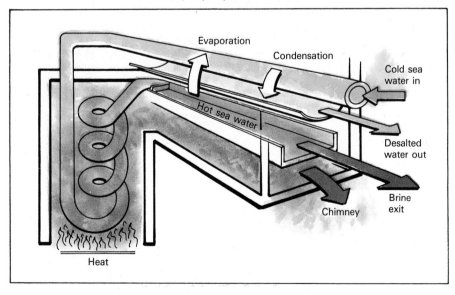

Southern California is facing increasingly severe and chronic water shortages, and there is increasing acrimony between farmers and city dwellers as they compete for limited supplies.

Basically this is a desert region, but the moderate, sunny climate is ideal for agriculture when the land is irrigated, and the region has also attracted millions of people to live and work. State and federal governments have supported both agriculture and urban growth by building dams and reservoirs in the mountains to the west and north. Even the Colorado River, on the other side of the mountains, has been tapped, and thousands of miles of canals have been built to bring water to the parched lands and cities. Providing abundant water succeeded in making the region exceedingly prosperous, in terms of both agriculture and cities. The price of this prosperity has been that population, and hence water demands in the region, has continued to grow so that today demand is essentially 100 percent of the average flow rates into reservoirs.

This situation makes shortages inevitable in southern California during droughts, which the climatic record shows are normal recurring events. How are the shortages to be managed? Who should be asked or required to cut back the most?

The provision of abundant water in the past has lead to water-wasteful ways. Farmers, who use about 85 percent of the total water, are charged as little as a penny per 1000 gallons (4000 liters). The all-but-free water has resulted in farmers' using water-wasting flood irrigation and growing crops that are extravagant in water use, such as rice and cotton. Also grown are low-value crops, such as alfalfa for cattle feed. Adding to the irony, the world-market price for rice and cotton is low because of oversupply; farmers make a profit on these crops only because of government-subsidized prices. In other words, taxpayers are paying the real costs of providing water and then paying above-market prices for the crops produced. Thus city dwellers say that farmers should be the ones to conserve.

However, city dwellers are not without blame either. They pay somewhat more than farmers do for water, but costs, in the range of 35 cents per 1000 gallons (9 cents per 1000 liters) are still nominal. Thus, water use in southern California is more than double the national norm because water is used extensively to water lawns and gardens and fill swimming pools.

Farmers claim that, since they are producing food for city people to eat, the city dwellers are the real users. Also, farmers point out that city people are effectively irrigating lawns simply to have green grass to look at whereas farmers are supporting the economy with crops.

What are the values involved in this dispute? How should the government allocate the water, and what means should be used to promote conservation? Even with conservation, can unlimited growth be supported? Is the government obligated to provide water to more and more people moving into a water-short area? If not, what means should be used to limit growth?

(In the spring of 1992, as rains seemed to end this round of drought, municipalities in southern California immediately began authorizing more water hookups, which had been restricted during the drought!)

p. 255). However, this is an option only for coastal cities. For inland cities and for irrigation, the costs of desalinization plus those of transporting the water are prohibitive.

USING LESS WATER

We have seen that, in the United States and other countries where water resources have been developed, water consumption has grown to very high levels (Table 11–2). Yet, water shortages are generally seen in terms of needing still more. Gaining increasing popularity, however, is the alternative concept: We could balance our demands with existing supplies by using much less—in other words, through water conservation and water recycling.

Where cities and farms are utilizing water from the same source (as is the case in southern California), there is obvious competition and antagonism between farmers and city dwellers over allocation of limited supplies (see Ethics Box above). There is room for conservation on both sides, however.

Irrigation

In the United States about 7 times more water is used for irrigation than for residential consumption. Where irrigation water is applied by traditional flood or center pivot systems, about 60 percent is wasted in evaporation, percolation, or runoff. This loss can be substantially reduced by installing **drip irrigation** systems, networks of plastic pipes with pinholes that

FIGURE 11–15
Drip irrigation. Irrigation is the most consumptive water use. Drip irrigation offers a conservative method of applying water, dripping it on each plant through a system of plastic pipes. (Lowell Georgia/Photo Researchers.)

drip water at the base of each plant (Fig. 11–15). Such systems have the added benefit of retarding salinization (Chapter 9). Although drip irrigation is being used more and more, the bulk of irrigation is still done by flood or center-pivot methods because they are less costly to install and maintain.

The low cost of flood and center-pivot systems, coupled with nominal or no charges for the water, encourages farmers to irrigate crops, such as corn and cotton, for which there are chronic surpluses but which, given guaranteed price supports by the government, never fail to turn a profit. Likewise, farmers find it profitable to irrigate alfalfa, which is then used for cattle feed to grow additional beef that most nutritionists agree we tend to overconsume. In effect, we are overdrawing water supplies and suffering the numerous environmental and ecological consequences to grow crops that are not really needed. The economics and irony of this are explored further in the Ethics Box on page 256.

Municipal Systems

Where cities and farmers are in competition for the same water supplies, shortages might be mitigated by having farmers use drip systems. However, many inland cities are overdrawing groundwater without competition from farming. For such cities there is no choice; they must either make better use of water or run out of it.

Water consumption in modern homes averages around 100 gallons of water per person per day. This volume includes, in addition to the water used for drinking and cooking, that used for flushing toilets (3–5 gallons per flush), taking showers (2–3 gallons per minute), doing laundry (20–30 gallons per wash), and so on. If watering lawns and filling pools are included, the per person volume is in the range of

300 gallons per day. As a point of contrast, consider the opposite extreme. Hikers backpacking in water-short regions readily discover that they can get by on less than 1 gallon per day for drinking, cooking, and essential washing. Obviously, they are doing without flushing toilets, showering, or washing clothes, which we would not recommend for city living. Still the contrast tells us that we could get by without hardship on considerably less than the current average of 100 gallons per day.

Numerous cities faced with water shortages or attempting to accommodate more growth without developing new supplies are promoting water conservation through "public awareness" programs. Such programs encourage people to take shorter showers, turn off the water while brushing teeth, and so on. In addition, water-saving devices such as toilets that use only 1.5 gallons per flush, dishwashers and laundry machines that use less water, and shower heads that deliver a finer spray are promoted. Gaining popularity in the Southwest is **xeroscaping,** landscaping with drought-resistant plants that need no watering. Such structural changes from water saving toilets through xeroscaping, create a continuing savings in water consumption without constantly having to remember or alter habits.

Also, gray water recycling systems are being adopted in some water-short areas. Gray water, the slightly dirtied water from sinks, showers, bath and laundry tubs, and so on, is collected in a holding tank and used for such things as flushing toilets, watering lawns, and washing cars. Many residents in southern California have adopted the habit of standing in a plastic tub while taking a shower. They then use the water collected for watering gardens and other purposes. A number of cities are using treated wastewater (sewage water) for irrigation, both to conserve water and to reduce pollution of receiving waters (see Chapters 12 and 13).

Most industries could be required to purify and reuse their wash water rather than just passing it through once. On a broader scale, a number of cities are investigating systems for purifying and reusing wastewater, and a few cities have already begun using such systems. Even traditional sewage treatment produces a relatively clean effluent (see Chapter 13). With the addition of a process such as desalination (see In Perspective Box, p. 255), wastewater could be made cleaner than water from any natural source.

If the idea of reusing sewage water turns you off, recall that *all* water is recycled by nature. There is hardly a molecule of water you drink that has not been through organisms—including humans—numerous times.

Tucson, Arizona, is an example of a desert city

The Death of the Aral Sea

In the 1930s, economic planners sitting in thick-walled stone buildings in Moscow set in motion a chain of events that has almost killed an entire sea.

In order to create vast cotton fields in the drylands of Soviet Central Asia, the planners had long irrigation canals dug, fed by the waters of two rivers, the Amu Daria and the Syr Daria, that flow into the inland Aral Sea. In the statistics of the central planners, the project was a huge success. The cotton harvests grew until the Soviet Union became the world's second largest cotton exporter, after China.

But the statistics did not show the effect of diverting most of the river flow into the Aral Sea.

By 1989, the sea was receiving only one-eighth the level of water as in 1960. Its water level had dropped by 47 feet, more than a quarter, and its volume had shrunk by two-thirds. Once the size of North America's Lake Huron, its total area diminished by 44 percent.

The shores of the sea receded, leaving fishing villages tens of miles from the shore. A new desert was created around the sea, with salt strewn in massive dust storms across a vast area.

At the same time, Moscow's demands for cotton cultivation were met with saturation use of pesticides, fertilizers, and herbicides that flowed into the rivers and canals. The population around the sea, deprived of clean drinking water and living on poisoned soil, experienced rising rates of disease and infant mortality.

"The problem of the Aral Sea is very simple," says Igor Zonn, a specialist on the subject, "The Aral Sea will be dead, not soon and not completely, but it will be dead."

Dr. Zonn and fellow Russian scientist Nikita Glazorsky, head of the Institute of Geology and until recently deputy environment minister of Russia, have been working for a long time to try to save the Aral Sea. . . .

The Russian scientists say a technical solution for the sea is already well-formulated. Much of the water is now wasted because of evaporation and drainage out of unlined irrigation canals and primitive irrigation technology.

At least half the 120 cubic kilometer flow of the two rivers could be saved by rebuilding the canals and introducing new irrigation systems. It would be enough to begin to stabilize the Aral Sea at its present level, though not enough to restore it. At the same time there must be a program of health care and a con-

certed effort to shift the economy away from cotton cultivation, they say.

But such measures require resources and a political will that is not present. The breakup of the Soviet Union and its replacement by a loose commonwealth has given the Russian government an opportunity to dump the problem into the laps of the five former Soviet Central Asian states that form the Aral Sea's water basin. This is the source of anger for Central Asians who see the cotton monoculture imposed by Moscow as a classic example of colonial-style exploitation.

"We are left face to face with the Aral Sea problem," Kazakhstan president Nursultan Nazarbayev said . . . "Meanwhile 97 percent of the cotton of the Central Asia and Kazakhstan is taken out to the European part of the Commonwealth of Independent States where the employment of 10 million workers depends on cotton use."

Russia has an ethical responsibility to help, agrees Dr. Glazorsky. "All of us live in this system, and we must solve this problem together."

that has implemented conservation and reuse to sustain water supplies. City ordinances require water-conserving toilets and showerheads, and the city shares the cost of putting these devices in older homes. Parks and golf courses are watered with treated wastewater. The major impetus behind such changes is simply coming to recognize the value of water, a topic we shall return to in the final section of this chapter.

In the face of growing population, neither developing more water resources nor conservation is an ultimate solution; such steps can only buy time. The ultimate question we must face is: With each additional person creating a certain additional demand for water, how much population growth can be accepted locally, regionally, or globally without causing increasing cost and hardship for everyone?

EFFECTS CAUSED BY CHANGING LAND USE

All the land area from which water drains into a particular stream or river is known as the **watershed** of that stream or river (Fig. 11–16). If the watershed is covered by natural forest with rich topsoil, most of the precipitation will infiltrate and recharge the

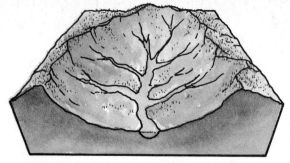

FIGURE 11–16
A *watershed* is all the land area that drains into a particular stream or river. The colored area depicts the watershed for this stream.

groundwater. In such a forested area, only exceptionally heavy or prolonged storms are sufficient to saturate the ground and cause significant runoff. Thus, a steam draining a forested watershed doesn't flood because water percolates into the groundwater. Such a stream continues to flow during dry periods because the groundwater continues to "leak" out through springs and seeps at a more or less constant rate. The stream is able to support a rich aquatic ecosystem as well as much of the surrounding terrestrial ecosystem.

With deforestation, rainfall quickly compacts the soil, runoff increases enormously, and infiltration decreases. The increase in runoff causes erosion, siltation of downstream waterways, and increased

flooding. For example, extensive flooding in Bangladesh is now common because of deforestation of the Himalayan foothills in India and Nepal. (This is just one example of how the actions of one country can have an environmental impact on another country.) The flooding in Bangladesh is only half the problem, however. With less infiltration, water tables drop, springs go dry, and the flow of streams and rivers is diminished. Available groundwater is also obviously less. Thus water shortages during dry periods are intensified.

An even more profound change in land characteristics—one we are all very familiar with—is caused by urban and suburban development. Runoff is increased enormously whenever hard, impervious surfaces such as roadways, parking lots, and rooftops are built. Even the soil of suburban lawns is generally compacted so much that the infiltration-runoff ratio is decreased significantly.

The change in stream flow that occurs with development has been well documented (Fig. 11–17). The change from a more-or-less constant volume to one that alternates between flooding or near flooding when it rains and near or total dryness when it doesn't is conspicuous. The increased flow during rains occurs because runoff enters the stream channels almost immediately. We often see stormwater from roadways and parking lots going down storm drains, which funnel the water into the nearest stream. Water flow between rains decreases because

FIGURE 11–17
Curves are for similar storms on Brays Bayou in Houston, Texas, before, during, and after development. Note the increasing height of the surge occurring with the storm and the decreasing volume of flow that occurs later in the cycle. Because of increasing runoff from new development in the watershed, established areas experience flooding where none occurred before. (Melissa Hayes English/Photo Researchers; V. Van Sickle, in *Effects of Watershed Changes on Stream Flow*, W. Moore & C. Morgan, eds. Austin: U. of Texas Press, 1969.)

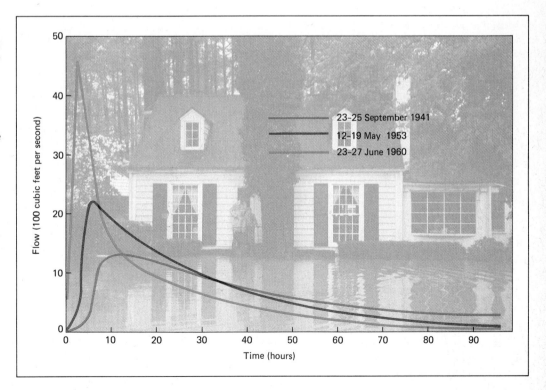

23–25 September 1941
12–19 May 1953
23–27 June 1960

Flow (100 cubic feet per second)

Time (hours)

FIGURE 11–18
The game stream channel during and shortly after an average rainstorm. (a) Because runoff from the streets and parking lots is funneled directly into streams they tend to flood with nearly every rain. (b) Because of lowered watertable resulting from decreased infiltration, the stream becomes dry shortly after the rain. (Photograph by BJN.)

the lower infiltration causes the water table to drop and many springs that previously fed the stream to go dry (Fig. 11–18).

Most streams in urban and suburban areas that used to flow quietly year-round and supported a diverse biota are now little more than open storm drains alternating between surges of water when it rains and dry, dead beds when it doesn't. Indeed, many such streams have been covered and simply incorporated into the storm-drain system. In urban areas, vast arrays of streams and tributaries no longer exist except as underground storm drains.

The surges from increasing runoff have a number of other effects as well.

Flooding The increased potential for flooding is self-evident. Countless communities, many of them expensive new suburban developments, have experi-

enced flooding with increasing frequency and severity as development has paved more and more of the upstream watershed.

Streambank Erosion Even short of flooding, the surges of water vastly accelerate erosion of the stream banks, undercutting trees and causing them to topple into the stream, which diverts water against and over the banks and causes even more erosion. Finer materials are washed away, but rocks, stones, and coarse sand are deposited. These deposits clog the channel, cause the water level to rise and result in even more bank erosion. Further loads of sediment may come from badly placed storm-drain outlets, which erode gullies down sides of the valley. The result is that the stream or river channel gets wider and shallower. Indeed, a stream channel may become completely filled such that the water is diverted and simply floods the valley floor (Fig. 11–19). This results in topsoil erosion and the death of many trees because what soil is left is waterlogged. Gradually, a narrow, tree-lined stream may be converted into a broad washway of drifts of sand and gravel.

Increased Pollution Earlier in this chapter we noted that runoff results in poor-quality water because the water tends to pick up anything that happens to be on the ground surface. In addition to the sediments from erosion, the following items are all abundant in urban runoff:

○ nutrients from lawn and garden fertilizer

○ insecticides and herbicides used on lawns and gardens

○ bacteria from fecal wastes of pets

○ road salt and other chemicals from surface treatments or spills

○ grime and toxic chemicals from settled vehicle exhaust and other air pollution

○ oil and grease picked up from road surfaces or disposed of in storm drains

○ trash and litter carelessly discarded on the ground

Indeed, urban runoff is now recognized as a major source of nonpoint pollution for many rivers and estuaries. Recall that the first source we learned about was runoff from agricultural lands. Today, the three main sources of water pollution in the United States are discharges from point sources, such as sewage treatment plants or industries, urban runoff, and agricultural runoff. Which is greatest in a given area obviously depends on which activities predominate in the watershed.

Turning to the other side of the infiltration-runoff ratio, decreased infiltration also exacerbates saltwater intrusion, land subsidence, and other problems related to falling water tables.

STORMWATER MANAGEMENT

The traditional approach to handling stormwater has been to get rid of it as quickly as possible. Parking lots, roadways, and lawns are graded to quickly shed water into storm drains, which funnel the water into the nearest convenient stream. If streambank erosion and flooding become too troublesome, streams are **channelized,** that is, the channel is dredged, straightened, or gently curved, and lined with rock or concrete to prevent bank erosion. The channelized stream carries both water and sediment more efficiently, and so the hazard of flooding in the area is greatly reduced. In addition to ecological damage, however, the channelized stream often transfers the flooding to a point farther downstream.

Fortunately, the failings of this traditional approach are now recognized, and new approaches are gradually being implemented. The modern ecological concept of stormwater management is to keep the stormwater near where it falls and preserve the natural infiltration-runoff ratio. A number of techniques have been developed to achieve this goal:

○ Dry wells and trenches—broad, shallow, rock-filled wells or trenches that receive runoff and then let it percolate into the ground.

○ Swales, which are depressions created to channel the flow of water, with barriers for holding excess runoff.

FIGURE 11–19
Sediment from upstream erosion fills and clogs stream channels, causing flooding and further erosion of the stream valley as the water is forced to find new pathways. In this photo, you can see how the stream channel has been nearly filled with fine gravel sediment, which eroded from upstream areas. (Photograph by BJN.)

○ Graded areas that allow would-be runoff to accumulate in broad shallow depressions and then infiltrate.

○ Parking lots paved with porous surfaces.

○ Retention ponds. Rather than being run directly into streams, storm drains are directed into artificial ponds that will hold the runoff and allow it to drain away slowly. Stormwater retention ponds may support pockets of wildlife (Fig. 11–20), or they may provide a practical source of water for nondrinking purposes, thus lessening the demand

---

ETHICS
Watershed Management

Dams are built and reservoirs are created in order to capture water supplies. What is the quality of water in the reservoir? Its quality is determined by the activities on the watershed that drains into the reservoir. If the watershed is forested, the major pathway of water into the reservoir will be infiltration into the forest topsoil, percolation into the groundwater, and seepage of the groundwater into the numerous tributaries of the streams that feed into the reservoir. Thus, most of the water in the reservoir, having been filtered through the forest soil and

roots and through the ground, has virtually the same quality as spring water.

As the watershed is deforested and developed, runoff increases. As noted in the text, runoff picks up all the pollutants on the ground surface—soil sediments, fertilizer, pesticides, pet and animal droppings, oil and grease, and any other chemicals used on or buried in the gound. Therefore, water quality in the reservoir can only deteriorate as development proceeds.

The objective of **watershed management** is to control activities in

the watershed to maintain water quality in the reservoir. Ideally, the water shed should be kept entirely forested. However, watersheds and reservoirs are frequently close to the cities they serve, and land in the watershed is a prime target for development as the cities grow. There are growing controversies between development interests and those wanting to protect the watershed for protection of water quality. How should such controversies be resolved?

FIGURE 11–20
Stormwater retention pond. Rather than letting excessive runoff from developed areas cause environmental damage, it may be funneled into a retention pond, from where it may either drain away slowly or infiltrate. The design of the standpipe allows the pond shown here to slowly drain away excess water while retaining enough to create a useful habitat for ducks and other organisms. (Photograph by BJN.)

on the municipal system. Large stormwater retention reservoirs can also provide recreational opportunities. During a storm the recreation area around the reservoir can store floodwater. Recreational equipment can be made so that it will not be damaged by the flooding water, and there is little inconvenience to people since most of them would leave at the time of the storm, whether the area was flooded or not.

○ Rooftops and parking lots designed to "pond" water and let it trickle away slowly.

Proper stormwater management can reduce the economic losses due to flooding, protect streams from excessive bank erosion, maintain a more uniform stream flow and thus the natural ecology of the stream or river valley, enhance infiltration and recharge of groundwater, and provide additional sources of (nonpotable) water.

Implementing Solutions

We have seen that vast opportunities exist for conserving and recycling water. The question is, How

can we get ourselves and others to implement these measures short of social or ecological disaster?

Water conservation has been promoted for years, and many communities have found it necessary to impose restrictions on watering lawns and washing cars in time of drought. Talk alone has a short-term and limited effect, however, and regulations get increasingly difficult to enforce and require a bureaucracy that becomes more and more unwieldy to manage. Few people can be talked into saving water when the costs or inconveniences of saving it outweigh the money saved. Yet most cities continue to provide water to consumers for only a tiny fraction of a penny per gallon, so that even extravagant use costs us relatively little. Indeed, our water bills reflect only the cost of treatment and piping; the water itself is considered free. None of the ecological costs, added pollution, or costs of depleting supplies are included. Therefore, there is little economic incentive to save water regardless of the virtues.

Many people point out that a direct way of promoting water conservation would be to raise the price. An "ounce" of economic incentive has more effect than many "pounds" of talk. Hardship on the poor can be avoided by having a sliding scale so that cost per gallon increases with the number of gallons used, and increasing costs can be phased in. For example, Arizona has passed a law that will gradually raise prices high enough to prevent the water table from dropping further. Having water cost more will also provide the incentive for retention and use of stormwater and for reuse of gray water. Higher water charges will pay for recycling of municipal waste-

water. In a related vein, many analysts advocate that water subsidies to farmers be phased out.

None of us relishes paying more for water. However, we must either be willing to accept these up-front costs or deal later with the hidden costs we shall ultimately pay for pursuing water-use policies that are nonsustainable.

 Review Questions

1. During evaporation and condensation, what happens to water molecules, and what energy changes are involved?
2. Trace three alternative pathways of a water molecule around the water cycle. What parts of the cycle are basically the same? What parts may differ from each other?
3. Define runoff, infiltration, surface water, capillary water, gravitational water, percolation, groundwater, leaching, water table, aquifer, recharge area, spring, seep.
4. Why do different regions receive different amounts of precipitation?
5. What are the major uses of water, and where and how do humans obtain water supplies?
6. What are the civic and ecological consequences of overdrawing surface waters?
7. What are or may be the civic and ecological consequences of overdrawing groundwater?

8. What are the pros and cons of expanding water supplies versus using less?
9. List several methods of water conservation, and describe the advantages and disadvantages of each in terms of cost, convenience, and amount of water saved.
10. What is the potential for reusing water, and what are the impediments?
11. How does development affect water quantity and quality in various parts of the water cycle?
12. What is meant by stormwater management? Why is it important?
13. What are the various techniques of stormwater management, and what are the pros and cons of each?
14. Are new policies called for to achieve sustainable water supplies? What policies would you support or promote?

Thinking Environmentally

1. The water cycle has the potential to provide sustainable supplies of high-quality water, but humans manage their water supplies in ways that are not sustainable. How can we change our habits to ensure sustainable water supplies without damaging the environment in the process?
2. For water in any part of your environment (natural or human-made), analyze how it is really part of the water cycle by describing where the water came from and where it will go.
3. Outline the many ways in which humans are altering the water cycle and the social and environmental impacts of this alteration.

4. Increasing numbers of people are moving to the arid Southwest, despite the fact that water supplies are already being overdrawn. If you were the governor of one of these states, what policies would you advocate regarding this situation?
5. Discuss the pros and cons of increasing the cost of water to consumers.
6. There are many proposals for additional housing developments around reservoirs that supply drinking water. How will such development affect water quality? What policies would you advocate regarding such development?

Finding a Balance Between Population, Soil, Water, and Agriculture

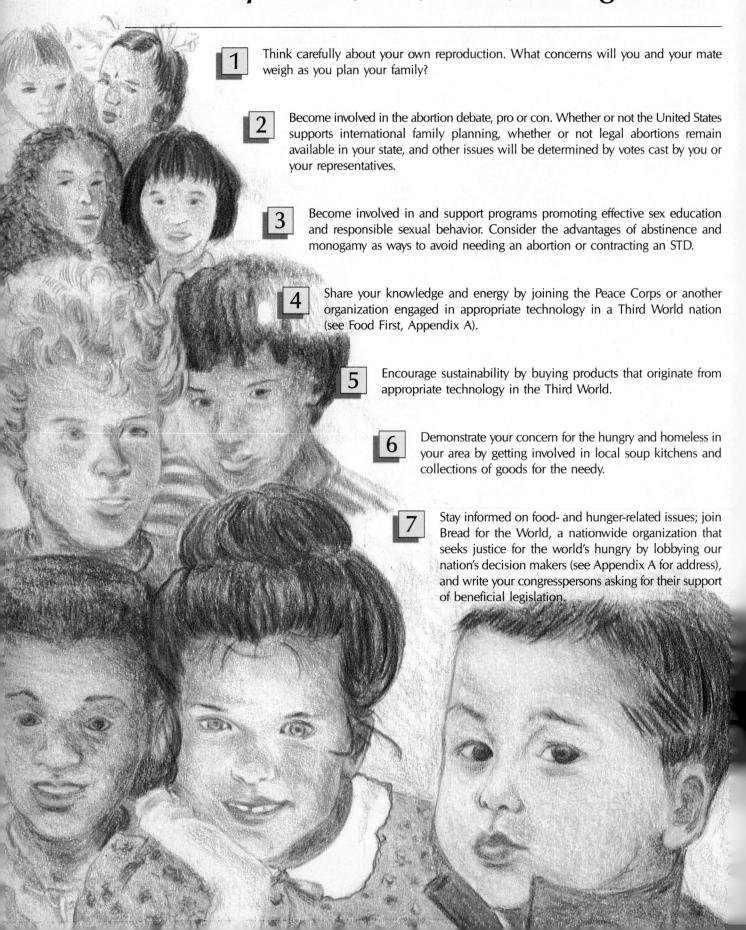

1. Think carefully about your own reproduction. What concerns will you and your mate weigh as you plan your family?

2. Become involved in the abortion debate, pro or con. Whether or not the United States supports international family planning, whether or not legal abortions remain available in your state, and other issues will be determined by votes cast by you or your representatives.

3. Become involved in and support programs promoting effective sex education and responsible sexual behavior. Consider the advantages of abstinence and monogamy as ways to avoid needing an abortion or contracting an STD.

4. Share your knowledge and energy by joining the Peace Corps or another organization engaged in appropriate technology in a Third World nation (see Food First, Appendix A).

5. Encourage sustainability by buying products that originate from appropriate technology in the Third World.

6. Demonstrate your concern for the hungry and homeless in your area by getting involved in local soup kitchens and collections of goods for the needy.

7. Stay informed on food- and hunger-related issues; join Bread for the World, a nationwide organization that seeks justice for the world's hungry by lobbying our nation's decision makers (see Appendix A for address), and write your congresspersons asking for their support of beneficial legislation.

Making a Difference

8 Produce some of your own food by planting a vegetable garden and raising animals; practice sustainable agriculture: avoid the use of herbicides and pesticides, compost kitchen and yard wastes, and build up a humus-rich soil. Any space—even a rooftop—with 6 hours of sunlight a day is enough.

9 Contact your county soil conservation or farm extension agent and ask for help in communicating to others the work being done to prevent erosion and runoff of farm chemicals.

10 Inventory the pesticides and herbicides in your home, and switch to either cultural control methods or "soft" pesticides based on natural products.

11 If you are concerned about the source of your food, consider buying organically grown foods. If they are not available where you shop, ask your grocer to stock them.

12 For your community/city, investigate: Where does the water come from and how is it treated? Are water supplies adequate or being overdrawn? How is stormwater managed in your area? What policies/actions are being considered to meet future needs?

13 Observe where stormwater from your own home and yard goes (does it mostly run off or infiltrate?) Install a system, perhaps just a barrel at a downspout, to capture and use stormwater for lawn/garden watering and car washing.

14 Examine your own eating habits and opt for a nutritionally balanced diet that includes less red meat and more fruit and vegetables.

Part Three

Pollution

An industrial plant sends a steady stream of pollutants into the air and the wind carries the pollutants away. If it didn't, there could be no field of flowers in the same scene. But where is "away"? The next county, or state, or nation? The upper atmosphere? Sadly, we are learning that there is no "away," as we face the complex problems of hazardous wastes, sewage, acid deposition, global warming and loss of the ozone shield. Must we simply accept the fact that there could be no economy without polluting, or that wherever there are people, they will have to send their natural wastes to the environment for disposal?

In this part we will take a hard look at the major forms of pollution. We will see that whether we are talking about air pollution, water pollution, pollution by toxic chemicals, or by human and animal excretions, there are responsible ways of dealing with the problems. It is possible to have people and economic enterprises without overburdening the earth, air and water with wastes. However, dealing responsibly with wastes is a deliberate and often costly matter, and it happens in the context of a human social and political system that can be slow to see the harm caused by pollution and reluctant to take costly action. As the chapters in this part unfold, look for the changes in public policy and individual lifestyles that will signal a successful move in the right direction—towards sustainability and the environmental revolution.

Sediments, Nutrients, and Eutrophication

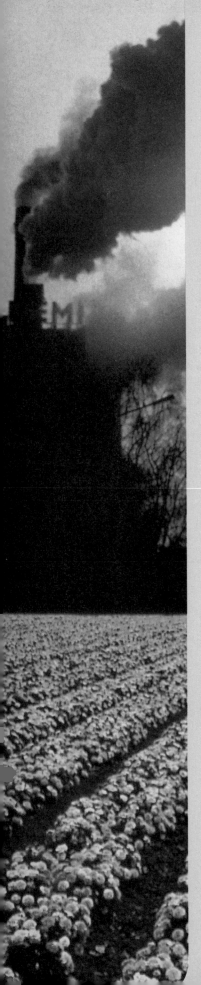

LEARNING OBJECTIVES

When you have finished studying this chapter, you should be able to:

1. Define pollution and pollutants and discuss the diversity of pollution problems.

2. Name and describe two categories of aquatic plants and contrast how the balance between them is altered by the nutrient content of the water.

3. Explain eutrophication, giving all the steps in the change from an oligotrophic to a fully eutrophic condition.

4. Contrast eutrophication in a body such as Chesapeake Bay with that occurring in shallow ponds and lakes.

5. Explain how overgrowth of aquatic plants leads to oxygen depletion and dieoff of fish and shellfish.

6. Contrast natural and cultural eutrophication.

7. Contrast methods of attempting to control eutrophication by attacking symptoms versus getting at the root cause.

8. Describe how soil sediments affect aquatic ecosystems and contribute to eutrophication.

9. Identify the major sources of sediment and discuss control strategies for each source.

10. Identify the major sources of nutrients leading to eutrophication and discuss control strategies for each source.

11. Describe two kinds of wetlands and how each plays a natural role in controlling eutrophication.

12. Describe how wetlands are being destroyed and the natural values lost as this destruction takes place.

Here in Part 3 we turn our attention to the study of pollution. Pollution is an everyday word, and everyone has some concept of what it means. Yet, a formal definition is in order. **Pollution**—of water, air, or soil—is the human-caused addition of any material or energy (heat) in amounts that cause undesired alterations. The material that causes the pollution is called a **pollutant.**

It is important to note the breadth and diversity of pollution. Any part of the environment may be affected, and virtually anything may be a pollutant. The only criterion is that the addition results in undesirable alterations. The impact of the undesirable alteration may be largely aesthetic—hazy air obscuring a distant view or the unsightliness of roadside litter, for instance. Or the impact may be biological, such as causing the dieoff of fish or forests. Or the impact may be on human health—toxic wastes contaminating water supplies, for example. Also the impact may range from very local—the contamination of an individual well, for instance—to global.

In all cases, pollution is rarely the result of wanton mistreatment of the environment; the additions that cause pollution are almost always the *byproducts* of otherwise worthy and essential activities, such as producing crops, creating comfortable homes, providing transportation, manufacturing products, harnessing the atom, and of our basic biological functions (excreting wastes). An overview of the categories of pollutants that result from various activities is shown in Figure 12–1.

Pollution problems have become more pressing over the years both because of increasing population and because of increasing per capita use of materials and energy. Also, many materials now widely used, such as aluminum cans, plastic packaging, and innumerable synthetic organic chemicals, are **nonbiodegradable.** That is, they resist attack and breakdown by detritus feeders and decomposers and consequently accumulate in the environment.

From this brief overview, it should be evident that the simple slogan, "Give a hoot; don't pollute" is a gross oversimplification of the problem. As long as we exist, there will be byproducts of our activities. Ultimately, ridding the environment of pollution will require basic changes in the way we generate power, provide transportation, manufacture products, and so on. The necessary changes will involve adopting means both of providing for our needs in ways that produce less-troublesome byproducts and of recycling byproducts.

No one technique is applicable to all pollution problems. Instead, to rid the environment of pollution, it is necessary in each case to:

1. identify the material or materials that are causing the pollution;

2. identify the sources of those polluting materials; and

3. develop and implement strategies for controlling the pollution from those sources.

Therefore, this and the following four chapters will each focus on particular categories of pollution. In this chapter, we focus on *eutrophication* (pronounced, *yoo-tro-fuh-kay-shun*), which basically refers to an overenrichment with nutrients. Numerous undesirable alterations in aquatic ecosystems take place as a result of nutrient oversupply. Our objective is to understand the range and severity of these alterations, how they occur, and potential control strategies.

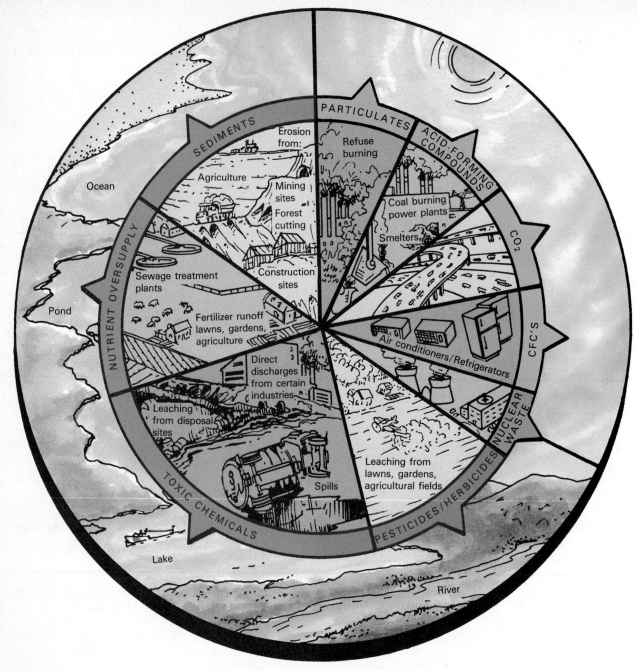

FIGURE 12–1
Pollution is the byproduct of otherwise worthy human endeavors. Major categories of pollution and the activities that cause them are shown here.

The Process of Eutrophication

THE CASE OF CHESAPEAKE BAY

The Chesapeake Bay (Fig. 12–2) is North America's largest estuary and prior to the 1970s was its most productive, yielding many millions of pounds of fish and shellfish and supporting vast flocks of waterfowl. Most of the food chains supporting this rich bounty had their origin in the seagrasses, half a million acres (200 thousand hectares) of underwater "grass" growing on the bottom 3–6 feet (1–2 m) beneath the surface. The beds of seagrass provided not only food but also spawning habitats and shelter for young fish and shellfish and dissolved oxygen for them to breathe.

In the early 1970s, the seagrasses in all the major rivers and subestuaries leading into the bay started dying. By 1975, the dieback was dramatic. By 1980,

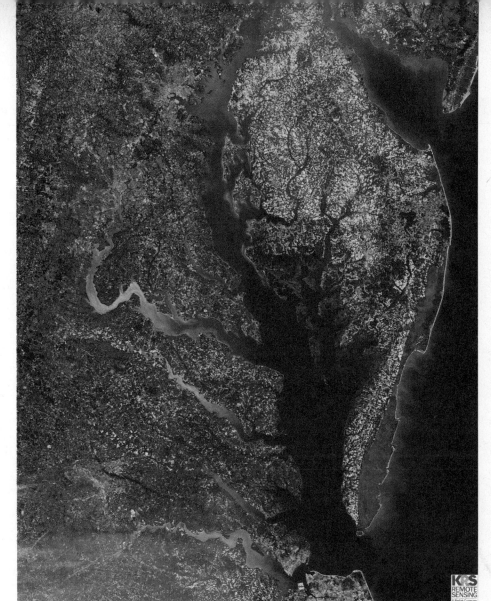

FIGURE 12–2
Photograph of Chesapeake Bay area taken by the *Earth Resources Satellite* at an altitude of 550 miles (870 km). The gray areas at the upper left are Washington, D.C., and Baltimore, Maryland. Light tan and reddish rectangles are agricultural fields. The green speckles and patches are remaining forest. The rust color on the lower east side of the bay is remaining tidal wetlands. Note the extent of urbanization and agriculture in the watershed. Both processes are sources of nutrients and sediments. The tan color of the three rivers in the lower left is a result of sediments eroding after a rain. (Satellite imagery provided by EOSOT, Landham, Maryland. © Chesapeake Bay Foundation, 1987.)

the grasses were gone except in the main stem of the lower bay. Populations of the fish, shellfish, and waterfowl that had depended on the grasses soon declined also. Even more devastating, the bottom waters in deep areas of the bay became depleted of dissolved oxygen, causing huge numbers of fish and shellfish to be suffocated. What caused the dieoff of seagrasses and the depletion of dissolved oxygen in the Chesapeake?

A team of scientists from the University of Maryland and the Virginia Institute of Marine Science, supported by grants from the Environmental Protection Agency, investigated the problem. Toxic chemicals from industry were ruled out because, while a problem in certain locations, they could not be responsible for a dieback throughout the bay. Herbicides used on farmlands were suspected, but tests showed that they did not reach damaging levels except in small ditches and streams receiving drainage directly from farm

fields. Then investigations turned to light, and this proved to be the key. The waters of the Chesapeake had become increasingly **turbid** (murky or cloudy), and the cloudiness was persisting over extended periods of time. The increased turbidity was cutting off the light required for photosynthesis, and the seagrasses were dying as a result. What was causing the increased turbidity? It was *sediments* (mainly clay particles in suspension) and *phytoplankton*.

With the loss of the seagrasses, dissolved oxygen was no longer being supplied by their photosynthesis. Even more harmful, bacterial decomposers, feeding on the abundance of dead seagrass, and detritus from phytoplankton were consuming the dissolved oxygen, thus making it unavailable to fish and shellfish.

The Chesapeake Bay has fallen prey to eutrophication. The process of eutrophication, which we shall describe in detail, is not unique to the Chesapeake.

In the last 40 years, many thousands of ponds, small lakes, and even some large lakes and certain rivers have suffered this fate, and the problem is continuing to spread. However, there are means of checking the problem, and remarkable recovery has occurred in some cases as a result.

Let us investigate the causes of this problem as well as the means of preventing or even perhaps reversing it.

TWO KINDS OF AQUATIC PLANTS

To understand more thoroughly the eutrophication that has occurred in Chesapeake Bay and other such bodies of water, we need to consider two distinct life forms of aquatic plants: *benthic plants* and *phytoplankton*.

Benthic plants (from *benthos*, "deep") are aquatic plants that grow attached to or rooted in the bottom. All common aquarium plants and seagrasses are examples (Fig. 12–3a). As shown in the figure, benthic plants may be divided into two categories: **submerged aquatic vegetation (SAV),** which generally grows totally under water, and **emergent vegetation,** which grows with the lower parts in water but the upper parts emerging from the water. The most important point to remember for understanding eutrophication is that benthic vegetation thrives in *nutrient-poor* water. This is because benthic plants take in mineral nutrients through their roots from the bottom sediments, just as land plants do. SAVs, however, require clear water so that sufficient light for their photosynthesis can penetrate, except when the water is relatively shallow, in which case they may reach the surface.

FIGURE 12–3
Two basic categories of aquatic plant life. (a) Benthic, or bottom-rooted, plants. These are subdivided into submerged aquatic vegetation (SAV) and emergent vegetation. (b) Phytoplankton, various species of plants that exist either as single cells or as small groups or filaments of cells that float freely in the water. Benthic plants withdraw nutrients from sediment and hence do well in nutrient-poor water, while phytoplankton depend on nutrients dissolved in the water.

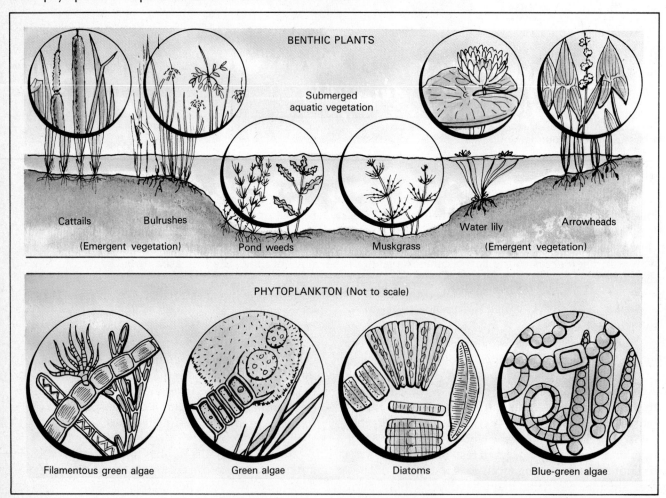

BENTHIC PLANTS

Submerged aquatic vegetation

Cattails   Bulrushes

(Emergent vegetation)   Pond weeds   Muskgrass   Water lily   Arrowheads

(Emergent vegetation)

PHYTOPLANKTON (Not to scale)

Filamentous green algae   Green algae   Diatoms   Blue-green algae

The depth to which adequate light for photosynthesis can penetrate is known as the **euphotic zone.** In very clear water, this depth may be nearly 90 feet (30 m). However, as water becomes more turbid, the euphotic zone is reduced; in extreme situations, it may be reduced to a matter of a few inches (1 in = 2.5 cm).

Phytoplankton (from *phyto*, plant, and *plankton*, floating) consists of numerous species of algae that grow as microscopic single cells or small groups, or "threads," of cells that maintain themselves near or on the water surface (Fig. 12–3b). Since phytoplankton are near or on the surface, turbid water is of less consequence to them than to benthic plants. Indeed, high populations of phytoplankton are a major cause of turbidity, as they absorb the light needed for photosynthesis. In extreme situations, water may become literally pea-soup green (or tea-colored, depending on the species involved), and a scum of phytoplankton may float on the surface and absorb essentially all the light. Phytoplankton reach such density only in *nutrient-rich* water because, not being connected to the bottom, they must absorb their nutrients from the water. Lack of nutrients in the water limits the growth of phytoplankton.

With this information, can you see how the balance between submerged benthic plants and phytoplankton may be shifted when nutrients are added to the water? As we shall see in the following section, adding nutrients is exactly what humans have done as an unwitting byproduct of other activities.

UPSETTING THE BALANCE BY NUTRIENT ENRICHMENT

The Oligotrophic Condition

Natural bodies of water, where their watershed has been undisturbed by human activities, are generally **oligotrophic,** meaning the water is *nutrient-poor.* This is because, as we have learned, natural ecosystems are very efficient in holding and recycling nutrients on the land. Very little erosion or leaching of nutrients occurs from forested watersheds, for example. The oligotrophic condition limits the growth of phytoplankton, but benthic plants may thrive to a depth of 45 feet (15 m) or so, the usual extent of the euphotic zone.

In addition to providing food and habitat, benthic plants, particularly the SAV, also aid in maintaining a high level of dissolved oxygen in deeper water. Oxygen from the atmosphere is very slow to dissolve in and mix through the water, but the oxygen released from the photosynthesis of SAV dissolves directly into the water. Thus, an oligotrophic body of water is characterized by being nutrient-poor but oxygen-rich from top to bottom (Fig. 12-4a). It may support a diverse ecosystem of fish and shellfish. Because the water is so clear, oligotrophic bodies of water are prized for their aesthetic and recreational qualities.

The Eutrophic Condition

As an oligotrophic body of water is enriched with nutrients, numerous changes are set into motion.

FIGURE 12–4
Dissolved oxygen levels typical of (a) oligotrophic and (b) eutrophic conditions.

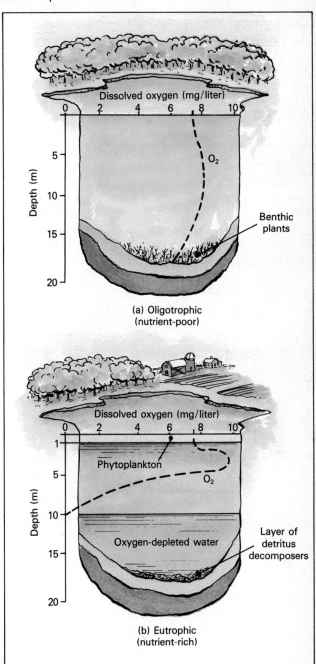

(a) Oligotrophic (nutrient-poor)

(b) Eutrophic (nutrient-rich)

First, the nutrient enrichment allows the rapid growth and multiplication of phytoplankton, resulting in increasing turbidity of the water. The increasing turbidity shades out the submerged benthic plants, interrupting food chains and causing loss of habitat. The shading effect may be exacerbated by sediments; clay particles in particular may remain in suspension for long periods of time, adding to the turbidity. (Muddy-looking water is the effect of these particles in suspension.) Dissolved oxygen is lost too. Phytoplankton do produce oxygen in photosynthesis, as do all green plants. Because these plants grow near the surface, however, the surface water becomes supersaturated with oxygen, and the excess oxygen escapes to the atmosphere. On a calm, sunny day, you can often observe bubbles of oxygen entrapped by filamentous algae being released at the surface. Thus, photosynthesis by phytoplankton does not replenish the dissolved oxygen of deeper water.

Still more is involved. Phytoplankton has remarkably high growth and reproduction rates. Under optimal conditions, phytoplankton can double its biomass every 24 hours, a capacity far beyond that of benthic plants. Thus, phytoplankton soon reaches a maximum population density, and continuing growth and reproduction are balanced by dieoff. Dead phytoplankton cells gradually settle out, resulting in heavy deposits of detritus on the bottom.

As decomposers, mainly bacteria, feed on the detritus, they consume oxygen in their respiration. This process causes the depletion of dissolved oxygen in the deeper layers of water and the consequent suffocation of the fish and shellfish that inhabit those waters (Fig. 12–4b). The process does not kill the bacteria, however. When dissolved oxygen is no longer available, they have the capacity to shift to alternative pathways of cell metabolism that do not require oxygen (see p. 60). Continuing to thrive, but using oxygen whenever it is available, such bacteria can deplete and hold the dissolved oxygen at zero as long as there is detritus to support their growth.

By the same token, organic matter added directly to a body of water may lead to depletion of dissolved oxygen. Such additions occur with the discharge of raw sewage and organic wastes from certain industries (such as food-processing plants and paper mills). Indeed, the potential impact of such wastes is commonly measured in terms of their **biochemical oxygen demand (BOD).** This is the amount of oxygen that will be absorbed, or "demanded," while the wastes are being digested or oxidized. Clearly, if the BOD of the wastes exceeds the amount of dissolved oxygen in the receiving water, oxygen depletion will occur. In rivers, dissolved oxygen is commonly depleted below the point of waste discharge. Further downstream, recovery may occur as the organic material is oxidized and additional dissolved oxygen enters from the atmosphere (Fig. 12–5). In the United States, pollution control laws now require such wastes to be treated before discharge to reduce the BOD. How this is done for sewage wastes is considered in Chapter 13.

In summary, a **eutrophic** body of water is one characterized by *nutrient-rich* water that supports abundant growth of algae and perhaps other aquatic plants at the surface. Under the surface layer, plant growth is greatly diminished or absent because of

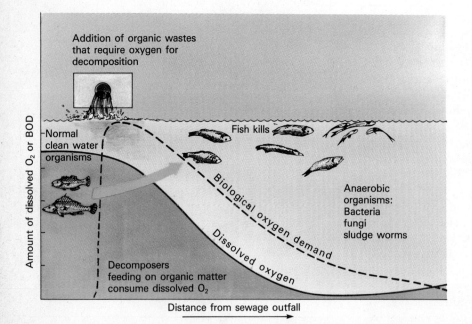

FIGURE 12–5
Dissolved oxygen and biological oxygen demand (BOD). Fish and other aquatic organisms depend on dissolved oxygen for their respiration. Additions of organic matter create an additional BOD because bacteria (decomposers) that feed on the organic matter consume oxygen as well. Thus, additions of organic wastes to waterways may result in fishkills through depletion of dissolved oxygen. After the organic wastes are fully decomposed, recovery of dissolved oxygen will occur as atmospheric oxygen slowly dissolves in the water.

shading; on the bottom is an accumulation of detritus. Dissolved oxygen is high at the surface because of photosynthesis of the phytoplankton but declines to near zero or zero at the bottom because of oxygen consumption by decomposers.

Eutrophic lakes have been called "dead," but, biologically, this is a misnomer because the total production of biomass by phytoplankton may greatly outpace that of the previous benthic plants. In turn, the phytoplankton may support large populations of certain fish that are adapted to feed on it and avoid the oxygen-depleted deep water. For example, in Chesapeake Bay, populations of bay anchovy and menhaden, which are filter feeders consuming phytoplankton and other microscopic organisms, are at all-time high levels. Since they are small and oily, however, these species are not good for eating or sport fishing. It is more correct, then, to view eutrophication as an *ecosystem disturbance* in which a diverse system based on submerged benthic vegetation changes to a simple system based on phytoplankton and emergent benthic vegetation. However, the eutrophic system is less than appealing for swimming, boating, and sport fishing and may even have the foul odors of dead organisms, which are generated by decomposition in the absence of oxygen. Therefore, it may be considered "dead" in terms of these values. Also, if the lake is a source of drinking water, its value may be greatly impaired because phytoplankton and mats of vegetation rapidly clog water filters and may cause a foul taste.

Eutrophication of Shallow Lakes and Ponds

In lakes and ponds where the water depth is 3 feet (about 1 meter) or less, eutrophication takes a somewhat different course, but the results are the same. Submerged aquatic vegetation may grow to a height of a meter or so. Therefore, in these shallow waters, the SAV may reach the surface, where it continues to have access to abundant light for photosynthesis. When such lakes and ponds are enriched with nutrients, it is often SAV growth that is stimulated. The vegetation grows rapidly, sprawling over and often totally covering the surface with dense mats of vegetation that make boating, fishing, or swimming impossible (Fig. 12-6). Whatever vegetation is below gets shaded out. As this vegetation dies and sinks to the bottom, it creates a BOD that often depletes the water of dissolved oxygen, causing the death of aquatic organisms other than bacteria.

Natural versus Cultural Eutrophication

Even oligotrophic lakes may be subject to periodic bursts of phytoplankton growth called algal blooms (see In Perspective p. 277) and over many thousands

FIGURE 12–6
Eutrophication in shallow lakes and ponds. In shallow water sufficient light continues to reaches the submerged aquatic vegetation. Hence, oversupply of nutrients stimulates its growth such that it reaches the surface (as seen in the foreground of this photo) and forms mats over the surface (as seen in the background). (Photograph by BJN.)

of years, all lakes are destined to a fate of natural eutrophication. Over such time spans, all lakes will eventually be filled and enriched with the sediments and nutrients that inevitably erode and leach from the land, even though the erosion and leaching rates are very, very slow. What humans have done is to vastly accelerate this process. We have caused changes in a few decades that normally would occur only over several thousand years. This accelerated process is referred to as **cultural eutrophication.**

There is always a tendency to blame pollution problems on industries dumping toxic compounds. Eutrophication is due to what are generally considered benign substances, however: fertilizer nutrients and in some cases soil particles. That such benign materials can cause such a major problem emphasizes a point made in Chapter 4: *Changing any natural factor can grossly upset the balance of an ecosystem.* Figure 12–7 provides a summary of cultural eutrophication.

COMBATING EUTROPHICATION

There are two general approaches to combating the problem of eutrophication. One is to *attack the symptoms:* the growth of vegetation or the lack of dissolved oxygen or both; the other is to *get at the root cause:* excessive inputs of nutrients and sediments. Attacking the symptoms continues to be pursued in some cases, but it is of dubious merit and cost-prohibitive in large lakes. We should be aware of these shortcomings so that we are not tempted to repeat mistakes.

Attacking the Symptoms

Chemical treatments Herbicides have been and continue to be used in agriculture to suppress the growth

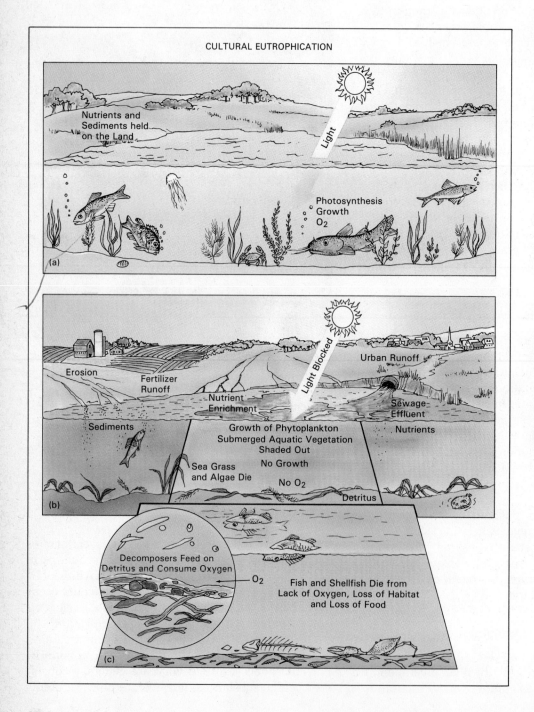

CULTURAL EUTROPHICATION

(a)
Nutrients and Sediments held on the Land

Light

Photosynthesis
Growth
O₂

(b)
Erosion
Fertilizer Runoff
Nutrient Enrichment
Sediments
Light Blocked
Urban Runoff
Sewage Effluent
Nutrients
Growth of Phytoplankton
Submerged Aquatic Vegetation Shaded Out
No Growth
Sea Grass and Algae Die
No O₂
Detritus

(c)
Decomposers Feed on Detritus and Consume Oxygen
O₂
Fish and Shellfish Die from Lack of Oxygen, Loss of Habitat and Loss of Food

FIGURE 12–7
Summary of eutrophication. (a) Natural bodies of water are generally nutrient-poor because nutrients are held on the land by natural ecosystems. Submerged aquatic vegetation getting nutrients from bottom sediments provides habitat, food, and dissolved oxygen for fish and shellfish. (b) Nutrients from sewage effluents and fertilizer runoff stimulate the growth of phytoplankton, which shade out the submerged aquatic vegetation and thus deprive fish and shellfish of habitat, food, and dissolved oxygen. The shading is compounded by sediments from erosion. (c) Decomposition of dead algae consumes additional oxygen, causing fish and shellfish to suffocate.

Algal Bloom and Bust

An algal "bloom" is not a flower! The term refers to a concentration of phytoplankton algae in surface waters. Algal blooms typically occur in the spring and cause the water to lose its normal clarity and take on a brownish or greenish cast. Quite often an algal bloom will appear suddenly in a body of water, last for a week or two, and gradually subside. Blooms that are found in eutrophic bodies of water can last for months, however, and create conditions that are highly undesirable until the algae undergo a decline. What factors are responsible for algal blooms? In answering this question, we will describe events in a typical lake in the north temperate zone.

The explanation really begins during the winter, when temperatures and light are at their seasonal low point. The algae and other organisms from the previous year have died, become detritus, and dropped into the lower depths of the water, often all the way to the bottom. Bacteria in the water column and in the sediments gradually decompose the detritus, and as they do the elements once present in the organisms (most importantly, nitrogen and phosphorus) are released as soluble, inorganic nutrients. Because freshwater reaches its maximum density at 4°C (39°F), water bodies in colder climates in the winter will be stratified, with the coldest water and ice on top. Thus the nutrients accumulate in the deeper water layers all during the late fall, winter, and early spring.

As spring develops, solar energy heats the water and provides increasing amounts of visible light for photosynthesis. Ice thaws, and the upper layers of water, which intercept the greatest amount of solar radiation, gradually warm. As they

warm from 0° to 4°C (32°–39°F) they become denser and sink, mixing with the deeper water. Wind action promotes vertical mixing of the water column, and soon the entire body of water is thoroughly mixed throughout. This situation is called the "spring turnover" and is the catalyst for two important events in the water body. The first is the reaeration of the water. As bacteria decompose detritus, they remove oxygen from the water, sometimes to the point of exhaustion of the oxygen. The spring turnover brings the deeper water layers to the surface, where the water comes into contact with atmospheric oxygen and is thoroughly reaerated. This spring reaeration is important to all aerobic life in the lake, because it charges the water with oxygen before the lake once again becomes stratified under summer conditions.

The second event catalyzed by the spring turnover is the spring bloom. Nutrients regenerated in deeper water layers are mixed throughout the water column, and the combination of nutrients and increasing light promotes the growth of many species of phytoplankton algae at once. In a short time, exponential growth of the algae leads to the densities referred to as blooms. Blooms occur in oligotrophic as well as eutrophic lakes, but densities are not as great in the former. Eventually, the bloom is limited, as all of the inorganic nutrients in the surface waters are removed by the phytoplankton. In the meantime, zooplankton, which increase more slowly than phytoplankton, are taking advantage of the high numbers of their phytoplankton food and beginning to eat down the bloom, providing another check to the phytoplankton numbers. The bloom turns into a bust!

The warming of the surface waters

eventually leads to summer stratification, with warmer water overlying the colder water. Some turnover of nutrients occurs in the surface layers as a result of food chain and bacterial activities, promoting continued summer algal growth but at much lower densities than during the spring bloom. Much of the algal biomass gradually sinks out of the upper lighted layers into the bottom water (along with fecal pellets and other organic detritus). Once again the process of bacterial decomposition occurs, and nutrients are regenerated in the cooler bottom waters and sediments.

You can probably guess what happens in the fall, as the upper layers cool and gradually mix with the deeper water and the "fall overturn" occurs. Once again the water is reaerated, and another bloom occurs. This time, however, light and temperature are diminishing, and the bloom is not usually as remarkable as the spring bloom and fairly quickly subsides.

There are many exceptions to this basic process. Very shallow lakes and ponds, for example, do not become stratified; they often are dominated by rooted vegetation, which competes successfully against the phytoplankton and prevents large blooms. Also, many eutrophic bodies of water continually receive nutrients from cultural sources, and phytoplankton blooms are sustained throughout the summer by these sources. However, it should be apparent that blooms are the result of the combination of nutrients and light, and if the blooms are undesirable, the only sure way to bring about control is to get at the source of the nutrients.

of unwanted plants. An obvious extrapolation of this concept was to believe that unwanted algae and benthic vegetation could also be eliminated by chemical treatments, and thousands of tons of chemicals were spread on U.S. ponds and lakes in the 1960s and 1970s. The results were less than inspiring. Planktonic algae—especially blue-green species, which are the most obnoxious—are among the most resistant of all organisms. Therefore, amounts of chemicals sufficient to kill them also have severe effects on virtually all other aquatic organisms. When concentrations of the herbicides dissipate, the algae are among the first species to reappear.

To date, no chemical has been found that will selectively kill algae and not harm other aquatic plants and animals. Nevertheless, copper sulfate is currently being used to control algae growth in some water-supply reservoirs where algae otherwise impart a bad taste to the water and cause excessive clogging of filters. Above trace amounts, however, copper is known to be highly toxic to all organisms. Therefore, you should remain skeptical of the long-term effects of this practice. Is it sustainable?

Aeration A mechanical aeration system in a eutrophic lake or pond will keep the dissolved oxygen high and at least prevent the fishkills due to suffocation. In addition, aeration may also reduce the levels of phosphate, which is frequently a limiting nutrient, by causing it to change to insoluble forms. Aeration does not cause any undesirable side effects. However, the costs of aerating a million plus acre body of water such as Chesapeake Bay would be too absurd to even contemplate.

Harvesting algae The idea of harvesting the overgrowth of aquatic vegetation comes to many people as a potential solution. However, the only feasible way to harvest phytoplankton is by filtering, and attempts to filter large amounts of phytoplankton from water are precluded by the fact that the algae quickly clog filters. In addition, the logistics, much less the costs, of filtering all the water of a body such as Lake Erie or Chesapeake Bay quickly puts filtering out of the realm of possibility.

In shallow lakes or ponds where eutrophication is resulting from the overgrowth of benthic vegetation, harvesting the mats of vegetation from the surface has somewhat more practicality. In some cases, residents have gotten together to remove the vegetation by hand. A raft made by lashing a few planks across two canoes, as shown in Figure 12–8, is remarkably effective for such an effort. Also, commercial aquatic weed harvesters are now on the market. A lake's appearance may be greatly improved by removing the unsightly mats of vegetation and other litter, and the harvested vegetation makes good organic fertilizer and mulch (see Chapter 9). However, even harvesting has a limited effect. The vegetation soon grows back because roots are left in the nutrient-rich sediments.

Controlling Inputs

Real control of cultural eutrophication requires decreasing the inputs of nutrients. Control of how much sediment enters a body of water is also crucial because sediment invariably brings along bound nutrients. Sediments add to turbidity and gradually

FIGURE 12–8
To avoid the use of chemical herbicides to control aquatic vegetation, these residents in Columbia, Maryland, resorted to harvesting by hand. (Photograph by Bruce Fink.)

fill the body of water. Because controlling sediments is a necessary aspect of controlling eutrophication, we now expand our discussion to consider sediments as well as nutrients.

Controlling Eutrophication

Combating cultural eutrophication requires first identifying the nutrient and sediment sources and then developing and implementing a control strategy for each source.

SOURCES OF SEDIMENTS

The source of all sediments is *soil erosion*, which occurs from croplands, overgrazed rangelands, and deforested slopes. Gully and streambank erosion result from increased runoff. Additional severe erosion may occur from construction and open mining sites.

Approximately 1.5 million acres (0.6 million hectares) of U.S. land are affected each year by development of housing tracts, highways, and other construction. The subsoil exposed by construction activity has little capacity for infiltration, and the slopes created are frequently steep and unstable. The runoff and resulting erosion are often severe. Water-gullied embankments along highways under construction are probably familiar to everyone (Fig. 12–9). Losses of 1000 tons of soil per acre are not uncommon, and this number in some cases can be as high as 10 000 tons per acre. The soil lost from a construction site during one year may exceed what would be lost over 20 000 to 40 000 years under natural conditions.

Mining creates similar situations. In the United States about 4 million acres (1.6 million hectares) have been disturbed by surface mining. Worse, mining activities increase each year as we turn to coal to meet our energy needs. Erosion from mining sites is similar to that from construction sites, but mines are generally larger and remain open for longer periods of time. A federal law requiring reclamation of strip-mined areas was passed in 1977, but it has been unevenly enforced and attempts to weaken it persist. In the meantime, some 0.8 million hectares of old mined soils still remain exposed to erosion.

Finally, there are any number of small miscellaneous sources of sediment. Notice the bare patches around schools, homes, and shopping centers and along highways where the earth has been denuded or where reseeding after construction did not take. Such areas often do not revegetate by themselves but remain open and erode indefinitely. These patches

FIGURE 12–9
The gullies in the embankment of this highway under construction attest to severe erosion. Such construction activities may be the most significant source of sediments entering waterways. (Photograph by BJN.)

may not be very large individually, but taken together they contribute significantly to the sediment problem.

Which sediment source is most prevalent in a given location will depend on the activities in the area.

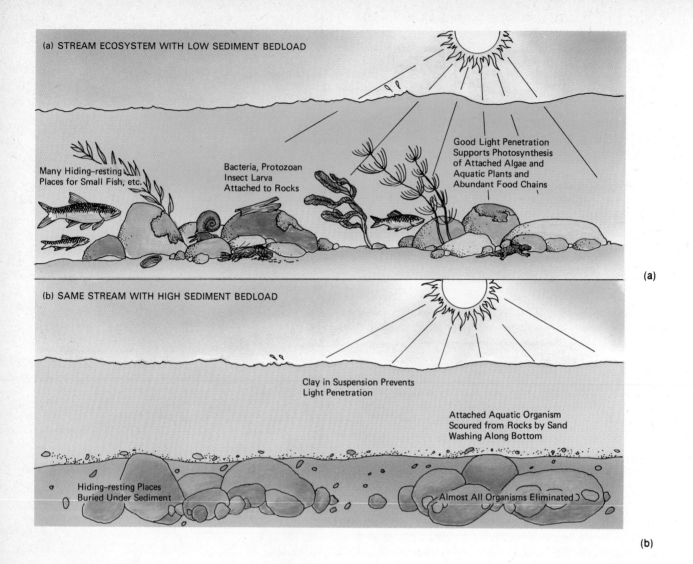

(a) STREAM ECOSYSTEM WITH LOW SEDIMENT BEDLOAD

Many Hiding–resting Places for Small Fish, etc.

Bacteria, Protozoan Insect Larva Attached to Rocks

Good Light Penetration Supports Photosynthesis of Attached Algae and Aquatic Plants and Abundant Food Chains

(a)

(b) SAME STREAM WITH HIGH SEDIMENT BEDLOAD

Clay in Suspension Prevents Light Penetration

Attached Aquatic Organism Scoured from Rocks by Sand Washing Along Bottom

Hiding–resting Places Buried Under Sediment

Almost All Organisms Eliminated

(b)

(c)

FIGURE 12–10
Negative impact of sediments on the aquatic ecosystems of streams and rivers. (a) The ecosystem of a stream that is not subjected to a large sediment bedload. (b) The changes that occur when there are large sediment inputs. (c) A river channel choked with sediment from upstream erosion. The sandbars seen here constitute the bedload; they shift and move with high water, preventing reestablishment of aquatic vegetation. Platte River at Lexington, Nebraska. (Charles R. Belluky/Photo Researchers.)

IMPACTS OF SEDIMENTS ON STREAMS AND RIVERS

In addition to contributing to eutrophication, sediments have a number of other environmental and ecological impacts. When erosion is slight, streams and rivers draining an area run clear. They support algae and other aquatic plants that attach to rocks or root in the bottom. These producers, plus miscellaneous detritus from fallen leaves and so on, support a complex food web of bacteria, protozoans, worms, insect larvae, snails, fish, crayfish, and other organisms. These organisms keep themselves from being carried downstream by attaching to rocks or, in the case of fish, seeking shelter behind or under rocks. Even fish that maintain their position by active swimming occasionally need such shelter to rest.

Eroded soil particles entering the waterway have an array of impacts. Sand, silt, clay, and humus are quickly separated from one another by the flowing water and are carried at different rates. Clay and organic particles are carried in suspension, making the water look muddy and reducing light penetration and photosynthesis. As this material settles, it coats everything and continues to block photosynthesis. It also kills the animal organisms by clogging their gills and feeding structures. Eggs of fish and other aquatic organisms are particularly vulnerable to smothering by sediment.

Equally destructive is the **bedload,** the sand and silt, which is not readily carried in suspension but is gradually washed along the bottom. As particles roll and tumble along, they scour organisms from the rocks. They also bury and smother the bottom life and fill in the hiding and resting places for fish and crayfish. Aquatic plants and other organisms are prevented from reestablishing themselves because the bottom is a constantly shifting bed of sand (Fig. 12–10). In addition, the coarser sediment settles in and clogs the channel, exacerbating flooding and streambank erosion, as discussed in Chapter 11.

Sediments do not receive the attention the news media give to hazardous wastes and certain other pollution problems, but erosion is so widespread throughout the world that few streams and rivers escape the harsh impact of excessive sediment loads. The Environmental Protection Agency ranks sediments as the number one problem for U.S. streams and rivers, and this is no doubt true for many other countries as well. Sediment also causes serious economic problems. Irrigation and shipping channels are in constant need for dredging as they fill with sediment. Many millions of cubic meters of water-storage capacity in reservoirs are lost each year because of sedimentation (Fig. 12–11). The U.S. Soil Conservation Service estimates that the impacts of sediments

FIGURE 12–11
Soil eroding from the land may settle out in reservoirs, gradually diminishing water-storage capacity, as seen here. The sediment-filled portions (flat area extending upstream into the background) now support wetland vegetation. This filling has occurred in just the past 35 years since the reservoir was created; the muddy water at the mouth of the stream shows that the process is continuing. (Photograph is a section of Liberty Reservoir, Baltimore, MD, by Richard Adelberg, Jr., Owings Mill, MD.)

cost the United States over $6 billion each year. Along with the costs of dredging, there are the problems and costs of disposing of the dredged material.

The magnitude of the problem is illustrated by Baltimore Harbor in Maryland. Baltimore Harbor needed to be dredged, but the project was held up for 15 years because of controversy over what to do with the dredged material. In this case, the material was a "mucky ooze" of fine clay and water plus a generous mixture of sewage sludge and industrial chemical wastes that had accumulated in the harbor over the years. The solution finally accepted was to create an offshore island of the dredged material.

SOURCES OF NUTRIENTS

Nutrient ions tend to bind and cling to clay and humus particles. Therefore, any source of erosion and sediments is also a source of nutrients. Also, any applied fertilizer, especially inorganic fertilizer, is subject to at least some leaching. Therefore, runoff from croplands, lawns, and gardens is a major source of nutrients. Of course, abundant use of fertilizer on soils with poor nutrient-holding capacity result in more leaching.

Nutrients are also contained in animal excrements. Thus runoff from cattle feedlots, dairy barns, horse stables, and other locations where animals are kept are a third major source of nutrients. The density of pets in urban and suburban neighborhoods is at least 100 times higher than the densities of wild animals of similar size in natural ecosystems. Therefore, runoff of pet excrements from lawns, streets, and sidewalks is a fourth source of nutrients.

Nor can humans be excluded from any list of nutrient sources. The density of humans living in urban areas is at least 1000 times higher than the density of populations of wild animals of similar size in natural ecosystems. Such a high density creates a great load of nutrients from human excrements. Most human sewage in developed countries is collected and treated. However, sewage treatment, as it presently exists in most towns and cities, does not remove nutrients. They are discharged with the water into natural waterways. Thus, discharges from sewage-treatment plants is a fifth major source of nutrients. Closely related is the nutrient-rich seepage from individual septic systems (see Chapter 13).

Phosphate-containing detergents are another source. Phosphate is generally the limiting nutrient in freshwater ecosystems. This means that increasing or decreasing the phosphate level alone will influence eutrophication. In areas where phosphate-containing laundry and dishwashing detergents and water softeners are used, the amount of phosphate going down the drain, through the sewage-treatment plant, and into the waterway may double or triple the amount from excrements alone.

The above sources, which are summarized in Figure 12–12, are often put into three categories: agricultural runoff, urban and suburban runoff, and sewage-treatment plants. Which source is most significant will obviously differ from area to area, depending on the number of farms, farming activities, and population centers in the watershed. In the Chesapeake Bay watershed, for example, it is estimated that contributions of nutrients from farm runoff, urban runoff, and sewage-treatment plants are about equal. In small lakes and ponds, often a single source is responsible. For example, countless developments of summer cottages surrounding small lakes have been constructed in recent decades. A large percentage of these lakes have become eutrophic as a result of the drainage from septic fields of the cottages.

In summary, natural terrestrial ecosystems are quite efficient in recycling nutrients. Recycling not

FIGURE 12–12
The major sources of eutrophication-causing nutrients.

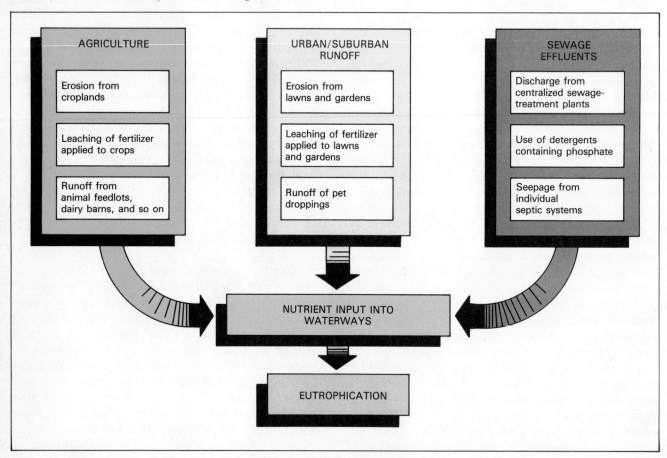

FIGURE 12–13
Tidal wetlands as seen here are immersed by high tides but drain with low tides. Such wetlands play a major role in purifying the water of adjacent estuaries or ocean, because sediments and nutrients entering with the water are largely filtered or absorbed by the wetlands as the water recedes. (C. C. Lockwood/ Earth Scenes.)

only sustains the ecosystem but also serves to keep waterways oligotrophic. In contrast, humans have created systems that are extremely inefficient in recycling nutrients. Thus existing and applied nutrients (inorganic fertilizer) readily enter waterways. Likewise we have constructed sewage-disposal systems that may function well in terms of protecting human health but make no pretense of recycling nutrients. In addition, deforestation and overgrazing diminish an area's capacity to recycle nutrients, and the nutrients are carried with erosion or leaching into waterways. Eutrophication of waterways as well as degradation of soils is the result.

Destruction of Wetlands and Eutrophication

Wetlands are land areas that are naturally covered by shallow water at certain times and are more-or-less drained at other times. Depending on the depth and permanence of water, wetlands are divided into marshes, swamps, and bogs, and they may be either fresh or salt water. Also, wetlands are divided into *tidal* and *nontidal*. **Tidal wetlands** are the often broad expanses of marsh grasses and reeds along coasts and estuaries, where the ground is covered by high tides but drained at low tide (Fig. 12–13). **Nontidal wetlands** are inland wetlands not affected by tides. The plant community of wetlands is typically dense stands of species of grasses, reeds and other kinds of emergent plants, and trees that are adapted to the periodic flooding.

Nontidal wetlands collect water during wet periods and allow it to percolate into the groundwater. Thus, they are important in reducing runoff and flooding and maintaining the water table. Along rivers, they receive and hold overflow and thus further reduce downstream flooding. The dense vegetation

of wetlands plays an important role in filtering sediments and absorbing nutrients from floodwaters. Hence wetlands are important in reducing eutrophication of downstream lakes and estuaries. Because they are thus enriched with nutrients, nontidal wetlands are tremendously productive ecosystems, supporting waterfowl and other wildlife.

Tidal wetlands play a similar role for bays and estuaries. Sediments and nutrients are carried onto the wetlands with rising tides. With the falling tide, water drains through the sieve of grasses and other vegetation, leaving most of the sediments and nutrients behind. The same occurs with wave action. Thus tidal wetlands filter out nutrients and sediments and help maintain an estuary in an oligotrophic condition. As do nontidal wetlands, tidal wetlands utilize the nutrients to become highly productive ecosystems that support many species of waterfowl and other wildlife. In addition, much of the productivity of fish and shellfish in the estuary is supported by biomass production in the surrounding wetlands.

A common human attitude toward wetlands, however, is that they are too wet to plow or build on and too dry to go boating on. Consequently, they have been viewed as "wasteland" good only for changing to "better" uses. Nontidal wetlands have been extensively channeled and drained for conversion to agriculture or filled for conversion to residential and commercial properties. Do the effects on runoff, flooding, groundwater, and eutrophication resulting from this short-sighted practice need to be enumerated?

Tidal wetlands have also been extensively dredged, filled, and bulkheaded—a process that entails dredging to a depth of a meter or so, using the dredged material to build up the other portion, and stabilizing the edge with a wall or bulkhead. The advantage in creating more usable bayfront property is

Once covering 7 million acres of southeast Florida, the everglades is a shallow subtropical wetland wilderness of sawgrass and tree islands blending into mangrove forests as it nears the sea. At the north of the everglades is Lake Okeechobee, the third largest freshwater lake in the United States. The lake is fed by the Kissimmee River and drains into the everglades to the south. A hundred years ago, the system absorbed the abundant rainfall of the summer wet season and allowed the slow flow of water through the sawgrass prairies to the sea. In the winter dry season, alligator holes sustained the underwater wildlife, while great concentrations of wading birds harvested fish in the shrinking wet areas. Wildlife was abundant, adapted to the cycle of summer rains and winter droughts.

Early settlers and state officials regarded the everglades as a wasteland—without value—and began over 100 years ago to drain the wetlands and convert them into "useful" land. By 1970, the everglades was laced by a system of 2000 miles of canals, Lake Okeechobee was diked to a height of 20 feet, and the Kissimmee River was channelized to a straight white-walled canal 48 miles long. The changes were designed to create agricultural and urban land, control flooding, and provide water for the growing population (Florida is now the fourth largest state in the United States). Fully half of the original everglades has now been converted to other land uses, and only one-fifth of the original everglades is under protection as the Everglades National Park (ENP). The ENP is considered an international treasure; the United Nations has designated it as an international biosphere reserve.

Two major eutrophication problems face the everglades system: (1) 200 000 acres of wetlands in the north were converted to dairy farms, which now drain into Lake Okeechobee. Nutrients from the dairy farms have stimulated the growth of masses of blue-green algae and water weeds, and the lake is now considered to be hypereutrophic—the most extreme phase of eutrophication. (2) South of the lake, 700 000 acres of glades have been drained to raise sugarcane, and the nutrients from fertilizers used on the sugarcane are draining directly into the ENP. The influx of nutrients is converting the sawgrass prairies to a monoculture of cattails, threatening to bring on a collapse of the fragile glades ecosystem.

Overriding these eutrophication concerns, the freshwater that flows southward is now almost totally controlled by the system of canals and control dams and is subject to a battle. On one side is the continued demand for water for agriculture (2 billion gallons per day) and urban areas (more than 3 billion gallons per day) in southern Florida; on the other side is the need for water in the ENP in a pattern that approximates the original wet and dry seasons before the canals were built. The tendency has been to treat the everglades as a reservoir, taking water for human needs when the supply is low and channeling excess water into the glades when rainfall is excessive. The result has been disastrous for the wildlife of the everglades, which is now wetter than normal in the rainy season and drier than normal in the dry season. The combination of impacts on the most conspicuous wildlife—the wading birds—has reduced the population 95 percent since 1870. Most of the

other wildlife has suffered serious declines also.

Agriculture and tourism are the basis of Florida's economy, and they are clearly dependent on the freshwater that would normally flow south through the everglades system. The unique wildlife of southern Florida is equally dependent on the water, to be supplied in concert with their adaptation to the seasonal cycle. The wildlife is also dependent on an ecosystem based on sawgrass, not one based on cattails. This battle over water and nutrients is now well recognized at the state and federal levels. Congress has mandated that a minimum amount of water be delivered to the ENP during times of low flow, and the state adopted in 1983 the "Save Our Everglades" initiative. This initiative has the goal of assuring "that the Everglades of the year 2000 looks and functions more like it did in 1900 than it does today." Key elements of this strategic planning initiative include cleaning up Lake Okeechobee, restoring a more natural flow of water into the ENP, and restoring major parts of the system like the Kissimmee River and drained lands adjacent to remaining wetlands. The price tag for the initiative runs into the hundreds of millions of dollars, and to date only a few elements of the plan have been put into place.

It has taken the people of Florida a hundred years to recognize the value of their unique wetland system, and they have a giant task in front of them if they are to maintain the megalopolis of southeastern Florida and the agricultural industry as well as the remnants of the natural system that has been there for centuries. How do you think they will do?

FIGURE 12–14
Bulkheaded shoreline. The natural shoreline of estuaries is a gentle slope covered by grasses and emergent vegetation gradually receding into the water. Much of this shoreline has now been dredged, filled, and bulkheaded to make it more usable by humans. Lost in the process are habitat, food production, and the filtering function which formerly removed nutrients and sediments. Additionally, waves smacking against the bulkhead keep sediments in suspension, aggravating the problem of eutrophication. (Photograph by BJN.)

obvious, but what are the tradeoffs? Not only are the productivity and cleansing capacity of the wetlands lost, but now waves smack against the bulkheads and create turbulence that stirs up sediments and keeps them in suspension for days and weeks at a time, cutting off light, photosynthesis, and growth of the submerged aquatic vegetation (Fig. 12–14).

CONTROLLING NUTRIENTS AND SEDIMENTS

From our discussion of sediment and nutrient sources, much of what needs to be done to control them is probably self-evident. Therefore, the following is just a cursory overview of remedial practices being implemented.

Best Management Practices on Farms, Lawns, and Gardens

Best management practices is a catch-all term that includes all the methods of soil conservation discussed in Chapter 9. Keeping the ground covered with vegetation or mulch to prevent erosion; strip cropping; using clover or other legumes for natural nitrogen addition; and using compost, manure, and other organic fertilizer in place of inorganic fertilizers top the list.

Where properties are adjacent to streams, rivers, or lakes, owners should plant buffer strips of trees between their fields and the waterways in order to catch and reabsorb the nutrient-rich leachate. The need for replanting deforested or overgrazed areas is self-evident. Where animal wastes from feedlots, dairy barns, or horse stables drain directly into waterways, ponds should be constructed to intercept the nutrient-rich runoff. Such water can then be recycled

as irrigation water, returning the nutrients to the soil (Fig. 12–15). Rather than letting animals wade into streams to drink and incidentally excrete wastes, farmers should provide water troughs so that animals will drink and excrete their wastes on the land.

When the Clean Water Act was reauthorized in 1987, a new section (section 319) was added requiring states to develop management programs to address nonpoint sources of pollution, that is, runoff from farms and from urban and suburban areas. Getting farmers to comply with best management practices is a relatively straightforward task. Farmers are relatively few in number, the practices will generally benefit their own economic interests, and the state can make other benefits contingent on the farmers' adopting such practices. Getting millions of individual homeowners to comply is another matter. Students taking a course like this and spreading the word may be the only solution.

Sediment Control on Construction and Mining Sites

Several techniques are now commonly used to reduce the loss of soil from construction sites. Rather than leaving the entire site bare and exposed to erosion for the duration of construction, contractors bring much of the site to final grade and restabilized it with grass immediately. Even temporary piles of dirt may be stabilized by a straw-mulch cover until needed. These techniques are particularly appropriate in highway construction.

Sediment from erosion may be caught in a **sediment trap,** which is essentially a pond into which runoff from a construction site is channeled. As sediment-laden water enters the pond, its velocity slows, sediment settles, and sediment-free water flows out

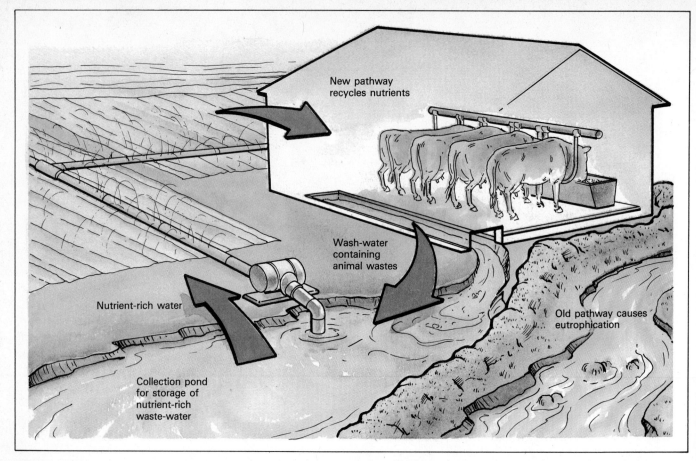

FIGURE 12–15
A collection pond for dairy barn washings. When washings from animal facilities are flushed directly into natural waterways, they may contribute significantly to eutrophication. This may be avoided by collecting the flushings in ponds from which both the water and the nutrients may be recycled. Many farmers use such flushings for irrigation.

Labels in figure:
New pathway recycles nutrients
Wash-water containing animal wastes
Old pathway causes eutrophication
Nutrient-rich water
Collection pond for storage of nutrient-rich waste-water

over a rock dam or through a standpipe (Fig. 12–16). After construction is completed, the sediment trap may be converted to a stormwater-retention reservoir. On small sites, the lower perimeter of the site may be diked with bales of straw, which filter the runoff and remove sediment. Or, a plastic fence may be used to catch water and direct it to a sediment trap. The same techniques may be applied to mining sites.

Many developers view sediment control as just an added expense that they want to reduce as much as possible. Erosion control on construction sites must therefore be promoted through legislative action and legal enforcement. Under Maryland state law, sediment-control regulations are added to other building codes and construction requirements so that developers must submit sediment-control plans along with other plans for site development. These plans are reviewed by the government and modified if necessary before a building permit is issued. Once construction starts, the site is inspected periodically to be sure sediment traps or other devices are installed and

maintained properly. Stormwater-control legislation works similarly.

Inspection and enforcement are frequently the weak link. It is not uncommon to find sediment traps that are nonfunctional for lack of proper installation or maintenance. This is an issue where local citizen action can be effective. Save Our Streams and Trout Unlimited are two citizen organizations whose members take it upon themselves to act as "inspectors" overseeing construction projects in their areas. When they find uncontrolled erosion threatening waterways, they report it to the authorities. If the authorities fail to take action, the organizations report the situation to the local media. Media publicity frequently brings action when all else fails.

Regarding the numerous small patches of eroding earth noted before, the best, and in many cases only, solution is for a few people to get together and do some raking, mulching, and reseeding. Government bureaucracies tend to be too large and cumbersome to deal with such small areas effectively. The

(a)

(b)

(c)

FIGURE 12–16
Sediment control at a construction site. (a) Runoff carrying sediment flows into a "pond" constructed at the lower edge of the site. Sediment settles (foreground), while water flows over the rock dam (where the person is standing). (b,c) Construction of a larger pond with a standpipe overflow will function both as a sediment trap during construction and as a permanent stormwater reservoir. (Photographs by BJN.)

Soil Conservation Service, which maintains an office in every county of the United States, does not have the work force to do the work but is generally enthusiastic about providing advice and technical assistance and sometimes seed or seedlings.

Preservation of Wetlands

The need to preserve wetlands is obvious. However, the economic interests and pressures for their development are tremendous. Despite their recognized ecological value and the fact that over half (54 percent) of the nontidal wetlands and even greater portions of tidal wetlands of most estuaries in the United States have already been destroyed, wetlands are still being converted to development. Developing laws and regulations that will provide suitable protection is an ongoing issue at the current time. The problems and controversies involved are addressed in Chapter 19.

Banning Phosphate Detergents

When the connection between eutrophication and phosphate detergents was recognized in the 1960s, the detergent industry developed a number of substitutes. Nevertheless, for economic reasons, it has continued to market phosphate detergents preferentially. It has proved necessary to legislatively ban the sale of phosphate detergents in order to cause

TABLE 12–1

Progress Toward Banning Phosphate from Detergents	
States with a ban on phosphate-containing detergents	
Indiana	North Carolina
Maryland	Pennsylvania
Michigan	Vermont
Minnesota	Virginia
New York	Wisconsin
States with a ban on phosphate-containing detergents in some areas	
Idaho	Ohio
Illinois	Oregon
Montana	Washington
Chain stores marketing only zero-phosphate detergents	
Kmart	Wal Mart

changeover to nonphosphate alternatives. Total or partial "phosphate bans" are now in effect in 16 states (Table 12–1).

Advanced Sewage Treatment

There has been much upgrading of sewage-treatment plants over the last two decades. However, most plants still do not remove nutrients from the wastewater. The need for advanced sewage treatment, including nutrient removal, speaks for itself. The process of sewage treatment and how nutrient-removal may be achieved are the subject of Chapter 13.

RECOVERY

The good news is that cultural eutrophication is reversible. Experience has shown that, when nutrient and sediment inputs are curtailed, existing quantities will gradually stabilize or be flushed from the system. Thus bodies of water may eventually make a substantial if not total recovery. Lake Washington provides one example (see In Perspective Box below). Controlling inputs of sediments and nutrients will be a major step toward sustaining or restoring the health of our waterways.

IN PERSPECTIVE
Lake Washington Recovery

One of the best ways to establish the connection between nutrients and cultural eutrophication would be to add massive amounts of nutrients to a body of water and then withdraw the nutrient input to see if the system recovered. Although no one intended it as a test of cultural eutrophication, just such an experiment was conducted on a large lake in the Seattle area—Lake Washington. The "experiment" has conclusively shown the crucial role of phosphorus in eutrophication of lakes, and it has made it clear that cultural eutrophication can be reversed.

Lake Washington is a 33.8-mi² lake east of Seattle and within commuting distance of greater metropolitan Seattle. The population began to spread outward from Seattle around the lake in the 1940s and 1950s, and by 1960 there were 11 sewage-treatment plants located close to the lake, sending effluents into the lake at a rate of 20 million gallons per day. The sewage was given state-of-the-art secondary treatment before being released to the lake or tributaries. At the same time, phosphorus-based detergents were becoming widely marketed and used by homeowners. Residents began to notice changes in the clarity of the water and in the fish populations during the 1950s, and it was obvious that

the lake was rapidly deteriorating in quality. In particular, blooms of phytoplankton dominated by blue-green algae began to occur every summer, creating unpleasant smells and masses of floating algae that washed up on the shores of the lake. These algae are symptomatic of high phosphorus loading in a lake system. They have the ability to fix atmospheric nitrogen and so can outcompete other algal species that take their nitrogen from solution.

Citizen concern led to the creation of a system that diverted the treatment plant effluents from the lake into tidally flushed Puget Sound. The diversion was phased in gradually between 1963 and 1968. The lake began to respond positively even before the diversion was complete. The water increased in clarity as algal blooms began to diminish, and, by 1975, the lake had finished responding to the diversion. Summer blooms of blue-green algae had completely ceased, and the total algal population of the lake had diminished by 90 percent or more.

This "experiment" in cultural eutrophication was closely followed by limnologist Tom Edmondson and co-workers from the University of Washington. Edmondson's work has demonstrated beyond question the key role of phosphorus in eutrophication. Before the diversion began,

the lake was receiving 220 tons of phosphorus per year from all sources. Under normal conditions, phosphorus input is balanced by deposition to the sediments. Edmondson demonstrated that more phosphorus was coming into the lake than was being deposited, with the resulting blooms of blue-green algae. By the end of the diversion, phosphorus input was reduced to less than 40 tons per year, well within the ability of the lake to handle by deposition to the sediments.

There are several morals to this story, the most important being that cultural eutrophication can be reversed by paying attention to nutrient inputs. Also, we now know never to allow sewage-treatment plant effluent to enter a lake unless it receives tertiary treatment to remove phosphorus. As a result of the work of Edmondson and other limnologists, lake management is now seen as a matter of controlling the phosphorus loading into the lake from all sources: sewage, lawn fertilizers, farms, street runoff, and failing septic systems. Finally, we see that with the right balance of nutrients, blue-green algal blooms can be prevented and the process of eutrophication can at least be put back on its natural schedule.

Review Questions

1. Name five general categories of human activity and give the kinds of pollution that result from each.
2. Define eutrophication.
3. What are the two categories of aquatic plants, and what is the effect of nutrient enrichment of the water on each?
4. How does eutrophication of a body such as Chesapeake Bay differ from eutrophication of shallow lakes and ponds?
5. How does overgrowth of aquatic plants lead to the dieoff of fish and shellfish?
6. What are three approaches to attacking the symptoms of eutrophication, and what are the limitations of each?
7. What must be done to correct the root cause of eutrophication?
8. Why do sediments as well as nutrients need to be controlled?
9. What are three major sources of sediments?
10. What techniques may be used to control sediment from each source?
11. What are three major sources of nutrients?
12. Give three subcategories of nutrients for each of the three main categories of nutrient sources.
13. What are strategies for controlling the nutrients from each subcategory noted in question 12?
14. What is meant by best management practices?
15. What are two kinds of wetlands?
16. How do wetlands provide a natural control for eutrophication?
17. What are humans doing to wetlands?
18. What tradeoffs are involved in developing wetlands for agriculture or building?

 ## Thinking Environmentally

1. A large number of fish are suddenly floating dead on a lake. You are called in to investigate the problem. You find an abundance of aquatic plant life and no evidence of toxic dumping. Suggest a reason for the fishkill.
2. A local developer plans to turn a major section of tidal wetlands on a productive estuary into a summer community and marina. Discuss the probable environmental impacts and tradeoffs involved in this plan.
3. A number of regions in the United States have banned the use of phosphate-containing detergents in recent years. What harm is caused by such detergents, and what is hoped to be achieved by such bans?
4. Describe how planting trees on an eroding hillside may protect aquatic life in an estuary many miles away.
5. What changes in our society are necessary to reduce eutrophication significantly?

13

Sewage Pollution and Rediscovering the Nutrient Cycle

LEARNING OBJECTIVES

When you have finished studying this chapter, you should be able to:

1. Describe the health hazards associated with untreated sewage.
2. Trace the development of early sewage-waste treatment systems.
3. List and describe the four categories of pollutants in raw sewage.
4. Describe the processes of preliminary and primary sewage treatment.
5. Compare two types of systems that may be used in secondary treatment.
6. Outline possible methods for disinfecting sewage treatment effluent.
7. Explain when tertiary treatment is needed and how it may be done.
8. Discuss irrigation with effluents as an alternative to tertiary treatment.
9. Evaluate the pros and cons associated with each of the three possible sludge-treatment options.
10. Describe two alternative systems for treating sewage.
11. Cite the goals and accomplishments of The Clean Water Act of 1972.
12. Describe impediments to using sewage waste as a resource.

The first principle of ecosystem sustainability is that nutrients must be recycled. In Chapter 12 we studied the many ways in which humans fall short in respecting this principle and the eutrophication problems that result. One of the major aspects of this problem involves our disposal of sewage wastes. As opposed to a cycle, we have created a one-way flow such that nutrients withdrawn from the soil by crop plants and cattle fodder are eventually discharged into waterways along with sewage effluents.

Sustainability of our human system will ultimately depend on our developing means to recycle the nutrients contained in sewage effluents. There are precautions that must be taken in recycling wastes, however; the most important is to safeguard human health. Our objectives in this chapter are to examine traditional methods of sewage treatment and to see how they may be amended or what alternative methods may be used in order to reestablish nutrient cycling while not compromising public health.

Sewage Handling and Treatment

HEALTH HAZARD OF UNTREATED SEWAGE

Untreated sewage is a major public health hazard because it is a vector in the spread of many infectious diseases. The excrement from humans and other animals infected with certain **pathogens** (disease-causing bacteria, viruses, or other parasitic organisms) contains large numbers of these organisms or their eggs (Table 13–1). Even after an infected person or animal recovers, it may continue to harbor low populations of the pathogen and thus pass the parasite. In other words, the once-infected person or animal, although no longer sick, continues to act as a carrier. If sewage wastes from infected individuals contaminate drinking water, food, or water used for swimming or bathing, the parasites may gain access to and infect other individuals (Fig. 13–1).

In general, pathogenic organisms survive only

TABLE 13–1

Pathogens Carried by Sewage	
Disease	**Infectious Agent**
Typhoid fever	*Salmonella typhi* (bacterium)
Cholera	*Vibrio cholerae* (bacterium)
Salmonellosis	*Salmonella* species (bacteria)
Diarrhea	*Escherichia coli*, *Campylobacter* species (bacteria)
Infectious hepatitis	Hepatitis A virus
Poliomyelitis	Poliovirus
Dysentery	*Shigella* species (bacteria) *Entamoeba histolytica* (protozoan)
Giardiasis	*Giardia intestinalis* (protozoan)
Numerous parasitic diseases	(Roundworms, flatworms)

FIGURE 13–1
A scene along the Ganges River in India. In many places in the Third World the same waterways are used simultaneously for drinking, washing, and disposal of sewage. A high incidence of disease, infant and childhood mortality, and parasites is the result. (T. Stoddard/Katz/ Woodfin Camp & Associates.)

a few days outside a host, and the number of organisms entering the body is an important factor in determining subsequent infection. When host populations are sparse, relatively little transfer of pathogenic organisms occurs because contamination levels remain low and considerable time elapses between elimination by one host and contact by the next. As host populations become denser, however, the reverse is true. This poses a particular problem for humans. Humans tend to live and work in high-density urban situations; thus, we make ourselves very vulnerable to the spread of pathogens. As long as the population is healthy, pathogens may remain below the threshold level of causing infection in healthy individuals. The greatest risk of dense populations is that, if an individual gets sick, the disease can spread with great rapidity if there is contamination of water or food supplies with untreated sewage.

Before the connection between disease and sewage-carried pathogens was recognized in the mid-1800s, disastrous epidemics were common in cities. For example, epidemics of typhoid fever that killed thousands of people were common in cities until the turn of this century. Today, public-health measures that prevent this disease cycle have been adopted throughout developed countries. These measures involve (1) purification and disinfection of public water supplies with chlorine or other agents (Chapter 11); (2) personal hygiene and sanitation, especially in relation to preparation and handling of food (it is recommended that certain foods prone to passing pathogens, for example, pork and chicken, always be well cooked); and (3) sanitary collection and treatment of sewage wastes. There are a variety of other measures as well. For example, oyster beds contaminated with raw sewage are closed to harvesting. Implicit in all measures is monitoring for sewage contamination (see In Perspective Box, p. 293).

Many people attribute good health in a population to modern medicine, but good health is more a result of disease prevention through public-health measures. In Chapter 11 we noted that 1.2 billion people do not have access to treated drinking water. However, an even greater number live where there is poor (or no) sewage collection or treatment. This is true of large sections of Third World cities that have grown rapidly in recent years (Chapter 6) as well as of rural areas. For example, 30 percent of Mexico City, 50 percent of Bangkok and similar percentages of numerous other Third World cities lack sewage-collection systems. In many cities where collection systems exist, raw sewage is still discharged into rivers. In countries of the former USSR, it is estimated that 20 percent of the sewage is still discharged in the raw state and another 50 percent receives only rudimentary treatment.

This low treatment rate means that large portions of the world's population are chronically infected with various parasites and are vulnerable to epidemics of any and all diseases spread via the sewage vector. In 1990 an outbreak of cholera in Peru killed several thousand people as it spread through Latin America because of unsanitary conditions. The high infant mortality rate in many Third World countries is one tragic consequence of the lack of control over water-borne diseases.

As we address sewage treatment, therefore, we must remain aware of the need to protect against disease as well as the ultimate goal of nutrient recycling.

BACKGROUND

Prior to the late 1800s, the general means of disposing of human excrement was the outdoor privy. Seepage from the privy not infrequently contaminated drinking water and caused disease, especially in places where privies and wells were located near one another. In the late 1800s, Louis Pasteur and other scientists showed that sewage-borne bacteria were responsible for many infectious diseases. This important discovery led to intensive efforts to rid cities of wastes as expediently as possible. Cities already had drain systems for stormwater, but using these for human wastes had been prohibited. With the urgency of the situation, however, minds quickly changed. The flush toilet was introduced, and sewers were tapped into storm drains. Thus, western societies initiated the system of flushing sewage wastes into natural waterways with no prior treatment.

The results of this practice should be obvious. Many streams and rivers became so overloaded with biochemical-oxygen-demanding detritus that dissolved oxygen was totally depleted, suffocating all aquatic organisms except for those bacteria and a few other organisms not requiring oxygen. Effectively, streams and rivers became open sewers emitting foul odors, lacking fish and waterfowl, and having great potential for spreading disease.

To alleviate the problem of polluted rivers and streams, people began to realize the need to develop facilities for treating the wastewater and to install separate systems for stormwater and sewage. Facilities were designed to remove pollutants from wastewater, and the first treatment plants in the United States were built around 1900. However, the combined volumes of sewage and stormwater soon proved impossible to handle. During heavy rains, wastewater would overflow the treatment plant and carry raw sewage into the receiving waterway. Gradually, regulations were passed requiring developers

In spite of general improvement in sewage collection and treatment, there are and always will be various failures in the system. Therefore, an important aspect of public health is the monitoring of water supplies and other bodies of water where there is human contact. It is worth understanding how this monitoring is done.

It would be exceedingly difficult, time-consuming, and costly to test for each specific pathogen that might be present. Therefore an indirect method called the **fecal coliform test** has been developed. This test is based on the fact that huge populations of a bacterium called *E. coli (Escherichia coli)* normally inhabit the lower intestinal tract of humans and other animals, and large numbers of the bacterium are excreted with fecal material. *E. coli* is found in nature only when it enters an ecosystem from animal excrement. It is not a pathogen; in fact, our bowels would not function properly without it. However, since *E. coli* is invariably part of fecal wastes, its presence indicates a persistent source of raw sewage and, consequently, the potential presence of other sewage-borne pathogens. Thus, *E. coli* is referred to as an **indicator organism.** The presence of *E. coli* indicates that water is contaminated with fecal wastes and pathogens may be present. The absence of *E. coli* indicates that water is free from such pathogens.

The fecal coliform test, one technique of which is shown in the figures, detects and counts the number of coliform bacteria in a sample of water. Thus, the test shows how much sewage pollution is present and the relative degree of hazard. For example, to be safe for drinking, water should have an average *E. coli* count of no more than 1 *E. coli* per 100 mL (about 0.4 cups) of water. Water with as many as 200 *E. coli* per 100 mL is still considered safe for swimming. Beyond that level, a

river may be posted as polluted, and swimming and other direct contact should be avoided. By comparison, raw sewage (99.9 percent water:0.1 percent wastes) has *E. coli* counts in the millions.

When water is too polluted for a desired use, two possible approaches exist: disinfection or reduction of sewage sources. Drinking-water supplies and swimming pools are generally disinfected with chlorine. However, there is obviously no way of disinfecting natural bodies of water without also killing everything else in the system. Therefore, there is no substitute for continuing and improving the quality of sewage treatment.

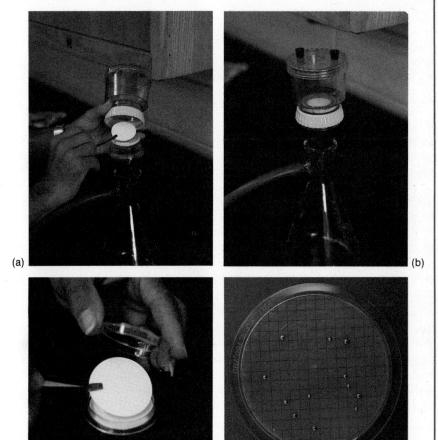

Testing water for sewage contamination—the millipore technique. (a) A millipore filter disc is placed in the filter apparatus. (b) A sample of the water being tested is drawn through the filter and any bacteria present are entrapped on the filter disc. (c) The filter disc is then placed in a petri dish on a special medium that supports the growth of bacteria and will impart a particular color to fecal coliform bacteria. The dish is then incubated for 24 hr at 38° C, during which time each bacterium on the disc will multiply to form a colony visible to the naked eye. (d) Fecal coliform bacteria indicating sewage contamination are identifiable as the colonies with a metallic green sheen as seen here. (Photographs by BJN, Bob Hudson and George Waclawiw.)

to install separate systems: **storm drains** for collecting and draining runoff from precipitation and **sanitary sewers** that receive all the wastewater from sinks, tubs, and toilets in homes and other buildings. (Note the distinction in terms; it is incorrect to speak of storm drains as sewers.)

Even in the developed world progress has been extremely uneven, however, and growing populations continually add to the burden. While tremendous progress has been made, cases of raw sewage overflowing with stormwater into waterways still abound. In other words, 100 years later we are still playing "catch-up," even in developed countries.

The situation is much worse in the Third World, where considerable portions of the population still have no efficient collection much less treatment of sewage. Especially in poor areas of mushrooming Third World cities, it is not uncommon to find raw sewage littering the ground, overflowing gutters and streams, and being discharged directly into rivers. Many of the people living in these areas must use these badly polluted waters for bathing, laundering, and even drinking (Fig. 13–1.)

SEWAGE TREATMENT

What Is Raw Sewage?

A sewer system brings all tub, sink, and toilet drains in homes and other buildings together just as the twigs and branches of a tree lead to the trunk. At the base of the "trunk," the total combination of all that goes into this collection system comes out—as *raw sewage* or *raw wastewater*. Because we use such large amounts of water to flush away small amounts of waste and often just run the water with no waste at all, **raw sewage** is about 1000 parts water for every 1 part of waste—99.9 percent water to 0.1 percent wastes. With the addition of stormwater, raw sewage is diluted still more. Low as their concentration may be, however, the wastes in raw sewage are highly significant. They are divided into four categories:

Debris and Grit Debris includes rags, plastic bags, and other objects flushed down toilets or washing in through storm drains in places where they are still connected to sewers. Grit is coarse sand and gravel, and also enters mainly through storm drains.

Particulate Organic Material Particulate organic material includes visible particles of organic matter, originating from food wastes from home garbage disposal units as well as fecal matter and bits of paper from toilets. Particulate organic material also includes living bacteria and other microorganisms that have begun to digest the waste, and possibly pathogenic

organisms. Particulate organic material will settle, although slowly, in still water.

Colloidal and Dissolved Organic Material Colloidal organic material originates from the same sources as particulate organic material; the main distinction is particle size. Whereas the visible particles described above will settle in still water, colloidal particles are so fine that they will not settle, at least not within any reasonable time period. Bacteria and other microorganisms, including pathogens, are also present in this category. In addition, there is dissolved organic material from soaps, detergents, shampoos, and other cleaning and washing agents that will not settle at all.

Dissolved Inorganic Material Dissolved inorganic material includes mainly nutrients such as nitrogen, phosphorus, and potassium from excretory wastes plus phosphate from detergents and water softeners.

These are the four categories of pollutants in "ideal" raw sewage. In addition, variable amounts of pesticides, heavy metals, and other toxic compounds are found in sewage because people pour unused portions of such materials down sink, tub, or toilet drains. Also, industries may discharge various toxic wastes into sewers. We shall consider the problems caused by "nonideal" sewage and how to cope with them at the end of the chapter. First, though, let us consider the treatment and potential uses of ideal sewage, that is, sewage that is not contaminated with such materials.

Wastewater-Treatment Steps

A sewage-treatment plant must be capable of removing all four types of pollutants if there is to be complete treatment. Debris and grit are removed by *preliminary treatment*. Particulate organic material is removed by *primary treatment*. Colloidal and dissolved organic material is removed by *secondary treatment*. Dissolved inorganic material is removed by *tertiary treatment*, which may also be called *advanced treatment*.

These levels of treatment have been added to treatment systems in steps as circumstances have demanded. Consequently there exist cities and communities that still discharge raw sewage, others that go only as far as primary treatment and then discharge the water, others that carry it through secondary treatment before discharge, and some that go all the way through tertiary treatment. In the United States and other developed countries, most cities now have secondary treatment, and increasing numbers of cities have or are installing tertiary treatment systems. As we noted earlier, much of the less-developed world lacks adequate sewers for waste collection, to say nothing of treatment facilities.

FIGURE 13–2
Bar screen. Channelling wastewater through a bar screen is the first step of preliminary treatment. The screen traps large pieces of debris, which are then removed from the screen by a mechanical rake. (Washington Suburban Sanitary Commission photograph.)

Preliminary Treatment (*Removal of debris and grit*) Because debris and grit would damage or clog pumps and later treatment processes, their removal is a necessary beginning step and is termed **preliminary treatment.** Debris is removed by letting raw sewage flow through a **bar screen,** a row of bars mounted about 1 inch (2.5 cm) apart (Fig. 13–2). Debris is mechanically raked from the screen and taken to an incinerator. After it passes through the bar screen, the water flows through a **grit-settling tank,** a swimming pool–like tank, where its velocity is slowed just enough to permit the grit to settle (Fig. 13–3). The settled grit is mechanically removed from these tanks and taken to landfills.

FIGURE 13–3
Grit-settling tank, the second step of preliminary treatment. Here the velocity of the water is slowed to about 1–2 feet (0.5 m) per second, a speed that allows sand and other coarse grit to settle to the bottom while the water with other pollutants flows over the top edge (foreground). The grit is removed by mechanical plows scraping the bottom. (Photograph by BJN.)

(a)

Barrier blocks overflow of scum

Clarified water

Rotating plow moves settled material to collection trough

Particulate organic material settles

Raw sludge

Raw sewage from preliminary treatment

(b)

FIGURE 13–4
Primary clarifiers used for primary treatment. (a) The water enters these tanks at the center and exits over the wire at the edge. The slow velocity (about 5–10 feet, or 2 m, per hour) of flow through the tanks permits the particulate organic material to settle while oil and grease rise. The settled organic material is pumped from the bottom, and the oil and grease are skimmed from the surface. These combined materials constitute raw sludge. (b) Cross section of clarifier. (Photograph by BJN; Courtesy of Walker Process Division of C.B.I.)

(a)

(b)

FIGURE 13–5
Trickling filters for secondary treatment. (a) The water from primary clarifiers is sprinkled onto and trickles through a bed of rocks about 6–8 feet (2–3 m) deep. (b) Various bacteria and other detritus feeders adhering to the rocks consume and digest the organic material in the water as it trickles by. The water is collected at the bottom of the filters. (Photographs by BJN.)

Primary Treatment (*Removal of particulate organic material*) Following preliminary treatment, the water moves on to **primary treatment,** where it flows very slowly through large tanks called **primary clarifiers** (Fig. 13–4). Because it flows extremely slowly through these tanks, the water is nearly motionless for several hours. This permits the particulate organic material, about 30 to 50 percent of the total organic material, to settle to the bottom, from where it can be removed. At the same time, fatty or oily material floats to the top and is skimmed off. All the material removed, both particulate organic material and fatty material, is known as **raw sludge;** we shall consider its treatment and disposal shortly.

Note that primary treatment involves nothing more complicated than putting polluted water in a "bucket," letting material settle, and pouring off the water. Nevertheless, it removes the particulate organic matter at minimal cost.

Secondary Treatment (*Removal of colloidal and dissolved organic material*) **Secondary Treatment** is also called **biological treatment** because it uses organisms, natural decomposers, and detritus feeders, to break colloidal and dissolved organic material down to carbon dioxide and water. An oxygen-rich environment is required for the organisms of secondary treatment to work. Either of two types of systems may be used: *trickling-filters* or *activated-sludge systems.*

In a **trickling-filter system,** the water exiting from primary treatment is trickled onto and allowed to percolate through a bed of fist-sized rocks 6–8 feet (2–3 m) deep (Fig. 13–5). The spaces between the rocks provide for good aeration. As in a natural stream, this environment supports a complex ecosystem of bacteria, protozoans, rotifers (organisms that consume protozoans), various small worms, and other detritus feeders attached to the rocks (Fig. 13–6). The organic material in the water (principally colloidal and dissolved organic material at this point), including pathogenic organisms, is absorbed and digested by these organisms as it trickles by. Clumps of organisms that occasionally break free and wash from the trickling filters are removed by passing the water through secondary clarifiers, tanks that work in the same way as primary clarifiers. Through primary treatment and a trickling-filter system, 85 to 90 percent of the total organic material is removed from the wastewater.

In the alternative type of secondary treatment, called an **activated-sludge system,** water from primary treatment enters a long tank that could hold several tractor-trailer trucks parked end to end (Fig. 13–7a). A mixture of detritus-feeding organisms, referred to as **activated sludge,** is added to the water

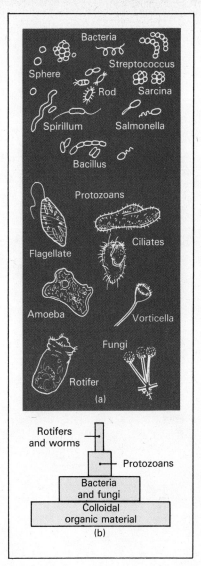

FIGURE 13–6
(a) Some of the organisms active in both trickling-filter and activated-sludge secondary-treatment systems. (b) These organisms represent a number of trophic levels through which the biomass of organic material entering the system is reduced by up to 90 percent.

as it enters the tank, and the water is vigorously aerated as it moves through the tank, creating an oxygen-rich environment ideal for the growth of the organisms. As the organisms feed, the biomass of organic material, including pathogens, is reduced.

From the aeration tank, the water goes into a secondary clarifier tank where the organisms, usually clumped on bits of detritus, settle relatively efficiently. The settled organisms become the activated sludge, which is pumped back into the entrance of the aeration tank (Fig. 13–7b,c). Thus, the detritus-consuming organisms are continuously recycled in the system, while water, now with 90 to 95 percent

(a)

FIGURE 13–7
(a) Aeration tank used in activated-sludge treatment. Wastewater from the primary clarifiers moves through the tank and is vigorously aerated by air forced up from the tubes along the bottom. (b) In the oxygen-rich environment of the aeration tank, microorganisms consume colloidal and dissolved organic material. Organisms (activated sludge) settle out in the secondary clarifier and are returned to the aeration tank while the clarified water flows on. (c) Aeration tank in operation. The rust-colored pipe in the foreground is the inlet pipe for activated sludge being recycled from the secondary clarifier. (Photographs by BJN.)

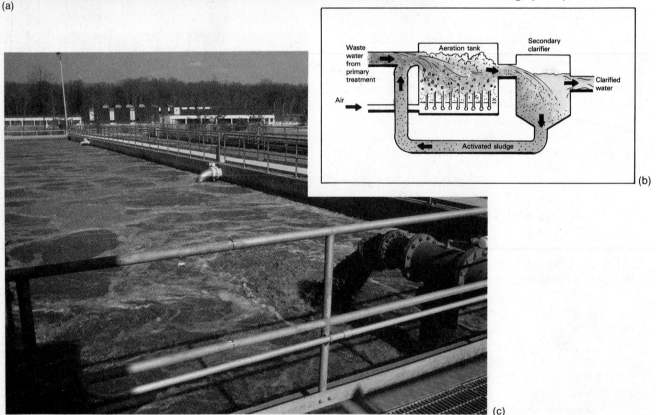

(b)

(c)

of all the organic material removed, flows on. Excess activated sludge is generally combined and treated with the raw sludge from the primary clarifiers.

When existing sewage-treatment plants are upgraded nowadays, the activated-sludge system is the one most commonly chosen because space is generally at a premium. Being a forced-air system, this method of secondary treatment permits more water to be processed in less space than trickling filters. The disadvantage is that large amounts of energy are required for the aeration pumps, making the operation much more costly than the trickling-filter systems, which operate by gravity.

The final effluent from secondary treatment both looks and smells relatively clean (a very slight cloudiness of colloidal organic material remains). To preclude any remaining risk of pathogens, however, it is customary to disinfect the water before discharge.

Disinfection Whenever there is any level of sewage treatment at all, the effluent is generally disinfected as it is discharged in order to kill pathogens. The most widely used disinfecting agent is chlorine gas since it is both effective and relatively inexpensive. This treatment also introduces chlorine into natural waterways, however, and even minute levels of chlorine can harm aquatic animals. The hatching of trout eggs, for instance, and development of the embryos are affected by the presence of chlorine. Also, chlorine reacts spontaneously with organic compounds to some extent to form **chlorinated hydrocarbons,** organic molecules with chlorine atoms attached. Many of these compounds are toxic and nonbiodegradable, and some have been identified as compounds that cause cancer, abnormal development, and reproductive problems.

Because of these two negative side effects, disinfecting agents other than chlorine are coming into use. One is ozone, which is extremely effective in killing microorganisms and in the process breaks down to oxygen gas, which improves water quality. However, ozone is unstable and hence explosive. Therefore it must be generated at the point of use, a step that demands considerable capital investment and energy. A second disinfection method is to pass the effluent though an array of ultraviolet lights mounted in the water. (Standard fluorescent lights without the white coating emit ultraviolet.) The ultraviolet radiation kills microorganisms but does not otherwise affect the water. In a third method, chemicals that react with chlorine to form inactive, harmless compounds are added after chlorine disinfection.

In 1972, about 5 percent of U.S. sewage was still being discharged raw into waterways and another 25 percent received only primary treatment. A major goal of the Clean Water Act of 1972 was to rid the nation's waterways of sewage pollution. Bringing all sewage facilities up to the level of secondary treatment was adopted as the means to accomplish this end. Only later did researchers recognize that nutrients alone, which still remain in the water after secondary treatment, are enough to cause eutrophication. Also, disinfection with chlorine was found to have the undesirable environmental side effects noted.

Tertiary or Advanced Treatment (*Removal of dissolved inorganic materials*) Where effluents from sewage-treatment plants are a factor in causing overenrichment and eutrophication, there is an obvious need to extend sewage treatment to remove one or more of the nutrients or to do something else with nutrient-rich effluent. There are a number of methods available for removing one or more nutrients from the water.

Which method is chosen depends on future use of the water, if any. In a water-short region, there may be economic justification for totally purifying the water by a process of desalinization and recycling it (see In Perspective Box, p. 255). If controlling eutrophication in the receiving waterway is the only goal of tertiary treatment, only the limiting nutrient, generally phosphate, need be removed. In addition to banning the use of phosphate detergents to reduce inputs, phosphate can be removed by mixing lime (calcium carbonate) into the water. Calcium combines chemically with phosphate to form an insoluble precipitate of calcium phosphate, which may be removed by another settling process. Where nitrogen is the nutrient causing the problem, the water can be run through chemical or biological processes to remove it. Biological processes involve using bacteria that convert the ammonium and nitrate ions present in the sewage to nitrogen gas, as occurs in the natural nitrogen cycle (see Fig. 3–17).

Since Washington, D.C., implemented tertiary treatment (nitrogen and phosphate removal) a few years ago, the Potomac River has cleared up considerably. Aquatic vegetation and waterfowl are returning, a hopeful sign for Chesapeake Bay if other cities follow suit.

Irrigating with Effluents and Biological Tertiary Treatment As an alternative to tertiary treatment, an increasing number of communities are using the nutrient-rich effluent for irrigation. For example, the secondary-treatment effluent from St. Petersburg, Florida, was causing eutrophication in Tampa Bay. Now St. Petersburg uses it to irrigate 4000 acres (1600 hectares) of urban open space, from parks and residential lawns to a golf course. Revenues from the water sales help offset operating costs. Bakersfield, California, receives $300 000 a year income from a 5000-acre (2000-hectare) farm irrigated with its treated effluent. Claton County, Georgia, is irrigating 2500 acres of woodland with partially treated sewage. Hundreds of other similar projects are under way around the country.

Using treated sewage water for irrigation is not feasible in every case, however. There are obvious climatic restrictions. Also, irrigation is a consumptive use; water shortages may not permit it. Finally, there may not be sufficient land nearby to absorb the effluents. Piping and pumping costs preclude moving the water more than a few miles or to significantly higher elevations.

Where irrigation is not feasible, aquatic or **wetland systems** may be developed to remove nutrients biologically. For example, in the 1960s and 1970s much of the land around Orlando, Florida, which was originally wetlands, was drained and converted to cattle pasture. At the time, Orlando was discharging

FIGURE 13–8
Orlando Easterly Wetlands Reclamation Project, Florida. Formerly cattle pasture, this is part of a mixed marsh that was created to remove nutrients from the wastewater of Orlando and to reestablish wildlife habitat. (Photograph courtesy of Post, Buckley, Schak & Jernigan, Inc.)

into the James River 13 million gallons (50 million L) per day of nutrient-rich effluent following secondary treatment. Through the Orlando Easterly Wetlands Reclamation Project, 1200 acres of pastureland has been converted back to wetlands. The project involved scooping soil from, and building berms around, pastures to create a chain of shallow lakes and ponds and replanting 1.2 million wetland plants ranging from bulrushes and cattails to various trees. The effluent entering the upper end now percolates through the wetland for about 30 days before entering the James River virtually pure. Thus the project has re-created a wildlife habitat, a step that protects biodiversity and at the same time provides tertiary treatment of sewage effluents and protects downstream water quality (Fig. 13–8). Wetland systems can be designed for any size, small as well as large. Thus they are becoming an increasingly popular alternative for small communities.

More contained artificial systems may also be designed for biological removal of nutrients from wastewater. For example, ponds filled with water hyacinths are being used in Mississippi and a number of other locations (Fig. 13–9a). These plants, which

FIGURE 13–9
(a) Artificial ponds with water hyacinths may be used for advanced sewage treatment. (Photograph by N. D. Vietmeyers, National Space Technology Laboratory.) (b) Water hyacinth plant. "Bulbs" at bases of leaves are air bladders that enable the plant to float on the surface while the roots dangle down and withdraw nutrients from the water solution. (Photograph by BJN.)

(a)   (b)

Coastal cities have a reputation for dragging their heels on sewage treatment. With the enormous potential of an ocean for tidal mixing and flushing, they have become accustomed to using the dilution solution for disposing of their sewage. The city of Boston, Massachusetts, presents an intriguing case in point. For decades, Boston sewage has received only primary treatment before discharge into Boston Harbor. Making matters worse, scum and sludge from the treatment process were also released to the harbor, with the result that Boston Harbor had such a well-deserved reputation for gross pollution that George Bush used it to demolish the environmental record of Michael Dukakis in the 1988 presidential campaign.

Under federal government pressure for years to clean up the harbor, the state established the Massachusetts Water Resources Authority (MWRA) in 1984 to oversee the cleanup. The MWRA mapped out a plan for the cleanup that conforms with the federally mandated full implementation by 1999. One element of that plan was put in place in December 1991: The flow of raw sludge was stopped and all sludge now goes to a pelletizing plant to be turned into fertilizer (see p. 305). A key component of the plan is the construction of a 9.5-mile-long pipe that will carry treated effluent well out of Boston Harbor and into Mas-

sachusetts Bay. The pipe will begin carrying effluent from a new primary-treatment plant in 1996, and by December 1999, the secondary-treatment phase of the plant will be in full operation and the program will be in full implementation.

The proposed 9.5-mile outfall pipe, which is already under construction, has stirred up a firestorm of criticism from citizens and advocacy groups on Cape Cod. Their concern centers on the impact of the effluent on water quality in Cape Cod Bay, which adjoins Massachusetts Bay to the southeast. The opponents would prefer that the outfall remain in Boston Harbor, which has already been affected by pollution and will probably always be so. They claim that because of the location of the outfall pipe, coastal water circulation may tend to carry the high-nutrient effluent into Cape Cod Bay. The Cape Codders cite numerous studies that show that nitrogen acts as a limiting factor in many coastal marine systems, and they fear that the added nitrogen from the Boston treatment plant will create conditions leading to blooms of algae, especially those that cause red tide. Red tides have been a recurring problem in recent years up and down the northeastern coast, often closing down shellfish harvesting because of the dangers of paralytic shellfish poisoning.

Resolution of this conflict may be

difficult. The MWRA claims that not only will the new treatment system obviously benefit Boston Harbor, it will also improve water quality in Massachusetts and Cape Cod bays. They point out that the system as currently operating is already polluting those bays, since the effluent is eventually carried out of the harbor by tidal action. Substantial efforts by leading marine scientists are under way to assess the potential impact of the pipe, but Cape Cod residents fear that the studies (funded by the MWRA) will be biased. Their request to the MWRA for $575 000 for an independent review of the ongoing studies was recently rejected. MWRA board members felt that the request was really a form of "blackmail" on the part of the Cape Cod advocacy groups. Adding to the complexity of the conflict, Susan Tierney, the state's environmental chief and also chairwoman of the MWRA, spoke in favor of the request.

On the surface, this conflict appears to be another case of NIMBY (Not In My Back Yard). One municipality is attempting to resolve its environmental problem in an apparently effective fashion, but there is the risk that the solution will create another problem many miles away as a result. What do you think should be done?

float on the surface and have an extensive root system dangling down into the water, are extremely efficient at absorbing nutrients (Fig. 13–9b). Such systems containing cattails and other "reeds" are also very efficient in removing nutrients. The plant material produced may be used for weaving mats and baskets or for cattle feed, or it may be fermented to produce alcohol or other products. Additional trophic levels, including fish, shellfish, and waterfowl, may be added to the system. Climatic factors and available space are the two factors that most restrict their development.

The capability of wetland systems in removing nutrients from wastewater is *not* to say that effluents may simply be discharged into natural wetlands. Indeed, it has been found that, where natural wetlands receive such effluents, the wetlands are significantly degraded and may never return to their original natural state. Re-created and artificial systems are distinct from natural wetlands, despite the fact that they may be highly productive and offer great potential.

A summary of water treatment procedures is shown in Figure 13–10. Despite the availability of all these levels of sewage treatment, coastal cities con-

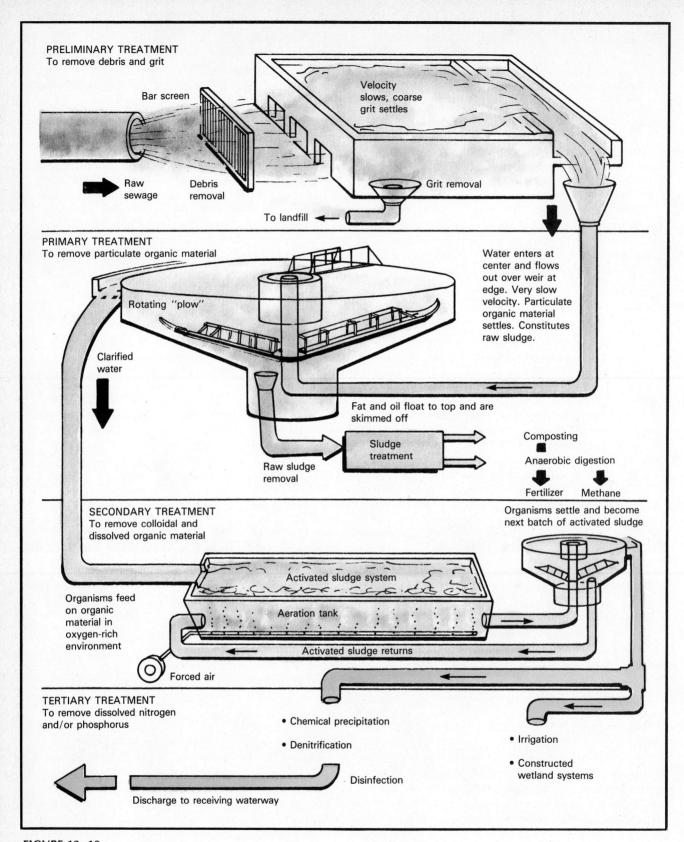

FIGURE 13–10
A summary of wastewater treatment.

tinue to depend on nonecological "solutions" to their sewage problems (see Ethics Box, p. 301).

Sludge-Treatment Options

Recall that particulate organic material, which is 30 to 50 percent of the total organic material in sewage, settles out in the primary clarifiers (primary treatment), and this settled material is called raw sludge. Pumped from the bottom of tanks, raw sludge is a black, foul-smelling, syrupy liquid consisting of 97–98 percent water, 2–3 percent organic material, and potentially including pathogens. Before this raw sludge is treated, excess activated sludge and sludges generated in tertiary treatment may be added to it.

Treatment and disposal of raw sludge are two more major aspects of sewage treatment. Past and still-persisting methods of disposal are to press out a major fraction of the water and incinerate or landfill the remainder. Until recently, a number of major U.S. cities on the Atlantic Coast were barging raw sludge to sea and dumping it. The ecological failing of these techniques is self-evident.

With suitable treatment, raw sludge can be converted to a *nutrient-rich humus* that can be used as organic fertilizer. In nature, detritus is broken down and converted to humus by the activity of soil organisms. Likewise, sludge treatment involves letting bacteria and other detritus feeders work on the sludge either in the absence of air (*anaerobic digestion*) or in the presence of air (*composting*). A third alternative is *pasteurization and drying*.

Anaerobic Digestion In **anaerobic digestion,** the raw sludge is put into large, airtight tanks called **sludge digesters.** In the absence of air, bacteria digest the material either by fermentation or by anaerobic respiration. A major byproduct of both processes is *biogas*. **Biogas** is about two-thirds methane; the remainder is carbon dioxide and various foul-smelling organic compounds that give sewage its characteristic odor. Natural gas, widely used for heating and cooking, is nearly pure methane. Because of its methane content, biogas is flammable and can be used for fuel as it is. In fact, it is commonly burned to heat the sludge digesters since the bacteria working on the sludge do best when maintained at about 100°F (38°C). Also, biogas can be refined and used to supplement natural gas supplies.

After 4 to 6 weeks, anaerobic digestion is more-or-less complete, and what remains is called **treated sludge.** It consists of the remaining organic material,

IN PERSPECTIVE
Water Conservation and Sewage Treatment

Some people see a potential conflict between the need for water conservation and sewage treatment. The concern is: With water conservation, will there be enough water going down drains to carry the wastes?

The concern is unwarranted. Since raw sewage is only 0.1 percent pollutants, reducing its water content by half would make it only 0.2 percent pollutants. It would be 99.8 percent rather than 99.9 percent water—an insignificant difference in terms of affecting the flow through pipes.

There would be a significant positive effect on sewage treatment, however. The positive effect would occur because the settling process of primary treatment and the biological processes of secondary treatment are time-dependent. The more

time the wastewater can spend going through these stages the more material will be removed. If water volume were cut by half, the remaining volume of wastewater could spend twice as much time going through the primary settling tanks and biological processes, thus considerably improving the degree of removal and the quality of the final effluent. Conversely, increasing amounts of water from increasing numbers of hookups cause the quality of treatment to decline, as increasing volumes of water are forced through existing systems faster.

Thus, in addition to saving water, water conservation may be promoted as a means toward improving the performance of sewage-treatment plants or avoiding the need to expand sewage-treatment facilities.

Another factor that affects the quality of the final effluent is how much "stuff" is put down the drain in the first place. The use of sink-drain garbage disposal units to dispose of food wastes has at least doubled the load of organic material (both particulate and colloidal) to be removed and treated by the sewage-treatment plant, and there is the additional water used to flush the wastes through the disposal unit as well. Home composting of food wastes as described in the In Perspective Box, page 195, is a more ecological alternative.

In conclusion, we can all have a positive impact on the sewage treatment of our cities and communities by practicing water conservation and not disposing of garbage down the drain.

FIGURE 13–11
Use of treated sludge. The treated sludge remaining after anaerobic digestion is a humus-rich, nutrient-rich liquid that is an excellent soil conditioner. The vehicle shown here is specially designed for applying sludge to soils. (Courtesy of Ag-Chem Equipment Co. Inc., Minneapolis, Minnesota.)

which is now a relatively stable, nutrient-rich, humus-like material in water suspension. Pathogens have been largely if not entirely eliminated, so they no longer present any significant health hazard.

Treated sludge makes an excellent organic fertilizer. It may be applied directly to lawns and agricultural fields in the liquid state in which it comes from the digesters, providing the benefit of both the humus and the nutrient-rich water (Fig. 13–11). Alternatively, much of the water may be removed by means of filter presses, leaving the organic material as a semisolid **sludge cake** (Fig. 13–12). The sludge cake is easy to stockpile, distribute, and spread on fields, with traditional manure spreaders. Since a large share of the nutrients are in water solution, however, much of the nutrient value goes with the filtered water, which is generally put back into the wastewater stream.

Use of treated sludge as an organic fertilizer—or "soil conditioner," as it is often called—is gaining increasing favor among farmers. At first there was difficulty getting farmers to accept the program, as they considered the sludge to be "a little free fertilizer and a lot of bad smell." Now, however, as more and more farmers recognize the additional benefits in improving soil, the demand for sludge cake exceeds the supply in many areas.

To cite just one success story, since 1979, a program operated by the Madison (Wisconsin) Metropolitan Sewerage District has recycled over 100000 tons (dry weight) of sewage sludge to 27500 acres (11000 hectares) of agricultural land. This program is estimated to have saved over $1.2 million in fertilizer costs.

Composting For **composting**, raw sludge is mixed with wood chips or other water-absorbing material to improve aeration and then piled in **windrows**, long

(a)

(b)

FIGURE 13–12
Dewatering treated sludge. After digestion of raw sludge, the remaining solid material may be removed from the water by means of filter presses seen here. (a) The treated sludge (98 percent water) is run between canvas belts going over and between rollers such that much of the water is pressed out. (b) The resulting "sludge cake" is a semisolid humus-like material that may be used as an organic fertilizer. (Photographs by BJN.)

narrow piles that allow air circulation and convenient turning with machinery. Bacteria and other decomposers break down the organic material to rich, humus-like material (see In Perspective Box, p. 195). Pathogens lose out in the competition. As long as the piles are kept well aerated, the obnoxious odors typical of anerobic respiration are negligible. After 6 to 8 weeks of composting, the resulting humus is screened out of the wood chips. The chips may be reused, and the humus is ready for application to soil.

A technique that is gaining increasing favor is the **co-composting** of sewage sludge and waste paper, a major constituent of refuse that presents another disposal problem (Chapter 20). Shredded paper absorbs the excess water from the sludge and facilitates aeration. The nutrients in the sludge enhance the ability of the decomposers to digest the paper. The combination of sewage sludge and paper ends up as a humus-like material or compost that may be used as a soil conditioner. This co-compost material is gaining

FIGURE 13–13
The Ocean Arks International greenhouse system, shown here, purifies raw sewage using a number of plants and animals living in a sequence of tanks through which the water passes. (Dann Blackwood.)

favor among farmers, landscapers, and homeowners in communities where it is available, often free of charge.

Pasteurization and Drying Raw sludge may be filter-pressed and the resulting sludge cake put through drying ovens that operate like oversized laundry dryers. In the dryers, the sludge is **pasteurized,** which means that heat is used to kill any pathogens (exactly the same process that makes milk safe to drink). The product is dry, odorless organic pellets. Milwaukee, Wisconsin, which has a particularly rich sludge resulting from the brewing industry, has been using this process for over 60 years. The city bags and sells the pellets throughout the country as an organic fertilizer under the trade name Milorganite℠. In 1991 Boston, Massachusetts, started up a pasteurization and drying facility that is now the nation's largest producer of organic fertilizer. Before the plant opened, Boston and surrounding communities had been dumping some 500 000 gallons of raw sludge every day into Boston Harbor, giving the harbor the reputation of being the nation's most polluted waterway.

Pasteurization and drying of raw sludge is still not a money-making proposition, however. It is estimated that the Boston facility will spend about $750 to produce a dry ton of pellets, which will sell in the range of $50 to $100. Much of the cost is energy for the drying process. The real cost comparison, however, must be in terms of the value of cleaning up polluted waterways.

Alternative Systems for Treating Raw Sewage

Large cities that already have sewage-collection systems and treatment plants may be locked into the options described above. However, small communities in need of a new sewage-treatment system have an opportunity to design an ecological system from the start. One such system is the *overland flow system* described in In Perspective Box, page 306. Another system, one that holds promise where space is available and the wastewater volume is not great, is the *oxidation pond system.* Here the raw sewage is directed into a series of shallow ponds, where algae and other vegetation consume the nutrients and then settle to the bottom. After passage through the series of ponds, the effluent has lost most of the remaining nutrient and particulate organic material.

John Todd and other researchers at Ocean Arks International, Falmouth, Massachusetts, have developed a greenhouse aquatic system for sewage treatment that is not limited by climate as open wetland systems may be (Fig. 13–13). Raw sewage first flows through a series of tanks, where bacteria consume the

The Overland Flow Wastewater Treatment System

An alternative to traditional wastewater treatment is the **overland flow system.** A facility utilizing this system was put into operation in Emmitsburg, Maryland, in 1989.

Raw wastewater, about 1 million gallons (4 million L) per day, from all the homes and commercial establishments of the community is first put through a pond where grit and particulate organic material settle. The water is then irrigated onto the long, narrow fields you see in the photograph. These fields have about a foot (0.3 m) of rich topsoil supporting a crop of reed canary grass, a forage grass that has a voracious appetite for nitrogen and other nutrients. Below the topsoil, the subsoil is compacted clay, which is impermeable to water, and slopes gently downward away from the irrigation pipe. The wastewater applied to one side of the field thus percolates through the topsoil, across the field, and into a collecting gutter at the opposite side. During this passage, natural organisms in the topsoil break down and utilize the organic wastes and maintain the richness and aggregate structure of the topsoil in the process, while the grass absorbs the nutrients. The water exiting into the collecting gutter is clear and nearly nutrient-free. It is collected into another reservoir and spray-irrigated onto forage crops so that none of the nutrients go to waste. The canary grass is periodically mowed and becomes feed for cattle. Thus, the nutrients make a complete cycle from wastewater to grass to beef to humans to wastewater and again back to the soil.

You may ask how a small community in western Maryland happened to install a state-of-the-art ecological method of wastewater treatment. First, the town did need a new wastewater treatment facility to accommodate population growth. However, it was largely the ecological thinking of one farmer in the area, Richard Waybright, who persuaded the town officials to go ecological. He volunteered some two hundred acres (about 80 hectares) of his own land to be the final recipient of the water and the nutrients. His forward thinking has created a situation in which everyone wins. He gets free water and nutrients for irrigation, Emmitsburg meets the standards for sewage treatment that will accommodate growth with a low-cost, low-maintenance system, and everyone benefits by not having the nutrients go into Chesapeake Bay, where they would cause more eutrophication.

Overland flow wastewater treatment system. Wastewater is being irrigated onto fields. Note the lush growth of grass benefiting from the nutrients. (Photograph by BJN.)

organic material (both particulate and dissolved), algae absorb the nutrients, snails eat the algae, and so on up the food chain. The system includes clams, several species of fish, and 120 species of plants, each performing a particular role. The end of the process yields, in addition to the organisms harvested for food, virtually pure water that with disinfection might be recycled for the municipal supply if needed.

The greenhouse system is quite economical to operate because all the energy required for nutrient removal comes from sunlight via photosynthesis. The major constraint of the system is the space requirement. The system can process only so much water only so fast, and additional volumes require additional space. Of course skilled management is another must.

Despite these limiting factors, the Ocean Arks system offers fantastic potential for small towns or

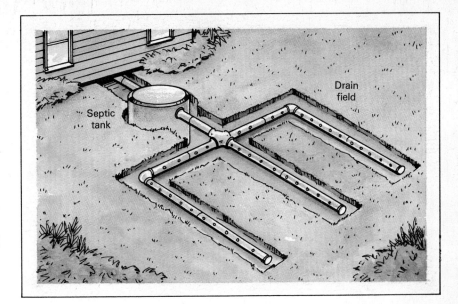

FIGURE 13–14
Individual sewage treatment. Septic tank and drain field. All the pipes and the tank are normally buried underground. They are shown uncovered here only for illustration. (USDA—Soil Conservation Service.)

villages, especially in the Third World in situations where no sewage treatment facilities exist at present. You can readily see how the system has the potential of providing food, employment, and raw materials (such as reeds for weaving) at the same time it is improving public health by taking care of the sewage problem.

Individual Septic Systems

Countless homes in rural areas are not connected to a municipal sewer system. Instead, they have individual septic systems consisting of a septic tank and a drain field (Fig. 13–14). Wastewater flows into the tank, where particulate organic material settles to the bottom. The tank acts like a primary clarifier in a municipal system. Water containing colloidal and dissolved organic material as well as the dissolved nutrients flows into the drain field and gradually percolates into the soil. Organic material that settles in the tank is digested by bacteria, but accumulations must still be pumped out every 2 to 3 years. Soil bacteria decompose the colloidal and dissolved organic material that comes through the drain field. Some people establish successful vegetable gardens over septic drain fields, thus exercising the sound principle of recycling the nutrients. The water, of course, rejoins the groundwater and helps to maintain the local hydrological cycle.

A properly maintained septic system may function indefinitely. However, colloidal organic material frequently enters the soil faster than it decomposes, gradually clogging the soil pores and forcing raw sewage to the surface, where it causes objectionable odors, contaminates surface water, and is a general health hazard. If the lot is not large enough to relocate the drain field, little can be done about this problem except to try to get connected to a municipal system as soon as possible. In many areas of the country, expensive suburban developments using individual septic systems have become quite obnoxious as a result of this problem. However, wherever soil has a structure that allows good absorption and aeration—characteristics that enhance the rate of bacterial decomposition—septic systems continue to be a viable treatment option.

An alternative for the handling of personal wastes right in the home is a **composting toilet.** An example is the Clivus Multrum developed in Sweden (Fig. 13–15). The Multrum receives only personal excrements and food wastes—no water other than urine. These wastes pass through a series of chambers as they decompose, and after 2 to 4 years they arrive at the final chamber as a stable, nutrient-rich humus that is suitable for application on lawns and gardens. The home must have other means for disposing of bath and other gray water. Since such water is not contaminated by human excrements, however, disposing of its presents few problems. Usually it is used for watering lawns and gardens. At a cost of about $2500, composting toilets are competitive with conventional septic systems but not with hookup to municipal sewers.

Taking Stock

As we noted at the beginning of this chapter, the disease hazards of untreated sewage were pointed out by Pasteur and other scientists in the late 1800s,

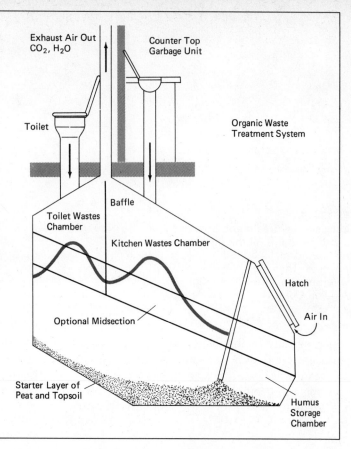

FIGURE 13–15
Clivus Multrum, a dry waste-treatment system. Sewage and food wastes deposited in the Clivus Multrum decompose aerobically. A dry, nutrient-rich compost is produced in 3 to 4 years. (Courtesy of Clivus Multrum, W. Pa., Inc.)

and the technology for primary and secondary (trickling-filter) treatment was developed at that time. There was no great rush, however, to keep up with ever-increasing amounts of sewage even in developed countries, much less in less-developed countries.

PROGRESS

In the period following World War II, rapidly expanding population and development in the United States led to the proliferation of signs reading "No Swimming! Polluted Water!" posted along formerly clean rivers and beaches. Much of the pollution was due to overflow of raw sewage from overloaded systems. A major turning point occurred as an outraged public finally brought enough pressure on Congress to pass the **Clean Water Act of 1972.** In addition to setting goals and schedules for cleanup, this act provided federal money for new interceptor sewers to collect wastes, for separating storm and sewage systems, and for constructing, expanding, and upgrading sewage-treatment plants. Nearly $50 billion has been spent on these efforts since the act was passed. Now most sewage in the United States is treated at least through the secondary level. Problems remain,

but many streams, rivers, harbors, lakes, and other waterways are notably cleaner now than they were in the late 1960s and early 1970s.

The major problem areas remaining in the United States are dealing with the nutrient-rich effluent from the secondary process and with sludge. What are the impediments to moving ahead with the many ecological options we have discussed?

IMPEDIMENTS TO PROGRESS

Contamination with Toxic Materials

The most serious impediment to using sewage sludge and nutrient-rich water for agricultural purposes is their contamination with toxic metals such as lead, mercury, cadmium, chromium, and nonbiodegradable organics. The conventional steps of sewage treatment do not affect such chemicals. Indeed, such chemicals may even poison the organisms used in secondary treatment and thus impede the process. Heavy metals tend to bind to organic material and therefore end up both in the treated sludge and in solution, making the effluent unfit for irrigation. Much of the sludge generated in the United States is still unfit for agricultural uses for this reason.

The source of these toxic materials is twofold. Most industries, particularly small shops, are connected into the same sewer systems as everyone else. Therefore, they may flush wastes containing these chemicals into the system. The fault lies not just with industry, however. Individuals at home or at work use cleaning fluids, pesticides, paints and other coatings, photographic chemicals, and so on, which commonly contain one or more toxic materials, and they pour unused or spent portions down the drain. Also, lead and copper may leach from water pipes.

Under the Clean Water Act, industries are now required either to pretreat their wastewater to remove toxic chemicals before discharging it into municipal sewer systems or to find alternative means of disposal. As standards for and enforcement of pretreatment have become more rigorous, the quality of sewage sludge has improved, and it is now in many cases usable. However, household wastes remain a problem.

Public Opinion and Lack of Understanding

Perhaps the biggest impediment to upgrading sewage treatment or developing use of byproducts is public opinion. Many people feel that what goes down the drain should be "out of sight, out of mind." People fail to make any connection between spotless, sanitary bathrooms and pollution problems in waterways. Combined water and sewer bills, which include the costs for sewage treatment, are only $5 to $10 a month per household. For another $2 to $3 per month, we could have state-of-the-art sewage treatment. Yet, many people feel adamantly that sewer charges should not be increased, and municipal referenda to raise money for sewage-treatment facilities fail again and again.

In addition, some people fail to understand the distinction between treated products and raw sewage. Consequently, they have an unrealistic fear that the treated products present a health threat. Such people are so adamantly against the use of sludge or wastewater for agricultural purposes that they will bring court action against individuals or municipalities to block their using it.

To overcome these problems, we need greater public understanding of (1) the ecological principal of recycling nutrients, (2) how failure to recycle nutrients causes degradation of soils and pollution of waterways, and (3) how modern sewage-treatment methods and recycling of byproducts can prevent soil degradation and water pollution and still be safe. As the public gains this understanding, we can expect progress to hasten. A major educational campaign in adopting sustainable solutions is in order.

 Review Questions _____

1. Why does raw sewage present a disease hazard?
2. When the disease hazards of sewage were recognized in the 1800s, what steps were first taken and what were the effects on natural waterways?
3. What is the water content of, and what are the four categories of, pollutants present in raw sewage?
4. What are the five steps of conventional wastewater treatment?
5. What is done, and what is removed in preliminary treatment?
6. What is done, and what is removed in primary treatment?
7. Why is secondary treatment also called biological treatment?
8. What are the two alternative systems that may be used in secondary treatment?
9. What is done, and what is removed in each of the systems named in question 8?
10. What problems may arise from the use of chlorine for disinfection? What are the alternatives?
11. What category of pollutants remains in the water

following secondary treatment and disinfection? What harm might it do?
12. What are two alternative methods for removing nutrients from wastewater?
13. What are the advantages of using treated wastewater for irrigation? What are the limitations of doing so?.
14. What is raw sludge, and where does it come from?
15. What are three alternative methods of treating raw sludge, and what end product(s) may be produced from each method?
16. Name and describe two alternative systems for handling sewage wastes.
17. What are the goals and accomplishments of the Clean Water Act of 1972?
18. What factors currently hinder the use of sewage as a resource?
19. How do less-developed countries compare with developed nations in terms of sewage collection and treatment?

Thinking Environmentally

1. Suppose a new community that will be comprised of several thousand people is going to be built in Arizona (warm, dry climate). You are called in as a consultant to design a complete sewage system including collection, treatment, and use/disposal of byproducts. Write an essay describing the system you recommend and giving a rationale for the choices involved.

2. Suppose your city/community has been disposing of raw sewage sludge in landfills but is now proposing to compost it and use the resulting material on city park-lands. The proposal is meeting considerable resistance from the public. Write a "Let-

ter to the Editor" describing the environmental advantages of this proposal.

3. Some people favor the use of oceans as a dumping ground for sewage. Do you support or oppose this alternative? Defend your position.

4. Many Americans will continue to live in rural locations where hookups to centralized sewage collection/treatment systems is impractical. What kinds of sewage systems would you recommend for such people, particularly in regions where soil drainage is poor and or there is a high frequency of heavy rains? Give a rationale for your recommendations.

14

Pollution from Hazardous Chemicals

LEARNING OBJECTIVES

When you have finished studying this chapter, you should be able to:

1. List and define the four categories of hazardous chemicals recognized by the EPA.

2. List eight ways in which a chemical may enter the environment at various times in its "life span."

3. Name two classes of chemicals that pose a particularly severe long-term toxic risk despite low levels of environmental contamination.

4. Define bioaccumulation and biomagnification and describe how they pertain to the toxic risk of the chemicals named in objective 3.

5. Describe how hazardous chemical wastes were disposed of before the early 1970s, including three methods of land disposal.

6. Name the two laws governing disposal of chemical wastes passed in the early 1970s and describe how these laws affected disposal.

7. Discuss shortcomings of land disposal methods that became evident in the 1970s and the magnitude of the problem.

8. Name the legislation that pertains to the cleanup of abandoned toxic waste sites and describe how cleanup is being implemented under this legislation.

9. Name and describe the major features of the legislation regulating disposal of hazardous wastes today.

10. Name and describe the major features and shortcomings of the legislation that is intended to ensure that municipal drinking water is safe.

11. Define what is meant by groundwater remediation and describe two techniques for accomplishing it.

12. Discuss current and future trends in the management and disposal of hazardous chemical wastes.

In Chapter 12 we noted that the second way humans have diverged from the principle of recycling materials is in the production of increasing amounts and kinds of *nonbiodegradable* materials, some of which may be highly toxic. Our first objective in this chapter is to examine the nature of these materials and how past management (or mismanagement) of them has led to a variety of pollution problems. As a second objective, we shall examine how these problems are being addressed currently and may be addressed in the future.

Overview of Chemical Hazards

Virtually everything we use—from the toothbrush and shampoo we use in the morning to the TV set we watch in the evening—is a product of chemistry. While improving life in many respects, however, many chemicals also present a variety of hazards. The U.S. Environmental Protection Agency categorizes chemical hazards as follows:

Ignitability Substances that catch fire readily (examples: gasoline and alcohol)

Corrosivity Substances that corrode storage tanks and equipment (example: acids) .

Reactivity Substances that are chemically unstable and may explode or create toxic fumes when mixed with water (examples: explosives, elemental phosphorus [not phosphate] and concentrated sulfuric acid)

Toxicity Substances that are injurious to health when ingested or inhaled (examples: chlorine, ammonia, pesticides, and formaldehyde)

Any material having one or more of these attributes is a **hazardous material,** or HAZMAT for short. (**Radioactive materials,** probably the most hazardous of all are treated as an entirely separate category and are discussed in Chapter 22.)

The hazard a material presents is frequently implicit in the use. Gasoline would not work as a fuel if it were not ignitable, for example. Often, however, the hazard is an undesirable side effect, such as the toxic fumes given off by many cleaning solvents. Most people agree that the advantages we gain from modern chemicals far outweigh the risks. However, this favorable tradeoff is achieved and maintained only insofar as we take suitable precautions to reduce the risks. This is true both at the level of the individual and at the societal level. For example, it is an individual user's responsibility to read the labels on various products and use them only as recommended, taking the precautions listed. At the societal level, it is government regulations that require such information be put on the label.

Use is only one event, so to speak, in a material's "life span." Implicit in your use of shampoo, for example, is that raw materials were obtained and various chemicals were produced to make both the shampoo and its container. Inevitably, there are chemical wastes and byproducts in the production process. In addition, consider the risks of accidents occurring in the manufacturing process and in the transportation of the raw materials, the finished product, or the wastes. What happens to the spent shampoo you rinse down the drain? What happens to the container you throw into the trash, still holding a last few drops or perhaps most of the shampoo when you don't like the brand? Now expand your thinking to include the hundreds of thousands of products used by billions of people, and perhaps you can begin to appreciate the magnitude of the situation.

Human and environmental risks are present at each stage of the life span of any manufactured product (Fig. 14–1). Many of these risks are beyond the control of the individual. Therefore, in this area of chemical hazards, as in other areas, we generally take

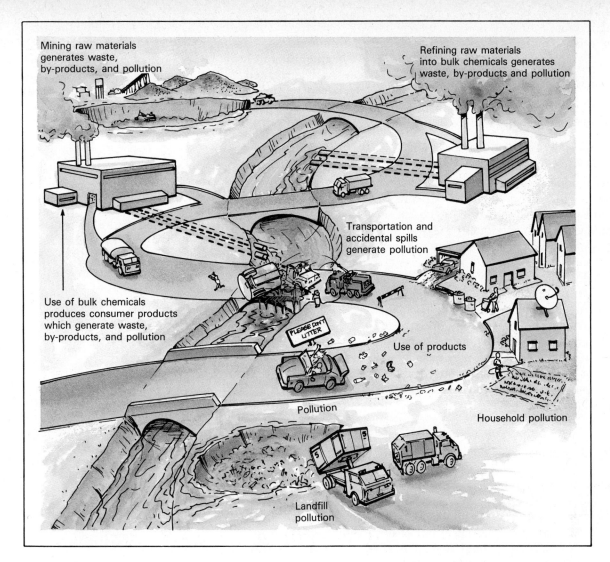

FIGURE 14–1
Life span of a chemical. The life span begins with the obtaining of raw materials and ends with final discard of the used product. At each step or in transportation between steps, wastes, byproducts, or the product itself may enter the environment, causing pollution and creating various risks to human and environmental health.

Within the illustration:

Mining raw materials generates waste, by-products, and pollution

Refining raw materials into bulk chemicals generates waste, by-products and pollution

Transportation and accidental spills generate pollution

Use of bulk chemicals produces consumer products which generate waste, by-products, and pollution

PLEASE DON'T LITTER

Use of products

Household pollution

Pollution

Landfill pollution

for granted that government has responsibility in protecting human and environmental health. Indeed, there are numerous government regulations requiring the chemical industry and transporters of chemicals to maintain certain safety standards. Much of our understanding of the long-term risks of toxic chemicals entering the environment has been gained only within the last 30 years and is still accumulating. Therefore, we are still in the midst of developing regulations and methods for managing toxic materials. This area will be the main focus of the rest of this chapter. The question of what levels of risk are acceptable is deferred until Chapter 17.

Toxic Chemicals and Their Threat

SOURCES OF CHEMICALS ENTERING THE ENVIRONMENT

There are many stages at which chemicals may enter the environment:

1. Intended use may involve introducing the entirety of the product into the environment—application of pesticides, fertilizers, and road salts, for example

2. Intended use may entail a fraction of the material going into the environment—the evaporation of solvents from paints and adhesives, for example

3. Disposal of waste byproducts generated in the course of industrial production

4. Disposal of spent solvents, cleaning fluids, washwater, lubricants, and so on, with whatever contaminants they may contain

5. Disposal of unused portions and amounts remaining in emptied containers

6. Leaks from storage tanks

7. Accidental spills or releases during production or use

8. Accidental spills or releases during transportation

These introductions into the environment occur at every level, from major industrial plants to small shops and individual homes. While single events involving large amounts of one chemical may constitute a disaster and make headlines, the total amount of all chemicals gradually entering the environment from millions of consumers and households is much greater than the amount from any single disaster.

Fortunately, a large portion of the chemicals introduced into the environment, even though they may be highly toxic in acute doses (high-level, short-term exposures), are gradually broken down and assimilated by natural processes. Therefore, once these chemicals are diluted sufficiently, they pose no long-term human or environmental risk. Indeed, until recently it was generally assumed that "dilution was a solution to pollution."

CHEMICALS PRESENTING LONG-TERM RISK

There are two major classes of chemicals for which the "dilution solution" fails, however: (1) heavy metals and their compounds and (2) nonbiodegradable synthetic organics. Far from "disappearing" into the environment, these chemicals tend to be absorbed from the environment and concentrated by organisms, including humans, till they may reach lethal doses. This concentration process obviously poses another dimension of risk to human and environmental health—a long-term risk from low, even minute levels.

Heavy Metals

The most problematic heavy metals are lead, mercury, arsenic, cadmium, tin, chromium, zinc, and copper. These metals are widely used in industry, particularly in metal-working or metal-plating shops and in such products as batteries and electronics. Because heavy metal compounds have brilliant color, they are used in paint pigments, glazes, inks, and dyes. They are also used in certain pesticides and medicines. Thus heavy metals may enter the environment wherever any of these products are produced, used, and ultimately discarded.

Heavy metals are extremely toxic because, as ions or in certain compounds, they are soluble in water and may be readily absorbed into the body, where they tend to combine with and inhibit the functioning of particular vital enzymes. Very small amounts can have severe physiological or neurological consequences. The mental retardation caused by lead poisoning and the insanity and crippling birth defects caused by mercury are particularly well known.

Nonbiodegradable Synthetic Organics

Synthetic organic compounds are the basis for all plastics, synthetic fibers, synthetic rubber, modern paintlike coatings, solvents, pesticides, wood preservatives, and innumerable other products. Being nonbiodegradable is a major part of their utility: We wouldn't want fungi and bacteria attacking and rotting our tires, paints, and so on.

These compounds are toxic because they are similar enough to natural organic compounds to be absorbed into the body. There they interact with particular enzymes, but their nonbiodegradability prevents them from being broken down or processed further. The result is that they upset the system. When a person ingests a sufficiently high dose, the effect may be acute poisoning and death. With low doses over extended periods, however, the effects are insidious and can be mutagenic (mutation-causing), carcinogenic (cancer-causing), or teratogenic (birth-defect-causing). They may cause serious liver and kidney dysfunction, sterility, and numerous other physiological and neurological problems.

A particularly troublesome class of synthetic organics is the **halogenated hydrocarbons,** organic compounds in which one or more of the hydrogen atoms have been replaced by an atom of chlorine, bromine, fluorine, or iodine. These four elements are classed as halogens, hence the name *halogenated hydrocarbons.* (Fig. 14–2). Of the halogenated hydrocarbons, the chlorinated hydrocarbons (also called organic chlorides) are by far the most common. These compounds are widely used in plastics (polyvinyl chloride), pesticides (DDT, kepone, and mirex), solvents (carbon tetrachlophenol), electrical insulation (polychlorinated biphenyls, which are the infamous PCBs always in the news), flame retardants (TRIS), and many other products. Additional chlorinated hydrocarbons and their health effects are listed in Table 14–1.

BIOACCUMULATION AND BIOMAGNIFICATION

The trait that makes heavy metals and nonbiodegradable synthetic organics particularly hazardous is their tendency to *accumulate* in organisms. Because of accumulation, small, seemingly harmless amounts re-

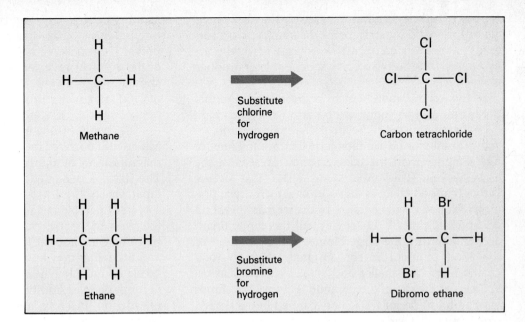

FIGURE 14–2
Halogenated hydrocarbons are organic compounds in which a halogen (Cl, Br, F, or I) has been substituted for hydrogen (H).

ceived over a long period of time may reach toxic levels. This phenomenon—referred to as **bioaccumulation**—can be understood as follows.

You are familiar with the concept of a filter. As water passes through a filter, impurities too large to pass through the pores accumulate on the filter. Basically, organisms act as a filter for heavy metals and synthetic organics. Heavy metals enter the body dissolved in water, but once bound to enzymes, the met-

als are removed from solution. Synthetic organics are highly soluble in lipids (fat or fatty compounds) but sparingly soluble in water. As they pass through cell membranes, which are lipid, they come out of water solution and enter into the lipids of the body. Thus, traces of heavy metals and synthetic organics that are absorbed with food or water are trapped and held by the body's enzymes and lipids, while the water and water-soluble wastes are passed in the urine. Since

TABLE 14–1

Examples of Toxic Synthetic Organic Compounds Frequently Found in Chemical Wastes

Chemical	Known Health Effects*							
	Mutations	Carcinogenic	Birth Defects	Still Births	Nervous Disorders	Liver Disease	Kidney Disease	Lung Disease
Benzene	X	X	X	X				
Dichlorobenzene	X			X	X	X		
Hexachlorobenzene	X	X	X	X	X			
Chloroform	X	X	X		X			
Carbon tetrachloride	X		X	X	X	X		
Chloroethylene (vinyl chloride)	X	X			X	X		X
Dichloroethylene	X	X		X	X	X	X	
Tetrachloroethylene		X			X	X	X	
Trichloroethylene	X	X			X	X		
Heptachlor	X	X		X	X	X		
Polychlorinated biphenyls (PCBs)	X	X	X	X	X	X		
Tetachlorodibenzo dioxin	X	X	X	X	X	X		
Toluene	X			X	X			
Chlorotoluene	X	X						
Xylene			X	X	X			

* Determined from tests on experimental animals.

Source: Adapted from S. Epstein, L. Brown, and C. Pope. *Hazardous Waste in America.* Copyright © 1982 by Samuel S. Epstein, M.D., Lester O. Brown, and Carl Pope.
Reprinted with permission of Sierra Club Books.

the body has no mechanism to excrete the heavy metals or synthetic organics or to metabolize them further, trace levels consumed over time gradually accumulate in the body and may sooner or later produce toxic effects.

Bioaccumulation, which occurs in the individual organism, may be compounded through a food chain such that organisms at the top of the chain accumulate concentrations of a contaminant that are millions of times higher than the concentration present in the environment. This multiplying effect that occurs through a food chain is called **biomagnification.** Biomagnification occurs because of the biomass pyramid (Chapters 2 and 3). Effectively, all the contaminant accumulated by the large biomass at the bottom of the food pyramid is concentrated, through food chains, into the smaller and smaller biomass of organisms at the top of the pyramid. Figure 14–3 shows biomagnification for DDT, now banned in the United States but still a widely used chlorinated hydrocarbon pesticide in the Third World.

One of the most distressing aspects of bioaccumulation and biomagnification is that there are no warning symptoms until contaminant concentrations in an organism or organisms are high enough to cause problems. Then, it is too late to do much about it.

As is often the case, bioaccumulation and biomagnification went unrecognized until serious problems brought the phenomena to light. In the 1960s

serious diebacks in the populations of many species of predatory birds, including Bald Eagles and ospreys, were observed. Investigations into the cause of the diebacks revealed the biomagnification of DDT shown in Figure 14–3. This finding led to the banning of DDT and many related pesticides in most developed countries, as described in Chapter 10.

A tragic episode in the early 1970s known as the Minamata disease revealed the potential for biomagnification of mercury and other heavy metals. The disease is named for a small fishing village in Japan where the episode occurred. In the mid-1950s, cats in Minamata began to show spastic movements, followed by partial paralysis, coma, and death. At first this was thought to be a peculiar disease of cats, and little attention was paid to it. However, concern escalated quickly when the same symptoms began to occur in people. Additional symptoms such as mental retardation, insanity, and birth defects were also observed. Scientists and health experts eventually diagnosed the cause as acute mercury poisoning.

A chemical company near the village was discharging wastes containing mercury into a river that flowed into the bay where the Minamata villagers fished. The mercury, which settled with detritus, was first absorbed and bioaccumulated by bacteria and then biomagnified as it passed up the food chain through fish to cats and humans. Cats had suffered first and most severely because they fed almost ex-

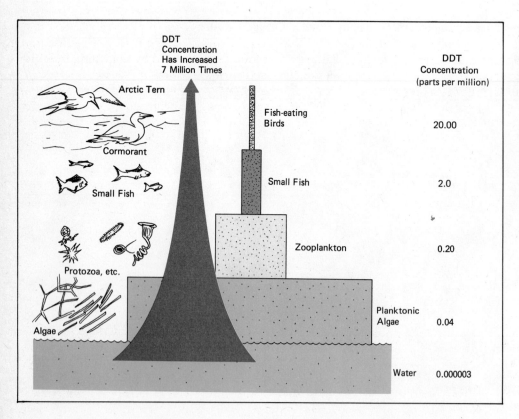

FIGURE 14–3
Biomagnification. Each successive consumer in a food chain receives a more contaminated food supply and accumulates the contaminant to yet a higher level. Scientists have observed that the concentration of the pesticide DDT was magnified almost 7 million times as it passed through the food chain shown here.

FIGURE 14–4
In the 1960s the Cuyahoga River carried such a load of flammable pollutants from Cleveland's Industries that this fire resulted. Incidents such as this led to the passage of the Clean Water Act of 1972 which now prevents such indiscriminate discharge of pollutants into waterways. (© 1992 The Plain Dealer Publishing Company.)

clusively on the remains of fish. By the time the situation was brought under control, some 50 people had died and 150 had suffered serious bone and nerve damage. Even now, the tragedy lives on in the crippled bodies and retarded minds of Minamata descendants.

INTERACTION BETWEEN POLLUTANTS AND PATHOGENS

Complicating the situation are interactions between pollutants and pathogens. For example, unusually high numbers of dead dolphins and other sea mammals were found washing up on beaches of the Atlantic Coast and other regions in 1991. Analysis of the dead animals showed that their tissues had bioaccumulated significant but sublethal levels of various chlorinated hydrocarbons, but the immediate cause of death seemed to be pathogens. Is there any relationship between these two factors, or is the presence of the pollutants only incidental? Many investigators contend that pollutants weaken the immune system, and consequently the organism may fall victim to the pathogens. It is extremely difficult to prove this hypothesis, but it creates another order of concern.

Management and Mismanagement of Hazardous Wastes

Historically, chemical wastes of all kinds have been disposed of as expediently as possible. From the dawn of the Industrial Age up to the very recent past,

it was common practice to exhaust all combustion fumes up smokestacks, vent all evaporating materials and solvents into the air, and flush all waste liquids and contaminated wash water into sewer systems or directly into natural waterways. Many human health problems occurred, but they were either not recognized as being caused by the pollution or accepted as the "price of progress." Indeed, much of our understanding regarding human health effects of hazardous materials is derived from those uncontrolled exposures. For example, the expression "mad as a hatter" comes from the fact that people who made hats in the 1800s frequently became insane. The insanity, it was later found, was caused by poisoning from the mercury used in the production process.

In the 1950s, as production expanded and synthetic organics came into widespread use in the developed countries, many streams and rivers became essentially open chemical sewers, as well as sewers for human waste. These waters were not only devoid of life; they were themselves hazardous. For example, in the 1960s the Cuyahoga River, which flows through Cleveland, Ohio, carried so much flammable material that it actually caught fire and destroyed seven bridges before it burned itself out (Fig. 14–4). Worsening pollution and increasing recognition of adverse health effects finally created a degree of public outrage that pushed Congress to pass the Clean Air Act of 1970 and the Clean Water Act of 1972. These acts set standards for allowable emissions into air and water and timetables for reaching those standards.

As a result of this legislation and subsequent amendments, industry has spent billions of dollars on pollution-control equipment. As a result, direct discharge of wastes into air and water has been greatly reduced. Although there is still a considerable distance to go, water quality in many areas has improved dramatically, and fish have returned to numerous formerly polluted waterways. (Many less-developed countries either have not passed such legislation or else existing legislation is poorly enforced. Consequently, uncontrolled discharge of toxic ma-

terials into air and water still occurs in many regions of the world.)

As industries removed pollutants from air and water waste streams, however, they were confronted with a new problem: What do we do with them now? Unfortunately, the writers of the Clean Air and Clean Water acts had not considered this problem, and hence the legislation offered no further guidance. The one major alternative that remained open and unregulated was land disposal. Thus, indiscriminate discharge into air and water was effectively traded for indiscriminate disposal on or in the land. With your understanding of the water cycle (Chapter 11), you can see how land disposal enormously increases the potential for groundwater contamination.

FIGURE 14–5
Deep-well injection, a technique used for disposal of large amounts of liquid wastes. The supposition is that toxic wastes may be drained into dry, porous strata inside the earth, where they may reside harmlessly "forever." The theory and precautionary safety measures are listed on the left. Prior to the 1980s, these measures frequently were not taken. Even when they are, the potential for failure remains (right). (Adapted with permission from *Environmental Action*, 1525 New Hampshire Ave. N.W., Washington, D.C. 20036.)

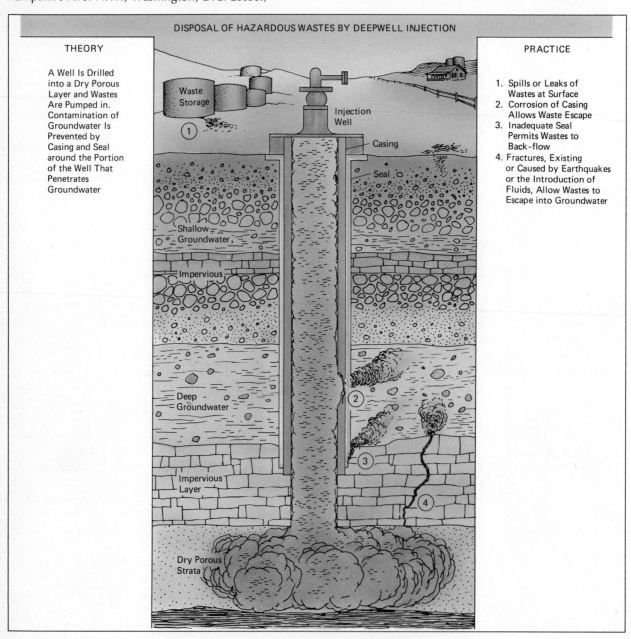

DISPOSAL OF HAZARDOUS WASTES BY DEEPWELL INJECTION

THEORY

A Well Is Drilled into a Dry Porous Layer and Wastes Are Pumped in. Contamination of Groundwater Is Prevented by Casing and Seal around the Portion of the Well That Penetrates Groundwater

Waste Storage

Injection Well

Casing

Seal

Shallow Groundwater

Impervious

Deep Groundwater

Impervious Layer

Dry Porous Strata

PRACTICE

1. Spills or Leaks of Wastes at Surface
2. Corrosion of Casing Allows Waste Escape
3. Inadequate Seal Permits Wastes to Back-flow
4. Fractures, Existing or Caused by Earthquakes or the Introduction of Fluids, Allow Wastes to Escape into Groundwater

METHODS OF LAND DISPOSAL

In the early 1970s there were three primary land-disposal methods: (1) deep-well injection, (2) surface impoundments, and (3) landfills. With conscientious implementation of safeguards, each of these methods has some merit. Without adequate regulations or enforcement, however, contamination of groundwater is virtually inevitable.

Deep-Well Injection

Deep-well injection involves drilling a "well" into dry, porous material below groundwater (Fig. 14–5). In theory, hazardous waste liquids pumped into the well soak into the porous material and remain isolated indefinitely. However, it is almost impossible to guarantee that fractures in the impermeable layer will not eventually permit wastes to escape and contaminate groundwater. Indeed, the introduction of wastes may produce enough stress to cause such fractures. Also, there are a number of additional ways in which wastes can escape into groundwater, as Figure 14–5 shows.

Surface Impoundments

Surface impoundments are simple excavated depressions into which liquid wastes are drained and held. They were the least expensive and hence most widely used way to dispose of large amounts of water carrying relatively small amounts of chemical wastes. As waste is discharged into the "pond," solid wastes settle and accumulate while water evaporates. (Fig. 14–6). If the pond bottom is well sealed and if evaporation equals input, impoundments may receive wastes indefinitely. However, inadequate seals may allow wastes to percolate into groundwater, exceptional storms may cause overflows, and volatile materials can evaporate into the atmosphere, adding to

FIGURE 14–6
Surface impoundments, an inexpensive technique for disposal of large amounts of lightly contaminated liquid wastes. The supposition is that only water leaves the impoundment, by evaporation, while wastes remain and accumulate in the impoundment indefinitely. Prior to the 1980s, plastic liners were not used and ponds were not always dug into clay. Even when these steps are taken, potential for failure remains, as the Practice column shows.

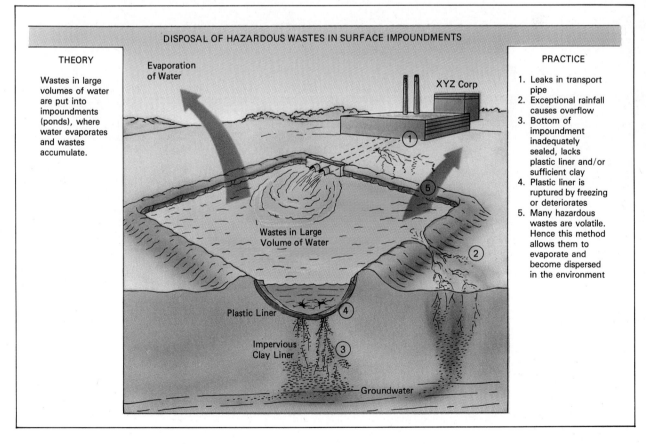

air pollution problems and eventually coming down with rainfall to contaminate water in other locations.

Landfills

When hazardous wastes are in a concentrated liquid or solid form, they are commonly put into drums and buried in landfills. If a landfill is properly lined, supplied with a system to remove leachate (material that may percolate out the bottom), and properly capped, it may be reasonably safe and is referred to as a secure landfill (Fig. 14–7). The various barriers are subject to damage or deterioration, however, and many experts feel it is only a question of time before contents will leach from even the most "secure" landfills.

Because land disposal was not regulated, however, there were numerous instances where even the most rudimentary precautions were not taken. There

(a)

FIGURE 14–7
Landfilling, a technique widely used for disposal of concentrated chemical wastes. (a) An operating hazardous waste landfill in Alabama. (Alon Reininger/Contact/Woodfin Camp Associates.) (b) Precautionary measures to make this method safe are listed in the Theory column. Before the 1980s, these measures frequently were not taken. Even when they are taken, potentials for failure remain. (Adapted from an illustration by Rick Farrell, copyright. All rights reserved.)

(b)

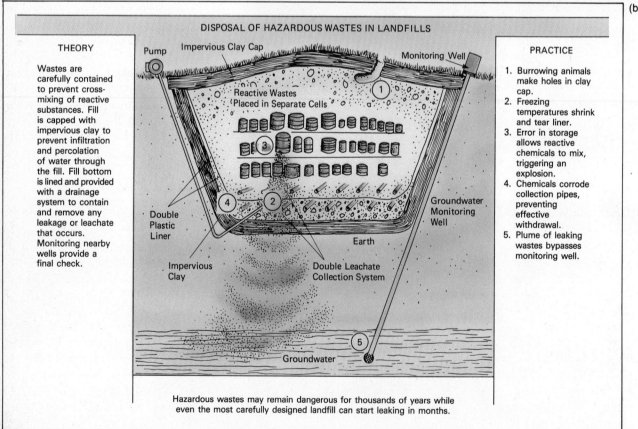

DISPOSAL OF HAZARDOUS WASTES IN LANDFILLS

THEORY

Wastes are carefully contained to prevent cross-mixing of reactive substances. Fill is capped with impervious clay to prevent infiltration and percolation of water through the fill. Fill bottom is lined and provided with a drainage system to contain and remove any leakage or leachate that occurs. Monitoring nearby wells provide a final check.

PRACTICE

1. Burrowing animals make holes in clay cap.
2. Freezing temperatures shrink and tear liner.
3. Error in storage allows reactive chemicals to mix, triggering an explosion.
4. Chemicals corrode collection pipes, preventing effective withdrawal.
5. Plume of leaking wastes bypasses monitoring well.

Pump · Impervious Clay Cap · Monitoring Well · Reactive Wastes (Placed in Separate Cells) · Double Plastic Liner · Impervious Clay · Double Leachate Collection System · Earth · Groundwater Monitoring Well · Groundwater

Hazardous wastes may remain dangerous for thousands of years while even the most carefully designed landfill can start leaking in months.

FIGURE 14–8
Midnight dumping at "Valley of the Drums" near Louisville, Kentucky about 1975. Thousands of drums of waste chemicals, many of them toxic, were simply unloaded at this site and left to "rot" seriously threatening the surrounding environment, waterways and aquifers. Incidents such as this led to the passage of the Resources Conservation and Recovery Act of 1976 which outlaws such indiscriminate disposal. (Van Bucher/Photo Researchers.)

were many cases where deep wells were injecting wastes directly into ground water. Abandoned quarries were used as landfills with no additional precautions being taken, and surface impoundments frequently had no seals or liners whatsoever.

Even worse, considerable amounts of waste failed to get to any disposal facility at all. Instead these materials ended up either being dumped in vacant lots and abandoned warehouses or accumulating on the site of the generator (the company where the wastes originate).

Midnight Dumping and On-Site Accumulation

Following passage of the Clean Air and Clean Water acts, the need for disposing of waste chemicals created the brand new business of chemical waste disposal. For a fee, disposal companies would haul away chemical wastes and dispose of them. Many reputable businesses entered the field, but, again in the absence of regulations, there were also disreputable operators. As stacks of drums filled with hazardous wastes "mysteriously" appeared in abandoned warehouses, vacant lots, or municipal landfills, it became clear that some operators were simply pocketing the fee and then unloading the wastes in any available location, frequently under cover of darkness (hence the term **midnight dumping,** Fig. 14–8).

Authorities trying to locate the individuals responsible found that they had gone out of business and were nowhere to be found. Some companies simply stored wastes on their own properties and then went out of business, abandoning the property and the wastes (Fig. 14–9). As drums containing hazardous chemicals corroded and leaked, there was great danger of reactive chemicals combining and causing explosions and fires. By far the greatest long-term hazard, however, is that toxic chemicals from any form of insecure land disposal may leach into groundwater.

Scope of the Problem

The mounting problem of unregulated land disposal of hazardous wastes was brought vividly to the public attention by the episode at Love Canal, which has become a battle cry of the hazardous waste problem.

Love Canal was an abandoned canal bed near Niagara Falls, New York. In the 1930s and 1940s, it served as a convenient burial site for thousands of drums of waste chemicals—over 20 000 tons in all. After the canal was filled and covered with earth, the land was eventually transferred to the city of Niagara Falls, and homes and a school were built on the edge of what had been the old canal. The area of covered chemicals became a playground. Over the years, children attending the school and coming in contact with

"black gooey stuff" oozing out of the ground began having unusual health problems, ranging from chemical burns and skin rashes to severe physiological and nervous disorders. Even more alarming, residents began to note that an unusually high number of miscarriages and birth defects were occurring. The situation climaxed in 1978 when health authorities identified the black ooze as a potent mixture of numerous chlorinated hydrocarbons known to cause birth defects and other disorders in experimental animals.

Some $3 billion in health claims were filed against the city of Niagara Falls, several hundred times its annual operating budget, and nearly 600 families demanded relocation at state expense. Eventually the state did purchase about 100 homes (Fig. 14–10), but most of the residents only gained further frustration because their claims could not be proved. The difficulty is that it is essentially impossible to prove the cause of any particular disorder when multiple factors are involved. Furthermore, we simply do not have sufficient understanding regarding the long-term health effects of most bioaccumulating synthetic organic chemicals.

Investigation into who was responsible for the Love Canal disaster eventually revealed that the fault lay more with the bureaucratic processes that allowed

development next to and on top of the site than with negligent disposal practices on the part of the chemical company. Nevertheless, the media attention given Love Canal did focus the public's attention on what was being done with toxic wastes and the hazards involved.

As both government and independent researchers began surveying the extent of the problem, bad disposal practices were found to be rampant. The World Resources Institute estimated that in the United States in the early 1980s, there existed 75000 active industrial landfill sites, along with 180000 surface impoundments and 200 other special facilities that were or could be possible sources of groundwater contamination. In most cases, the contaminated area was relatively small, 200 acres (80 hectares) or less, but in total the problem was immense and affected virtually the entire country. As studies and tests proceeded, thousands of individual wells, some in every state, and some major municipal wells were closed because of being contaminated with toxic chemicals.

Many cases of private-well contamination

IN PERSPECTIVE
Grass Roots in a Toxic Wasteland

Given the magnitude of the toxic waste problem in the United States, it is tempting to think that things might be better in a society characterized by central planning on the part of the regime in power, that environmental concerns might more easily be addressed at the highest levels of power. Recent events have put such thinking to rest. Now that the shrouds of secrecy have been lifted from the former Communist countries of eastern Europe and the republics of the former USSR, it is apparent that central planning has been responsible for the worst kinds of environmental pollution imaginable. Pollutants are emitted from the stacks of industry and power plants with no controls, ruining thousands of square miles of forests and creating untold health problems; untreated sewage from cities fouls the rivers, destroying fish and rendering the water unfit even for industrial uses; heavy metals and toxic chemicals pour untreated into the Baltic Sea, turning the bottom into a marine desert.

The truth is, the leaders of Communist regimes almost completely ignored environmental concerns in the interest of promoting economic production. In the words of Murray Feshback and Alfred Friendly, Jr. in their book *Ecocide in the U.S.S.R.,* "No other great industrial civilization so systematically and so long poisoned its air, land, water and people. . . . And no advanced society faced such a bleak political and economic reckoning with so few resources to invest toward recovery." The people of eastern Europe and the former USSR are now faced with both a bankrupt environment and a bankrupt economy, and they are looking to the Western democracies for help in both areas.

In environmental affairs in the United States and other democratic societies, grass roots action has always preceded change in public policy. The progress in solving the hazardous waste problem—the laws passed by Congress and the implementation and enforcement of those laws, which you are going to read about in the following part of this chapter—has been a direct result of grass roots pressure both on recalcitrant leaders and through electing new leaders with environmental values. All the evidence indicates that without this grass roots pressure—citizens demanding a clean, healthful environment—political leaders fall into promoting various economic and business concerns for the benefit of their own "in-group" and ignore environmental responsibilities.

How are the environmental grass roots in the countries of eastern Europe and the former USSR? Grass roots movements are never encouraged in totalitarian societies; they are seen as subversive elements. After the nuclear accident at Chernobyl in 1986, environmentalism achieved some legitimacy in the USSR, grass roots movements were tolerated, and some protests were successful in bringing about limited environmental changes. However, now that political processes are open, environmental grass roots movements have declined as people have become more concerned with basic economic survival. There is currently a fatalism in the response of people living in the polluted former Communist lands. They know their environment is unhealthy, but they still feel powerless to do anything about it. Such is the continuing legacy of Communism.

Thus, experience demonstrates that it is in liberal democracies, where citizens have the freedom to organize and to demand a better environment, that pollution control and prevention regulations are passed and implemented. Of course the freedom to organize and act is only half the answer; people must take advantage of that freedom. We can only hope that the newly freed people of the former Communist nations and other peoples of the world including ourselves will take advantage of freedom and act on behalf of creating a better environment.

Sources: R. Liroff. "Eastern Europe: Restoring a Damaged Environment." *EPA Journal* 16; no. 3 (1990); R. Brandt. "Soviet Environment Slips Down the Agenda." *Science,* 255 (January 3, 1992).

---

caused by careless disposal were discovered only after people experienced "unexplainable" illnesses over prolonged periods. While these incidents did not receive the media attention of Love Canal, they were nevertheless devastating to the people involved.

Another problem uncovered in the aftermath of Love Canal was that tests showed that drinking water supplies in many cities were tainted with toxic synthetic organic chemicals. In most cases, it was deemed that the trace levels detected were not dangerous, and so no action was taken. However, just how much constitutes a safe versus a harmful level is a subject of considerable controversy.

Even more important than the damage already done was the recognition that this is an ongoing problem. The Environmental Protection Agency estimates that close to 150 million tons of hazardous wastes are generated each year in the United States (close to two-thirds of a ton per person). In the late 1970s, as much as 90 percent was being disposed of improperly. Even in new sites specifically designated for toxic wastes, the EPA found that monitoring systems were inadequate to detect leakage.

CLEANUP OF HAZARDOUS WASTE SITES

There are four major aspects to the hazardous waste problem.

1. We need to: ensure that drinking water currently being consumed is safe,
2. Clean up thousands of existing sites from which toxic materials are already or may soon be leaching into groundwater and thus contaminating future drinking water,
3. **Remediate** contaminated groundwater (that is, return groundwater to its original uncontaminated state), and
4. Provide proper management and disposal of hazardous wastes being generated now as well as those that will be generated in the future.

The U.S. Congress has responded to these needs by passing a number of laws under which considerable progress has been made in each of the four areas.

Ensuring Safe Drinking Water

To protect the public from the risk of toxic chemicals contaminating drinking water supplies, Congress passed the **Safe Drinking Water Act of 1974.** Under this act, the EPA sets standards regarding allowable levels of various toxic chemicals, and municipal water supplies are monitored for these chemicals on a reg-

ular and continuing basis. If specified toxic chemicals are found, the water supply is closed until adequate purification procedures or other alternatives are adopted. The law also ordered the closing of all deep injection wells except those that can be proved safe. Thus, there are only a handful of such wells still operating in the United States, and those few have passed extremely rigorous geological tests.

There are still shortcomings to the law, however. There is no provision for systematic monitoring of private wells. Therefore contamination may not be recognized until people experience "unexplained" illness or report "funnytasting" or "funnysmelling" water and have the water tested on their own initiative. (Your county public health department will provide this service, albeit for a charge.) If dangerous levels of contamination are found in a private well, the well may be closed but there is no compensation. The owner must accept the burden of getting water from an alternative source or installing purification measures that may be prohibitively expensive. In extreme cases, owners have been known to simply abandon their homes (Fig. 14–11). If low levels of contamination deemed "safe" are found in a private well, no action is taken, and the owners are left on their own to decide what to do. Likewise, there is no provision for monitoring bottled water, which many people have turned to drinking to avoid presumed contamination of municipal water supplies. Ironically, in 1991 it came to light that numerous samples of bottled water contained higher levels of contamination than did municipal supplies.

Cleaning Up Existing Sites (Superfund)

A major federal program aimed at protecting groundwater from leakage from abandoned chemical waste sites was initiated by the **Comprehensive Environmental Response, Compensation, and Liability Act of 1980,** popularly known as **Superfund.** Through a tax on chemical raw materials, this legislation provides a fund for the identification of such sites, protection of groundwater near the site, remediation of groundwater if it has been contaminated, and cleanup of the site. Initially funded at $1.6 billion for the period 1980–1985, it was re-funded at $8.6 billion in 1986–1990. In 1990 it was funded for another 5 years at $9.1 billion. It is estimated that the fund will go on for at least another 30 years and another $300 billion. Such is the magnitude of the problem; Superfund is one of the EPA's largest ongoing programs.

After a slow start and considerable controversy on how to proceed, much of it political, the Superfund program has now reached a standard of performance that most environmentalists rate as a success. The basic strategy is as follows:

FIGURE 14–11
Leaching of toxic wastes from landfills and contamination of water supplies has forced the abandonment of properties in many areas. (Photograph by Thomas Busier.)

Protection of public health from groundwater contamination is the highest priority. Therefore, as sites are identified, they are first analyzed in terms of current and potential threat to groundwater supplies. The waste is sampled to determine its characteristics, and groundwater around the site is sampled to determine whether or not contamination is occurring. If it is determined that no immediate threat exists, nothing more may be done. If a threat to human health does exist, the most expedient measures are taken immediately to protect the public. These measures may include digging a deep trench and installing a concrete dike around the site and recapping it with impervious layers of clay and plastic to prevent infiltration. Thus the wastes are isolated, at least for the short term. If the situation has gone past the threat stage and contaminated groundwater is or will be reaching wells, remediation procedures, discussed below, are begun immediately.

The worst sites (those presenting the most immediate and severe threats) are put on a **National Priorities List (NPL)** and scheduled for total cleanup. A site on this list is reevaluated in terms of the most cost-effective method of cleanup, and finally cleanup begins. One procedure is to excavate the wastes and contaminated soil and incinerate them in a kiln assembled on the site (Fig. 14–12). Another method, used where there is only contaminated soil, is to drill a ring of injection wells around the site wells and a suction well in the center. Water containing a harmless detergent is injected into the injection wells and drawn to the suction well, cleansing the soil along the way. The withdrawn water is treated to remove the pollutants and is then reused for injection.

As of 1991, the EPA had logged approximately 34000 sites, and assessments had been completed on about 32700. Of the assessed sites 20500 were deemed to require no further action, 11000 were awaiting further investigation, and nearly 12000 were placed on the NPL. Of the NPL sites, 773 were in the analysis or design phase, cleanup was under way at 310, and cleanup had been completed at 63. (That leaves only 10854 sites to go!) Cost for cleanup is averaging $26 million per site. At this rate, it will cost about $300 billion to clean up the remaining 10854 sites currently

FIGURE 14–12
Toxic Waste Mobile Incinerator. The Environmental Protection Agency is cleaning up some Superfund sites by running contaminated materials through this incinerator which is brought to and erected on the site. (Tim Lynch/Liaison International.)

An Example of a Site Put on the National Priorities List

The following is but one example of the now more than 12 000 sites on EPA's NPL listing. Described are the conditions at the time of listing in 1987. Residents in the affected area have since been provided with a safe drinking water supply, but as of 1992 this site was still in the process of being evaluated to determine the best method of cleanup.

Conditions at listing (January 1987): The Obee Road Site consists of a plume of contaminated ground water in the vicinity of Obee Road in the eastern section of Hutchinson, Reno County, Kansas. The Kansas Department of Health and Environment (KDHE) has been investigating the area since July 1983. At that time, the State detected volatile organic chemicals, including benzene, trans-1,2-dichloroethylene, chlorobenzene, 1,1-dichloroethylene, tetrachloroethylene, trichloroethylene, vinyl chloride, and tolu-

ene, in wells drawing on a shallow alluvial aquifer. An estimated 1,900 residents of suburban Obeeville obtained drinking water from private wells in the aquifer. Hutchinson has connected the homes of the Obeeville residents to the Reno County Rural Water District #4.

Preliminary work by the State has tentatively identified a source of the contamination as the former Hutchinson City landfill, which is located at the eastern edge of what is now the Hutchinson Municipal Airport. Before closing in about 1973, the landfill accepted unknown quantities of liquid wastes and sludges from local industries, as well as solvents from small metal-finishing operations at local aircraft plants. Also, the Department of Defense (DOD) may have disposed of solvents at the landfill. DOD owned or maintained the airport until 1963.

Another possible source of con-

tamination is individual septic tank systems. Commercial septic tank cleaners commonly contain trichloroethylene, dichloromethane, and benzene.

Status (April 1987): Water District #4 is providing an alternate water supply to the Obee school system, which was drawing water from a contaminated well.

Further investigation is needed to define the extent of the problem and identify those responsible. KDHE is preparing an application to EPA for funds to conduct a remedial investigation/feasibility study.

Source: Hazardous waste site listed under the EPA Comprehensive Environmental Response, Compensation, and Liability Act of 1980 (CERCLA)("Superfund")

on the NPL. However, additional sites are being added to the list at the rate of about 100 per year.

It may seem that we have barely scratched the surface of this problem. Nevertheless, the EPA claims that "the net result of the Superfund cleanup work at NPL sites has been to reduce the potential risks from exposure to hazardous waste to more than 23.5 million of the 41 million people who live within four miles of these sites."

Groundwater Remediation

It has frequently been said that contaminated groundwater is effectively lost forever because there is no way to purify an aquifer and it may be hundreds of years before wastes are flushed out. However, technology for **groundwater remediation** has been developed in recent years and is expanding rapidly. The technique involves drilling wells, pumping out the contaminated groundwater, purifying it, and injecting the purified water back into the ground or discharging it into surface waters. (Figure 14–13).

In many cases, the soil or water is contaminated with toxic organic compounds that are biodegradable.

The problem is that they do not degrade because of lack of organisms or oxygen or both. In a process called **bioremediation,** oxygen and organisms are injected into contaminated zones. The organisms feed on and eliminate the pollutants as in secondary sewage treatment and then die when the pollutants are gone.

Bioremediation may also be used in place of detergents to decontaminate soil. It is a rapidly expanding and developing aspect of waste cleanup applicable to leaking storage tanks and spills as well as to waste disposal sites.

MANAGEMENT OF WASTES BEING PRODUCED NOW

The Clean Water Act

The Clean Water Act of 1972 and its various amendments remain the cornerstone legislation protecting surface waters from pollution. As noted previously, it is under this act that the EPA sets standards for allowable levels of pollutants discharged into surface waters. However, both setting and enforcing the

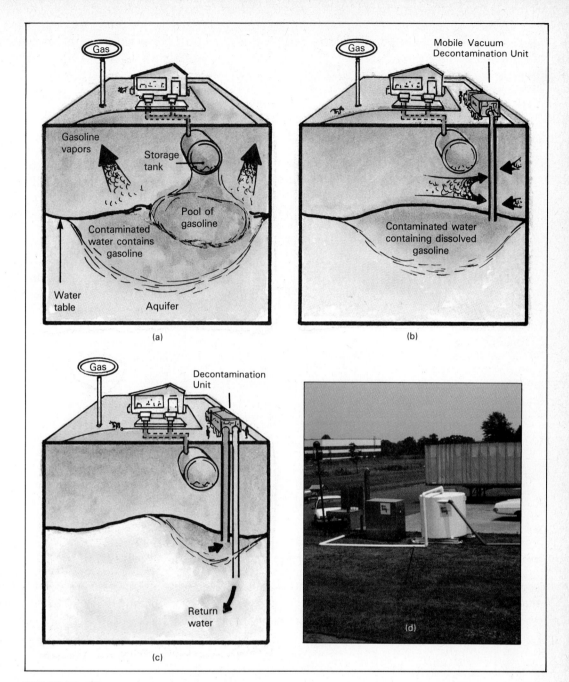

FIGURE 14–13
Groundwater remediation. (a) Typical subsurface contamination from a leaking fuel tank at a gas station. (b) After the leak has been repaired, the vacuum extraction process causes gasoline and residual hydrocarbons in the soil and on the water table to evaporate and then removes the vapors, preventing further contamination of the groundwater. (c) Contaminated groundwater is pumped out, treated, and returned to the ground.

standards involves a difficult balancing act between the ideal of no pollution on the one hand and economic practicality on the other. A company could meet a demand for immediate stopping of all pollution only by shutting down and putting all its employees out of work. To avoid this, the Clean Water act allows the granting of **interim permits,** which ef-

fectively say: You can go on polluting for the time being, but you must reduce discharges by X amount by Y date. A **compliance schedule** is negotiated between the company and the EPA or the state equivalent. Fines may be levied for noncompliance, but the schedule may be renegotiated and extensions granted.

Unfortunately, there is considerable opportunity for abuse of interim permits. Also, small firms, homes, and farms are exempt from regulation. Thus, progress notwithstanding, there is still considerable discharge of toxic wastes, and many waterways remain polluted.

The Resources Conservation and Recovery Act

To control the indiscriminate land disposal of hazardous wastes, Congress passed the **Resource Conservation and Recovery Act of 1976 (RCRA),** commonly spoken of as "reck'-ra." Along with its subsequent amendments, RCRA is the cornerstone legislation for management and disposal of wastes currently being generated.

One main feature of this act is "cradle-to-grave" tracking of all hazardous wastes. The generator must fill out a form detailing the exact kind and amount of waste generated. Persons transporting the waste and operating the disposal facility must each sign the form, vouching that amounts of waste transferred are accurate, and copies go to the EPA. All phases are subject to unannounced inspections. The generator is responsible for any waste "lost" along the way or any inaccuracies in reporting. You can see how this provision of RCRA curtails midnight dumping and ensures that wastes get to a designated disposal facility.

Another feature of RCRA is that all landfills and other facilities receiving hazardous wastes must meet standards for construction, operation, and monitoring. As a result, thousands of facilities in the United States have been closed down because they could not meet standards. Many, if not most, of the substandard facilities pose the same threats to groundwater and require the same remediation procedures as described above for Superfund sites. The only difference is that here the users and operators of the facilities are known and are being made to foot the cleanup bill.

Amendments to RCRA passed in 1984 addressed the mounting problem of leaking underground storage tanks (USTs). Hence these amendments are often referred to as **UST legislation.** One consequence of our automotive-based transportation system is millions of underground fuel storage tanks at service stations. Putting such tanks underground greatly diminishes the risk of explosions and fires, but it also hides leaks. Underground storage tanks have a life expectancy of about 20 years before they may spring leaks. Many thousands of tanks are crossing this threshold each year and springing leaks. Without monitoring, leaks went undetected until nearby residents began to see fuel-tainted water flowing from their faucets. UST legislation requires strict monitoring of fuel supplies so that leaks may be detected early. When leaks are detected, remediation must begin within 72 hours, using the technique illustrated in Figure 14–13.

Accidents and Emergency Response

As modern society uses increasing amounts and kinds of hazardous materials, the stage is set for accidents to become widescale disasters. Transport is an area particularly prone to accidents. Over the course of a year, about 4 billion tons of hazardous materials are transported by road, rail, and water in the United States. Numerous accidents involving rail and truck tankers filled with hazardous chemicals have required the evacuation of residents.

Management and response are improving. The **Department of Transportation Regulations (DOT Regs)** specify kinds of containers and methods of packing to be used in the transport of various hazardous materials. The regulations are intended to reduce the risk of spills, fires, and any poisonous fumes that could be generated in case of an accident.

DOT Regs require that every individual container and the outside of a truck or railcar must carry a standard placard identifying the hazards (flammability, corrosiveness, potential for poisonous fumes, and so on) of the material inside. You may see HAZMAT signs on highways restricting truckers with hazardous materials to particular routes or lanes. Increasingly, fire and police departments have special emergency response teams. These are people who are specifically trained to know the potential dangers involved with each kind of hazardous material and how to respond to each to protect both themselves and the public.

Toxic Substances Control Act (TSCA)

In the past, new synthetic organic compounds were introduced for a specific purpose without any testing of potential side effects. For example, in the 1960s a new compound known as TRIS was found to be a very effective flame retardant and was widely used in children's sleepware. It was only later discovered that TRIS is a potent carcinogen. Treated sleepware was immediately withdrawn from the market. It is not known how many children (if any) got cancer from TRIS, but the warning from this and other such cases is obvious.

Congress responded by passing the Toxic Substances Control Act of 1979, which requires any new chemical to be tested for carcinogenic properties. Depending on the outcome of tests, uses may be restricted or a product may be kept off the market altogether.

Laws applying to hazardous wastes are summarized in Figure 14–14.

Labels in figure:

CAA (Clean Air Act) Limits discharges into the air

OSHA (Occupational Safety and Health Act) Protects workers' health and safety

DOT (Department of Transportation Regulations) Assures safe transport

RCRA (Resource Conservation and Recovery Act) Assures that wastes get to suitable disposal facilities

Disposal facility

HAZARDOUS WASTE

TOXIC WASTE

TSCA (Toxic Substance Control Act) Requires new chemicals to be shown safe for specific uses

CWA (Clean Water Act) Limits discharges into waterways

Superfund: Provides for cleanup of abandoned hazardous waste sites

SDWA (Safe Drinking Water Act) Sets standards for drinking water

FIGURE 14–14
Summary of the major laws pertaining to the protection of workers, the public, and the environment from hazardous materials.

How Are We Doing?

Considering that we started to tackle the hazardous material mess only 20–25 years ago, progress has really been quite remarkable, and all those who played a role in bringing about corrective measures deserve tremendous credit. Certainly risks to public and environmental health have been considerably reduced as a result. Yet, shortcomings and loopholes remain.

Considerable amounts of hazardous wastes escape regulation by RCRA. The major loophole is that "small operations," those that produce less than 100 kilograms (220 lbs) of hazardous waste per month, are exempt, as are homeowners and farmers. You can see why this is the case. Attempting to regulate *everyone* would be an exorbitant expense and would require a police state besides. Yet, the total amounts disposed of by these unregulated parties is immense. Items such as batteries and unemptied pesticide containers go into the trash and end up in municipal landfills. Therefore, municipal landfills are effectively toxic waste dumps. Yet they are not regulated as such, and few have the necessary protective measures.

Another area of concern is the legally permitted discharges that are still occurring under the Clean Water and Air acts. Previously, the law required only that chemicals in discharges be below certain concentrations; the total amounts (concentration × volume) were not calculated. Compiling and totaling these data for the first time in 1989, the Environmental Protection Agency came to the disturbing realization that, every year, *13 billion pounds* (6 billion kg) of toxic chemicals are being legally discharged into the environment without control. Of this total, 2.2 billion pounds (1 billion kg) goes into the air; the rest goes into the water and onto the land. This amounts to over 44 pounds (20 kg) per person each year. The Clean Air Act Amendments of 1990 address the problem of emissions into the air, but the other discharges still await attention.

A considerable number of companies are working to circumvent the regulations rather than to cooperate with them. A new kind of midnight dumping has recently come to light. Shipments of toxic wastes falsely labeled as products or raw materials are going to less-developed countries, where people who need income, unaware and unsuspecting of their hazards, accept them for disposal without safeguards and for token payments (see In Perspective Box, p. 330). Numerous firms have moved their entire production facilities to less-developed countries, where requirements for pollution control, safety precautions, worker protection, and so on are either weaker than in the United States or nonexistent.

In 1991 it came to light that a number of waste-handling companies were negotiating with Native Americans to place disposal facilities on their reservations, where they would not be subject to EPA regulations that apply to the rest of the country. (In a number of respects, Native American reservations are governed as independent countries.) This follows a trend of locating waste facilities in areas dominated by ethnic minorities who lack the political power to fight back. (See Ethics Box, p. 331.)

FUTURE MANAGEMENT

It is widely recognized that, even at its best, land disposal is not a sustainable solution to the toxic waste problem because the lifetime of the wastes will inevitably be longer than the lifetime of protective barriers. Also, with the tightening of EPA regulations, facilities that are permitted to receive toxic wastes are few and far between, a situation that leads to exorbitant hauling distances. In short, regulations have driven up costs of land disposal such that companies are increasingly turning to alternatives.

IN PERSPECTIVE
Exporting Toxic Waste

In 1988, ships from Italy made repeated visits to Koko, a small Nigerian fishing village. As a result of a deal between an Italian businessman and some misguided Nigerian entrepreneurs, the ships left behind a total of 3800 tons of highly toxic wastes, including 150 tons of PCBs. When the Nigerian government learned of the deal, they acted decisively: The village's residents were evacuated, the Nigerians involved were punished, and the wastes were removed by Italy after the Nigerian ambassador was recalled from Rome. The ship bearing the wastes from Nigeria, the *Karin B.*, tried to dump the waste in five other European countries before returning to Italy.

This incident is considered to be only the tip of an iceberg of commercial traffic in hazardous waste. The legal complexities and high costs of disposing of hazardous waste in the developed nations provide strong incentives to waste-disposal firms to export the wastes. Impoverished countries are tempted to accept the wastes in return for hard currency. As an example, following reunification of Germany, it was discovered that the oppressive East German regime had been accepting 5 million tons of hazardous waste annually from Western countries, especially West Germany. The Germans are now learning that "there is no away," as they are faced with the costs of cleaning up the toxic waste dumps. No precise figures on this toxic trade are available (because much of this trade is illegal), but there are reports of hundreds of shipments and millions of tons of toxic wastes to developing countries.

Many developing nations are incensed about this overt exploitation of their poverty, and to date 81 countries have passed bans against importing toxic wastes. The problem has been addressed by the United Nations Environment Program; an international treaty (the Basel Accord) was developed and signed in March 1989, by 53 nations. The treaty falls short of an outright ban. Instead, it establishes a system of "informed consent" whereby both exporting and importing nations must approve shipments. As part of the accord, the proposed importing waste facility must be "environmentally adequate."

Critics point out that the United States was primarily responsible for watering down the treaty; representatives of the Bush Administration effectively lobbied against stipulations that would cut back the trade in toxic waste, claiming that the commerce represented a legitimate activity as long as receiving countries were agreeable. In particular, the United States threatened not to sign the treaty if it prevented nations with stricter environmental standards from shipping toxic wastes to nations with more lenient standards. Currently, the EPA receives notifications of the intent to export hazardous wastes from generators within the United States. These applications are usually for shipment to Canada or Mexico, but an increasing number are directed toward Third World nations.

Most observers see an outright ban as unacceptable to too many nations, so the major goal of environmentalists is to establish the principle that nations should insist on the same or better standards for dealing with hazardous wastes in any nation receiving those wastes as in the nation exporting the wastes—a sort of international golden rule, which says, in effect, do not do to other nations what you would not do to your own people. It will be an encouraging sign if nations can reach an equitable agreement on this serious international concern.

Sources: French, Hilary, "A Most Deadly Trade," *WorldWatch* Vol 3 no 4 (July/Aug 1990), pp. 11–17. Uva, M.D. and J. Bloom, "Exporting Pollution: The International Waste Trade," *Environment* vol 31 no 5 (June 1989) pp. 45, 43–44.

Item: The largest commercial hazardous waste landfill in the United States is located in Emelle, Alabama. This landfill receives wastes from Superfund sites and every state in the continental United States.

Item: Kettleman, California, has been selected as the site for the state's first commercial hazardous waste incinerator.

Item: A Choctaw reservation in Philadelphia, Mississippi, was targeted to become the home of a 466-acre hazardous waste landfill.

These three places have two things in common: They are considered to be appropriate locations for hazardous waste disposal, and they are predominantly populated by people of color. African Americans make up 90 percent of Emelle's population; Latinos represent 95 percent of Kettleman's population; and the Choctaw reservation is entirely Native American. At issue is the question of environmental racism.

Several recent studies have documented the fact that all across the United States, waste sites and other hazardous facilities are more likely than not to be located in towns and neighborhoods where the majority is nonwhite. If it is also true that these same towns and neighborhoods are less affluent, this only makes a stronger case for addressing the obvious injustices. The wastes involved are primarily generated by affluent industries and the affluent majority, but somehow the wastes tend to end up well away from where they were generated and in the backyards of people of color. It seems fair to assume that the siting of hazardous facilities is a matter of political power, and those with the power would like to have the sites well away from their own backyards.

Environmental racism has gotten the attention of the EPA. EPA Administrator William K. Reilly created the Environmental Equity Workgroup, made up of 40 professionals from within the agency. The workgroup submitted its report in February 1992 and concluded that although there is a lack of data on actual environmental health effects by race, differences in exposure to some environmental pollutants are correlated with race and socioeconomic factors. The workgroup recommended a number of steps to be taken by the EPA to address concerns of environmental racism.

Within the nonwhite communities there are signs of a groundswell of grass roots concern about environmental racism and equity. The success of minorities in pushing for civil rights in recent decades has set a precedent that is beginning to empower people of color in seizing the initiative in this newly emerging arena. As a start, activists are turning their attention to cleaning up their own communities. Calling themselves the Toxic Avengers, a group of African American and Latino students in Brooklyn, New York, has taken on projects like battling the Radiac Research Corporation's toxic waste facility in the neighborhood, starting a recycling program, and conducting neighborhood workshops on fighting pollution. Watchdog is a grass roots, multiracial, working-class group in Los Angeles that has successfully worked toward amending the regulations of the South Coast Air Quality Management District so that the regulations reflect special concerns of minorities and low-income workers. The Good Road Coalition is an alliance of grass roots groups that successfully blocked the proposal by a Connecticut company to build a 6000-acre landfill on the Rosebud Sioux reservation in South Dakota.

What is still uncertain about the environmental justice movement is the response of the white majority in the United States and in particular the very organizations that promote environmentalism. Will the nongovernmental organizations join with people of color in fighting this apparent manifestation of racism? And, will businesses and local political bodies be willing to address race-related environmental inequalities? What do you think?

Source: EPA Journal, 18, no. 1 (1992).

Incineration

Most synthetic organic compounds, like natural organic compounds, are flammable. Even flame-resistant chlorinated hydrocarbons can be broken down with oxygen to carbon dioxide, water, and harmless chlorine compounds. It is simply a matter of keeping the chemicals at a high enough temperature for a long enough period of time.

The kilns used for producing cement have the conditions required for breakdown of synthetic organic wastes. Consequently many cement companies have recently adopted a second business—disposing of hazardous wastes. Organic wastes are mixed with the regular fuel oil and fed into the kiln, thus contributing fuel value in the process of being destroyed (Fig. 14–15). Any resultant toxic ash is incorporated into the cement, which ultimately provides a safe "container."

Biodegradation

Synthetic organic compounds are notorious for being nonbiodegradable. However, scientists are discov-

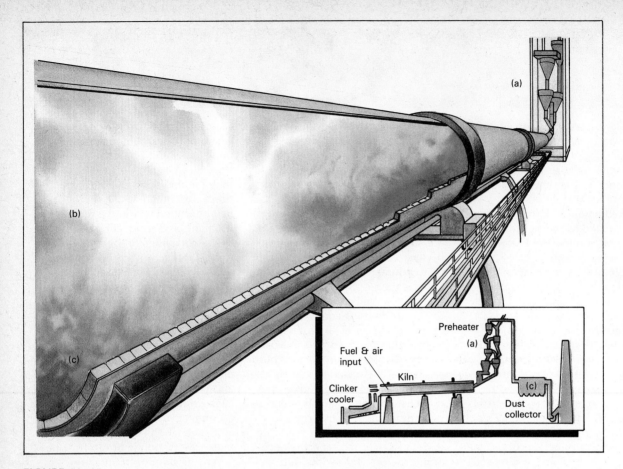

(a)

(b)

(c)

Preheater
(a)

Fuel & air
input

Kiln

Clinker
cooler

(c)

Dust
collector

FIGURE 14–15
Cement kilns may be used for the destruction of hazardous wastes. A cement kiln is a huge,
rotating "pipe," typically 15 ft in diameter and 230 ft long, mounted on an incline. Solid wastes
fed in with the raw materials (a) are fully incinerated and reacted with cement compounds as
they gradually tumble toward the combustion. Flammable, liquid wastes added with the fuel
and air (b) impart fuel value as they are burned. Waste dust is trapped (c) and recycled into the
kiln. (Redrawn with permission. Southdown, Inc. Houston, TX 77002)

ering that some bacteria will break down these compounds, particularly if suitable conditions are provided. If this approach proves fruitful, someday systems similar to those used in secondary sewage
treatment may be used to dispose of toxic organic
wastes.

Pollution Avoidance—The Solution for a Sustainable Society

Gustave Speth, president of the World Resources Institute, points out that, in addressing problems of pollution, our inclination has been to go for "tailpipe"
solutions—adding something on at the end of the
process to contain or dispose of the pollutants. The
United States has never encouraged development of
techniques that would prevent much of the pollution
in the first place, a strategy known as **pollution avoidance** or **pollution prevention** (as opposed to pollution
control). Thus, we produce five times more pollutants
for every dollar of product than do the Japanese.

Do not make the mistake of thinking that tail-

pipe solutions are used because they are cheap, because they're not. U.S. industry is currently spending
close to $100 billion a year on pollution control and
waste disposal. The Congressional Office of Technological Assessment estimates that output of polluting byproducts could be cut by half with current
technologies and by another 25 percent with development of new technologies. Costs of pollution control would be reduced accordingly. Gradually companies are beginning to make this discovery. For example Carrier, Inc., a maker of air conditioners,
installed equipment for more precise metal cutting.
According to a spokesperson for the company, the
new equipment greatly reduces cutting wastes, eliminates the need for a toxic cutting lubricant, and leads
to the production of a better product, all results that
make the company more competitive. Clairol
switched from water to foam balls for flushing pipes
in hair-product production, reducing waste water by
70 percent and saving $250 thousand per year. Other
techniques being adopted include purifying and recycling rather than disposing of spent solvents; sub-

stitution of water-based inks and paints for those based on synthetic organic solvents; inks and dyes based on biodegradable organic compounds rather than on heavy metals; use of abrasives rather than solvents for paint removal; use of water and biodegradable detergents rather than synthetic organic solvents for cleaning. In total, industries stand to save many billions of dollars by adopting these pollution-avoidance measures. As a welcome side effect, the measures will make the companies more competitive in world markets as well as reduce pollution.

The concept of pollution avoidance is extremely applicable to the individual consumer. So far as you are able to reduce or avoid using products containing toxic chemicals, you are preventing those amounts of chemicals from going into the environment. You are also reducing the byproducts resulting from producing those chemicals. Increasingly, companies are gradually beginning to produce and market "green products," a term used for products that are more environmentally benign than their traditional counterparts. How fast and to what degree these green products displace or replace traditional products will in large part depend on how we behave as consumers. In other words, consumerism could be an extremely potent force—if consumers did not buy products made with or containing toxic materials, such products would not be produced. (See Making a Difference following Part Three for more specifics.)

 ## Review Questions

1. The Environmental Protection Agency divides hazardous chemicals into what four categories?
2. Considering the total "life span" of a chemical, at what stages may it enter the environment?
3. What are the two classes of chemicals that pose the most serious long-term toxic risk?
4. Why do the two classes of chemicals mentioned in question 4 pose a long-term risk?
5. What is meant by bioaccumulation and biomagnification?
6. How were chemical wastes generally disposed of before 1970?
7. What three methods of land disposal were used, and what were the shortcomings of these methods?
8. What two laws pertaining to the disposal of hazardous wastes were passed in the early 1970s?
9. In what ways did the two laws noted in question 9 alleviate the problem of environmental contamination, and in what ways did they make it worse?

10. What three laws were passed to correct the problems that developed from unregulated land disposal of hazardous wastes?
11. What actions are mandated in the Safe Drinking Water Act that protect against municipal water supplies being contaminated?
12. How does Superfund legislation provide for the cleanup of abandoned chemical waste sites?
13. What actions are mandated by RCRA that prevent midnight dumping and disposal of wastes in unsafe facilities?
14. How can contaminated groundwater be cleaned up?
15. What is being done to protect us from transportation accidents involving shipments of hazardous chemicals?
16. What methods of hazardous waste disposal may be used (more) in the future?
17. What is meant by pollution avoidance?

 ## Thinking Environmentally

1. Suppose there is a proposal to build an incineration facility near your community. It is proposed that hazardous wastes currently being landfilled at the same location will be disposed of in the new facility. Would you support or oppose the proposal? Give a rationale for your position.
2. Our knowledge regarding bioaccumulation of toxic chemicals and its long-term health effects in humans is very limited. Why?
3. Prior to the 1970s, it was not illegal to dispose of hazardous chemicals in unlined pits, and many companies did so. Should they be held responsible today for the contamination those wastes are caus-

ing, or should the government (taxpayers) pay for the cleanup? Give a rationale for your position.
4. Under RCRA, the company that originally produces a chemical waste remains responsible for cleanup even though some other party may have committed the improper disposal that caused a problem. Is this fair? Give a rationale for your position.
5. Should the export of hazardous wastes to Third World countries or Native American reservations be prevented? Propose legislation mandating steps or procedures that would do so.

15

Air Pollution and Its Control

LEARNING OBJECTIVES

When you have finished studying this chapter, you should be able to:

1. Describe the natural cleansing processes that take place in the air.

2. Trace the origins of industrial smog and photochemical smog.

3. Understand the impetus for air pollution legislation and the basic control methods employed.

4. Name eight major air pollutants.

5. Describe three ways air pollutants affect human health.

6. Describe the effects of pollutants on crops, forests, and materials.

7. List the primary pollutant products of combustion.

8. List the secondary pollutants and describe the processes by which they are formed.

9. Understand the "command-and-control" strategy of pollution regulation.

10. Describe the programs of the Clean Air Act of 1990.

11. Analyze the costs and benefits associated with air pollution controls.

12. Discuss steps that could still be taken to improve air quality and also improve our way of life.

13. Describe the major sources of indoor air pollution and the extent to which public policy addresses them.

As we learned in Chapter 3, the atmosphere surrounding the earth consists of a mixture of gases and particles. This mixture has always been changing, but with the advent of the Industrial Revolution, it has been changing more rapidly than ever, and the effects on natural ecosystems and human health have proved to be dramatic and serious. Some of the changes are regional, affecting the air in the vicinity of the polluting sources. Other changes extend into the upper atmosphere and are essentially global in their distribution. In this chapter, we look at the lower atmospheric problems. Chapter 16 deals with more global changes.

Background of the Air Pollution Problem

Air pollutants are substances in the atmosphere that have harmful effects. Organisms are able to deal with certain levels of pollutants without suffering ill effects. The pollutant level below which no ill effects are observed is called the **threshold level.** Above the threshold level, we begin to see the effects of pollutants. However, the effect caused by a pollutant depends on both its concentration and the length of time of exposure. Higher levels may be tolerated if the exposure time is short. Thus, for any given pollutant, the threshold level is high for shorter exposures but gets lower as exposure time increases (Fig. 15–1). It is not the absolute amount of pollutant but rather the *dose* that is important. As we learned in Chapter 2,

FIGURE 15–1
The threshold level diminishes with increasing exposure time. The threshold level differs for each pollutant. Also, different species and individuals may have very different threshold levels to a given pollutant, and the threshold level may differ depending on the presence or absence of other pollutants or factors that may be causing stress.

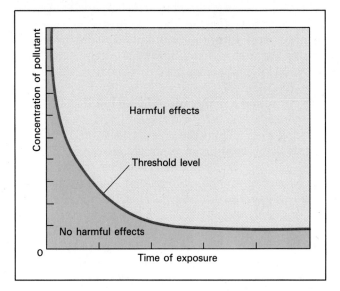

dose is defined as concentration multiplied by time of exposure.

Three factors determine the level of air pollution:

○ Amount of the pollutants put into the air
○ Amount of space into which the pollutants are dispersed
○ Mechanisms that remove pollutants from the air

Volcanoes, fires, and dust storms have sent smoke and other pollutants into the atmosphere for millions of years. Coniferous trees and other plants emit volatile organic compounds into the air around them. However, there are mechanisms in the biosphere that remove, assimilate, and recycle these natural pollutants. First, the pollutants disperse and dilute in the atmosphere. Then, as shown in Figure 15–2, a naturally occurring cleanser, the hydroxyl radical (OH), oxidizes many of them to products that are harmless or that can be brought down to earth by precipitation. Microorganisms in the soil further convert some of these products into harmless compounds. The chemistry of all these cleansing reactions is complex and is still being investigated, but we do know that through such processes natural pollutants are held below threshold levels (except in the immediate area of a source, such as around an erupting volcano).

With the discovery of fire, humans began adding to these natural pollutants. Down through the centuries, venting combustion and other fumes into the atmosphere remained the natural way to avoid their obviously noxious effects. With the Industrial Revolution of the 1800s came crowded cities and the use of coal for heating and energy, and air pollution began in earnest. In *Hard Times*, Charles Dickens commented on a typical scene: "Coketown lay shrouded in a haze of its own, which appeared impervious to the sun's rays. You only knew the town was there because there could be no such sulky blotch upon the prospect without a town." This shrouding haze became known as **industrial smog,** an irritating grayish mixture of soot, sulfurous compounds and water

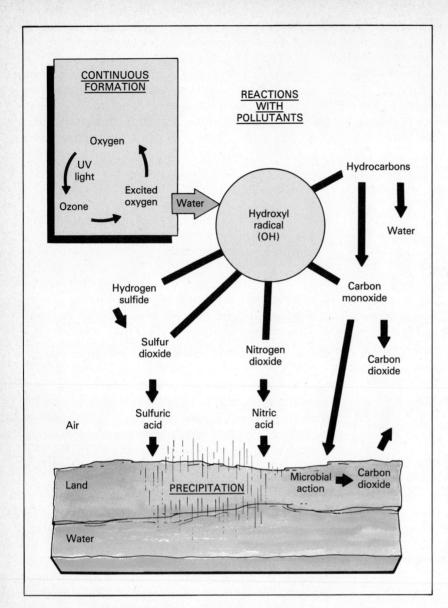

FIGURE 15-2
Atmospheric cleansing by the hydroxyl radical and by soil microbes. The hydroxyl radical is continuously formed in the atmosphere and reacts readily with most pollutants, converting them to substances that are either less harmful or brought to the earth in precipitation. Soil microbes are also able to convert some pollutants to less toxic forms.

vapor. This kind of smog continues to be found wherever industries are concentrated and coal is the primary energy source.

Until recently, air pollution in cities and near certain industrial sources was considered a local problem, a natural outcome of economic growth and human technology. A short distance away from pollution sources, air quality generally remained good. In the 1950s, however, with the mushrooming use of cars for commuting (see Chapter 24), whole metropolitan areas became enshrouded in a brownish haze on a daily basis (Fig. 15–3). This haze is called **photochemical smog** because sunlight is involved in its formation, as we shall see later.

Certain weather conditions intensify smog levels, the most significant being a *temperature inversion*. Under normal conditions, daytime air temperature is highest near the ground because sunlight strikes the earth and the heat absorbed radiates out to the air near the surface (Fig. 15–4a). The warm air near the ground rises, carrying pollutants upward and dispersing them at higher altitudes. At night, when the sun is no longer heating the earth, the currents cease. This condition of cooler air below and warmer air above is called a **temperature inversion** (Fig. 15–4b). Inversions are usually very short-lived, as the next morning's sun begins the process anew, and any pollutants that accumulated overnight are carried up and away. During cloudy weather, however, the sun may not be strong enough to break up the inversion for hours or even days. Or a mass of high-pressure air may move in and sit above the cool surface air, trapping it.

When such "long-term" temperature inversions occur, pollutants can build up to dangerous levels, prompting local health officials to urge people with

FIGURE 15–3
A typical episode of photochemical smog. (a) Early in the morning, the air is clear.
(b) Midmorning of the same day, the air is very hazy with smog. The haze is the
result of pollutants from the exhaust of rush-hour traffic reacting in the atmosphere.
The reactions are promoted by sunlight. These photographs are close to the same
view over Los Angeles. ((a) Tom McHugh/Photo Researchers; (b) Georg Gerster/
Photo Researchers.)

FIGURE 15–4
A temperature inversion may cause episodes of high concentrations of air pollutants.
(a) Normally air temperatures are highest at ground level and decrease at higher
elevations. Since the warmer air rises, pollutants are carried upward and diluted in
the air above. (b) A temperature inversion is a situation in which a layer of warmer
air overlies cooler air at ground level. This blocks the normal updrafts and causes
pollutants to accumulate like cigarette smoke in a closed room.

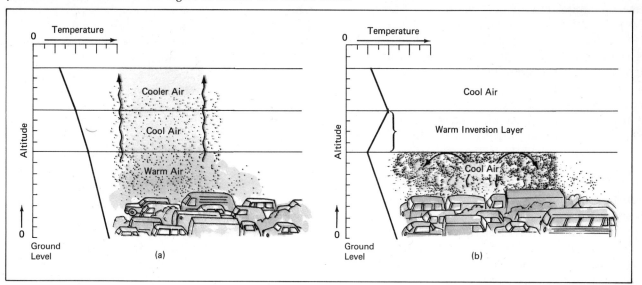

FIGURE 15–5
Air pollution in this Puerto Rican city has caused the decline and death of the palm trees lining the streets of the city. (Peter Arnold, Inc.)

By the 1950s and 1960s, it was obvious that human-made pollutants were overloading the natural cleansing processes. The unrestricted discharge of pollutants into the atmosphere could no longer be tolerated.

Under grass-roots pressure from citizens, the U.S. Congress passed the **Clean Air Act of 1970.** Together with amendments passed in 1977 and 1990, this law, administered by the Environmental Protection Agency (EPA), represents the foundation of U.S. air pollution control efforts. It calls for identifying the most widespread pollutants, setting **ambient standards**—levels that need to be achieved to protect environmental and human health—and establishing control methods and timetables to meet these goals.

Four stages are involved in meeting these mandates:

○ Identifying the pollutants.

○ Demonstrating which pollutants are responsible for particular adverse health and/or environmental effects, so that reasonable standards may be set. (If standards are set below threshold levels, vast sums may be spent on control without benefit.)

○ Determining pollutant sources.

○ Developing and implementing suitable controls.

All phases are ongoing processes. As new information regarding adverse effects of pollutants is discovered, standards may be made more stringent, and new strategies for control must be developed and implemented.

As a result of progress under the Clean Air Act, air quality in most cities in the United States is markedly better now than it was in the mid-1900s. However, not all problems have been solved, and elsewhere in the world, air pollution has become a major source of forest and crop damage and a rising threat to human health. Eastern Europe, the republics of the former USSR and many Third World cities are experiencing air pollution reminiscent of the early Industrial Revolution in England.

breathing problems to stay indoors. For many people, smog causes headaches, nausea, and eye and throat irritation and aggravates preexisting respiratory conditions such as asthma and emphysema. In some industrial cities under severe temperature inversions, air pollution reached such high levels that mortalities increased significantly. These cases became known as *air pollution disasters.* For example, London experienced repeated episodes of inversion-related disasters in the mid-1900s, with pollution-related deaths reaching 4000 in one episode.

The ill effects of air pollution are not limited to people living in cities. In recent years, many species of trees and other vegetation in cities have begun to die back (Fig. 15-5), and farmers near cities have suffered damage or even total destruction of their crops because of air pollution. Pollution-induced damage became so routine that it forced the complete abandonment of citrus growing in some parts of California and vegetable growing in certain areas of New Jersey, which were among the most productive regions in the country. A conspicuous acceleration in the rate of metal corrosion and deterioration of rubber, fabrics, and other materials was also noted.

Major Air Pollutants and Their Effects

MAJOR POLLUTANTS

The following eight pollutants or pollutant categories have been identified as most widespread and serious:

1. **Suspended particulate matter (SPM).** This is a complex mixture of solid particles and aerosols (liquid particles) suspended in the air. We see

FIGURE 15–6
A fossil fuel power plant near Boston sends out a plume of pollutant particles and gases. Since the stack lacks electrostatic precipitators, the plume contains substantial amounts of suspended particulate matter. (Ray Pfortner/ Peter Arnold, Inc.)

these particles as dust, smoke, or haze (Fig. 15–6). Particulates may carry any or all of the other pollutants dissolved in or adhering to their surfaces.

2. **Volatile organic compounds (VOC).** These include materials such as gasoline, paint solvents, and organic cleaning solutions, which evaporate and enter the air in a vapor state, as well as fragments of molecules resulting from incomplete oxidation of fuels and wastes.

3. **Carbon monoxide (CO).** An invisible gas that is highly poisonous to air-breathing animals.

IN PERSPECTIVE
Mexico City—Life in a Gas Chamber

"I'm getting out of here one of these days," said one resident of Mexico City. "The ecologists are right. We live in a gas chamber." By all accounts, Mexico City residents have the worst air in the world. The city exceeded safety limits for ozone for 310 days in 1988, with values sometimes more than 300 percent over norms set by the World Health Organization. The city's 3 million vehicles (a large proportion of which are old and in a state of disrepair) and 35 000 industries pour more than 5 million tons of pollutants into the air every day. The city, because it is nestled in a natural bowl, experiences thermal inversions every day during the winter. One study showed that 40 percent of young children suffer chronic respiratory sickness during the winter months; in January 1989, the government closed schools for a month to lessen

the exposure of school children to city air. Coughs and conjunctivitis are normal to city dwellers. Writer Carlos Fuentes has nicknamed the metropolis "Makesicko City!"

Mexico City's problems are indicative of a general Third World dilemma. Jobs and industrial activity, vital for a nation's economy, depend on the use of motor vehicles and the industrial plants ringing the city. Many automobile owners are absolutely dependent on their cars but can ill afford to keep them in a state of repair that reduces exhaust emissions. Other owners often see cars as status symbols, and will acquire one as soon as possible—often a clunker discarded from the developed world. The industries are not regulated as they are in the United States and lack the technology and capital it would take to control emissions.

The Mexican government has initiated an air-monitoring system and, more recently, mandated emissions tests and catalytic converters for automobiles. The city has adopted a "Days of No Driving" policy, where drivers are given decals that permit them to drive only on specific days. New air pollution laws are beginning to be enforced: One major refinery responsible for 4 percent of the air-quality problem was shut down by President Salinas in the spring of 1991, and 80 other industries were temporarily or permanently closed for violating the new laws. Officials predict that the air will get better by the mid-1990s as a result of these measures. In the meantime, Mexico City's 20 million inhabitants are part of an unintentional experiment in demonstrating the health effects of severe air pollution.

4. **Nitrogen oxides (NO_x).** Several nitrogen-oxygen compounds, all gases. They are converted to nitric acid in the atmosphere and are a major source of acid deposition (Chapter 16).

5. **Sulfur oxides, mainly sulfur dioxide (SO_2).** Sulfur dioxide is a gas poisonous to both plants and animals. It is converted to sulfuric acid in the atmosphere and is the major source of acid deposition.

6. **Lead and other heavy metals.** Lead, in particular, is very dangerous at low concentrations and can lead to brain damage and death.

7. **Ozone and other photochemical oxidants.** You probably know that ozone in the upper atmosphere must be preserved to shield us from ultraviolet radiation (Chapter 16). However, ozone is also highly toxic to both plants and animals. Therefore, ground-level ozone is a serious pollutant. This case emphasizes the fact that "a pollutant is a chemical out of place."

8. **Air toxics and radon.** The air toxics include carcinogenic chemicals, radioactive materials, and other toxic chemicals (e.g., asbestos, vinyl chloride, benzene) that are emitted as pollutants but are not included in the above list of conventional pollutants. Radon is a radioactive gas produced by natural processes within the earth. All radioactive substances have the potential to be damaging to any living matter they contact.

ADVERSE EFFECTS OF AIR POLLUTION ON HUMANS, PLANTS, AND MATERIALS

It is important to recognize that air pollution is not a single entity but an alphabet soup of the above materials mixed with the normal constituents of air. Further, the amount of each pollutant present varies greatly depending on proximity to the source and various conditions of wind and weather. As a result, we are exposed to a mixture that varies in makeup and concentration from day to day, even from hour to hour, and from place to place. Consequently, the effects we feel or observe are rarely if ever the effects of a single pollutant; they are the combined effect of the whole mixture of pollutants acting over the total life span, and frequently the effects are synergistic.

For example, both plants and animals may be stressed by pollution such that they become more vulnerable to other environmental factors such as drought or attack by parasites and diseases. Given the complexity of this situation, it is extremely difficult to determine the role of any particular pollutant in causing the observed result. Nevertheless, some significant progress has been made in linking cause and effect.

Effects on Human Health

Humans breathe 14 kg of air into their lungs each day. Although some of the pollution symptoms people suffer involve the moist surfaces of the eyes, nose, and throat, the major site of air pollutant impact is the lungs (Fig. 15–7). Three categories of impact can be distinguished: (1) **chronic:** pollutants cause gradual deterioration of a variety of physiological functions over a period of years; (2) **acute:** pollutants bring on life-threatening reactions within a period of hours or days; (3) **carcinogenic:** pollutants initiate changes in cells that lead to uncontrolled growth and division—cancer.

Almost everyone living in areas of urban air pollution suffers from chronic effects. Long-term exposure to sulfur dioxide can lead to bronchitis (inflammation of the bronchi); chronic inhalation of ozone and particulates can lead to fibrosis of the lungs, a scarring that permanently impairs lung function. Carbon monoxide reduces the capacity of the blood to carry oxygen, and low levels of carbon monoxide over a long period of time can contribute to heart disease. Chronic exposure to nitrogen oxides is known to impair the immune system, leaving the lungs open to attack by bacteria and viruses.

The chronic effects of air pollution are reflected in the morbidity and mortality data from Poland, a nation with perhaps the worst air in the world. Life expectancy for men is lower than it was 20 years ago; 30 to 45 percent of students are below international norms for height, weight, and other health indicators. In Upper Silesia, where there is a heavy concentration of industry, respiratory disease is 47 percent higher than in the rest of Poland (all of which is high by U.S. standards), there are 15 percent more cases of circulatory disease, and 100 percent more deaths from liver disease.

In some cases, air pollution has reached levels that caused death. Although these acute effects are attributed to air pollution, it should be noted that these deaths occur among people already suffering from severe respiratory or heart disease or both. Although the gases present in air pollution are known to be lethal in high concentrations, such concentrations are not reached in outside air. Therefore, deaths attributed to air pollution are not the direct result of simple poisoning. However, intense air pollution puts an additional stress on the body, and if a person is already in a weakened condition (e.g., elderly, asthmatic), this additional stress may be fatal.

The heavy metal and organic constituents of air pollution include many chemicals known to be carcinogenic in high doses. According to industrial reporting required by the EPA, 107 000 tons of carcinogens were released into the air in the United States

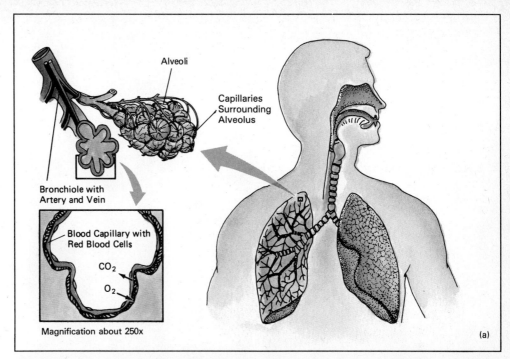

(a)

(b)

FIGURE 15–7
(a) In the lungs, air passages branch and rebranch and finally end in millions of tiny sacs called alveoli. Alveoli are surrounded by blood capillaries. As blood passes through these capillaries, oxygen from the air we inhale diffuses from the alveoli into the blood, and carbon dioxide diffuses in the reverse direction and leaves the body in the air we exhale. (b) On left, normal lung tissue; on right, lung tissue from a person who suffered from emphysema, a chronic lung disease in which some of the structure of the lungs has broken down. Cigarette smoking and heavy air pollution are associated with the development of emphysema and other chronic lung diseases. (A. Glauberman/Photo Researchers.)

in 1987, and this is undoubtedly an underestimate. The presence of trace amounts of these chemicals in air may be responsible for a significant portion of the cancer observed in humans. Once an air pollutant is shown by laboratory studies to have mutagenic or carcinogenic properties, the Clean Air Act of 1990 says it may not be emitted if there is a lifetime risk of cancer greater than 1 in 1 million to the most exposed individual in the population.

In some cases, exposure to a pollutant can be linked directly to cancer and other problems by way of epidemiological evidence. One pollution factor that clearly and indisputably is correlated with cancer and other lung diseases is cigarette smoking (Fig. 15–8). According to former Surgeon General Everett Koop, cigarette smoking is the major single cause of death due to cancer in the United States, responsible for 30 percent of all cancer deaths. Studies have shown that

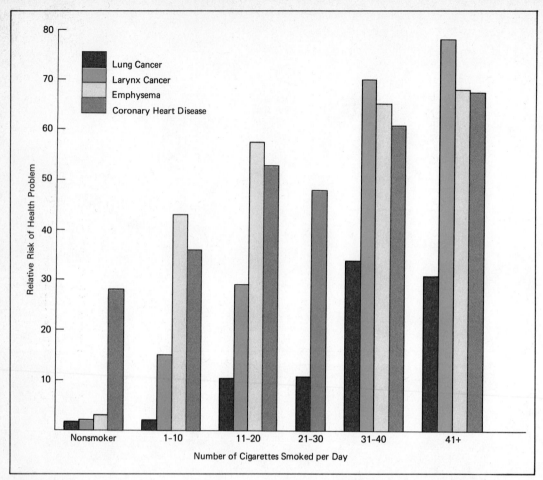

FIGURE 15-8
Cigarette smoking is strongly correlated with the development of cancer and of many chronic diseases.

smokers living in polluted air experience a much higher incidence of lung disease than smokers living in clean air—a synergistic effect where the two factors combine to produce an effect greater than the simple sum of the two. Certain diseases typically associated with occupational air pollution show the same synergistic relationship with smoking. For example, black lung disease is seen almost exclusively among coal miners who are also smokers. Lung disease among those exposed to asbestos predominates in smokers.

One of the heavy metal pollutants, lead, deserves special attention. Lead poisoning has been recognized for several decades as a cause of mental retardation. Researchers once thought that ingestion of peeling paint chips that contain lead was the main source of lead contamination. In the early 1980s, however, elevated lead levels were shown to be much more widespread than previously expected, and they were present in adults as well as children. Learning disabilities in children and high blood pressure in adults were found to be correlated with high levels

of lead in the blood. The major source of this widespread contamination was traced to leaded gasoline. The lead in exhaust fumes may be inhaled directly or may settle on food, water, or any number of items that are put in the mouth. This knowledge led the EPA to mandate the phase out of leaded gasoline, which was completed at the end of 1988. The result has been a dramatic reduction in lead concentrations in the environment.

Effects on Agriculture and Forests

How pollutants affect vegetation is determined by growing plants in chambers where they can be subjected to any desired concentration of pollutants and the results compared with field observations. Open-top chambers set up in the field enable plants in one chamber to receive filtered air, while plants in an adjacent chamber receive unfiltered air, and pollutants are monitored. Through such experiments it is possible to determine which pollutants cause damage and to extrapolate from this information a broader picture of the effects of air pollution on agriculture, forests, and ecosystems in general.

Experiments show that plants are considerably

more sensitive to gaseous air pollutants than humans are. Before emissions controls, it was common to see wide areas of totally barren land or severely damaged vegetation downwind from smelters or coal-burning power plants (Fig. 15–9). The pollutant responsible was usually sulfur dioxide. The dieback of vegetation in large urban areas and damage to crops, orchards, and forests downwind of the urban centers was determined to be mainly due to ozone and other photochemical oxidants.

The open-chamber experiments also show that plants in clean (filtered) air often grow larger than plants in unfiltered air. This suggests that existing levels of pollution in air are responsible for a general reduction of growth without conspicuous signs of damage or abnormality. Ozone is by far the most significant factor in this effect. A recent assessment of crop production in the United States estimated that without ozone pollution, crop yields would increase as follows: corn, 3 percent; wheat, 8 percent; soybeans, 17 percent; and peanuts, 30 percent. This represents about $5 billion worth of agricultural productivity, which is about 10 percent of total farm income. These figures represent an average across the United States. The effects in badly polluted areas are far more severe.

The negative impact of air pollution on wild plants and forest trees may be even greater than for agricultural crops. Open-chamber experiments in the Blue Ridge Mountains in northwestern Virginia have shown that the growth of various wild plants was substantially reduced by ozone even though the ozone level remained below standards set for human health. Significant damage to economically valuable ponderosa and Jeffrey pines is occurring along the entire western slope foothills of the Sierra Nevada in California. In the San Bernardino Mountains of California, an area that receives air pollution from Los Angeles, the rate of tree growth has been reduced by 75 percent. As the next chapter shows, acid deposition from air pollution also has a great impact on the growth of forest trees downwind of major urban areas.

Forests under pollution stress are more susceptible to damage by insects and other pathogens than are unstressed forests. For example, death of ponderosa and Jeffrey pines in California is generally attributed to western pine beetles, which invade pollution-weakened trees. Even normally innocuous insects may cause mortality when combined with pollution stress. If widespread pollution worsens, reductions in tree growth and survival could occur with disastrous suddenness as threshold levels of more and more species are exceeded.

Even without drastic dieoffs, the decrease in primary production must ultimately affect the rest of the

FIGURE 15–9
The countryside around this Butte, Montana smelter is devasted by the toxic pollutants from the industrial processes. (Joern Gerdis/Photo Researchers.)

ecosystem, including soils, because they depend on primary production. As sensitive species die out, they are replaced by more-resistant species in the process of ecological succession. Where this will lead is uncertain, but numerous foresters and ecologists have little doubt that large-scale biological changes are already under way in the landscape as a result of current levels of air pollution.

Effects on Materials and Aesthetics

Walls, windows, and all other surfaces turn gray and dingy as particulates settle on them. Paint and fabrics deteriorate more rapidly, and the sidewalls of tires and other rubber products become hard and checkered with cracks because of oxidation by ozone. Corrosion of metals is dramatically increased by sulfur dioxide and acids derived from sulfur and nitrogen oxides, as are weathering and deterioration of stonework (Fig. 15–10). These and other effects of air pollutants on materials increase the costs for cleaning or replacement by hundreds of millions of dollars a year. Many of the materials damaged are irreplacable.

Besides being a matter of health, a clear blue sky and good visibility—in contrast to the haze of smog— have significant aesthetic value and a psychological

FIGURE 15–10
The corrosive effects of acids from air pollutants are dissolving away the features of many monuments and statues, as seen in the faces of these statues in Brooklyn. (Ray Pfortner/Peter Arnold, Inc.)

impact on people. Can a value be put on this? Many of us spend thousands of dollars and hundreds of hours commuting long distances to work and back so that we can live in a less-polluted environment. Ironically, the resulting traffic and congestion cause much of the very pollution we are trying to escape. Too many of us have come to accept pollution as the necessary cost of "progress." Is it? Can we consider redefining progress in terms of improving human and environmental health rather than just increasing our gross national product? Environmentalists believe we can, but it requires political action and comes with an up-front economic cost, as we shall see.

Pollutant Sources and Control Strategies

In large measure, air pollutants are direct and indirect byproducts of the burning of coal, gasoline, other liquid fuels and refuse (waste paper, plaster, and so on). These fuels and wastes are organic compounds. With complete combustion, the byproducts of burning them are carbon dioxide and water vapor, as this equation for the combustion of methane shows:

$$CH_4 + 2O_2 \rightarrow CO_2 + 2H_2O$$

However, oxidation is seldom complete, and substances far more complex than methane are involved.

PRIMARY POLLUTANTS

The first six of the major pollutants listed above are called **primary pollutants** because they are the direct products of combustion and evaporation.

When fuels and wastes are burned, particles consisting mainly of carbon are emitted into the air; these are the particulates seen as soot and smoke. In addition, there are various unburned fragments or fuel molecules. These are the VOC emissions. Incompletely oxidized carbon is carbon monoxide, in contrast to completely oxidized carbon, which is carbon dioxide. Combustion takes place in the air, which is primarily 78 percent nitrogen and 21 percent oxygen. At high combustion temperatures, some of the nitrogen gas is oxidized to form the gas nitric oxide (NO). In the air, nitric oxide immediately reacts with additional oxygen to form nitrogen dioxide (NO_2) or nitrogen tetroxide (N_2O_4) or both. These compounds are collectively referred to as the nitrogen oxides. Nitrogen dioxide absorbs light and is largely responsible for the brownish color of photochemical smog.

In addition to their organic matter content, fuels and refuse contain impurities and additives, and these substances are also emitted into the air during burning. Coal, for example, contains from 0.2 to 5.5 percent sulfur. In combustion, this sulfur is oxidized, giving rise to the gas sulfur dioxide. Coal may contain heavy metal impurities, and of course refuse contains an endless array of "impurities."

According to EPA data, U.S. 1989 emissions of these five primary pollutants—particulates, VOCs, CO, NO_x, and SO_2—amounted to 128 million metric tons. By comparison, in 1970, when the first Clean Air Act became law, these same five pollutants totaled 194 million metric tons. The relative amounts emitted in the United States, and their major sources, are shown in Figure 15–11. The EPA requires the constant monitoring of ambient pollutant concentrations

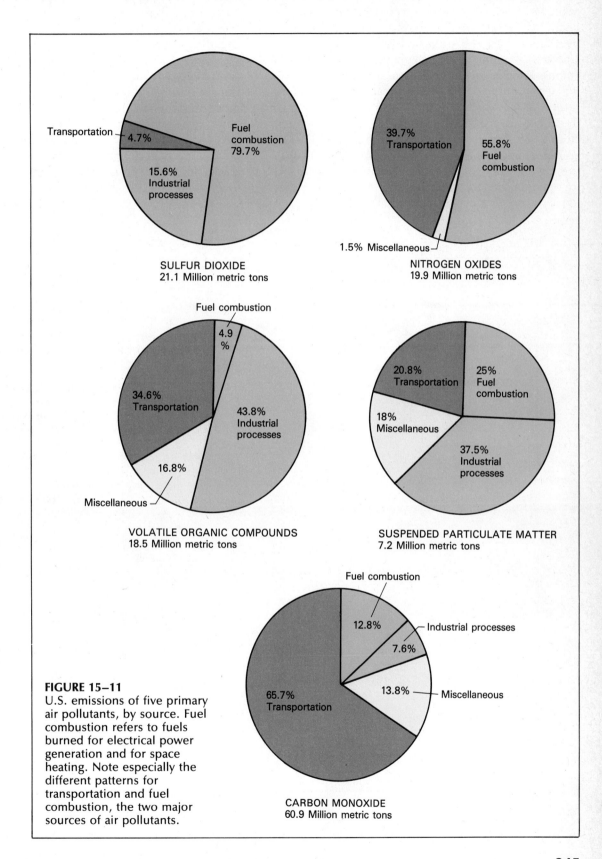

FIGURE 15–11
U.S. emissions of five primary air pollutants, by source. Fuel combustion refers to fuels burned for electrical power generation and for space heating. Note especially the different patterns for transportation and fuel combustion, the two major sources of air pollutants.

SULFUR DIOXIDE
21.1 Million metric tons

NITROGEN OXIDES
19.9 Million metric tons

VOLATILE ORGANIC COMPOUNDS
18.5 Million metric tons

SUSPENDED PARTICULATE MATTER
7.2 Million metric tons

CARBON MONOXIDE
60.9 Million metric tons

in selected regions. Trends in the ambient levels of five primary pollutants and of the secondary pollutant ozone are shown in Figure 15–12.

The sixth primary pollutant in our list, lead and other heavy metals, we treat separately because the quantities emitted are far less than levels for the first five. Before the EPA-directed phaseout, lead was added to gasoline as an inexpensive way to prevent engine knocking. Emitted with the exhaust, it remained airborne and traveled great distances before settling as far away as the glaciers of Greenland. Since the phaseout, concentrations of lead in the air in U.S. cities have shown a remarkable decline (Fig. 15–12). At the same time, lead in children's blood has dropped greatly. Lead concentrations in Greenland ice have also declined significantly. All these declines

indicate that lead restrictions in the United States and a few other Northern Hemisphere nations have had a global impact. However, lead is still used in some manufacturing processes in the United States and is subject to few restrictions elsewhere in the world.

As with lead, the concentration of air toxics and radon in the air are minute compared with those of the other primary pollutants. Some of the air toxic compounds—benzene, for example—originate with transportations fuels. Most, however, are traced to industries and small businesses. Radon, on the other hand, results from the spontaneous breakdown of fissionable material in rocks and soils. It escapes naturally to the surface and enters into buildings through cracks in foundations and basement floors, sometimes collecting in buildings.

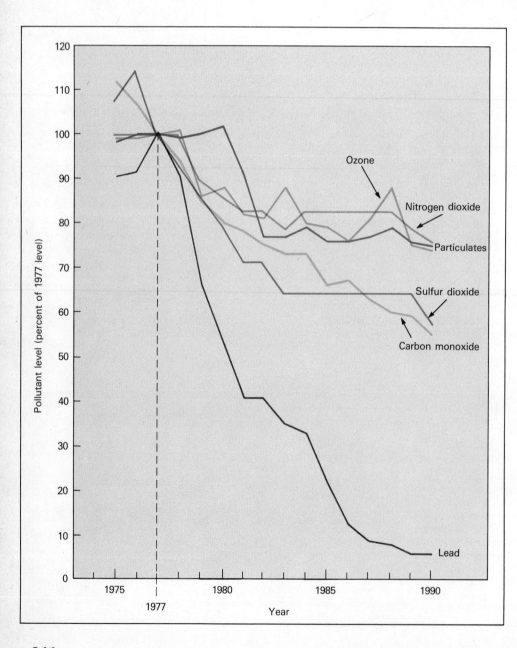

FIGURE 15–12
Trends in ambient air pollutants in selected regions of the United States, using 1977 as a base of 100. Nitrogen oxides have not improved because of a lack of attention to this pollutant. Ozone improved, but gains have been erased as increasing numbers of vehicles and numbers of kilometers driven offset lower emissions per car. Apparent improvement in sulfur dioxide has been at the expense of increased acid rain because of the tall stacks that are used by power plants. Real improvements have been made in lead by phasing it out of gasoline.

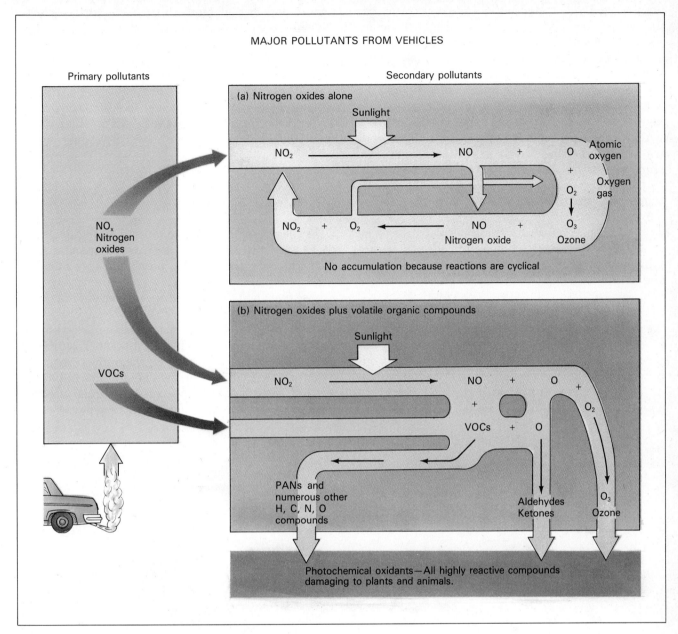

MAJOR POLLUTANTS FROM VEHICLES

Primary pollutants

Secondary pollutants

(a) Nitrogen oxides alone

Sunlight

NO_2 ⟶ NO + O   Atomic oxygen

+ O_2 Oxygen gas

NO_2 + O_2 ← NO + O_3
Nitrogen oxide   Ozone

No accumulation because reactions are cyclical

NO_x Nitrogen oxides

VOCs

(b) Nitrogen oxides plus volatile organic compounds

Sunlight

NO_2 ⟶ NO + O

+   + O_2

VOCs + O O_3

PANs and numerous other H, C, N, O compounds

Aldehydes Ketones

O_3 Ozone

Photochemical oxidants—All highly reactive compounds damaging to plants and animals.

FIGURE 15–13
Formation of ozone and other photochemical oxidants. Ozone is the most injurious.
(a) Nitrogen oxides, by themselves, would not reach damaging levels because
reactions involving them are cyclic. (b) When VOCs are also present, however,
reactions occur that lead to the accumulation of numerous damaging compounds,
most significantly ozone.

SECONDARY POLLUTANTS

Some of the primary pollutants may undergo further reactions in the atmosphere and produce additional undesirable compounds. These latter products are called **secondary pollutants.**

Ozone and numerous reactive organic compounds are formed as a result of chemical reactions between nitrogen oxides and volatile organic carbons,

with sunlight providing the energy necessary to cause the reactions to occur. Since sunlight provides the energy, these products are collectively known as **photochemical oxidants.**

The major reactions involved are shown in Figure 15–13. Light energy is absorbed by nitrogen dioxide and causes it to split to form nitric oxide and atomic oxygen. The atomic oxygen rapidly combines with oxygen gas to form ozone. If other factors are not involved, ozone and nitric oxide react to reform nitrogen dioxide and oxygen gas, and there is no appreciable accumulation of ozone (Fig. 15–13a).

When volatile organic compounds are present, however, the nitric oxide reacts with them, and these reactions result in several serious problems. First, the reaction between nitric oxide and VOCs leads to highly reactive and damaging compounds known as peroxyacetyl nitrates, or PANs (Fig. 15–13b). Second, with the nitric oxide tied up in this way, the ozone tends to accumulate. Also, numerous aldehyde and ketone compounds are produced as volatile organic compounds are oxidized by atomic oxygen, and these are also noxious.

In a sense, sulfuric and nitric acids can also be considered secondary pollutants, since they are formed as a result of the reaction of sulfur dioxide and nitrogen oxides with atmospheric moisture (Chapter 16).

Particulates take on additional potency when they adsorb numerous other pollutants on their surface. *Ad*sorption means that the pollutants simply adhere to a surface, like flies sticking to flypaper as opposed to "soaking in" (*ab*sorption). When breathed in, many of these contaminated particulates are fine enough to penetrate deeply into the lungs, releasing the adsorbed pollutants on the moist lung surfaces, where they often remain trapped for life.

Sources of air pollution are summarized in Figure 15–14. Note that industrial processes are the major source of particulates, whereas transportation and fuel combustion account for the lion's share of the other pollutants. As we shall see, strategies for control of air pollutants are different for these different sources.

SETTING STANDARDS

The Clean Air Act of 1970 mandated setting standards for four of the primary pollutants—SPM, sulfur dioxide, carbon monoxide, and nitrogen oxides—and for the secondary pollutant ozone because at the time these five were the pollutants recognized as the most widespread and objectionable. These five are now known as **criteria pollutants.** Lead was added to the list of criteria pollutants in recent years. The **primary standard** for each is based on the highest level that can be tolerated by humans without noticing any ill effects, minus a 10 to 50 percent margin of safety (Table 15–1). It is significant that air pollution stan-

FIGURE 15–14
The prime sources of the major air pollutants.

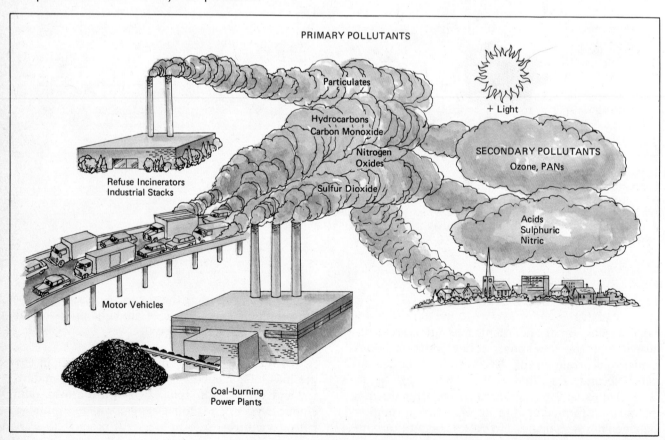

TABLE 15–1

National Ambient Air Quality Standards for Criteria Pollutants

Pollutant	Averaging Time[a]	Primary Standard
PM-10 particulates[b]	1 year	50 µg/m³
	24 hours	150 µg/m³
Sulfur dioxide	1 year	0.03 ppm
	24 hours	0.14 ppm
	3 hours	0.5 ppm
Carbon monoxide	8 hours	9 ppm
	1 hour	35 ppm
Nitrogen oxides	1 year	0.05 ppm
Ozone	1 hour	0.12 ppm
Lead	3 months	1.5 µg/m³

[a] Averaging time is the time period over which concentrations are measured and averaged.

[b] PM-10 is the particulate fraction having a diameter smaller than or equal to 10 micrometers. It replaces SPM as the standard. A µg (microgram) is one-millionth of a gram.

Source: EPA 1990 Annual Report on Air Quality in New England, July, 1991.

dards are set according to human health criteria and not according to impacts on other species or on atmospheric chemistry.

In the 1980s, the EPA added a new standard for particulates based on information that smaller particulate matter (less than 10 micrometers in diameter, called PM-10) has the greatest health effect because of its capacity to be inhaled. In addition, **National Emission Standards for Hazardous Air Pollutants** (NESHAPS) have been issued for eight toxic substances: arsenic, asbestos, beryllium, benzene, coke oven emissions, mercury, radionuclides, and vinyl chloride. The **Clean Air Act of 1990** greatly extended this section of the EPA's regulatory work by specifically naming 189 toxic air pollutants.

CONTROL STRATEGIES

The basic strategy of the 1970 Clean Air Act was to regulate air pollution to the point where the criteria pollutants remained below the primary standard levels. This strategy is referred to as a **"command-and-control"** approach because industry was given regulations to achieve a set limit on each pollutant, to be accomplished by specific control equipment. The assumption was that significant improvements could be made in human and environmental health by reducing the output of pollutants. If your area was in violation for a given pollutant, a local government agency would go after the source(s) and mandate reductions in emissions until you were in compliance.

Unfortunately, this strategy proved difficult to implement. Most of the regulatory responsibility fell on the states and cities, which were often unable or unwilling to enforce control. Many areas violated the standards. Even after 25 years of air pollution control, 96 cities still consistently fail to meet the ozone standards; 41 do not meet the carbon monoxide standards; 72 fail to meet the SPM standards. To be sure, total pollutants have been reduced by some 34 percent, during a time when population and economic activity have increased, but we are still sending out huge quantities of pollutants into the atmosphere. (There has been one outstanding accomplishment: the 97 percent reduction in lead emissions in the past 20 years.)

The Clean Air Act of 1990 addresses these failures by targeting specific pollutants more directly and putting more teeth into achieving compliance, such as the imposition of sanctions. However, it provides more flexibility than the command-and-control approach by allowing polluters to choose the most cost-effective way to accomplish the goals, and it uses a market system to allocate pollution among different utilities (see Chapter 16).

Particulates

Prior to the 1970s, the major sources of particulates were from the open burning of refuse and industrial stacks. The Clean Air Act of 1970 mandated the phaseout of open burning of refuse and required particulates from industrial stacks be reduced to "no visible emissions."

The alternative generally taken for refuse disposal was landfilling, a solution that has created its own set of environmental problems, discussed in Chapter 20. To reduce stack emissions, many industries were required to install filters, electrostatic precipitators, and other devices as shown in Figure 15–15. However, the solid wastes removed from exhaust gases frequently contain heavy metals and other toxic substances, discussed in Chapter 14. These measures have resulted in a marked reduction of particulates since the 1970s (Fig. 15–12). However, particulates continue to be released from steel mills, power plants, cement plants, smelters, construction sites, diesel engines, and so on. Wood-burning stoves and other wood and grass fires also contribute to the SPM load, making regulation even more difficult.

Under the Clean Air Act of 1990, the 72 regions of the United States that have failed to attain the required levels must submit attainment plans. These plans must be based on **reasonably available control technology** (RACT) measures, and must convince the EPA that the standards will be reached by the end of 1994.

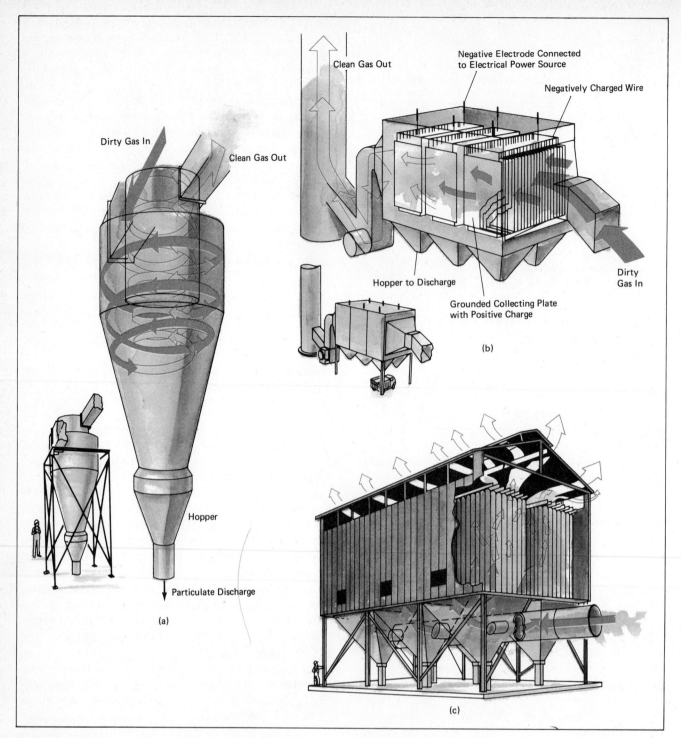

FIGURE 15-15

Devices to remove particulates from exhaust gases. (a) Cyclone precipitator. Particles are removed by centrifugal force as exhausts are swirled. (b) Electrostatic precipitator. Particles are electrically charged, then attracted to plates of the opposite charge. (c) Bag house. Exhaust gases are forced through giant vacuum cleaner bags. None of these devices removes very fine particles or polluting gases.

Pollutants from Motor Vehicles

Cars, trucks, and buses release nearly half of the pollutants that foul our air. Vehicle exhaust sends out VOCs, carbon monoxide, and the nitrogen oxides that lead to ground-level ozone and PANs. Additional VOCs come from evaporation of gasoline and oil vapors from the fuel tank and engine systems.

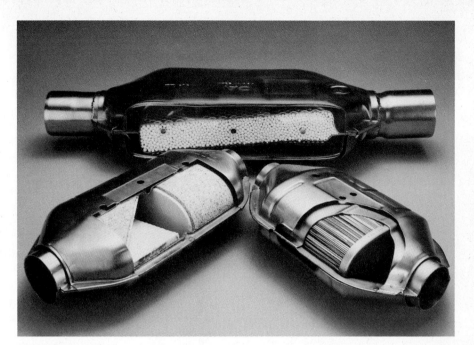

FIGURE 15–16
Cutaways of three kinds of catalytic converters: (top–ceramic pellets, left–metal foil plates, right–ceramic monolith substrates). The catalysts are platinum, palladium and rhodium, metals that promote the oxidation of residual VOCs and carbon monoxide to CO_2 and water. Catalytic converters are now required on all automobiles. (GM/ AC Rochester.)

The Clean Air Act of 1970 mandated a 90 percent reduction in these emissions by 1975. This timing proved to be unrealistic, but enough improvements have been made over the years that a new car today actually emits 95 percent less pollutants than pre-1970 cars with no controls. This is fortunate because vehicle driving in the United States has been increasing much faster than the population; between 1970 and 1990, the number of vehicle miles has doubled, from 1 trillion to 2 trillion miles per year. It is hard to imagine what the air would be like without the improvements mandated by the Clean Air Act.

The reductions in auto emissions have been achieved with a considerable array of pollution control devices, including computerized control of fuel mixture and ignition timing, which give more complete combustion of fuel and decrease VOC emissions. However, the most significant control device on cars remains the **catalytic converter** (Fig. 15–16). As the exhaust passes through this device, a chemical catalyst consisting of platinum-coated beads oxidizes most of the VOCs to carbon dioxide and water. The catalytic converter also oxidizes carbon monoxide to carbon dioxide. It does nothing, however, to control nitrogen oxides. Also, continuing failure to meet standards in many regions of the United States and the aesthetically poor air quality in many cities made it clear that further legislation was in order.

In preparing for the Clean Air Act of 1990, the EPA considered three options to improve the air in our cities: (1) Tighten up emissions standards still more; (2) encourage cleaner-burning fuel; (3) get people to drive less. The new act addresses the first two but makes only a token move in the direction of less driving. Here are the highlights of the motor vehicle and fuels sections:

1. New cars sold in 1994 and thereafter must emit 30 percent less VOCs and 60 percent less nitrogen oxides than 1990 cars. Emissions-control equipment must function properly (as represented in warranties), to 100 000 miles; 1990 cars need work only for 50 000 miles. Canisters in the gas tank will be required to capture vapors during refueling. It is estimated that these measures will add less than $200 to the cost of a new car. Buses and trucks must meet more stringent standards. Also, EPA is given authority to control emissions from all non-road engines that contribute to air pollution, specifically, lawn and garden equipment, motorboats, and farm equipment.

2. Starting in 1992 in the 41 cities with continuing carbon monoxide problems, oxygen will be added to gasoline in the form of alcohols and ethers to stimulate more complete combustion and cut down on carbon monoxide emissions. Starting in 1995, specially clean fuels will be required for the nine cities with the worst air pollution. These measures will result in higher gasoline prices.

3. Employers in cities with high smog levels will be required to take steps to increase the number of their employees who carpool or use mass transit.

Sulfur Dioxide and Acids

Measurements of sulfur dioxide in city air indicate that there has been great success in controlling this pollutant (Fig. 15–12). These results are highly misleading, however, because control strategies have led to other problems that are as bad if not worse. The major source of sulfur dioxide is coal-burning electric power plants. A power plant may burn up to 10 000 tons of coal per day. Contaminated with 3 percent sulfur, this results in daily emissions nearing 1000 tons of sulfur dioxide.

Rather than reducing these emissions, these industries lobbied for and obtained permission under the Clean Air Act to use a "dilution solution." They shut down small, inefficient plants in cities and built huge facilities in rural areas, with very tall smokestacks that ejected pollutants above the inversion layer. The idea was that if pollutants were ejected into the upper atmosphere, they would dilute and effectively disappear. Far from disappearing, however, the sulfur dioxide from the tall stacks converts to sulfuric acid, which has been shown to be the major source of acid precipitation (Chapter 16).

Thus, sulfur dioxide improvements in city air quality have been at the expense of declining rural air quality, and reducing sulfur dioxide at ground level has been at the expense of greatly worsening the acid deposition problem. The Clean Air Act of 1990 specifically addresses this problem by requiring a reduction of 10 million tons (roughly half) of sulfur dioxide emissions from coal-burning power plants by the year 2000.

Ozone

It is now well understood that the best way to reduce ozone levels is to reduce emissions of the volatile organic compounds. The steps outlined above for motor vehicles will address the source of about half of VOC emissions. **Point sources** (industries) account for 30 percent of the VOC emissions, and **area sources** (numerous small emitters such as dry cleaners, print shops, household products) represent the remaining 20 percent. RACT measures have already been mandated for many point sources, and much progress has been made through EPA, state, and local regulatory efforts to reduce emissions from these sources. Under the Clean Air Act of 1990, regions with continuing violation of ozone standards will be required to further control VOC emissions from both point and area sources. The degree of control and the timing vary according to the seriousness of the violation, but the means of control will range from prohibition of some consumer products to annual fees paid by point sources ($5000 per ton of VOC), to sanctions such as withholding of federal highway funds and other grants. The latter is a means of forcing recalcitrant states to address their air pollution problems.

Air Toxics

By EPA estimates, the total amount of toxic substances emitted into the air in the United States is over 6 million metric tons annually. Cancer and other adverse health effects, environmental contamination, and catastrophic chemical accidents are the major concerns associated with this category of pollutants.

Under the Clean Air Act of 1990, Congress identified 189 toxic pollutants and directed the EPA to identify major sources of these pollutants and develop **"maximum achievable control technology"** (MACT) standards. The setting of these standards includes both control technologies and options for substituting nontoxic chemicals, giving industry some flexibility in meeting goals. State and local air pollution authorities will be responsible for seeing that industrial plants achieve those goals. This program should reduce emissions by at least 20 percent by 2005, at an estimated cost to industry of $7 billion.

Other provisions of the act include a renewed emphasis on calculating risks to public health from air toxics from all sources (for example, radon in buildings), and more attention given to the problem of accidental releases.

TAKING STOCK

The Cost of Controlling Air Pollution

Without question, measures taken to reduce air pollution carry an economic cost. The United States invests heavily in pollution control, a clear signal that human health and environmental quality rank highly in the development and implementation of public policy. There is strong citizen support for this; according to a *New York Times* poll in March, 1990, 75 percent of Americans agreed that environmental improvements should be made regardless of cost. Other polls indicate that most Americans are ready to pay more for cleaner gasoline. Currently, pollution control costs are estimated at $125 billion per year. About one-fourth of that is accounted for by air pollution. The EPA estimates that, when fully implemented in 2005, the new Clean Air Act will cost an additional $25 billion. Although this is an enormous sum, it represents less than 1 percent of the estimated $7 trillion economy for that year.

Some critics—especially economists—have charged that air pollution controls are not cost-effective, that is, the benefits are not nearly as great as the costs. They see lost opportunities for growth and tend to disregard *avoided costs*. We have noted that with existing levels of air pollution, damage to human health, crops, forests, materials, and aesthetic values is substantial. What would these damages and losses be without pollution control? Economists who have factored in these avoided costs conclude that each dollar spent on pollution control has given us a return of two to three dollars savings, to say nothing of avoided misery and suffering. In addition, pollution control is now a major industry. Providing over 2 million jobs, it is a significant part of the national economy.

Future Directions

In the early 1970s, many cities were experiencing more than 100 days a year of pollution in the "unhealthful" range or above; Los Angeles was having on the order of 300 such days per year. The original goal of air pollution control was to achieve "good" air quality on all days by 1975. It hardly needs to be said that this goal was not accomplished; however, marked progress has occurred for five of the primary air pollutants (Fig. 15–12). Certainly air quality is very much better than it would be if controls had not been initiated.

However, the backsliding of the last few years is very troubling. Basically, pollution control devices and greater fuel efficiency reduced a car's emissions by 95 percent from 1969 to 1989. As old, polluting cars were replaced by new "cleaner" ones over this period, emissions levels decreased despite increasing numbers of cars and miles driven. Now, with most cars being cleaner models, this reduction-by-replacement has leveled out, and the increasing number of cars is becoming the dominant factor, bringing the decline in vehicle-related pollutants to a halt. The new Clean Air Act addresses the problem primarily by mandating increased emissions controls and cleaner-burning fuels. Whether these measures will be suc-cessful in further reducing air pollution remains to be seen.

A more general approach that would benefit the whole country would be to increase the fuel efficiency standard for cars, which is now locked at 27.5 miles per gallon (mpg). Intense lobbying by the automobile and petroleum industries prevented inclusion of greater fuel efficiencies in the Clean Air Act of 1990. Technologies are available that could bring average mileage per gallon to 60 by the year 2000. For example, two-stroke engines (which fire on every thrust of the piston instead of every other thrust as in the current four-stroke engine) cut engine weight in half; lightweight cars with smaller engines give mileages in the 80 to 100 mpg range. Increasing the fuel efficiency of vehicles would proportionately reduce emissions. It would also address the problem of the carbon dioxide greenhouse effect (Chapter 16) and the problem of future crude oil shortages (Chapter 21).

California, which by all measures has the greatest problem with photochemical smog and vehicular pollutants, has taken a completely different tack to reduce the pollution from vehicles. Starting in 1998, California law requires that 1 percent of all vehicles sold there must be "emission-free"—in other words, electrical. The percentage rises to 5 percent by 2001. Similar laws have been enacted in nine northeastern states and the District of Columbia. At the 1992 North American International Auto Show in Detroit, automakers displayed a range of electrical cars that are close to the production stage (Fig. 15–17). The cars are much lighter than conventional cars and are definitely limited in range (100 to 150 miles before needing a recharging). They also lack such amenities as air conditioning and other power accessories. However, the electrical cars may well be the wave of the future. You will recognize, of course, that switching from gasoline to electrical power basically transfers

FIGURE 15–17
The Ford Allegheny Ecostar Van, one of the many electric cars being developed for the U.S. market in anticipation of a greater demand for pollution-free vehicles in pollution-impacted cities. (Courtesy Ford Motor Company.)

FIGURE 15–18
Indoor air pollution. Pollutants originate from many sources and can accumulate to unhealthy levels, leading to the problem of "sick building syndrome" (eye, nose, and throat irritations, nausea, irritability, fatigue).

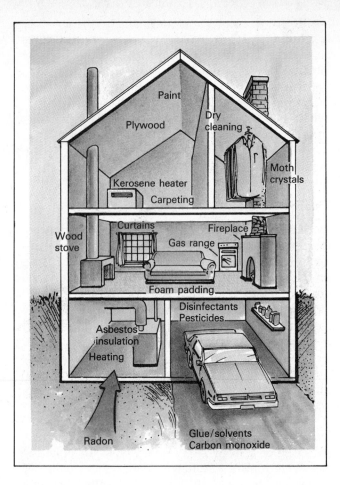

the site of pollution emission from the moving vehicles to the power plants (Chapter 21).

It can be argued that all this emphasis on how vehicles burn fuel is just tinkering with the mechanics of the problem. To achieve lasting progress, we need to address the fundamental way we have organized society and industry. Consider: We pour billions of dollars into expanding and improving our highways; we build malls by every major highway corridor; we build homes in bedroom communities scores of miles from the major employment centers; we put little into improving our mass transportation systems. All of these developments encourage more driving instead of less. In effect, we are placing ourselves in deeper bondage to urban sprawl and hours spent in daily commutes. It would make great sense if, in the coming years, we were to fashion legislation and policies that would guide new development and redevelopment of depreciated urban areas so as to *reduce* average commuting distances and times and facilitate greater use of mass transit systems. At the same time, development of efficient mass transit systems and more fuel-efficient vehicles should be made a matter

IN PERSPECTIVE
Portland Takes a Right Turn

The American lifestyle has come to mean spending hours in the automobile each day going to and from work. Vehicle miles per year in the United States have increased much more rapidly than population, and too often the only response of state and city governments is to build more lanes on the expressways.

Portland, Oregon, had its share of expressways 20 years ago, but took a different tack when faced with the prospect of more and more commuters on the roadways. In response to an Oregon land-use law, Portland threw away its plans for more expressways and instead built a light rail system. This system now carries the equivalent of two lanes of traffic on all the roads feeding

into downtown Portland (figure at right). The result? Smoggy days have declined from 100 to 0 per year, the downtown area has added 30 000 jobs with no increase in automobile traffic, and Portland's economy has prospered.

This has clearly been a situation where everyone wins. The Portland solution seems so sensible that you have to ask, Why is it the exception and not the rule?

Metropolitan Area Express (MAX) in Portland, Oregon. This public transportation system has made a major contribution to the clean air and continued economic success of downtown Portland. (Courtesy Tri-Met, Portland, OR.)

of national priority. These measures would do much to move our society in the direction of sustainability.

Indoor Air Pollution

After recognizing the many pollutants in outside air, you may be inclined to remain indoors to escape the hazards. However, air inside the home and workplace often contains much higher levels of hazardous pollutants than outdoor air. In a recent report, the National Academy of Sciences called indoor air pollution an issue that "has been largely overlooked" and a matter "of immediate and great concern." The overall indoor air pollution problem is threefold. First, increasing numbers and kinds of products and equipment used in homes and offices give off potentially hazardous fumes. Second, buildings have become increasingly well-insulated and sealed; hence, pollutants are trapped inside, where they accumulate to potentially dangerous levels. Third, people are exposed more to indoor pollution than to outdoor pollution. The average person spends 90 percent of his or her time indoors, and the people who spend the most time inside are those most vulnerable to the harmful effects of pollution: small children, pregnant women, the elderly, and the chronically ill.

SOURCES OF INDOOR POLLUTION

The sources of indoor air pollution are numerous (Fig. 15–18). They include:

○ Formaldehyde and other synthetic organic compounds given off from plywood, particleboard, foam rubber, "plastic" upholstery, and no-iron cotton sheets and pillowcases.

○ A wide host of compounds from foods burned on the stove or oven.

○ Incomplete combustion and impurities from fuel-fired heating systems such as gas or oil furnaces, kerosene heaters, and woodstoves.

○ Fumes from household cleaners and other "strong" cleansing agents.

○ Fumes from glues and hobby materials.

○ Pesticides.

○ Air fresheners and disinfectants. Most air fresheners work by either dulling the sense of smell so you don't notice noxious odors or by introducing "high-intensity" smells that cover up odors.

○ Aerosol sprays of all sorts.

○ Radon. As warm air escapes from the top of a house creating a partial vacuum, radon may be drawn through the basement floor and, trapped in the house, may then reach hazardous levels.

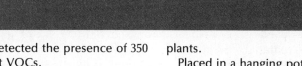

IN PERSPECTIVE

Plants for Your Health

In 1989 the EPA submitted a report to Congress stating that "indoor air pollution represents a major portion of the public's exposure to air pollution and may pose serious acute and chronic health risks. [Further] indoor air pollution is among the nation's most important environmental health problems." Carpets, furniture coverings, wallboard, and other home furnishings are made from plastics and artificial fibers that give off VOCs. Cleaners, insecticides, glues, air fresheners, hair spray, shoe and nail polish, magic markers, oil-based paints, and any number of other commonly used products add to the pollutant load. An EPA study of one Washington, D.C., nursing home detected the presence of 350 different VOCs.

In the early 1970s, Bill Wolverton, an environmental engineer working for NASA was given the problem: How can air be kept clean and healthful in a spaceship? He started testing the effectiveness of house plants as "bioregenerative" agents. They proved even more effective than he had dared hope. Starting with "dangerous" levels of various VOCs, Wolverton found that some plants proved capable of reducing the pollution to nondetectable levels within 24 hours. Two of the most effective plants were spider plants and philodendrons, which are also among the easiest-to-grow house plants.

Placed in a hanging pot by a window, these plants create a cascading curtain of green. They tolerate almost any light conditions. Watering every week or two will suffice. They are resistant to pests and they don't produce blossoms, which may be allergenic to some people. The plants are easily propagated, philodendrons from cuttings and spider plants from the numerous plantlets that are produced on runners. They are among the most common and least expensive plants wherever house plants are sold. They are, indeed, plants for your health!

Smoking in public is becoming a socially unacceptable habit, forcing many smokers to retreat to the restrooms or specially designated "smoking areas" in order to satisfy their needs. Almost all states have laws restricting smoking in public places, and federal law prohibits smoking on all domestic airline flights. Whereas once it was considered simply a nuisance to have to breathe secondhand smoke (sidestream smoke, as it is called), evidence has shown that consistent breathing of sidestream smoke can lead to some of the same consequences experienced by smokers. In one study, it was found that nonsmoking wives of men who smoke are twice as likely to die of lung cancer as wives of men who do not smoke. Some workers (bartenders and musicians, for instance) involuntarily inhale the equivalent of 10 or more cigarettes per day. All of this

new knowledge has put smokers on the defensive, as nonsmokers adopt slogans like "Your right to smoke stops where my nose begins," and "If you smoke, don't exhale!"

Smoking, of course, has its defenders. Smokers feel they have a right to indulge in their habit and would like nonsmokers to understand that smokers have a physical addiction that is quite difficult to break. Some smokers have become belligerent about the restrictions and are willing to take their rights into court. However, court challenges to the restrictions on smoking have not gone well for smokers. For example, a federal court upheld a ruling that prohibited a firefighter from smoking, whether on or off the job. The judges ruled that smoking was not protected by the constitutional right to privacy. Civil libertarians object to such restrictions on the basis that a person has a right to

smoke; others counter with arguments that individual liberty can be limited if there is a rational purpose—in this case, firefighters must be in top condition and are subjected to hazards in the course of their work that smoking could aggravate.

The battles over smoking often take place in the home. Is it hospitable to invite smokers to dinner but ask that they refrain from smoking? How can one respond to a child who puts pressure on a parent not to smoke by refusing a good-night kiss?

In the United States, nonsmokers are in the majority, and they seem to be gaining the upper hand in the controversy over public smoking. Have we gone too far, or should we do more to curb sidestream smoke? What do you think?

The EPA has estimated that between 5000 and 20 000 cancers per year are caused by radon.

○ Asbestos. Asbestos is a natural mineral that has fiberlike crystals. It is mined from the earth and was once used as a heat-insulating and fire-retarding material. Steam-heating pipes were wrapped in asbestos, ceilings in many public buildings were covered with it, and it was used in ironingboard covers, paints, and roofing materials. In the 1960s, research workers determined that the inhalation of asbestos fibers is associated with a unique form of lung cancer that develops as many as 20 to 30 years after exposure. EPA began regulating asbestos in the mid-1970s and initiated intensive campaigns to remove it from schools. However, this program has moved forward slowly, and many buildings still contain asbestos. Also, many asbestos products remain in use.

○ Smoking. Smoking carries a much higher health risk than the average exposure to any of the above materials, and it may act synergistically to increase the risk of exposure to other indoor (and outdoor) pollutants. Smoking has also been shown to in-

crease the health risks of nonsmokers subjected to breathing "secondhand" smoke.

PUBLIC POLICY AND INDOOR POLLUTION

The EPA recently announced a ban on practically all uses of asbestos, to be completed by 1997. In the meantime, the EPA has published guidelines for controlling old asbestos in public buildings, with an emphasis on schools. The guidelines caution school administrators against assuming that removal is the only option. In most cases, asbestos can be sprayed with a sealant (encapsulation) or enclosed, at costs much lower than removal costs and with far less risk of exposure to the public.

Smoking is essentially a form of portable air pollution. The smoker is the primary recipient of the pollution, but the effects are spread to all those who breathe the smoke-clouded air and, in the case of pregnant women, to the unborn fetus. Worldwatch Institute's William Chandler maintains that "Tobacco causes more death and suffering among adults than any other toxic material in the environment." The Surgeon General has issued repeated warnings against smoking since 1964, and public policy has taken those warnings seriously by requiring warning labels on smoking materials, banning cigarette advertising on television, promoting a smoke-free workplace, requiring nonsmokers' areas in restaurants, and banning smoking on all domestic airline flights. Since the warnings began, the U.S. smoking population has gradually dropped from 40 percent to 26 percent. Unfortunately, another trend has appeared—people are likely to start smoking at younger ages.

Our approach to tobacco is strange. On the one hand, we try to keep people from suffering the serious health effects of smoking, and on the other hand, we subsidize the tobacco industry and encourage the overseas sales of U.S. tobacco products. Cigarettes have even been part of the Food For Peace Program! Although the links between cigarette smoking and health effects are now firmly established, the tobacco industry has managed to avoid liability for the deaths caused by smoking. However, as long as there are smokers, there will be a population whose illness and mortality will continue to signal to us what to expect if we fail to keep air pollution under control.

 Review Questions _____

1. What natural cleansing processes take place in the air?
2. Describe the origin of industrial smog and photochemical smog.
3. Discuss the four basic stages in improving air quality as mandated by the Clean Air Acts of 1963 and 1970.
4. Name the air pollutants considered most widespread and serious.
5. What impact does air pollution have on human health?
6. Describe the negative effects of pollutants on crops, forests, and other materials. Which pollutants are mainly responsible for these effects?
7. Which primary pollutants are the products of combustion?

8. Name the secondary pollutants of combustion. By what processes are these formed?
9. Describe how the Clean Air Act of 1970 used the "command-and-control" strategy of pollution regulation.
10. Discuss several highlights of the Clean Air Act of 1990 that address air pollutants from motor vehicles.
11. In what ways does this new law require industries and small businesses to control the emission of VOCs and air toxics?
12. What further steps could be taken to improve the quality of our air?
13. Name the major sources of indoor air pollution. Discuss the legislative regulations that address each.

 Thinking Environmentally _____

1. Describe the major sources of air pollution in your community. What is the possibility of a severe air pollution problem occurring where you live?
2. The cost of reducing sulfur dioxide emissions from many of our utility plants would require a significant boost in electricity rates to cover incurred costs. Should consumers be required to pay for these pollution-reducing efforts? How should these costs be met?
3. How might traditional patterns of life change if cleaner air is not achieved? Write a short essay describing the possibilities.
4. Motor vehicles release close to half of the pollutants that dirty our air. What alternatives might be introduced to encourage a decrease in our use of

automobiles and other vehicles with internal combustion engines?
5. Lead paint, asbestos, tobacco, and many other products have been linked to adverse effects on human health. Research one such case that has been brought into the courts. Describe the alleged injustice, trial proceedings, and outcome. Was the decision reached a fair one? Why, or why not?
6. Who pays for the health-care costs of those who become ill from smoking? If you were the Surgeon General of the United States, what further steps would you advocate in regard to smoking regulations, or would you back off? Defend your answer.

16

Major Atmospheric Changes

LEARNING OBJECTIVES

When you have finished studying this chapter, you should be able to:

1. Describe air pollutants in the context of atmospheric structure.

2. Understand pH, acids, and bases.

3. Discuss the two major acidic pollutants.

4. Describe the effects of acid deposition on aquatic ecosystems, forests, and human artifacts.

5. List the major strategies for controlling acid emissions and evaluate their effectiveness.

6. Understand the politics of acid deposition and list the provisions of Title IV of the Clean Air Act of 1990.

7. Describe how the greenhouse gases maintain heat in the atmosphere.

8. List the different greenhouse gases and evaluate their contribution to present and future global warming.

9. Describe the most significant possible impacts of future global warming.

10. Suggest steps that should be taken to stabilize atmospheric concentrations of the greenhouse gases.

358

11. Describe the stratospheric ozone shield, including how it is formed and broken down.

12. Describe how CFCs affect the stratospheric ozone concentration.

13. Understand how the Antarctic ozone "hole" is formed.

14. Evaluate the political and economic steps being taken to protect the ozone shield.

Thousands of lakes on our planet are lifeless; tens of thousands more are threatened; forests are dying back. The climate shows signs of warming, threatening the world with unprecedented droughts and other climatic shifts. The public is being warned of rising rates of skin cancer because of increased exposure to ultraviolet radiation. All these situations have been identified as **anthropogenic,** meaning they come from human activities. In effect, air pollution from the human system is now affecting the entire biosphere. Before we address these problems, however, it would be helpful to consider the basic structure of the atmosphere.

The atmosphere is a collection of gases that gravity holds in a thin envelope around the earth. The lowest layer, the **troposphere,** extends up 10 miles (15 km) and (except for local temperature inversions) gets colder with altitude (Fig. 16–1). The troposphere is well mixed vertically; pollutants can reach the top within a few days. This layer contains practically all of the water vapor and clouds; it is the site and source of our weather. Substances entering the troposphere may be washed back to the earth's surface by precipitation.

Capping the troposphere is the **tropopause,** a region of temperature reversal. Above the tropopause is the **stratosphere,** a layer where temperature increases with altitude to about 40 miles. The temperature increases here primarily because the stratosphere contains ozone (O_3), a form of oxygen that

FIGURE 16–1
Structure and temperature profile of the atmosphere. Weather and most pollutants are found in the troposphere. The ozone shield is located in the stratosphere.

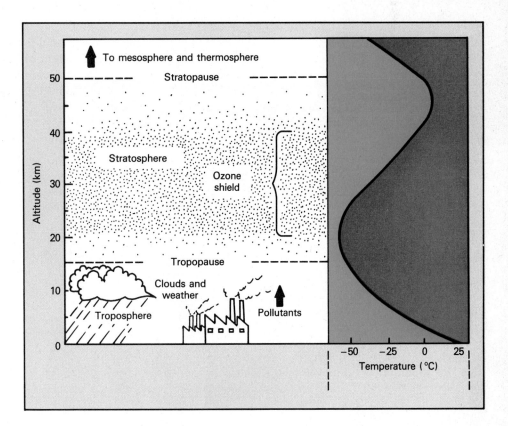

absorbs high energy radiation emitted by the sun. Because there is little vertical mixing in the stratosphere and no precipitation from it, substances that enter it can remain there for a long time.

Beyond the stratosphere are two more layers, the mesosphere and the chemosphere, where ozone concentration declines, and only small amounts of oxygen and nitrogen are found. Because none of the reactions we are concerned with in this chapter occur in the mesosphere or chemosphere, we shall not discuss those two layers.

In this chapter we shall examine three phenomena involving major atmospheric changes—acid deposition, global warming, and ozone depletion. Keep in mind the characteristics of the troposphere and stratosphere as we proceed (Table 16–1).

Acid Deposition

Acid deposition refers to any precipitation—rain, fog, mist, or snow—that is more acidic than normal. It also refers to the fallout of dry acidic particles. Broad areas of North America, as well as most of Europe and other industrialized regions of the world, are regularly experiencing precipitation that is between 10 and 1000 times more acidic than normal. This is affecting ecosystems in diverse ways, as illustrated in Figure 16–2.

To understand the full extent of this problem, first we must understand some principles regarding the nature of acids and how we measure their concentration.

TABLE 16–1

Characteristics of Troposphere and Atmosphere	
Troposphere	**Stratosphere**
Extent: ground level to 10 mi	Extent: 10 mi to 40 mi
Temperature normally decreases with altitude	Temperature increases with altitude
Much vertical mixing	Little vertical mixing, very slow diffusion exchange of gases with troposphere
Substances entering may be washed back to earth	Substances entering remain here unless attacked by light or other chemicals
All weather and climate take place here	Isolated from the troposphere by the tropopause layer

ACIDS AND BASES

Acidic properties (sour taste, corrosiveness) are due to the presence of hydrogen ions (H^+, a hydrogen atom without its electron), which are highly reactive.

Therefore, an **acid** is any chemical that releases hydrogen ions when dissolved in water. Chemical formulas of a few common acids are shown in Table 16–2. Note that all of them ionize (the components separate) to give hydrogen ions plus a negative ion.

TABLE 16–2

Common Acids and Bases						
Acid	**Formula**	**Yields**	**H^+ Ion(s)**	**Plus**	**Negative Ion**	
Hydrochloric acid	HCl	$\rightarrow$	H^+	$+$	Cl^-	Chloride
Sulfuric acid	H_2SO_4	$\rightarrow$	$2H^+$	$+$	SO_4^{2-}	Sulfate
Nitric acid	HNO_3	$\rightarrow$	H^+	$+$	NO_3^-	Nitrate
Phosphoric acid	H_3PO_4	$\rightarrow$	$3H^+$	$+$	PO_4^{3-}	Phosphate
Acetic acid	CH_3COOH	$\rightarrow$	H^+	$+$	CH_3COO^-	Acetate
Carbonic acid	H_2CO_3	$\rightarrow$	H^+	$+$	HCO_3^-	Bicarbonate
Base	**Formula**	**Yields**	**OH^- Ion(s)**	**Plus**	**Positive Ion**	
Sodium hydroxide	$NaOH$	$\rightarrow$	OH^-	$+$	Na^+	Sodium ion
Potassium hydroxide	KOH	$\rightarrow$	OH^-	$+$	K^+	Potassium ion
Calcium hydroxide	$Ca(OH)_2$	$\rightarrow$	$2OH^-$	$+$	Ca^{2+}	Calcium ion
Ammonium hydroxide	NH_4OH	$\rightarrow$	OH^-	$+$	NH_4^+	Ammonium ion

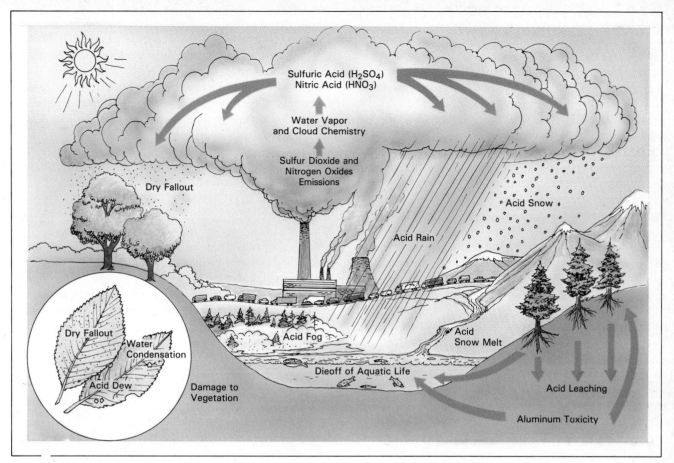

FIGURE 16–2
Acid deposition. Emissions of sulfur dioxide and nitrogen oxides react with the hydroxyl radical and water vapor in the atmosphere to form their respective acids, which come back down either as dry acid deposition or mixed with water, causing the precipitation to be abnormally acidic. Various effects of acid deposition are noted.

The higher the concentration of hydrogen ions in a solution, the more acidic the solution is.

The bitter taste and caustic properties of all alkaline, or basic, solutions are due to the presence of hydroxyl ions (OH^-, oxygen-hydrogen groups with an extra electron). Hence, a **base** is any chemical that releases OH^- ions when dissolved in water (Table 16–2).

The concentration of hydrogen ions is expressed as **pH.** The pH scale goes from 0 (highly acidic) through 7 (neutral) to 14 (highly basic) (Fig. 16–3). The numbers on the scale stand for the negative logarithm (powers of 10) of the hydrogen ion concentration expressed in grams per liter. For example, to say that a solution has a pH of 1 means that the concentration of hydrogen ions in the solution is 10^{-1} g/

FIGURE 16–3
The pH scale.

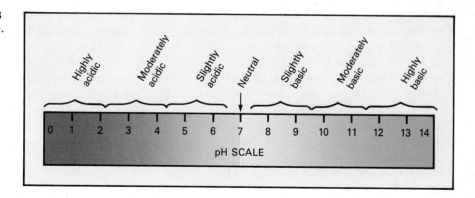

L (= 0.1 g/L); pH = 2 means that the hydrogen ion concentration is 10^{-2} g/L, and so on.

At pH = 7, the hydrogen ion concentration is 10^{-7} (0.0000001) g/L, but here the OH^- concentration is also 10^{-7} g/L. This is the neutral point, where small but equal amounts of H^+ and OH^- are present in pure water. The pH numbers above 7 continue to express the negative exponent of hydrogen ion concentration, but they also represent an increase in OH^- concentration. Because solutions above pH = 7 contain higher concentrations of OH^- than of H^+, they are referred to as basic solutions.

We use the same negative logarithm arrangement to express the concentration of hydroxyl ions: **pOH.** You may already know from your chemistry courses that there is a reciprocal relationship between pH and pOH. As the pH of a given solution goes up, its pOH goes down, and vice versa. In fact, we can state this as a rule: for any aqueous solution pH + pOH = 14. For example, pH = 13 means that the hydrogen ion concentration is 10^{-13} (a decimal followed by 12 zeros and a one) grams per liter. However, the OH^- concentration of this same solution is 10^{-1} grams per liter, and so the pOH = 1. Recall that at neutrality, equal amounts of H^+ and OH^- are present; pH = 7 and pOH = 7 (and again pH + pOH = 14).

Since numbers on the pH scale represent powers of 10, there is a *tenfold difference* between each unit and the next. For example, pH 5 is ten times more acidic (has ten times more H^+ ions) than pH 6; pH 4 is ten times more acidic than pH 5, and so on.

One easy way to measure pH is with indicator paper, which is available from any laboratory supply house. This paper contain pigments that change color when they are wetted with an acidic or a basic solution. The pH is determined by dipping a strip of indicator paper in the solution and matching the color of the wet paper with a color chart provided with the paper. Where accuracy and precision are important, however, it is necessary to use electronic instruments for measuring pH.

EXTENT AND POTENCY OF ACID DEPOSITION

In the absence of any pollution, rainfall is normally slightly acidic, pH 5.6, because carbon dioxide in the air readily dissolves in and combines with water to produce an acid called carbonic acid (Table 16–2). *Acid precipitation*, then, is any precipitation with a pH of 5.5 or less. Because dry acidic particles are also found in the atmosphere, the combination of precipitation and dry particle fallout is called *acid deposition.*

Unfortunately, acid precipitation is now the norm over most of the industrialized world. As Figure 16–4 shows, for instance, the pH of rain and snowfall over a large portion of the eastern United States and part of eastern Canada is typically about 4.5. Many areas in this region regularly receive precipitation having a pH of 4.0, and occasionally as low as 3.0. Fogs and dews can be even more acidic; in mountain forests east of Los Angeles, scientists found pH 2.8 fog water dripping from pine needles, almost 1000 times more acidic than normal.

THE SOURCE OF ACID DEPOSITION

Chemical analysis of acid precipitation reveals the presence of two acids, sulfuric acid (H_2SO_4) and nitric acid (HNO_3), at a ratio of about two to one. We saw in Chapter 15 that burning fuels produce sulfur dioxide and nitrogen oxides, and so the source of the acid deposition problem begins to become clear. These oxides enter the troposphere in large quantities from both anthropogenic and natural sources. Once in the troposphere, they are oxidized by the **hydroxyl radical** (an OH group with an electron missing) to sulfuric and nitric acids, which dissolve readily in water or adsorb on particles and are brought down to earth in acid deposition.

We must recognize that natural sources contribute substantial quantities of pollutants: 50–70 million tons per year of sulfur (from volcanoes, sea spray, and microbial processes) and 30–40 million tons per year of nitrogen oxides (from lightning, biomass burning, and microbial processes). Anthropogenic sources are estimated at 100–130 million tons per year of sulfur dioxide and 25–35 million tons per year of nitrogen oxides. The vital difference between these two sources is that anthropogenic oxides are strongly concentrated in industrialized regions, whereas the emissions from natural sources are well spread out over the globe. Even more important, levels of the anthropogenic oxides have increased at least fourfold since 1900, while levels of the natural emissions have remained fairly constant. As Figure 15–11 indicates, 21 million tons of sulfur dioxide are released into the air annually in the United States, 80 percent from fuel combustion (mostly from coal-burning power plants). Some 20 million tons of nitrogen oxides are released annually, 40 percent traced to transportation emissions and 56 percent to fuel combustion at fixed sites. For the eastern United States, the source of over 50 percent of the acid deposition has been identified as the tall stacks of 50 huge coal-burning power plants (Fig. 16–5). Recall from Chapter 15 that these tall stacks were built to alleviate the sulfur dioxide pollution at ground level. The unfortunate result of the "dilution solution" is that emitting sulfur dioxide and

FIGURE 16–4
Regions receiving acid deposition. Monitoring the pH of precipitation now reveals that acid deposition is occurring over most of the eastern United States and Canada. It is especially severe in the northeastern United States and eastern Canada.

FIGURE 16–5
(a) Standard smoke stacks of this coal-burning power plant were replaced by new 1000 foot (330-m) stacks to aid in the dispersion of pollutants into the atmosphere. The taller stacks alleviate local air pollution problems but create more widespread distribution of acid-generating pollutants. (b) Locations of the 50 largest sulfur dioxide emitters, all of which are utility coal-burning power plants. These facilities account for over 50% of the acid deposition falling on the eastern United States. (Grapes/Michaud/Photo Researchers.)

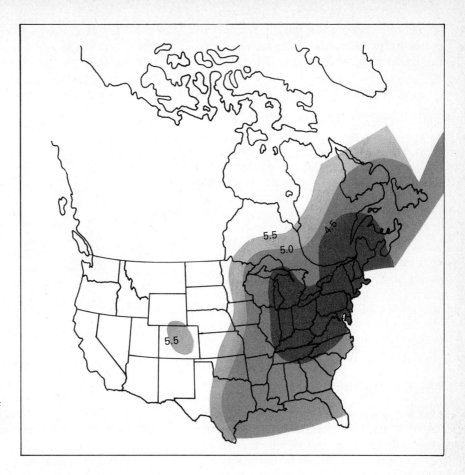

(a) (b)

nitrogen oxides high in the air simply provides more opportunity for them to convert to acids and spread hundreds of miles from the source.

THE EFFECTS OF ACID DEPOSITION

Acid deposition has been recognized as a problem in and around industrial centers for over 100 years, but effects on ecosystems were not noted until about 35 years ago, when anglers started noticing precipitous declines in fish populations in many lakes in Sweden, Ontario, and the Adirondack Mountains of upper New York State. Scientists in Sweden were the first to identify the cause as increased acidity of the lake water and to link this increased acidity with precipitation having an abnormally low pH. Since that time, while the ecological damage has continued to spread, studies have revealed many ways in which acid deposition affects and may destroy ecosystems.

Effects on Aquatic Ecosystems

The pH of an environment is extremely critical because it affects the functioning of virtually all enzymes, hormones, and other proteins in the bodies of all organisms living in the environment. Most freshwater lakes, ponds, and streams have a natural pH in the range of 6–7, and organisms are adapted accordingly. The eggs, sperm, and developing young of these organisms are especially sensitive to changes in pH. Most are severely stressed and many die if the environmental pH shifts as little as one unit from the optimum.

As such aquatic ecosystems are acidified, there is a rapid dieoff of virtually all higher organisms either because the acidified water kills them or because it keeps them from reproducing. As Figure 16–2 shows, acid deposition may leach aluminum and various heavy metals from the soil as the water percolates through. Normally, the presence of these elements in the soil does not pose a problem because they are bound in insoluble mineral compounds and therefore are not absorbed by organisms. As these compounds are dissolved by low-pH water, however, they may be absorbed and are highly toxic to both plants and animals. For example, mercury tends to accumulate in fish as lake waters become more acidic. Mercury levels are so high in the Great Lakes that many bordering states advise against eating sport fish caught in these bodies of water.

In Norway and Sweden the fish have died in at least 6500 lakes and 7 Atlantic salmon rivers. In Ontario, Canada, approximately 1200 lakes are now dead, and in the Adirondacks, a favorite recreational region for New Yorkers, more than 200 lakes are without fish, and many are devoid of all life save a few bacteria. The appearance of such lakes is deceiving. From the surface they are clear and blue, the outward signs of a healthy oligotrophic condition. However, a view under the surface is eerie. In spite of ample light shimmering through the clear water, there is not a sign of life; these waters are barren.

In addition to fish, acidified water affects the other organisms normally nourished by aquatic ecosystems: loons and other waterfowl, insect-eating birds, and many mammals.

During the 1980s, the Environmental Protection Agency conducted the National Acid Precipitation Assessment Program (NAPAP), at a cost of $600 million. Part of this program involved the National Surface Water Survey (NSWS), where samples were collected from 2311 lakes (representative of a total of 28300 lakes) and 500 streams (representative of 64300 streams). Acid deposition had acidified an estimated 1180 lakes and 4670 streams, mainly in the Adirondacks, the mid-Atlantic states, New England, northern Florida, and the upper Midwest.

As Figure 16–4 indicates, wide regions of the United States receive roughly equal amounts of acid deposition, and yet not all areas have acidified lakes. Many areas remain apparently healthy, while others have acidified to the point of becoming lifeless. How is this possible? The key lies in understanding the concept of buffering capacity. A system may be protected from pH change, despite addition of acid, by a **buffer**—a substance that, when present in a solution, has a large capacity to absorb hydrogen ions and thus hold the pH relatively constant.

Lakes and streams receive their water from rain and melted snow as they drain through soils in their drainage basins. If the soils are in regions of limestone ($CaCO_3$) rock, the lakes will contain dissolved limestone. Limestone is a natural buffer (Fig. 16–6); its presence protects lakes from the effects of acid deposition in many areas of the North American continent. The regions found in the NSWS survey to be sensitive to acid deposition are mostly in regions of granitic rock, which does not yield any chemical compounds that are good buffers.

Any buffer has limited capacity. Limestone, for instance, is used up by the buffering reaction and so is no longer available to react with more added H^+ (Fig. 16–6). Ecosystems that have already acidified and collapsed are those that had very little buffering capacity. Those that remain healthy have greater buffering capacity. However, many of these still-healthy lakes are gradually losing their buffering capacity as more and more acid continues to be deposited. In time, many healthy lakes will join the growing number of dead ones if acid deposition is not curbed.

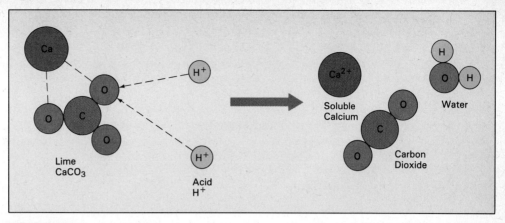

FIGURE 16–6
Buffering. Acids may be neutralized by certain nonbasic compounds called *buffers*. A buffer such as limestone (calcium carbonate) reacts with the hydrogen ions as shown. Hence the pH remains close to neutral despite the additional acid. Note, however, that the buffer is consumed by the acid. Limestone is the most widespread natural buffer.

Effects on Forests

Along with dying lakes, decline of forests has also become conspicuous. From the Green Mountains of Vermont to the San Bernadino Mountains of California, the dieback of forest trees is causing great concern. Forest decline is a very serious problem in major parts of Europe. Spruce and other coniferous trees seem especially vulnerable to the effects of air pollution. Commonly, the affected trees lose needles and often fall prey to insect and disease attack before they succumb. Most scientists are convinced that the combination of acid deposition and high ground-level ozone concentration is responsible for this dieback, since it is found in regions receiving heavy doses of these pollutants. Exact cause and effect has been difficult to trace, but recent experimental evidence shows that dieback involves periodic ozone damage to leaf surfaces and damage to root functions that results once acid deposition has upset soil chemistry. It is the latter that seems the more serious, as the soils become permanently changed by the constant input of acid pollutants from the atmosphere.

The effects of sulfuric and nitric acid deposition on soils are complex. Nutrients are leached from soil and humus and are washed away. Nitrate, from nitric acid, is readily taken up by roots and can stimulate unbalanced growth. Aluminum, a very common element in soil, is released into solution by acids and can disturb the normal ion ratios in soil water as well as directly inhibit plant root activities. Soil-buffering capacity may be used up, leading to rapid changes in soil chemistry. In a region of high acid deposition centered on the border between eastern Germany and Czechoslovakia, the woods died at an alarming rate. A 1980 study showed that 60 percent of the fir trees were healthy; two years later 98 percent were dead or dying. This rapid collapse could have been caused by exhaustion of the soil's buffering capacity.

As changes occur in tree populations, you can surmise the effect on wildlife populations. If a forest ecosystem collapses suddenly, the ramifying effects on soil erosion, sedimentation of waterways, flooding, and deterioration of water supplies could be quite serious. At the very least, we can expect a succession in which the dying trees are replaced by acid-loving species, but the variety of the latter type of plants is very limited. Most are mosses, ferns, and other scrubby plants that are economically worthless even for grazing.

Effects on Humans and Their Artifacts

One of the more noticeable effects of acid deposition is the deterioration of artifacts. Limestone and marble (which is a form of limestone) are favored materials for the outside of buildings and monuments (collectively called **artifacts**). The reaction between acid and limestone is causing these structures to weather and erode at a tremendously accelerated pace. Monuments and buildings that have stood for hundreds or even thousands of years with little change are now dissolving and crumbling away, as Figure 15–10 shows. The corrosion of buildings, monuments, and outdoor equipment by acid deposition costs billions of dollars for replacement and repair each year in the United States.

While the decay of such artifacts is a tragic loss

in itself, it should also stand as a grim reminder of how we are dissolving away the buffering capacity of ecosystems. In addition, some officials are concerned that acid deposition's mobilization of aluminum and other toxic elements may result in contamination of both surface and groundwater supplies. Increased acidity of water also mobilizes lead from the pipes used in some old plumbing systems and from the solder used in modern copper systems.

COPING WITH ACID DEPOSITION

As the song says, "What goes up, must come down." The sulfur and nitrogen oxides pumped into the troposphere at the rate of 41 million tons per year in the United States come down as acid deposition, generally to the east of their origin because of the way weather systems flow. The deposits cross national boundaries: Canada receives half of its acid deposition from the United States; Scandanavia gets most of its from Great Britain and other western European nations. There is now a broad consensus in the United States (with the notable exception of the utility industry, which is responsible for much of the problem) that something must be done about acid deposition.

Ways to Reduce Acid-forming Emissions

Scientists calculate that a 50 percent reduction in present acid-causing emissions would effectively prevent further acidification of the environment. This reduction would not correct the already bad situations, but natural buffering processes are estimated to be capable of preventing further deterioration. Because we know that about 50 percent of the acid-producing emissions comes from the tall stacks of coal-burning plants that generate electricity, control strategies center on these sources. Six main strategies have been proposed: (1) fuel switching, (2) coal washing, (3) fluidized bed combustion, (4) scrubbers, (5) alternative power plants, and (6) reductions in electricity consumption.

Economic factors do not favor the first two strategies. Low-sulfur coal exists in the western United States, but the cost of transporting 200 million tons of it per year to eastern power plants would be prohibitive. Oil is more expensive than coal, and switching to it as a fuel for generating electricity would make us even more dependent on foreign oil than we are now. Coal washing to remove sulfur is both economically and environmentally costly; large amounts of polluted water would be the outcome.

In fluidized bed combustion, coal is burned in a mixture of sand and lime; the mixture is kept churn-

ing (fluid) by forced air coming from underneath. The sulfur in the coal combines with the lime in the course of combustion and is removed with the ash. When new plants are built, this may be the preferred method of emissions control, but to install this process in existing plants would entail tearing down and rebuilding a major portion of each plant.

Scrubbers are "liquid filters." They involve putting the exhaust fumes through a spray of water containing lime (Fig. 16–7). The sulfur dioxide reacts with the lime and is precipitated as calcium sulfate ($CaSO_4$). Scrubbers have been required for all major power plants and smelters built since 1977, but the law has not required retrofitting the large, older plants in the U.S. Midwest that are the source of much of the acid rain plaguing the East and Canada. Plants in Europe and Canada and a few in the United States have demonstrated that scrubbers can be added onto existing power plants in a relatively short time; they are highly effective in controlling emissions, and they are not prohibitively expensive.

Strategies 5 and 6 of our list require major shifts in U.S. energy policy. Nuclear power represents the only current alternative technology for generating electricity and accounts for 21 percent of electrical power generation in the United States. However, as Chapter 22 explains, the future of nuclear energy is in serious question because of concerns about safety and the nuclear waste problem. Reducing electricity consumption makes very good sense but will involve major changes in many sectors of economic and social life in the nation, as explained in Chapter 21.

Political Developments

The decision to do something about acid deposition in the United States has been discouragingly slow in coming. The seriousness of the problem and the involvement of coal-burning power plants were clearly evident by the end of the 1970s. Indeed, in 1981 the National Academy of Sciences, after studying the problem, concluded that the circumstantial evidence linking power plant emissions to acid deposition was "overwhelming." Additional research and studies completed since that time have only added to the evidence. Yet no significant action occurred until 1990.

The problem has been one of different regional interests. Western Pennsylvania and the states of the Ohio River Valley, where older coal-burning power plants produce most of the electrical power, argued that controlling their sulfur dioxide emissions would make electricity in the region unaffordable. They also maintained that acid deposition was not well understood and that the problem was overemphasized. A coalition of politicians from these states, producers of

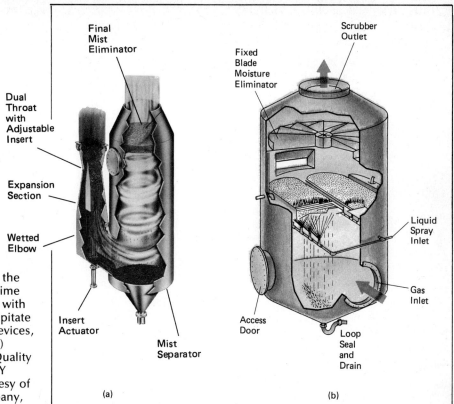

FIGURE 16–7
Scrubbers. Sulfur dioxide may be removed from flue gases by passing the furnace exhaust through a spray of lime and water. The sulfur dioxide reacts with the lime, and a calcium sulfate precipitate is removed. Two designs for such devices, called *scrubbers*, are shown here. (a) (Courtesy of FMC Corporation-Air Quality Control.) (b) A cutaway view of a SLY IMPINJET(TM) Gas Scrubber. (Courtesy of The W. W. Sly Manufacturing Company, Cleveland, Ohio.)

Figure labels (a):
Final Mist Eliminator
Dual Throat with Adjustable Insert
Expansion Section
Wetted Elbow
Insert Actuator
Mist Separator

Figure labels (b):
Scrubber Outlet
Fixed Blade Moisture Eliminator
Liquid Spray Inlet
Gas Inlet
Access Door
Loop Seal and Drain

high-sulfur coal, and the electrical power industry effectively blocked all attempts through the 1980s at passing legislation that would take action on acid deposition.

On the other side of the question were New York and New England, as well as most of the environmental and scientific community, who argued that it was both possible and necessary to address acid deposition and that the best way to do it was to control sulfur dioxide emissions. Cheering them on and applying diplomatic pressure were the Canadians. However, the Reagan administration of the 1980s took the position that the issues were not clear enough for action and promoted NAPAP to subject the problem to more study, a favorite delaying tactic of recalcitrant politicians.

Title IV of the Clean Air Act of 1990

The outcome is now history. President Bush promised to control acid deposition during the 1988 election campaign that brought him to power, and Congress finally passed the Clean Air Act of 1990. **Title IV** of this act is the first law in our history to address the acid deposition problem, and it does so by mandating reductions in both sulfur dioxide and nitrogen oxide levels. The major provisions of Title IV are:

1. By 2000, total sulfur dioxide emissions must be reduced 10 million tons below 1980 levels. This is the 50 percent reduction called for by scientists.

2. This reduction will be implemented gradually. One hundred and ten of the larger coal-burning power plants in the midwestern and eastern states are to install emissions-control devices by 1995. By 2000, a total of 1000 utility plants will have to have such devices. The utilities are required to install equipment that closely monitors their emissions of acid-generating gases.

3. In a major departure from the command-and-control approach described in Chapter 15, Title IV authorizes the EPA to use a free-market approach to regulation. Basically, each plant is granted "allowances" based on formulas in the legislation. The penalty for exceeding allowances is severe: a $2000 per ton fee and the requirement that the utility must compensate for the excess emissions the next year. A utility may not emit more sulfur dioxide than allowances permit, but if it emits less, the difference in allowance may be sold. Another utility may purchase the first plant's allowance difference in place of reducing its own emissions. Allowance trading is expected to become a major market activity; anyone may buy and sell allowances, including environmental groups and pri-

vate citizens. It is expected that this free-market approach will cut the utilities' compliance costs by $1 billion per year.

4. Beyond 1995, new utilities will not receive allowances; they must buy into the system by purchasing existing allowances. Thus, there will be a finite number of allowances in existence.

5. Nitrogen oxide emissions must be reduced by 2 million tons by the year 2000. This is to be accomplished by regulating the boilers used by the utilities and by mandating continuous monitoring of emissions.

In light of the passage of the Clean Air Act of 1990, Canada and the United States have signed a treaty whereby Canada will cut its sulfur dioxide emissions in half and cap them at 3.2 million tons by the year 2000. The economic fallout of Title IV will not be clear for a decade or more, but the immediate beneficiaries will be New York, New England, and eastern Canada. The ultimate beneficiaries will be the remaining healthy aquatic and forested ecosystems that will now be protected from future acid deposition. It is hoped that ecosystems already harmed will be able to recover from current damage.

Global Warming

Forty U.S. Senators recently wrote a letter to President Bush in which they stated that **global warming** represents "the single most important environmental issue of the century." Why were they writing to the president about an issue that he certainly is aware of? What is the evidence that their statement is true? Are we now on the threshold of a major human-caused change in the climate? If so, why is that a problem?

THE EARTH AS A GREENHOUSE

You are familiar with the way the interior of a car heats up when the car is sitting in the sun with the windows closed. This heating occurs because sunlight comes in through the windows and is absorbed by the seats and other interior objects. In being absorbed, the light energy is converted to heat energy, which causes the objects to become hot and to give off heat energy in the form of infrared radiation. Unlike sunlight, infrared radiation is blocked by glass and so cannot leave the car. The energy thus trapped causes the interior temperature to rise. This is the

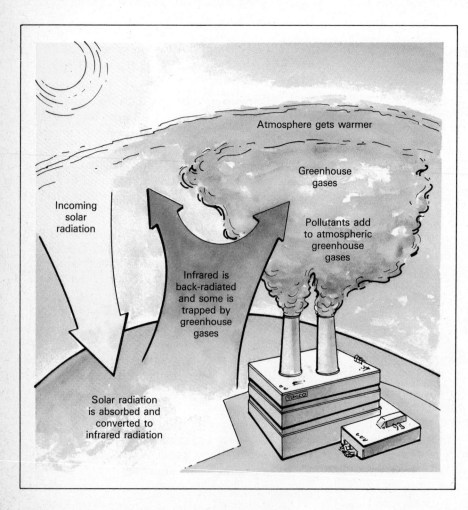

FIGURE 16–8
The greenhouse effect. Solar radiation is absorbed and converted to infrared radiation. As this radiates back through the atmosphere to outer space, some is absorbed by the greenhouse gases and insulates the earth, raising the temperature in the troposphere. Various pollutants add to the greenhouse gas content of the atmosphere, and more infrared radiation is trapped, leading to global warming.

same phenomenon that keeps a greenhouse warmer than the surrounding environment.

On a global scale, carbon dioxide, water vapor, and other trace gases in the atmosphere play a role analogous to the glass in a greenhouse. Therefore they are called **greenhouse gases.** Light energy comes through the atmosphere, is absorbed by the earth and converted to heat energy at the earth's surface. The energy, now in the form of infrared energy, radiates back upward through the atmosphere and into outer space. The greenhouse gases naturally present in the troposphere absorb some of the infrared radiation; other gases (N_2, O_2) in the troposphere do not. (Only the gases in the troposphere are important to our present discussion because the stratosphere is thermally isolated from the lower atmosphere and thus from the earth.) The greenhouse gases are like a heat blanket, insulating the earth but allowing the heat to escape eventually (Fig. 16–8). Without this insulation, average surface temperatures on earth would be 33°C colder and life as we know it would be impossible.

Our global climate is dependent on the concentrations of greenhouse gases. If these concentrations increase or decrease, our climate will change accordingly. Indeed, geological evidence indicates that the earth has undergone major climatic changes, fluctuating between ice ages and warmings over time spans of tens to hundreds of thousands of years. During the height of the most recent ice age, 18000 years ago, global temperatures were 3°–5°C colder than they are today. An analysis of gas bubbles trapped in ice from glaciers laid down during the ice age indicates that carbon dioxide concentrations then were 60 percent lower than those of today—a clear suggestion of cause and effect.

THE CARBON DIOXIDE STORY

The first alarm sounded in 1938. In an article entitled "The Artificial Production of Carbon Dioxide and Its Influence on Temperature," scientist G. Callendar reasoned that human use of fossil fuels had the potential to increase atmospheric carbon dioxide concentrations and if that happened, the climate could change. This suggestion was largely forgotten, however, until 1958, when C. Keeling began measuring carbon dioxide levels in Mauna Loa, Hawaii. The measurements continue to be recorded, and they reveal a striking increase in atmospheric levels (Fig. 16–9). Carbon dioxide concentrations increased exponentially until the energy crisis in the mid-1970s

FIGURE 16–9
Atmospheric carbon dioxide concentration. Carbon dioxide concentration in the atmosphere fluctuates between winter and summer because of seasonal variation in photosynthesis. The average concentration is gradually increasing owing to human activities, namely, burning fossil fuels and deforestation.

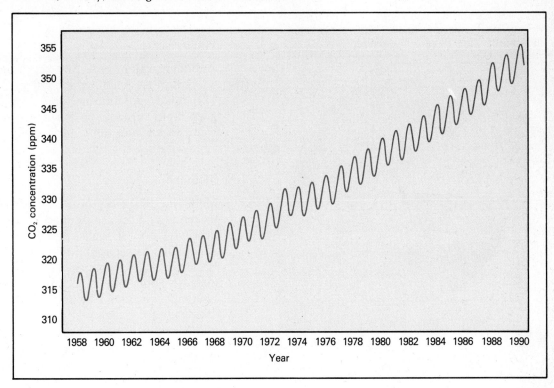

and linearly since then. The data also reveal an annual oscillation of 5 ppm, a reflection of photosynthesis and respiration in terrestrial ecosystems in the Northern Hemisphere. When respiration predominates (late fall through spring), levels rise; when photosynthesis predominates (late spring through early fall, levels fall). The most striking aspect, however, is the steady rise in carbon dioxide levels, which is continuing at 1.5 ppm/year. Carbon dioxide levels are now (1993) up to 360 ppm, 25 percent higher than they were before the Industrial Revolution. Our insulating blanket is thicker, and there is every reason to expect that it will have a warming effect.

As Callendar suggested long ago, the obvious place to look for the source of the increasing carbon dioxide levels is our use of fossil fuels. Every kilogram of fossil fuel (coal, liquid fuels derived from crude oil, and natural gas) burned results in the production of about 3 kg of carbon dioxide. (The mass triples as each carbon atom in the fuel picks up two oxygen atoms in the course of burning and becoming CO_2.) Currently, nearly 6 billion tons of fossil fuel carbon is burned each year, adding about 18 billion tons of carbon dioxide to the atmosphere (Fig. 16–10).

If all the carbon dioxide emitted from burning fossil fuels accumulated in the atmosphere, the concentration would rise by at least 3 ppm per year, not 1.5 ppm. Fortunately, the oceans are undersaturated with carbon dioxide, and there is now broad agreement that the oceans serve as a *sink* for much of the

carbon dioxide emitted. There are limitations to the ocean's ability to absorb carbon dioxide, however, because only the upper 300 m of the ocean exchanges gases with the atmosphere. (The deep oceans do mix with the upper layers, but only very slowly—a mixing time of about 1500 years.)

The seasonal swings obvious in Figure 16–9 show that the biota can influence atmospheric carbon dioxide levels. Can some of the "lost" carbon dioxide be traced to net biological uptake? Can we hope that the forests will take up increasing amounts of carbon dioxide as our emissions continue to rise? In reality, the forests are now serving as an additional *source* of carbon dioxide, not a sink! The reason: Forests are being cut and burned at a rate of 2 percent per year. It is estimated that the burning of forest trees is adding 1–2 billion tons annually to the 6 billion tons of carbon dioxide-carbon already coming from industrial processes. The net loss of forests, a serious biological concern in itself, therefore is also a cause for alarm when studied in the context of global warming.

OTHER GREENHOUSE GASES

Several other gases also absorb infrared radiation and add to the insulating effect of carbon dioxide (Table 16–3). Some of these gases are generated from anthropogenic sources and are increasing in concentration, raising the concern that future warming will extend well beyond the calculated effects of carbon dioxide alone.

Water Vapor

Although water vapor can trap infrared energy, its concentration in the troposphere is quite variable. Through evaporation and precipitation, water undergoes rapid turnover in the lower atmosphere, and water vapor does not tend to accumulate over time.

Methane

Methane (CH_4) is a product of microbial fermentative reactions and is also emitted from coal mines, gas pipelines, and oil wells. Because methane is generated in the stomachs of ruminants, animal husbandry is thought to be responsible for much of the methane increase appearing in the troposphere. This gas is gradually destroyed by reaction with the hydroxyl radical, but it is being added faster than it is being broken down. The concentration of methane has doubled since the Industrial Revolution, as revealed in ice corings.

FIGURE 16–10
Worldwide carbon dioxide emissions from fossil fuel burning. Total emissions in 1992 were approximately 18 billion metric tons of carbon dioxide.

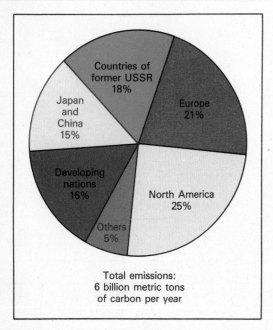

Total emissions:
6 billion metric tons
of carbon per year

TABLE 16–3

Gas	Avg. concentration 100 yr Ago (ppb)[a]	Approx. current concentration (ppb)	Avg. residence time in atmosphere (yr)
Anthropogenic Greenhouse Gases in the Atmosphere			
Carbon dioxide (CO_2)	290 000	360 000	100
Methane (CH_4)	900	1700	10
Nitrous oxide (N_2O)	285	310	170
Chlorofluorocarbons and other halocarbons	0	3	60–100

[a] Parts per billion

Nitrous Oxide

Nitrous oxide (N_2O) levels are also on the increase. Sources of nitrous oxide include biomass burning and use of chemical fertilizers; lesser quantities come from burning of fossil fuel. This is a particularly unwelcome emission because its long residence time (170 years) will be a problem in both the troposphere—where it contributes to warming—and the stratosphere—where it contributes to the destruction of ozone.

Chlorofluorocarbons and Other Halocarbons

These gases are entirely anthropogenic in origin and are increasing more rapidly than any other greenhouse gas. Like nitrous oxide, halocarbons are long-lived and contribute both to global warming in the troposphere and ozone destruction in the stratosphere. They are used as refrigerants, solvents, and fire retardants and have a much greater capacity (10000 times) for absorbing infrared radiation than does carbon dioxide.

Together these other greenhouse gases are estimated to trap about 60 percent as much infrared radiation as carbon dioxide. At the rate their tropospheric concentrations are rising, their impact will equal that of carbon dioxide within a decade or two.

AMOUNT OF WARMING AND ITS PROBABLE EFFECTS

It is certain that levels of carbon dioxide and the other greenhouse gases are increasing in the troposphere as a result of human activities. The greenhouse effect is also well established. How can we be *certain*, however, that the increase in greenhouse gases will bring on global warming? The short answer is that we can't. Many variables affect global temperatures, including reflection of solar radiation by cloud cover, changes in the sun's intensity, and levels of airborne particulate matter from volcanic activity. But the more informed answer to the question of future global warming is that all the evidence examined so far points to the *very high probability* that, as levels of greenhouse gases increase in the troposphere, global temperatures will indeed rise and the climate will undergo major changes as a consequence.

The Contributions of Modeling

Powerful computers have made it possible to include a great number of variables in the construction of models of global warming. One model, put together by a group of NASA scientists headed by James Hansen, electrified the scientific community in 1981 with its ability to track known temperature changes and link them to past and future carbon dioxide levels and global temperature changes. Hansen's model showed that the combination of carbon dioxide and volcanic emissions was responsible for most of the observed temperature changes in the last century. Results since 1981 have only confirmed the basic thrust of this model: We are living in a time of significant global warming, and what we are experiencing is consistent with what the NASA model predicts for the future (Fig. 16–11). A warming trend of more than 0.7°C over the last century has coincided with an increase of 30 percent in carbon dioxide concentration over the same period.

A number of such models have been used to predict the future. All agree that, if the concentration of greenhouse gases were to double, the earth would warm up between 1° and 5°C, with most estimates around 3°C. This doesn't sound like much, but consider that temperatures were only 5°C cooler during the last ice age, 18000 years ago. At that time, the ice sheets stretched across North America from New York through the Great Lakes states and all of Canada, with thicknesses up to a mile.

Impacts of Future Warming

Rising global temperatures are linked to two major impacts: **regional climatic changes** and **a rise in sea**

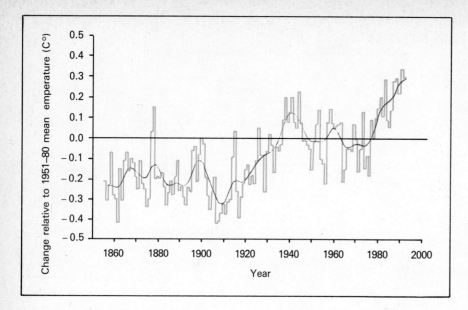

FIGURE 16–11
Annual surface temperature for the world, 1856–1991. Baseline comparison with 1951–1980 average. Solid line represents running average of 5 years. Warming trend for the past 17 years is conspicuous.

level. Both of these effects will likely occur in a matter of decades, requiring rapid and costly adjustments and involving major disruptions of natural ecosystems. The warming associated with a doubling of greenhouse gas levels is likely to be more pronounced in polar regions, as much as 10°C, and less pronounced in equatorial regions, 1°–2°C.

Warming will affect rainfall and agriculture seriously. The present-day difference in temperature between the poles and the equator is a major driving force for atmospheric circulation. Greater heating at the poles than at the equator will reduce this force, changing atmospheric circulation patterns as well as rainfall distribution patterns. Some regions of the world are likely to see an increase in rainfall; other areas, a decrease, as shown in Figure 16–12.

North Africa, which is largely desert at present, is likely to profit by increasing rainfall. However, the United States and Canada are likely to be losers. The central portion of North America is a major "bread-

FIGURE 16–12
Regions of the world that are likely to become wetter or drier as a result of the greenhouse effect. Predictions are based on computer models and are speculative.

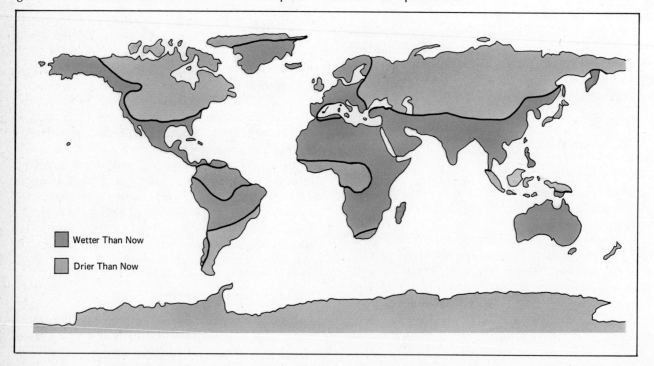

Wetter Than Now

Drier Than Now

basket" of the world, producing huge amounts of wheat and corn for world trade. Rainfall, often marginal for these crops, is likely to become much less. Quoting Walter Orr Roberts, former director of the National Center for Atmospheric Research,

The Dust Bowl of the middle 1930s in the United States was the greatest climatic disaster in the history of our nation. . . . [However] the Dust Bowl of the 30s may seem like children's play in comparison to the Dust Bowl of the 2040s. Because of the effects of the warming . . . natural rainfall may decline by as much as 40%, and the summers will be hotter, increasing the evaporation of soil moisture. The soils will desiccate, and the winds will lift them to the skies.*

Nor can irrigation be expected to provide much relief. Recall from Chapter 11 that the water table is already being drawn down to support agriculture through much of the region; by 2040, and probably considerably before then, most of the groundwater for the region will be exhausted.

The greatest difficulty, however, is not knowing what to expect. Already farmers lose an average of one in five crops because of unfavorable weather. As the climate shifts, the vagaries of weather will become more pronounced, and crop losses are likely to increase.

With global warming, sea level will rise because of two factors: thermal expansion as ocean waters warm and melting of glaciers and icefields. Sea level is already on the rise, at the rate of 1–1.5 mm/yr. Most of this rise is attributed to the global warming of the last century.

An enormous amount of water is stored in the world's remaining ice—enough to raise sea level by 75 m. Oceanographers and climatologists are not in agreement about the magnitude of sea level rise; most projections suggest a rise between 0.5 and 1.5 m during the next century. Even the lowest estimate will flood many coastal areas and make them much more prone to storm damage, forcing people to abandon properties and migrate inland. The highest estimate would cause utter disaster for most coastal cities, which are home to half of the world's population and its business and commerce. Are inland cities and communities ready to accommodate the billions of people that will be displaced? Are we ready to build dikes or modify all ports to accommodate the higher sea level? Note that this estimate of 0.5–1.5 m does not extend beyond the next century; the impact will certainly be greater beyond the year 2100.

* Roberts, Walter Orr, "It is Time to Prepare for Global Climate Changes," *Conservation Foundation Letter*, April, 1983.

COPING WITH GLOBAL WARMING

Is Global Warming Here?

There is so much natural variation in weather from year to year that trends are not always apparent. However, the unusually hot summer of 1988 and the accompanying drought in the United States (nearly 40 percent of the U.S. corn crop was lost) did much to draw attention to the possibility that we may already be experiencing global warming. Indeed, six of the hottest years on record occurred during the 1980s; the 1990 global temperature was the warmest ever recorded, and 1991 was the second warmest. How much evidence do we need before we agree with greenhouse scientist James Hansen when he claims that we are now well into global warming?

Energy Scenarios

Clearly, the potential for global warming is tied mainly to carbon dioxide emissions, which in turn are

FIGURE 16–13
Warming effects of different energy use scenarios. No growth implies continued output of carbon dioxide at present levels. Current energy use is increasing at 2 percent per year as indicated by the middle line. If global production of carbon dioxide increases at a rate of 4 percent per year, as occurred during the 1960s and 1970s, the temperature change would be dramatic.

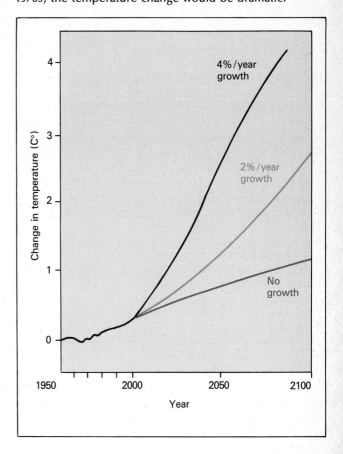

tied to fossil fuel consumption. Many scenarios have been drawn up to evaluate the warming impacts of different energy choices (Fig. 16–13). However, all scenarios agree that, unless the nations of the world cut back on emissions, we can expect the concentrations of greenhouse gases to double during the twenty-first century, most probably by 2050.

Political Developments

The world's industries and transportation networks are so locked into the use of fossil fuels that massive emissions of carbon dioxide and other greenhouse gases seem bound to continue for the foreseeable future. However, it is certainly possible to lessen the rate of addition and eventually bring about a sustainable balance. The means for accomplishing this lie in the international scientific and political arenas, and there has been an intensification of activity on these fronts in recent years.

A variety of steps have been suggested to combat global warming. The appropriate goal is *stabilization of the greenhouse gas content of the atmosphere.* Here are some of the suggestions:

1. Place a worldwide cap on carbon dioxide emissions by limiting the use of fossil fuels in industry and transportation.

2. Encourage the development of nuclear power, but deal effectively with the environmental problems.

3. Speed up international agreements to completely phase out the chlorofluorocarbons before the year 2000.

4. Stop the loss of tropical forests, and encourage tree planting over vast areas now suffering from deforestation.

5. Make energy conservation rules much more stringent (tighten building codes to require more in-

---

ETHICS
Staking a Claim on the Moral High Ground

The United States has rightly laid claim to the "moral high ground" in the matter of protecting the ozone layer. This country was the first to ban CFCs in spray cans and took the lead in bringing international accord on the Montreal Protocol. It is possible that our example may have motivated other nations to follow our lead in this matter.

Putting these actions in the context of morality means that we are concerned about doing what is right. Whether this is an ongoing concern or just an isolated case of enlightened self-interest played out on the international stage, remains to be determined. The reason? Global warming looms before us as a problem that is enormously more difficult than the CFC problem. If addressed effectively this larger problem will require major policy changes with far-reaching consequences. Taking action on ozone-depleting chemicals is child's play compared with what lies ahead if we are to act on global warming. Scientists estimate that it would take a 50 to 80 percent cut in the emissions of

greenhouse gases to stabilize the global climate. As the world's leading producer of greenhouse gas emissions, the United States is in a rather conspicuous and pivotal position. What we do will carry a great deal of weight in determining the outcome of negotiations.

The dilemma is a familiar one: short-term interests versus long-term gains. In the short run, our economy is geared to growth, and achieving a steady-state or—horrors—cutting back is extremely unpopular politically and economically. There is a long run, though, that young people can expect to see a good deal of, even if those in power will not live to see it. Twenty-first-century Americans as well as the rest of the world will bless us if we make the right choices. But time is running out, as the pressures of population growth and continuing development force increasingly greater energy production on top of an already damaging level.

To occupy the moral high ground, we have to deal with an existing confrontation between the industrial

world and the Third World. The industrial nations currently account for more than 70 percent of the greenhouse gas emissions. These have been called "luxury emissions" because they largely come from a way of life that is unavailable to the majority of the world. Most of the Third World contributions to greenhouse gases are in the form of methane from cows and rice paddies and carbon dioxide from burning forests—referred to as "survival emissions." The pathway to the moral high ground will involve a willingness to undergo a reduction in luxury emissions and a transfer of technology so that the Third World can reduce its survival emissions and also achieve the goals of economic development. Many of the industrialized nations (Japan, Australia, many European countries) have started out on that pathway. However, the world is watching while the United States decides where to position itself.

---

sulation, use energy-efficient lighting, and so forth).

6. Reduce the amount of fuels used in transportation by raising mileage standards, encouraging car pooling, and stimulating mass transit in urban areas.

Climatologist Stephen Schneider advocates the **tie-in strategy:** "Society should pursue those actions that provide widely agreed societal benefits even if the predicted change does not materialize." Thus, investing in more efficient use and production of energy reduces acid deposition, makes good economic sense, lowers the harmful health effects of air pollution, and lowers our dependency on foreign oil—and, of course, reduces carbon dioxide emissions.

People who are skeptical of global warming have achieved some prominence with suggestions that sun spots or other variations in solar activity are responsible for the observed warming. Their views were not taken seriously by a recent panel of 200 scientists (International Panel on Climate Change), who concluded that global warming is going to happen. It will involve at least a 2.5°C temperature rise by 2050 and a sea level rise between 8 and 29 cm by 2030, according to these scientists. Major shifts in the climate can be expected.

Unfortunately, although most skeptics are neither scientists nor climatologists, their views have had an impact on the political scene in the United States. Citing the uncertainties of global warming science, the Bush administration has been advocating a "cautious approach" that has left the United States isolated from the other industrialized nations. At an international conference convened to draw up a treaty for the 1992 U.N. Conference on Environment and Development, the United States refused to agree to a limit on carbon dioxide emissions, as did a group of Third World nations concerned that such a limit might stifle their economic growth. The proposal on the table for the 1992 U.N. conference is an agreement to stabilize greenhouse gas emissions at 1990 levels by the year 2000 or 2005. Eighteen European nations, Australia, and Japan have agreed to the U.N. conference proposal. Such a limit would clearly slow global warming.

One option given serious consideration by some Cornucopian thinkers is to do nothing about global warming. It isn't hurting anyone now and may never happen anyway. According to this view, there are such uncertainties surrounding the issue that we should not risk the negative economic impacts entailed in curbing emissions. Even if the predictions do come true, we shall have adequate time to adapt to the climate changes and rising sea levels, they argue. We can build dikes around our coastal cities as the Dutch have done. A warmer climate might not be so bad for people in the north. In fact, this "greenhouse hysteria" is another example of environmentalist terrorism (like bashing nuclear power plants, crying out about radon gas, and so on).

In response to these ideas, our purpose in this text is to bring you what we think is the most current and responsible reflection of scientific knowledge. It should be evident that we the authors are not in accord with the Cornucopian ideas reflected here or with those described elsewhere in the text. We hope that when you encounter such ideas, you will recognize their self-serving and one-dimensional nature. We believe that global warming is indeed the most challenging environmental issue facing our society today and that our response to it will reflect how serious we are about being stewards of the planet.

Depletion of the Ozone Shield

NATURE AND IMPORTANCE OF THE SHIELD

Radiation from the sun includes ultraviolet (UV) radiation along with visible light. *Ultraviolet radiation* is like visible light radiation except that UV wavelengths are slightly shorter than the wavelengths of violet light, which are the shortest wavelengths the human eye can see (Fig. 16–14). Ultraviolet radiation contains more energy than does visible light. On penetrating the atmosphere and being absorbed by biological tissues, UV radiation destroys protein and DNA molecules (this is what occurs when you get a sunburn). If the full amount of ultraviolet radiation falling on the stratosphere came through to the earth's surface, it is doubtful that any life could survive; plants and animals alike would simply be "cooked." Even the small amount (less than 1 percent) that does come through is responsible for all the sunburns and more than 200000 cases of skin cancer per year in the United States, as well as untold damage to plant crops and other life forms.

We are spared more damaging effects from ultraviolet rays because most of it (over 99 percent) is absorbed by ozone in the stratosphere (Fig. 16–1). Thus, ozone in the stratosphere is commonly referred to as the **ozone shield.**

The ozone we are speaking of here is the same molecule (O_3) described in Chapter 15 as a serious air pollutant. Recall that one definition of pollution is a chemical *out of place*. Ozone is out of place in the troposphere and thus is a pollutant there. In the stratosphere, however, ozone belongs; it is not a pollutant up there.

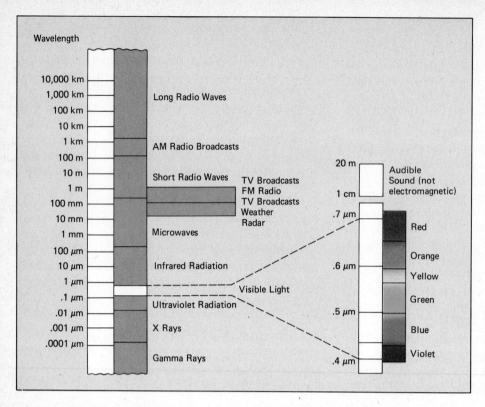

FIGURE 16–14
Ultraviolet, visible light, infrared, and many other forms of radiation are just different wavelengths of the electromagnetic spectrum.

IN PERSPECTIVE
Atmospheric Trouble from Air Traffic

Researchers have recently found unusually high concentrations of nitrogen oxides at altitudes between 5 and 9 miles above sea level. The source of these oxides is mainly air traffic: The normal cruising altitude for commercial airlines is 5–8 miles (9–13 km). This height is close enough to the tropopause to enable the gases to drift into the stratosphere. Recent work has shown that, in the stratosphere, nitrogen dioxide may break down ozone, adding to the impact of the CFC emissions. Further, nitrogen oxides and water vapor contribute to the formation of the stratospheric clouds implicated in the formation of the Antarctic ozone hole. Studies indicate that ozone depletion over the Northern Hemisphere may be twice as high as what would occur

with CFCs alone. Measurements above Europe have shown a 3 percent decrease in ozone between 6 and 10 miles and an 8 percent loss between 10 and 13 miles. These data point to the conclusion that much of the measured ozone losses above the temperate zone in the Northern Hemisphere is traceable to air traffic.

Of course, the bulk of the nitrogen oxides emitted from aircraft remain in the high troposphere, where they contribute to tropospheric ozone formation and global warming. Substances released so high in the troposphere tend to remain there at least 100 times longer than those released at ground level, due to less active cloud formation and precipitation at those heights. When these substances finally do

come down, they contribute to acid deposition. Thus, one activity—high-altitude air traffic—leads to ozone layer depletion, global warming, *and* acid deposition!

What can be done? Cruising altitudes should be kept below the tropopause, which can be as low as 6 miles during the winter. Plans to develop supersonic aircraft that cruise at higher than 9 miles should be scrapped forever. And air traffic should be subjected to the same scrutiny as all other forms of transportation, with attention given to the development of jet engines that emit only low levels of nitrogen oxides and to limitations on the volume of traffic. (From "Climate: Air Traffic Emissions," *Environment*, November 1991.)

FORMATION AND BREAKDOWN OF THE SHIELD

Ozone Formation

The need to maintain the ozone shield needs no elaboration. However, there are anthropogenic pollutants that are causing it to break down. Ozone in the stratosphere is a product of ultraviolet radiation acting on oxygen (O_2) molecules. The high-energy UV radiation causes some O_2 to split apart into free O atoms, and these atoms combine with O_2 to form O_3, as shown in Figure 16–15. All the O_2 is not converted to O_3, however, because free O atoms may also combine with O_3 molecules to form two O_2 molecules. Thus, the amount of ozone in the stratosphere is not static; it represents an equilibrium between these two reactions. The presence of other chemicals in the stratosphere can influence this equilibrium, as we have learned in recent years.

Chlorofluorocarbons in the Atmosphere

Chlorofluorocarbons (CFCs) are one type of halogenated hydrocarbon (Chapter 14). They are nonreactive, nonflammable, nontoxic molecules in which both chlorine and fluorine atoms replace some of the hydrogens. At room temperature, CFCs are gases under normal (atmospheric) pressure but they liquify under modest pressure, giving off heat in the process and becoming cold. When they revaporize, they reabsorb the heat and become hot. These attributes have led to their widespread use (several million tons per year) for the following purposes:

1. Chlorofluorocarbons are used in virtually all refrigerators, air conditioners, and heat pumps as the heat-transfer fluid. As these machines break down or are ultimately scrapped, their CFCs generally escape into the atmosphere.

2. A second major use is in the production of plastic

FIGURE 16–15
The ozone shield. (a) Ozone (O_3) in the stratosphere absorbs ultraviolet radiation (UV) from the sun. Without this protection, UV radiation could destroy most life on earth. (b) Ozone is formed in the stratosphere when UV radiation causes oxygen molecules to split into free oxygen atoms that may then combine with another oxygen molecule to form ozone. However, a free oxygen atom may also combine with an ozone molecule to produce two oxygen molecules. Consequently an equilibrium between oxygen and ozone is established and maintained.

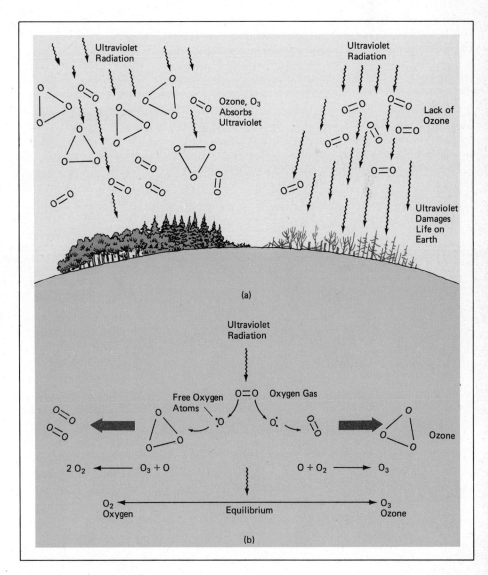

foams. Chlorofluorocarbons are mixed into liquid plastic under pressure (they are soluble in organic materials). When the pressure is released, the CFC gas causes the plastic to foam just as the carbon dioxide in a soda causes foaming when the pressure is released. After the foam is made, the CFCs escape to the air.

3. A third major use is in the electronics industry for cleaning computer parts, which must be meticulously clean. Again, spent CFCs escape into the air.

4. Finally, CFCs are still used in most countries other than the United States as the pressurizing agent for aerosol cans, which, of course, release them into the air.

In 1972 chemist Sherwood Rowland read a report describing the presence of CFCs in the atmosphere. Rowland and his colleague Mario Molina reasoned that, although CFCs would be stable in the trophosphere, in the stratosphere they would be subjected to UV radiation, which would break them apart. This reaction would release free chlorine atoms, which could attack stratospheric ozone to form chlorine monoxide (ClO) and O_2 (Fig. 16–16). Even worse, the chlorine is a **catalyst**: a chemical that promotes a chemical reaction without itself being used up by the reaction. Every chlorine atom may cause the breakdown of 100000 molecules of ozone, unless there are other chemicals in the atmosphere that bind the chlorine. Thus, CFCs are damaging in that they act as transport agents to get chlorine atoms into the stratosphere. The damage can persist because the chlorine atoms are removed from the stratosphere very slowly, and CFCs are continuously moving up into the stratosphere.

Rowland and Molina published a classic paper in 1974 that concluded that CFCs were damaging the stratospheric ozone layer through the release of chlorine atoms, and the result would be increased UV radiation and more skin cancer. The EPA was convinced that CFCs were a threat, and in 1978 banned their use in aerosol cans. Manufacturers in the United States quickly switched to nondamaging substitutes such as butane, and things were quiet for several years. Use of CFCs in applications other than aerosols continued, and critics demanded more convincing evidence of their harmfulness.

The Ozone "Hole"

In the fall of 1985, a British team of atmospheric scientists working in Antarctica reported a gaping "hole" in the ozone shield over the South Pole (Fig. 16–17). In an area the size of the United States, ozone levels were 50 percent lower than normal. The hole would have been discovered earlier by NASA satellites monitoring ozone levels, except that computers screening data were programmed to reject data showing a drop of 30 percent or more as instrument "hiccups." It had been assumed that the loss of ozone, if it occurred, would be slow, gradual, and uniform over the whole earth. This hole came as a surprise, and if it had occurred anywhere but over the pole, the results in UV damage would have been catastrophic. News of the ozone hole stimulated an enormous scientific research effort. The hole has reappeared every fall (spring in the Antarctic) and has been intensifying—exactly what would be expected if increasing levels of CFC were responsible. The recurring hole over the Antarctic may cause a significant destruction of marine phytoplankton. If it does, this loss will have a severe impact on virtually all Antarctic wildlife, from penguins to whales, because phytoplankton productivity is the basis of the food chains there.

A unique set of conditions is responsible for the hole. Basically, the coming of the Antarctic winter in June creates a vortex (like a whirlpool) in the strato-

FIGURE 16–16
Destruction of the ozone shield. Chlorfluorocarbons introduced into the troposphere work their way into the stratosphere, where they are attacked by ultraviolet radiation and release free chlorine atoms, which catalyze the breakdown of ozone. Not all the ozone is destroyed because it keeps being produced, but the oxygen-ozone balance is shifted to a lower ozone concentration.

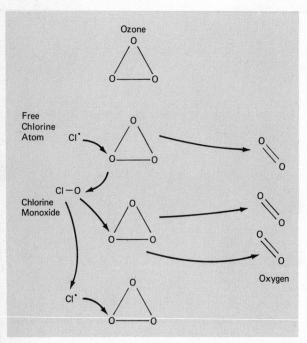

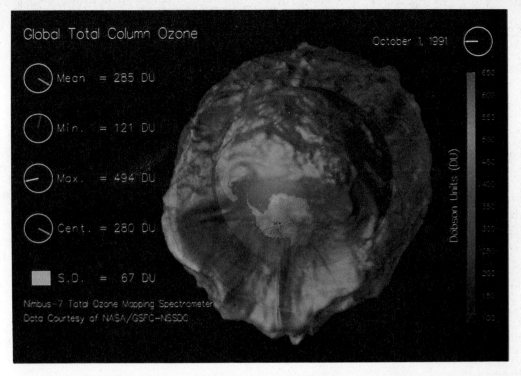

FIGURE 16–17
Computer-based visualization of stratospheric ozone in the Southern Hemisphere, October 1, 1991. Data source is NASA's Nimbus-7 spacecraft. The visualization clearly shows the pronounced ozone depletion (the ozone hole) that occurs during the Antarctic spring as a result of the winter vortex and the ozone depleting chemical reactions occurring there. (NASA/IBM/Treinish.)

sphere that confines stratospheric gases within a ring of circulating air. The extremely cold temperatures of the Antarctic winter cause the small amounts of moisture and other chemicals present in the stratosphere to form clouds. In the summer, when the clouds are not present, gases such as nitrogen dioxide and methane react with chlorine monoxide and chlorine to "trap" the chlorine, keeping ozone depletion low. When the clouds are present, though, in winter, cloud particles provide surfaces on which chemical reactions occur and release molecular chlorine (Cl_2) from its chemical traps. When sunlight returns to the Antarctic in the spring, the warmth of the sun breaks up the clouds and UV light attacks molecular chlorine, releasing free chlorine and initiating the chlorine monoxide catalytic cycle, which rapidly destroys ozone. By November, the beginning of the Antarctic summer, the vortex breaks down, and ozone-rich air returns to the Antarctic. By this time, however, ozone-poor air is spread all over the Southern Hemisphere. Shifting patches of ozone-depleted air have caused UV radiation increases of 20 percent above normal in Australia, where television stations now report daily UV readings and warnings for Australians to stay out of the sun. Scientists in Chile have reported increases in skin allergies and sunburn among schoolchildren, and cataracts in sheep, all traced to increases in UV radiation.

Early in 1992, researchers monitoring the stratosphere over the Arctic found record-high levels of chlorine monoxide. They concluded that the level of Arctic ozone is poised for a fall if a polar vortex forms.

If an ozone hole develops over the Arctic, it will be far more serious than the Antarctic hole because ozone-depleted air will extend outward over highly populated regions of North America, Europe, and Asia.

Ozone losses have not been confined to the polar regions, although they are most spectacular there. Recent reports from EPA and the U.N. Environment Program reveal losses during the 1980s of 3 to 5 percent all across the temperate and tropical zones of both hemispheres. That is triple the losses that occurred in the 1970s. The depletion is now known to be present during the summer, when UV radiation is strongest and can do the most damage to people and crops. A further loss of at least 3 percent is expected during the 1990s. The EPA has calculated that the 1980's losses will have caused 12 million people in the United States to develop skin cancers over their lifetime, and that 93000 of these cancers are expected to be fatal.

COMING TO GRIPS WITH OZONE DEPLETION

International Agreements

Under its Environmental Program, the United Nations convened a meeting in Montreal, Canada, in 1987, to address ozone depletion. Member nations reached an agreement, known as the **Montreal Protocol,** to scale CFC production back 50 percent by 2000; to date, 68 nations (including the United States) have signed the treaty.

The 1987 Montreal Protocol and its 1990 amendment are a remarkable and encouraging development in human affairs. Most of the nations of the world have agreed to take steps that are economically costly in order to protect a global resource—the ozone shield. This agreement was reached on the basis of research from the scientific sector, and it was a response not to an immediate problem but to one that is expected to occur in the future. The protocol is a statement that today's generation actually has some concern about future generations.

There are some important lessons from the protocol that should not be missed as the nations of the world begin to fashion an international policy on global warming:

1. It is significant that such broad-based cooperation can be forthcoming from the collection of nations that are most accustomed to putting national self-interest at the top of their agendas.

2. It is quite apparent that scientists must continue to play a role in the crafting of policy. It was their work that first drew attention to the ozone losses and the global warming complex. Scientists have continued to be outstanding advocates for action based on knowledge that they have accumulated. The development of consensus among the scientists has been likewise remarkable.

3. Governments must be bold enough to act even while scientific certainty is lacking. By its very nature, the scientific method cannot provide certainty short of an actual observation. At some point, delaying a decision becomes an immoral act. The costs of such delay may be enormous.

4. Leadership by the United States was crucial in bringing about the Montreal Protocol; such leadership may be just as essential in moving to a treaty on global warming.

5. The United Nations, through its Environment Program, played a crucial role in the ozone accord. It will likely be equally important in the global warming negotiations.

(From R. Benedick, "Lessons from the Ozone Hole," *EPA Journal*, March/April 1990.)

The protocol was written before CFCs were so clearly implicated in driving the ozone destruction and before the threat to Arctic and temperate zone ozone was recognized, however. Because ozone losses during the late 1980s were greater than expected, an amendment to the protocol was adopted in June 1990. The amendment requires protocol nations to completely phase out the major chemicals destroying the ozone layer by 2000 in developed countries and by 2010 in developing countries. There is concern that even this may not be strict enough. The problem is that, even with an immediate total ban on new CFC production, there is so much of these chemicals already present in existing refrigerators and air conditioners that normal breakdown of the units will increase atmospheric CFC levels for some years. Still, an immediate total ban is the best we can do, and for this reason many people are calling for one.

Action in the United States

The United States is by far the leader in production and use of CFCs and other ozone-depleting chemicals, with du Pont Chemical Company being the major producer. To its credit, du Pont pledged in 1988 to phase out production by 2000. In late 1991, a spokesperson announced that, in response to new data on ozone loss, the company would accelerate its phaseout by 3 to 5 years. Company scientists are well along in their development of suitable substitutes. Many of the large corporate users of CFCs have announced goals for a complete phaseout by 1994 (AT&T, IBM, Northern Telcom, for example). Clearly, much can be done in the private sphere to bring us quickly to a total ban.

The Clean Air Act of 1990 also addresses this problem, in **Title VI**, "Protecting Stratospheric Ozone." Title VI is a comprehensive program that restricts production, use, emissions, and disposal of a whole family of chemicals that have been identified as ozone-depleting. For example, it calls for a phaseout schedule for the hydrochlorofluorocarbons (HCFCs), a family of chemicals that will be used as less damaging substitutes for CFCs until nonchlorine substitutes are available. The act also regulates the servicing of refrigeration and air-conditioning units. The EPA's hand has been substantially strengthened by Congress, and we can hope that the agency will continue to be aggressive in taking effective action to protect the ozone layer.

There is a provision in Title VI requiring the president to accelerate the phaseout schedule for ozone-depleting chemicals if new data warrant it. Fol-lowing the recent news from atmospheric scientists that the Arctic appears poised for the development of an ozone hole, President Bush announced in Feb-ruary 1992 a speedup of the phaseout deadline. Ac-cording to his directive, U.S. chemical companies must halt all CFC production by December 31, 1995.

 Review Questions

1. What are the important characteristics of the stratosphere? Of the troposphere?
2. What is the difference between an acid and a base?
3. What is the pH scale? The pOH scale? How are they related?
4. What two major acids are involved in acid dep-osition? Where does each come from?
5. How can a shift in environmental pH affect aquatic ecosystems?
6. In what other ecosystems can acid deposition be observed? What are its effects?
7. Discuss several strategies for controlling acid dep-osition. Which are considered by ecologists to be the most effective?
8. What provisions were included in the Clean Air Act of 1990 to address the problem of acid dep-osition?

9. Describe the heat-trapping effects of carbon diox-ide.
10. Which of the greenhouse gases are the most sig-nificant contributors to global warming?
11. What are four possible impacts of rising global temperature?
12. What steps could be taken to stabilize the green-house gas content of the atmosphere?
13. How is the ozone shield formed? What causes its breakdown?
14. Describe four sources of CFCs entering the strat-osphere.
15. How do CFCs affect the concentration of ozone in the stratosphere?
16. What causes the formation of an ozone "hole"?
17. Discuss several efforts that are currently under way to protect our ozone shield.

 Thinking Environmentally

1. What arguments have the utility industries used to delay action on acid deposition? Evaluate the validity of their concerns.
2. Compile a list of ways you are producing carbon dioxide. What steps could you take to decrease this production?
3. In your opinion, does the greenhouse effect exist? Defend your stance.

4. How would a ban on chlorofluorocarbon produc-tion change the way you live? Give your reasoning for or against such a ban.
5. The Swedish government has made efforts to at-tack the symptoms of acid deposition directly by trying to neutralize acidic lakes with lime. Discuss the pros and cons of this method.

17

Risks and Economics of Pollution

LEARNING OBJECTIVES

When you have finished studying this chapter, you should be able to:

1. Trace the origins of cost-benefit analysis.

2. Explain how cost-benefit analysis addresses externalities.

3. Understand how pollution-control costs are generated and what the magnitude of those costs is nationally.

4. List the most important benefits of environmental regulation.

5. Evaluate the cost effectiveness of pollution control.

6. Discuss shadow pricing and its application to human life and the environment.

7. Understand some of the complexities in comparing costs and benefits.

8. Define hazard, risk, and risk analysis.

9. Describe the four steps in the process of risk analysis at the EPA.

10. Evaluate the discrepancy between the public's perception of a risk and scientists' evaluation of that risk.

11. Describe the elements of the outrage component of public risk perception.

12. Understand the connection between risk perception and public policy.

13. Explain the factors that determine how regulatory decisions are made.

14. Evaluate comparative risk analysis as a new strategy for the EPA.

Chapters 12–16 show the many serious pollution problems we face. These chapters also show that rising public concern has led to a network of laws and agencies to deal with our interactions with the environment. The political process in the United States has led to the establishment of public policies that are administered by regulatory agencies such as the Environmental Protection Agency and the Food and Drug Administration. Environmental public policy is now a major factor in decision making at all levels of government, from the approvals for new building lots to international affairs. Environmental regulation represents a large cost outlay for businesses and governments alike, causing some concern that we might be spending too much to regulate problems that are not as serious as some environmentalists say. How clean does our air need to be? How much risk to human and environmental health is really present in our air, water, and food? Have we come to the point where environmental concerns are being overemphasized and our economic growth stifled?

This chapter focuses on public policy and environmental concerns as they have been influenced by the disciplines of economics and the natural sciences. Our aim is to address two important questions. The first—*How can we measure the cost-effectiveness of pollution regulation?*—leads us into cost-benefit analysis, a major tool of the new discipline of environmental economics. The second—*How can we be sure that our public policies are providing adequate protection for humans and the environment?*—leads us to risk analysis, which brings scientific judgment to bear on the regulatory process. These are highly important matters, for they reflect the way in which our political process is responding to environmental problems.

Cost-Benefit Analysis

In the first decade of strong environmental regulation—the 1970s—most environmental policies were developed with little consideration of economics. Early policies were typical command-and-control responses to air and water pollution, focusing on controlling emissions from cars and factories, releases from sewage outfalls and other point sources, and pesticides being sprayed over large areas. The policies undoubtedly had a significant economic impact on businesses, consumers, and the workforce. Concerned that the U.S. economy in general and business in particular might be overregulated and thus unduly restricted, President Ronald Reagan issued Executive Order 12291 in February 1981. This order required all executive departments and agencies to support every new *major* regulation with cost-benefit analysis. To qualify as *major*, a regulation must meet at least one of three requirements: it must (1) impose annual costs of at least $100 million, or (2) cause a significant increase in costs or prices for some sector of the economy or geographic region, or (3) have a significant adverse effect on competition, investment, productivity, employment, innovation, or the ability of U.S. firms to compete with foreign firms.

The order had the effect of applying an economic tool to many decisions on public policies dealing with the environment. Let us take a closer look at this process.

COSTS VERSUS BENEFITS

A **cost-benefit analysis** compares the estimated costs of a project with the benefits that will be achieved. Such analysis is often used as a means of rationally deciding whether to proceed with a given project. All costs and benefits are given monetary values and compared. Such comparison is commonly referred to as the **benefit-cost** (or cost-benefit) ratio. A favorable ratio for a project means that the benefits outweigh the costs. Such a project is said to be **cost-effective,** and there is economic justification for proceeding with it. Analysis usually involves considering several options for accomplishing the project and selecting the option with the best benefit-cost ratio. If costs are projected to outweigh benefits, the project may be revised, dropped, or shelved for later consideration.

Cost-benefit analysis in environmental issues is intended to build efficiency into policy so that society does not have to pay more than necessary for a given level of environmental quality. If the analysis is done properly, it will take into consideration *all* of the costs and benefits associated with a regulatory option. In so doing, cost-benefit analysis addresses the problem of *externalities*.

In the language of economics, an **externality** is an effect of the business process not included in the usual calculations of profit and loss. For example, when a business pollutes the air or water, this pollution imposes a cost on society in the form of poor health or the need to treat water before using it. This is an *external bad*. As another example, when workers improve job performance as a result of experience and learning, this improvement is not credited on the company's ledgers. It is considered as an *external good*.

Amenities such as clean air, uncontaminated groundwater, and a healthy ozone layer are not privately owned and therefore can be degraded without the need to compensate any owners. In the absence of regulatory controls, there are no direct costs to a business for degrading these amenities (in other words, they are externalities). Therefore there are no incentives to refrain from polluting them.

Because cost-benefit analysis includes *all* of the costs and benefits of a project or a regulation, it effectively brings the externalities into the economic accounting—if it is done properly.

THE COSTS OF ENVIRONMENTAL REGULATIONS

The costs of pollution control include the price of purchasing, installing, operating, and maintaining pollution control equipment and implementing a control strategy. Even the banning of an offensive product costs money because jobs are lost, new products must be developed, and machinery may have to be scrapped. In some instances, a pollution-control measure may result in the discovery of a less expensive way of doing something. However, such money-saving controls are relatively rare. In most cases, any form of pollution control costs money. Thus, the effect of most regulations is to prevent an external bad by imposing economic costs that are ultimately shared by government, business, and consumers.

Pollution-control costs generally increase exponentially with the level of control to be achieved (Fig. 17–1a). That is, a partial reduction in pollution may

be achieved by a few relatively inexpensive measures. However, further reductions generally require increasingly expensive measures, and 100 percent control is likely to be impossible at any cost. Because of this exponential relationship, regulatory control often has to focus on stimulating new ways of reducing

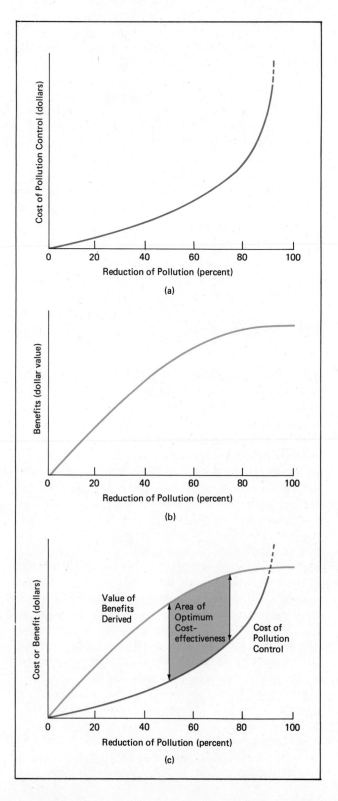

FIGURE 17–1
The cost-benefit ratio of pollution reduction. (a) The cost of pollution control increases exponentially with the degree of control to be achieved. (b) However, additional benefits to be derived from pollution control tend to level off and become negligible as pollutants are reduced to near or below threshold levels. (c) When the curves for costs and benefits are compared, we see that the optimum cost-effectiveness is achieved at less than 100 percent control. Expenditures to achieve maximum reduction may yield little if any additional benefit and hence may be cost-ineffective.

384

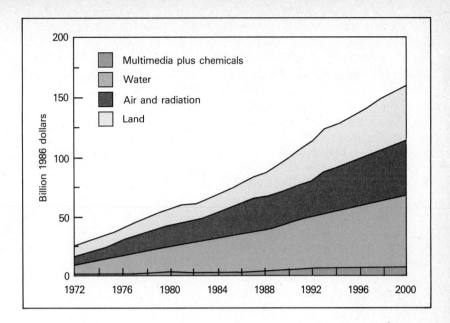

FIGURE 17–2
Total yearly costs of pollution control in the United States, assuming full implementation of existing regulatory laws. (From EPA, *Environmental Investments*, 1990.)

pollution. Indeed, the costs of pollution control represent a powerful incentive to make substitutions, to recycle materials, or to redesign industrial processes.

A 1990 EPA report on the costs of a clean environment considered costs from all pollution sources and for all environments (air, radiation, land, water, chemicals, and so on) and projected those costs to the year 2000 (Fig. 17–2). For 1993, the projected costs are $125 billion, approximately 2.2% of our gross national product. Thus, the costs of improving environmental quality represent a major economic outlay. What benefits have we received in return?

THE BENEFITS OF ENVIRONMENTAL REGULATION

The benefits of regulatory policies are seldom as easy to calculate as the costs. Estimating benefits is often a matter of estimating the costs of damages that *would* occur if the regulations were not imposed. For example, the projected environmental damage that would be brought on by a given level of sulfur dioxide emissions from a coal-fired power plant (an external bad) becomes a benefit (an external good) when a regulatory action cuts those emissions in half. Benefits include such things as improved public health, reduced corrosion and deterioration of materials, reduced damage to natural resources, preservation of aesthetic and spiritual benefits, increased opportunities for outdoor recreation, and continued opportunities to use the environment in the future. The dollar value of these benefits is derived by estimating, for example, the reduction in health-care costs, the reduction in maintenance and replacement costs, and the economic value generated by the enhanced rec-

reational activity. Examples of potential benefits are listed in Table 17–1.

The relationship between pollution reduction and benefit value is very different from the relationship between reduction and costs shown in Figure 17–1a. Significant benefits are frequently achieved by modest degrees of cleanup. As cleanup approaches

TABLE 17–1

Benefits That May be Gained by Reduction and Prevention of Pollution

1. Improved human health
   Reduction and prevention of pollution-related illnesses
   Reduction of worker stress caused by pollution
   Increased worker productivity
2. Improved agriculture and forest production
   Reduction of pollution-related damage
   More vigorous growth by removal of pollution stress
   Higher farm profits benefiting all agriculture-related industries
3. Enhanced commercial and/or sport fishing
   Increased value of fish and shellfish harvests
   Increased sales of boats, motors, tackle, and bait
   Enhancement of businesses serving fishermen
4. Enhancement of recreational opportunities
   Direct uses such as swimming and boating
   Indirect uses such as observing wildlife
   Enhancement of businesses serving vacationers
5. Extended lifetime of materials and cleaning
   Reduction of corrosive effects of pollution extending the lifetime of metals, textiles, rubber, paint, and other coatings
   Cleaning costs reduced
6. Enhancement of real estate values
   Real-estate values depressed in polluted areas
   Reduction of pollution will enhance them

100 percent, little if any additional benefits may be realized (Fig. 17–1b). This follows from the fact that organisms can often tolerate a threshold level of pollution without ill effect (see Fig. 15–1). Therefore, reducing a pollutant to below threshold levels will not yield an observable improvement.

COST EFFECTIVENESS

Figure 17–1c makes it clear that, with modest degrees of cleanup, benefits outweigh costs. As cleanup efforts move toward the 100 percent mark, however, the lines cross, and costs exceed the value of benefits. Consequently, while it is tempting to argue that we should strive for 100 percent control, demanding upwards of 90 percent control may involve enormous costs with little or no added benefit. At the point when control of a particular pollutant reaches 90 percent, it makes more sense to allocate dollars and effort to other projects where greater benefits may be achieved for the money spent. Optimum cost effectiveness is achieved at the pollution-reduction point where the benefit curve is the greatest distance above the cost curve.

What has been the result of cost-benefit analyses to date? Air and surface-water pollution reached critical levels in many areas of the United States in the late 1960s, and since that time huge sums of money have been spent on pollution abatement. Cost-benefit analysis shows that, overall, these expenditures have more than paid for themselves in decreased health-care costs and enhanced environmental quality. Does this mean that further expenditures of pollution control will prove equally cost-effective? Or are we at the point where further expenditures will yield little if any benefit, and the money will be effectively wasted? At the very least, industry, many economists, and government officials now demand more documentation of presumed benefits before consenting to further expenditures.

Some observers believe that this demand represents real questions about the cost effectiveness of additional expenditures for environmental protection. Others believe that it represents a more sophisticated method of protecting economic self-interests at the expense of environmental quality and society at large. To understand these views, we must take a more detailed look at cost-benefit analysis.

Problems in Performing Cost-Benefit Analysis

The concept behind cost-benefit analysis is relatively straightforward. Simply estimate the costs that may be incurred and the value of benefits that may be derived, and compare the two numbers. The difficulty lies in obtaining realistic estimates and in making objective comparisons.

ESTIMATING COSTS

In most cases, pollution-control technologies and strategies are understood and available. Thus, equipment, labor, and maintenance costs can be estimated fairly accurately. Unforeseen problems that increase costs may occur, but as technology advances and becomes more reliable, experience is gained, lower-cost alternatives frequently emerge, and such unforeseen increases are negligible. The costs of pollution control are likely to be highest at the time they are initiated and then decrease as time passes (Fig. 17–3a). The importance of this trend will become evident when we consider the time span over which costs and benefits are compared.

ESTIMATING BENEFITS

The value of some benefits can also be estimated fairly accurately. For example, it is well recognized that air pollution episodes cause increases in the number of people seeking medical attention. Since the medical attention provided has a known dollar value, eliminating air pollution episodes provides a health benefit of that value. As another example, consider a polluted lake that is upgraded to the point where it will again support water recreation. The benefits of this are estimated by assigning a value of $3 to $5 to each anticipated swimmer-day. This figure is based on the fact that most people will pay this price for admission to a pool.

Many benefits however, are difficult to estimate. Accurate cost-benefit analysis depends on assigning monetary values to every benefit, but how can a dollar value be put on maintaining the integrity of a coastal wetlands, for example, or on the enjoyment of breathing cleaner air? The answer: Find out how much people are willing to pay to maintain these benefits. How can this be done if there is no free market for the benefits? Again, economists have an answer: **shadow pricing.** Shadow pricing involves asking people what

they *might* pay for a particular benefit, if it were up to them. For example, in assessing benefits of cleaner air in the Los Angeles basin, homeowners were asked to place a value on improving their air quality from

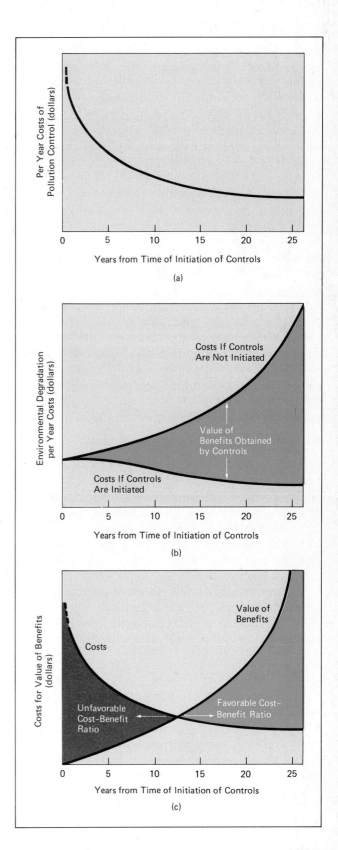

FIGURE 17–3
Evaluation of the optimum cost effectiveness of pollution control expenditures changes with time. (a) Pollution-control strategies generally demand high initial costs. Costs generally decline as those strategies are absorbed into the overall economy. (b) Benefits may be negligible in the short term, but they increase and continue to accrue as environmental and human health recover from the impacts of pollution or are spared increasing degradation. (c) When these two curves are compared, we see that what may appear as cost-ineffective expenditures in the short term (5–10 years) may, in fact, be very cost-effective expenditures in the long term.

poor to *fair*. The average response (in 1977) was $30 a month. Total benefit was then calculated from the number of households in the basin times the average value.

The Value of Human Life

Shadow pricing becomes difficult when the analysis has to place a value on human life. Many of the pollutants to which people are exposed are hazardous; they exact a toll on health and mortality. To estimate the benefits of regulating such pollutants, it is necessary to calculate how many lives will be saved or how many people will enjoy better health. Finding a value for these benefits is fraught with ethical difficulties.

One approach compares how much we pay people to perform hazardous jobs with how much we pay for the same kind of work without the hazards. Another approach asks hypothetical questions, such as, What is the minimum pay you will demand to accept a risk of 1 in 50 that you will get cancer from working with chemical A? Such approaches have resulted in a range of values for human life that covers two orders of magnitude, from tens of thousands of dollars to $10 million. Any cost-benefit analysis that must factor in risk to human life requires that human lives be valued somewhere in this very broad range. Obviously, the outcome of the analysis is determined by the value selected.

Nonhuman Environmental Components

How does shadow pricing work for the nonhuman components of natural environments—for a population of rare wildflowers, for example, or a wilderness site? Such entities depend entirely on how willing people are to pay for their preservation. Again, monetary values must be assigned to their existence, and again the outcome will be strongly influenced by a very subjective element in the analysis.

It is encouraging that more and more people recognize the importance of nonhuman environmental components and clearly value them highly, even if it is difficult to express their value in dollars.

PROBLEMS IN COMPARING COSTS AND BENEFITS

Even after valid cost and benefit estimates are obtained, the comparison is often a complicated matter. We have seen that, during the initial stages of control, costs are high, and observable benefits are usually few or none. As time passes, however, costs generally moderate while benefits increase and accumulate.

Consequently, whether benefits outweigh costs or vice versa depends on whether one takes a short-term or a long-term view. A situation that appears to be cost-ineffective in the short term may prove extremely cost-effective in the long term (Fig. 17–3c). This is particularly true for pollution problems like acid rain or groundwater contamination from toxic wastes. In these instances, the consequences of delaying control may seriously affect large geographic areas and many millions of people, and they may be irreversible.

Those who bear the costs of pollution control and those who receive most of the benefits are frequently different groups of people. For example, industry and its shareholders may bear the costs of curtailing effluents into a river, while people who enjoy sports fishing gain the benefits. Obviously, the two parties are more than likely to reach different conclusions regarding whether benefits outweigh costs. Thus, a spokesman from the American Petroleum Institute called some recent environmental regulations unwise and inefficient and claimed that U.S. industries would be hurt by them. The Clean Air Act of 1990, groundwater protection laws, and the ongoing resistance to opening up the Arctic National Wildlife Refuge to oil exploration were specifically criticized.

This problem of who pays and who benefits is very complex when pollutants produced in one state or country have their greatest negative impact in another state or country. This is particularly true of acid deposition.

PROGRESS AND PUBLIC SUPPORT

The U.S. public continues to place high priority on environmental concerns. A recent survey found that four out of five people in this nation are convinced that pollution affects the quality of their lives, and three out of four believe that current laws are too weak. Another survey indicated that 75 percent of the U.S. public believe that environmental improvements should be made regardless of cost.

As we have seen, the cost of controlling pollution is high and will continue to rise. What benefits do we receive in exchange for this cost? EPA Deputy Administrator F. Henry Habicht summarized the substantial changes that have been wrought in the past 20 years of the EPA's existence:

Because of EPA's efforts to implement national laws, air emissions from cars, power plants, and large industrial facilities have been curtailed sharply; hundreds of primary and secondary waste-water treatment facilities have been constructed; ocean-dumping of wastes has been virtually eliminated; land disposal of untreated hazardous wastes has largely stopped; hundreds of hazardous waste sites

One of President Bush's concessions to business concerns has been the creation of the **Council on Competitiveness,** headed by Vice President Dan Quayle. The council reviews all new regulatory rules to determine their potential impact on the business community and proposes changes in the rules if it wants to. The agency that originated the regulatory rule must then either accept the council's changes, negotiate a compromise with the council, or precipitate a political battle.

The executive director of the council is Allan Hubbard, co-owner of World-Wide Chemical and also a substantial stockholder in PSI Resources, Inc. (a power company). Both of these companies participate in lobbying efforts to weaken Clean Air Act provisions. In spite of his clear connections to two polluting industries, Hubbard has received a waiver of the federal conflict-of-interest laws from Vice President Quayle.

The council drew public attention in 1991 because of the major revision of a wetlands manual produced by scientists from four federal agencies. Finding it too easy for land to qualify as wetlands under the procedure recommended in the manual, the council rewrote the guidelines in a way that removed protection

from half of the existing wetlands in the United States—more than 20 million hectares. This change by the council will enable land developers to move in on land that is currently protected by law. EPA Director William K. Reilly has criticized the council's changes and reminded them of the president's campaign promise to hold the line on wetlands. As this book goes to press (September 1992), however, the council has not backed down on the wetlands redefinitions.

The council has also been at work on the new Clean Air Act regulations. In its review of proposed regulations from the EPA, the council wrote in about 100 changes. One of the changes was to reject a request by the EPA that the federal government be allowed to enforce air pollution laws if a local government does not. If this change is allowed to stand, the government and citizens' groups would be prevented from suing a company when a local government does not enforce air pollution laws. Another change allows industry not to count its worst days of pollution when measuring smokestack emissions. Both companies in which Hubbard has financial ties will benefit greatly from the relaxation of air pollution laws.

Congress has begun to scrutinize

the activities of the council. Quayle's relaxation of conflict-of-interest concerns has been challenged by the House Government Operations Committee in the light of a directive from the president that government officials should avoid even the appearance of a conflict of interest. An inquiry was held on whether Hubbard's governmental actions benefited his financial interests, with the result that Hubbard has agreed to put his holdings into a blind trust (an arrangement that requires a redistribution of investments that is unknown to the holder).

In effect, the Council on Competitiveness represents what has been called a *"Shadow Government,"* that is, one operating outside the normal channels envisioned by Congress and the Constitution. The council is turning back important proposed federal regulations at the request of affected business interests. Its continued activity represents a powerful threat to our nation's ability to respond to the increasingly serious national and global environmental problems we have been documenting in this book.

---

have been identified and 52 have been cleaned up; and the production and use of substances like asbestos, DDT, PCB's, and leaded gasoline have been banned. In the aggregate, actions like these have had a measurable, positive effect on environmental quality in this country, and they have set an example for other countries around the world.

Do these benefits outweigh the costs? To answer this question, consider the phaseout of leaded gasoline as just one example. This process, occurring over an 8-year period, has cost about $3.6 billion, according to a cost-benefit report by the EPA. Benefits were valued at over $50 billion, $42 billion of this for avoided medical costs.

Executive Order 12291 is still in effect. A cost-benefit analysis accompanies every new regulatory rule from the EPA and other federal agencies. Even if a rule is not classified as *major,* the accompanying documentation must demonstrate by cost-benefit analysis that the rule is *not* major. According to an EPA spokesman, the final decision on any regulatory rule is made on the basis of the legislative statute that generated the rule and not strictly on cost-benefit considerations. Some statutes are more amenable than others to cost-benefit analysis, and some prohibit the process as a basis of decision. When a given rule is proposed, comments are solicited from all interested parties, and those comments are addressed when the final rule is issued. The cost-benefit analysis may become a point of contention and lead to changes in the rule, but the legislative statute is ultimately the basis of decision.

In summary, cost-benefit analysis and the economic considerations represented by the process are firmly entrenched in U.S. public policy. The process is not in the driver's seat, but it does influence regulatory decision making at many points.

Risk Analysis, Perception, and Management

All of us would like to live long and healthy lives until, perhaps at the age of 85 or 90, we keel over in the strawberry patch and quietly expire. When confronted with the news that our days may be cut short by the effects of radiation, pesticides, second-hand smoke, or chemicals in our drinking water, we be-

FIGURE 17–4
Hazards are an unavoidable part of life, as shown by these photos. Some are a matter of lifestyle choices, and some are a consequence of where we live or work. All hazards carry a risk that something unpleasant will happen.

come indignant. We expect someone in authority to keep us from dying young. The high priority we place on environmental health is largely a reflection of this concern for *hazards* coming at us from the air we breathe, the food we eat, and the water we drink. It is this concern at the grass roots level that is largely responsible for the billions of dollars spent on environmental protection.

In fact, our lives in this late-twentieth-century technological society are honeycombed with hazards (Fig. 17–4). In the context of environmental science, a **hazard** is defined as anything that can cause (1) injury, disease, or death to humans; (2) damage to personal or public property; or (3) deterioration or destruction of environmental components. Many hazards are a matter of personal choices we make everyday. We may eat too much, get too little exercise, drive cars, sunbathe, use addictive and harmful drugs, or choose hazardous occupations. Why do we subject ourselves to these hazards? Basically, because we derive some real or perceived *benefit* from them. Wanting the benefits, we are willing to take the *risk* that the hazard will not harm us. Other hazards, however, are a result not of personal choices but of choices made by other people: pesticides used by farmers, chemicals emitted by power plants, radiation from nuclear testing. Still other hazards are the result of natural causes, and no one does any choosing: radiation from radon, earthquakes, lightning, floods, and so on.

Each hazard in our lives carries with it a finite risk that something unpleasant will happen to us. Here, a **risk** is defined as the probability of suffering injury, disease, death, or other loss as a result of exposure to a hazard. Our actions each day involve choices that may subject us to higher risks (driving too fast in order to get somewhere on time) or lower risks (passing up that temptation to jaywalk).

In the best of all possible worlds, we might want to know exactly what the risks are of our individual actions—that is, we would like some evaluation of the risks we are subject to. This knowledge, it is hoped, would enable us to make informed choices as we consider the benefits and risks of the hazards around us. We might learn, for example, that 82 million people go swimming each year, and 2600 of them drown while swimming. The risk of drowning, then, is 32 in a million. **Risk analysis**—which is what we just did—is *the process of evaluating the risks associated with a particular hazard before taking some action*. To reduce the risk of drowning, for example, we might decide not to go swimming at a beach where no lifeguard is on duty, or we might choose to swim in a calm lagoon rather than in the ocean. Table 17–2 lists some everyday risks, expressed as the probability of dying from a given hazard. For example, the table indicates that the annual risk of dying from smoking

TABLE 17–2

Some Commonplace Hazards, Ranked According to the Degree of Risk

Hazardous Action	Annual Risk[a]	
Cigarette smoking, 1 pk./day	3.6 per	1 000
All cancers	2.8 per	1 000
Mountaineering (mountaineers)	6 per	10 000
Motor vehicle accident (total)	2.4 per	10 000
Police killed in line of duty	2.2 per	10 000
Air pollution, eastern U.S.	2 per	10 000
Home accidents	1.1 per	10 000
Frequent-flying professor	5 per	100 000
Alcohol, light drinker	2 per	100 000
Sea-level background radiation	2 per	100 000
Four tablespoons peanut butter/day	8 per	1 000 000
Electrocution	5.3 per	1 000 000
Drinking water containing EPA limit of chloroform	6 per	10 000 000

Source: R. Wilson and E. A. C. Crouch. "Risk Assessment and Comparisons: An Introduction." *Science*, 236 [1987], 267. Copyright 1987 by the AAAS.

[a] Probability of dying.

one pack of cigarettes a day is 3.6 per 1000, meaning that in the course of a year, 3.6 out of every 1000 people who smoke a pack a day will die from smoking-related diseases.

Not very many people actually make informed choices about the hazards in their lives on the basis of such risk analysis. However, risk analysis has become an important process in the development of public policy and is being regarded as a major way of applying science to the hard problems of environmental regulation.

RISK ANALYSIS BY THE EPA

Risk analysis began at the EPA in the mid-1970s as a way of addressing the cancer risks associated with pesticides and toxic chemicals. Since then, the process has continued to focus largely on risks to human health. As currently performed at the EPA, there are four steps in risk analysis: *hazard assessment, dose-response assessment, exposure assessment,* and *risk characterization*. One important limitation to the process is the fact that the analyst must work with the available information, which never seems to be sufficient. The final estimate of risk therefore includes a measure of the imprecision that always accompanies the use of imperfect information.

Hazard Assessment: Which Chemicals Cause Cancer?

Hazard assessment is the process of examining evidence linking a potential hazard to its harmful effects.

In the case of accidents, the linkage is obvious. The use of cars, for example, involves a certain number of crashes and deaths. In these cases, *historical data*, such as the annual highway death toll, are very useful for calculating risks. In other cases the linkage is not so clear because there is a time delay between first exposure and final outcome. For example, establishing a linkage between exposure to certain chemicals and the development of cancer some years later is often difficult. In cases where linkage is not obvious, the data may come from two sources: epidemiological studies and animal tests. An **epidemiological study** is one that tracks how a sickness spreads through a community. Thus, in our example of finding a link between cancer and exposure to some chemical, an epidemiological study would look at all the people exposed to the chemical being examined and determine whether this population has more cancer than the general population. Data from such studies are considered to be the best data for risk analysis and have resulted in scores of chemicals being labeled as *known* human carcinogens.

The second data source, animal testing, is used when we want to find out *now* what might happen many years in the future. For example, we do not want to wait 20 years to find out that a new food additive causes cancer, so we accept evidence from **animal testing** (Fig. 17–5). A test involving several hundred animals (usually mice) takes about 3 years and costs upwards of $250000. If a significant number of the animals develop tumors after being fed the substance being tested, the results indicate that the substance is either a *possible* or *probable* human carcinogen, depending on the strength of the results. Two objections have been raised about animal testing: (1) rodents and humans may have very different responses to a given chemical; and (2) the doses used on the animals are often unrealistically high. A point that supports the value of animal testing is that all chemicals shown by epidemiological studies to be human carcinogens are also carcinogenic to test animals.

Hazard assessment is the important first step in risk analysis. Whether it involves an analysis of accident data, epidemiological studies, or animal testing, hazard assessment tells us that we *may* have a problem that requires regulation.

Dose-Response and Exposure Assessment: How Much for How Long?

When animal tests show a linkage between exposure to a chemical and an ill effect, the next step is to analyze the relationship between the concentrations of chemicals in the test (the **dose**) and both the incidence

FIGURE 17–5
Laboratory mice are routinely used to test the potential of a chemical to cause cancer. This is an important source of information used to assess the presence of a hazard in food, cosmetics or the workplace. (Courtesy the Jackson Laboratory.)

and the severity of the response in the test animals. From this information, projections are made about the number of cancers that may develop in humans who are exposed to different doses of the chemical. This process is **dose-response assessment.**

The next task is **exposure assessment.** This procedure involves identifying human groups already exposed to the chemical, how that exposure came about, and the doses and length of time of the exposure.

Risk Characterization: How Many Will Die?

The final step, **risk characterization,** is to pull together all the information gathered in the first three steps in order to determine the risk and its accompanying uncertainties. More commonly, risk is expressed as the probability of a fatal outcome due to the hazard, as was done in Table 17–2 for our everyday risks.

The EPA expresses cancer risk as "upper-

TABLE 17–3

Risks to Ecology and Human Welfare, as Developed by the EPA's Science Advisory Board (not ranked within categories)		
High Risk	**Medium Risk**	**Lower Risk**
Habitat alteration and destruction	Herbicides and pesticides	Oil spills
Species extinction and loss of biodiversity	Toxics, nutrients, BOD, turbidity in surface waters	Groundwater pollution
Stratospheric ozone depletion	Acid deposition	Radionuclides
Global climate change	Airborne toxics	Acid runoff to surface waters
		Thermal pollution

Source: Reducing Risk: Setting Priorities and Strategies for Environmental Protection. USEPA Science Advisory Board, September 1990.

bound, lifetime risks," meaning the top of the range of probabilities of the risk, calculated over a lifetime. The new Clean Air Act directs the EPA to regulate chemicals that have a cancer risk of greater than one in a million ($>1 \times 10^6$) for the people who are subject to the highest doses. That is the same standard employed by the Food and Drug Administration for regulating chemicals in food, drugs, and cosmetics. Because people often have difficulty conceptualizing such data, risks may be expressed in different ways, such as reduction in life expectancy caused by engaging in a risky activity. Thus, calculations show that smoking one cigarette reduces life expectancy by 5 minutes, which is just about the amount of time it takes to smoke the cigarette!

Risks to the Natural Environment

In September 1990, EPA Director William K. Reilly received a report from a Science Advisory Board (SAB), which had been convened at his request and asked to evaluate different environmental risks in light of the most recent scientific data. The board consisted of 39 scientists and other experts from academia, state governments, industry, and public interest groups. Their report, entitled *Reducing Risk: Setting Priorities and Strategies for Environmental Protection*, has intensified interest in the use of science and risk analysis in setting environmental public policy.

The SAB separated risks into two categories: **health risks,** which has been the EPA's major concern since its origin, and **ecological risks.** It is significant that the SAB recommended strongly that the EPA give at least equal attention to ecological risks and health risks. The board divided ecological risks into high, medium and low-risk categories, as shown in Table 17–3. In performing the analyses, the board took care to weigh the *temporal* dimensions of the hazards—the length of time over which a problem is generated, understood, and corrected—and the *spatial* dimensions—the extent of geographical area affected by the problem. Thus, such problems as global warm-

ing and ozone depletion, which have long-term global implications, were ranked much higher than, say, oil spills, which have a more localized, short-term impact.

One of the most important problems raised by the scientists was the significant difference between their evaluation of risks and that of the public. Table 17–4 indicates the top 11 environmental problems from the SAB study, compared with the top 28 concerns of the public, as indicated in a Roper poll conducted in 1990. Clearly, the public perception of risks is quite different from that of the scientists. This discrepancy deserves a further look.

RISK PERCEPTION

The U.S. public is increasingly concerned about environmental problems; most of that concern can be traced to the fear of hazards that pose a risk to human life and health. People *perceive* that their lives are more hazardous than ever before, but this is not true. In fact, our society is freer from hazards than it has ever been, as evidenced by increased longevity. Why, then, do people protest against nuclear power plants, waste sites, and pesticide residues in food when, according to experts, these hazards pose extremely small risks? The answer lies in people's **risk perceptions**—their intuitive judgments about risks. In short, people's perceptions are not consistent with the reality of the situation.

Hazard versus Outrage

The reason for the inconsistency between public perception and actual risk calculations, according to Peter Sandman of Rutgers University, is that public perception of risks is more a matter of outrage than hazard. Sandman holds that, while the term *hazard* primarily expresses concern for fatalities only, the term *outrage* expresses a number of additional concerns:

TABLE 17–4

Public Concerns vs the EPA's Top 11 Risks

The EPA's Top 11 (not in rank order)	Public Concerns (in rank order)[a]
Ecological Risks	1. Active hazardous waste sites
Global climate change	2. Abandoned hazardous waste sites
Stratospheric ozone depletion	3. Water pollution from industrial wastes
Habitat alteration	4. *Occupational exposure to toxic chemicals*
Species extinction and biodiversity loss	5. Oil spills
	6. *Destruction of the ozone layer*
Health Risks	7. Nuclear power plant accidents
Criteria air pollutants (e.g., smog)	8. Industrial accidents releasing pollutants
Toxic air pollutants (e.g., benzene)	9. Radiation from radioactive wastes
Radon	10. *Air pollution from factories*
Indoor air pollution	11. Leaking underground storage tanks
Drinking water contamination	12. Coastal water contamination
Occupational exposure to chemicals	13. Solid waste and litter
Application of pesticides	14. *Pesticide risks to farm workers*
Stratospheric ozone depletion	15. Water pollution from agricultural runoff
	16. Water pollution from sewage plants
	17. *Air pollution from vehicles*
	18. Pesticide residues in foods
	19. *Greenhouse effect*
	20. *Drinking water contamination*
	21. Destruction of wetlands
	22. Acid rain
	23. Water pollution from city runoff
	24. Nonhazardous waste sites
	25. Biotechnology
	26. *Indoor air pollution*
	27. Radiation from X-rays
	28. *Radon in homes*

Source: "Counting on Science at EPA." *Science* 249 [1990], 616. L. Roberts. Copyright 1990 by the AAAS.

Items in italics also appear on the EPA's list.

---

1. Lack of familiarity with a technology: how nuclear power is produced and how toxic chemicals are handled, for example.

2. Involuntariness of risks: Research has shown that people who have a choice in the matter will accept risks roughly 1000 times as great than when they have no choice.

3. Memorability of hazards: Accidents involving many deaths (Bhopal) or failures of technology (Three Mile Island) are thoroughly imprinted into public awareness by media coverage and are not quickly forgotten.

4. Overselling of "safety": The public becomes suspicious when scientists or public relations people play up the benefits of a technology and play down the hazards.

5. Morality: Some risks have the appearance of being morally wrong. If it is wrong to foul a river, the notion that the benefits of cleaning it are not worth the costs is unacceptable. You should obey a moral imperative regardless of the costs.

6. Control: People are much more accepting of a risk if they are in control of the elements of that risk, as in automobile driving, indoor air pollution, and radon.

7. Fairness: The benefits and the risks should be connected. If the benefits go to someone else, why should you accept any risk?

Public perception of risks is strongly influenced by the media, which are far better at communicating the outrage elements of a risk than the hazard. Public concern over oil spills rose precipitously following the Exxon Valdez spill in Alaska, an accident that received extensive media coverage and had a high "outrage quotient." Cigarette smoking, however, which causes *500000 deaths a year in the United States alone*, receives minimal media attention because it is not "news" and there's no outrage factor. Indeed, it has been suggested that if all the year's smoking fatalities occurred on one day, the media would have a field day and smoking would be banned the next day!

Risk Perception and Public Policy

The most serious problem with the discrepancy between experts and the public is that, generally speaking, it is *public concern* that drives *public policy*, rather than cost-benefit analysis or risk analysis conducted by scientists. The EPA's funding priorities are set largely by Congress, which reflects public concern. How important is this problem?

If public outrage is the primary impetus for public policy, some serious risks will get less attention than they deserve. In particular, risks to the environment are perceived as much less important than they really are, because of the public's preoccupation with risks to human health. As Table 17–4 indicates, two major ecological risks on the EPA's list—habitat alteration and species extinction and biodiversity loss—do not even show up on the list of public concerns. And, only 3 of the public's top 10 concerns are on the EPA's list.

This difference of opinion points to the importance of *risk communication*, a task that definitely should not be left to the media. The value of ecosystems and their connections to human health and welfare need far greater emphasis in the public consciousness, and this responsibility falls to the scientific and educational communities as well as governmental agencies.

It is true that the public's concern for more than the probabilities of fatalities might have merit. Public outrage must be heard, understood, and given a reasonable response. It may not be the best source of public policy, but it reflects certain values and concerns that could easily be omitted by an "objective" risk analysis. The fact is, subjective judgments are going to play a role at every step in the risk-analysis process, from hazard assessment to perception to risk management.

RISK MANAGEMENT

Regulatory Decisions

The analysis of hazards and risks is a task that primarily falls to the scientific community. The task is incomplete, however, if no further consideration is given to the information gathered. Risk management naturally follows, and this is the responsibility of lawmakers and administrators. It is EPA policy to separate risk analysis and risk management within the agency. **Risk management** involves (1) *a thorough review of the information available pertaining to the hazard in question and the risk characterization of that hazard* and (2) *a decision on whether the weight of evidence justifies a regulatory decision.* Without doubt, public opinion can play a powerful role in determining the outcome. In general, however, a regulatory decision will hinge on one or more of the following considerations:

1. **Cost-benefit analysis.** As we learned earlier, a comparison of costs and benefits, if done fairly, can in many cases make a regulatory decision very clear-cut.
2. **Risk-benefit analysis.** A decision can be made on the benefits versus the risks of being subjected to a particular hazard, especially where those benefits cannot be easily expressed in monetary values. The use of medical X-rays is a good example. X-rays carry a calculable risk of cancer, but the benefit derived from the X-ray of, say, a broken bone is much greater than the small cancer risk involved.
3. **Public preferences.** As we have seen, people have a much greater tolerance for risks that they feel are under their own control or are voluntarily accepted.

Risk management has been thoroughly incorporated into the EPA's policy-making process for at least 10 years. Its most common use has been in the design of regulations. The process has enabled the agency to target appropriate hazards for regulation and to determine where to aim the regulations—that is, at the source, at the point of use, or at the disposal of the hazardous agent. As was done for toxic chemicals and cancer, a given risk level may be adopted as a standard against which policy makers can measure new risks. The EPA's struggles with Alar (see In Perspective Box, p. 396) illustrate how this standard can be applied, as well as several other risk management problems.

Comparative Risk Analysis

One of the advantages of risk analysis is that it provides a common language and a common procedure for comparing hazards. The basic process is understandable, even if it is fraught with its own set of hazards (as in the use of subjective judgments at many points). Conceivably, the process of risk analysis may help society to choose wisely among a broad range of policy options available for reducing risks. The 1990 report of the SAB—*Reducing Risk*—addressed another potential use of risk analysis, however: **setting environmental priorities.** Indeed, much of their work was directed toward this goal.

The group separated into different fields (human health, ecology) and defined different types of risks by which problems could be compared: cancer and noncancer health risks, impacts on materials, ecological effects, and economic impacts. They ad-

Alar (trade name for the pesticide *daminozide*) has been used by applegrowers since the late 1960s. It preserves the appearance of red apples and allows them to ripen slowly. In March 1989, the Natural Resources Defense Council (NRDC), a private environmental organization, announced the results of a 2-year study on pesticide residues in food and drew special attention to Alar. The report severely criticized the EPA for disregarding the safety of children in its tolerance of pesticide residues in some foods. The problem, said the NRDC, is that children consume much more fruit and juices than do adults, and they eat more relative to their body weight. Yet the EPA had ignored these facts in calculating exposure and dosages as they performed risk analysis. In particular, the NRDC study claimed that Alar would cause the future death of 6000 of today's children. Food industry experts denied that there was a problem with

Alar, and the EPA disagreed with the NRDC's calculations. The issue was immediately picked up by CBS, and a great public outcry ensued, resulting in the voluntary removal of Alar by its manufacturers, Uniroyal, and a cancellation procedure by the EPA.

The NRDC's risk analysis led to a 25-fold greater risk calculation than did the EPA's, a discrepancy that some experts maintain is not at all unusual, given the uncertainties and assumptions made in the calculation process. According to actuarial tables, of the 22 million schoolchildren alive today, 25 percent or 5.5 million, will eventually get cancer even if they never eat apples. Six thousand excess cancers above 5.5 million is a miniscule addition to the total risk of cancer, raising it to 25.025 percent. However, there is another way to look at these NRDC results. The U.S. population is 240 million. Divide this number by 6000 and you get 40000. So 6000 excess cancers means that one person in

40000 will get cancer because of Alar. Since the EPA must regulate chemical residues when the cancer risk is greater than 1 in a million, the cancer risk from Alar clearly indicates the need for the EPA to act.

One unfortunate casualty in this episode was public confidence in food safety. In addition, applegrowers suffered a temporary decline in sales of red apples and have sued the NRDC and CBS for those losses. More recent EPA reviews of animal test results have lowered the calculated risk, but the problem is moot now that Alar has been withdrawn and the apple industry is getting along quite well without it. The Alar episode does clearly point out the enormous importance of public perception of risks and the role of the media in exploiting the outrage component of public risk perception.

mitted that these risk types were not directly comparable but agreed that information could be collected within each category and rankings constructed according to how serious the board members rated the separate hazards. One result of this ranking was shown in Table 17–3. As a result of their deliberations, one strong recommendation of the board was that the EPA needs to refocus its policy orientation more toward the natural environment. Given the direct links between public concern and public policy, this refocusing will require convincing the public of the great value of natural ecosystems. To do so, the weight of scientific evidence must be brought to bear on the major environmental problems—which means strong governmental support of research. And the EPA will have to improve its communication with the public and the media, a large task in itself.

The EPA has expressly stated the objective of cultivating public confidence in the general goals of our scientific and environmental leadership, so that the U.S. public will become willing to delegate much

of the responsibility for risk analysis and management to those most capable of doing it. It is hard to imagine a more important assignment. Faced with global warming, ozone layer depletion, acid deposition, the loss of habitats and biodiversity, the EPA has both the opportunity and the responsibility to be out in front on the issues. As one significant response to this task, the EPA established in late 1991 the Office of Environmental Education, with the broad goals of increasing environmental literacy and encouraging young people to pursue careers in mathematics, science, engineering, and other fields important in future environmental improvement.

Conclusion

We have addressed two important components of our political process in this chapter, one—**cost-benefit** analysis—giving an economic perspective, and the other—**risk analysis**—bringing science more directly to bear on regulatory decisions affecting the environment. In a democratic society, however, public policies on environmental issues should be an expression of more than economics and science. They should represent the broad commitment of all segments of society to the stewardship of the environment. The discrepancies revealed in Table 17–4 indicate the need for a higher level of environmental literacy, so that more of the public can respond to such problems as habitat loss, endangered species, and the other concerns that rank highest in the views of the experts. These are costly problems to solve, but if we fail to address them, we shall lose out on a very large scale. Ultimately all of society will receive the benefits of environmental protection and will suffer the costs and risks of environmental degradation.

 Review Questions _____

1. Define the term *cost-benefit analysis* as it relates to environmental regulation.
2. How does cost-benefit analysis address externalities?
3. List specific costs and benefits of pollution control.
4. Discuss the cost effectiveness of pollution control.
5. What difficulties may be encountered when trying to apply shadow pricing to human life and the environment?
6. Describe several problems that may arise when comparing cost and benefit estimates.
7. Define risk analysis.
8. Discuss the four steps of risk analysis used by the EPA.
9. Why do scientists and the public often come to very different conclusions regarding cost-benefit analysis?
10. What factors have been known to generate public outrage?
11. Discuss the relationship between public risk perception and public policy.
12. What considerations are combined to form a regulatory decision?
13. Describe the EPA's new strategy of comparative risk analysis.

 Thinking Environmentally _____

1. Consider Table 17–4. Which would you rank as the top three ecological and health risks? Which of the seven public outrage factors might be playing a role in your rankings?

2. Imagine that you have been appointed to a risk-benefit analysis board. Explain why you approve or disapprove of widespread use of:

   genetic engineering
   drugs to slow the aging process
   nuclear power plants
   the abortion pill

3. Suppose your town wanted to spray trees to rid them of a deadly pest. How would you use cost-benefit analysis to determine whether this is a good idea?

4. What costs of our industrial processes are now treated as externalities? How could we internalize these costs?

5. Consider the complexities and limitations of cost-benefit analysis. Should we continue to support its use to determine public policy? How great a role should it play? Support your answers.

Pollution

1 Reduce your contribution to air pollution by amending your lifestyle to drive fewer miles: arrange to live near your workplace, car pool, use public transport, bicycle to work or school, avoid unnecessary trips.

2 Keep informed on how the Clean Air Act of 1990 is being implemented and enforced. If there are attempts to weaken it, protest to your congresspersons. Ask them to promote new standards to increase the average vehicle fuel efficiencies.

3 Be an advocate of energy conservation on all fronts, in supporting the "tie-in strategy" to combat global warming; insist on full U.S. participation in global efforts to curb greenhouse gas emissions.

4 Adopt ways to use less fuel and electricity both at home and where you work (turn thermostats back, turn off lights when leaving a room, wear sweaters, use energy-efficient light bulbs, etc.).

Making a Difference

5   Survey the lakes and ponds in your area and determine whether cultural eutrophication is occurring. If you find cases, follow up by consulting a local environment group and get involved in combating the problem.

6   Use phosphate-free detergents for all washing. Information will be printed on the container (often in fine print!).

7   Explore sewage treatment in your community. To what extent is the sewage treated? Where does the effluent go? The sludge? Are there problems, alternatives that are ecologically more sound? Consult local environmental groups for help.

8   Read labels and become informed about the potentially hazardous materials you use in the home and workplace. Wherever possible, use nontoxic substitutes. If toxic substances are used, insure that they are disposed of responsibly.

9   Encourage your community to set up a hazardous waste collection center where you can take leftover hazardous materials.

10   Properly maintain all fuel-burning equipment to burn efficiently (oil burner, gas heater, lawnmower, outboard motor, automobile, etc.).

11   Refuse to become a smoking victim: If you are a smoker, stop; insist that others do not smoke in your presence or in your home.

12   Make sure that anyone servicing a refrigeration or air-conditioning unit for you will recapture and recycle the CFCs if the system needs to be opened.

13   Monitor hazards in your own life and take steps to reduce behavior ranked more risky by the experts. Be discriminating as you hear about hazards and risks; especially be aware of the media's tendency to exploit the *outrage* element of risks.

14   Be aware of the pressures put on regulators by business concerns to weaken regulatory rules, as with the Council on Competitiveness, and insist on fairness and consistency in the application of laws passed by Congress.

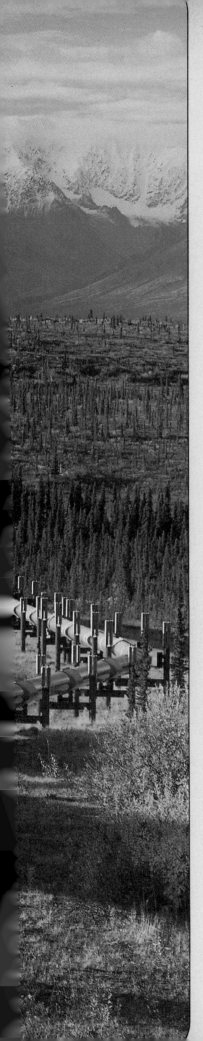

Part Four

Resources: Biota, Refuse, Energy, and Land

The Alaska oil pipeline snakes its way down through tundra and forest to the port of Valdez, an intrusion into a beautiful and fragile wilderness that is symbolic of the dilemma facing our society. We may love the beauty of wild ecosystems and species, and turn to them for enjoyment and rest, but we also must look on much of the natural world as resources for exploitation. Can we have it both ways? Can we hope to preserve nature while we also make use of it?

In this Part we turn our attention to the resources that make it possible for some of us in the developed world to enjoy our late-20th-century lifestyle, and for those in the Third World to move slowly along the path of economic development. These resources—fossil fuels, fisheries, forests, and the like—are not evenly distributed on the earth, and their extraction and use represents a great part of the economic exchange between peoples. Wars are fought over such resources, international treaties are forged to regulate access and exchange—yet the most important actors in the picture are the individuals who through their lifestyle choices are the consumers of resources. The good news is that consumers can become stewards and bring the attitude of care and concern to the living environment that sustains us and the resources buried in the earth as minerals. Trash can be recycled, energy can be conserved, solar options can be adopted, land can be set aside for preservation. This is what the environmental revolution is all about—the transition to a sustainable society.

18

Wild Species: Diversity and Protection

LEARNING OBJECTIVES

When you have finished studying this chapter, you should be able to:

1. Distinguish between instrumental and intrinsic value in assigning worth to natural species.

2. Explain how natural species are important in agriculture, forestry, aquaculture and animal husbandry; medicine; recreation, science, and aesthetic enjoyment; and commercial trade.

3. Discuss the attempts to assign intrinsic value to wild species.

4. Define biodiversity and estimate the extent of current biodiversity.

5. Document the extent of biodiversity losses, both known and unknown.

6. Analyze the ways in which habitat conversion, fragmentation, and simplification affect biodiversity.

7. Explain how human population growth, pollution, and overuse of wild species are involved in biodiversity loss.

8. Describe the different ways exotic species can impact biodiversity.

9. Explore the possible consequences of species extinctions.

10. Describe the relationship between game animal management and biodiversity.

11. Trace the origins of the Endangered Species Act and describe its role in preserving species in the United States.

402

About 1.4 million species of plants, animals, and microbes have been examined, named, and classified, but scientists estimate that there are between 3 and 30 million additional species that have not yet been systematically explored. These natural species of living things, collectively referred to as **biota,** are responsible for the structure and maintenance of all ecosystems. These species and the ecosystems they form represent a form of wealth—**biological wealth**—that sustains human life and economic activity. It is as if the natural world were an enormous bank account. Like money in a bank account, biological wealth is capable of paying vital, life-sustaining dividends indefinitely, but only as long as the capital is maintained.

Our early ancestors were thoroughly integrated into natural ecosystems as hunter-gatherers. Their survival depended on learning the ways of the other animals and identifying plants that could be eaten. Some 10 000 years ago, humans began learning to select certain plant and animal species from the natural biota and to propagate them, and the natural world has never been the same. Over time, vast areas of forests, savannahs, and plains were converted to fields and pastures as the human population grew and human culture flourished. In the process, many living species were exploited to extinction, and others disappeared as their habitats underwent development. At least 500 plant and animal species have become extinct in the United States alone, and thousands more are at risk. We have been drawing down our biological capital, with unknown consequences.

Now, living in cities and suburbs in the industrialized world and getting all our food from supermarkets, our connections to nature seem remote. That is an illusion. Our interactions with the natural world have changed, but we are still dependent on biological wealth. Our counterparts in the Third World are not so insulated from the natural world; their dependence is much more immediate as they draw sustenance and income directly from forests, grasslands, and fisheries. However, these people too are drawing down their biological capital, with consequences that are obvious and grave.

To gain insight into this set of problems, we shall in this chapter give attention to wild species. We shall consider their importance to us, examine the perspective of biodiversity, and then look at the public policies that seek to protect wild species. In the next chapter our focus will broaden to an examination of human impact on whole ecosystems, with special emphasis on human patterns of use (and misuse) of those systems and their living components. Dividing these topics into two chapters is a somewhat arbitrary distinction; it will be obvious through the entire discussion that protecting wild species also means preserving their ecosystems and vice versa.

Wild Nature

A FIERCE GREEN FIRE

Aldo Leopold was a pioneer in the field of conservation and environmental ethics. In his 1949 classic *A Sand County Almanac,* Leopold tells the story of a wilderness trip in the southwestern United States with some friends. They were eating lunch on a hillside when they noticed an animal crossing a small river below them. They watched as the animal, evidently a she-wolf, was met on the other side of the river by a half-dozen grown pups, who tumbled all over her with their greetings.

In those days we had never heard of passing up a chance to kill a wolf. In a second we were pumping lead into the pack, but with more excitement than accuracy: how to aim a steep downhill shot is always confusing. When our rifles were empty, the old wolf was down, and a pup was dragging a leg into impassible slide-rocks.

We reached the old wolf in time to watch a fierce green fire dying in her eyes. I realized then, and have known ever since, that there was something new to me in those eyes—something known only to her and to the mountain. I was young then, and full of trigger itch; I thought that because fewer wolves meant more deer, that no wolves would mean hunter's paradise. But after seeing the green fire, I sensed that neither the wolf nor the mountain agreed with such a view.

This incident not only changed Leopold's *attitude* toward shooting wolves, it changed his *valuation* of wolves and other wild species. He began to understand the importance of wolves in keeping deer herds from overeating their food supply, and in time Leopold articulated an ethic that stressed the *value* of natural species and the land. By value, he meant more than economic worth; he meant that wild species and their habitats have a right to exist, and protecting that right is a matter of morality. The fierce green fire in the old wolf's eyes was symbolic of wild nature, and when it went out, Leopold saw that something very precious was gone.

INSTRUMENTAL VERSUS INTRINSIC VALUE

It was not so long ago that hunters on horseback would ride out to the vast herds of buffalo roaming the North American prairies and shoot them by the thousands, often taking only the tongues for markets back east. The passenger pigeons that darkened the skies in huge flocks were ruthlessly killed at their roosts to fill a lively demand for their meat, until the species was gone (Fig. 18–1). Plume hunters decimated egrets and other shorebirds to satisfy the demands of women's fashions in the late 1800s. Appalled at this wanton destruction, nineteenth-century naturalists called for an end to the slaughter, and the U.S. public began to be sensitized to the losses that were occurring. People began to see natural species as worthy of preservation, and naturalists began to look for ways to justify their calls for conserving nature. There was an emerging sense that species should not be hunted to extinction. But why? Were these early conservationists just concerned that there might not be any animals left to hunt or trees left to chop down?

Their problem then, and our problem now, is to establish that wild species have some *value* that makes it essential that they be preserved. If we can identify that value, then we shall find it much easier to justify the action we must take to preserve wild species.

Philosophers who have addressed this problem inform us that there are two kinds of value to be considered. The first is **instrumental value.** A species or individual organism has instrumental value if its existence or use benefits some other entity. This kind of value is usually *anthropocentric;* that is, the beneficiaries are human beings. Clearly, there are many

FIGURE 18–1
Clouds of passenger pigeons darkened American skies during the 18th and 19th centuries, but relentless hunting pressure so decimated the species that it became extinct in the early 20th century. (Asa C. Thoresen/Photo Researchers.)

species of plants and animals that have instrumental value to humans and will tend to be preserved (or conserved, as we would say) so that we can continue to enjoy the value that we derive from them.

The second kind of value we must consider is **intrinsic value.** We assign intrinsic value to something when we agree that it has value for its own sake; that is, it does not have to be useful to us to possess value. How do we know that something has intrinsic value? That is a philosophical question, and it comes down to a matter of moral reasoning. People often disagree about intrinsic value, as illustrated by the animal rights view (see Ethics Box, p. 405).

As we study the problem of species loss, we shall see that there are many people who argue that no species on earth (except *Homo sapiens*) has any intrinsic value. However, we shall also see that if there is no recognition of the instrinsic value of species, it is difficult to justify preserving many that are appar-

Activists who support the rights of animals have attracted much public attention. They have publicized their cause by drawing attention to a number of particularly objectionable forms of cruelty to animals (such as leg-hold steel traps and pharmaceutical eye tests on rabbits). In Chapter 4 we considered animal rights in the light of hunting and the place of individual animals in natural ecosystems. Here let us look at animal rights in the context of *value*.

Animal rights proponents differ from people who are concerned about preserving animal species in one main way. The former value the individual animal, whereas the latter value the whole population as a species. An animal rightist would argue that the individual animal has intrinsic value and that it is not valid for one animal (human beings) to uphold intrinsic value for individuals of its own species but not for individuals of other species. Animal rights and human rights are given parallel status on the basis that animals (at least, higher animals) have "perception, memory, a sense of the future,

an emotional life together with feelings of pleasure and pain, and the ability to initiate action in pursuit of their desires and goals," as articulated by ethicist Tom Regan.

It follows that, if something has intrinsic value, it deserves respect, meaning that how we treat it becomes a matter of justice. Animal rightists point to the inconsistencies in our dealings with animals: We may not feed poisoned food to squirrels and pigeons in the park, but we may shoot them out of the trees; we may not bludgeon our pet dog if it fails to please us, but we may subject the eyes of rabbits to chemicals until the rabbits are blinded. In short, we recognize animal rights in some situations but deny them in others.

Philosophers point out that, to be entirely consistent as an animal rightest, one should be against zoos, hunting, trapping, eating animals, and performing experiments on animals. It is permissible to kill an animal only in self-defense, in the same way we would defend ourselves against a human intruder. The

use of animals for laboratory dissection is also wrong, primarily because it is assumed that the animals were wrongfully caught and killed.

Most people reject this viewpoint as being improper sentimentalism. Animals are not equal to people in terms of their intrinsic value, and although we do accord to individual animals some rights (the right not to be treated with cruelty, for example), there is no philosophical imperative to assume they have equal rights with humans. As journalist Richard Coniff pointed out in a recent *Audubon Magazine* article, there is a contradiction in arguing on the one hand that humans are merely a part of nature and should not have any more rights than other animals and on the other hand that humans have a moral obligation to give up practices that are entirely consistent with our animal nature— namely, to use other species to satisfy our needs for food and clothing.

Can we have a high moral obligation and at the same time have no more rights or value than any other animals? What do you think?

ently insignificant or very local in distribution. Although there are problems in establishing intrinsic value for species, there is growing support at the grass roots level in favor of preserving species that may have no usefulness to humans and may never be seen by anyone except a few naturalists or systematists.

Value of Natural Species

The value of natural species can be categorized into five areas:

○ Sources of agriculture, forestry, aquaculture, and animal husbandry

○ Sources for medicine

○ Recreational, aesthetic, and scientific value

○ Commercial value

○ Intrinsic value

The first four categories mostly reflect instrumental value. In the case of aesthetic and scientific value, it could be argued that these sometimes represent intrinsic value.

SOURCES FOR AGRICULTURE, FORESTRY, AQUACULTURE, AND ANIMAL HUSBANDRY

Since most of our food comes from agriculture, we tend to believe that it is independent of natural biota. This is not true. Recall that in nature both plants and animals are continuously subjected to the rigors of natural selection. Only the fittest survive. Consequently, wild populations have numerous traits for parasite resistance, competitiveness, tolerance to adverse conditions, and other aspects of **vigor**. In addition, as we learned in Chapter 5, the gene pools of wild populations generally harbor the variations that enable adaptation to changing conditions.

Conversely, populations grown for many generations under the "pampered" conditions of agriculture tend to lose these traits for vigor because they are selected for production, not vigor. For example, a high-producing plant that lacks drought resistance is irrigated, and its drought resistance is ignored. Also, in the process of breeding plants for maximum production, virtually all genetic variation is eliminated. Indeed, the cultivated population is commonly called a **cultivar** (for *culti*vated *vari*ety), indicating that it is a highly selected strain of the original species with a *minimum* of genetic variation. Cultivars, when provided with optimal water and fertilizer, do give outstanding production under the specific climatic conditions to which they are attuned. With their minimum genetic variation, however, they have virtually no capacity to adapt. If climatic conditions change from what a cultivar is adapted to, its production may drop to nil and it will be unable to adapt to the new conditions because its gene pool lacks the necessary variation.

To maintain vigor in cultivars and to adapt them to different climatic conditions, plant breeders comb wild populations of related species for the desired traits. When found, these traits are introduced into the cultivar through crossbreeding or genetic engineering. Note, though, that any such improving trait comes from a related wild population, that is, from natural biota. If natural biota with such wild populations are lost, the options for continued improvements in food plants will be greatly reduced.

Also, tremendous potential for developing new agricultural cultivars will be lost. From the hundreds of thousands of plant species existing in nature, humans have used perhaps 7000 in all, and modern agriculture has tended to focus on only about 20. This limited diversity in agriculture makes it ill-suited to production under many environmental conditions. For example, we tend to think of arid regions as being unproductive without irrigation. However, there are many wild species belonging to the bean family that produce abundantly under dry conditions. Scientists estimate that there are 75 000 plant species with edible parts that might be brought into cultivation. Many of these could increase production in environments that are less than ideal. The winged bean, native of New Guinea, is a good example (Fig. 18–2). This plant is a veritable supermarket, with every part edible: pods, flowers, stems, roots, and leaves. Recently introduced to many Third World countries, it has already made a significant contribution to improving nutrition. Loss of biological diversity undercuts such future opportunities.

Another area in which biodiversity has instrumental value to humans is pest control. In Chapter 10 we discussed the tremendous and invaluable opportunities to control pests through introducing natural enemies and increasing genetic resistance. Natural enemies and genes for increasing resistance can come only from natural biota. Destroying natural biota will destroy such opportunities.

Since we select species from nature for animal husbandry, forestry, and aquaculture, essentially all the same arguments can be made in connection with those important enterprises.

To use our concept of biological wealth, we can

FIGURE 18–2
The winged bean, a climbing legume with edible pods, seeds, leaves, and roots. This tropical species is an example of the great potential of wild species for human use. (Courtesy ECHO, Educational Concerns for Hunger Organization.)

look at natural biota as a bank in which are deposited the gene pools of all the species involved. As long as natural biota are preserved, we have a rich endowment of genes in the bank, which we can draw upon as needed. Thus natural biota are frequently referred to as a **genetic bank.** Depleting this bank cannot help but deplete our future.

SOURCES FOR MEDICINE

This genetic bank also serves medicine, as the following example illustrates. For thousands of years, the indigenous people of the island of Madagascar used an obscure plant, the rosy periwinkle, in their folk medicine. This plant grows only on Madagascar, and if it had become extinct before 1960 hardly anyone outside Madagascar would have cared. In the 1960s, however, scientists extracted from this plant two chemicals, which they called "vincristine" and "vinblastine," having medicinal properties. These chemicals have revolutionized the treatment of childhood leukemia and Hodgkin's disease. For example, before their discovery, leukemia was almost always fatal in children; today, with vincristine treatment, there is a 95 percent chance of remission. These two drugs now represent a $100 million a year industry.

The rosy periwinkle is just one of hundreds of such stories. The venom from a Brazilian pit viper (a poisonous snake) led to the development of the drug Capoten, used to control high blood pressure. An extract from the bark of the Pacific yew has proved to be so valuable for treating cancer that the American Cancer Society has asked that this tree be listed under the Endangered Species Act so that its habitats will be preserved. Indeed, 25 percent of pharmaceuticals in the United States contain ingredients originally derived from native plants, representing $8 billion of annual revenue for drug companies and better health and longevity for countless people. It is likely that the search for such chemicals has barely scratched the surface.

RECREATIONAL, AESTHETIC, AND SCIENTIFIC VALUE

The species in natural ecosystems also provide the foundation for numerous recreational and aesthetic interests, ranging from sport fishing and hunting to hiking, camping, bird watching, photography, and so on (Fig. 18–3). Interests may range from casual aesthetic enjoyment to serious scientific study. Virtually all our knowledge and understanding of evolution and ecology have come from studying wild species and the ecosystems in which they live. Pleasure and satisfaction may even be indirect. For instance, one may never see a whale, but knowing that whales and similar exciting animals exist provides a certain aesthetic pleasure. The great popularity of nature films attests to this. Further, knowing that the earth and its biosphere continue to support and maintain such wildlife provides a sense of well-being.

Recreational and aesthetic values constitute a very important source of support for maintaining wild species. These activities involve a great number of people and represent a huge economic enterprise. To cite one example, 84 percent of the Canadian people take part in some form of wildlife-related recreation. In the United States, at least 50 million people do some recreational hunting or fishing each year, spending over $38 billion to do so. Very likely, the broadest public support for preserving wild species and habitats is traced to the aesthetic enjoyment people derive from them.

COMMERCIAL VALUE

Recreational or aesthetic activities support commercial interests: sporting goods stores, tourist and travel accommodations, and so on. **Ecotourism**—where tourists visit a place in order to observe unique ecological sites—represents the largest income-generating enterprise in many Third World countries. As the amount of leisure time available to people increases, more and more money will be spent on recreation. Since some percentage of these recreation dollars will be spent on activities related to the natural environment, any degradation of that environment affects commercial interests. Examples abound of businesses folding because a lake or beach, for example, has become polluted and is no longer suitable for fishing or swimming.

In addition to the indirect support resulting from recreation, natural species support a number of commercial interests directly. Commercial fishing, logging, and the trade in exotic "pets" (uncommon species of fish, reptiles, mammals, birds, and plants) are the most conspicuous examples. In one recent study, the annual contribution of wild plant and animal species to the U.S. economy was calculated to be more than $200 billion, which is about 4.5 percent of the gross domestic product. The percent contribution of wild plant and animal species to the economy is undoubtedly much higher in Third World countries, where economic pressures to exploit natural resources are intense. Of course, it is these commercial activities that are in many cases leading to the decline and extinction of wild species.

FIGURE 18–3
Natural biota provide numerous recreational, aesthetic, and scientific values, a few of which are depicted here.

INTRINSIC VALUE

The usefulness (instrumental value) of many species is apparent. What about those species that have no obvious value to anyone—probably the majority of plant and animal species, many of which are rare or inconspicuous in the environment? Some observers believe that the most successful strategy for preserving all wild species is to emphasize the intrinsic value of species rather than unknown or uncertain ecolog-

ical and economic instrumental values. Thus, we should recognize that the extinction of a species per se is an irretrievable loss of something of value.

Many observers who view wild species as having a basic right to existence claim that humans have no right to terminate a species that has existed for thousands or millions of years and represents a unique set of biological characteristics. They argue that long-established existence carries with it a right to continued existence. Some support this view by arguing that there is value in every living thing and that one kind of living thing (human) has no greater value than any other. This argument, however, can lead to some difficulties, such as having to defend the rights of pathogens and parasites. In a more common viewpoint, ethicists maintain that humans, because we have the ability to make moral judgments, have a special responsibility toward the natural world, and that responsibility includes concern for other species.

Some ethicists find their basis for intrinsic value in religion. For example, Old Testament writings express God's concern for wild species when He created them. Jewish and Christian scholars maintain that by declaring His creation good and giving it His blessing, God was saying that all wild things have intrinsic value and therefore deserve moral consideration. The Islamic Qur'ān proclaims that the environment is the creation of Allah and should be protected because it praises the Creator. This ethical concern for wild species underlies many religious traditions and represents a potentially powerful force for preserving biodiversity. What is now needed is to convince people that they have a responsibility to make the transition from belief to action and work towards saving what is left of the natural world.

Thus we see that even if species have no demonstrable use to humans, it can be argued that they have a right to continue to exist. Only rarely (for parasites and pathogens) can we claim that there is any moral justification for driving other species to extinction.

Biodiversity

We saw in Chapter 5 that natural selection operating over time leads to **speciation**—the creation of new species—as well as **extinction**—the disappearance of species. Over geological time, the net balance of these processes has favored the gradual accumulation of more and more species—in other words, biodiversity. The concept of biodiversity usually includes the genetic diversity within species and also the diversity of ecosystems and habitats. Its main focus, however, is the species.

How much biodiversity is there? No one knows. The only certainties are that (1) 1.4 million species have been described and (2) many more than that exist. Most people are completely unaware of the great diversity of species within any given taxonomic category. Groups especially rich in species are the flowering plants (220 000 species) and the insects

(750 000 species), but even less diverse groups, such as birds or ferns, are rich with species that are unknown to most people. Taxonomists are aware that their work in finding and describing new species is incomplete. Groups that are conspicuous or important commercially, such as birds, mammals, fish and trees, are much more fully explored and described than, say, insects, very small invertebrates like mites and soil nematodes, fungi, and bacteria.

Estimates of the number of species on earth today are based on recent work in the tropical rainforests, which hold more living species than all other habitats combined. The upper bound of the estimate keeps rising as taxonomists explore the rainforests more and more. According to Harvard biologist E. O. Wilson, the number could be as high as 100 million species! Whatever the number, the planet's biodiversity represents an amazing and diverse storehouse of biological wealth.

THE DECLINE OF BIODIVERSITY

Losses Known and Unknown

A recent inventory of the wild plant and animal species in the 50 states suggests that at least 9 000 species are at risk of being lost. At least 500 species native to the United States are known to have become extinct since the early days of colonization. Over 100 of these are vertebrates (Fig. 18–4). Currently, the U.S. Fish and Wildlife Service and the National Marine Fisheries Service list almost 600 species as "endangered" or "threatened," categories that elicit protection and recovery plans under the Endangered Species Act (Table 18–1). Thousands more are awaiting evaluation so that their status can be verified and plans made for their protection.

The species named on this long list are only part of the problem. Nationwide declines in nonlisted wild species are also occurring. Commercial landings of fish are down 42 percent since 1982; waterfowl populations have shown an overall decline of 30 percent since 1969; many U.S. songbird species, such as the Kentucky Warbler, Wood Thrush, and Scarlet Tanager, have been declining and have disappeared entirely from some local regions. The North American Breeding Bird Survey conducts a census of breeding birds in 2000 areas of the United States, and their latest results show that populations of 20 species of songbirds declined significantly between 1978 and 1987.

Worldwide, the loss of biodiversity is even greater. Most of the extinctions of the past several hundred years have occurred on oceanic islands, where small land masses limit the size of populations

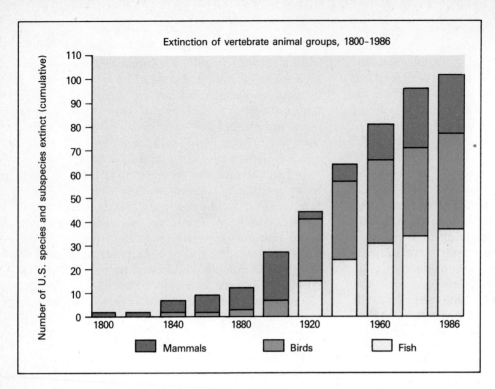

FIGURE 18–4
Extinction of vertebrate animal groups in the United States, 1800–1986. The cumulative extinction of mammals, birds, and fish is shown. (From *Environmental Trends*, 1989. Washington, D.C.: Council on Environmental Quality, p. 108.)

and human intrusions are most severe. Recently, attention has focused on the tropics, where biodiversity is at a peak. Here the rate of loss of species is estimated from information on the richness of species in tropical habitats. This richness is almost unimaginable. E. O. Wilson identified 43 species of ants from a single tree in a Peruvian rainforest, a level of diversity equal to the entire ant fauna of the British Isles. Other scientists found 300 tree species in a single 2.5-acre

(1-hectare) plot and as many as 10 000 insect species on a single tree in Peru.

Assuming there are 2 million species in the tropical forests (probably a conservative estimate) and a clearance rate of 1.8 percent per year for those forests, Wilson calculated that tropical deforestation is responsible for the loss of 4 000 species a year. Other scientists have projected that as many as 750 000 species will be lost by the end of the twentieth century!

TABLE 18–1

1990 Federal listings for threatened and endangered U.S. plant and animal species

Category	Endangered	Threatened	Total	Species with Recovery Plans
Mammals	53	8	61	29
Birds	74	11	85	69
Reptiles	16	17	33	25
Amphibians	6	5	11	6
Fishes	54	33	87	44
Snails	3	6	9	7
Clams	37	2	39	29
Crustaceans	8	2	10	5
Insects	11	9	20	12
Arachnids	3	0	3	0
Plants	180	60	240	120
Total	445	153	598	351
Total endangered U.S. species	445	(265 animals, 180 plants)		
Total threatened U.S. species	153	(93 animals, 60 plants)		
Total listed U.S. species	598	(358 animals, 240 plants)		

Source: Department of the Interior, U.S. Fish and Wildlife Service, *Endangered Special Technical Bulletin* 15(11):16 (December 1990).

These calculations have led some workers to use the term *holocaust* in describing the dimensions of the problem. Not all scientists agree with these calculations, however (see In Perspective Box below). They point to the great number of assumptions that must be made and the lack of good data on which to base those assumptions. The most important conclusion we can draw is that many species are in decline and some are becoming extinct. No one knows how many, but the loss is real, and it represents a continuing depletion of the biological wealth of our planet.

Physical Alteration of Habitats

The greatest loss of biodiversity is caused by the physical alteration of habitats:

1. **Conversion.** Natural areas are converted to farms, housing subdivisions, shopping malls, marinas, and industrial centers. When, for example, a forest

is cleared, it is not just the trees that are destroyed. Every other plant and animal that occupies that destroyed ecosystem either permanently or temporarily (migrating birds) also suffers. The idea that this wildlife will simply move "next door" and continue to live in an undisturbed section is erroneous. As we saw in Chapter 4, population balances lead to each area having all the wildlife it can support. Any loss of natural habitat can result in only one thing: a proportional reduction in all populations that require that habitat. Thus, the decline in songbird populations cited above has been traced to a combination of the loss of winter forest habitat in Central and South America and the increasing fragmentation of summer forest habitat in North America.

2. **Fragmentation.** A certain minimum area is required to support a *critical number* of individuals of any natural population. This minimum area must be large enough to compensate for years of

IN PERSPECTIVE
The Dangers of Crying Wolf

Biodiversity is under assault worldwide, and a number of prominent biologists have gone on record documenting a catastrophic loss of species and the dire consequences. Current losses are estimated to be as high as 40 000 species a year in the tropical forests alone. At current rates of clearing land, some scientists estimate that we will lose one-fourth of all species within 50 years. If there are 30 million species, as some taxonomists suspect, this means the loss of 7.5 million species! In light of these possibilities, calls are going out for such responses as stopping the development of all remaining natural lands in the developed world and giving huge sums of money to the poor nations so that they can quickly bring population growth down and convert their agriculture to sustainable, high-yield use of currently existing croplands.

This sounding of the alarm, according to some critics, has become a dogma among many biologists and

is based on some highly questionable assumptions. Three key assumptions lie behind the "mega-extinction" scenario. One is that habitat losses always mean species losses and often in direct proportion. Forests cleared are not necessarily forests gone forever, however; many are converted to second-growth forests, which support many of the original species.

A second assumption is that the larger a geographical area, the larger the number of species it holds. In many large areas, however, the number of species levels off even though the geographical area is large. So there may be no net loss of species even though some of an area is lost.

The third questionable assumption is that the number of species in existence is far larger than the number of species we know about; hence estimated extinction numbers are similarly high. Thus, the catastrophic extinction rates predicted by biologists are based

largely on estimated numbers of species that no one has ever seen. Also, there is a disturbing lack of agreement among observers who calculate the number of existing species, often an order of magnitude or greater.

Support for the mega-extinction scenario is so strong that those who challenge it from within the biological research community risk their reputations and the normally cordial relationships expected from colleagues. Challengers do exist despite the large stakes, however, and these critics are calling for their colleagues to examine their data and base their predictions only on supportable evidence. They also ask that only realistic public policy alternatives be advocated.

The problem is, if the dire predictions are not shown to be true, some of the real problems—tropical deforestation and wetlands loss, for instance—will also get less than the attention they deserve.

adverse weather. That is, more area will be required during a dry year than during a normal year. If development reduces the habitat to a point where it cannot support the critical number during an adverse year, the entire population will perish. Similarly, development (such as a highway) that fragments a territory and prevents migration between the two fragments will cause a population to perish if neither area can adequately support the critical number. Also, reducing the size of a habitat creates a greater proportion of edges, a situation that favors some species but may be detrimental to others. For example, the Kirtland's Warbler is an endangered species highly dependent on large patches of second-growth jack pines. It is endangered because its habitat has been greatly fragmented, creating edges that favor the Brown-headed Cowbird, a nest parasite that can invade the forest and lay its eggs in the nest of the rare warbler.

3. **Simplification.** Human use of habitats often simplifies them. We might, for example, remove fallen logs and dead trees from woodlands for firewood, thus diminishing an important microhabitat on which several species depend. When a forest is managed for production of a few or one species of tree, tree diversity obviously declines, and with it so does the diversity of a cluster of plant and animal species dependent on the less favored kinds of tree. Streams are sometimes "channelized" by clearing the streambed of fallen trees and riffles and even straightening out the stream by dredging. Such alterations inevitably bring on a loss of diversity of fish and stream invertebrates.

The Population Factor

Past losses in biodiversity can be attributed to the expansion of the human population over the globe. Continuing human population growth will bring on continued alteration of natural ecosystems and the inevitable loss of more wild species. The losses will be greatest where human population growth is highest: in the Third World (where biodiversity is also greatest). People's desire for a better way of life, the desperate poverty of rural populations, the global market for timber and other natural resources—these are powerful forces that will continue to draw down biological wealth.

The key to holding down the loss in biodiversity lies in bringing human population growth to a halt. There is an inverse relationship between human population size and the survival of species worldwide, illustrated in Figure 18–5. If the human population increases to 12 billion, as some demographers believe

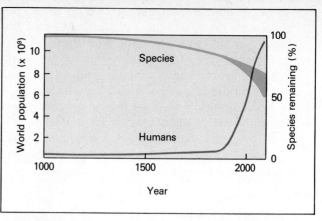

FIGURE 18–5
The inverse relationship between human population size and species survival worldwide. Uncertainty about the extent of species becoming extinct is reflected in the width of the species curve. (From M.E. Soulé, "Conservation: Tactics for a Constant Crisis," *Science*, 253 [1991], 744. Copyright 1991 by the AAAS.)

it will, the consequences for the natural world are enormous.

Pollution

Another major factor causing loss of biodiversity is pollution. Of course, pollution is a form of habitat destruction or alteration. We make it a separate category in this discussion, however, because the cause is not a direct attack on a natural ecosystem but an unintended "side effect" of other activities.

Despite the side-effect nature of pollution, its consequences may be just as severe as those caused by deliberate conversions. Forest dieback is caused by acid deposition and air pollution; species dieoff on Chesapeake Bay is due to sediments and nutrients; DDT devastates wild bird populations; the list is endless. Some scientists project that global warming from the greenhouse effect may be the greatest catastrophe to hit natural biota in 60 million years. (The fossil record reveals that a massive episode of extinctions of plants and animals occurred at that time. The probable cause was the earth being hit by a major asteroid, but there is still controversy regarding this hypothesis.)

The reason for projecting catastrophic effects from global warming is that most species adapt only very slowly and hence can adapt only to very gradual changes. The greenhouse effect may cause more warming in the next 50 years than what would normally occur in the next 1000 years, a 40-fold increase in the rate of change. It is probable that numerous species will be unable to adapt so rapidly. Nor can they migrate quickly enough. Trees, for example, can spread their seeds only a few miles each generation. Seeds landing this distance from the parent must grow to mature trees and shed seeds before the population can move the next few miles, and so on. Scientists speculate that the rate of climatic warming will far outpace the ability of most tree species to migrate northward, thus trapping them in inhospitable climates. Every species that dies out doubtlessly will take others with it. If the forest trees can't migrate, neither can the rest of the wildlife that depend on them for food and habitat.

Exotic Species

An **exotic species** is one introduced into an area from somewhere else, often a different continent. Because the exotic species is not native to the new area, it is often unsuccessful in establishing a viable population and quietly disappears. This is the fate of many pet birds, reptiles, and fish that escape or are deliberately released. Occasionally, however, an introduced species finds the new environment very much to its liking. As we saw in Chapter 10, most of the insect pests and plant parasites that plague agricultural production were accidentally introduced. Exotic species are major agents in driving native species to extinction.

The transplantation of species by humans has occurred throughout history, to the point where most people are not aware of the distinction between native and exotic species living in their lands. Literally hundreds of weeds and forage plants were brought by the European colonists to the Americas; most of the common field, lawn, and roadside plant species in eastern North America are exotics. Among the animals, the most notorious have been the house mouse and the Norway rat; others include the wild boar, donkey, horse, nutria, and even the red fox, which was brought to provide better fox hunting for the early colonial equestrians. One of the most destructive exotics is the house cat. The 60 million domestic and feral cats in the United States are very efficient at catching small mammals and birds. One recent study of cat predation in the British Isles showed that cats kill 20 million birds a year in Great Britain.

It might be supposed that introduced species would add to the biodiversity of a region, but many introductions have exactly the opposite effect. The new species are often very successful predators that eliminate native species not adapted to their presence. For example, the brown tree snake was introduced to the island of Guam as a stowaway on cargo ships during World War II (Fig. 18–6). This snake is mildly poisonous, grows to a length of 13 feet and eats almost anything. In little over 40 years, the snakes have eliminated most of the island's birds; 6

FIGURE 18–6
The brown tree snake, accidentally introduced onto Guam, has decimated bird life on the island. Often lacking in natural predators, islands are especially vulnerable to the harmful effects of exotic species. (Klaus Uhlenhut/Animals, Animals.)

of Guam's 18 native species of birds have become extinct, and populations of the remaining birds are so decimated that survivors are rarely seen or heard. Efforts at controlling the snakes have proved futile. There are no natural predators on the island, and the snake population has reached such high density that they have invaded houses in search of prey (such as puppies and cats). At present, wildlife officials are concentrating their efforts on preventing the snakes from spreading to other islands in the Marianas chain.

Other ecological impacts occur when an exotic species is introduced, as documented in Chapter 4, the most common being competition for resources. An Asian weed, hydrilla, was brought to Florida from India by a man selling plants for home aquariums. Since 1950 it has spread throughout lakes and reservoirs in the South, and federal agencies are spending $5 million a year in attempts to control it. The weed chokes intakes of power plants and water works, and because of its tendency to spread over the water surface, causes fish and invertebrates to die by preventing oxygen from reaching the deep water.

One of the most recent invading exotics is the zebra mussel (see In Perspective Box, p. 414), which came to the United States in 1986 in the ballast water of ocean-going vessels. The zebra mussel is expected to cause $4 billion in damage in the Great Lakes alone

Invasion of the Zebra Mussels

Since the early 1800s, the Great Lakes have become host to at least 115 exotic species of plants, fish, algae, and molluscs. A few of the invaders were deliberately introduced (silver and Chinook salmon, for example), but most came in by accident. Ships account for the largest percentage of the exotics, especially through their ballast water.

In 1986, a ship took in ballast water from the Elbe or Rhine River in Europe and released the water in the St. Clair River near Detroit when it took on cargo. In this ballast water there must have been a sizable quantity of the planktonic larvae of *Dreissena polymorpha,* the zebra mussel. The larvae settled to the bottom and soon grew to adult size (about 1.5 inches, or 4 cm, long) and began to reproduce—with a vengeance! The mussel has now been sighted in all of the Great Lakes and will undoubtedly spread to New York and mid-Western lakes, the Mississippi River drainage and into the Southeast. The latest report shows that zebra mussels have entered the Hudson River.

The mussels have found the biological conditions in the Great Lakes ideal. Mussel densities greater than 30 000 individuals per square yard have been found. They attach to any hard surface with strong

threads (called byssal threads) and feed on plankton carried to them by normal currents. Unfortunately, they particularly like hard surfaces where water flow is continuous and so have settled in great densities on the water intake pipes of municipal water supplies and thermal power stations. There they pile up on top of each other and eventually block the pipes. Repairs in Detroit and Monroe, Michigan, and Cleveland, Ohio, have already cost $300 million; rough estimates of the damage for the entire Great Lakes have run to $4 billion over the next decade.

As severe as this intake-pipe problem is, a greater fear is that the mussels will bring about catastrophic changes in the ecology of the Great Lakes. Where they reach high densities, they are able to filter virtually all of the larger plankton from the water. Indeed, in the eutrophic European waters to which they are native, the mussels have a positive impact on the water as they filter out the dense plankton and detritus. Food chains that support the Great Lakes fisheries are plankton-based, however, and it is feared that the mussels will intercept crucial quantities of the plankton and thus cut back greatly on energy flow to the higher trophic levels (the desirable fish species). The mussels

also may cause the demise of native bivalve populations through competition for food and because of their habit of attaching to the shells of the other bivalves. There are several fish species that prey on the zebra mussel in Europe and keep numbers from reaching pest levels, but similar species are lacking in North America and no one believes that importing yet another exotic species is a good idea.

In response to this invasion, Congress, in October 1990, passed the Non-indigenous Aquatic Nuisance Prevention and Control Act—too late to stop the zebra mussels but intended to prevent further invasions by exotics brought in via ballast water by prohibiting ballast water discharge. The act also funds research and control programs directed toward minimizing the impact of the zebra mussels. In the meantime, biologists are waiting for the zebra mussels to reach their peak and, it is hoped, settle down to an equilibrium level where their impact is diminished. Because they are so dependent on plankton, high densities of mussels will probably continue to occur only in the most eutrophic regions of the Great Lakes: Lake St. Clair, Lake Erie, and possibly Green Bay and Saginaw Bay.

over the next decade; its impact on native species is likely to be devastating.

Overuse

It is obvious that killing whales, fish, trees, or any other species at a rate exceeding the species' reproductive rate will lead to the ultimate extinction of the species. Yet, deliberate overuse is another major assault against wild species. Such overuse is driven by a combination of economic greed, ignorance, and insensitivity.

Some consumers are willing to pay exorbitant

prices for such things as furniture made from tropical hardwoods like mahogany and for exotic pets, furs from wild animals, and innumerable other "luxuries," including polar bear rugs, baskets made from elephant and rhinoceros feet, ivory-handled knives, and reptile-skin shoes and handbags. For example, some Indonesian and South American parrots sell for up to $5 000 in the United States. A panda-skin rug can bring $25 000. The gall bladder of the North American black bear is valued as a folk medicine in Korea, where a single gall bladder can fetch as much as $2000. Such prices create a powerful economic incentive to exploit the species involved.

FIGURE 18–7
"Products" from endangered species confiscated in the breakup of a black market ring. Even though trade in such products is illegal and even though the animals are protected, poaching and black market trade are still causing merciless slaughter for often trivial products. (U.S. Fish and Wildlife Service photo. Steve Hillebrand.)

The prospect of extinction does not curtail the activities of exploiters because the prospect of a huge immediate profit outweighs other considerations. Even when the species is protected, the economic incentive is such that poaching and black market trade continue (Fig. 18–7). The World Wildlife Federation reports that poaching and black market trade in endangered species have reached epidemic proportions and continue despite efforts to control them.

It is easy to place the blame on the economic greed of the persons doing the killing. However, equal blame must be shared by the consumers who offer the "reward." It is their ignorance or insensitivity to the fact that their money is fostering the extinction of invaluable wildlife that is fueling the situation. Unfortunately, the situation shows no sign of abatement.

Particularly severe is the growing fad for exotic "pets": fish, reptiles, birds, and house plants. In virtually every case, these species are gathered, trapped, or hunted in the wild. Keeping these animals or plants for pets may seem acceptable since there is no intent to kill them. However, for every living specimen that makes it to the buyer, several are killed or die in the process of capture, transportation, and marketing. Few survive in captivity for any length of time. And even when these plants and animals do survive, they are still removed from the natural breeding population. In terms of maintaining the species, they are no better than dead. Numerous species of tropical fish, birds, reptiles, and plants are thus headed toward extinction because of exploitation by the pet trade.

Even more severe examples of overuse are the overcutting of forests for firewood, the overcultiva-

tion of land, and overgrazing. These practices not only deplete the resource in question, they set into motion a cycle of erosion and desertification with effects far beyond the exploited area. Figure 18–8 provides a summary of the major factors causing the decline in biodiversity.

CONSEQUENCES OF LOSS OF BIODIVERSITY

What will happen as more and more rare or unknown species pass from the earth because of our activities? A few people will mourn their passing, but as biologist Norman Myers put it, the sun will continue to come up every morning, and the newspaper will appear on our doorstep. Currently, we seem to be getting away with it, as evidenced by the fact that many species have already become extinct as a consequence of human activities, and life goes on. However, we really do not know what we are losing when we lose species. Some ecologists have likened biodiversity loss to an airplane flight, where we continually pull out rivets as the plane cruises along. How many rivets can we pull out? Is this a wise activity? So far, we have gotten away with driving species to extinction, but the world is certainly less beautiful and less sustainable as a result.

We might someday lose what ecologists call a **keystone species,** a species whose role is absolutely vital for the survival of many other species in an ecosystem. For example, a group of insects called orchid bees play a vital role in tropical forests by pollinating trees. These bees travel great distances and are able to pollinate trees and other plants that are often widely separated from each other. If these bees dis-

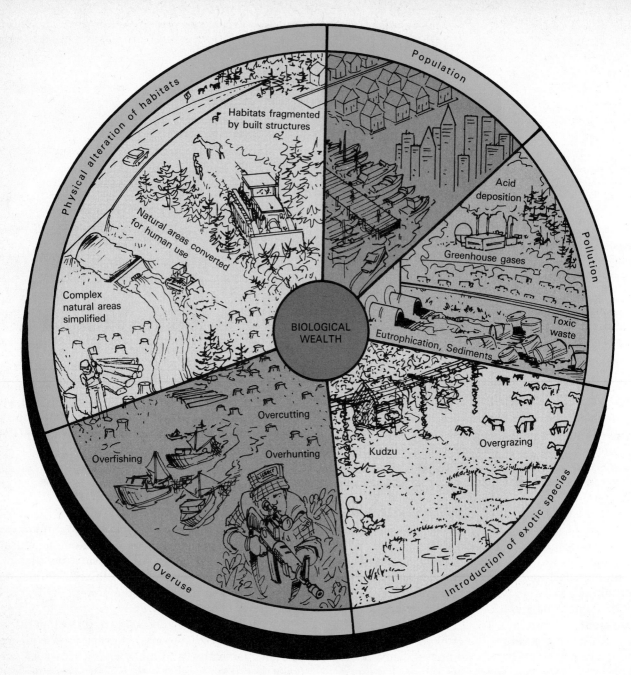

FIGURE 18–8
A summary of the ways in which biological wealth is being threatened, causing a decline in biodiversity.

appear, many other plants, including major forest trees, will eventually follow. Or, perhaps the keystone species in an ecosystem is a predator that keeps herbivore populations under control.

Furthermore, biodiversity will continue to decline because we shall continue to lose the natural habitats in which wild species live. This loss is definitely costly because natural ecosystems provide vital services to human societies. Recreational, aesthetic, and commercial losses will be inevitable. The loss of wild species, therefore, will bring certain and unwelcome consequences, because such loss is linked

directly to the degradation or disappearance of ecosystems (our focus in Chapter 19).

Saving Wild Species

Support is growing in the United States in favor of preserving wild species. This support is coming from a broad constituency of people who are united in their desire to staunch the loss of biodiversity. Let us turn our attention to some of those actions.

GAME ANIMALS IN THE UNITED STATES

Game animals are those traditionally hunted for sport or meat. In the early days of the United States, there were no restrictions on hunting, and a number of species were hunted to near extinction (bison, wild turkey) or extinction (great auk, passenger pigeon). As game animals became scarce in the face of increasing hunting pressure, regulations came into effect. State governments backed up by the federal government began to set regulations and hire wardens for enforcement.

One success story is the wild turkey. A favorite game species, this bird was hunted to the brink of extinction but was making a slow comeback by the 1930s as a result of hunting restrictions. At that time there was a total population of about 30 000 individuals in a few scattered states. After World War II state and federal programs addressed the need for protecting turkey habitats. The birds were reintroduced into areas they had once inhabited, and hunting quotas were strictly limited. The turkey is now found in 49 states and has risen to a total population of 4.5 million as a result of these measures.

Using hunting fees as the revenue source, state wildlife managers enhance the habitats supporting important game species and provide special areas for hunting. They monitor populations and adjust seasons and bag limits accordingly. Game preserves, parks, and other areas where hunting is prohibited are maintained to protect both habitat and certain breeding populations.

Common game animals, such as deer, rabbits, and squirrels, are well adapted to the rural field and woods environment. Being thus adapted to the humanized environment and protected from overhunting, populations of these common game animals are being maintained. However, some problems are emerging:

1. The number of animals killed on roadways now far exceeds the number killed by hunters. Increasing numbers of animals on roadways as rural areas are developed are a serious hazard to motorists, with scores of people killed annually from collisions with deer and moose.

2. Opossums, skunks, raccoons, and deer are thriving in highly urbanized areas, creating different hazards. For example, in 1992 a rabies epidemic among "urban" raccoons was a public health risk in several states in the eastern United States.

3. Some game animals have no predators except hunters and tend to reach population densities that push them into suburban habitats where they cannot be effectively hunted. The white-tailed deer, for example, has become a pest to gardeners and fruit nurseries and also poses a public health risk because it is often infested with Lyme disease ticks.

THE ENDANGERED SPECIES ACT

In colonial days, huge flocks of snowy egrets inhabited coastal wetlands and marshes of the southeastern United States. In the 1800s, when women's fashion turned to fancy hats adorned with feathers, egrets and other birds were hunted for their plumage. By the late 1800s, egrets were almost extinct. In 1886, the newly formed Audubon Society began a press campaign to shame "feather wearers" and end this "terrible folly." The campaign caught on, and gradually attitudes changed and laws followed.

Florida and Texas were first to pass laws protecting plumed birds. Then, in 1900, Congress passed the **Lacey Act,** which forbids interstate commerce in illegally killed wildlife, making it more difficult for hunters to sell their kill. Since then, numerous wildlife refuges have been established to protect the birds' breeding habitats (Fig. 18–9). With millions of people visiting these refuges and seeing these birds in their natural locales, attitudes have changed significantly.

FIGURE 18–9
A tricolor heron stabs at its prey in Everglades National Park. Herons and egrets were brought close to extinction in the late 19th century by plume hunters. Such wanton killing is a thing of the past, but habitat destruction is causing heron and egret populations to decline once again. (E. R. Degginger/Animals, Animals.)

Today the thought of hunting these birds would be abhorrent to most of us, even if official protection were removed. Thus protected, egret populations were able to recover substantially. However, their numbers are headed down once again since wetland habitats outside the very limited refuges continue to be destroyed and degraded by pollution.

Congress took a major step to protect species from extinction when it passed the **Endangered Species Act** of 1973 (most recently reauthorized in 1988). As we learned at the beginning of Chapter 4, an **endangered species** is defined as one that has been reduced to the point where it is in imminent danger of becoming extinct if protection is not provided. The act also provides for the protection of **threatened species,** species judged to be in jeopardy but not on the

---

IN PERSPECTIVE
A Beach for the Birds

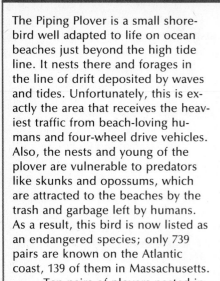

The Piping Plover is a small shorebird well adapted to life on ocean beaches just beyond the high tide line. It nests there and forages in the line of drift deposited by waves and tides. Unfortunately, this is exactly the area that receives the heaviest traffic from beach-loving humans and four-wheel drive vehicles. Also, the nests and young of the plover are vulnerable to predators like skunks and opossums, which are attracted to the beaches by the trash and garbage left by humans. As a result, this bird is now listed as an endangered species; only 739 pairs are known on the Atlantic coast, 139 of them in Massachusetts.

Ten pairs of plovers nested in 1990 on Plum Island, a part of the Parker River National Wildlife Refuge north of Boston. Because parts of the beach have been closed to protect the birds, these 10 pairs successfully raised 14 young in 1990, and refuge biologists believe that the population has the potential to increase. Accordingly, refuge manager Jack Filio closed the entire 6 miles (10 km) of beach between April and early July 1991 in order to allow the piping plovers to nest undisturbed. The closure met with angry protests from local business people and politicians concerned about the impact of the decision on local tourism.

"Nobody argues the point that endangered species should be protected, but at what cost to the community?" asked a spokesman.

Protestors accused the refuge manager of putting animals before people and of being unwilling to compromise. Filio pointed out that the primary purpose of the refuge is the protection of wildlife, especially endangered or threatened species. Recreational use is secondary to these concerns.

In spite of the closure, only 5 pairs nested in 1991, but they successfully fledged 13 young. The beach was reopened for recreation in phases as chicks in various nests reached the age of 35 days, when they could fly. (Refuge personnel believed that the decrease in number of nesting pairs was a consequence of winter storm erosion, which removed much of the best upper beach habitats over long stretches of the beach.)

Events at Plum Island are symptomatic of a broader problem. The U.S. Fish and Wildlife Service has been severely criticized by some members of Congress and a General Accounting Office report for allowing incompatible and harmful uses on some of the refuges it manages. The Parker River refuge has clearly taken a stand in favor of the wildlife it is charged with protecting.

The piping plover is an endangered species that breeds on outer beaches just above the highwater line. It is showing remarkable recovery in response to measures that protect it from human and predator intrusion. (Perry D. Slocum/Animals, Animals.)

---

brink of extinction. When a species is officially recognized as being either endangered or threatened, the law provides for substantial fines for any killing, trapping, uprooting (plants), or commerce of it or its parts. (Because the United States is a major market, making such commerce illegal protects endangered species of other countries as well as those in the United States.) The act requires the U.S. Fish and Wildlife Service to draft recovery plans for protected species. Habitats must be mapped, and a program for preservation and management of critical habitats must be designed, such that the species can rebuild its population. As Table 18–1 indicates, almost 600 species are currently listed for protection under the act, and recovery plans are in place for 350 of those.

There are several shortcomings in the Endangered Species Act. A major one is that protection is not provided until a species is officially listed as endangered or threatened by the Fish and Wildlife Service. Most ecologists are extremely concerned by the slow pace at which species are being added to the official list. Hundreds of species in the United States—over 400 in Hawaii alone—may either become extinct or else have their habitats so reduced that extinction is inevitable before they can be listed.

The listing of a species is sometimes a controversial project, as is shown by the case of the Spotted Owl. This species has become a focal point in the battle to save some of the remaining old-growth forests of the Pacific Northwest. The owl is found only in these forests and has dwindled to a population between 6000 and 8000. In June 1990, the Fish and Wildlife Service listed the owl as a threatened species. The listing prompted the service to set aside 6.9 million acres (2.8 million hectares) of the old-growth forests, enough to guarantee the owl protection into the future. Neither environmentalists nor members of the timber industry are pleased with the action. Calling the plan "a legal lynching of an entire region by an out-of-control federal agency," the timber industry claims that 33000 jobs will be lost if the forests are protected from cutting. (The service admitted that jobs will be lost but stated that many of these jobs would be lost anyway because of other market forces.) The Wilderness Society claims that the acreage is insufficient to ensure the owl's survival and that political pressure from the timber industry prevented the agency from setting aside a much larger area.

A second major problem of the Act is that, despite official protection, poaching and illegal trade remain rampant because there is insufficient funding provided in the Act to create adequate enforcement. This problem is worsening as increasing numbers of people in the United States are unable to find adequate employment and turn to illicit activities.

A few species have gained exceptional public attention, and heroic efforts have been mounted to save them. Efforts to save the Whooping Crane, for example, included virtually full-time monitoring and protection of the single remaining flock, which for years numbered only in the teens. To obtain a higher reproduction and recruitment rate, eggs were collected and artificially incubated (when eggs are continually removed from a nest, the female can produce up to 14 eggs!). The chicks hatched in incubation and were then placed in nests of related Sandhill Cranes to be raised by them as foster parents (Fig. 18–10). This effort seems to have paid off; the Whooping Crane is now at least holding its own.

Similar efforts to save the California Condor have recently met with some success. In a controversial move, wildlife officials trapped the five remaining wild condors in the mid-1980s. With successful breeding in captivity in the Los Angeles Zoo, the population is now over 50 individuals. Early in 1992, a pair of the condors was released back into the wilds.

Most large zoos in the United States are beginning to take an active role in breeding other endangered animals, such as the rhinoceros and elephant. Seeds are being collected from wild plants and stored in seed banks maintained by international cooperative efforts among universities and governments. Many ecologists believe that this is the only way left to save endangered species. However, for every an-

FIGURE 18–10
A 10-day-old Whooping Crane chick being reared in captivity by Sandhill Crane foster parents. (Courtesy of U.S. Fish and Wildlife Service.)

imal or plant that receives enough attention to be saved from extinction by breeding a few captive individuals, hundreds of lesser-known species will doubtlessly become extinct as habitat destruction and other factors take their toll.

The few individuals maintained and bred in captivity are a very limited representation of the total genetic diversity that existed in the wild population. All zoos can do is make a concerted effort to enhance genetic diversity by exchanging individual animals and frozen sperm for artificial insemination.

The Endangered Species Act is a formal recognition of the importance of preserving wild species regardless of economic importance—in effect, imparting *intrinsic value* to wild species. Species listed have legal rights to protection under the law. The act is something of a last resort for wild species, but it embodies an encouraging attitude toward nature that has now become public policy. Actions taken under auspices of the act have demonstrated a firm com-

mitment to preserve wild species, even if it involves some economic losses. Funding is currently about $10 million a year, which is far less than needed to accomplish full listing and recovery, but it is a major step in the right direction.

FURTHER STEPS

It is the continuing loss of habitat worldwide that is largely responsible for the current trend in loss of biodiversity. Strategies to address this loss must focus on preserving the natural ecosystems that sustain wild species. In the next chapter, we shall turn our attention to those ecosystems—how we depend on them, what we are doing to them, and what we must do to resist the forces that seem to be leading to a future where all that is left of wild nature is what we have managed to preserve in parks, preserves, and zoos.

 Review Questions _____

1. Define *instrumental value* and *intrinsic value* as they relate to determining the worth of natural species.
2. Describe ways in which we humans of today are still dependent on natural species.
3. What is biodiversity? What are the best current estimates of its extent?
4. Describe both the known and estimated losses in biodiversity.
5. How do habitat conversion, fragmentation, and simplification affect biodiversity?
6. How do human population growth, pollution and overuse affect biodiversity?

7. What impacts can exotic species have on biodiversity?
8. What are three possible consequences of species extinctions?
9. In what ways can intrinsic value be assigned to wild species?
10. Discuss the negative and positive effects of game animal management on biodiversity.
11. How does the Endangered Species Act preserve rare species in the United States? What are the shortcomings of this act?

 Thinking Environmentally _____

1. Choose an endangered or threatened animal or plant species and research what is currently being done to preserve it. What dangers is it subject to? Write a protection plan for it.
2. You have been given the task of maintaining or increasing biodiversity on a small island. What steps will you take?
3. Some people argue that each individual animal has an intrinsic right to survival. Should this right extend to plants and microorganisms? Justify your answer.

4. Monies for enforcing the Endangered Species Act are limited. How should we decide which species are the most important to save?
5. Is it right to use animals for teaching and research? Defend your view.
6. Do all species have a right to exist? What about the anopheles mosquito, which transmits malaria? Tigers that kill people in India? Bacteria that cause typhoid fever? Defend your position.

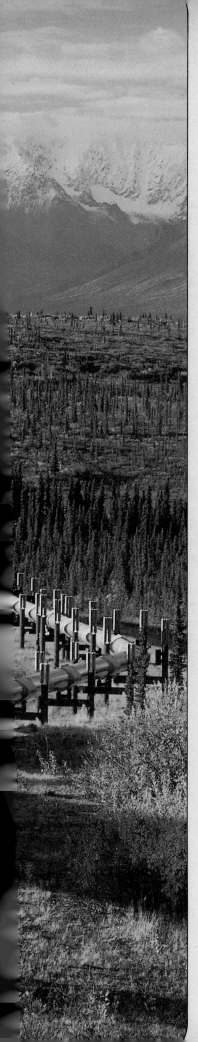

19

Ecosystems As Resources

LEARNING OBJECTIVES

When you have finished studying this chapter, you should be able to:

1. Describe the most important natural services performed by ecosystems.

2. Explain the tragedy of the commons and give an example of how it operates today.

3. Distinguish between conservation and preservation.

4. Explore the concept of the maximum sustainable yield and the difficulties of applying it.

5. Describe the recent activities in the United States directed toward restoring natural systems.

6. Understand why forests are cleared and the consequences of clearing.

7. Describe the problems caused by rainforest destruction, particularly recent developments in Brazil.

8. Trace recent trends in exploitation of the ocean fisheries.

9. Understand how and why whales are killed, the current moratorium on killing them, and the emergence of whale watching.

10. Describe the recent history of wetlands losses and the current problem of defining wetlands.

11. Describe the ways in which public lands are protected in the United States.

12. Understand the commercial use of forests in the United States and describe some problems with the management of those forests.

13. Evaluate land trusts as a means of protecting natural lands.

In Chapter 18 we focused on natural species as biological wealth capable of sustaining human life and economic activity if used wisely. As noted in Chapter 2, however, each species can be sustained only in the context of an overall

421

ecosystem. Thus preservation of ecosystems is necessary for the preservation of species. In addition, ecosystems provide many essential natural services for maintenance of the biosphere as they process energy and recycle nutrients and water.

It takes no great insight into world affairs to observe that natural ecosystems have not been faring well; one has only to read the papers or look around. Much of this text has documented case after case where natural systems have been either lost to other uses or else degraded by pollution and unwise use. As we traced the loss of biodiversity in Chapter 18, we observed that the most vital way to save wild species is to preserve the ecosystems in which they are found— in itself an important reason to maintain natural ecosystems. Populations of wild species continue to decline, however, because ecosystems are also in decline. Beyond the loss of species is the degradation or loss of the natural services ecosystems provide.

To continue our banking analogy from the previous chapter, we are spending the principal of our biological wealth, exchanging it for short-term economic gain, a process that is unsustainable. On the surface, spending principal seems to be so foolish that it defies reason. However, there are strong forces at work that make such exploitation attractive.

In this chapter we look at major ecosystems that sustain human life and economy, examine the gains and losses as we convert them to other uses, and look in particular at those ecosystems that are in the deepest trouble.

Biological Systems in a Global Perspective

MAJOR SYSTEMS AND THEIR SERVICES

As we learned in the early chapters of this book, the earth is occupied by ecosystems that vary greatly in species composition but exhibit common functions such as energy flow and the cycling of matter. The major terrestrial and aquatic ecosystem types—the biomes—reflect the response of biota to different climatic conditions. We can divide the earth's area into several broad categories for purposes of this chapter (Table 19–1): **forests and woodlands, grasslands and savannahs, croplands, wetlands,** and **desertlands and tundra.** We can separate the oceanic ecosystems into **coastal ocean and bays, coral reefs,** and **open ocean.** These eight systems provide all of our food; much of our fuel; wood for lumber and paper; raw materials for fabrics, leather, and furs; oils and alcohols; and much else.

Ecosystems also perform a number of valuable **natural services** as they process energy and circulate matter in the normal course of their functioning (Fig. 19–1). Some of the services are general and pertain to essentially all ecosystems; others are specific to the categories indicated. Some of the most important are:

1. **Maintenance of the hydrological cycle.** Plants absorb water from soils and release it through transpiration, returning the water to the atmosphere. Forests, grasslands, and wetlands maintain a favorable distribution of water, absorbing it when it is abundant and releasing it gradually. In doing so, they prevent much flooding.

2. **Modification of the climate.** Because plants absorb considerable amounts of solar radiation and release water vapor through transpiration, ecosystems moderate temperature and help to maintain an even climate.

TABLE 19–1

Natural Ecosystems on the Earth's Surface		
Ecosystem	Area (million mi²)	Percent of Land Total
Forests and woodlands	18.2	31.7
Savannahs and grasslands	11.6	20.2
Croplands	6.2	10.7
Wetlands	2.0	3.5
Tundra, desertlands	19.6	34.0
Total land area	57.6	
Coastal ocean and bays	8.4	
Coral reefs	0.2	
Open ocean	164	
Total ocean area	172.6	

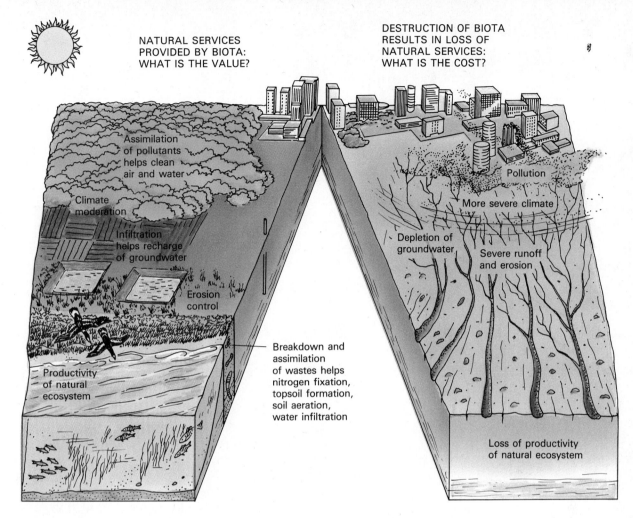

NATURAL SERVICES PROVIDED BY BIOTA: WHAT IS THE VALUE?

Assimilation of pollutants helps clean air and water

Climate moderation

Infiltration helps recharge of groundwater

Erosion control

Productivity of natural ecosystem

Breakdown and assimilation of wastes helps nitrogen fixation, topsoil formation, soil aeration, water infiltration

DESTRUCTION OF BIOTA RESULTS IN LOSS OF NATURAL SERVICES: WHAT IS THE COST?

Pollution

More severe climate

Depletion of groundwater

Severe runoff and erosion

Loss of productivity of natural ecosystem

FIGURE 19–1
Natural ecosystems perform invaluable natural services, including air and water purification, maintenance of chemical cycles, erosion control and soil building, climate control, and pest management.

3. **Absorption of pollutants.** Wetlands are particularly significant in stabilizing sediments, absorbing excess nutrients, and thus preventing eutrophication.

4. **Transformation of toxic chemicals.** Microbes in all ecosystems transform many toxic organic and inorganic chemicals into harmless products.

5. **Erosion control and soil building.** Plant cover and ground litter absorb the potentially destructive impact of rainfall and prevent soil breakup; plant roots bind the soil. Forests are especially crucial in preventing erosion on hilly terrain. All of the plants, small animals, and microorganisms in terrestrial systems contribute to soil formation.

6. **Pest management.** Natural ecosystems contain a diverse array of insect predators. Even small patches of natural habitat, such as hedgerows and woodlots, contribute to the control of pests in adjacent agricultural lands.

7. **Maintenance of the oxygen and nitrogen cycles.** Photosynthesis continuously regenerates oxygen, and soil microbes maintain soil fertility through nitrogen fixation. They also prevent the buildup of potentially harmful nitrogenous compounds (for example, ammonium and nitrite).

8. **Carbon storage and maintenance of the carbon cycle.** The global carbon cycle is maintained by energy flow within natural ecosystems. Over 500 billion metric tons of carbon is stored in the standing biomass of forests, more than is found in the entire atmosphere. Much more is present in the organic matter of soils.

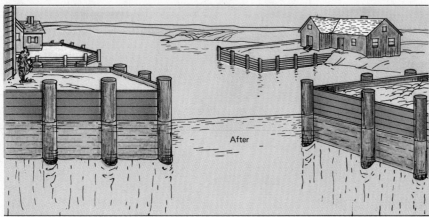

FIGURE 19–2
Tidal wetlands provide a number of valuable services, worth more than $100 000 a year for just one acre. Especially important are water purification and fish propagation. These services are lost when the wetland is bulkheaded and converted to vacation homes.

All natural and altered ecosystems (provided their functional integrity is maintained) perform some or all of these natural services, free of charge. We tend to take these services for granted until the ecosystems and thus the services are lost. For example, deforestation in India is largely responsible for the massive siltation and flooding that are causing human tragedy and suffering in Bangladesh. Loss of wetlands, which provided sediment and nutrient control, is a major factor in the eutrophication of Chesapeake Bay (Chapter 12).

To put it in monetary terms, scientists have calculated that it would cost more than $100 000 a year to artificially duplicate the water purification and fish propagation capacity provided by a single acre of natural tidal wetland (Fig. 19–2). Even if such sums were available for artificially carrying out these processes, the energy expenditures involved (producing and burning the requisite fuels) would probably lead to a net increase in pollution rather than a decrease. Thus, there is no real way we can compensate for the losses of natural services that will be incurred as natural ecosystems are destroyed. We shall simply have to suffer the consequent deterioration in quality of life.

ECOSYSTEMS AS NATURAL RESOURCES

If the natural services performed by ecosystems are so valuable, why are we still draining wetlands and cutting down old-growth forests? The answer is revealing: *A natural area will receive protection only if the value a society assigns to its natural functions is higher than the value the society assigns to exploiting its natural resources.* When we refer to natural ecosystems and the biota in them as **natural resources,** we are consciously placing them in an economic setting and losing sight of their ecological importance. A resource is expected to produce something of economic value for its owner. Thus, even though an acre of wetland provides valuable services for Chesapeake Bay, the owner of that acre will benefit much more by draining or filling it in and selling to a builder. Its conversion removes the acre as a functional wetland and thus represents a *loss* in natural services but an economic *gain* to its owner.

Some natural ecosystems are maintained in a natural or seminatural state because that is how they provide the greatest economic value for their owners. For example, most of the state of Maine is owned by

With the exception of the open oceans, all natural ecosystems come under the sovereignty of the different nation-states. Some natural lands are remote from human habitations, but the vast majority are found where humans can get access to them and to the resources they hold. Indeed, it was usually the products and services of the natural lands that drew humans to them in the first place and nurtured the increases in prosperity and population that followed.

Now, in the late twentieth century, essentially all natural lands are impacted by human exploitation. That exploitation is thoroughly managed in industrialized nations. In the United States, public lands are managed by federal and state agencies, such as the National Forest Service and state Fish and Game departments. Private lands are managed by the owners, with certain constraints placed on them by governmental authorities.

In many less developed nations, however, natural resources are exploited with little restraint, often by poor and needy people trying to subsist on what they can extract from the land or the sea. Concerns for conservation and preservation are usually swept aside when basic survival needs must be met.

In all cases, though, whether rich nation or poor, public land or private, natural lands are exploited and managed by *people*—not by agencies. For this reason, people's *attitudes* about nature are crucial. Some attitudes lead inexorably toward the destruction of nature and the tragedy of the commons: the same sad result regardless of whether the motive is short-term profit taking or basic survival in the face of dire human need. There is no wisdom in pursuing this course, no recognition of the rights and needs of future human generations, let alone the rights of the natural world.

There is one attitude, however, that is capable of bringing dignity to the human exploitation of nature—the attitude of the **steward.** A steward is one to whom a trust has been given; something has been placed under his or her care. Stewardship of natural lands and seas is desperately needed in this crowded and hungry world. It is an attitude that recognizes that humans do not own the earth, an attitude that sees beyond the present to the needs and rights of those yet to be born. For some people, stewardship is an attitude that stems from a recognition that the land and the wild things belong to their creator, God, who has

given dominion over His creation to humankind but who always holds His stewards accountable for their care. For other people, stewardship is a matter of wisdom that stems from a deep understanding of the natural world and the necessary limits of human use of that world.

It is encouraging that very often those who are given responsibility for some part of the natural world end up really *caring for* it. Those who work closest to nature often not only love their work—they also love the natural world they work in. It is common to find exploiters who have become caretakers or foresters and rangers who have risked their jobs to bring misuse to the attention of their bosses or the public.

Stewardly care of the natural world is also a possibility for *you,* as you examine your own attitudes and interactions with the environment. We all have opportunities to exercise stewardly care. Indeed, it may be true that unless the attitude of stewardship is widely appreciated and practiced in a society, and eventually on a global scale, human tenure of the earth in the twentieth century will be seen as the age of folly, a time when untold biological wealth was squandered beyond the point of no return.

---

private corporations that periodically harvest the timber for lumber and for paper manufacturing. However, if Maine experienced a population explosion and corporations could sell their land to developers for more money than can be gained from timber harvest, forested lands would quickly become house lots.

Finally, some natural ecosystems either are publically owned (state and federal lands) or cannot be owned (ocean ecosystems). These ecosystems are still considered natural resources and may be subject to economically motivated exploitation. Obviously, wise exploitation of these systems will maintain the natural services they perform. However, exploitation

is not always wise. We now address some of the issues surrounding human exploitation of natural ecosystems.

PATTERNS OF USE OF NATURAL ECOSYSTEMS

Tragedy of the Commons

Where a resource is owned by many people in common or by no one, it is known as a **common pool resource,** or a **commons.** Examples of natural resource commons are many: federal grasslands where private ranchers graze their livestock, coastal and open-ocean fisheries, groundwater, nationally owned woodlands and forests in the Third World, and so forth. Exploitation of such common pool resources presents some serious problems and will often lead to the eventual ruin of the resource—a phenomenon that is called the *tragedy of the commons.*

As described by biologist Garrett Hardin in a now classic essay by the same title, the original "commons" was pastureland in England provided free by the king to anyone who wished to graze cattle. Herdsmen were quick to realize that whoever grazed the most cattle stood to benefit most. Even if he realized that the commons was being overgrazed, any herdsman who withdrew his cattle simply sacrificed his own profits while others went on using the commons. His loss became their gain, and the commons was overgrazed in any case. Consequently, herdsmen would add to their herds until the commons was totally destroyed. They were locked into a system that led to their ruin.

Harvesting soft-shelled clams in New England provides a modern example. The clam flats in any coastal town are effectively a commons belonging to the townspeople; commercial access to them is limited to town residents. Although diggers may recognize that the clams are being overharvested, any digger who curtails his or her own take diminishes personal income while competitors do not. His or her loss becomes their gain, and the clams are depleted in any case. Thus, while the size and number of the clams diminish, indicating overharvesting, digging continues. Higher prices brought on by the shortage of clams continue to make the digging profitable despite declining harvests, until finally there are so few clams to harvest that the industry collapses. Coastal Massachusetts towns experience regular cycles of scarcity and recovery in the soft-shelled clam industry, a clear sign that the tragedy of the commons is at work.

Obviously, the tragedy of the commons can be avoided, but not without some limitation of freedom of access. One arrangement that usually mitigates the tragedy is *private ownership.* When a renewable natural resource is privately owned, access to it is restricted, and, in theory, it will be exploited in a manner that guarantees a continuing harvest for its owner(s). This theory does not hold, however, when an owner maximizes immediate income and then moves on. Where private ownership is unworkable, the alternative is that *access to the commons be regulated.* Regulation should allow for (1) protection so that the benefits derived from the commons can be sustained, (2) fairness in access rights, and (3) common consent of the regulated.

Conservation and Preservation

The beauty of natural biota is that it is a **renewable resource.** It has the capacity to replenish itself through reproduction despite certain quantities being taken, and this renewal can go on indefinitely. Recall from Chapter 4 that every species has the biotic potential to increase its numbers and that, in a balanced ecosystem, the excess numbers fall prey to parasites, predators, and other factors of environmental resistance. It is difficult to find fault with activities that effectively put some of this excess population to human use. The tragedy occurs when users (hunters, fishermen, loggers, and so on) take more than the excess and deplete the breeding populations, threatening or causing extinction of the species.

Conservation of natural biota, then, does not, or at least should not, imply no use by humans whatsoever, although it may sometimes be temporarily expedient in a management program to allow a certain species to recover its population size. Rather, the aim of conservation is to *manage or regulate use* so that it does not exceed the capacity of the species or system to renew itself.

Preservation is often confused with conservation. The objective of preservation of species and ecosystems is to *ensure their continuity regardless of their potential utility.* Effective preservation often precludes making use of the species or ecosystems in question. For example, it is not possible to maintain old-growth forests and at the same time harvest the trees. Thus a second-growth forest can be *conserved* (trees cut but at a rate that allows recovery) but an old-growth forest must be *preserved* (no cutting).

Maximum Sustainable Yield

The central question in managing a commons is: *How much continuous use can be sustained without undercutting the capacity of the species or system to renew itself?* The maximum use a system can sustain without impairing its ability to renew itself is called the **maximum sustainable yield** (MSY). This is *the highest possible rate of use that the system can match with its own rate of replacement or maintenance.*

The concept of maximum sustainable yield applies to more than just preservation of natural biota. It is also the central question in maintaining parks, air quality, water quality and quantity, soils, and, indeed, the entire biosphere. *Use* can refer to cutting of timber, hunting, fishing, number of park visitations, discharge of pollutants into air or water, and so on. Natural systems can withstand a certain amount of use (or abuse in terms of pollution) and still remain viable. However, a point exists where increasing use

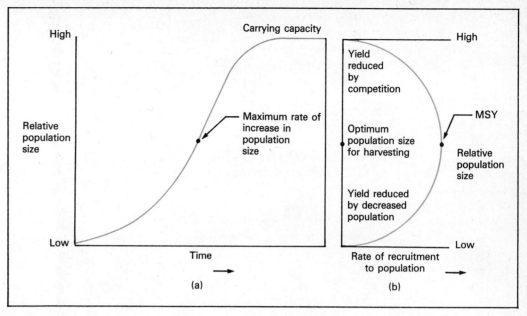

FIGURE 19–3
Maximum sustainable yield occurs not at the maximum population level but rather at an *optimal* population level. (a) The logistic curve of population size in relation to carrying capacity. The optimal population level is where the rate of population increase is at a maximum, which is well below the carrying capacity. (b) Here recruitment is plotted against population size, showing the effects of competition and decreased population levels. The maximum sustainable yield occurs where the population is at the optimal level, where population increase rate is at a maximum.

begins to destroy regenerative capacity. Just short of that point is the maximum sustainable yield.

An important consideration in determining MSY is the carrying capacity of the ecosystem, the maximum population the ecosystem can support on a sustainable basis. If population size is well below the carrying capacity of the ecosystem (Fig. 19–3a), then allowing a population to grow will increase the number of reproductive individuals and thus the yield that can be harvested. However, as population size approaches carrying capacity, new individuals must compete with older individuals for food and living space. As a result, recruitment may fall drastically (Fig. 19–3b). When a population is at or near carrying capacity, production—and hence sustainable yield—can be increased by thinning the population so that competition is reduced and optimal growth and reproductive rates are achieved. Thus, MSY is obtained with optimal, not maximum, population size.

The matter is further complicated by the fact that carrying capacity and hence optimal population are not constant. They may vary from year to year as weather fluctuates. Replacement may also vary from year to year since some years are particularly favorable to reproduction and recruitment while others are

not. Of course human impacts such as pollution and other forms of habitat alteration adversely affect reproductive rates, recruitment, carrying capacity, and consequently sustainable yields (Fig. 19–4).

Managing natural populations to achieve MSY is fraught with difficulties. Accurate estimates must constantly be made of the population size and the recruitment rate. The usual approach is to use the estimated MSY to set a fixed quota. If the data on population size and recruitment are not accurate, it is easy to overestimate the MSY, especially when there are economic pressures to maintain a high harvest quota. Later in the chapter we shall see how the MSY approach has failed to prevent the overharvesting of many marine fish species.

In summary, to achieve the objectives of conservation, we must be aware of (1) the concept of MSY and (2) social and economic factors that cause overuse and other forms of environmental degradation that diminish MSY. We must then establish and obey regulations that are necessary to protect natural resources:

○ Natural resources cannot be treated as a commons because, wherever they have been, the tragedy of

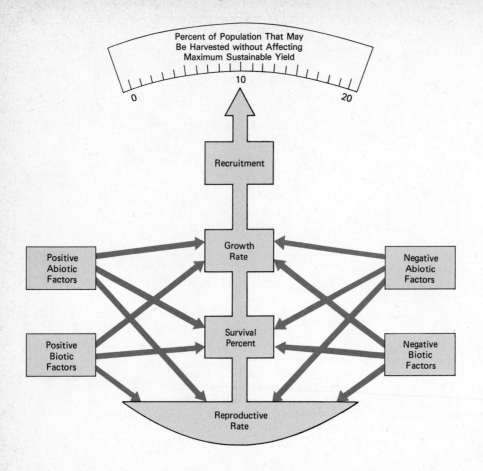

FIGURE 19–4
The percentage of the population that can be harvested each year without depleting the stock depends on recruitment rates, which vary from year to year and are influenced by many factors.

the commons inevitably resulted. Biota must be put under an authority that is responsible for the sustainability of the biota and can regulate its use.

○ Regulations must be enforced.

○ Economic incentives that promote violation of regulations must be eliminated.

○ Suitable habitats must be preserved.

○ Habitats must be protected from pollution.

There are situations where exploitation and degradation have gone too far, and whole communities and species have been driven into a decline. In these cases, natural services and potential uses can be restored only if the habitats are restored. In recent years, the need for restoration of endangered species and damaged ecosystems has become clear, and a new subdiscipline with that objective has appeared, namely, **restoration ecology.**

Restoration

As we have documented throughout this text, there is a global trend of destruction and degradation of natural ecosystems, accompanied by the decline and disappearance of thousands of wild species. Biodi-

versity in all of its manifestations is on the decline. At the same time, global recognition of this trend and its potentially disastrous consequences is on the rise so that today there is a growing commitment to restore natural systems that have been lost or damaged. A great increase in restoration activity has occurred during the last 20 years, spurred on by federal and state programs and the growing science of restoration ecology.

The intent of ecosystem restoration is simple: repair the damage so that normal functioning returns and native flora and fauna are once again present. Because of the complexity of natural ecosystems, however, the task of restoration is often difficult in practice. Restoration of a degraded or altered ecosystem, for example, is not simply a matter of "letting nature take its course." Often soils have been disturbed, pollutants have accumulated, important species have disappeared, and other—often exotic—species have achieved dominance. For these reasons, a thorough knowledge of ecosystem and species ecology is essential for success.

The ecological problems that can be ameliorated by restoration include those resulting from soil erosion, surface strip-mining, draining wetlands, coastal damage, deforestation, overgrazing, desertification, and lake eutrophication. One of the greatest challenges to restoration is **derelict lands,** lands so degraded that they have lost the entire ecosystem that once occupied them. In the case of strip-mined lands, everything down to the mined ore deposits has been removed (Fig. 19–5). Restoration must begin with trucking in soil healthy enough to support the desired ecosystem. The next stage is adding suitable plant species by sowing seeds or planting seedlings. In many cases, it is sufficient to establish an early successional stage that will in time develop into the desired climax ecosystem, as described in Chapter 4.

It is now possible to breed many endangered species in captivity and reintroduce them to the wild; successful examples are the Peregrine Falcon in North America and the Arabian oryx in the Arabian peninsula. Restoration of the Arabian oryx succeeded because the species was preserved in the San Diego Wild Animal Park; the last wild oryx was shot in 1972 by trophy hunters. Restorationists brought in 21 animals from San Diego and established natural herds within large enclosures, then released them to the wild. They now range over 2000 square miles of desert, watched over by Bedouin trackers.

As the values of natural ecosystems are recognized, efforts to restore damaged or lost ones are going to become increasingly important in the future. John Berger, executive director of Restoring the Earth (a private organization dedicated to the task of restoration), states that "the business of restoration is likely to become a multibillion dollar global enterprise" because of its capacity to ameliorate many of our worst environmental problems.

Ecosystems under Pressure

The decline of biodiversity is directly linked to the welfare of all the earth's ecosystems. Although human activities affect virtually all ecosystems, some are under more pressure than others.

FORESTS AND WOODLANDS

Forests are the normal ecosystems in regions with adequate year-round rainfall to sustain tree growth. They are also the most productive systems the land can support. As we have seen, forests and woodlands (ecosystems with mixed trees and grasses) perform a number of vital natural services, and they provide us with lumber, the raw material for making paper, and fuel for cooking and heating. In spite of all these valuable uses, the major threat to the world's forests is not simply exploitation but rather total removal.

FIGURE 19–5
An aerial view of a surface coal mine in Clearfield, Pennsylvania. Strip-mining removes everything down to the ore deposits and leaves the land in a derelict state. Restoration of such land is a challenge to our ecological knowledge and our sense of values. (Grapes/Michaud/Photo Researchers.)

Worldwide, one-third of the area originally covered by forests and woodlands is now devoid of trees. Most of the deforestation in the developed countries has been halted, but it is proceeding at an accelerating pace in the Third World. Why are forests being cleared? The most direct answer is that even though forests are very productive systems, it has always been difficult for humans to exploit them for food. Most of the energy in forests goes to detritus and decomposer food webs, not to the grazing food web (Chapter 2). In contrast, grasslands have short food chains, where herbaceous growth supports large herbivores that can yield meat and other animal products. Alternatively, natural grasses can be replaced with cultivated ones (grains) and used directly. Thus, the forests have always been an obstacle to conven-

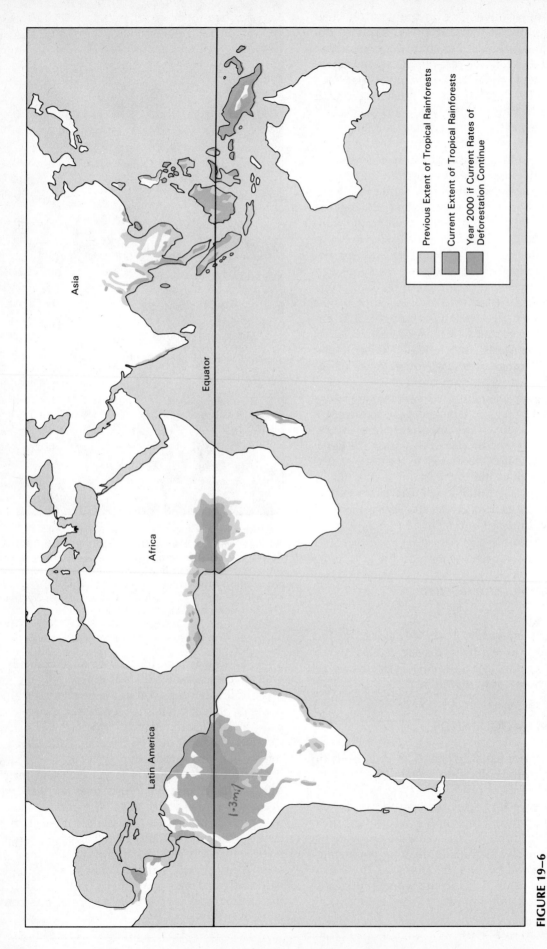

FIGURE 19–6

Tropical rainforests originally covered most of the equatorial regions of South America, Africa, and Indonesia. Their extent was already greatly reduced when recent exploitation and clearing for agriculture accelerated greatly. Currently, 40 million acres are being lost each year to deforestation.

Legend:
- Previous Extent of Tropical Rainforests
- Current Extent of Tropical Rainforests
- Year 2000 if Current Rates of Deforestation Continue

tional animal husbandry and agriculture. The first task our European ancestors faced when they came to the New World was to clear the forests so they could raise crops. Once the forests were cleared, continual grazing or plowing effectively prevented their return.

Clearing a forest has immediate and severe consequences for the land:

1. The overall productivity of the area is reduced
2. The standing stock of nutrients and biomass, once stored in the trees and leaf litter, is enormously reduced.
3. Biodiversity is greatly diminished.
4. The soil is more prone to erosion and drying.
5. The hydrological cycle is changed, as water drains off the land instead of being released by tree transpiration or percolating into groundwater.

Tropical Rainforests

Tropical rainforests are of greatest concern. Covering a broad equatorial band across South America (mainly Brazil), Africa (mainly Zaire), and Indonesia, they are the habitat for millions of plant and animal species, the majority of which are still unidentified. Climatologists reason that rainforests are also crucial in maintaining the earth's climate. At the very least, their destruction is contributing significantly to the loading of the atmosphere with carbon dioxide, which is currently warming the climate. Yet rainforests are being destroyed at a phenomenal rate—almost 40 million acres a year, according to a 1991 report by the Food and Agricultural Organization of the United Nations (Fig. 19–6). This is 50 percent greater than the deforestation rate of the early 1980s.

Destruction is being caused by a number of factors, all of which come down to the fact that the countries involved are poor and have rapid population growth. The huge numbers of young people in the newest generation cannot find jobs or live on the land that barely supports their parents. Therefore, many Third World governments are encouraging the colonization of forested lands, which begins with deforestation. For example, Indonesia has embarked on a program of resettlement and intends to convert 20 percent of its remaining forests to agricultural production. Brazil grants 250 acres (100 hectares) of land and provides subsidies of food, money, and education to peasants who are willing to migrate to the Amazonian rainforests. Some 7 million Brazilians have moved into the rainforests since 1970, clearing the land for small-scale agriculture. Unfortunately many tropical soils are unsuitable for agriculture and become exhausted of nutrients within a few years.

Settlers then move either back to the cities or on to new land, often selling their lots to cattle ranchers.

The problem is exacerbated by the governments involved because they have huge debts (over $113 billion in the case of Brazil). To raise money to pay the interest on their loans, they are selling the logging rights, generally to multinational companies, which come in and wastefully destroy the forests to obtain their hardwood logs for furniture and make no effort to replant or restore the forest. These companies clearly see the forest as a commons from which to get as much as they can while they can. They have no vested interest in maintaining sustainable yield.

Agricultural rights are sold to wealthy ranchers, who clear the forest and replant it in grass for grazing cattle to supply meat to industrial nations. Over the past 10 years, more than 3 percent of the Latin American rainforests has been converted to pasturelands, another instance of the global economy driving exploitation in poor nations.

Brazil is a main focal point of concern over the rainforests. The Amazonian rainforest of Brazil covers 1.3 million square miles and is the world's largest tropical rainforest. During the 1970s and 1980s, Brazil embarked on several ambitious schemes designed to move the nation rapidly into the ranks of the industrialized nations. Loans from the World Bank and other international agencies provided capital to launch two major projects: Grande Carajás and Polonoreste. The government is also developing hydroelectric power in the water-rich Amazon basin; dozens of dams are planned, and those already built have flooded thousands of square miles of rainforest.

The Grande Carajás project is a huge industrial enterprise expected to involve an area in eastern Brazil the size of France and Great Britain. Its cost is projected at $62 billion. Besides the clearing needed to develop roads, towns, and industrial sites, the project has led to great losses of forests because of the need for charcoal to fuel iron-ore smelters.

Polonoreste is a development scheme built around a highway into the Amazonian rainforest, BR364. As settlers and ranchers have moved out from the highway, the forest in the region has declined from 97 percent cover to 80 percent. The fires of burning rainforests from these and other projects were vividly evident from NASA's *Discovery* satellite mission in 1988, alerting the world in a dramatic way to the extent of rainforest destruction (Fig. 19–7).

In the last several years, there has been an international outcry against rainforest destruction. Brazilian officials see a certain hypocrisy in this outcry because it comes mainly from industrialized nations, who have long since thoroughly developed their own forests and prairies. Brazilians also point to the enormous contributions industrialized nations have made

Acre is Brazil's westernmost state and in many ways brings the old U.S. Wild West to mind. Along with the adjoining state of Amazonas, it is a vast, sparsely settled frontier that supports large cattle ranches whose owners have turned rainforests into pastures. Like the old U.S. West, the Brazilian west has had its share of lawlessness, most apparent these days in the struggle over land ownership, which has led to thousands of deaths. It is a classic conflict: poor forest-dwellers who make their living from products of the forests versus wealthy ranchers who cut and burn the forests so that the land will grow grass for their cattle. The conflict has spawned its heroes, none better known than Chico Mendes. His story is proof that one person with a vision and the will to turn that vision into reality can make a difference in world affairs.

Francisco (Chico) Mendes was a *serenguero*—a rubber tapper—one of hundreds of thousands of Brazilians who harvest the latex rubber and Brazil nuts from rainforest trees. Realizing that the serengueros were fast losing their livelihood to the onslaughts of the cattle ranchers as the rainforest went up in smoke, Mendes worked tirelessly to organize the serengueros against continued cutting of the forests. Soon he was not only a serious threat to the ranchers but also an international figure who in 1987 successfully testified in opposition to U.S. support of the Inter-American Development Bank (which was granting loans to Brazil to build roads into Acre's forests). How serious a threat Mendes was to the ranchers became clear on December 22, 1988, when, in spite of police protection, he was fatally shot as he stepped outside his house. Two years later, a rancher and his son were convicted of the killing and sentenced to 19 years in prison.

The legacy of Chico Mendes lives on, though. A key element in his crusade was **extractive reserves**—forestlands that would remain uncleared so that the serengueros could continue their way of life. Since Mendes's death, the Brazilian government has set aside 14 extractive reserves in Amazonia, covering 7½ million acres (3 million hectares) of forests in four Brazilian states. The rubber tappers' union founded by Mendes has become a vital national force, with its eye on protecting many more millions of acres. The economy of the rubber tappers is a fragile one, however, and the two major rainforest products—rubber and Brazil nuts—face serious competition from foreign plantation-grown products. Nevertheless, the concept of Mendes's extractive reserves is spreading throughout the tropics. And the success of the rubber tappers has ignited hope for other impoverished Third World grass roots movements.

Judson Valentim is another Brazilian who has brought relief to the rainforests, albeit in a way very different from that of Chico Mendes: Valentim is providing aid to the group normally identified as the enemy—the ranchers. He is an agronomist with the Brazilian Institute of Ranching and Agriculture who was given the task of doing something about the failure of grasslands that had been converted from rainforest. Typically, pastures would support 20 cows on 25 acres (10 hectares) of land in the first year or two after clearing. Within 5 years, however, land quality would decline as a result of erosion, drying and hardening of the soil, and fertility loss, so that the same area might support no more than 5 cows. This decline led to more rainforest clearing in order to graze a constant number of cattle. Valentim reasoned that if the land already cleared could be restored to a productive state, the pressures on the rainforest would diminish.

In the 1970s, some landowners attempted to establish rubber plantations in northern Brazil. This enterprise eventually failed, but one aspect lived on. Kudzu, that scourge of the U.S. South had been brought in to cover the ground between rubber trees. The kudzu quickly spread (no surprise!) to other areas of cleared land, and Valentim began experimenting with the plant as a source of cattle feed. The results were highly successful. Kudzu produces three times the forage of grasses, does well during the dry season, overgrows other weeds, is resistant to pests, and, because it is a legume, restores fertility to the soil by nitrogen fixation. Cattle raised on kudzu are thriving, gaining as much as 450 pounds (200 kg) a year more than grass-fed cattle and producing more milk. Land planted to kudzu is now supporting 25 cows on 25 acres, better than freshly cleared pasture.

The ranchers of Acre adopted kudzu with enthusiasm. Valentim started a consulting service and now travels about introducing kudzu techniques and seeds to new areas. The use of kudzu is now law in Acre. Ranchers may graze their cattle only in previously disturbed areas and only if they plant kudzu, restrictions aimed at preventing a new round of deforestation. Valentim believes that kudzu pastures will support twice the number of cattle currently in Acre without any more forest clearing. The government is watching the experiment closely, but all indicators point to kudzu and Valentim's work as a substantial step toward saving the remaining Brazilian rainforests.

FIGURE 19–7
A 1988 photo from the space shuttle Discovery shows smoke from burning rainforests so extensive that it obscures all land features in the Amazon River basin of South America. (NASA.)

to global warming and ozone layer depletion, certainly two of the most pressing global environmental problems. They see attempts to stop the development projects as infringements on their national sovereignty.

The Brazilian government has recognized, however, that there are serious problems with their colonization and industrial schemes. The subsidies to cattle ranchers have been halted, and the rate of deforestation has consequently declined significantly, perhaps as much as 80 percent from 1990 to 1992. The government has placed 7½ million acres (3 million hectares) of rainforest into **extractive reserves**—land that is protected for native peoples who live in the forest and other Brazilians who gather latex rubber and Brazil nuts from mature forest trees. Also, international loan agencies have become more environmentally conscious and now require environmental impact assessments to accompany loan requests. The wealthier Western countries are considering support for a $1.6 billion project that would buy land and protect it, and they are exploring other kinds of work for the ranchers, farmers, prospectors, and others who are currently exploiting the Amazon.

All of these developments suggest that global awareness of the importance of the tropical rainforests has reached the level of effective action, and the high rate of deforestation of the 1980s may soon diminish.

OCEAN FISHERIES

The open ocean has traditionally been considered an international commons, and by the end of the 1960s numerous species and areas were being seriously depleted by overfishing. In 1977, a number of nations, including the United States, extended their limits of jurisdiction from the previous 3–12 miles to 200 miles offshore. Since many prime fishing grounds are located between 12 and 200 miles from shores, this action effectively removed the fish from the international commons and placed them under the authority of a particular nation, allowing regulation of fishing. As a result, some fishing areas recovered, until nationally based fishing fleets expanded to exploit the fisheries. (**Fishery** refers either to a limited marine area or to a group of fish species being exploited.)

The total recorded catch from marine fisheries has increased remarkably since 1950, when it was just

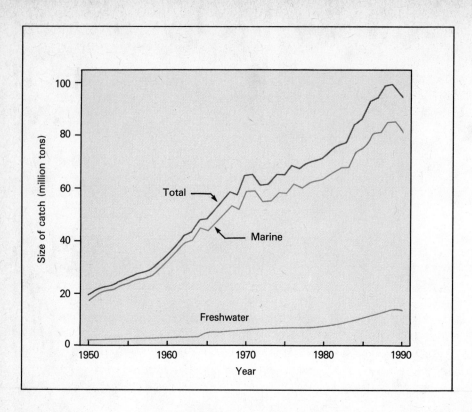

FIGURE 19–8
The global fish catch, 1950–1990.
(From World Resources 1990–91.
Copyright 1991, Oxford University
Press.)

20 million metric tons (Fig. 19–8). It is now over 80 million metric tons and has shown a sharp increase since 1983. These figures are a measure of the sea's productive potential, but in a sense they obscure what is really happening. Many species and areas continue to be overfished, and when this happens, fishermen turn to new areas and formerly less desirable species. For example, 70 percent of the rise during the 1980s is attributed to increased catch of just four species not previously exploited; many other species yielded decreasing harvests. By all measures, the world's fisheries have reached the limit of sustainable yields, and most observers believe a significant downturn is coming.

Events on George's Bank, New England's richest fishing ground, are indicative of the trends. Cod, haddock, and flounder were the mainstay of the fishing industry for centuries. In the early 1960s, these species amounted to two-thirds of the fish population. Overfishing has resulted in a decline in these species and a rise in the so-called rough species, dogfish and skates, which now constitute 75 percent of the fish population on the bank. There is concern that the dominance of these rough species may be permanent; they may be usurping the ecological role of the more desirable species, such that even if quotas of cod, haddock, and flounder are drastically reduced, these more desirable fish may not come back.

The decline of the prized species on George's Bank is directly traceable to the unrestrained free-market approach to the fishery. Promoted by recent federal administrations, it is a classic case of the tragedy of the commons. The National Marine Fisheries Service (NMFS) has jurisdiction over the coastal fisheries, and no one thinks they are doing a good job. A recent government-sponsored report (the Chandler Report) revealed that the agency has suffered severe cuts in budget and personnel and is now faced with too much to do and not enough resources to do it. In response to a suit by the Conservation Law Foundation against NMFS, the service has agreed to mount a program to restore the cod, haddock, and flounder fisheries by 1997. It is likely that simply lowering the quotas on the desirable fish will not be enough to restore them; action against the rough species may be required.

Drift-netting on the High Seas

New fishing technology is leading to even greater overfishing. High-tech radar and depth-finders, sometimes aided by aircraft, locate fish. Larger vessels go to sea for weeks, freezing their catch until the holds are filled. The most devastating innovation, however, is drift-netting. Gill nets up to 30 feet in depth are reeled off the back of the ship by the mile; a single boat will hold 30 miles of net. In theory, the nets catch only fish that are of a diameter similar to that of the net openings. However, the nets have a tendency to collapse around larger fish, diving birds, sea turtles, dolphins, porpoises, seals, and whales, and millions of these animals are killed each year. The practice has been described as "strip-mining life from the sea." Drift-nets are operated by the Japanese, Korean, Taiwanese, and French; they are extremely effective in catching upper-ocean species such as tuna and flying squid. The nets were used in offshore New Zealand waters for several years before they were discovered; New Zealand and 21 island nations of the Pacific then banned the drift-netters from their waters. In response to initiatives from these nations, the United Nations passed a resolution in December 1989 banning the use of drift-nets everywhere on the high seas by 1992. Under threats of trade sanctions from the United States, Japan agreed to the moratorium. However, the Taiwanese, who have hundreds of vessels in the industry, threaten to continue the practice, and unfortunately, there are no teeth in the resolution to force them to do otherwise. Again, the tragedy of the commons is being played out, and the last players are those with the most devastating technology and the fewest reservations against using it.

International Whaling

Whales, found in the open ocean as well as coastal waters, are suffering from overexploitation that was finally halted in the late 1980s. Whales once were harvested for their oil but are now prized for their meat, considered a delicacy in Japan and a few other countries. An organization of nations with whaling interests, the International Whaling Commission (IWC), decided in 1974 to regulate whaling according to the principle of maximum sustainable yield. Whenever a species of whale dropped below the optimal population for MSY, the IWC instituted a hunting ban in order to allow the population to recover. At that time, three species (right whale, bowhead whale, and blue whale) were in serious decline and were immediately protected. Because of difficulties in obtaining reliable data on and enforcing catch limits, the IWC took more drastic action by placing a moratorium on the harvesting of all whales in 1986.

The International Union for Conservation of Nature (IUCN) publishes the up-to-date status of whale populations in its *Red Data Book*. Table 19–2 lists the status of nine whale species of commercial interest. Populations range from an estimated remaining 600 blue whales to approximately 650 000 minke whales. Although reliable data are still hard to acquire, many of the whale species appear to be recovering. The bowhead whale, for example, has just been upgraded from "endangered" to "vulnerable" because of rising population numbers.

There is heavy pressure from three members of the IWC–Japan, Norway, and Iceland—to reopen whaling, and their interests focus on the minke

TABLE 19–2

Status of Large Whale Species

Endangered	Vulnerable	Insufficient Knowledge
Blue whale	Bowhead whale	Minke whale
Northern right whale	Fin whale	Sperm whale
	Humpback whale	
	Sei whale	
	Southern right whale	

Source: International Union for Conservation of Nature. *Red Data Book.* 1991.

whale, which is judged by some scientists to be numerous enough to absorb a sustainable exploitation. Currently each of these three nations takes up to 300 minke whales per year, for "scientific purposes." Of course, the meat is sold once the "science" has been satisfied. At the end of 1991, Iceland announced its intention to withdraw from the IWC, claiming that the commission had been "taken over by radical nations" who want to halt all whaling for any purposes. Indeed, there are strong pressures from many organizations to prohibit whaling for ethical reasons.

One reason for the rising interest in protecting whales is the opportunity many people have had to observe whales first–hand, since whale-watching has become an important tourist enterprise in coastal areas (Fig. 19–9). Stellwagen Bank, within easy reach of boats from Boston, Cape Ann, and Cape Cod, Massachusetts, has become the center of a whale-watching industry estimated to generate $17 million annually. From spring through fall, scores of boats venture offshore daily to watch the whales that congregate over the bank. Many of the humpback whales seem to enjoy entertaining the visitors and often frolic alongside the boats for hours.

FIGURE 19–9
A humpback whale entertains a boatload of whalewatchers in the Gulf of Maine off the Massachusetts coast. Whalewatching has replaced whaling in New England waters that were once famous as a whaling center. (Jan Robert Factor/Photo Researchers.)

Besides its obvious aesthetic and entertainment values, whale-watching is of scientific value. Whale-watching tour boats usually carry a biologist along who identifies the whale species and interprets the experience for the visitors. The biologists are often associated with groups such as the Cetacean Research Unit, which has studied the whales of Stellwagen Bank since 1979 and has published many papers on the humpback whale.

Freshwater and Estuarine Fisheries

Exploitation of fish and shellfish that live in fresh water and estuaries is generally controlled by regulations similar to those used to control hunting. The greatest threats to these fisheries now are pollution, sedimentation, eutrophication, and toxic chemicals. In some areas of the Great Lakes and the Hudson and Mississippi rivers, some game fish are so contaminated with toxic chemicals, gained through bioaccumulation, that commercial fishing is banned and people fishing for sport are advised not to eat their catch. Notably, a few rivers, such as the Willamette in Oregon and the Detroit in Michigan, which were little more than open sewers in the 1960s have been cleaned up and support fishing again. For the first time in 200 years, salmon have begun to spawn in the Connecticut River, a tribute to pollution control efforts and the building of fish ladders on the many dams on the river.

WETLANDS

Among the most productive ecosystems known, wetlands provide crucial habitat for many species of birds, mammals, and fish of commercial or recreational importance. These and other natural services they perform—protection of water supplies and flood protection, for example—place wetlands very high in any scheme of economic valuation.

There exits a great diversity of wetland types, from the extensive saltmarshes of the eastern U.S. and Gulf coasts, to the great Everglades of Florida, to marshes and shrub and tree swamps of inland areas. For a long time they were all considered wastelands—sources of mosquitoes, obstacles to development, and places to dump trash and industrial debris. People believed that the best way to deal with such wastelands was to drain them and put the resulting dry land to use. Most of the loss was to agricultural land since, drained of their water, wetland soils are very productive. In addition, coastal wetlands were often filled to provide valuable land for homesites, marinas, and city expansion (see Fig. 19–2). Since the founding of the United States in 1776, over 120 million acres of the nation's wetlands have been lost, which is approximately half of the area they originally occupied.

Fortunately, the great natural value of wetlands has now been well established, and the scale of wetland loss has been much reduced thanks to federal and state protection. This reduction was accomplished mainly during the 1980s by:

1. legislation removing incentives that encouraged wetlands destruction (1982 Coastal Barrier Resources Act, 1985 Food Security Act, 1986 Tax Reform Act);
2. provision of incentives for farmers to protect wetlands (1990 Farm Act);
3. provision of funds for wetlands acquisition (Emergency Wetlands Resources Act of 1986); and
4. provision of funds and incentives for the restoration of drained wetlands (1985 Food Security Act).

In 1988, at the request of the EPA, the Conservation Foundation sponsored the National Wetlands Policy Forum, where major public policy concerns were aired. At its close, the forum recommended that "the nation establish a national wetlands protection policy to achieve *no overall net loss* of the nation's remaining wetlands." It was this call for no net loss that George Bush echoed in his run for the presidency in 1988, as he presented himself as the "environmental candidate."

However, the Bush administration found a way to get around that pledge. The **Federal Manual for Identifying and Delineating Wetlands,** established in 1989 by the EPA, was revised in 1991 by redefining wetlands in a way that removed up to 30 million acres (12 million hectares) nationwide from protection. Essentially, an area has to be wet at the surface for more than 21 consecutive days a year in order to qualify as a wetland, as opposed to the EPA's prior definition of 7 consecutive days. In Chapter 17 (Ethics Box, p. 389) we noted that the Council on Competitiveness actively promoted this change in response to pressures from developers and the farm lobby. As we go to press, the 1991 version of the manual is still under review and may be revised again in response to heavy criticism from the environmental and scientific communities. One additional ray of hope exists: At stake here is only *federal* protection of the wetlands; many states have their own wetlands protection laws that will not be affected by the new 21-day definition.

PUBLIC LANDS IN THE UNITED STATES

As we saw in Chapter 18, in order to save wild species, we must protect their habitats. The time has passed when we could justify ecosystem losses because there were suitable substitute habitats just over the hill. With the rising human population, industrial expansion, and pressure to convert natural resources to economic gain, however, there will always be reasons to exploit natural ecosystems. The last resort for many species and ecosystems is protection by law in the form of national parks, wildlife refuges, and reserves. Worldwide, some 4500 areas have received this kind of protection, representing 3.2 percent of the landmass. However, in the Third World, many of these are "paper parks," where exploitation continues and protection is given only lip service.

The United States is unique among the nations of the world in setting aside a major proportion of its landmass for public ownership. Nearly 40 percent of the nation's land is publicly owned and managed by state and federal agencies for a variety of purposes. The distribution of public lands, shown in Figure 19–10, is greatly skewed toward the western U.S. and Alaska, a consequence of historical settlement and land distribution policies. Although most of the East

FIGURE 19–10
Distribution of public lands in the United States. (Environmental Trends, 1989. Council on Environmental Quality.)

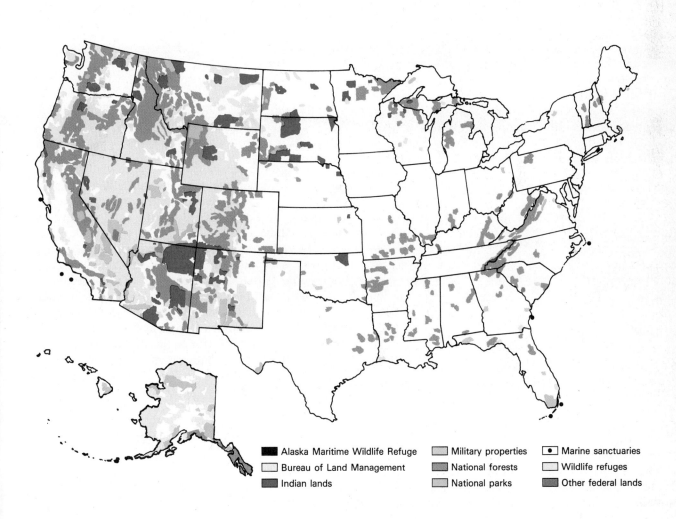

■ Alaska Maritime Wildlife Refuge	▨ Military properties	⊡ Marine sanctuaries
▢ Bureau of Land Management	▨ National forests	▢ Wildlife refuges
▨ Indian lands	▨ National parks	▨ Other federal lands

and Midwest is in private hands, there are still functioning natural ecosystems on much of those lands.

Land given the greatest protection (preservation) is designated Wilderness. Authorized by the Wilderness Act of 1964, it includes 90 million acres at 465 locations. The act provides for the permanent protection of these undeveloped and unexploited areas so that natural ecological processes can operate freely. Permanent structures, roads, and motor vehicles and other mechanized transport are prohibited. Timber harvest is excluded. Some livestock grazing and mineral development are allowed where prior use existed; hiking and other similar activities are allowed.

The National Parks (administered by the National Park Service) and National Wildlife Refuges (administered by the Fish and Wildlife Service) provide the next level of protection to 170 million acres. Here the intention is to protect areas of great scenic or unique ecological significance, protect important wildlife species, and provide public access to view these scenic wonders. These dual goals of protection and providing public access are often conflicting because the parks and refuges are extremely popular, sometimes drawing so many visitors that protection can be threatened by those who want to see and experience the natural sites.

Increasingly, agencies are beginning to understand the need to manage natural sites as part of larger ecosystems. For example, the Great Smoky Mountains National Park is now part of the Southern Appalachian Man and the Biosphere Cooperative (Fig. 19–11). Private and public land managers now have a decision-making body that can aid in the larger task of protecting natural resources. When black bears leave the park, for example, they become fair game for hunters. State wildlife managers and park officials now cooperate in setting reasonable limits on the number that can be taken.

This cooperative approach is important for the continued maintenance of biodiversity, since so much of the nation's natural lands remain outside of protected areas. It may also help to restrict development up to the borders of the parks and refuges, avoiding the situation where the refuge is a fairly small island in a sea of developed landscape.

National Forests

There is no doubt that U.S. forests represent an enormously important natural resource, capable of providing habitat for countless wild species as well as providing natural services and products. The United States contains about 740 million acres of forests, of which about two-thirds is managed for commercial timber harvest. Almost three-fourths of the managed commercial forestlands is in the East and privately

FIGURE 19–11
Southern Appalachian Biosphere Reserve, showing the relationship between the Great Smoky Mountains National Park and the broader land area of the cooperative. (Environmental Quality 1990. Council on Environmental Quality.)

owned; the remainder is mainly in the West and administered by a number of governmental agencies, primarily the National Forest Service and the Bureau of Land Management.

Deforestation is no longer a problem in the United States. Although we have cut all but 5 percent of the forests that were here when the first colonists arrived, second-growth forests have regenerated wherever forestlands have been protected from conversion to croplands and houselots. There are more trees in the United States today than there were in 1920. In the East, second-growth forests have aged to the point where some look almost as they did before their first cutting, except for the lack of the American Chestnut.

Also, there has been active reforestation (tree planting) of some cut lands. However, most of the reforested land has been planted with a single commercially valuable species. Many observers point out that this is effectively a monoculture, involving almost as great a loss in biodiversity as if the forest had been converted to cornfield.

Recently the U.S. Forest Service has been criticized for condoning unnecessarily aggressive logging in the national forests, particularly where some old-

The Spotted Owl controversy pits jobs against the preservation of the old-growth forests of the West. Timber interests maintain that preservation will result in 20000 lost jobs. In their view, the controversy comes down to this: We should continue to cut the old-growth forests for the sake of keeping loggers employed.

General Motors and other major businesses facing hard times lay off thousands of workers, and no one questions their need or right to do so. Clearly, the survival of General Motors is more important than keeping all their workers employed. It is assumed that those laid off will find other employment. According to conventional wisdom, corporate America must survive if the economy is to have a chance of recovering from economic bad times.

In a real sense, ecosystems are the corporations that sustain the economy of the biosphere (see Box figure.). If we want those ecosystems to survive and recover, we may have to tighten our belt and withdraw some of the work force engaged in exploiting them. The maintenance of these systems is obviously more important than some jobs. Why can't laid-off loggers seek other employment the way laid-off

Forested ecosystems can be viewed as nature's corporations, providing lumber and firewood for the economy. They can only do so if they are maintained sustainably, however, and not exploited just to provide jobs for loggers. (Bill Hubbell/Woodfin Camp & Associates.)

autoworkers, steelworkers, or computer engineers do? Why should a natural ecosystem be "bankrupted" just to maintain the *temporary* employment of a few (*temporary* because the loggers will be out of a job in a few years when the old-growth forests are finally all cut)?

Clearly, it is short-sighted to as-

sume that ecosystems are there just to provide jobs and that the jobs are more important than the ecosystems. When the old-growth forests are gone, we shall have lost more than just loggers' jobs—we shall have lost a priceless heritage and a major part of the natural world that provides us with vital services.

growth (virgin) stands remain. A battle is raging over old-growth forests in Oregon and Washington and protection of the Spotted Owl, a threatened species (see In Perspective Box, p. 439). On one side are the loggers, who claim that the owl does perfectly well in second-growth forests and who are eager to get access to the 300–1000 year old trees because these trees provide the greatest profit. On the other side is the Fish and Wildlife Service and environmental groups, who maintain that the Spotted Owl can exist only in old-growth forests. In the middle is the National Forest Service, being pressured by Western politicians to proceed with logging in order to preserve jobs and the states' economies. There is more than the Spotted Owl at stake; old-growth forests are becoming an "endangered species" in their own right,

a national treasure. As mentioned above, only 5 percent of U.S. forestlands are old-growth areas. If logging continues, this last 5 percent will soon be gone forever.

The Forest Service is also criticized for its general management policies. Government foresters select tracts of forest they judge ready for harvesting and then lease the tracts to private companies who log and sell the timber. In theory the Forest Service should generate a profit from the sale of leases; in fact, they sell the leases too cheaply and use the funds to provide access roads and other services. Independent auditors estimate that some $5.6 billion in potential federal revenue has been lost over the past decade because of these subsidized sales. In other words, the national forests are losing money, and

U.S. taxpayers are heavily subsidizing the logging industry. Of course, the loggers are happy with this arrangement. Pressure to phase out the money-losing sales is mounting in Congress, and it is likely that the Forest Service will have to change its practices in the near future.

PRIVATE LAND TRUSTS

With so much land in private hands, the future of natural ecosystems in the United States is always uncertain. This is especially true for land having special appeal for recreation and aesthetic enjoyment because here the potential for development is great. Often landowners and townspeople want to protect natural areas from development but are wary of turning the land over to a governmental authority. One very creative option is the **private land trust,** a non-

profit organization that will accept either outright gifts of land or **easements**—arrangements where the landowner gives up development rights into the future but retains ownership. The land trust may also purchase land to protect it from development. The land trust movement is growing; there were 429 land trusts in the United States in 1980, and now there are over 900, as reported by the *Land Trust Alliance,* an umbrella organization in Washington serving the local and regional land trusts. Local and regional land trusts protect about 2 million acres of land, and large national land trusts such as *The Nature Conservancy* protect an additional 3.5 million acres.

Land trusts are proving to be a vital link in ecosystem preservation. The oldest trust is *The Trustees of Reservations* in Massachusetts, founded in 1891 and now guardian of 17 000 acres throughout the state. The Trustees, as the organization is often called, has received some ecologically prime and scenic land

IN PERSPECTIVE
Environmental Backlash

"We are sick to death of environmentalism, and so we will destroy it," says Ron Arnold, executive vice president of the Center for Defense of Free Enterprise (CDFE). "Environmentalism is the new paganism. Trees are worshipped and humans sacrificed at its altar." Arnold is a fundraiser for conservative causes, and now that communism is all but dead, conservatives need a new enemy. Arnold and his partner, Alan Gottlieb, would like to convince them that environmentalists are their new enemy.

Whenever use of federal lands is restricted or wetlands given protection, people who want to use those lands for profit have formed lobbying organizations. Their rallying cry is "wise use," by which they mean that natural ecosystems should be used (by them!) and not simply preserved. They want to see environmental legislation dismantled and push forcefully the concept that private property owners should be able to do what they want with their land. The Public Lands Council,

which represents 31 000 cattle and sheep ranchers, the Western States Public Lands Coalition, which fights stricter mining laws, and the Blue Ribbon Coalition, which wants more access to public lands for off-road vehicles, are three of the largest wise-use lobbying groups. Their main targets are environmental advocacy groups such as the Sierra Club, Nature Conservancy, Greenpeace, and Wildlife Defense Fund. Environmental groups now consider the wise-use movement a serious threat, pointing out that behind the groups are big corporations who have an interest in seeing public lands opened up for more exploitation, including Exxon, Honda, and Kawasaki.

The tactics of CDFE are mass mailings of strongly anti-environmentalist literature, with a plea for funds to support their cause. From 1986 to 1991 they raised several million dollars on behalf of organizations that fight environmentalists. Arnold charges $3000 a day to speak or help organize groups in support of the

wise-use concept. Arnold and Gottlieb have published their own book, *The Wise-Use Agenda,* in which they advocate opening up the national parks to logging, drilling for oil on wildlife refuges, and systematically cutting all of the remaining old-growth forests.

In perspective, this movement was to be expected. People and corporations are almost always going to act in their own self-interest. In this case, that self-interest lies in freedom to use natural lands belonging to the public. The current recession that is plaguing the U.S. economy will surely prompt more of this kind of thing. Arnold and Gottlieb have a good thing going for themselves. It is unfortunate, however, that they have been successful in convincing some people that environmentalism is a worthy enemy and that government-imposed environmental protection is a threat to basic human freedoms. The success they've had means that the message of environmental science has not penetrated deeply enough into our society.

through the years and now maintains many of its properties for public use compatible with preservation. The land trusts are serving the common desires of landowners and rural dwellers to preserve the sense of place that links the present to the past. At the same time, the undeveloped land remains in its natural state, sustaining natural populations and promising to do so into the future.

Conclusion

Ecosystems everywhere are being exploited for human needs and profit. In addition to all the examples we just looked at, other areas in trouble could be mentioned: coral reefs swept clean of fish, rangelands overgrazed, mangrove swamps cleared for shrimp farms, rivers overdrawn for irrigation water. The list is endless. Our purpose in this chapter has been to highlight some of the most crucial problems and to indicate steps being taken to correct them. It is certain that greater pressures will be put on natural ecosystems as the human population continues to rise. These pressures must be met with increasingly effective protective measures if we want to continue to enjoy the fruits of these ecosystems. The problems are most difficult in the Third World, where poverty forces people to take from nature in order to survive. If natural areas are to be preserved, the needs of people must be met in ways that do not involve destroying ecosystems. People must be provided with alternatives to exploitation, a process requiring both wise leadership and effective international aid.

Sustainability—that crucial concept once again—should be the goal of all our interactions with natural systems. The sooner we recognize the wisdom of that basic approach, the better our chances of having more than some remnants of nature left by the time the human population levels off sometime in the twenty-first century.

 Review Questions

1. What are the five major categories of terrestrial ecosystems? What are the three major categories of oceanic ecosystems?
2. Describe the natural services that all ecosystems perform.
3. What is the tragedy of the commons? Give two examples of commons and how they are mistreated.
4. Compare and contrast the terms *conservation* and *preservation*.
5. What does *multiple sustainable yield* mean? What factors complicate its application?
6. Describe how a derelict land area is restored.
7. What are five major consequences of deforestation?
8. Why is tropical deforestation considered one of the world's most serious environmental problems?
9. How does the tragedy of the commons apply to commercial ocean fishing?
10. What three countries are pressuring the IWC to reopen commercial whaling, and what is their rationale for resuming the killing?
11. What four important protection steps for wetlands in the United States were taken in the 1980s?
12. Contrast the two levels of ecosystem protection given to public lands in the United States.
13. What two practices of the National Forest Service have come under recent public scrutiny?
14. How do land trusts work, and what roles do they play in preserving natural lands?

 Thinking Environmentally

1. What natural resource commons do you share with those around you? Are you exploiting or conserving these commons?
2. Does our National Park system encompass enough area, or should it be expanded? Either way, defend your answer. If you believe the system should be expanded, propose some new areas to be added.
3. What incentives and assistance could the United States offer Brazil to keep the tropical rainforests from further harmful development?
4. Research the preservation efforts and effectiveness of one of the following conservation groups: The Sierra Club, The Appalachian Mountain Club, The Nature Conservancy, The Audubon Society.
5. Identify and study an area in your community where destruction or degradation of natural land or wetland is an issue. What if anything is being done to protect this land?

20

Converting Trash to Resources

LEARNING OBJECTIVES

When you have finished studying this chapter, you should be able to:

1. Sketch the historical background of refuse disposal.

2. Name the main components of municipal solid waste today and describe how each is disposed of.

3. Describe four problems that stem from landfilling with refuse.

4. Describe the features that new landfills must have in order to prevent problems.

5. Discuss the pros and cons of incineration.

6. Describe some of the costs and limitations of landfills.

7. Describe ways in which the total volume of refuse might be reduced.

8. Name six factors that make it difficult to recycle refuse and describe how each can be overcome.

9. Give examples of laws that might be passed to promote recycling.

10. Explain what composting is and describe the components of refuse that might be composted.

11. Describe how refuse may be converted to energy.

12. Discuss the concept of integrated waste management.

On your next trip to the mall, take a good look at all the products piled high on the shelves and hanging from the racks of the scores of retailers there. Everything you see will eventually end up as trash. These goods are there to be purchased, used, and then thrown away; the economy depends on it. Little of the materials used to make all these products is ever reused, a clear violation of the first principle of sustainability: that ecosystems dispose of wastes and replenish nutrients by recycling all elements. Common sense dictates that a throwaway society is not sustainable, but it appears that we have had to prove this fact experimentally. The results: We are on the brink of a crisis as we produce more trash than ever before and are rapidly running out of places to put it. Yet, solutions that might avert a full-blown crisis are available and are being implemented. In this chapter, we look at the dimensions of the crisis and discuss some potential solutions.

The Solid Waste Crisis

So far in this book we have talked about animal feedlot wastes (Chapter 12), sewage wastes (Chapter 13), and industrial wastes (Chapters 14 and 15). A fourth category of wastes, the focus of this chapter, is **municipal solid waste** (MSW), defined as the total of all the materials thrown away from homes and commercial establishments (commonly called trash, refuse, or garbage).

Over the years, the amount of municipal solid waste generated in the United States has grown steadily, in part because of increasing population but more so because of changing lifestyles and the increasing use of disposable materials and excessive packaging (Fig. 20–1). It now amounts to somewhat over 4 pounds (2 kg) per person per day. At the current U.S. population of 250 million, that is enough

waste to fill 75 000 garbage trucks each day, a total of 185 million tons (166 million metric tons) per year.

Studies show that the refuse generated by municipalities is roughly composed of:

Paper, paperboard	41%	Food wastes	8%
Yard waste	18%	Glass	8%
Metals	9%	Other	7%
Plastics	9%		

However, the proportions vary greatly depending on the generator (commercial versus residential), the neighborhood (affluent versus poor), and the time of year (during certain seasons, yard wastes, such as grass clippings and raked leaves, add to the solid waste burden, often equaling all the other categories combined).

Traditionally, local governments have assumed

FIGURE 20–1
Output of solid wastes in the United States. While the population increased by 22 percent between 1970 and 1990, the amount of MSW collected increased by 60 percent and the per capita waste output increased by 30 percent. (Data from J. E. Young, "Discarding the Throwaway Society," Worldwatch Paper 101. © 1991 by Worldwatch Institute.)

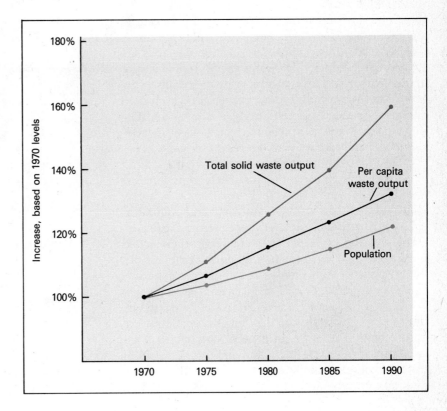

the responsibility for collecting and disposing of MSW. The local jurisdiction may own the trucks and employ workers, or it may contract with a private firm to provide the collection service. Collected MSW is then disposed of in a variety of ways, as discussed below. Alternatively, some municipalities have opted for putting all trash collection and disposal in the private sector. The collectors bill each home by volume of trash. This system allows competition between collectors and gives homeowners a strong incentive to reduce trash volume.

Until the 1960s most MSW was disposed of in open, burning dumps. The waste was burned to reduce volume and lengthen the life span of the dump site, but refuse does not burn well. Smoldering dumps produced clouds of smoke that could be seen from miles away, smelled bad, and created a breeding ground for flies and rats. Some cities turned to incinerators, which are huge furnaces in which high temperatures allow the waste to burn more completely than in open dumps. Without air pollution controls, however, incinerators were also prime sources of air pollution. Public objection and air pollution laws forced the phaseout of open dumps and many incinerators during the 1960s and early 1970s. Open dumps were then converted to *landfills*.

CURRENT DISPOSAL PROCESSES

At present in the United States, three-fourths of municipal solid waste is disposed of in landfills and the remaining one-fourth is equally divided between incineration and recycling (Fig. 20–2). The pattern is different in countries where population densities are higher. High-density Japan, for instance, incinerates about half of its trash and recycles over half of the rest. Many Western European nations (also high-density areas unable to devote valuable land to landfills) deposit less than half of their municipal waste in landfills and incinerate most of the rest.

In a **landfill,** the waste is put on or in the ground and covered with earth. Because there is no burning and because each day's fill is covered with a few inches of earth, air pollution and vermin populations are kept down. Unfortunately, aside from those concerns and the minimizing of cost, no other factors were given real consideration when the first landfills were opened. Municipal waste managers generally had no understanding of or interest in ecology, the water cycle, or what products would be generated by decomposing wastes, and they had no regulations to guide them. Therefore, in general, any cheap, conveniently located piece of land on the outskirts of town became the site for a landfill. This site was frequently a natural gully or ravine, an abandoned stone

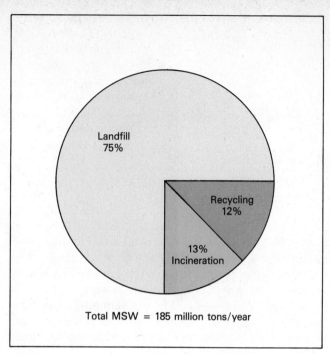

FIGURE 20–2
Current processes of MSW disposal in the United States.

quarry, a section of wetlands, or a previous dump. Once the municipality acquired the land, dumping commenced with no precautions taken. The plan usually was that after the site was full, it would be covered with earth and converted to a park or playground. Thus, landfilling was originally thought of as a means of upgrading "wasteland" to a higher use as well as a means of disposing of trash.

PROBLEMS OF LANDFILLS

With the understanding you have gained from previous chapters, you should be able to predict the consequences of landfilling:

○ Leachate generation and groundwater contamination

○ Methane production

○ Incomplete decomposition

○ Settling

Leachate Generation and Groundwater Contamination

The most serious problem by far is groundwater contamination. Recall that as water percolates through any material, various chemicals in the material may dissolve in the water and get carried along, a process called leaching. The water with various pollutants in it is called **leachate**. As water percolates through MSW, a noxious leachate is generated, which consists of residues of decomposing organic matter combined with iron, mercury, lead, zinc, and other metals from rusting cans, discarded batteries, and appliances—generously spiced with paints, pesticides, cleaning fluids, newspaper inks, and other chemicals. The nature of the landfill site and the absence of precautionary measures noted above funnel this ''witch's brew'' directly into groundwater aquifers.

All states have some municipal landfills that either are or soon will be contaminating groundwater, but Florida is in a real crisis. The state is topographically flat and hence has vast areas of wetlands. Most of the land is only a few feet above sea level and rests on water-saturated limestone. No matter where Florida's landfills were located, they were either in wetlands or just a few feet above the water table. Since residents rely on groundwater for 92 percent of their fresh water, you can guess the result: more than 200 municipal landfill sites on the Superfund list. Recall from Chapter 14 that Superfund is the federal program to clean up sites that are in imminent danger of jeopardizing human health through groundwater contamination. It will cost between $10 million and $100 million to clean up each site. So much for cheap waste disposal!

Methane Production

Because 67 percent of MSW is organic materials, it is potentially subject to natural decomposition. However, buried wastes do not have access to oxygen. Therefore, their decomposition is anaerobic, and a major byproduct of this process is biogas, which is about two-thirds methane, a highly flammable gas (p. 303). Produced deep in a landfill, methane may seep horizontally through the earth, enter basements, and cause explosions as it accumulates and is ignited. Over 20 homes at distances up to 1000 feet from landfills have been destroyed, and some deaths have occurred as a result of such explosions. Also, methane seeping to the surface kills vegetation by poisoning the roots. Without vegetation, erosion occurs, exposing the unsightly waste. A number of cities have exploited the problem by installing ''gas wells'' in landfills. The wells tap the biogas, and the methane is purified and used as fuel. There are now 70 commercial landfill gas facilities in the United States; the largest, in Sunnyvale, California, generates enough electricity to power 100 000 homes.

Incomplete Decomposition

The plastic components of MSW cannot be decomposed because of their chemical composition. For this reason, much emphasis has been placed on developing biodegradable plastics. There are serious questions about the degradability of these plastics, however. The term *biodegradation* refers to the complete breakdown of carbon compounds to carbon dioxide and water. All the purported biodegradable plastics do is disintegrate into a fine polymer powder that still resists microbial breakdown.

Recent research on landfills has shown that even materials always assumed to be biodegradable—newspapers, wood, and so on—are degraded only slowly if at all. Newspapers 30 years old were recovered in a readable state; layers of telephone directories were found marking each year, practically intact. Since paper materials are 40 percent of MSW, this is a serious matter. The reason paper and other organics decompose so slowly is the lack of suitable amounts of moisture; the more water percolating through a landfill, the better the biodegradation of paper materials. However, the more percolation there is, the more toxic leachate is produced!

Settling

Finally, waste settles as it compacts. Luckily, this eventuality was recognized from the beginning, and so buildings have never been put on landfills. Settling presents a problem where landfills have been converted to playgrounds and golf courses, though, because it creates shallow depressions that collect and hold water. This process converts some of the land back to a ''wetland,'' albeit usually one with noxious leachate seeping to the surface.

IMPROVING LANDFILLS: TRYING TO FIX A WRONG ANSWER

Recognizing the above problems, the Environmental Protection Agency has upgraded siting and construction requirements for new landfills. Under current regulations:

○ New landfills are sited on high ground well above the water table. Often the top of an existing hill is bulldozed off to supply a source of cover dirt and at the same time create a floor that is above the water table.

○ The floor is first contoured so that water will drain into a tile leachate-collection system. The floor is then covered with at least 12 inches of impervious clay or a plastic liner or both. On top of this is a layer of coarse gravel and a layer of porous earth. With this design, any leachate percolating through the fill will encounter the gravel layer and then move through that layer into the leachate collection system. The clay layer or plastic liner prevents leachate from ever entering the groundwater. Collected leachate can be treated as necessary.

○ Layer upon layer of refuse is positioned such that the fill is built up in the shape of a pyramid. Finally it is capped with a layer of clay and a layer of topsoil and then seeded. This clay-topsoil cap and the pyramidal shape help the landfill to shed water. In this way, water infiltration into the fill is minimized, and so less leachate is formed.

○ Finally, the entire site is surrounded by a series of groundwater monitoring wells that are checked periodically, and such checking must go on indefinitely.

These design features are summarized in Figure 20–3.

The regulations do protect groundwater, but the landfill pyramids may well last as long as the Egyptian pyramids (though they are not likely to become tourist attractions!). And if they do break down, they become a threat to the groundwater—therefore, the need for monitoring remains. As trash keeps coming, more and more landfills are created, leading only to more problems in the future. It is clear that we are "fixing a wrong answer."

INCINERATION: ANOTHER WRONG ANSWER

Currently, 128 incinerators are operating in the United States, burning about 25 million tons of waste annually—13 percent of the waste stream. Incineration of municipal solid waste has some advantages over landfilling. Its primary appeal is reduction of volume; incinerators can reduce trash volume by 80 to 90 percent, and thus greatly extend the life of a landfill (which is still required to receive the ash). Also, most newer incinerators are designed to generate electricity, which is sold to offset some of the costs of disposal.

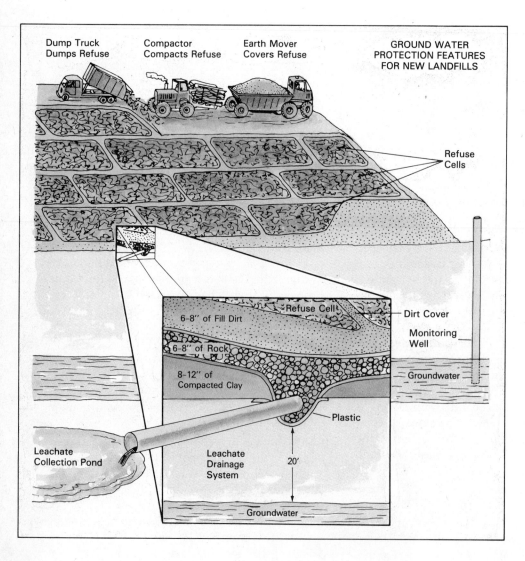

Dump Truck Dumps Refuse
Compactor Compacts Refuse
Earth Mover Covers Refuse
GROUND WATER PROTECTION FEATURES FOR NEW LANDFILLS
Refuse Cells
Refuse Cell
6-8" of Fill Dirt
Dirt Cover
Monitoring Well
6-8" of Rock
8-12" of Compacted Clay
Groundwater
Plastic
Leachate Collection Pond
Leachate Drainage System
20'
Groundwater

FIGURE 20–3
Features of a modern landfill with environmental safeguards. The landfill is sited on a high location well above the water table. The bottom is sealed with compacted clay or a plastic liner or both overlaid by a rock or gravel layer with pipes to drain the leachate. Refuse is built up in layers as each day's refuse is covered with earth so that the completed fill has a pyramid-shape that sheds water. The fill is provided with groundwater monitoring wells.

Incineration has some serious drawbacks, however. Trash does not burn cleanly. Despite being equipped with air pollution control devices, incinerator stacks emit toxic fumes into the air as burning oxidizes and vaporizes the assortment of metals, plastics, and hazardous materials that inevitably end up as municipal waste. Incinerators are very expensive to build, and their siting has the same problem as the landfills: No one wants to live near one. Incinerator ash is loaded with metals and other hazardous substances and must be disposed of in secure landfills. The worst criticism of the incineration option is that, even if the incinerator generates electricity, the process wastes both energy and materials when compared with recycling. Indeed, incinerators compete directly with recycling for burnable materials such as newspapers and represent a major impediment to recycling in many municipalities. For these reasons, incineration represents a wrong direction in waste management.

COSTS OF MUNICIPAL SOLID WASTE DISPOSAL

The costs of disposing of MSW are becoming prohibitive. Increasing costs are not just a result of the new design features of landfills. More and more, they reflect the expenses of acquiring a site and providing transportation. **Tipping fees**—the costs assessed at the disposal site—now exceed $100 a ton at some landfills, and the waste collector must recover this cost as well as transportation costs to the site. Tipping fees at incinerators are no better; one trash-to-energy incinerator in Saugus, Massachusetts, recently had to increase its tipping fee from $22 to $80 per ton to cover the cost of adding a mandated stack scrubber.

No one wants a landfill in her or his backyard, and with spreading urbanization there are few areas near cities not already dotted with high-priced suburban homes. Any site selection, then, is met with protests and often legal suits declaring, "Anywhere

---

---

but here!'' The legal costs incurred in overcoming these objections, if they are overcome, are often as expensive as all other costs combined. Or, the process leads to selection of a very distant site, which involves inordinate hauling expenses. There are also limits of state lines; no state wants to act as the dump site for another state (see In Perspective Box, p. 447).

With new landfills being held back by costs and legal objections, the sad fact is that most MSW is still going into old landfills with inadequate safeguards. Of the 6000 municipal landfills in the United States, which receive 75 percent (over 130 million tons per year) of our MSW per year, 75 percent are unlined, 95 percent do not have leachate collection systems, and 75 percent do not monitor groundwater.

About 1200 old landfills are scheduled to close by 1997 either because they have reached capacity or because of environmental problems. New landfills are being constructed at less than half this rate, however. In the 1960s, many environmentalists thought that our throwaway society would meet limits in the form of shortages of resource materials. It is ironic to note that the actual limit is turning out to be space to dump the garbage.

Running out of space is beside the point, however, since even if new sites could be obtained, the system is not sustainable. Fortunately, there is a better way.

Solutions

REDUCING WASTE VOLUME

The best strategy of all is to reduce waste at its source. We noted earlier that the increased amount of waste produced over the years is largely a result of changing lifestyles, notably the growing use of disposable products and excessive packaging. This may be changing. To mention just two recent developments, concerned over the mass of disposable diapers in MSW, many families are switching to cloth diapers, and environmentally concerned consumers have successfully pressured some producers to reduce packaging. The manufacturers of compact discs have agreed to downsize their unnecessarily large packages, for example.

An option that thus far has received too little attention is the potential for reducing waste volume by keeping products in use longer. Reusing items in their existing capacity is the most efficient form of recycling. The use of returnable versus nonreturnable beverage containers is a prime case in point.

Returnable versus Nonreturnable Bottles

Before the 1960s, most soft drinks and beer were marketed by local bottlers and breweries in returnable bottles that required a deposit. Trucks delivered filled bottles to retailers and picked up empties to be cleaned and refilled. This procedure is efficient when the distance between producer and retailer is relatively short. As the distance increases, however, transportation costs become prohibitive because the consumer pays for hauling the bottles as well as for the beverage. In the late 1950s, distributors, bent on expanding markets and growth, observed that transportation costs could be greatly reduced if they used lightweight containers that could be thrown out rather than shipped back. Thus, no-deposit, nonreturnable bottles and cans were introduced. The throwaway container is also an obvious winner for its manufacturers, who profit by each bottle or can they produce.

Through massive advertising campaigns promoting national brands and the convenience of throwaways, a handful of national distributors gained dominance during the 1950s and 1960s, and countless local breweries and bottlers were driven out of business. At the same time, bottle and can manufacturing grew into a multibillion-dollar industry.

The average person drinks about a quart of liquid each day. For 250 million Americans, this daily consumption amounts to some 1.3 million barrels of liquid. That a significant portion of this volume should be packaged in single-serving containers that are used once and then thrown away is bizarre. It is difficult to imagine a more costly, wasteful way to distribute fluids.

Beverages in nonreturnable containers and those in returnable containers appear to be priced competitively on the market shelf, but this equality evaporates when you look at the hidden costs of single-use containers. Nonreturnable containers constitute 6 percent of the solid waste stream in the United States and about 50 percent of the nonburnable portion; they also constitute about 90 percent of the nonbiodegradable portion of roadside litter. Broken bottles along the road are responsible for innumerable cuts and other injuries, not to mention flat tires. Both the mining of the materials they are composed of and the manufacturing process create pollution. All of these are hidden costs that do not appear on the price tag, but we pay them with taxes for litter cleanup, our injuries, flat tires, environmental degradation, and so on.

In an attempt to reverse the trend, environmental and consumer groups have promoted **bottle bills**—laws that facilitate the recycling or reuse of beverage containers. Such bills generally call for a deposit on all beverage containers—both returnables and throwaways. Retailers are required to accept the used containers and pass them along for recycling or reuse.

Bottle bills have been proposed in virtually every state legislature over the last decade. In every case, however, the proposals have met with fierce opposition from the beverage and container industries and certain other special interest groups. The reason for their opposition is obvious—economic loss—but the arguments they put forth are more subtle. The container industry contends that bottle bills will result in loss of jobs and higher beverage costs for the consumer. They also claim that consumers will not return the bottles and litter will not decline.

In most cases, the industry's well-financed lobbying efforts have successfully defeated bottle bills. However, some states—10 as of 1992—have adopted bottle bills despite industry opposition (Table 20–1). Their experience has proved the beverage and bottle industry's arguments false. More jobs are gained than lost, costs to the consumer have not risen, a high percentage of bottles are returned, and there is a marked reduction in can and bottle litter. In some cases, local breweries and bottlers are making a comeback, thus improving the local economy.

A final measure of the success of bottle bills is continued public approval. Despite industry efforts to repeal bottle bills, no state that has one has repealed it. As this text goes to press, a national bottle bill has been introduced in Congress. Opponents are arguing that the bill will threaten the newly won successes in curbside recycling, with some justification. Beverage containers represent the most important source of revenue in curbside recycling. However, as curbside recycling currently reaches only 15 percent of the U.S. population, a national bottle bill will recover a much greater proportion of beverage containers (states with bottle bills report 80 to 97 percent rates of return of containers).

Other Measures

Whenever items are reused rather than thrown away, the effect is a reduction in waste and better conservation of resources. In this respect, it is encouraging to see the growing popularity of yard sales, flea markets, and other "not new" markets (Fig. 20–4).

TABLE 20–1

States That Have Bottle Bills

State	Year Passed	State	Year Passed
Oregon	1972	Iowa	1978
Vermont	1973	Massachusetts	1978
Maine	1976	Delaware	1982
Michigan	1976	New York	1983
Connecticut	1978	California	1991

FIGURE 20–4
One way to recycle. The yard sale has become a Saturday morning staple in communities all over the country. (George E. Jones III/Photo Research.)

Consideration of our domestic wastes and their disposal emphasizes what a mammoth stream of materials flows in one direction from our resource base to disposal sites. Just as natural ecosystems depend on recycling nutrients, the continuance of a technological society will ultimately depend on our learning to recycle or reuse not only nutrients but virtually all other kinds of materials as well.

THE RECYCLING SOLUTION

In addition to reuse, recycling is another obvious solution to the solid waste crisis. More than 75 percent of MSW is recyclable material. Of course, many people have been advocating recycling for a long time now, and various groups and individuals have been recycling paper, glass, and aluminum cans on a small scale for decades. What has prevented recycling from being implemented on a large scale? There are some real impediments to large-scale recycling that must be overcome. However, once these problems are recognized and understood, they can be and are being solved. Indeed, recycling is now a tremendous growth industry with prospects for a bright future.

Impediments to Recycling

The problems that must be overcome are:

Sorting We are used to the convenience of throwing all our refuse into a single container and handling it in one bulk mass. For recycling, the various constituents must be separated either in the home or after collection.

Lack of Standards Sorting is made even more difficult by lack of standards. That is, several kinds of plastic

ETHICS
Recycling—The Right Thing to Do?

The town meeting is an institution in Massachusetts. Held once a year, it is a forum for decision making on the town's budget and all sorts of ancillary business (such as deciding on a leash law, changing the zoning requirements, or accepting new streets). Originally, the whole town was expected to attend; nowadays, 10 percent of the voting population is considered a good turnout. Microphones are set up in the auditorium, and anyone in the town can express her or his opinion on matters on the docket.

In recent years, Massachusetts towns have taken a strong turn toward fiscal stringency, a symptom of the larger economic recession in the United States. Many school programs have been radically cut, town personnel have been let go, and many items on town budgets have been voted down. In the face of such stringency, it is encouraging that recycling programs have gone against the cutback trend in almost every town and city in Boston's Northshore area. By 1992, most Northshore towns had some kind of recycling program. Very often, the

program was proposed by a nonpaid town official acting on behalf of grass roots citizens' groups. Arguments pro and con were aired at the town meeting, and, in town after town, recycling was adopted by overwhelming votes even though the programs do not pay for themselves; it still costs money to recycle.

The city of Beverly is one example. Beverly has curbside recycling, and the town picks up 50 tons of recyclables a week, at a cost of $30 a ton. This compares unfavorably with the current tipping fee of $22 a ton at a regional trash-to-energy incinerator. When asked why the townspeople were in favor of recycling, one Beverly town official stated that they're doing it because "it's the right thing to do!"

One town—Topsfield—recently celebrated its twentieth year of a recycling program, clearly a leader in the trend. Not all towns have jumped on the bandwagon, however. Danvers, for example, has held back because "it's just not cost-effective," according to the town public works director.

However, the handwriting is not only on the wall, it's now a matter of a state **solid waste master plan** issued in 1990. Leaf waste, large appliances, and tires were phased out of landfill dumping statewide in 1991; the ban extended to other yard waste, metals, and glass in 1992, and by 1994, recyclable paper and plastics will be banned from landfills. Obviously, the recycling trend is not simply a matter of virtuous decision making on the part of the townspeople; they see the state mandates coming and are acting wisely. However, it should be noted that recycling in Massachusetts began as a grass roots movement. It has reached the state level and is now working back down to the grass roots to catch those towns that have been dragging their feet.

Recycling is one of the most obvious ways for people to demonstrate their concern for the environment. And perhaps the reason it got started is the most basic one: Because it's the right thing to do. Do you agree?

or grades of paper may be used in similar products or even in the same product.

Reprocessing There must be companies capable of receiving the materials collected and converting them into salable materials. Otherwise it is off to the landfill after sorting.

Marketing There must be industrial or consumer markets to buy the products made from recycled material. Otherwise the manufacturing company goes bankrupt, and the products become refuse before they are even sold.

Separation between Government and Private Enterprise In general, refuse collection is done by local governments, and governments are reluctant to (and probably shouldn't) get into the business of producing and marketing materials, which is the realm of private enterprise. Conversely, companies engaged in production like to deal with clean, uniform raw materials, which trash is not. Therefore, with few exceptions, they have been less than eager to deal with refuse. Lack of cooperation between local governments and the private sector frequently impedes recycling.

Vested Interests in the Status Quo Tremendous profits can be maintained indefinitely in manufacturing and selling bottles, cans, and other items that are used only once and then discarded. The vested interests who profit from the throwaway habit have been a potent force against implementing any form of recycling.

Hidden Costs Since refuse disposal is usually financed out of tax revenues, people generally do not realize how much they are paying. The costs of cleaning up a hazardous site or monitoring such a site forever to check for groundwater contamination are not tallied into the costs of disposal. With costs thus hidden, refuse disposal may seem like a free (and care-free) service. The costs of alternatives seem expensive by comparison, even though the long-term costs would be less.

Addressing the Problems

Problems should not be taken as an excuse for inaction, however; rather, they should be taken as an opportunity to develop creative solutions. Thousands of communities in the United States (15 percent of the population) are overcoming the aforementioned obstacles in one way or another and entering into curbside recycling. Let us look at a few of the basic ideas being tried.

Government-Business Partnerships Companies that will provide full-service recycling—that is, collection through processing, including production of certain products from recycled materials—are forming and growing rapidly. Governments are forming partnerships with such companies. Basically the company is contracted to collect and recycle a certain minimum percentage of the municipality's waste stream. In return, the government gives the company certain guarantees such as exclusive collection and marketing rights of certain recycled materials in their area. Further, the government may agree to purchase certain amounts of recycled paper, compost, and plastic. Finally the government agrees to continue to dispose of a certain quantity of material that cannot be recycled at the present time. Can you see the need for these guarantees? Without them the company could be forced out of business, to the detriment of both parties.

Sorting Sorting is best done at the source (homes), although it may be done after collection. Sorting at the source requires the cooperation of a large portion of the population, but it is relatively inexpensive since the work is "volunteer." A system that is gaining acceptance in a number of communities is the issuance of color-coded containers for plastic, metals, glass, paper, yard wastes, and "other." A trailer with colored bins is drawn behind the regular trash truck, and workers dump containers in the respective bins. Unsorted refuse continues to go into the regular trash truck for traditional disposal. The sorted refuse is then transported to a facility that processes it for further distribution to businesses that deal in recycled materials (see In Perspective Box, p. 453).

Alternatively, unsorted trash can be picked up by regular trucks and separated after collection. Refuse separation facilities have been built and are in operation. The general scheme for one such facility is shown in Figure 20–5. However, such equipment is very costly to purchase, and operation and maintenance costs are also high. The payback from the sale of recycled materials comes nowhere close to offsetting these costs. The major savings comes from not having to pay tipping fees at landfills or incinerators. When the savings in fees are added to proceeds from the sale of recycled materials, the total can more than offset the maintenance costs.

In Third World countries many poor people make their living by picking through dumps and reselling "garbage" (Fig. 6–6). This is a sign of their desperate poverty, however, not a recommended solution to the garbage problem.

Reprocessing and Profits There is an abundance of alternatives for reprocessing various components of

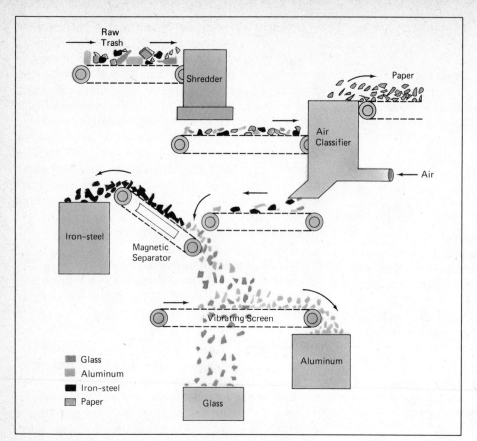

FIGURE 20–5
Schematic flow diagram for the separation of MSW after collection. Separation can be achieved, but is it superior to separation at the source, as in curbside recycling? Does the value of the separated materials justify the costs of separation?

Labels in figure: Raw Trash, Shredder, Air Classifier, Paper, Air, Iron-steel, Magnetic Separator, Vibrating Screen, Aluminum, Glass

Legend: Glass, Aluminum, Iron-steel, Paper

refuse, and people are coming up with new ideas and techniques all the time. A few of the major established techniques follow.

○ Paper can be repulped and reprocessed into recycled paper, cardboard, and other paper products; finely ground and sold as cellulose insulation; or shredded and composted (see below).

○ Glass can be crushed, remelted, and made into new containers or crushed and used as a substitute for gravel or sand in construction materials such as concrete and asphalt.

○ Some forms of plastic can be remelted and fabricated into "synthetic lumber" (Fig. 20–6). Such lumber, since it is not biodegradable, has potential for use in fence, sign, and guardrail posts, docks, decks, and other outdoor uses.

FIGURE 20–6
Waste plastics can be fabricated into "synthetic lumber," which has great potential for outdoor uses, such as the deck boards of this dock, because it is nonbiodegradable. This deck was provided by the Chicago Park District as part of a 1988 Pilot Recycling Program. The deck material was manufactured from approximately 84 500 plastic milk bottles by Eaglebrook Profiles, Chicago, IL. (Photograph by BJN.)

Currently, most recycling is on a town-by-town basis. Either the towns or their waste contractors must find markets for the recycled goods: cans, bottles, newspapers, and plastics. This problem has kept many municipalities from getting into recycling. The solution? A regionalized materials recycling facility (MRF), referred to in the trade as a "murf." Here's how the state-owned MRF in Springfield, Massachusetts, works:

Basic sorting takes place when waste is collected, either by curbside collection or by town recycling stations (sites where townspeople can bring wastes to be recycled). The waste is then trucked to the MRF and handled on three tracks—one for metal cans and glass containers, another for paper products, and a third for plastics. The materials are moved through the facility by escalators and conveyor belts, tended by workers who inspect and do further sorting. The objective of the process is to prepare materials for the recycled goods market. Glass is sorted by color, cleaned, crushed into small pebbles, and then shipped to glass companies, where it replaces the raw materials that go into glass manufacture—sand and

soda ash—and saves substantially on energy costs. Cans are sorted, flattened, and sent either to de-tinning plants or to aluminum processing facilities. Paper is sorted, baled, and sent to reprocessing mills. Plastics are sorted into four categories, depending on color and type of plastic polymer, and then sold.

The facility's clear advantages are its economy of scale and its ability to produce a high-quality end product for the recycled materials market. Towns know where to bring their waste, and they quickly become familiar with the requirements for initial sorting during collection and transfer.

Currently, there is only one MRF in Massachusetts, built with state funds at a cost of $5 million and operated by the state Department of Environmental Protection. Several more are on the drawing board. After one year of operation, the Springfield MRF was declared a success. It took in 42 886 tons of materials from 91 towns representing a population of 750 000. The process has diverted 22 to 50 percent of the waste stream (depending on the community) from landfills and incinerators, and saved the towns millions of dollars in disposal costs.

The facility has attracted recycled wastes from Connecticut and New York while those states are setting up their own regional recycling centers.

There have been some unanticipated problems at the Springfield facility during its first year of operation. Several towns have reduced their trash so much that they are not fulfilling their contractural obligations to a regional incinerator. Landfill revenues have declined, forcing landfill operators to lower their fees in order to attract more haulers. These appear to be temporary problems that will be worked out in time, however. The state is committed to recycling almost half of municipal solid waste, and incinerating most of the rest, by the year 2000. In theory, landfills will operate mainly as recipients of incinerator ash and materials that can't be either recycled or burned.

Very likely, the future of regional recycling will be in the private sector. In order for this to work, the recycling market will have to be substantially strengthened. However, all indications point to such facilities as the wave of the future.

○ Metals can be remelted and refabricated. Making aluminum from scrap aluminum saves up to 90 percent of the energy required to make aluminum from virgin ore.

○ Food wastes and yard wastes (leaves, grass, and plant trimmings) can be composted to produce a humus soil conditioner.

○ Textiles can be shredded and used to strengthen recycled paper products.

○ Old tires can be remelted or shredded and made into a number of other products.

In addition, literally hundreds of new processes are being developed and commercialized to make refuse components into more valuable end products.

Thus recycling is becoming increasingly profitable. Consequently, the profit potential of recycling is bringing many new companies into the field despite the vested interests of some old established companies in maintaining the status quo.

Promoting Recycling through Mandate

A number of measures are being adopted by various state and local governments that mandate or at least support recycling. These include:

Mandatory Recycling Laws A number of states have passed mandatory recycling laws. Such laws require that each county, under threat of loss of state funds, recycle a certain percentage of its refuse by a certain

FIGURE 20–7
Composting refuse. Keeping material moist but well aerated results in odor-free decomposition of paper, food, yard, and other organic wastes into humuslike material. Photograph shows windrow of refuse being turned and fluffed by machinery to aid decomposition. (Courtesy Wildcat Manufacturing Co., Inc. Freeman, SD.)

date. Massachusetts, for example, has adopted a goal of recycling 46 percent of all of its MSW by 2000.

Banning the Disposal of Certain Items in Landfills
Yard wastes are a good first candidate here because they take up considerable volume and can easily be collected separately, composted into humus, and applied to park lands. To date, Florida, Illinois, Minnesota, New Jersey, Pennsylvania, and Wisconsin have banned yard wastes from landfills. Of course, items that present a toxic or explosive hazard, such as car batteries, are already generally banned.

Mandating Government Purchase of Recycled Materials A state can require that all its agencies buy a certain percentage of recycled paper. Requiring highway departments to use plastic signposts and parks departments to use compost can extend recycling beyond paper.

Mandating the Use of Recycled and Reusable Materials in Packaging One-third of landfill space is taken up by packaging materials. Great savings—at the source and by recycling—can come from requiring that all packaging be reusable or made (at least partly) of recycled materials. A bill to that effect has been introduced in the Massachusetts legislature, calculated to triple recycling rates in the state by the year

2000 if enacted. The province of Ontario has already passed a similar initiative aimed at soft drink packaging.

COMPOSTING

One way of treating some forms of refuse that is rapidly growing in popularity is composting. Recall that composting involves the natural biological decomposition (rotting) of organic matter in the presence of air (see Chapter 9, In Perspective Box, p. 195). The end product is a residue of humuslike material, which can be used as an organic fertilizer. Composting was one method of treating sewage sludge described in Chapter 13. Likewise, since refuse is usually 60 to 80 percent organic matter (paper and food wastes)—or more when yard wastes are added—it may be treated by composting (Fig. 20–7).

A number of companies have entered into the business of selling equipment or of building and running facilities for composting refuse. Glass, metals, and plastics may be removed either before or after composting and recycled as desired. Also, raw sewage sludge may be mixed with the refuse to achieve a synergistic composting of both simultaneously. Paper helps to dewater the sewage sludge and provides better aeration, and the sludge supports better decomposition. There is a good market for compost from landscaping firms, and it may be used on city parks or agricultural fields. It is important, however, that wastes to be composted be free of toxic household products and heavy metals. Otherwise, the compost is unacceptable for agricultural use.

REFUSE-TO-ENERGY CONVERSION

Because it has a high organic content, refuse, including the plastics portion, can be burned. When burned, unsorted MSW releases about half as much energy as coal, pound for pound. We noted earlier in this chapter that incineration of MSW releases toxic fumes into the air and keeps recyclable materials (75 percent of MSW) from being recycled. Despite these serious drawbacks, a number of municipalities across the United States have built plants to incinerate MSW and generate electricity at the same time—almost three-fourths of the currently operating incinerators. Sale of the electricity offsets some of the costs of MSW

disposal, and the incinerators are usually able to compete successfully with landfills for MSW.

Incinerator ash must be landfilled, but because ash makes up only about 10 to 20 percent of the original volume, the life of the landfill will be extended about five- to tenfold. More important, since the incinerated material is not subject to further decomposition and settling, it can sometimes be used as fill dirt in construction sites, road beds, and so forth. However, the heavy metals in the ash may preclude this kind of use in many areas.

A refuse-to-energy plant that went into operation in Baltimore, Maryland, in 1984 is shown in Figure 20–8. This plant is capable of consuming 2000 tons of unsorted MSW per day. Steam produced in boilers drives a 60 000-kilowatt generator that produces enough power to service about 60 000 homes. Exhaust gases from the combustion are processed through electrostatic precipitators to remove some of the objectionable air pollutants.

The major drawback of converting refuse to energy is that it largely precludes future options for recycling and composting.

FIGURE 20–8
Conversion of municipal solid waste to energy. This diagram represents a refuse-to-energy plant in Baltimore, Maryland. The facility can incinerate 2000 tons of unseparated MSW per day to generate 60 megawatts of on-line power, enough to supply 60 000 homes.

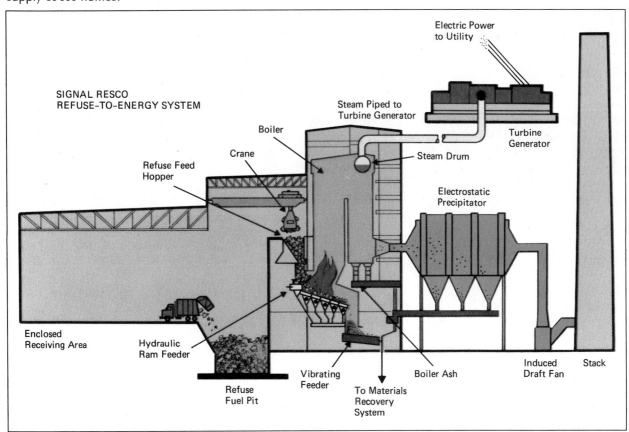

INTEGRATED WASTE MANAGEMENT

It is important to remember that it is not necessary to fasten on a single method of waste handling. Almost any combination of recycling, composting, and reducing waste volume may be used. Further, recycling can be introduced gradually, pursuing a number of options while phasing out landfilling. This system of having several alternatives in operation at the same time is called **integrated waste management.** Balancing the interests of all parties involved obviously requires skilled managers. However, the days where each town could go its own way are in the past; regional and state management of solid waste is becoming the rule. Public policy is moving in the direction of integrated waste management under regulations at the state and even the federal level. Given the dimensions of the solid waste crisis, this increased attention to management is a welcome development.

It is also important to remember, however, that true management of MSW begins in the home. Lifestyle changes have been responsible for the increase in per capita MSW; it would be entirely possible to bring about a decrease in per capita MSW through lifestyle changes. The growing use of recycling will undoubtedly draw the attention of consumers to choices they make in the purchase and use of materials. Those who want to be part of the solution to the solid waste crisis will find many ways to reduce the waste they generate. Our society can make real progress toward sustainability in the MSW arena only through a combination of public policy and personal lifestyle changes that reject the throwaway mentality.

 Review Questions

1. List six components and their percent composition in MSW. What level of government is responsible for disposal of these wastes?
2. Trace the historical development of refuse disposal. What method of disposal is now most common?
3. What are the major costs and limitations of landfilling?
4. Outline the EPA's latest regulations for the construction of new landfills.
5. What are the advantages and disadvantages of incineration?
6. Why is landfilling not a sustainable option for solid waste disposal?

7. How can waste volume be reduced?
8. What seven impediments to recycling have prevented its large-scale implementation?
9. How are communities overcoming these obstacles to recycling?
10. What laws have been adopted by state and local governments to support recycling?
11. Which waste materials may be composted?
12. Discuss the pros and cons of converting refuse to energy.
13. What is integrated waste management?

 Thinking Environmentally

1. Compile a list of all the plastic items you used and threw away this week. Consider how you could reduce the length of this list.
2. If you live near the ocean or a large lake, walk along the beach for 20 minutes and collect all of the items that are clearly not natural. Where did these come from? What is their volume per 100 yards of beach?
3. How and where does your school dispose of solid waste? Is a recycling program in place? How well does it work?

4. Why do you suppose the United States creates so much more waste than other countries?
5. Does your state have a bottle bill? Is it effective? If you live in a state without such a bill, explore the politics that have prevented the bill from being adopted.
6. Suppose your town planned to build an incinerator near your home. You are concerned about your family's health but also about the rising costs of solid waste disposal. Explain your decision for or against the incinerator.

Energy Resources: The Rise and Fall of Fossil Fuels

OUTLINE

LEARNING OBJECTIVES

When you have finished studying this chapter, you should be able to:

1. Describe how the major sources of energy used in the United States have changed from 1800 to the present.

2. Describe how electrical power is generated, why it is called a secondary energy source, and what primary energy sources are used to generate it.

3. Tell what environmental problems may be associated with the use of electrical power.

4. List the four major categories of energy use and name the primary energy sources currently used to supply each.

5. Name the three major fossil fuels and describe their origin.

6. For crude oil, distinguish between estimated reserves, proven reserves, and production, and describe the reasons behind increase or decrease in the figures for the amounts of each.

7. Contrast U.S. oil production and consumption before 1970 with that after 1970 and explain what was done to resolve the disparity.

8. Describe what brought on the "oil crisis" of the 1970s, what occurred to make supplies available again, and what actions were taken to resolve the underlying problem.

9. Describe what brought on the collapse of oil prices in 1986 and the impacts this had on exploration, U.S. production, and conservation.

10. Explain how the 1991 Persian Gulf War was related to controlling oil supplies and prices and how the United Nations forces' winning the war has not changed the situation.

11. Identify alternative fossil fuels that the United States might exploit to supplement declining reserves of crude oil, and explain the advantages and disadvantages of each.

12. List the basic avenues toward resolving the long-term energy problem, and compare economic and environmental advantages and disadvantages of each.

In Chapter 3 we learned that natural ecosystems use sunlight as their energy source—a source that is nonpolluting and nondepletable. Therein we saw the second principle for sustainability. Humans remain dependent on the sun and photosynthesis for all agricultural, forage, and forest production. However, technological development has been synonymous with development of additional sources of energy to run all kinds of machinery and processes used in production, transportation, heating, cooling, lighting finally to disposing of wastes. Try to imagine what life would be like without energy in addition to sunlight. The importance of supplementary energy sources speaks for itself.

By far the most widely used supplementary energy source is fossil fuels—coal, crude oil, and natural gas—and use of these fuels is continuing to grow. This situation is bound to change in the next 50 years, however, for two reasons: Major reserves of these fuels are becoming depleted, and the environmental impacts of using fossil fuels, particularly global warming, will force cutbacks.

We have a choice of two pathways into the future. We can ignore the depletion and environmental impact issues and pursue our present course until shortages and environmental problems cause such disruptions that changes will be forced on us. Or, we can start now to reduce our dependence on fossil fuels and develop alternatives that will enable a smooth transition to sustainable energy sources. Which of these paths we follow will depend largely on what you and your generation decide.

The objective of this chapter is to provide an overview of our current energy situation. In particular, we shall focus on fossil fuel resources, which provide most of our current energy demands, and the potential (or lack of potential) of those resources to meet future demands.

FIGURE 21–1
The development of energy sources has in large part supported the development of civilization.

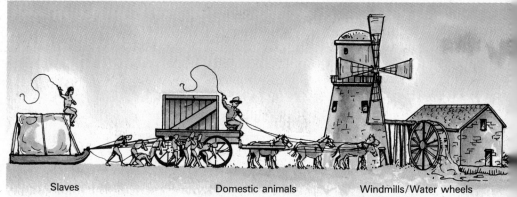

Slaves   Domestic animals   Windmills/Water wheels

Part Four Biota, Refuse, Energy, and Land

Energy Sources and Uses in the United States

Throughout human history, the advance of technological civilization has been tied to the development of energy sources (Fig. 21–1). In early times (and still persisting in less-developed regions), the major energy source was human labor, doing things by hand. Some people lived in relative luxury by exploiting the labor of others—slaves, indentured servants, and minimally paid workers. Human labor was supplemented to some extent with domestic animals for agriculture and transportation, water power and wind power for milling grain, and a few other uses. However, the limitations of these power sources are self-evident.

By the early 1700s designs for many kinds of machinery had already been conceived. The limiting factor was a power source to run them. The breakthrough that launched the Industrial Revolution was the development of the steam engine in the late 1700s. Steam engines operate on the principle of boiling water to produce high-pressure steam, which is used to push a piston back and forth in a cylinder. Through a crankshaft, the piston powers the drive wheel of the machinery. Steam engines rapidly become the power source for steamships, steam shovels, steam tractors, steam locomotives, and stationary engines to run sawmills, textile mills, and virtually all other industrial plants.

The first major fuel for steam engines was firewood. Then, as demands for energy increased and firewood around industrial centers became scarce, coal was substituted. By the end of the 1800s, coal had become the dominant fuel and remained so into the 1940s. In addition to being used as fuel for steam

engines, coal was widely used for heating, cooking, and industrial processes. In the 1920s, coal provided 80 percent of all energy used in the United States, with similar percentages in other industrializing countries.

While coal and steam engines powered the Industrial Revolution, they were far from ideal. The smoke and fumes from the numerous coal fires made air pollution in cities far worse than anything seen today. Writers have recorded that often you could not see the distance of a city block because of the smoke. Coal is also notoriously hazardous to mine and dirty to handle, and burning coal requires the disposal of large quantities of ash. As for steam engines, because of the size and bulk of the boiler, the engines were heavy and cumbersome to operate (Fig. 21–2). Often the fire had to be started several hours before the engine was put into operation, in order to heat the boiler sufficiently.

In the late 1800s, simultaneous development of three things—the internal combustion engine, well-drilling technology, and the capacity to refine crude oil into gasoline and other liquids fuels—combined to provide an alternative to steam power. Replacement of coal-fired steam engines and furnaces with gasoline and diesel engines and fuel oil furnaces was an immense step forward in convenience. Also, air quality was greatly improved because cities were gradually rid of the smoke and soot from burning coal. (It was only in the 1960s with the tremendous proliferation of cars that pollution from gasoline engines became a problem.) Further, the gasoline internal combustion engine provides a valuable power-to-weight advantage that allowed rapid advances in technology. A 100-horsepower gasoline engine weighs but a tiny fraction of a 100-horsepower steam engine and its boiler, and jet engines have an even

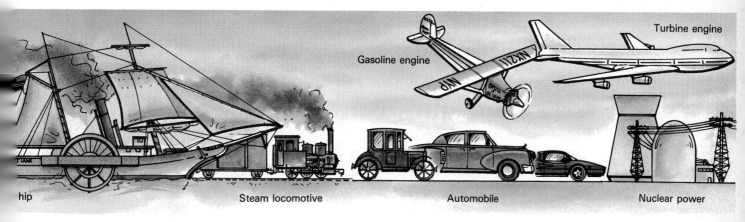

ship   Steam locomotive   Gasoline engine   Turbine engine   Automobile   Nuclear power

FIGURE 21–2
Colossal steam-driven tractors marked the beginning of modern agriculture. Today tractors half the size do 10 times the work and are much easier to operate. (Lowell Georgia/Photo Researchers.)

greater power-to-weight ratio. Automobiles and other forms of transportation would be cumbersome, to say the least, without this power-to-weight advantage, and airplanes would be impossible.

Replacement of one energy technology with another is a very gradual process, however, because it is most cost-effective to use existing machinery until it wears out before replacing it. It was the late 1940s before crude oil surpassed coal to become the dominant energy source for the United States, but its use has continued to grow. Crude oil currently provides about 41 percent of the total U.S. energy demand. The story is similar throughout the rest of the world, although the timing of events differs. We have noted that many poor regions of the world remain dependent on firewood. Coal is still the dominant fuel used in the countries of Eastern Europe and in China. Development throughout the world, so far as it has occurred since the 1940s, has been largely predicated on technologies that consume gasoline and other fuels refined from crude oil. Thus, oil is the mainstay not only for the United States but for most other countries as well, both developed and less-developed.

Coal has not passed from the picture even in the United States, however; it has become a major energy source for the generation of electrical power, as we shall discuss shortly. In this role, coal currently provides 23 percent of the total U.S. energy demand.

Natural gas, the third primary fossil fuel, is found in association with oil or in the process of looking for oil. Being largely methane, natural gas burns more cleanly than oil. Thus, in terms of pollution, it is a more desirable fuel. Despite its obvious fuel potential, however, at first there was no practical way to transport it from wells to consumers. Any gas re-

leased from oil fields was (and in many parts of the world still is) flared—that is, vented and burned in the atmosphere, a tremendous waste of valuable fuel. Gradually the United States constructed a network of pipelines that are continuous from wells to consumers. With the completion of these pipelines, the use of natural gas for heating, cooking, and industrial processes escalated rapidly because of its cleanliness, convenience (no storage bins or tanks are required on the premises), and relatively low cost. Currently natural gas provides 24 percent of U.S. energy demand.

Thus, three fossil fuels—crude oil, coal, and natural gas—provide over 88 percent of U.S. energy. The other 12 percent is provided mostly by nuclear power and water power, which, along with coal, are used for the generation of electrical power. This changing pattern of energy sources in the United States is shown in Figure 21–3.

Again the picture is similar for other countries of the world, although the percentages differ somewhat depending on the energy resources of any given country.

ELECTRICAL POWER PRODUCTION

In 1831, an English scientist named Michael Faraday discovered that passing a coil of wire through a magnetic field causes a flow of electrons—an electrical current—in the wire. An electric generator is basically a coil of wire that rotates in a magnetic field, or the magnetic field is rotated within a stationary coil (Fig. 21–4). However, our old friends, the First and Second Laws of Thermodynamics (Chapter 3), prevent us from getting something for nothing or even breaking

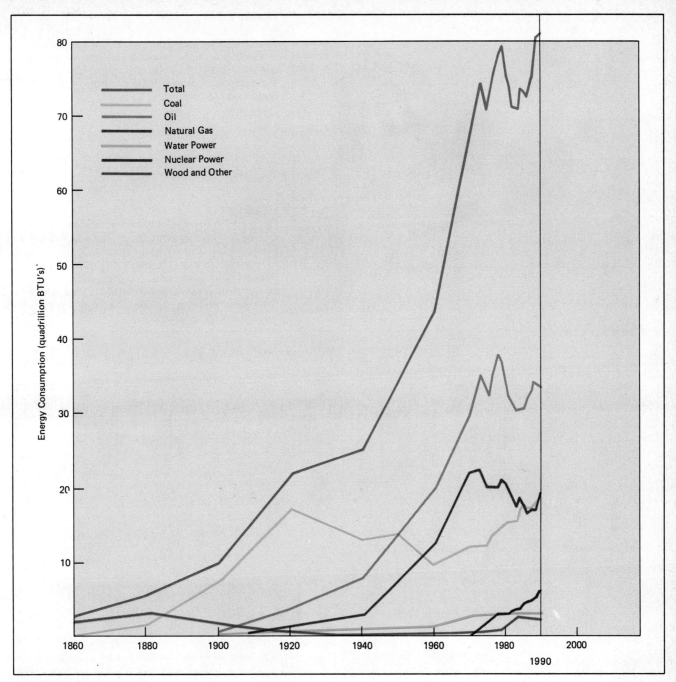

FIGURE 21–3

Energy consumption in the United States, 1860–1990, total consumption and major primary sources. Note how the mix of primary sources has changed over the years and how the total amount of energy consumed has continued to grow. Note skyrocketing increase in the use of oil following World War II (1945) as the U.S. population shifted to commuting to and from work by car. Conspicuous decreases in consumption occurred in the 1970s because of factors described in the text, but in recent years consumption has again begun to increase. (Data from the U.S. Department of Energy.)

even. As the current is created, it creates a magnetic field that opposes the first magnetic field and resists the movement of the wire. Also, some of the electrical energy created is converted to heat and lost. The result is that an amount of energy from another source,

greater than the output of the generator, must be expended in turning the generator. Electrical power is called a **secondary energy source** because it depends on a **primary energy source** (coal or water power, for instance) to turn the generator.

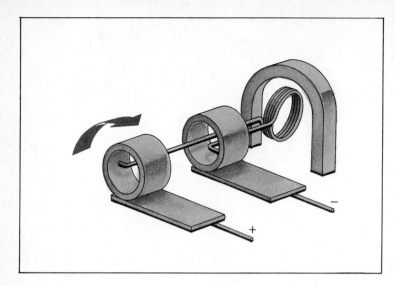

FIGURE 21–4
Principle of an electric generator. Rotating a coil of wire in a magnetic field induces a flow of electricity in the wire. In accordance with the laws of thermodynamics, more energy must go into turning the generator than is gotten out in electricity.

FIGURE 21–5
Electricity is produced commercially by driving generators with (a) steam turbines, (b) gas turbines, and (c) water turbines. Percentage of the U.S. electricity demand derived from each source is indicated. Small amounts of power are now also coming from solar wind energy. (d) Component of a steam turbine being assembled.

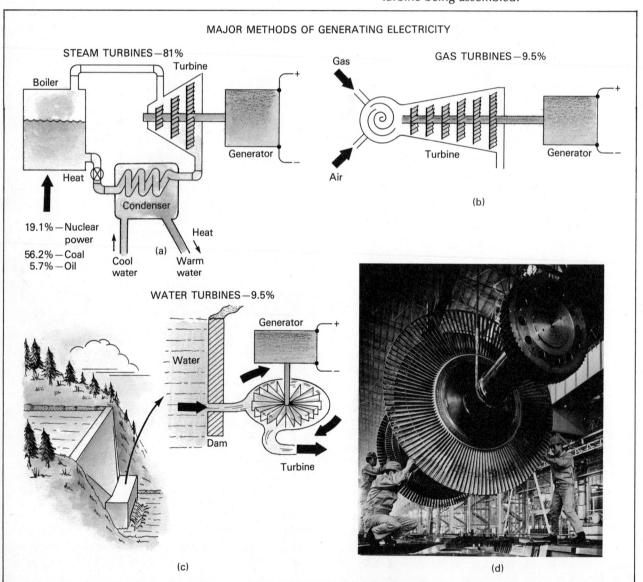

MAJOR METHODS OF GENERATING ELECTRICITY

STEAM TURBINES—81%

Boiler
Turbine
Generator

Heat
Condenser
19.1%—Nuclear power
56.2%—Coal
5.7%—Oil
Cool water
(a)
Warm water
Heat

GAS TURBINES—9.5%

Gas
Air
Turbine
Generator
(b)

WATER TURBINES—9.5%

Water
Generator
Dam
Turbine
(c)

(d)

The most widely used technique for generating electrical power is to use a primary energy source to boil water to create high-pressure steam, which drives a turbine—a sophisticated paddle wheel—coupled to the generator (Fig. 21–5a). The combined turbine and generator are called a **turbogenerator**. Any primary energy source can be used to boil the water. Coal, oil, and nuclear energy are most commonly used at present, but burning refuse, solar energy, and geothermal energy (heat from the earth's interior) may be more widely used in the future.

In addition to steam-driven turbines, gas and water (hydro-) turbines are also used. In a gas turbogenerator, the high-pressure gases produced by the combustion of a fuel, usually natural gas, drive the turbine directly (Fig. 21–5b). With water power, water under high pressure from behind the base of a dam or at the bottom of a pipe from the top of a waterfall is used to drive a hydroturbogenerator (Fig. 21–5c). Wind turbines are also coming into use.

Electrical power is often promoted as being the ultimate clean, nonpolluting energy source. It is true that, once generated, electricity involves essentially no pollution beyond heat losses. However, there may be any amount of pollution and any number of environmental impacts stemming from the primary energy source used to generate the power. Effectively what occurs with electrical power is that the problems are transferred from the *point of use* to the *point of generation*. For example, we have already discussed the problem of sulfur dioxide emanating from the tall stacks of coal-burning power plants as being the major source of acid rain, which affects much of the world (Chapter 16). Creating a dam and reservoir for

hydroelectric power involves the displacement of people, farmland, and wildlife and severely alters the course of the river being dammed, as discussed in Chapter 11. The hazards of radioactive wastes from nuclear power plants, discussed in Chapter 22, are well known. Additional impacts of electrical energy come from other parts of the fuel cycle, such as mining the coal or processing the uranium ore.

These problems take on even greater significance when the relatively low efficiency of electrical power production is considered. In order to drive a turbine, steam must pass from high pressure at the boiler side of the turbine to low pressure at the opposite side. This drop in pressure involves an increase in entropy and a necessary thermodynamic loss of heat. In fact, the most conspicuous feature of coal-burning or nuclear power plants is the huge cooling towers used to dissipate the lost heat into the environment (Fig. 21–6). This lost heat is *60 to 70 percent* of the energy released from the primary energy source. In other words, only 30 to 40 percent of the primary energy is converted to electricity. Then about 10 percent of the power generated (3–4 percent of the total) is lost to heat because of resistance along transmission lines from generators to homes.

An alternative to cooling towers is to pass water from a river, a lake, or the ocean over the condensing system (Fig. 21–5a) and discharge the heated water back into the environment. Thus waste heat is effectively discharged into the body of water. All the small organisms, both plant and animal, drawn with the water through the condensing system are effectively cooked, and the warm water added back to the receiving body may have a number of additional im-

FIGURE 21–6
Cooling towers. The most conspicuous feature of both coal-burning and nuclear power plants are these structures which are necessary to cool and recondense the steam from the turbines before returning it to the boiler. The tower creates a convection current which draws air over the condensing coils inside the tower. (Neal Palumbo/ Gamma Liaison.)

pacts on the aquatic ecosystem. Therefore, waste heat discharged into natural waterways is referred to as **thermal pollution**.

CONNECTING SOURCES TO USES

In addressing such questions as, Are energy resources sufficient? and Where shall we get additional energy?, we need to consider more than just the *source* of energy because some forms of energy lend themselves especially well to one use but not to another. For example, here in the United States we have developed an economy in which essentially all prime movers—all moving machinery, such as tractors and

bulldozers, as well as all cars, trucks, trains, and other transportation vehicles—depend on liquid fuels. Nuclear power, however, is suitable only for the generation of electrical power. Consequently, whether you are a supporter of nuclear power or an opponent, it is important to recognize that any amount of nuclear power will do little to alleviate the demand for liquid fuels, which come from crude oil. Of course, this situation might change with the development of electric vehicles. However, just the substitution of electric cars in the United States, to say nothing of all the other prime movers, would involve a $15 trillion (1 trillion = 1000 billion) investment.

Thus, it is important to understand how different energy sources are connected to particular end uses. Energy uses can be placed into four major categories:

○ Transportation
○ Industrial processes
○ Commercial and residential uses (mainly space heating and cooling, water heating and lighting)
○ Electric power generation

The major pathways from sources to end uses are shown in Figure 21–7. As noted above, the trans-

FIGURE 21–7
Energy flow from primary sources to end uses. This is a highly simplified diagram showing only major pathways, but it emphasizes that, given current technology, primary sources are connected to end uses in specific ways. Transportation is almost totally dependent on liquid fuels refined from crude oil, while coal and nuclear energy are useful only in producing electricity, almost none of which is used in transportation. Note the large portion of energy wasted at each conversion step. (Data from *Statistical Abstracts of the United States*, U.S. Department of Commerce, 1991.)

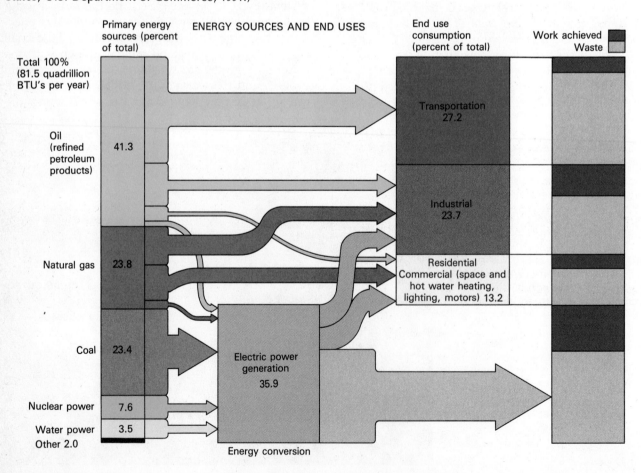

portation category, which includes all prime movers, is virtually 100 percent dependent on crude oil. Nuclear power, coal, and water power, however, are exclusive to the production of electrical power. Natural gas and oil are more versatile energy sources, but it makes sense to use them as efficiently as possible. For example, using natural gas to generate electrical power and then using that electricity for space heating is extremely inefficient because of the 60–70 percent thermodynamic loss that occurs in the production of the electricity. The natural gas can be used more efficiently in a modern furnace, in which better than 90 percent of the energy content of the gas goes into heating the home. (The other 10 percent goes up the chimney.)

Note in Figure 21–7 how much consumed energy goes directly to waste heat rather than to perform useful work. Some of this waste is the inevitable thermodynamic loss. In most cases, however, the losses are much greater than they need be. In other words, the opportunity exists to greatly increase the efficiency of energy use. Obtaining greater efficiency is a major part of energy conservation and is discussed in more detail later in this chapter.

Thus, the questions we posed at the beginning of this section—Are energy resources sufficient? Where shall we get additional energy?—require looking at more than just energy per se. For each end use, we must ask: Are energy resources *for that use* sufficient? Similarly, in considering where we might get additional energy, we need to ask, What form of energy will best meet our needs in this particular situation? Or might our needs be met by conservation in this case? More attention paid to developing an energy source that is appropriate to the existing technology may save trillions of dollars in investments. In the case of cars, for example, money may be more effectively spent on developing technologies to produce large quantities of hydrogen gas, which can be burned in place of gasoline in existing vehicles, rather than on switching to electric vehicles.

It is the failure of energy policy makers to pay adequate attention to the connection between energy sources and end uses that has brought us to the current energy problem, which promises to develop into a crisis in the near future if current trends are not altered.

The Energy Problem (or Crisis): Declining Reserves of Crude Oil

The main energy problem confronting the United States and most of the rest of the world today is that virtually all transportation depends on liquid fuels refined from crude oil; however, we can foresee world supplies of crude oil becoming severely limiting within the next 20 to 30 years. The objective of this section therefore is to explain the inevitability of depletion and the problems it is creating for us.

To understand the limits of oil reserves, we must understand some basic concepts regarding their nature and exploitation.

FORMATION OF FOSSIL FUELS

The reason coal, crude oil, and natural gas are called *fossil fuels* is that all three once were living matter (Fig. 21–8). Early in the earth's biological history, photosynthesis outpaced the activity of consumers and decomposers. Consequently, large amounts of organic matter accumulated, especially on the bottoms of shallow seas and swamps. Gradually, this material was buried under sediments eroding from the land and, over millions of years, was converted to coal, crude oil, and natural gas. (Which of these three was formed depended on the specific conditions and time involved.)

Formation of additional fossil fuels by natural processes may be continuing to this day. However, one fact precludes any notion of supplies being replenished as we use them: We are using fossil fuels outrageously faster than they ever formed. About 1000 years' worth of natural accumulations was required to produce the amount of crude oil that the world now uses in one day. Because supplies are finite, there is no question that sooner or later we shall run out of these fuels. The pertinent question is: How long will reserves last?

EXPLORATION, RESERVES, AND PRODUCTION

The science of geology provides understanding regarding the location and extent of ancient shallow seas. On the basis of this understanding and experience, geologists make educated guesses as to where oil or natural gas may be located and the amounts that may be found. These *educated guesses* are the world's **estimated reserves**, which of course may be far off the mark. There is no way to determine whether estimated reserves actually exist or not except by the next step: exploratory drilling.

If exploratory drilling strikes oil, further drilling is conducted to determine the extent and depth of the **oil field**. From this further drilling, a fairly accurate estimate can be made regarding how much oil can be cost-effectively recovered from the field at current prices. This amount now becomes **proven reserves**.

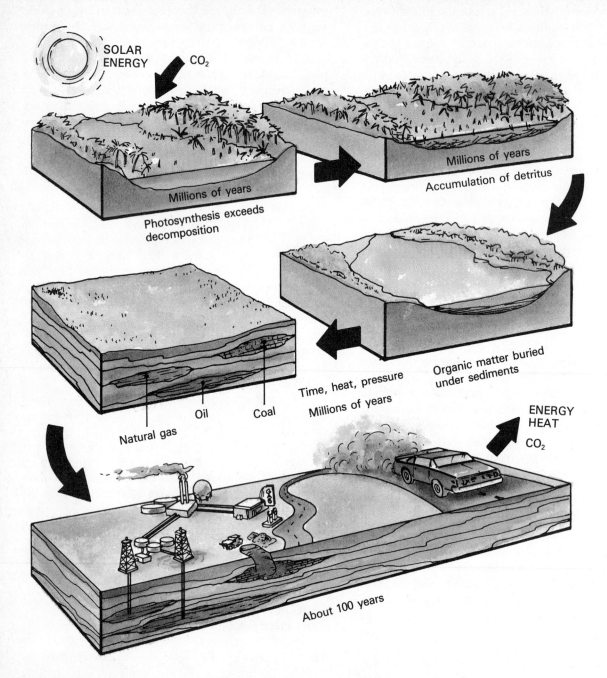

FIGURE 21–8
Energy flow through fossil fuels. Coal, oil, and natural gas are derived from
photosynthesis that occurred in early geological times. Deposits are limited, and as
they are used up the energy they contained will never be available to us again.

The final step, **production**, is the withdrawal of oil or gas from the field. Of course, *production* as used here is a euphemism. "It is production," in the words of ecologist Barry Commoner, "only in the sense that a boy robbing his mother's pantry can be said to be producing the jam supply."

    Production from a given field cannot proceed at any rate desired because crude oil is a viscous fluid

contained in sedimentary rock like water in a sponge. Initially the field may be under so much pressure that the first penetration produces a gusher. However, gushers are generally short-lived, and the oil then seeps slowly from the sedimentary rock into a well from where it is withdrawn. In some situations—as we have seen in Kuwait, for example—the natural entry of sea water, which is heavier than oil, into the oil field effectively floats the oil to the surface with little or no pumping. In either case, a general rule is that *maximum annual production is limited to about 10 percent of the remaining reserves.* Consider this rule in terms of maximum production from a field having 100 million barrels (one barrel equals 42 gallons) of recoverable reserves. The first year's maximum production will be 10 million barrels (10 percent of 100) leaving 90 million barrels. The second year's maximum production will be 9 million barrels (10 percent of 90), leaving 81. The third year's production will be 8.1 million barrels, and so on (Fig. 21–9).

Furthermore, only about 30 percent of the oil present in any sedimentary rock formation seeps form the rock into a well. Additional oil may be recovered by such secondary techniques as injecting steam into adjacent wells to force the oil into the producing well. However, these techniques obviously involve more cost, and the 10 percent rule still applies.

The size of reserves is often stated as: They will last X years at current rates of production. The implication is that the current production rates can be maintained for X years and then we shall suddenly run out. *This is absolutely false.* Barring discoveries of new reserves, maximum production will inevitably decline at the rate of about 10 percent per year. There will always be reserves remaining, however, and we can always, in theory, get 10 percent of what remains. Therefore, there is no running out as such; there is only a continuously diminishing production.

There is, however, an economic cutoff. When the value of the oil produced no longer justifies the costs of production, the field is shut down as unec-

onomical. Do you see how rising and falling oil prices can influence production on this basis? There was an economic boom in Texas from the mid-1970s to the mid-1980s as high oil prices justified the reopening of old oil fields. Then an economic bust occurred in the latter 1980s as the drop in oil prices caused the fields to be shut down again.

Finally, there is a *net energy yield* cutoff that even economics will not alter. Extracting oil from the ground requires the expenditure of a certain amount of energy to drill wells, run pumps, create steam, transport and refine the oil, and so on. The amount of energy obtained from the oil minus the amount spent in getting it is the **net energy yield**. There is obviously no point in continuing production when the net energy yield approaches zero.

Because of the 10 percent rule, you can see that constant production rates can be maintained only insofar as new discoveries (reserve additions) maintain proven reserves at a level of at least 10 times the annual production levels (reserve subtractions). If at any time new discoveries fail to keep pace with production and proven reserves fall below the 10-times factor, production will decrease accordingly.

DECLINING U.S. RESERVES AND INCREASING IMPORTATION OF CRUDE OIL

The failure to discover new oil, and hence declining reserves and production levels, are the crux of the energy problem in the United States. New oil discoveries began to fall short of production in the 1960s.

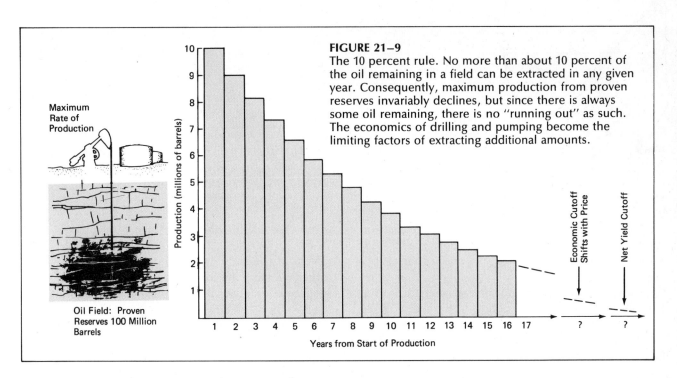

FIGURE 21–9
The 10 percent rule. No more than about 10 percent of the oil remaining in a field can be extracted in any given year. Consequently, maximum production from proven reserves invariably declines, but since there is always some oil remaining, there is no "running out" as such. The economics of drilling and pumping become the limiting factors of extracting additional amounts.

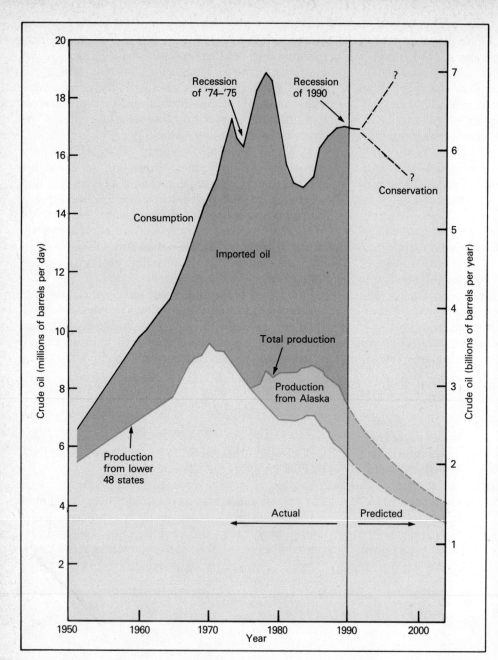

FIGURE 21–10
Oil production and consumption in the United States. Four stages can be seen: (1) Up to 1970, discovery of new reserves allowed production to parallel increasing consumption. (2) Early in the 1970s, lack of new discoveries caused production to turn down while consumption continued to climb, causing a vast increase in oil imports and bringing on the oil crisis of the 1970s. (3) In the late 1970s–early 1980s conservation decreased consumption, and higher prices stimulated production, which included bringing the Alaskan oil fields into production. Thus, dependence on foreign oil decreased considerably during this period. (4) In the late 1980s and on into the 1990s, consumption resumed a rapid increase while production resumed its decline because of diminishing reserves. (Data from *Statistical Abstracts of the United States*, U.S. Department of Commerce, 1988.)

The last major oil discovery for the United States was made in 1968 in Alaska. Thus, U.S. oil production peaked in 1970 and has been on a generally downward trend ever since. (There was a temporary reversal in the mid-1970s as the Alaskan oil field was brought into production.) Yet, in the early 1970s, the pace of oil consumption continued to increase, and the amounts involved were enormous—in 1977 U.S. oil consumption reached nearly 19 million barrels per day (Fig. 21–10).

To fill the "energy gap" between increasing consumption and falling production in the early 1970s, the United States became increasingly dependent on imported oil, primarily from the Arab countries of the Middle East. European countries and Japan did like-wise. Since imported oil cost only about $2.30 per barrel in the early 1970s and Middle Eastern reserves were more than adequate to meet the demand, this course seemed to present few problems.

THE OIL CRISIS OF THE 1970s

Thus, in the early 1970s the United States and most other Western nations were becoming increasingly dependent on oil imports from a small group of countries known as **OPEC**, the Organization of Petroleum Exporting Countries (Fig. 21–11). In 1973, OPEC, recognizing the dependence of industrialized nations on OPEC oil, decided to take advantage of the situation.

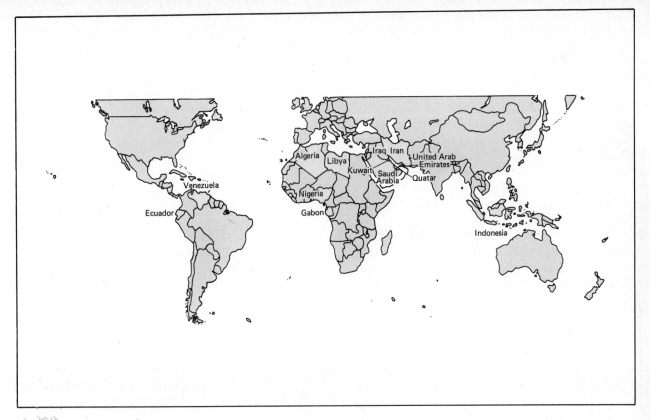

FIGURE 21-11
The nations in OPEC, the Organization of Petroleum Exporting Countries.

FIGURE 21-12
Gas "crisis" of 1978. Because of curtailment in imports of foreign oil, supplies fell below demand. The shortfall was only a few percent but the public perceived it as a crisis, panicked, and lined up, in some cases for miles, to get gas. What will happen when even more severe shortages occur in the future? (Robert McElroy/Woodfin Camp & Associates.)

They temporarily suspended oil exports, and this action threw the industrialized world into a crisis. People lined up for blocks and waited for hours to get gasoline at the few stations that had it to sell (Fig. 21-12). People were afraid to go anywhere, fearing that they would not have gas for emergencies. Motels, restaurants, and shopping malls became deserted.

The panic was much greater than what the shortage merited. To understand why this overreaction occurred, we must recognize that, compared with the amount of oil used (close to 20 billion barrels per year worldwide), relatively little is maintained in storage. We depend on a fairly continuous flow from wells to points of consumption. Consequently, if production is cut by just a few percent, supplies in storage are rapidly depleted, spot shortages occur, and, because our lifestyle is so dependent on vehicles for transportation, the situation is perceived as a crisis and panic occurs.

The reverse situation also occurs. If production moves ahead of consumption by just a few percent, available storage tanks fill to capacity, there is literally no place to put the excess oil, and the situation is referred to as an oil glut.

By curtailing supplies and causing the "oil crisis" of 1973, OPEC was able to get the rest of the world to pay $10.50 a barrel as opposed to the previous $2.30 a barrel. By continuing to limit oil production all through the 1970s, OPEC was able to keep supplies tight enough to force prices higher and higher. In the early 1980s, a barrel of oil cost $30 to $35, and the United States was paying about $60 billion per year for imported oil (Fig. 21–13). (All prices are costs at the time without adjustment for inflation.)

From a world perspective, then, running out of oil was not a real threat in the 1970s. However, rising oil prices demonstrated that, as long as we are dependent on OPEC oil, we remain economically and politically hostage to OPEC. Short of going to war, our options are limited. In 1991 that case was proved by the Persian Gulf War. Before studying the immediate causes of that war, let's fill in some more information leading up to it.

VICTIMS OF SUCCESS: SETTING THE STAGE FOR THE PERSIAN GULF WAR

Spurred by the desire to become energy-independent and even more by escalating fuel prices, the United States and other oil-dependent nations took significant steps toward independence in the 1970s and early 1980s. Major steps were:

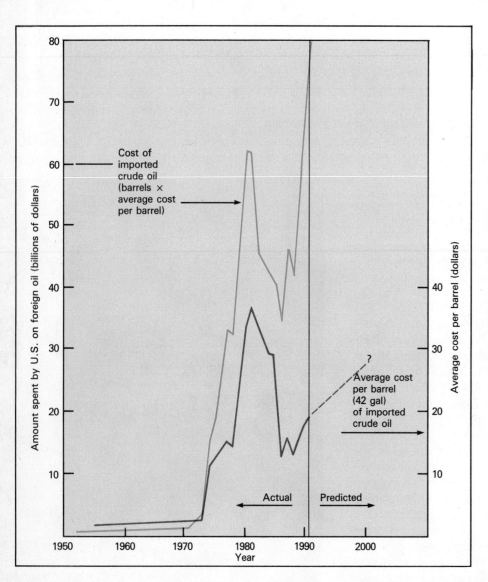

FIGURE 21–13
The cost of oil dependency. The red line shows the cost per barrel of imported oil (average for each year). The cost per barrel times the number of barrels imported gives the blue line, the amount the United States paid to foreign countries each year for oil. Since the mid-1970s imported foreign oil has been a major portion of the United States' balance-of-trade deficit, which contributes to inflation and economic weakness. With dependence on foreign oil again increasing, where are these figures likely to head in the future (dotted projections)? (Data from *Statistical Abstracts of the United States*, U.S. Department of Commerce, 1988.)

- Exploratory drilling was stepped up.
- Fields that had been closed down as uneconomical when oil was only $2.30 a barrel were reopened.
- Standards were set for automobile fuel efficiency. In 1973 cars averaged 13 miles per gallon (mpg). The government mandated stepped increases so that new cars would average 27.5 mpg by 1985. Also, mpg information was required in all car advertisements.
- Development of alternative energy sources was begun. The government supported research and development efforts and gave tax breaks to building owners who installed alternative energy systems.
- A strategic oil reserve was created. Thus far, the United States has "stockpiled" about 515 million barrels of oil (equivalent to 60 days of imports) in underground caverns in Louisiana.

It is noteworthy that, with the exception of the stockpile, all these steps have appeal only when the price of oil is high. If the price is low, the economic returns or savings do not justify the necessary investments.

Results of these major steps taken were not immediate, nor can quick returns be expected. For example, it takes at least 3 to 5 years to design and start production on a new car model. Then it is another 6 to 8 years before enough new models are sold and old ones retired to affect the overall "fleet." Through the 1970s, therefore, OPEC was able to manipulate world oil prices through the economic law of supply and demand. By restricting their own production and keeping world supplies tight, they were able to force prices higher and higher.

As we came into the 1980s, however, the efforts by oil-importing countries to become independent of OPEC began to pay off. The United States remained substantially dependent on foreign oil, but oil consumption was headed down and production was up because old fields had been reopened and the Alaskan field had been brought into production by the construction of the Alaskan pipeline (Fig. 21–10). Discovery and development of significant new fields in Mexico and the North Sea (between England and Scandinavia) made the world less dependent on OPEC oil. As a result of all these factors plus OPEC's inability to restrain its own production sufficiently to keep supplies tight, oil production exceeded consumption in the mid-1980s, and there was an oil glut.

What happens when supply exceeds demand? In 1986, world oil prices crashed from about $30 per barrel to close to $10. (From that time until the Persian Gulf War prices fluctuated in the range of $14 to $18 per barrel. Since the war they have fluctuated around

$20 per barrel.) This crash in oil prices was to the apparent benefit of consumers, industry, and oil-dependent Third World nations, but it hardly solved the underlying problem. Indeed, it set the stage for another crisis.

We noted above that high oil prices stimulated constructive responses. Conversely, the collapse in oil prices undercut those responses:

- Exploration, which became more costly as wells were drilled deeper and in more remote locations, was sharply curtailed. In 1982, there were 3105 drilling rigs operating in the United States; in 1992 there were 660.
- Production from old fields, which was costing around $10–$15 per barrel to pump, was terminated, causing economic devastation in Texas and Louisiana.
- Conservation efforts languished. Government standards for automobile fuel efficiency were decreased to 26 mpg with no intention to increase them further. The requirement to include mpg information in car advertisements was dropped.
- Tax incentives and other subsidies for development or installation of alternative energy sources were terminated, destroying many new businesses engaged in solar and wind energy. Grants for research and development of alternative energy sources (except for nuclear power) were sharply cut.

Thus, as a result of decreased oil prices in the late 1980s, U.S. oil production began to decline even more steeply than had been the case before, and consumption started to climb again as people bought less-fuel-efficient cars (Fig. 21–10). In particular, consumers started buying more "muscle cars,"—four-wheel-drive vehicles, light trucks, and vans, all of which get significantly lower mileage than fuel-efficient cars—and using them for everyday commuting (Fig. 21–14).

As we came into the 1990s, the U.S. dependence on foreign oil was at 45 percent—higher than in 1973 when the first crisis erupted—and growing. The situation for the rest of the non-OPEC world was not much better. Production from the major fields in Mexico, the North Sea, and the (then) USSR was declining, and consumption in both less-developed and developed countries was growing. Thus, OPEC was again in a position to manipulate world oil prices—if only it could get its own members to cooperate. The problem is that each OPEC member is caught in a dilemma similar to the tragedy of the commons. If one country cuts its production and others don't, its loss simply becomes their gain. On the other hand,

FIGURE 21-14
Powerful vehicles that have very low mpg ratings are becoming popular again with U.S. drivers. The wastefulness of driving such vehicles in city traffic is obvious. (Photograph by BJN.)

any country that increases production benefits, although the increase may result in lower oil prices. With all the OPEC nations squabbling among themselves, the stage was set for a confrontation.

Enter Saddam Hussein. If this head of Iraq could take over enough production capacity, which was close at hand in Kuwait and Saudi Arabia, he could manipulate production and thus world oil prices all by himself and rapidly become the richest, most powerful person on earth. In August 1990 he invaded Kuwait, which was producing about 6 million barrels per day, a level nearly equal to the amount of U.S. imports. Of course, the United States and the rest of the world were not about to let the rest of the scenario unfold. Thus, a huge military buildup and the Persian Gulf War of early 1991 followed. The war achieved the military goal of getting Saddam Hussein out of Kuwait, but he did succeed in knocking out Kuwait's production for better than a year by dynamiting and setting fire to nearly all of Kuwait's almost 700 wells.

WE WON THE WAR BUT THE PROBLEMS REMAIN

The Persian Gulf War did nothing to solve the basic problem confronting the United States and the rest of the non-OPEC world. U.S. dependence on foreign oil has continued to grow; it recently passed the 50 percent mark and, if this trend continues, will be 60–70 percent by the end of the decade. Many other countries, both developed and less-developed, are following a similar unenlightened path.

Through 1992 Saudi Arabia was pumping sufficient oil to maintain prices at about $20 per barrel. (This is hardly a sacrifice on the part of the Saudis. Middle Eastern reserves are such that they still cost

under a dollar per barrel to pump.) However, OPEC, or even Saudi Arabia by itself, is clearly in a position to restrain production and send oil prices soaring as happened in the early 1970s. Another tyrant like Saddam Hussein might appear. Of course either event could be solved by another military confrontation, but at what cost? When the cost of maintaining our military presence in the Middle East is added to the cost of oil, we are paying well over $100 per barrel, making oil the most expensive of all energy sources.

A most significant fact to note is that OPEC production cannot be maintained indefinitely in any case. As reserves are drawn down, production rates will meet a maximum and decline thereafter, just as they did in the United States. When will this occur? That depends on withdrawal rates and on the uncertain extent of OPEC reserves. However, most geologists estimate that it will occur within the next 15–20 years. When OPEC production does begin to decline, tight and decreasing oil supplies will be permanent, and we shall have to deal with the consequences.

Some people hope and even assume that vast additional reserves of crude oil will be discovered, but the likelihood of that happening is becoming increasingly remote. First, discoveries would have to be vast and frequent indeed. To support the world's current consumption level (about 20 billion barrels per year), we would have to discover another "Saudi Arabia" every few years. This is simply not occurring, nor is it likely to occur because of what geologists call the "Easter egg" argument.

When the people on an Easter egg hunt first begin searching for hidden eggs, the discovery rate is proportional to the searching effort. However, the searchers reach a point where, in spite of their continued efforts, fewer and fewer eggs are found. This situation leads to the logical conclusion that most of the eggs have already been found. The same argument pertains to the discovery of oil fields. The fact is that exploratory efforts since the mid-1970s have turned up remarkably little new oil. Indeed, multibillion-dollar investments in exploratory drilling in what were considered "likely locations" have continued to turn up nothing but dry holes. On the basis of this decline in returns from exploratory drilling, the U.S. Geological Survey has periodically lowered its forecasts of estimated reserves.

Finally, going for maximum discovery and production in the short term can have only one result: Reserves will be depleted more quickly, and production will decline more rapidly in the long term (Fig. 21–15). In the words of Earl Hayes, former chief scientist at the U.S. Bureau of Mines, "Talk of rising petroleum (and gas) production for long periods is both immoral and nonsensical. Whatever slight gain

FIGURE 21–15

Exploitation of a resource follows a bell-shaped curve. Production increases with increasing exploration and exploitation up to a point. Then, production invariably falls off as the resource is depleted. (a) Maximum effort put into exploration and exploitation leads to a higher peak production but then a more rapid falloff in production as the resource is exhausted more quickly. (b) Medium and (c) restrained levels of exploitation achieve lower levels of peak production, but then the resource remains available for a longer period of time. Which of these curves are we intent on following with respect to exploitation of the world's oil resources?

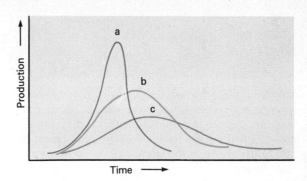

ETHICS

Trading Wilderness for Energy in the Far North

The search for new energy sources to satisfy the enormous energy appetite of the United States economy has turned northward in recent years. Discovery and development of the huge Prudhoe Bay oil field in Alaska and the building of the 800-mile Alaska pipeline have dramatized the stark contrasts between wilderness and modern technology. Scenes of caribou and grizzly bears are superimposed on oil rigs and the pipeline carrying oil south to Valdez. East of Prudhoe Bay is the Arctic National Wildlife Refuge (ANWR), the largest single wilderness area in the United States. ANWR is home to 180 000 caribou and several hundred native people whose subsistence way of life is tied to the caribou. ANWR is also the summer breeding ground for innumerable birds and other wildlife, but it is also believed to be on top of another huge oil field that industrial interests are anxious to exploit.

Twenty five hundred miles to the south and east of ANWR, three major river systems flow through a vast wilderness of lakes, bogs, and forests before emptying into the James and Hudson Bays. In addition to wildlife, this area of northern Canada is home to 10 000 Cree and Inuit people, 40 percent of whom still lead the way of life of their

ancestors living off the land. But, Hydro-Quebec sees vast potential in this wilderness for power generation. In what is called the James Bay project, they have already converted 4500 square miles of the wilderness into a network of diversions, dams, and reservoirs, and this is just one third the total project. Now the Crees are demanding in court a full review of environmental and human costs before Hydro-Quebec proceeds with the next two phases of development.

ANWR and James Bay promise to be major environmental battlegrounds of the 1990s. On one side are the energy and political interests who see values in terms of energy needs for a thriving economy and national security. On the other side are native people and environmentalists who see values in terms of the beauty and biodiversity of the northern wilderness, as well as the human rights of native peoples. Those favoring development claim that it will be done in an environmentally sound way and the native people will be compensated. Those favoring preservation are not satisfied with good intentions. They point to the Exxon Valdez disaster in which 11 million gallons of crude oil were spilled in the pristine waters of Prince Edward Sound, despite all

safeguards. They point out that present oil development in Alaska has introduced much of the worst of American culture to native peoples; a sense of lost identity is apparent as their traditional culture is swamped by consumer goods and pop culture. Even more environmentalists question development on the basis of its sustainability. The estimated oil reserve underlying ANWR is 3.5 billion barrels; this is about six months supply at the current rate of use in the United States. The power produced by the James Bay project will be mostly exported from Canada to the United States. Are there ways to have adequate power without purchasing it at the expense of wilderness?

How do we balance our need for energy, jobs, and economic growth, against caribou, wolves, and grizzly bears? How do we weigh the traditions and cultures of native peoples against the impacts of our own? What do you think?

(Sources include "Power in a Land of Remembrance," by Harry Thurston, Audubon, Nov-Dec 1991; and "Arctic Wars: The Fight for the Arctic National Wildlife Refuge, National Audubon Society film, 1990)

Economic and environmental constraints make it clear that the world economy cannot remain dependent on fossil fuel energy for the indefinite future. In time, the world will have to adjust to an economy based on sustainable energy sources. The sooner we begin to make this transition, the less disruptive the transition will be to economic and social institutions. The challenge confronting this generation is to shape energy policies that will both develop the renewable energy sources and at the same time make environmentally and economically sensible use of current fossil fuels. There is good evidence that natural gas may be the ideal transitional energy source to help us to get to a future sustainable energy economy.

The most abundant reserve fossil fuel, coal, is also the most highly polluting, and it fails to meet the need for transportation energy. Oil is currently the most versatile fossil fuel, but its supply is clearly limited and is subject to politically motivated manipulations. Natural gas is more abundant than oil, is more widely dispersed around the world, and is much less polluting than oil or coal. It is readily useful for electrical power generation and heating, but its use as a transportation fuel is still limited. Although it burns

cleanly and well in vehicle engines, natural gas must be stored in bulky pressurized tanks on board vehicles, and must be widely distributed and stored for routine consumer use. Nevertheless, the advantages of lower price (one-third that of gasoline) and much lower pollution are a strong incentive to increase the fleet of natural gas vehicles. They number in the hundreds of thousands now, but may well be in the millions within 10 years as a result of clean air laws.

Of the three fossil fuels, natural gas has shown the most rapid growth in use—up 87 percent in the last 20 years, and still increasing. It now constitutes one-fourth of all world fossil fuel use. This development is the result of recent widespread discoveries of natural gas fields and new technologies for generating power and heat from natural gas. In the United States, domestically produced gas is priced much lower and produces more energy than domestically produced oil. Current estimates of natural gas reserves in the United States indicate that there is enough to last 60 years at the current rate of use—and the U.S. has only 5 percent of the world's proven gas reserves. Worldwide, the data indicate that there is enough natural gas to double the

current world use in the next 20 to 30 years, and then to sustain that level for an additional 20 years.

Many scientists believe that the most likely sustainable energy system of the future is one based on hydrogen gas produced by solar energy (see Chapter 23). Hydrogen can run automobiles, generate electricity in limited areas, and provide the energy for heating systems. The beauty of natural gas is that it facilitates the transition to hydrogen far better than the other fossil fuels and is much less polluting. Hydrogen can be mixed with natural gas to stretch its supply, and systems based on natural gas use can be converted to hydrogen gas with minimal disruption and cost. Although many technological changes will be required, a solar/hydrogen economy will not require any major scientific breakthroughs. However, getting there from here will mean that our policymakers and engineers have to make a gradual break with the fossil fuel energy system of the present, and the evidence indicates that natural gas will make that break possible.

(*Source:* "Building a Bridge to Sustainable Energy," by Christopher Flavin, in *State of the World 1992*, Worldwatch Institute)

might be achieved for a very few years will be at the expense of the youth of today." And what are the tradeoffs for the environment? See Ethics Box page 473.

Options for the Future

Instead of pretending that oil shortages are history and continuing to increase consumption, we should be making every effort to accommodate ourselves to the decreasing supplies that will inevitably mark the future. Waiting until the crisis is upon us may well

be too late because, as experience from the 1970s shows, it takes many years for measures to have any significant effect.

There are three options: (1) getting "oil" from alternative fossil fuels, what we call juggling fossil fuels; (2) conservation; and (3) developing nonfossil fuel energy alternatives. It is not a matter of selecting one of these options and excluding the others. They can all make important contributions. The key thing will be to understand the pros and cons of each and put together a suitable mix. In the remainder of this chapter, we focus on alternative fossil fuels and to some extent on conservation and alternative energy sources.

JUGGLING FOSSIL FUELS

Natural Gas

While exploratory drilling in recent years has not yielded much additional oil, it has turned up considerable reserves of natural gas. With the installation of a tank for compressed gas in the trunk and some modifications of the engine fuel-intake system—costing about a thousand dollars—cars will run perfectly well on natural gas. Such cars are in widespread use in Buenos Aires, where service stations are equipped with compressed gas to refill the tank. Natural gas is a clean-burning fuel; hydrocarbon emissions are nil. Not to be using our natural gas surpluses for vehicles to supplement oil supplies seems shortsighted. To be sure, natural gas reserves are not sufficient to substitute for the total oil demand, and they could never be a permanent solution. However, they could provide an interim measure of considerable merit (see In Perspective Box, p. 474).

Coal

In contrast to its limited reserves of crude oil, the United States is exceptionally well endowed with coal (Fig. 21–16). Even tripling current rate of use, these reserves could supply this country's energy needs for over 150 years. As stated earlier in this chapter, however, the problem is the mismatch between coal and a fuel we can burn in vehicles.

There are chemical processes that will convert coal to a liquid or gas fuel. Such coal-derived fuels are referred to as **synthetic fuels** or **synfuels**. With both government and corporate support, a great deal of research went into upscaling these chemical processes to commercial production in the late 1970s and early 1980s. The projects were all abandoned, however, because they proved too expensive, at least for the present. Profitable production of synthetic fuels would involve prices of $60 to $70 per barrel.

Even more troublesome, however, are the environmental impacts of using coal. Most coal is in deposits that can be exploited practically only by strip-mining. In underground mining, at least 50 percent of the coal must stay in place to support the mine roof. In strip-mining, gigantic power shovels turn aside the rock and soil above the coal seam and then remove the coal (Fig. 21–17). It is evident that this procedure results in total destruction of the ecosystem. Although such areas may be reclaimed—that is, regraded and replanted—it takes many decades before an ecosystem anything like the original is reestablished. In arid areas of the West, water limits make

FIGURE 21–16
Major coal deposits in the United States: a solution or a potential ecological disaster? (U.S. Geological Survey.)

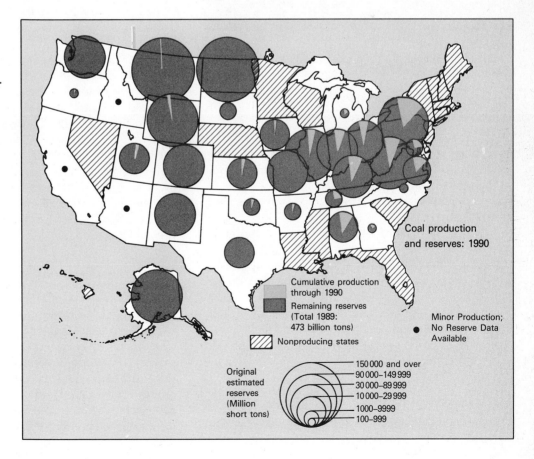

FIGURE 21–17
Strip-mining of coal. To keep electric power plants running, the United States is strip mining about a billion tons of coal per year. It is hard to conceive of a more ecologically destructive activity. (Patrick Phelan/The Stock Market.)

FIGURE 21–18
Oil shale. Self-sustaining combustion in high-grade "paper shale." Most shale will not burn but must be heated to remove oil-like material. (U.S. Department of Energy.)

it questionable whether the ecosystem could ever be reestablished. Consequently, strip-mined areas may be turned into permanent deserts. Furthermore, erosion and acid leaching from the disturbed earth may have numerous adverse effects on waterways and groundwater of the area.

The environmental hazards of coal are already well known from the nearly one billion tons mined annually in the United States for coal-burning power plants. The thought of doubling or tripling this impact to run vehicles on coal-derived fuel makes a mockery of all we have studied about environmental protection. In addition, converting coal to liquid fuel creates a number of polluting byproducts that would be difficult and expensive to control.

Oil Shales and Tar Sands

The United States has extensive deposits of **oil shales** in Colorado, Utah, and Wyoming. Oil shale is a fine sedimentary rock containing a mixture of solid, wax-like hydrocarbons called kerogen (Fig. 21–18). When shale is heated to about 600°C, the kerogen releases hydrocarbon vapors that can be recondensed to form a black, viscous, crude oil-like substance that can be refined into gasoline and other petroleum products. However, it requires about a ton of oil shale to yield little more than half a barrel of oil. The mining, transportation, and disposal of wastes necessitated by an operation producing, for instance, a million barrels a day (5 percent of U.S. demand) would be a Herculean task, to say nothing of its environmental impact. Perhaps to our environmental good fortune, oil shale, like oil from coal, has proved economically impractical for the present.

Tar sands are a sedimentary material containing bitumine, an extremely viscous, tarlike hydrocarbon. When tar sands are heated, the bitumine can be "melted out" and refined in the same way as crude oil. Northern Alberta, Canada, has the world's largest tar-sand deposits, and Canada is developing commercial production from them, the cost being competitive with current oil prices. The United States has smaller, less-rich tar-sand deposits, mostly in Utah, that might produce oil for $50–$60 per barrel. Again pollution and other forms of environmental impact are problems.

The Greenhouse Effect

A major drawback of all fossil fuels is their potential to cause global warming through the greenhouse effect described in Chapter 16. Because they are carbon-based fuels, they cannot be burned without the production of carbon dioxide. Different fossil fuels produce different amounts of carbon dioxide per unit of energy delivered. Of the top three—coal, oil, and gas—coal produces the most and natural gas produces the least, another argument for the use of natural gas. Synfuels would produce even more carbon dioxide than coal because there is a considerable energy loss entailed in converting coal to liquid fuels. Likewise, fuel derived from oil shales or tar sands would produce more carbon dioxide than oil because extracting oil from these sources involves a much lower net energy yield. An international treaty to cap carbon dioxide emissions, which the United Nations is striving to implement, might entirely preclude development of synfuels, oil shale, and tar sands as well as reduce oil consumption.

In the final analysis, if we are willing to go all out in juggling fossil fuels, we might keep running on them for some time, on the order of 100 years or more. However, even 100 years is a relatively short time in terms of human history. The choice to continue with fossil fuels certainly does not meet the criterion of being sustainable. Even for the short term, there are tremendous environmental costs associated with obtaining and burning fossil fuels.

CONSERVATION

Consider the discovery of a new oil field with a production potential of at least 6 million barrels a day. This is three times the size of the Alaskan field. Furthermore, assume that this new-found field is inexhaustible and that its exploitation will not adversely affect the environment. Of course, such an oil field sounds like a dream, but this is effectively what can be achieved through *conservation*.

As noted earlier in this chapter, this "conservation reserve" has already been tapped to some extend, especially as the average mpg of cars was doubled from 13 to 26 mpg. This by itself is already saving about 2 million barrels of crude oil per day. As a result, during the late 1970s and early 1980s the economy expanded without a corresponding rise in energy consumption for the first time in history. If the same relationship between energy and gross national product that existed in the early 1970s had persisted, we would now be using about 25 percent more energy than we are. Said another way, we are already obtaining about 25 percent of our energy from conservation, with an annual savings approaching $100 billion. In short, advances in conservation have been extremely cost-effective as well as environmentally benign.

A number of energy experts estimate that at least 80 percent of the "conservation reserve" remains to be tapped. Cars are currently available that get at least double the current average of 26 mpg, and Peugeot, Toyota, Volkswagen, and Volvo all have prototypes that get from 70 to 100 mpg. Renault has a prototype that attained a remarkable 124 mpg in test runs. All the car companies are waiting for government mandates, consumer demand, or fuel shortages before putting these models into production so that high sales will be guaranteed.

A concept known as *cogeneration*, described in the In Perspective Box, page 478, is another way of gaining large amounts of energy from the conservation reserve.

Other conservation measures are readily available but underutilized. Improved insulation and double-pane windows for homes and buildings would save on heating fuel. Substitution of fluorescent lights for conventional incandescent light bulbs would cut back on our electricity demands. Incandescent light bulbs are only 5 percent efficient, whereas fluorescent bulbs are between 20 and 25 percent efficient. By this switch, power required for lighting can be reduced by 75–80 percent. Similarly, more efficient refrigerators and other appliances could reduce energy demands by 50–80 percent. While these energy-conserving light bulbs and appliances are more expensive than conventional models, calculations show that they will more than pay for themselves in energy savings. Also, some utilities are subsidizing the adoption of fluorescent light bulbs, for example, as a cost-effective alternative to building additional power plants.

Finally there is the whole area of adjusting our personal habits and lifestyles to be more conserving, a topic addressed further in Chapter 24.

Many observers have pointed out that demanding more energy while failing to conserve is like demanding more water to fill a bathtub while leaving the drain open. To be sure, conservation by itself will not eliminate demands for energy, but it can make the demands much easier to meet regardless of what options are chosen to provide the primary energy.

DEVELOPMENT OF ALTERNATIVE ENERGY SOURCES

Two major pathways for developing energy sources as alternatives to fossil fuels lie before us. One is to pursue nuclear power; the other is to pursue solar energy. Of course nuclear power has been in use to some extent over the past 30 years, and various forms of solar energy have also been developed. In other words, both nuclear and solar alternatives are at a stage where they could be rapidly expanded to provide further energy needs if suitable public acceptance (in the case of nuclear power) or pressure and government support (in the case of solar energy) were provided. The following chapter focuses on the nuclear option, and Chapter 23 focuses on solar options so that we can make a more informed decision as to which of these alternatives to pursue.

There are many situations where a single-energy source can be put to use to produce both electrical and heat energy. This is called **cogeneration.** Cogeneration can be employed to conserve energy in two fundamental ways: 1) A power plant whose primary function is to generate electricity also can be used to heat surrounding homes and buildings. Consolidated Edison has been using steam from waste heat in New York City for years. 2) A facility whose primary function is to produce heat or mechanical energy can also generate electricity, which is used by the facility and can also be sold to the local utility. The system can work for all sorts of fuels and heat engines. This is by far the most common application of cogeneration.

The arrangement makes perfect sense. Every building and factory requires both electricity and heat. Traditionally, electricity and heat come from separate sources—the electricity from power lines brought in from some central power station and the heat from an on-site heating system, usually fueled by natural gas or oil. Ordinary power plants are at best 30 percent efficient in their use of energy. Combining the two in cogeneration can provide an overall effi-

ciency of fuel use as high as 80 percent. Besides providing savings on fuel, the cogenerating systems have at least 50 percent lower capital costs per kilowatt generated compared with conventional utilities. This combination of cost-saving makes cogeneration quite competitive with utility-produced power.

Let's consider an example. The Santa Barbara (California) County Hospital recently installed a cogeneration system based on natural gas as the fuel source. Natural gas drives a turbine with the capacity to generate 7 megawatts of electricity, much of which is used by the hospital. Boilers use the waste heat from the turbine to produce high-pressure steam, which provides all of the hospital's heating and cooling needs. Any steam not required by the hospital is directed back to the turbine to increase its power. Electrical power not used by the hospital is sold to the local utility. The system has the added advantage of protecting the hospital against wide-scale blackouts or brown-outs that may occur in the centralized utility system.

Enabling legislation known as PURPA (Public Utility Regulatory Policies Act of 1978) was essential for stimulating cogeneration. Under this

act, a facility with cogenerative capacity is entitled to use fuels not usually available to electricity generating plants (for example, natural gas), must produce above standard efficiency of fuel use and is entitled to sell cogenerated electricity to the utility at a fair price. After many challenges by the utilities (which considered the procedures an interference with their business), the Supreme Court decided in favor of PURPA, and the legislation is now firmly established.

Cogeneration is becoming increasingly popular in U.S. industries as a major strategy for cutting energy costs. Given the future predictions of a downturn in availability and major price increases for petroleum, pressures favoring energy conservation will increase and cogeneration will continue to grow. The Federal Energy Regulatory Commission believes that cogeneration may provide as much as 25 percent of our electricity by the year 2000.

(*Sources:* "Cogeneration—Efficient Energy Source," by E.L. Clark, in *Annual Review of Energy* 11:275, 1986; The Energy Sourcebook, edited by R. Howes and A. Fainberg, Am. Institute of Physica, 1991.)

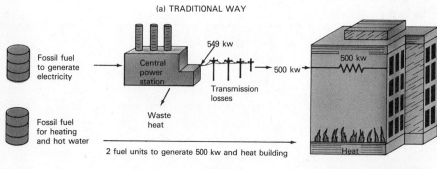

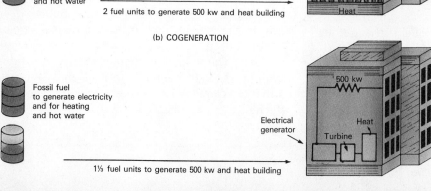

Cogeneration. (a) The traditional method of providing the energy needed for buildings. (b) In cogeneration, heating needs are provided by the heat lost from an on-site power-generating system.

Review Questions

1. How have the fuels to power homes, industry, and transportation changed from the beginning of the Industrial Revolution to the present?

2. Electricity is a nonpolluting form of energy. Why is it not substituted for polluting forms of energy?

3. What three primary sources of energy are used to generate most of the electrical power in the United States and how are they used?

4. What are the four major categories of energy use?

5. What is the most used primary source of energy for each of the four major categories of energy use?

6. What is the distinction between estimated reserves and proven reserves of crude oil, and what factors cause the amounts of each to change?

7. Draw a graph that will express the inevitable trend of maximum production per year in the absence of new discoveries (reserve additions). Explain why this trend will occur.

8. How does the price of oil influence the amount produced?

9. What are the trends in oil consumption, discovery of new reserves, and U.S. production leading up to 1970? Following 1970?

10. What was done by the United States in the early 1970s to resolve the disparity between oil production and consumption?

11. What events caused the sudden oil shortages of the mid-1970s and then the return to abundant but more expensive supplies?

12. What responses were made by the United States and other industrialized nations to the "oil crisis" of the 1970s, and what were the impacts of these responses on production and consumption?

13. Why did an oil glut occur in the mid-1980s, and what were its consequences in terms of oil prices and continuing pursuit of the responses listed in question 12?

14. How did the events listed in question 13 set the stage for the Persian Gulf War of 1991?

15. Does having won the Persian Gulf War of 1991 mean that energy problems are finally behind us? Explain what the war accomplished and what it failed to accomplish.

16. What three general avenues might be pursued toward resolving our long-term energy situation?

17. What are the three alternative fossil fuels that the United States might exploit to supplement declining reserves of crude oil, and what are the economic and environmental advantages and disadvantages of each?

18. Give examples of energy conservation.

19. What are some economic and environmental advantages of conservation regardless of the availability of alternative energy supplies?

Thinking Environmentally

1. Statistics show that less-developed countries use far less energy per capita than developed countries. Explain why this is so.

2. The United States and many other developed and less-developed countries are going all out to expand highway systems and produce more cars and trucks—in short, to expand a transportation system that depends on liquid fuels. Discuss the sustainability of this system. Suggest alternatives.

3. From 1980 to 1992 gasoline prices in the United States declined considerably relative to inflation. Predict what they will do in the next 12 years, and give a rationale for your prediction.

4. Suppose your region is facing power shortages— brown-outs. How would you propose solving the problem? Defend your proposed solution on both economic and environmental grounds.

5. List all the environmental impacts that occur during the "life span" of gasoline, that is, from exploratory drilling to consumption of gasoline in your car.

6. Develop what you feel should be the long-term policy for the United States to meet future transportation needs. Consider and balance the potentials and risks of continuing to increase imports, increasing exploration, developing synfuels or oil shale deposits, conservation, and alternative modes of transportation. Suggest laws, regulations, taxes, or subsidies that might be used to implement your proposal.

Nuclear Power: Promise and Problems

LEARNING OBJECTIVES

When you have finished studying this chapter, you should be able to:

1. Contrast the outlook for nuclear power in the 1960s with the present outlook.

2. Explain how nuclear power works.

3. Describe the general design of a nuclear reactor and a nuclear power plant.

4. List the environmental advantages and disadvantages of nuclear power in contrast to coal.

5. Describe the hazards of radiation from radioactive substances.

6. Discuss the problems associated with the disposal of radioactive wastes.

7. Explain what happened at Three Mile Island and Chernobyl and discuss the consequences of those accidents.

8. Discuss new efforts being made in nuclear power safety.

9. Give three economic reasons that have halted the construction and operation of many nuclear power plants.

10. Consider the advantages and drawbacks of breeder and fusion reactors.

11. Present four reasons for the current opposition to nuclear power.

As the industrialized nations continue their heavy use of fossil fuels, global temperatures are on the rise, giving increasingly strong support to the argument that our use of fossil fuels is bringing on global warming. The Gulf War of 1991 provided graphic evidence of how developed nations dependent on overseas sources of crude oil are willing to use military force to protect those sources. In light of these two serious energy-related problems, it seems prudent to do everything possible to develop alternative energy sources. Nuclear power is an alternative that does not contribute to global warming, and there is sufficient uranium to fuel nuclear reactors well into the twenty-first century, with the possibility of extending the nuclear fuel supply indefinitely through reprocessing technologies. Thirty-one nations now have nuclear power plants either in place or under construction, and in some of these countries, nuclear power generates more electricity than any other source. Is this the key to the global energy future, the best route to a sustainable relationship with our environment? Our objective in this chapter is to investigate the potential and the problems of nuclear power.

Nuclear Power: Dream or Delusion?

From the time fossil fuels were first used, geologists recognized that they would not last forever. Sooner or later other energy sources would be needed. The end of World War II, marked by the awesome power of the atom, was the time of decision. The U.S. government desperately wanted to show the world that the power of the atom could benefit humankind as well as destroy it and embarked on a course to lead the world into the "Nuclear Age."

It was anticipated that nuclear power could produce electricity in such large amounts and so cheaply that we would phase into an economy in which electricity would take over virtually all functions, includ-

FIGURE 22-1
(a) Changing fortunes of nuclear power in the United States. Since the early 1970s, when orders for plants reached a peak, few utilities have called for new plants and many have canceled earlier orders. (The last plant order not subsequently canceled was placed in 1974.) Nevertheless, the number of plants in service increased steadily as plants under construction were completed. Unless new orders are placed, 113 plants will be the peak.

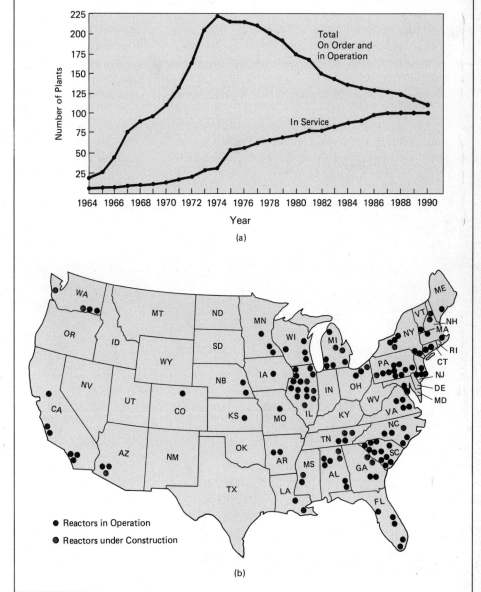

ing the generation of other fuels, at nominal costs. Thus, the government moved into research, development, and promotion of commercial nuclear power plants (along with continuing development of nuclear weaponry). Utilizing this research, companies such as General Electric and Westinghouse constructed nuclear power plants ordered and paid for by utility companies. The Nuclear Regulatory Commission (NRC), an agency in the Department of Energy, sets and enforces safety standards for the operation and maintenance of these plants.

In the 1960s and early 1970s, utility companies moved ahead with plans for numerous nuclear power plants (Fig. 22–1). By 1975, 53 plants were operating in the United States, producing about 9 percent of the nation's electricity, and another 170 plants were in various stages of planning or construction. Officials estimated that by 1990 several hundred plants would be on line, and by the turn of the century as many as 1000 would be operating. A number of other industrialized countries were in step with their own programs, and some less-developed nations were going nuclear by buying plants from industrialized nations.

Since 1975, however, the picture has changed dramatically. Utilities stopped ordering nuclear plants, and numerous existing orders were canceled. The last plant order in the United States not subsequently canceled was placed in 1974. In some cases construc-

tion was terminated even after billions of dollars had already been invested. Most striking, the Shoreham Nuclear Power plant on Long Island, New York, after being completed and licensed at a cost of $5.5 billion, was turned over to the state of New York in the summer of 1989 to be dismantled without ever having generated a single watt of power. The reason was that citizens and the state deemed that there was no possible way to evacuate people from the area should an accident occur. Similarly, just a few weeks earlier, California citizens voted to shut down the Rancho Seco Nuclear Power Plant located near Sacramento, which had a 15-year history of troubled operation.

At the end of 1991, there were only 110 functioning nuclear power plants in the United States plus another 3 under construction. It appears that nuclear power in the United States will peak at 113 operating plants in the mid-1990s, generating about 21 percent of our electricity, and then head downward as older plants are "decommissioned." Outside the United States, roughly 400 plants are either operating or under construction, mainly in France, the republics of the former USSR, Japan, Germany, Canada, and the United Kingdom. In general, the story abroad is similar to that in the United States—drastic downward revisions if not total moratoriums on construction of new plants. Only France and Japan remain fully committed to pushing forward with nuclear programs. Thus, France is now producing 70 percent of

FIGURE 22–2
Percentage of electricity generated by nuclear power for all countries that have nuclear power plants. (Reproduced with permission from the Annual Review of Energy, Vol. 15. © 1990 by Annual Reviews, Inc.)

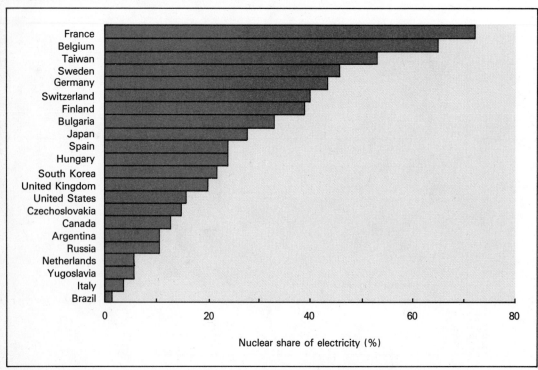

its electricity from nuclear power and has plans to go to more than 80 percent (Fig. 22–2).

After the catastrophic accident at Chernobyl in April 1986, it is not hard to see why nuclear power is being rethought. Yet, concerns over global warming and impending oil shortages are real. Although the United States has abundant coal reserves, coal-powered electricity generates more of the greenhouse gases than any other form. Since oil is so vital to transportation and home heating, it is unwise to use dwindling supplies for power generation. By turning our backs on nuclear power, are we throwing away our best opportunity for supplying future energy needs? Are the public's concerns over safety justified?

Very likely, public opinion and reaction will be the determining factors in whether we revive the nuclear dream or put it to rest. In order to react intelligently to this issue, we need a clear understanding of what nuclear power is and of its pros and cons.

HOW NUCLEAR POWER WORKS

The objective of nuclear power technology is to control nuclear reactions so that energy is released gradually as heat. As with fossil fuel–powered plants, the heat is used to boil water and produce steam, which then drives conventional turbogenerators.

The release of nuclear energy is a phenomenon completely different from the burning of fuels or any other chemical reactions we have discussed. To begin with, materials involved in chemical reactions remain unchanged at the atomic level, although the visible forms of these materials undergo great transformation as atoms rearrange to form different compounds. Nuclear energy, however, involves changes at the atomic level through one of two basic processes: *fission* or *fusion*. In the process known as **fission,** a large atom of one element is split to produce two smaller atoms of different elements (Fig. 22–3a). In the proc-

FIGURE 22–3
Nuclear energy is released from either (a) fission, the splitting of certain large atoms into smaller atoms, or (b) fusion, the "fusing" together of small atoms to form a larger atom. In both cases some of the mass of the starting atom(s) is converted to energy.

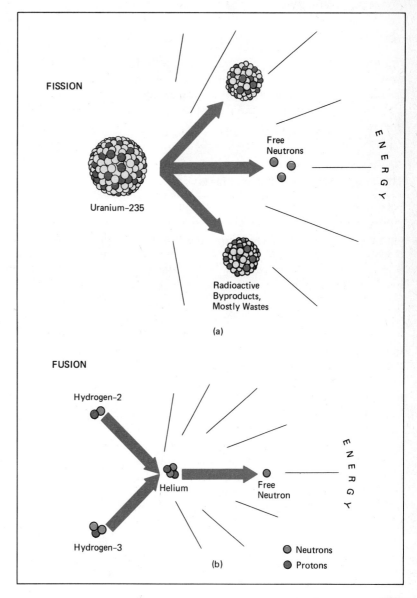

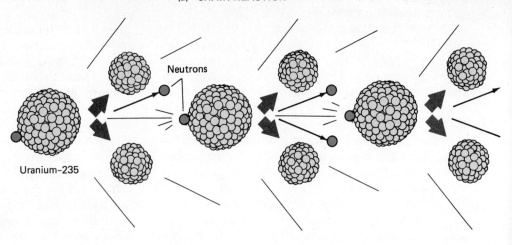

Neutrons

Uranium-235

ess known as **fusion,** two small atoms combine to form a larger atom of a different element (Fig. 22–3b). In both fission and fusion, the mass of the products is less than the mass of the starting material, and the lost mass is converted to energy in accordance with the law of mass-energy equivalence ($E = mc^2$) first described by Albert Einstein. The amount of energy released by this mass-to-energy conversion is tremendous. The sudden fission or fusion of a mere 1 kg of material releases the devastatingly explosive energy of a nuclear bomb.

The Fuel for Nuclear Power Plants

All current nuclear power plants utilize the fissioning (splitting) of uranium-235. The element uranium occurs naturally in various minerals in the earth's crust. It exists in two primary forms, or *isotopes*: uranium-238 (^{238}U) and uranium-235 (^{235}U). Like uranium, many other elements exist in more than one isotopic form. The **isotopes** of a given element contain different numbers of neutrons but the same number of protons and electrons. Some isotopes are unstable and release particles or rays, which are called **radioactive emissions,** to be discussed shortly.

The number that accompanies the chemical name or symbol of an element is called the **mass number** and represents the number of neutrons and protons in the nucleus of the atom. Since, by definition, all atoms of any given element must contain the same number of protons, variations in mass number represent variations in numbers of neutrons. Thus ^{238}U contains 92 protons, 92 electrons, and 146 neutrons, while ^{235}U contains 92 protons, 92 electrons, and 143 neutrons. While ^{238}U contains 3 more neutrons than ^{235}U, both isotopes contain the 92 protons that *define* the element uranium. Although all isotopes of a given element behave the same chemically, their other char-

acteristics may differ profoundly. In the case of uranium, ^{235}U atoms will fission but ^{238}U atoms will not.

It takes a neutron hitting the nucleus at just the right speed to cause ^{235}U to undergo fission. Since ^{235}U is an unstable isotope, a small but predictable number of the ^{235}U atoms in any sample undergo radioactive decay and release, among other things, neutrons. If one of these released neutrons moving at just the right speed hits another ^{235}U, the hit ^{235}U becomes ^{236}U, which is very unstable and fissions immediately into lighter atoms (**fission products**). The fission reaction gives off several more neutrons and releases a great deal of energy (Fig. 22–3a). Ordinarily these neutrons are traveling too fast to cause fission, but if they are slowed down and then strike another ^{235}U, they can cause fission to occur again. In this way, more neutrons and more energy are released, with the potential to repeat the process. A domino effect, known as a chain reaction, may thus occur (Fig. 22–4a).

A chain reaction does not occur in nature because ^{235}U atoms are too dispersed among other elements and more stable ^{238}U atoms, which absorb neutrons without fissioning. In fact, 99.3 percent of all uranium found in nature is ^{238}U; only 0.7 percent is ^{235}U. Hence when a ^{235}U atom spontaneously decays in nature, it seldom triggers fission in another atom, and the event goes unnoticed without the aid of radiation detectors such as Geiger counters.

To make nuclear "fuel," uranium ore is mined, purified, and *enriched*. **Enrichment** involves separating ^{235}U from ^{238}U to produce a material containing a higher concentration of ^{235}U. Since ^{238}U and ^{235}U are chemically identical, enrichment is based on their slight difference in mass. The technical difficulty of enrichment is the major hurdle that prevents less-developed nations from advancing their own nuclear capability.

When ^{235}U is highly enriched and placed in an appropriate mass, the spontaneous fissioning of an

FIGURE 22–4

(a) A simple chain reaction. When a uranium atom fissions, it releases two or three high-energy neutrons in addition to energy and split "halves." If another uranium-235 atom is struck by a high-energy neutron, it fissions and the process is repeated, causing a chain reaction. (b) A self-amplifying chain reaction leading to a nuclear explosion. Since two or three high-energy neutrons are produced by each fission, each fission may cause the fission of two or three additional atoms. Hence the entirety of a suitably concentrated mass of fissionable material may be caused to fission in a tiny fraction of a second, resulting in a nuclear explosion. (c) In a sustaining chain reaction the extra neutrons are absorbed in control rods so that amplification does not occur.

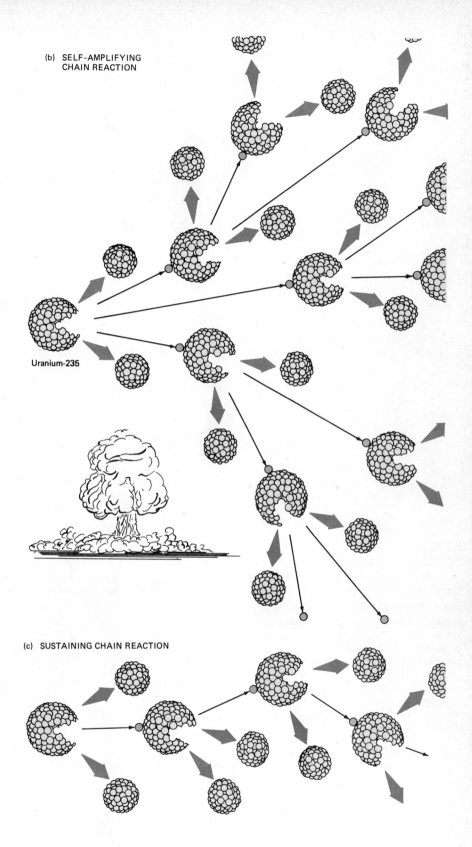

(b) SELF-AMPLIFYING CHAIN REACTION

Uranium-235

(c) SUSTAINING CHAIN REACTION

atom can trigger a chain reaction. In nuclear weapons, small masses of virtually pure uranium-235 or other fissionable material are forced together so that the two or three neutrons from a spontaneous fission cause two or three more atoms to fission; each of these triggers two or three more fissions, and so on. The whole mass fissions in a fraction of a second, releasing all the energy in one huge explosion (Fig. 22–4b).

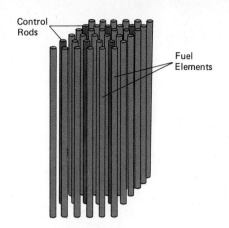

Control
Rods

Fuel
Elements

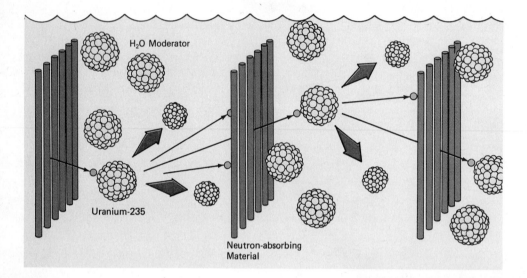

H₂O Moderator

Uranium-235

Neutron-absorbing
Material

(a)

FIGURE 22–5

(a) In the core of a nuclear reactor, a large mass of
uranium is created by placing uranium in adjacent
tubes, the fuel elements. The uranium is not sufficiently
concentrated to permit a nuclear explosion, but it will
sustain a chain reaction that will produce a tremendous
amount of heat. The rate of the chain reaction is
moderated by inserting or removing rods of neutron-
absorbing material (control rods) between the fuel
elements. The fuel and rods are surrounded by the
moderator fluid, pure water. (b) Technicians (Homer
Simpson and friend?) ready the reactor core housing to
receive uranium fuel elements in this nuclear reactor.
(Erich Hartmann/Magnum.)

(b)

The Nuclear Reactor

A nuclear reactor for a power plant is designed to
sustain a continuous chain reaction but not allow it
to amplify into a nuclear explosion (Fig. 22–4c). This
control is achieved by enriching the uranium to only
3 percent ^{235}U and 97 percent ^{238}U. This modest enrichment will not support the amplification of a chain
reaction into a nuclear explosion.

A chain reaction can be achieved in a nuclear

reactor only if a sufficient mass of enriched uranium is arranged in a suitable geometric pattern and is surrounded by a material called a *moderator*. The **moderator** slows down the fission neutrons so that they are traveling at the right speed to trigger another fission. In slowing down the neutrons, the moderator gains heat. In U.S. nuclear plants, the moderator is very pure water and the reactors are called light water reactors (LWRs). (The term *light water* distinguishes between ordinary water, H_2O, and the substance known as "heavy" water, D_2O). Other moderators employed in reactors of different design are graphite and heavy water.

To achieve the geometric pattern necessary for fission, the enriched uranium is made into pellets that are loaded into long metal tubes. The loaded tubes are called **fuel elements** or **fuel rods.** Many fuel rods are placed close together to form a reactor core inside a strong reactor vessel that holds the water, which serves as both moderator and heat-exchange fluid (coolant).

The chain reaction in the reactor core is controlled by rods of neutron-absorbing material, referred to as control rods, inserted between the fuel elements. The chain reaction is started and controlled by withdrawing and inserting the control rods as necessary. The nuclear reactor is simply this assembly of fuel elements, moderator-coolant, and movable control rods (Fig. 22–5). As the control rods are removed

and a chain reaction is initiated, the fuel rods and the moderator become intensely hot.

The Nuclear Power Plant

In a nuclear power plant, heat from the reactor is used to boil water to provide steam for driving conventional turbogenerators. One way to boil the water is to circulate it through the reactor. In most U.S. power plants, however, a double loop is employed. The moderator-coolant water is intensely heated by circulating it through the reactor, but it does not boil because the system is under very high pressure. This superheated water is circulated through a heat exchanger, where it boils other water to produce the steam used to drive the turbogenerator.

This double loop design isolates hazardous materials in the reactor from the rest of the power plant. However, it has one serious drawback. If the reaction vessel should break, the sudden loss of water from around the reactor, called a "loss-of-coolant accident" (LOCA), could result in the core overheating. The sudden loss of the moderator-coolant water would cause fission to cease, since the moderator is no longer present. However, even though fission stops, the fuel core can still overheat because 7 percent of the reactor heat comes from radioactive decays in the newly formed fission products. In time, this uncontrolled decay heat can melt the materials in the core,

FIGURE 22–6
Schematic diagram of a nuclear power plant.

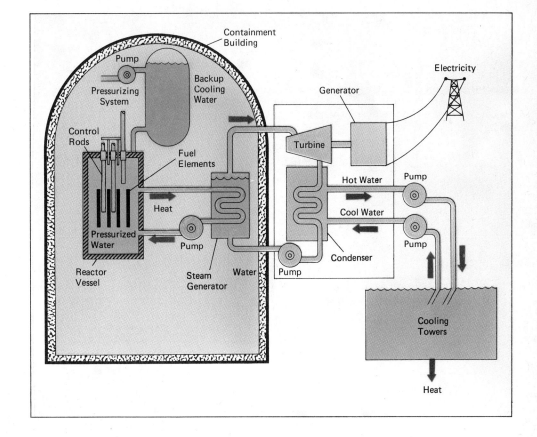

a situation called a **meltdown.** Then the molten material falling into the remaining water could cause a steam explosion. To guard against all this, there are backup cooling systems to keep the reactor immersed in water should leaks occur, and the entire assembly is housed in a thick concrete containment building (Fig. 22–6).

The fissioning of about one pound (0.5 kg) of uranium fuel releases energy equivalent to burning 50 tons of coal. Thus, one fueling of the reactor with about 60 tons of enriched uranium is sufficient to run the power plant for as long as 2 years. The spent fuel elements are then removed and replaced with new ones.

COMPARISONS OF NUCLEAR POWER AND COAL POWER

In canceling plans for nuclear power plants, we did not decide to do without electricity; effectively we opted to build coal-fired power plants instead. The United States does have abundant coal reserves, but is burning coal the course of action we wish to pursue? Nuclear power has some decided environmental advantages over coal-fired power (Fig. 22–7). Comparing a 1000-megawatt nuclear plant with a coal-fired plant of the same capacity for 1 year, we find:

○ Fuel needed—The coal plant consumes about 3.5 million tons of coal. If this is obtained by strip-mining, some environmental destruction and acid leaching will result. If the coal comes from deep mines, there will be human costs in the form of accidental deaths and impaired health. The nuclear plant requires about 30 tons of enriched uranium obtained from mining 7500 tons of ore, with much less harm to humans and the environment.

○ Carbon dioxide emission—The coal plant emits over 10 million tons of carbon dioxide into the atmosphere, contributing to global warming. The nuclear plant emits none.

FIGURE 22–7
The environmental impacts of nuclear power and coal-fired power.

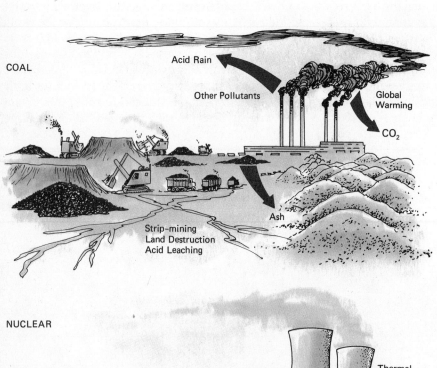

COAL

Acid Rain

Other Pollutants

Global Warming

CO_2

Ash

Strip-mining
Land Destruction
Acid Leaching

NUCLEAR

No Smokestacks

Thermal Pollution

Uranium Mining

Radioactive Wastes

○ Sulfur dioxide and other emissions—The coal plant emits over 400 000 tons of sulfur dioxide and other acid-forming pollutants. It also releases low levels of many radioactive chemicals found naturally in the coal. The nuclear power plant produces no acid-forming pollutants but will release low levels of radioactive waste gases.

○ Solid wastes—The coal plant produces about 100 000 tons of ash, requiring land disposal. The nuclear plant produces about 250 tons of radioactive wastes requiring safe storage and ultimate disposal.

○ Accidents—A worst-case accident in the coal plant could result in worker fatalities and a destructive fire, a situation common to many industries. Depending on the type of nuclear plant, accidents can range from minor emissions of radioactivity to catastrophic releases that can lead to widespread radiation sickness, scores of human deaths, untold numbers of cancers, and widespread, long-lasting environmental contamination.

Of course it is the radioactive wastes and releases and the potential for accidents that have led the public to reject nuclear power. Are these problems that cannot be overcome?

RADIOACTIVE MATERIALS AND THEIR HAZARDS

Assessing the hazards of nuclear power necessitates our understanding a little about radioactive substances and their danger.

When uranium or any other element fissions, the split halves are atoms of lighter elements: iodine, cesium, strontium, cobalt, or any of some 30 other elements. These newly formed atoms—called the *direct products* of the fission—are generally unstable isotopes of their respective elements. Unstable isotopes (called **radioisotopes**) gain stability by spontaneously ejecting subatomic particles or high-energy radiation, or both. The particles and radiation are collectively referred to as **radioactive emissions.** Any materials in and around the reactor may also be converted to unstable isotopes and become radioactive by absorbing neutrons from the fission process (Fig. 22–8). These *indirect products* of fission, along with the direct products, are the **radioactive wastes** of nuclear power. (Radioactive fallout from nuclear explosions also consists of these direct and indirect fission products.)

Radioactive emissions can penetrate biological tissue. They leave no visible mark, nor are they felt, but they are capable of breaking molecules within cells. In high doses, radiation may cause enough damage to prevent cell division. Thus radiation can be focused on a cancerous tumor to destroy it. However, if the whole body is exposed to such levels of radiation, a generalized blockage of cell division occurs that prevents the normal replacement or repair of blood, skin, and other tissues. This result is called radiation sickness and may lead to death a few days or months after exposure. Very high levels of radiation may totally destroy cells, causing immediate death.

In low doses, radiation may damage DNA, the

FIGURE 22–8
Radioactive wastes and radioactive emissions. Nuclear fission results in the production of numerous unstable isotopes, the radioactive wastes. They give off potentially damaging radiations until they regain a stable structure.

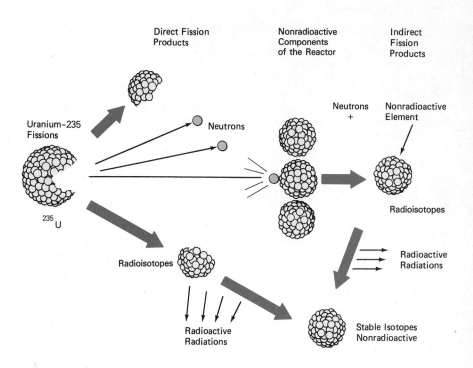

genetic material inside the cell. Cells with damaged (mutated) DNA may then begin dividing and growing out of control, forming malignant tumors. If the damaged DNA is in an egg or a sperm cell, the result may be birth defects in offspring. The effects of exposure to radiation may go unseen until many years after the exposure. Other effects include a weakening of the immune system, mental retardation, and the development of cataracts.

Health effects are directly related to the level of exposure, and many scientists believe that no dose is without some harm. Not surprisingly, a major concern regarding nuclear power is that it may inadvertently cause large numbers of the public to be exposed to low levels of radiation, thus elevating the risk of cancer and other disorders.

However, nuclear power is by no means the only source of such radiation. There is normal **background radiation** from radioactive materials, such as the uranium and radon gas that occur naturally in the earth's crust, and cosmic rays from outer space. For most people, background radiation is the major source of radiation exposure. In addition, we deliberately expose ourselves to radiation from medical and dental X-rays, by far the largest source of human-induced exposure and, for the average person, equal to one-fifth the exposure from background sources. Thus, the argument becomes one of *relative hazards*. Does or will the radiation from nuclear power significantly raise radiation levels and elevate the risks of developing cancer?

During normal operation of a nuclear power plant, the direct fission products remain within the fuel elements and the indirect products are maintained within the containment building that houses the reactor. No routine discharges of radioactive ma-

IN PERSPECTIVE
Radiation Phobia?

It is a familiar scene to TV viewers: Angry people with placards picket a nuclear power plant; some are dragged off by police for obstructing access to the plant. The protesters are convinced that nuclear power is dangerous and a threat to health, even when a plant is operating normally. Often the ranks of the protesters include medical doctors and research scientists along with the more familiar members of environmental protest organizations. What is it about nuclear power that ignites such strong feelings? David Willis, Professor Emeritus of Radiation Biology at Oregon State University, believes that many protesters are motivated by fear of something they do not understand—in fact, a radiation phobia. This fear is made worse by the way the media portray radiation. For example, the word *radiation* is very often preceded by the adjective *deadly*. Yet only a few people suffer ill effects from radiation, and those effects are almost always due to occupational or medical accidents involving carelessness in handling radiation sources.

Perhaps the fear occurs because

we are not able to perceive radiation by our normal senses of smell, taste, touch, vision, or hearing. We hear that there is radiation all around us in the form of cosmic rays—the background radiation—but there is nothing we can do about it, and life seems to go on. Willis points out that yes, ionizing radiation can induce cancer in humans, but the kinds of cancer induced are identical with cancers that arise spontaneously from all sorts of causes, mostly unknown. The higher the radiation dose, the greater the tendency to develop cancer. As the dose declines, however, as in small exposures to radiation that might be an occupational hazard for some lines of work, radiation effects simply disappear in the statistical data showing normal cancer development in a given population.

To illustrate this, Willis uses an analogy of an airplane descending into a fog bank as it comes in to land. Above the fog, the airplane is quite visible, but when it descends into the fog, it disappears from view even though it is still there. By analogy, the airplane represents the can-

cers induced from ionizing radiation. Its descent represents the effects of declining doses in fewer cancers induced. The fog bank represents the cancers normally occurring from a variety of causes, or spontaneously. Any effects of low levels of radiation are completely masked by the large fog of cancers naturally occurring. If some cancers are induced by low-level radiation, the number is insignificantly low compared with the natural rate.

Willis asserts that many opponents of nuclear power are afraid of something that actually carries a very low risk, especially when compared with other hazards in our environment. In spite of numerous studies, no linkage has been shown between the incidence of cancer and the presence of a nuclear facility. Willis has debated with nuclear opponents and often finds that they will not listen to a reasoned explanation of the health effects of ionizing radiation and the demonstratedly low risks involved with nuclear power. The result is apparent: Nuclear power is on the wane and may never recover.

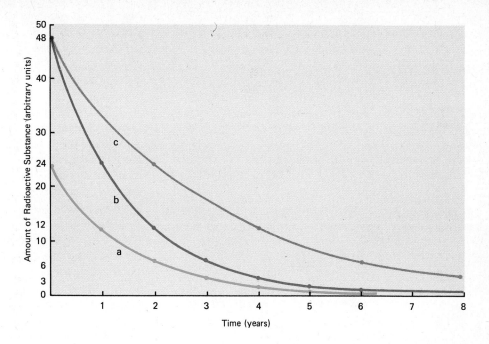

FIGURE 22–9
Radioactivity for any isotope declines as shown. Regardless of the starting amount, one-half decays during each successive half-life. (a) A substance with a half-life of 1 year starting with 24 units; (b) the same substance starting with 48 units; (c) a substance with a half-life of 2 years. The half-life for different isotopes may vary from less than one second to many thousands of years.

terials into the environment occur. Even very close to an operating nuclear power plant, radiation levels from the plant are lower than normal background levels. A radiation detector will pick up more radiation from the earth and concrete on a basement floor than it will when held within 150 yds of a nuclear power plant. Careful measurements have shown that public exposure to radiation from normal operations of a power plant is about 1 percent of natural background.

However, the main concern is not about *normal* operation. It is about disposal of radioactive wastes and the potential for catastrophic accidents, as occurred in Chernobyl in 1986.

Disposal of Radioactive Wastes

To understand the problems surrounding nuclear waste disposal, we must understand the concept of *radioactive decay*. As unstable isotopes eject particles and radiation, they become stable and cease to be radioactive. This process is known as **radioactive decay.** As long as radioactive materials are kept isolated from humans and other organisms, the decay process proceeds harmlessly.

The rate of radioactive decay is such that half of the starting amount of a given isotope will decay in a certain period of time. In the next equal period of time, half of the remainder (half of a half, which equals one-fourth of the original) decays, and so on, as shown in Figure 22–9. The time for half the amount of a radioactive isotope to decay is known as its **half-life.** The half-life of an isotope is always the same, regardless of the starting amount.

Note from Figure 22–9 that decay of a radioisotope is never 100 percent complete; radioactivity is reduced by only half in each half-life, and there is always an undecayed portion remaining. However, it is generally considered that radiation is reduced to insignificant levels after 10 half-lives.

Each particular radioactive isotope has a characteristic half-life. The half-lives for various isotopes range from a fraction of a second to many thousands of years. Uranium fissioning results in a heterogeneous mixture of radioisotopes, the most common of which are listed in Table 22–1.

Much of the radioactivity of fission wastes will dissipate in a period of months or a few years as the short-lived isotopes decay. To be safe, however, long-lived isotopes require isolation for up to 240 000 years (ten times the half-life of plutonium). Thus, the problem of nuclear waste disposal is twofold:

○ Short-term containment to allow the radioactive decay of short-lived isotopes. Thus, in 10 years fission wastes lose more than 97 percent of their radioactivity. Wastes can be handled much more easily and safely after this loss occurs.

○ Long-term containment (tens of thousands of years) to provide protection from the long-lived isotopes.

For short-term containment, the spent fuel is first stored in deep swimming pool–like tanks on the sites of nuclear power plants. The water in these tanks dissipates waste heat, which is still generated to some

TABLE 22–1

A Few of the Most Common Radioactive Isotopes Resulting from Uranium Fission and Their Half-Lives

Short-Lived Fission Products	Half-Life (days)
Strontium-89	54
Yttrium-91	59.5
Zirconium-95	65
Niobium-95	35
Molybdenum-99	2.8
Rubidium-103	39.8
Iodine-131	8.1
Xenon-133	15.3
Barium-140	12.8
Cerium-141	32.5
Praseodymium-143	13.9
Neodymium-147	11.1

Long-Lived Fission Products	Half-Life (years)
Krypton-85	10.27
Strontium-90	28
Rubidium-106	1.0
Cesium-137	30
Cerium-144	.8
Promethium-147	2.6

Addition Products of Neutron Bombardment	Half-Life (years)
Plutonium-239	24,000

degree, and acts as a shield for radiation. After a few years of decay, spent fuel may then be placed in dry casks in order to save space; here they are simply air-cooled. (Nuclear weapons facilities also maintain various tanks for short-term containment of radioactive wastes.)

The development of and commitment to nuclear power went ahead without ever fully addressing the issue of ultimate long-term containment. It was generally assumed by nuclear proponents that the long-lived wastes could be solidified, placed in sealed containers, and buried in deep stable rock formations (geologic burial) as the need for such containment became necessary (Fig. 22–10). In the meantime, the wastes from the world's commercial reactors have been accumulating at a rate of 9500 tons a year, reaching 103 000 tons at the end of 1992, all of which is stored on site at the power plants. At least one-fourth of these wastes are in the United States. Furthermore, because of neutron bombardment of the reactor walls, all nuclear power plants will eventually add to the stockpile of radioactive wastes. Scientists estimate that dismantling a decommissioned power plant will generate more nuclear waste than the plant produced during its active life.

Some of the worst failures in handling wastes have occurred at U.S. and former Soviet military facilities, in connection with the manufacture of nuclear weapons. Liquid high-level wastes stored at many facilities in the United States have leaked into the en-

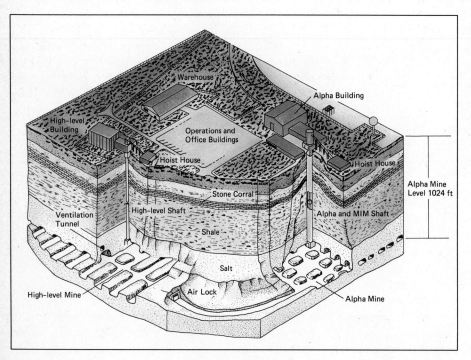

FIGURE 22–10
Disposal of radioactive wastes from nuclear power plants. These wastes must be isolated from the environment for thousands of years. Elaborate plans have been made for their disposal, but will the plans ensure that the wastes will stay where they are put?

vironment and contaminated wildlife, sediments, and soil (at Hanford, Washington; Fernald, Ohio; and Savannah River, South Carolina). Soviet military weapons facilities have been even more irresponsible, often dumping the wastes into the nearest body of water. Thus, standing for one hour on the shore of Lake Karachay, in the southern Ural Mountains, will kill a person from radiation poisoning within a week. The lake dried up one summer, and winds blew radioactive dust across the countryside, contaminating 41 000 people. This lake is considered to be the most polluted lake on earth, a legacy of the Cold War and an enormous continuing source of radioactive contamination.

The United States and most other countries using nuclear power have decided on geologic burial for ultimate consignment of nuclear wastes, but no nation has developed plans to the point of actually carrying out the burial. Many nuclear nations have not even been able to find a site potentially suitable for receiving the wastes. Where sites have been selected, many questions about safety have surfaced. The basic problem is that no rock formation can be guaranteed to remain stable and dry for tens of thousands of years. Everywhere scientists look, there is evidence of volcanic or earthquake activity or groundwater leaching within the last 10 000 years or so, which is to say that it may occur again in a similar

ETHICS

Showdown in the New West

In a scenario reminiscent of the Civil War, two western states have become the focus of a controversy over states rights. On one side is the federal government in the form of the Department of Energy (DOE), which is trying to solve the problem of what to do with nuclear wastes and thinks it has a reasonable solution. On the other side are politicians in the states designated as repository sites, who are responding to the voices of citizens of those states. At stake may be the whole future of nuclear energy in the United States.

The DOE has constructed a billion-dollar Waste Isolation Pilot Plant 2150 feet beneath the desert in southwestern New Mexico, 26 miles from the city of Carlsbad. The plant is on federal land and has passed all necessary safety reviews. It was ready to open for business in October 1991. "Business," in this case, was to become the repository for up to a million barrels of plutonium wastes from nuclear weapons plants and laboratories around the country. (No commercial nuclear wastes will go there.) The DOE wants to bring a few thousand barrels to the New Mexico site in order to demonstrate the usefulness of the facility, eager to prove that nuclear wastes can be

safely moved and placed in such a repository. The main objection to the New Mexico site is the prospect of accidents on the narrow roads in the area. There was talk of barricades in the streets to block the trucks if the DOE starts them rolling.

New Mexico sued the DOE on the grounds that it is in violation of federal law prohibiting the opening of such a repository without congressional approval. The state of Texas plus four environmental groups joined New Mexico in bringing suit, and in November 1991 a Federal District Court judge issued an injunction that indefinitely barred the DOE from opening the repository. It now appears that it will take an act of Congress to reverse the judicial decision, and in the meantime, the nuclear wastes continue to pile up.

Opposition to long-term storage of nuclear wastes is perhaps even stronger in Nevada, which has the dubious distinction of being selected for the nation's only repository for *commercial* nuclear wastes, at Yucca Mountain. The site is a barren ridge in the desert about 100 miles northwest of Las Vegas. In 1989, the Nevada legislature passed a bill making it unlawful for any agent or agency to store high-level

radioactive waste in the state. All sorts of surveys have been carried out to test the mood of Nevadans regarding long-term storage of wastes in their states. The results have been uniformly negative. Three-fourths of Nevadans agreed that the state should continue to do all it can to prevent establishment of the Yucca Mountain site. It is clear that people perceive the risks of such a site as enormous and unacceptable, in contrast to the opinions of scientists who have performed risk calculations for the site. In this case, the state's attempts to pose a legal blockade to the DOE have failed. The Supreme Court has ruled that Nevada must process DOE applications for permits to continue work on the site.

You can see the dilemma. Public and political support for any form of nuclear energy is at an all-time low, and people in these two western states are especially indignant at becoming the repository of nuclear wastes from around the nation. Yet we need to do something with this legacy of nuclear energy other than to pass the problem along to the next generation. What do you think should be done?

period of time. If such events did occur, the still-radioactive wastes could escape into the environment and contaminate water, air, or soil with consequent effects on humans and wildlife.

In the United States, efforts to locate a long-term containment facility, which have been going on for about 25 years, have been characterized by the term *NIMBY*, which stands for "not in my backyard." A number of states, under pressure from citizens, have

passed legislation categorically outlawing the disposal of nuclear wastes within their boundaries. In the meantime, the need to select and develop a long-term repository has become increasingly critical. With the current lack of public support for nuclear power, it may be extremely difficult for the United States or any other nation to establish an acceptable long-term nuclear waste facility (see Ethics Box, page 493).

At the end of 1987, Congress called a halt to debate and arbitrarily selected a remote site, Yucca Mountain in southwestern Nevada, to be the nation's nuclear dump. Not surprising, Nevadans have fought this selection, passing a law in 1989 that prohibits anyone from storing high-level radioactive waste in the state. The federal government has the power to override state prohibitions, though, and intensive study and evaluation of this site are now under way. To date, studies have shown that the storerooms will be 1000 feet (300 m) above current groundwater levels, presumably safe from groundwater invasion. However, earthquakes or volcanic activity could change conditions in a short time, and the site lies in a zone of fault lines and has a history of volcanic activity only 20 000 years ago. If plans go forward, the Yucca Mountain facility will begin receiving wastes from commercial and military facilities all over the country in 2010. It should not escape notice that this plan means that thousands of tons of radioactive wastes will be shipped by rail and truck through congested areas across the whole country. An even worse alternative is to let the problem remain unresolved. This is and will remain a serious problem for the foreseeable future just with existing nuclear power plants. Building more plants can only intensify it.

FIGURE 22–11
An aerial view of the Chernobyl reactor three days after the explosion (April 29, 1986). The Chernobyl explosion in the former USSR was the worst nuclear power accident ever, directly killing at least 33 people and putting countless numbers of people in the surrounding country at risk for future cancer deaths. (Shone/Gamma-Liaison.)

The Potential for Accidents

Prior to 1986, the scenario for a worst-case nuclear power plant disaster was a matter of speculation. Then at 1:24 A.M. local time on April 26, 1986, events at Chernobyl in the former USSR made the speculation a reality (Fig. 22–11). While conducting a test, electrical engineers unfamiliar with nuclear safety systems tried to operate the reactor at a prohibitively low percentage of its power. When they realized that this could lead to trouble, they tried to get power back up (instead of shutting down completely and starting over) by disabling major elements of the safety system and withdrawing almost all of the control rods. Power increased more rapidly than the operators could react, and a runaway reaction was achieved before control rods could be reinserted. Within seconds, the reactor went from 6 percent power to more than 30 times normal full power. A meltdown and fire occurred, a steam explosion blew the 1000-ton top off

A wrong-way drunk driver plowed head-on into a flatbed truck on Interstate 91 in Springfield, Massachusetts, early in the morning of December 16, 1991. The truck's diesel fuel burst into flames, and the fire consumed 12 wooden shipping crates on the trailer. Rescuers pulled the occupants from both vehicles in time to save their lives. What made this something more than a "normal" head-on collision was the fact that the truck was carrying 12 steel and lead casks of nuclear fuel destined for the Vermont Yankee nuclear power plant in Vernon, Vermont.

There were some anxious moments when state police and firefighters arrived on the scene and saw the symbol on the side of the truck showing that it carried a radioactive cargo; all but one of the casks withstood the impact and the fire intact. One suffered a small break, but none of the ceramic uranium dioxide pellets inside spilled. Even if they had, the pellets were harmless, because they had not yet been irradiated. If they had been spent nuclear fuel pellets, it would have been a different matter, but, as we have seen, spent nuclear fuel is not transported but kept on site. However, if the western nuclear repository sites ever get activated (see Ethics Box, p. 493), much more highly radioactive cargo will be on the highways.

Nuclear opponents were upset because the pertinent towns had not been notified that nuclear fuel was coming through. State Public Health officials replied that although the state issues permits for the movement of such cargos, federal regulations prohibit notification of cities and towns on the route because of a desire to avoid possible terrorist interception.

This accident is a stern reminder that human error must always be factored into any assessment of the risks of a technology. Plans to move any nuclear waste should be based on the assumption that the truck will have a head-on collision at any moment. The next test of fire could have far more serious consequences.

the reactor, and enormous quantities of radioactive materials were ejected thousands of feet into the atmosphere. Release of radioactivity continued for 10 days, until fires could be extinguished and the core cooled down with liquid nitrogen and carbon dioxide.

Only 2 of the engineers were killed by the explosion, but 31 of the personnel brought in to contain the aftermath died of radiation sickness within the next 6 months. It is sobering to note that responding to a nuclear accident may be a suicidal mission, yet the escape of radioactive material would have been considerably greater without the efforts of these people.

As the radioactive fallout settled, 135 000 people, everyone in a 20-mile radius around the plant, were evacuated and eventually relocated. Today the area remains uninhabitable because the soil is contaminated with radioactive compounds. Increased levels of radiation resulting from the disaster were detected around the globe, including in the United States.

Over a broad area in the republics of Belarus, Russia, and Ukraine, buildings and roadways were washed down to flush away radioactive dust. Many tons of foodstuffs caught in fallout areas in Europe were banned from market. Even with these precautionary measures, however, many people in or near the evacuation zone may have been exposed to radiation levels that could lead to cancers and birth defects in future years. Current estimates range from 14 000 to 475 000 cancer deaths worldwide from this accident.

Are we in danger of such an explosion occurring at U.S. nuclear power plants? Nuclear proponents argue that the answer is no because U.S. power plants have a number of design features that should make a repeat of Chernobyl impossible. The Chernobyl reactor used graphite as a moderator, rather than water as in light water reactors. LWRs are incapable of developing a power surge more than twice normal power, well within the designed capacity of the reactor vessel. In LWRs there are more backup systems to prevent core overheating, and reactors are housed in a thick concrete-walled containment building designed to withstand explosions such as the one that occurred in Chernobyl. The Russian reactor had no containment building.

However, LWRs are not immune to accidents, the most serious being a complete core meltdown as a result of total loss of coolant. This has never happened, but there was a close call at Three Mile Island.

On March 28, 1979, the Three Mile Island nuclear power plant near Harrisburg, Pennsylvania, suffered a partial meltdown as a result of a series of human and equipment failures. The steam generator (see Fig. 22–6) shut down automatically because of a lack of power in its feedwater pumps, and eventually a pressure valve on top of the generator opened in response to the pressure buildup. Unfortunately, the valve remained stuck in the open position and drained coolant water from the reactor vessel. Operators responded poorly to the emergency, shutting down the emergency cooling system at one point and shutting down the pumps in the reactor vessel. One design error compounded the problem: Instruments told operators that the reactor was full of water when actually it needed water badly. The core was uncovered for a time and suffered a partial meltdown, and a small amount of radioactive gas was released to the atmosphere.

The drama held the whole nation, particularly the 300 000 residents of Harrisburg poised for evacuation, in suspense for several days. The situation was eventually brought under control and no injuries occurred, but it could have been much worse if meltdown had been complete. The reactor was so badly damaged and so much radioactive contamination had occurred inside the containment building that the cleanup and repair process won't be complete until the mid-1990s and is proving to be as costly as building a new power plant.

Real costs of the accidents at Three Mile Island and especially Chernobyl must also be reckoned in terms of public trust. Public confidence in nuclear energy was declining in the 1970s but plummeted after the two accidents. The accidents pointed to human error as a highly significant factor in nuclear safety, and human error is something the public understands well. Nuclear proponents suffered a serious loss of credibility with Three Mile Island, but Chernobyl was their worst nightmare come true—a full catastrophe, just as the antinuclear movement had predicted might some day occur.

Safety and Nuclear Power

As a result of Three Mile Island and other lesser incidents in the United States, the Nuclear Regulatory Commission has upgraded safety standards not only in the technical design of nuclear power plants but also in maintenance procedures and in the training of operators. Thus, proponents contend, nuclear plants were designed to be safe in the beginning and now they are safer than ever. It is estimated that as a result of new procedures instituted after the accident at Three Mile Island, nuclear plants are now six times safer than before. Some proponents of nuclear power claim that we now have the technology to build *inherently safe* nuclear reactors, reactors designed in such ways that any accident would result in an au-

tomatic quenching of the chain reaction and suppression of decay heat. In reality, there is no such thing as an inherently safe reactor since the concept implies no release of radioactivity under any circumstances—an impossible expectation.

Instead, nuclear scientists are proposing a new generation of nuclear reactors with built-in *passive safety* features, rather than the *active safety* features found in current reactors. The distinction is important, and it represents a major change in the direction of thinking within the nuclear community. **Active safety** relies on operator-controlled actions, external power, electrical signals, and so forth. As accidents have shown, operators may override such safety factors, and electricity, valves, and meters can fail or give false information. **Passive safety,** on the other hand, involves engineering devices and structures that make it virtually impossible for the reactor to go beyond acceptable levels of power, temperature, and radioactive emissions.

In anticipation of a possible renewed interest in nuclear power, engineers are now designing LWRs that include passive safety features and much simpler, smaller power plants—the so-called advanced light water reactors (ALWRs). For example, one passive safety feature would be to position a cold-water reservoir so that, in the event of a LOCA, the water would drain by gravity to the reactor core (Fig. 22–12). In addition, the design will make it impossible for operators to inactivate the passive safety systems. Another feature being planned is to build reactors as modular units small enough to conduct decay heat outward into the soil; with heat conducted away like this, the reactors could not possibly experience a core meltdown. One such modular reactor has been built in Germany, and when tested against a LOCA its core suffered no damage.

The motive behind these new designs is restoration of the public's confidence in nuclear energy. Previously, nuclear proponents had emphasized the very low probabilities of accidents; as we have seen, though, improbable events can happen, and when they happen to nuclear power plants, the consequences are awesome. The new emphasis is on convincing the public of the fundamental safety of ALWR designs by demonstrations and explanations that nontechnical people can understand.

ECONOMIC PROBLEMS WITH NUCLEAR POWER

As Figure 22–1 shows, U.S. utilities were turning away from nuclear power considerably before the disaster at Chernobyl. The reasons are mainly economic.

First, increasing safety standards for the con-

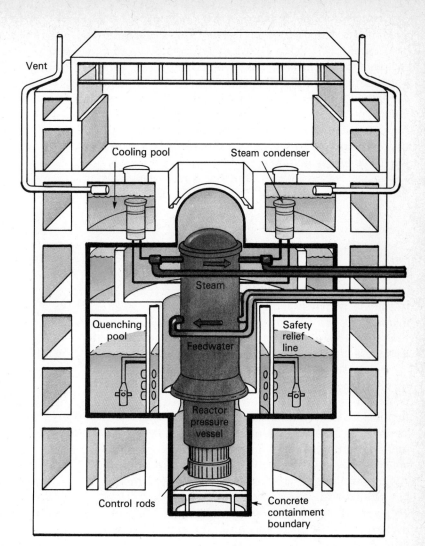

FIGURE 22–12
Advanced light water reactor showing passive safety features. The core is surrounded by three concentric structures: (1) a reactor pressure vessel, in which heat from the reactor directly boils water into steam; (2) a concrete chamber (outlined in black) and water pool, which contains and quenches steam vented from the reactor in an emergency; and (3) a concrete building, which acts as a secondary containment vessel and shield. Any excessive pressure in the reactor will automatically open valves that release steam into a quenching pool, reducing the pressure. Water from the quenching pool can, if necessary, flow downward to cool the core. Evaporation from a pool on top of the containment building limits the buildup of containment pressure. (From "Advanced Light-Water Reactors," by M. W. Golay and N. E. Todreas. Copyright © 1990 by Scientific American, Inc. All rights reserved.)

struction and operation of nuclear power plants have caused costs of nuclear power plants to increase at least fivefold even after inflation is considered. Second, public protests have frequently delayed construction or startup. Such delays increase costs still more because the utility is paying interest on its investment of several billion dollars even when the plant is not producing power. As these costs are passed on, consumers become even more disillusioned with nuclear power. Finally, safety systems may protect the public, but they do not prevent an accident from being financially ruinous to the utility. Since radioactivity prevents straightforward cleanup and repair, an accident, as Three Mile Island demonstrated, can convert a multibillion-dollar asset into a multibillion-dollar liability in a matter of minutes. Thus, nuclear power involves a financial risk that utility executives are reluctant to take.

Another factor that promises to increase the cost of nuclear-generated electricity is a shorter-than-expected lifetime for nuclear power plants. Originally it was thought that nuclear plants would have a lifetime of about 40 years. It now appears that their lifetime will be considerably less. Worldwide, 58 nuclear plants have been closed, with an average lifetime of 16 years. This shorter lifetime substantially increases the cost of the power produced because the cost of the plant must be repaid in the shorter period.

Plant lifetimes are shorter than originally expected because of a problem known as **embrittlement.** As well as maintaining a chain reaction, some of the neutrons from fission bombard the reactor vessel and other hardware. Gradually, this neutron bombardment causes the metals to become brittle such that they may crack under stress—for example, when emergency coolant waters are introduced in the event of a LOCA. When the reactor vessel becomes too brittle to be considered safe, the plant must be shut down or, in technological jargon, decommissioned.

Decommissioning can be extremely costly. By

the time this step becomes necessary, the power plant components will have accumulated so much radioactivity from neutron bombardment that the only safe course of action will be to seal off the entire containment building for an indefinite period of time and construct a new plant. Of the plants operating in the United States, it is estimated that at least 10 will be decommissioned in the 1990s and most of the rest in the 10 years thereafter.

MORE-ADVANCED REACTORS

Breeder Reactors

Uranium, especially ^{235}U, is not a highly abundant mineral on earth. At the height of nuclear optimism in the 1960s, when as many as 1000 plants were envisioned by the turn of the century, it was foreseen that shortages of ^{235}U would develop. Breeder reactors were seen as the solution to this problem.

Recall that when a ^{235}U atom fissions, two or three neutrons are ejected. Only one of these neutrons hitting another ^{235}U atom is required to sustain the chain reaction; the others are absorbed by something else. The breeder reactor is designed so that these extra neutrons are absorbed by the nonfissionable ^{238}U. When this occurs, the ^{238}U is converted to ^{239}Pu (plutonium), which is fissionable. The ^{239}Pu can be purified and used as a nuclear fuel just as ^{235}U is. Thus, the breeder converts nonfissionable ^{238}U into a useful nuclear fuel. Since there are generally two neutrons in addition to the one needed to sustain the chain reaction, the breeder may produce more fuel than it consumes. As 99.7 percent of all uranium is ^{238}U, converting this to ^{239}Pu through breeder reactors effectively increases the nuclear fuel reserves over 100-fold.

Breeder reactors present all of the problems and hazards of standard fission reactors plus a few more. If a meltdown occurred in a breeder, the consequences would be much more serious than in an ordinary fission reactor because of the large amounts of ^{239}Pu, which has the exceedingly long half-life of 24 000 years. In addition, because plutonium can be purified and fabricated into nuclear weapons more easily than ^{235}U, the potential for the diversion of breeder fuel to weapons production is greater. Hence, the safety and security precautions needed for breeder reactors are greater.

With its scaled-down nuclear program, the United States currently has enough uranium stockpiled to fuel all reactors that are operating or under construction through their lifetimes. Thus, there is no urgency for the United States to develop breeders, and construction of the prototype breeder at Clinch River, Tennessee, has been discontinued. However, it should be noted that in the United States and elsewhere, small breeder reactors are operated for military purposes. France is currently the only nation proceeding with development of commercial breeder reactors.

Fusion Reactors

The vast energy emitted by the sun and other stars comes from fusion (Fig. 22–3b). The sun, as well as other stars, is composed mostly of hydrogen. Solar energy is the result of fusion of this hydrogen into helium. Scientists have duplicated this process in the hydrogen bomb, but hydrogen bombs hardly constitute a useful release of energy. The aim of fusion technology is to carry out this fusion in a controlled manner in order to provide a practical heat source for boiling water to power steam turbogenerators.

Since hydrogen is an abundant element on earth (two atoms in every molecule of water) and helium is an inert, nonpolluting, nonradioactive gas, hydrogen fusion is promoted as the ultimate solution to all our energy problems—that is, pollution-free energy from a virtually inexhaustible resource, water. However, the dream is still a long way from reality. Indeed, a fusion plant has not yet been proven possible, much less a practical alternative.

At the present state of the art, fusion power is still an energy consumer rather than a producer. The problem is that it takes an extremely high temperature—some 100 million degrees Celsius—and pressure to get hydrogen atoms to fuse. In the hydrogen bomb, the temperature and pressure are achieved by using a fission bomb as an igniter—hardly a practical alternative to sustained, controlled fusion.

A major technical problem is how to contain the hydrogen while it is being heated to the tremendously high temperatures required for fusion. No material known can withstand these temperatures without vaporizing; however, two concepts are being tested. One is the Tokamak design, in which ionized hydrogen is contained within a magnetic field while being heated to the necessary temperature. The second is laser fusion, in which a tiny pellet of frozen hydrogen is dropped into a "bull's-eye" where it is hit simultaneously from all sides by powerful laser beams (Fig. 22–13). The laser beams simultaneously heat and pressurize the pellet to the point of fusion.

Some fusion has been achieved in both types of devices, and in late 1991 a European team working in Oxfordshire, England, reported a brief yield of a million watts of power, considered the most significant breakthrough to date. As yet, however, the break-even point has not been reached. More energy is required to run the magnets or lasers than is ob-

FIGURE 22–13
Laser fusion. In this experimental instrument at Lawrence Livermore Laboratory, 30 trillion watts of optical power are focused onto a tiny pellet of hydrogen smaller than a grain of sand located in the center of this vacuum chamber. For less than a billionth of a second the fusion fuel is heated and compressed to temperatures and densities like those found in the sun. (Alexander Tsiaras/Science Source/Photo Researchers.)

tained by the fusion. The most optimistic workers in the field believe that, with sufficient money for research—$10 billion—the break-even point might be reached in the late 1990s. Even if this goal is achieved, however, it is still a long way from a practical commercial fusion-reactor power plant. Developing, building, and testing such a plant would require at least another 20 to 30 years and many more billions of dollars. Additional plants would require additional years. Thus, fusion is, at best, a very long-term option. Many scientists believe that fusion power will always be the elusive pot of gold at the end of the rainbow.

Even if the break-even point is reached, fusion energy still promises to be neither clean nor an unlimited resource. Current designs use not regular hydrogen (^{1}H) but rather the isotopes deuterium (^{2}H) and tritium (^{3}H)—the *d-t* reaction. Fusion of deuterium alone (a *d-d* reaction) demands much greater tem-

peratures and pressures than the *d-t* reaction and has a low energy yield. Deuterium is a naturally occurring isotope that is nonradioactive and can be isolated in almost any desired amounts from the hydrogen in seawater. Tritium, however, is an unstable radioactive isotope that must be produced. Current plans call for the production of tritium by a breeding reaction where the element lithium is bombarded with neutrons. The neutrons will be produced by the fusion of deuterium and tritium. The overall reaction is:

^{2}H + ^{3}H → ^{4}He + n + Energy
(Deuterium   Tritium   Helium   Neutron   (fusion
isolated   reaction)
from water) ^{3}H + ^{4}He ← n + ^{6}Li
                    Tritium   Helium   Neutron   Lithium
                                                        (breeding
                                                        reaction)

Lithium is not an abundant element, and it could easily become the limiting factor in the wide-scale use of fusion reactors. Also, tritium is radioactive, hence hazardous. It is also gaseous and difficult to contain. As a result, fusion reactors could easily become a source of radioactive tritium leaking into the environment unless effective (and costly) designs prevent leaks. On the plus side, tritium has a short half-life and emits only weak beta particles (electrons), and the amounts needed in the reactor are so small as to be dangerous to the public only if a massive loss were to occur.

One distinct advantage of fusion over fission is the absence of spent fuel wastes, which we have identified as one of the major problems with current nuclear power. However, the reactor hardware will be embrittled and made highly radioactive by the constant bombardment from neutrons. Thus, there will be the cost of constantly replacing reactor components and the problem of disposing of components that have been made radioactive. This problem is considered to be very similar in magnitude to that of an equal-power fission reactor. Finally, fusion reactors promise to be the source of unprecedented thermal pollution. A steam turbogenerator is only 30 to 40 percent efficient and, if half the power produced must be fed back into the reactor to sustain the fusion process, the overall reactor is only 15 to 20 percent efficient. That is to say, 80 to 85 units of heat energy will be dissipated into the environment for every 15 to 20 units of electrical energy produced. At best, all this boils down to the fact that fusion power will be exceedingly expensive if it is achieved at all.

In the spring of 1989, there was great excitement regarding fusion power when two scientists announced that they had achieved the fusion of hydrogen atoms at room temperature, called "cold fusion." The claim was that when palladium electrodes were used to pass an electric current through heavy water (water in which the hydrogen atoms have been replaced by deuterium), some of the deuterium diffused into the electrodes and fused, which generated heat in addition to that from the electric current being used. However, no neutrons were detected in the reaction, and most scientists believe that the reactions would prove to be ordinary chemical reactions and not nuclear fusion. Attempts to reproduce the experiments in other laboratories around the world have failed, and the scientific community has generally rejected the claims of cold fusion.

Fusion research may continue in an effort to increase human understanding of physical processes. However, it seems likely that solar technologies will be producing power less expensively and with fewer risks before fusion reactors become a reality.

CONCLUSION: THE FUTURE OF NUCLEAR POWER

Worldwide, public opposition to nuclear energy is higher now than it has ever been and raises the question of whether nuclear energy has any future once the currently operating plants have lived out their life span. This opposition is based on several premises:

○ First, people have a general distrust of technology they do not understand, especially when that technology carries with it the potential for catastrophic accidents or the hidden but real capacity to induce cancer (see In Perspective Box, p. 490).

○ Second, some scientists and others who do understand the technology are skeptical of the way in which it is being managed. They are aware that the same agencies (in the United States, the NRC) involved with licensing and safety regulations are also strong proponents of the commercial nuclear industry. As an example of this concern, the U.S. General Accounting Office recently investigated allegations of the use of substandard parts in 72 U.S. nuclear power plants and concluded that NRC was negligent and in fact tried to prevent the information about the substandard parts from getting to the public.

○ Third, the nuclear industry has repeatedly presented nuclear energy as extremely safe, using arguments based on the low probabilities of accidents occurring. When accidents occur, as they have, probabilities become realities and the arguments are moot.

○ Fourth, there is the crucial problem of nuclear waste. All parties agree that the waste must be placed somewhere safe, but there the agreements end, and siting a long-term nuclear waste repository has become an apparently impossible political as well as technological problem in country after country.

Completely aside from these objections, there is the basic mismatch between nuclear power and the energy problem. As we have emphasized, the United States' main energy problem involves an eventual shortage of crude oil for transportation, yet nuclear power produces electricity, which is not used for transportation. If we were moving toward a totally electric economy that included even electric cars, nuclear-generated electricity could be substituted for oil-based fuels. Unfortunately, electric cars have not yet proved practical, and the outlook for them in the near future remains uncertain. Consequently, nuclear power simply competes with coal-fired power plants in meeting limited demands for electricity. The fact is that, given the escalated costs and additional financial risks of nuclear power plants, coal is cheaper, and the United States does have abundant coal reserves.

However, there are still the environmental problems of mining and burning coal, including acid precipitation and global warming. If these were factored into the price of coal, we would find its price considerably higher than it now appears. Of course, costs such as long-term confinement of nuclear wastes and decommissioning of power plants have yet to be factored into the cost of nuclear power.

At the bottom line, we find that one of the most touted energy options, nuclear power, is also one of the most feared technologies in use anywhere. The public simply does not trust the safety of the reactors and the plans for waste disposal. Opponents believe that the technology poses long-term risks to both human and environmental health and will in no way solve our most critical energy problem—fuel for transportation. Also, although the United States has relatively abundant reserves of uranium, this is not a renewable resource. Opting to meet the energy problem by all-out exploitation of this resource would assuredly lead to another energy resource crisis within 100 years. Over the course of history, it is a very short period.

If nuclear energy is to have a brighter future, it will be because we have found continued use of fossil fuels to be so damaging to the atmosphere that we have placed limits on their use—and it will be because we have learned how to build safe nuclear reactors and have found safe places to deposit the wastes.

 Review Questions

1. Compare the current outlook for nuclear power with that of the 1960s.
2. Describe how energy is produced in a nuclear reaction.
3. How do nuclear power plants and nuclear reactors work?
4. What are the advantages and disadvantages of nuclear power as compared with coal energy?
5. How are radioactive wastes produced, and what hazards are associated with these wastes?
6. What are the two stages of nuclear waste disposal?
7. What problems are associated with long-term containment of nuclear waste?

8. Describe what happened at Chernobyl and Three Mile Island.
9. Describe the features that might make a nuclear power plant inherently safe.
10. Discuss the economic reasons that have caused many utilities to opt for coal-burning rather than nuclear-powered plants.
11. How do breeder and fusion reactors work? Does either one offer promise for alleviating our oil shortage?
12. Does nuclear power address our shortfall in crude oil production? Can it?

 Thinking Environmentally

1. Discuss the advantages and disadvantages of nuclear power. Are we overly concerned or not concerned enough about nuclear accidents? Could an accident like Chernobyl happen in the United States?
2. The environmental impact of a nuclear accident is not limited to national boundaries. Should we encourage the establishment of international regulations for these power plants? Who would define the regulations? How would they be enforced?
3. Would you rather live next door to a coal-burning plant or a nuclear power plant? Defend your choice.

4. Many people feel that nuclear power is a fading dream. Discuss this statement from a historical perspective.
5. Alvin Weinberg, one of the scientists who developed the first fission reactor, believes that we have no choice other than to opt for nuclear power. In order to maintain ourselves at our present numbers and affluence, he believes that we must commit ourselves to nuclear power and the proper operation of these plants. Do you agree or disagree? Why?
6. What do you believe should be done with nuclear waste?

23

Solar and Other Renewable Energy Sources

OUTLINE

LEARNING OBJECTIVES

When you have finished studying this chapter, you should be able to:

1. List the three obstacles facing the use of solar energy and describe two approaches for overcoming these obstacles.

2. Describe how solar energy may be used for space and hot water heating.

3. Contrast active and passive solar heating systems and give the advantages and disadvantages of each.

4. Describe what has proved to be the most cost-effective method of overcoming the problem of heat storage.

5. Name and describe two methods that are proving practical for producing electrical power from sunlight.

6. Discuss why storage of solar-produced power is not an important issue, at least for the next 15 years.

7. Describe the main features of a system by which solar energy might be used to provide fuel for vehicles.

8. List four sources of energy that are considered indirect forms of solar energy.

9. Discuss the outlook for each energy source listed in objective 8 in terms of its potential to provide cost-effective power with acceptable environmental impacts.

10. Describe what is meant by geothermal energy, how it may be used, and its potential for further development.

11. Describe how tides and waves may be harnessed for the production of power and discuss the potential for their further development in the United States.

12. Contrast the de facto energy policy of the United States with a policy for achieving sustainable energy supplies.

502

Conventional wisdom among most government leaders, many energy experts, and the public at large is that we are stuck with dependence on fossil fuels and nuclear power. According to this "wisdom," our energy needs are such that we must continue to exploit increasing amounts of fossil fuels and develop more nuclear power, and tolerate the consequences. The consequences, as we have seen, are considerable: land devastated by strip-mining, coastal areas devastated by oil spills, polluted air, acid rain, global warming, military confrontations, and nuclear accidents. Further, this wisdom maintains that solar energy is a pie-in-the-sky dream that has little practicality in the real world.

Yet, in the past two decades, solar technologies have been making quiet advances. The point has been reached where in many situations sunlight can provide needed heat and electrical power at costs that are nearly comparable to those of traditional sources. When the hidden costs of traditional energy sources are considered—the costs of those consequences listed above, which are not included in our fuel bills—solar energy is much cheaper than traditional fuels. Best of all, it is sustainable. Over the next 30 years, the world could phase into solar-based energy supplies and attain the second principle of sustainability. The major obstacles in the way of this transition are making people aware of the opportunity and benefits of solar energy and overcoming the inertia of political leaders and special groups with vested interests in staying with traditional energy sources regardless of the consequences.

The objective of this chapter is to provide an overview of the potential of solar energy and certain other renewable energy sources. At the end we shall discuss an energy policy that can lead us toward adopting these sustainable energy resources.

Solar Energy

Solar energy originates with thermonuclear fusion reactions occurring in the sun. The solar energy reaching the earth is radiant energy ranging from ultraviolet light, which is largely screened out by ozone in the stratosphere, through visible light, to infrared light (heat). The most energy is in the visible light part of the spectrum (Fig. 23–1).

In Chapter 3, we learned that light, through photosynthesis, is the energy source for all major ecosystems. This use of solar energy, we learned, is a major principle of sustainability because it is nonpolluting and nondepletable—using solar energy will in no way shorten the life span of the sun; hence solar energy is also referred to as renewable energy. Finally, we learned that only 0.2–0.5 percent of the solar energy reaching the earth is trapped by photosynthesis—that is to say, all ecosystems together are driven by only a tiny fraction of the solar energy reaching the earth. A similarly small percentage of the solar energy reaching the earth could *sustainably* supply all the energy needs of human societies without altering the biosphere in any way.

If solar energy has these advantages, why are we not using it? The obstacles to using solar energy are in three areas. First its tremendous abundance comes from the fact that it falls more-or-less uniformly over the entire earth. Nowhere is it particularly intense. However, human activities usually require large amounts of energy at specific points, as in furnaces, engines, and motors. Second, solar energy is largely in the form of light, while most of our energy needs are for liquid fuels and electrical power. Third, solar energy is obviously not available during night hours or cloudy weather.

People with a vested interest in traditional energy sources frequently point to these obstacles and dismiss solar energy as impractical. However, all these obstacles mean is that problems exist in two areas: (1) economically collecting solar energy over large areas and converting it to forms that can be conveniently transported, stored, and used in existing equipment, and (2) designing facilities that can utilize diffuse sunlight. As indicated above, technology has advanced sufficiently that these problems have been largely overcome. In the following sections we shall look at the ways in which solar energy is beginning to be used as an economically practical as well as sustainable energy source.

SOLAR ENERGY FOR SPACE AND HOT WATER HEATING

About 25 percent of the total U.S. energy budget is used for heating buildings in cold weather and for

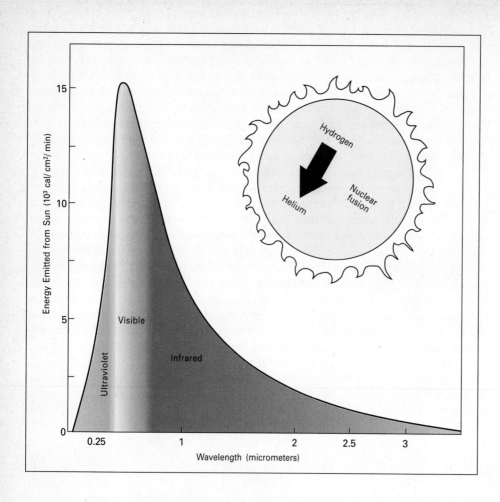

FIGURE 23–1
The solar energy spectrum. The greatest output of solar energy is in the visible light part of the spectrum.

providing hot water for washing. The temperatures involved here are modest, 68–72°F (20°–22°C) for space heating and 120–140°F (50°–60°C) for hot water. Using a gas or oil flame with a temperature of over 2000°F (1000°C) to provide this *low-temperature heat* is a thermodynamically inefficient use of energy. Using electrical power for this purpose is even more wasteful, like using a crane to lift a pillow. Solar advocates point out that we have been so conditioned to thinking of energy in terms of furnaces, heaters, and engines that we overlook the fact that solar energy is ideally suited for providing low-temperature heat because sunlight falling on any black surface is readily absorbed and converted to heat in the desired temperature range. You have probably experienced this in trying to walk barefoot on an asphalt pavement in the sun.

Therefore, complex collection or conversion equipment is not required when solar energy is to be used for low-temperature heat. A *flat-plate collector* will suffice. There are countless variations of **flat-plate collectors,** but all are basically composed of a black surface covered by a clear plastic or glass "window." The black surface absorbs sunlight and converts it to heat, and the window prevents the heat from escaping (Fig. 23–2); recall the greenhouse effect

described in Chapter 16. Air is heated by passing it between the window and the black surface; water is heated by passing it through tubes embedded in the surface. Thus, there is minimal cost in collection and conversion of solar energy to heat.

Beyond the collector, however, solar heating systems may be *active* or *passive* and may or may not include a means of heat storage. An **active solar heating system** is one that uses either pumps or blowers to circulate the air or water from the collector to the desired location. A **passive solar heating system** relies on natural convection currents (created when hot air or water rises) to move the air or water.

A diagram of an active system with storage is shown in Figure 23–3. Note the complexity of the system: pumps, blower, valves, and a great deal of piping. This complexity emphasizes the basic problem with active systems; while they work, all the plumbing is costly to install and troublesome to maintain. Also, if the energy required for manufacturing the components is considered, the net energy yield of such systems, over the lifetime of the system, is low at best and may be negative. Consequently, such active systems, which some people installed in the early 1970s, contributed to giving solar heating the undeserved reputation of being impractical. How-

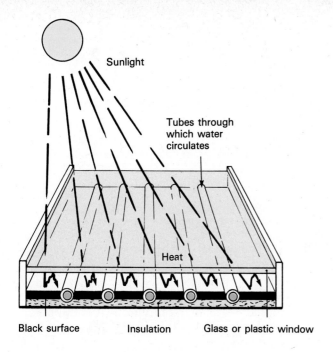

Sunlight

Tubes through
which water
circulates

Heat

Black surface   Insulation   Glass or plastic window

FIGURE 23–2
The principle of a flat-plate solar collector. Sunlight is converted to heat as it is absorbed by a black surface. A clear glass or plastic window over the surface allows the sunlight to enter but traps the heat. Air or water is heated by passing it over and through tubes embedded in the black surface.

ever, the problem is not with the concept of using solar energy but with the failure to follow what engineers call the KISS principle, which stands for *Keep It Simple, Stupid!*

Keeping it simple, in this case, means deleting the pumps, blower, valves, and most of the piping and going to a passive system in which convection currents do most of the work. Passive systems are relatively inexpensive and maintenance-free. Innumerable plans are available for passive solar heating of buildings, but the basic concept is that shown in Figure 23–4. With large, sun-facing windows, the building itself acts as the collector. In winter, sunlight beams through the window, heating the interior; at night, insulated drapes or shades are pulled to trap the heat inside. To avoid excessive heat load in the summer, an awning or overhang can shield the window from the high summer sun.

Early solar designs—even passive ones—included heat storage. A widely promoted idea involved circulating air through a "reservoir" of rocks, which readily absorb excess heat and then release it as the temperature drops (Fig. 23–5). However, in terms of the amount of heat stored and later used, the storage reservoir proved to be the most expensive component of the installation and contributed to the "impracticality" of solar heating.

The key to cost-effective storage has been found to be *improved insulation* of the space being heated.

FIGURE 23–3
An active solar heating system with heat storage. Such systems have generally proved to be cost-ineffective and fraught with maintenance problems. Note the amount of piping plus two pumps and a blower. (From B. Anderson, with M. Riordan, *The Solar Home Book*, Harrisville, N.H.: Brick House Publishing Co., 1976, p. 118.)

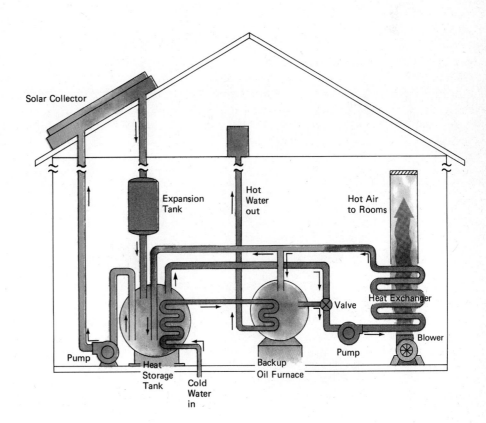

Solar Collector

Expansion Tank

Hot Water out

Hot Air to Rooms

Heat Exchanger

Valve

Blower

Pump

Pump

Heat Storage Tank

Cold Water in

Backup Oil Furnace

(a)
December 21
Insulation

N

(b)
June 21

FIGURE 23–4

Passive solar heating. In contrast to expensive and complex active solar systems, solar heating may be achieved by suitable architecture and orientation of the home at little or no additional cost. (a) The fundamental feature is large, sun-facing windows that permit sunlight to enter during winter months. Insulating drapes or shades are drawn to hold in the heat when the sun is not shining. (b) Suitable overhangs, awnings, or deciduous plantings will prevent excessive heating in the summer. (From B. Anderson, with M. Riordan, *The Solar Home Book*, Harrisville, N.H.: Brick House Publishing Co., 1976, p. 87.)

With good insulation plus thermally efficient doors and windows, heat that enters the building in the form of sunlight is stored in the air, furnishings, and interior walls of the building. There is no requirement for additional heat-storage facilities. (It is important to note that whatever kind of heating system is being used—solar or fossil fuel—improved insulation will increase its efficiency. Heating a poorly insulated building is like trying to keep a leaky bucket full of water. Good insulation also serves to keep a building

FIGURE 23–5

Passive solar heating may include heat storage. Rocks are the preferred heat-storage material because they readily absorb and give off heat, and they are inexpensive.

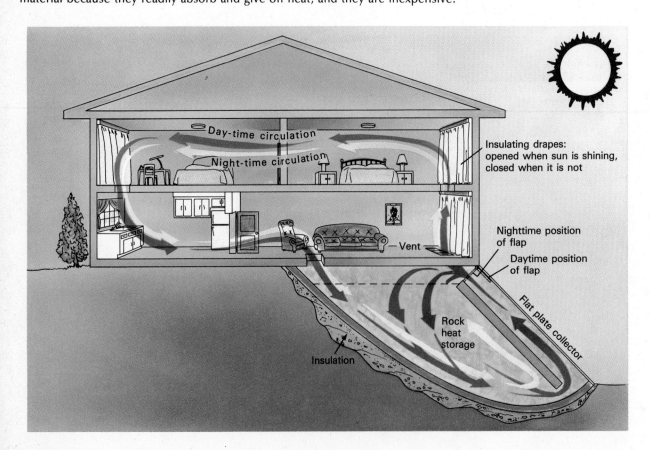

Day-time circulation

Night-time circulation

Insulating drapes: opened when sun is shining, closed when it is not

Vent

Nighttime position of flap

Daytime position of flap

Flat plate collector

Rock heat storage

Insulation

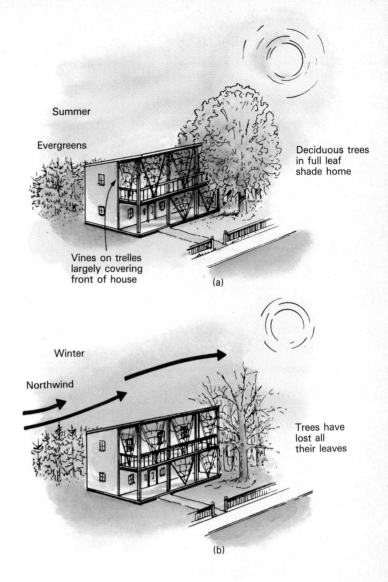

FIGURE 23–6
Landscaping may be an important adjunct to solar heating and cooling. (a) In summer, the house may be shaded with deciduous trees or vines. (b) In winter, leaves drop and the bare trees allow the house to benefit from sunlight. Evergreen trees on the opposite side protect and provide insulation from cold winds.

cool in summer, reducing the need for air conditioning.)

Along with improved insulation, appropriate landscaping can contribute to the heating and cooling efficiency of both solar and nonsolar designs. In particular, deciduous trees or vines on the sunny side of a building will block much of the excessive summer heat while letting the desired winter heat pass through (Fig. 23–6).

Another common criticism of solar heating, active or passive, is that a backup heating system is still required for periods of inclement weather. Again, good insulation is the answer to this criticism. People with well-insulated solar homes find that they have very little need for backup heating even in cold climates. When backup is needed, a small wood stove or gas heater suffices. In any case, the criticism concerning the need for backup heating misses the point because the objective of solar heating is to reduce our dependency on traditional fuels. If solar heating and improved insulation reduce the demand for conventional fuels by a mere 20 percent, this is still an ongoing yearly savings of 20 percent of both the fuel and its costs.

Savings can be much greater than 20 percent, however. The Center for Renewable Resources holds that in almost any climate, a well-designed passive solar home can reduce energy bills by 75 percent with an added construction cost of only 5 to 10 percent. What about existing homes? Nearly every homeowner can profit by improving insulation, and in innumerable situations, passive solar features can be retrofitted into existing homes (Fig. 23–7). Such additions, especially if owner-installed, may very quickly pay for themselves.

It is disappointing that, more than 20 years after recognizing the energy crisis and the virtues of solar energy, we are still building homes in traditional

(a)

FIGURE 23-7
Many homeowners could save on fuel bills by adding homemade solar collectors as shown here. (Photograph by BJN.)

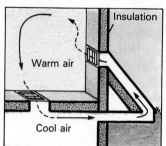

Insulation

Warm air

Cool air

(b)

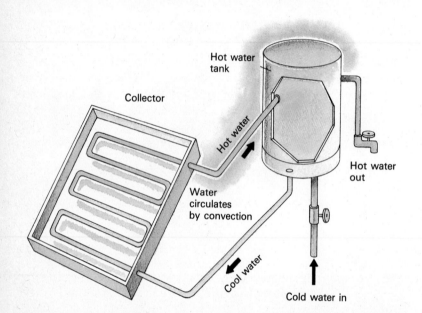

Hot water tank

Collector

Hot water

Hot water out

Water circulates by convection

Cool water

Cold water in

FIGURE 23-8
(a) Solar water heater. In nonfreezing climates, simple water convection systems may suffice. Where freezing occurs, an antifreeze fluid is circulated. (b) Solar panels supply 82 percent of the water heating required by this laundromat in Los Angeles. (Tom McHugh/Photo Researchers.)

(a)

(b)

LaunderLand
COIN-OP LAUNDRY

ways with traditional furnaces and ignoring the potential benefits of the sun. Insulation is somewhat improved, but it is often poorly installed. The Center for Renewable Resources concludes, "The chief barrier to more wide-spread use of passive solar designs is ignorance. Many builders and consumers are unaware of the potential benefits of passive solar design. . . . [Further, these benefits] are often ignored by policy makers and have received meager government support." One factor helping to maintain this "state of ignorance" has been intensive advertising campaigns by utility and oil companies purporting that solar energy is not practical or cost-effective at present. It should not escape notice that these companies make a profit on each unit of fossil-fuel energy regardless of its cost, where it comes from, or its future availability.

Much of the same holds true for solar water heating. About one-third to one-half of the total energy consumed in an average household is for heating water. In nonfreezing climates, passive systems suffice (Fig. 23–8). In climates where the temperature drops below freezing in the winter, the circulating fluid can be antifreeze. A solar water heater may not provide the full heat desired, requiring traditional backup heating. However, heating the water only halfway with solar energy will result in a 50 percent savings on fuel bills.

On a per capita basis, Cyprus is the world's largest solar energy user; 90 percent of the homes and a large portion of the hotels and apartment buildings have solar water heaters. In Israel, solar energy provides 65 percent of domestic hot water. In the United States, there are an estimated 800 000 solar hot water systems operating, but this represents only about one-half of 1 percent of the total.

In conclusion, there is tremendous opportunity to cost-effectively apply solar energy and improved insulation to space and water heating needs. The amounts of traditional fuels needed for these purposes could thereby be greatly reduced. In the United States in particular partial conversion to solar energy would make a large amount of natural gas available for fueling vehicles as noted in Chapter 21.

SOLAR PRODUCTION OF ELECTRICITY

Solar energy can also be used to produce electrical power, thus providing an alternative to coal and nuclear power. Currently, two methods are proving to be economically viable: *photovoltaic cells* and *solar trough collectors*.

Photovoltaic Cells

Photovoltaic cells, commonly called **solar cells,** were developed in the 1950s to provide power for space satellites. Their deceptively simple appearance belies a very sophisticated materials science and technology (Fig. 23–9). Each cell consists of two very thin layers of material. The lower layer has atoms with single electrons in the outer orbital; such electrons are easily lost. The upper layer has atoms lacking one electron from their outer orbital; such materials readily gain electrons. The kinetic energy of light striking this "sandwich" dislodges electrons from the lower layer, and they "fall" into the upper layer, creating an elec-

FIGURE 23–9
The thin wafer of material with wires attached is a photovoltaic cell. Converting light to electrical energy, this cell provides enough energy to run the small electric motor needed to turn these fan blades. (Photograph by BJN.)

tric potential between the two layers. This potential provides the electric current through the rest of the circuit, which connects the upper side through a motor or other electrical device back to the lower side. Thus, with no moving parts, solar cells convert light energy directly to electrical power. Since there are no moving parts, solar cells do not wear out, but their current life span is in the range of 20 years because of deterioration due to exposure to the weather.

So far, the largest solar cell that can be mass produced is about 2 inches (5 cm) in diameter. Such a cell in full sunlight will provide about the same power output as a standard flashlight battery. However, cells can be connected together to obtain any amount of power (Fig. 23–10). Nor do they need direct sunlight. They continue to produce power as light fades, although output falls off with input, as should be expected by the laws of thermodynamics. Current solar cells achieve a conversion efficiency of light energy to electrical energy of between 10 and 20 percent.

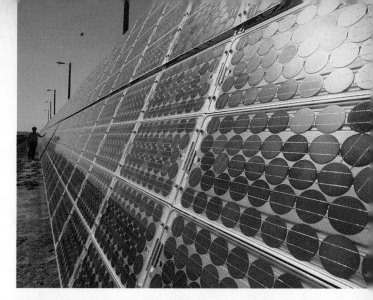

FIGURE 23–10
Photovoltaic cells can be wired together to obtain any amount of power desired. The array seen here is used to power an irrigation system on a farm in Nebraska. No backup or storage of power is needed in this situation because the plant's need for water corresponds to light intensity which causes evapotranspiration. (David Copp/ Woodfin Camp & Associates.)

ETHICS
Transfer of Energy Technology to the Developing World

The nations of the industrialized North have achieved their level of development using energy technologies based largely on fossil fuels (to a lesser extent nuclear). Their development took place during a time when fossil fuels were inexpensive; only as the expense of those fuels has risen have the nations of the North begun to get serious about technologies to make energy use more efficient. As we saw in Chapter 21, these nations still have a long way to go in conserving energy; just putting into place efficiency technologies already developed could save three-fourths of energy expenditures in the United States.

These same traditional fossil fuel–based technologies have been adopted by the developing nations of the South, except that they are lagging behind in efficiency. On the whole, the industrialized North is three times more efficient in energy use than the developing South. As we have seen in chapter 21, how-

ever, the fossil fuel–based energy pattern of the twentieth century will have to be largely replaced with renewable energy systems in the next century. If the rich nations are willing to play a significant role in helping the poor nations to develop, they should not promote the wasteful development path the North has taken. Instead, there is the opportunity to engage in some "leap-frogging" technology transfer. Thus, the North could put its development dollars into such technologies as electrification of rural areas via photovoltaics and efficient public transport systems for the cities.

The solar route is especially attractive for many of the climates in the Third World, where an estimated 2 billion people lack electricity. A model program is the work of Richard Hansen, whose nonprofit development group Enersol helped establish the Asociación de Desarrollo de Energía Solar (ADES) in the Dominican Republic. ADES has es-

tablished a revolving low-interest loan fund and the necessary equipment to help rural people electrify their homes. Rooftop photovoltaic units the size of two pizza boxes produce enough power for five electric lights, a radio, and a television set. Similar programs have been started in Sri Lanka and Zimbabwe.

It is entirely possible that if this strategy of development aid is followed, the nations of the South may end up pointing the way to a sustainable energy future for the rest of the world. The Rio Declaration signed at the Global Summit in June 1992 commits the industrial countries to greater levels of development aid to the poor countries and also stipulates that the development should be sustainable. As the details of this commitment get fleshed out, it will be interesting to see if solar energy receives the attention it deserves. If it does, there is much the donor countries could learn from the results. What do you think?

We have seen earlier in the text that principles of sustainability can be derived from an understanding of how natural ecosystems function. Throughout the text we have been examining the ways in which our human system can be brought into closer conformity with these principles as we look to the future. In a real sense, we are simply learning from nature as we adhere to basic ecological principles of sustainability. There are other ways we can learn from nature, and one example is seen in a new design for solar cells.

Late in 1991, scientists Michael Grätzel and Brian O'Regan published a paper in the prestigious British journal *Nature* describing a solar cell that functions somewhat like green plant chlorophyll. In fact, the new cell is even more efficient than chlorophyll at converting solar energy to electrical energy, achieving an efficiency of 12 percent. Unlike most current solar cells, which are based on silicon semiconductor materials, the Grätzel-O'Regan cell uses titanium oxide. The unique feature, however, is inclusion of a light-harvesting dye, which is a complex organic carbon molecule built around rubidium atoms. Chlorophyll is a complex organic carbon molecule built around magnesium atoms. Both molecules (called "antenna" molecules) have the capacity to become "excited" by photons of light; when they do, they emit electrons. In the new solar cell, the electrons are conducted to the titanium oxide and then transmitted as a current to the negative electrode of the solar cell. In green plants, electrons are conducted to other molecules and converted into chemical energy.

The new cells are more efficient than standard silicon-based solar cells, are easy to fabricate, and will cost much less to manufacture. The major uncertainty about the cells is their durability. However, observers note that the basic principle of the new cell is sound—no surprise, since it is based on a system that has been around for many millions of years! The invention is considered to be a significant advance in photovoltaics, with great potential for helping that technology become competitive with traditional energy sources.

This low efficiency is in one respect immaterial since the primary energy source, sunlight, is free. However, the cost of producing the number of cells needed for a given amount of power is not immaterial. As with any other new technology, the first cells produced were notoriously expensive. In the early 1970s photovoltaic power was calculated to cost about $30 a kilowatt-hour. With improvements in production techniques, costs have come down dramatically to about 30 cents per kilowatt-hour in 1991. This cost is still four to six times the cost of power from traditional sources. Yet even at this cost photovoltaic cells are economically practical for many uses. You are undoubtedly familiar with the widespread use of solar cells replacing batteries in pocket calculators and other such devices. Panels of solar cells are the most economical way of providing power at points that are far from utility lines: rural homes, irrigation pumps, traffic signals, radio transmitters, lighthouses, and offshore oil-drilling platforms, to name just a few. Rural electrification projects based on photovoltaic cells are beginning to spread throughout the Third World.

As costs have come down, production and sales of photovoltaic cells have grown steadily, from negligible levels in the early 1970s to 50 megawatts' (1 megawatt = 1 million watts) worth of cells in 1990, and sales are projected to double every 5 years. Perhaps this is a conservative estimate, because recent breakthroughs promise to cut costs by 50 to 80 percent again.

Thus, we seem to be close to the point where photovoltaic power plants may be producing power at the same cost as traditional power plants. The Pacific Gas and Electric Company, in conjunction with ARCO Solar, has a demonstration photovoltaic power plant near Bakersfield, California, which has a capacity of 6.5 megawatts, enough to power 2400 homes (Fig. 23–11a). As noted above, people in rural areas far removed from utility lines are already finding it more cost-effective to install photovoltaic panels on their roofs than to run long power lines. As costs of solar panels decline, this practice will undoubtedly become increasingly common.

Many people assume that storage batteries are a necessary part of solar electrical power. However, where power from conventional utilities is available, batteries are neither cost-effective nor necessary, at least for the time being. The reason is that about 70 percent of the overall demand for electrical power is during daytime hours. Thus, there are great savings to be had by using solar panels just for daytime needs and continuing to rely on conventional sources for nighttime demands. In particular, the demand for air

(a)

(b)

FIGURE 23–11
(a) World's first photovoltaic (solar cell) power plant, located near Bakersfield, California. The array of 220 34-foot (11-m) panels produces 6.5 megawatts at peak, enough for 2400 homes. (T. J. Florian/Rainbow.) (b) The Sunraycer built by General Motors is an experimental car powered by solar energy. Limitations in speed and maximum range of solar cars prevent them at this time from offering a viable alternative to internal combustion engines. (Sygma.)

conditioning, which consumes more power than any other single item, is well matched to energy from photovoltaic cells. In addition, solar-powered air conditioners could operate independently from the rest of the electrical system, thus avoiding the costs of interconnecting systems. The Japanese are currently developing solar-powered air conditioners to be marketed by 1995.

Experimentation with vehicles run on photovoltaic cells is under way. However, this program seems to have a long way to go to reach a practical solar commuting car. The fundamental problem is that the limited surface area of a vehicle—even ultraefficient designs—does not permit sufficient light absorption to provide more than very limited speeds or distances (Fig. 23–11b).

Solar Trough Collectors

Another method of producing electrical power from sunlight that is proving economically practical is the solar trough collector system (Fig. 23–12). Sunlight hitting the collector is reflected onto a pipe running down the center of the trough and heats oil or other fluid circulating through the pipe. The heated fluid is then used to boil water, producing steam to run a conventional turbogenerator. This solar trough concept was pioneered by an inventor named Charles Abbott in the 1930s (Fig. 23–13). Sixty years later it is finally coming into its own.

Luz International has built several solar trough facilities in southern California with a combined capacity of 350 megawatts, about one-third the capacity of a large nuclear power plant. The most recent Luz facility, which converts a remarkable 22 percent of incoming sunlight to electrical power, is producing power at a cost of 8 cents per kilowatt-hour, barely more than the cost at nuclear or coal-fired facilities. Indeed, one can make a sound argument that solar-produced power is already considerably cheaper because the hidden costs of air pollution, strip-mining, and nuclear waste disposal are not included in the

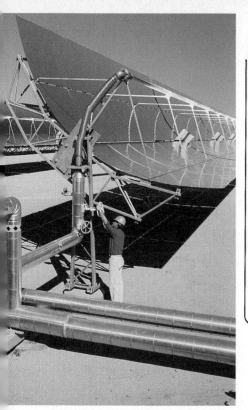

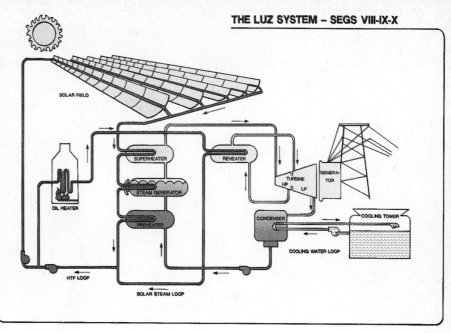

FIGURE 23–12
Several solar trough facilities are now in operation in southern California. (a) The curved reflector focuses sunlight on and heats oil in the pipe. (b) The heated oil is used to boil water and generate steam to drive a conventional turbogenerator. (Courtesy of Luz International.)

FIGURE 23–13
The solar trough concept of harnessing solar energy was pioneered by Charles Abbott in the 1930s. Shown here is Abbott, on right, with his ½-horsepower "solar boiler." This machine was used to provide power for the National Broadcasting Company to broadcast news of this event to the United States and Canada in 1936. (Smithsonian Institution Archives. Record Unit 7005, Box 189. Charles G. Abbott Papers, 1889–1973 and record of the Smithsonian Astrophysical Observatory.)

FIGURE 23–14
Georgetown University's Intercultural Center (Washington, D.C.) supports an array of 4400 photovoltaic modules providing a power output of 300 kilowatts under full sun. Such rooftop arrays may become commonplace in the future as production costs of photovoltaic cells are reduced. (U.S. Department of Energy.)

cost of power from traditional sources. Likewise, criticism that solar power requires an exorbitant amount of land to "harvest" the sunlight rings hollow when the thousands of acres that are devastated by strip-mining of coal each year or that might be made uninhabitable by a nuclear accident are considered. Also, photovoltaic systems can utilize existing rooftops so that additional land is not required (Fig. 23–14).

Because about two-thirds of the electrical demand occurs in daytime hours when industries and offices are in operation, the problem of sunlight not being available during the night is not as serious as might be supposed. For the next 15–25 years, all the solar power produced can be used directly for supplying the extra daytime demand. The amount of traditional fuels consumed in existing power plants could thereby be reduced on the order of 70 percent while still allowing them to provide the nighttime load. In the long run, we might envision the night-time load being carried by forms of indirect solar energy such as wind power or water power (discussed below).

Finally, when the need does arise, excess power generated at one time of the day may be stored by using it to pump water from a low-elevation reservoir to a reservoir at a higher elevation. When additional power is needed, the water is allowed to run back down through turbogenerators. In a similar manner, extra power may be used to pump air into underground caverns and allow the compressed air to drive turbines when needed.

Two additional methods of producing electrical power from sunlight are undergoing development. The first is what is commonly called a "power tower". An array of sun-tracking mirrors is used to focus the sunlight falling on several acres of land onto a boiler mounted on a tower in the center. The intense heat generates steam to drive a conventional turbogenerator. Southern California Edison, developers of the research facility shown in Figure 23–15, feels that this system may generate power even more economically than the solar trough system.

The second is solar ponds. An artificial pond is partially filled with brine (very salty water), and fresh water is placed over the brine. Because it is much denser than fresh water, the brine remains on the bottom, and little or no mixing occurs. Sunlight passes through the fresh water but is absorbed and converted to heat in the brine. The fresh water then acts as an insulating blanket and holds the heat in. The hot brine solution can be circulated through buildings for heating, or it can be converted to electrical power by vaporizing fluids with low boiling points and using the vapors to drive low-pressure turbogenerators. Since the pond also acts as a very effective heat-storage unit, it will supply power continuously. Israel has pioneered the development of solar ponds.

We observed in Figure 21–5 that about 75 percent of our electrical power is currently generated by coal-burning and nuclear power plants. Thus, development of solar electrical power will gradually reduce the need for coal and nuclear power. We can

FIGURE 23–15
A power tower for producing electrical power from sunlight. Sun-tracking mirrors are used to focus a broad area of sunlight onto a boiler mounted on the tower in the center. The steam produced is used to drive a conventional turbogenerator. This facility is in southern California. (Courtesy of Southern California Edison Co.)

also note that development of solar space and hot water heating will greatly reduce our demand for natural gas, which currently supplies most of these needs. However, we have yet to address the area that is our greatest concern: our dependency on crude oil as the source of fuels for vehicles and other prime movers. Solar energy has potential here as well.

SOLAR ENERGY TO PRODUCE FUEL FOR VEHICLES AND AS A SUBSTITUTE FOR NATURAL GAS

In considering energy sources to sustain our transportation system, we should first remember that conservation—increasing vehicle mileage—remains the option with the greatest potential for reducing demand for crude oil in the short term (10 years). Second, insofar as development of solar heating reduces the demand for natural gas in the residential and commercial sectors, this fuel may be transferred to the transportation sector (see p. 474). Third, electric vehicles might run on solar or any other source of energy used to generate the power. Battery technology remains a limiting factor for electric cars, however. Storage batteries are both very expensive and heavy relative to the amount of power stored. For example, General Motors' newest prototype electric car, which may be marketed within the next few years, carries 800 pounds (360 kg) of batteries to store the equivalent of only a little more than a gallon (4 L) of gasoline. Thus, even though the car is ultra-efficient, its range is still limited to about 100 miles (140 km) between rechargings. Perhaps a breakthrough in battery tech-

nology will improve the outlook for electric vehicles, but there are no indications for this occurring in the near future.

However, the potential also exists for using solar energy to produce hydrogen, which can be used as a fuel for existing vehicles. Hydrogen in pure form (H_2) is a clean-burning gas—the only waste product is water vapor—so pollution problems would be reduced as well:

$$2\,H_2 + O_2 \rightarrow 2\,H_2O + \text{energy}$$

The hurdle currently blocking the use of hydrogen as fuel involves the fact that there is essentially no free hydrogen on earth; it must be produced. Free hydrogen gas may be readily produced through the electrolysis of water—a process in which an electric current passed through water causes the water molecules to separate and re-form into hydrogen gas and oxygen gas (Fig. 23–16). However, our old friends, the laws of thermodynamics, demand that an energy equivalent greater than that gained by burning the hydrogen be put into the process. So far, limits of primary sources of energy have limited the production and use of large amounts of hydrogen.

An in-depth analysis by the World Resources Institute, "Solar Hydrogen: Moving Beyond Fossil Fuels," describes a way around the energy difficulty. The WRI's proposal is that we build arrays of solar trough or photovoltaic generating facilities in the deserts of the southwestern United States, where land is cheap and sunlight plentiful. The electrical power would be used to produce hydrogen gas from water

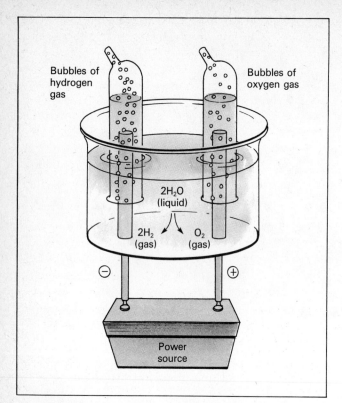

FIGURE 23–16
Solar production of hydrogen gas. An electric current passed through water causes the water molecules to break apart and form hydrogen gas and oxygen gas. The needed electrical power can be provided by solar energy. The gas can be stored, transported through pipelines, and burned as fuel in furnaces, stoves, and vehicles.

In the figure: Bubbles of hydrogen gas / Bubbles of oxygen gas / $2H_2O$ (liquid) / $2H_2$ (gas) / O_2 (gas) / Power source

by electrolysis. Conveniently, Texas is the hub of a network of natural gas pipelines that were originally constructed to transport natural gas from the Texas oil fields. Most of the pipelines throughout the nation are now underutilized because those fields have been largely depleted. The hydrogen gas produced in the Southwest could be transported throughout the nation by way of these pipelines.

Present vehicles could easily be adapted to run on hydrogen. The only changes needed would be to the fuel tank and the fuel intake system. Current service stations would dispense hydrogen gas into vehicle "fuel tanks," which would consist of materials that adsorb large amounts of hydrogen and release it only slowly upon being heated. This slow release is necessary to make the hydrogen-fueled vehicle safe from the hazard of an exploding fuel tank.

WRI's economic analysis projects that solar-produced hydrogen gas could be produced for a cost equivalent to gasoline costing $1.65 to $2.35 per gallon. This may not sound cheap, but it is less than the projected costs of synfuels (Chapter 21) and less than what a number of European nations and Japan are already paying. If the military costs of protecting our oil supplies in the Mideast are added in, this price is also less than what we are paying for imported oil. Most important of all, solar-produced hydrogen gas would be a sustainable energy supply. Germany is constructing what will be the world's first solar-hydrogen plant near Nüremberg. It will utilize photovoltaic cells.

Natural gas reserves promise to last considerably longer than crude oil reserves, and natural gas produces less pollution and carbon dioxide than do the other fossil fuels (Chapter 21). Therefore there is less urgency for finding a substitute for natural gas. However, when it is needed, hydrogen can be used as a substitute for that as well. Hydrogen can be distributed through existing gas mains and burned in existing gas appliances with relatively little modification of burners.

POWER FROM INDIRECT SOLAR ENERGY

Solar energy drives the water cycle, atmospheric circulation, and photosynthesis. Thus, water power, wind, and biomass are considered indirect forms of solar energy. Like direct solar energy, they may be used to provide substantial sources of energy. Indeed, humankind has been using these indirect forms of solar energy since ancient times.

Hydropower

Water power has been used for millennia by diverting water from natural streams or rivers over various kinds of paddle wheels or turbines. The power output of waterwheels is relatively meager, however, and so the trend for the last century or so has been to build high dams to create a substantial head of hydrostatic pressure (Fig. 23–17). The water under high pressure flows through the base of the dam and drives turbogenerators producing **hydroelectric power.** About 9.5 percent of the electrical power generated in the United States currently comes from hydroelectric dams, most of it from about 300 large dams concentrated in the Northwest and Southeast.

While hydroelectric power is basically a nonpolluting, renewable energy source, it still involves tremendous tradeoffs:

○ Dams have drowned out some of the most beautiful stretches of river in North America, as well as wildlife habitat, productive farmlands, forests, and areas of historic, archaeological, and geological value. Glen Canyon Dam [on the border between Arizona and Utah] drowned one of this world's most spectacular canyons. Tellico Dam in Tennes-

FIGURE 23–17
Hoover Dam. About 13.5 percent of the electrical power used in the United States comes from large hydroelectric dams such as this. Water flowing through the base of the dam drives turbines. (Lowell Georgia/Photo Researchers.)

see eliminated an ancient Cherokee village and the site of the oldest continuous habitation on the North American continent.*

○ The reservoir behind the Aswan High Dam in Egypt has caused the spread of a parasitic worm that causes a debilitating disease. The reservoir has also increased humidity over a widespread area, which is now causing the rapid deterioration of ancient monuments and artifacts that until now had stood virtually unchanged for many centuries.

○ Since water flow is regulated according to the need for power, dams play havoc downstream because water levels may go from near flood levels to virtual dryness and back to flood in a single day. Other ecological factors are also affected because sediments with nutrients settle in the reservoir and smaller amounts reach the river's mouth.

* B. Blackwelder, "Dams: A Change of Course," *National Parks*, 58 (July/Aug 1984), pp. 8–13.

Few sites conducive to large dams remain in the United States. There are already 65 000 U.S. dams; only 2 percent of the nation's rivers remain free-flowing, and federal surveys show that fish habitat is suffering in 68 percent of the nation's streams because of damming. Thus, proposals for more dams are embroiled in controversy over whether the projected benefits justify the ecological and sociological trade-offs. In 1991 the Environmental Protection Agency, after protracted disputes, finally denied permits for the proposed Two Forks Dam in Colorado on environmental grounds.

The James Bay project in northern Quebec is creating even more controversy. If completed—one-third has already been completed—the project will consist of 215 dams and dikes, as well as 9 river diversions, and will affect fish and wildlife throughout an area the size of New England, New York, and Pennsylvania. It will largely displace two tribes of Native Americans. The power produced will largely be exported to the United States.

Many less-developed countries have great potential for developing water power, but the tradeoffs noted above must be kept in mind.

Wind Power

Windmills also have been used throughout history. Until the 1940s, most farms in the United States used windmills for pumping water and generating small amounts of electricity. In the 1930s and 1940s, windmills fell into disuse as transmission lines brought abundant lower-cost power from central generating plants. However, with the energy crisis of the early 1970s, wind began to be reconsidered as a source of energy for generating electrical power on a large scale. A large number of different designs were tested, but what has proved most practical is the age-old concept of airplane-type propeller blades turning a generator geared to the shaft. (Wind driving a generator is more properly called a **wind turbine** rather than a "windmill.")

Economies of scale suggested that very large wind turbines should be most efficient. Thus in the 1970s the government funded the building of a number of huge wind turbines with blades as much as 300 feet (100 m) from tip to tip mounted on 200-foot (70-m) towers (Fig. 23–18). Such machines, in optimal wind conditions, could theoretically generate as much as 2.5 megawatts, enough to provide for about 2500 homes. Results from such machines have proved disappointing, however. The stress on materials is tremendous, and so the machines suffered from repeated breakdowns. The result is that development of gigantic wind turbines has been all but terminated.

On the other hand, more modestly sized wind

FIGURE 23–18
MOD 2 wind turbine in Washington State; blades, 300 feet from tip to tip; tower, 200 feet tall; capacity, 2.5 megawatts in winds 14–45 mph. The future of such wind turbines is in doubt, however, because severe stresses cause maintenance problems. (U.S. Department of Energy photograph.)

turbines, machines with blade diameter in the range of 50 feet (17 m) and capable of generating about 100 kilowatts, have proved practical. **Wind farms,** arrays of 50 to several thousand such machines, are now producing power in a number of locations around the world (Fig. 23–19). California, with 17 000 machines generating 1500 megawatts (supplanting the need for two nuclear power plants) is the world's largest producer of wind-generated power, but a number of European countries are not far behind. By the end of this decade, Britain, Denmark, the Netherlands, Italy, and Germany will have a combined wind-generating capacity of over 3000 megawatts. With gradual improvements in design and reliability, wind farms are now generating power for as little as 6 cents per kilowatt-hour, very competitive with traditional sources.

It is hardly surprising that this way of harnessing wind energy is growing rapidly. Electricity-producing wind turbines are now installed in 95 countries from the tropics to the Arctic, but still the potential has hardly been tapped. Most regions of the world have areas where winds are constant enough to make wind turbines practical. The American Wind Energy Association calculates that wind farms located throughout the Midwest could meet the electrical needs for the entire country, while the land beneath the turbines could still be used for farming. Wind farms in different locations connected to the already

FIGURE 23–19
Windfarm at Altamont Pass, an area east of San Francisco. Such arrays of wind turbines are proving to be an economically competitive way of generating electrical power. The land under the wind turbines may still be used for agriculture. (Tom McHugh/Photo Researchers.)

In addition to being a nonpolluting renewable energy source, photovoltaic, solar trough, and also wind generating facilities have another advantage: They can be installed quickly and added to the utility system in relatively small increments. To understand this advantage let's draw a comparison with a nuclear power plant.

A nuclear power plant can only be cost-effective in a large size—about 1000 megawatts capacity, enough to power a million average homes. Such a plant, even assuming public acceptance, requires in the order of 10 to 15 years from the time of the decision to build, to getting it into operation, and it costs in the order of 5 to 10 billion dollars. Therefore, a utility taking this route must borrow billions of dollars and keep that money tied up 10 to 15 years before a return on the investment is realized by selling power from the new plant. Then, if the 1000 megawatts of additional power is not needed when the new plant comes on line, you can see that the return on the investment may be meager indeed. In the late 1970s and early 1980s a number of utility companies went bankrupt or nearly so because they had huge amounts of money tied up in nuclear power plants and slack demand for the power. In other words, building large power plants (nuclear or coal) involves considerable financial risk on the part of the utility and consumers, as consumers are the ones who ultimately pay.

On the other hand, solar or wind facilities can be operational within a few months of the decision to build; they can be installed in small increments, and additional modules can be added as demand requires. The utility is not placed in the position of having to guess at what power demands will be in 10 or 15 years. Through solar or wind, a utility can add capacity essentially as it is needed, making relatively small investments at any given time and having those investments start paying back almost immediately. You can see that this involves much less financial risk both for the utility and for consumers.

existing electrical grid would provide backup for each other since the wind is invariably blowing somewhere. Wind farms would also provide a sustainable complement and backup to direct solar facilities.

Wind turbines are not entirely without problems. One can get tired of their appearance on the landscape and their continual whirring and whistling. Also, they are a significant hazard to birds. Location of wind farms on migratory routes could be disastrous for some avian populations.

Biomass Energy or Bioconversion

Biomass energy, or **bioconversion,** refers to the direct burning of any form of biomass—wood, waste paper, agricultural wastes, manure—or converting it to fuel (Fig. 23–20). Recall that certain microorganisms produce either alcohol or methane gas when they digest biomass in the absence of oxygen (Chapter 3). Effectively, using biomass energy means using the production from current photosynthesis as opposed to the ancient photosynthesis used in the burning of fossil fuels. Therefore, biomass energy is considered another form of indirect use of solar energy.

Where forests are ample relative to population, firewood can be a sustainable energy resource, and indeed this has been the case over much of human history. In the United States, wood stoves have enjoyed a tremendous resurgence in recent years—about 5 million homes rely entirely on wood for heating and another 20 million use it for partial heating, so much so that air pollution from wood stoves has become such a problem that some communities are finding it necessary to apply restrictions.

However, any thoughts of firewood being a large-scale energy source must be put in the context of maximum sustainable yields and preservation of the biodiversity of ecosystems discussed in earlier chapters. Too much of the Third World, where over a billion people still depend on firewood as their only source of fuel for cooking and warmth, is already suffering from the severe ecological effects of deforestation as well as the human deprivation caused by running out of wood (Chapter 9). Therefore, a blanket recommendation to burn wood because in theory it is renewable is environmentally irresponsible.

Similarly, proposals that trees or other crops be grown specifically for biomass energy—either for direct burning or for conversion to alcohol or methane—must be weighed against the impacts of soil ero-

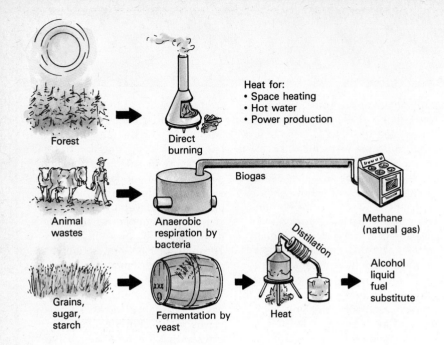

Heat for:
• Space heating
• Hot water
• Power production

Forest

Direct burning

Biogas

Animal wastes

Anaerobic respiration by bacteria

Methane (natural gas)

Distillation

Grains, sugar, starch

Fermentation by yeast

Heat

Alcohol liquid fuel substitute

FIGURE 23–20
Bioconversion. Biomass may be burned directly or converted to gas (methane) or liquid (alcohol) fuel.

sion, fertilizers, and pesticides that are generally entailed in such production. Also, energy crops will invariably compete with food crops because either may be grown on the same land.

Where energy can be produced as a byproduct of waste disposal, however, biomass energy may have some merit. For example, many sawmills and woodworking companies are now burning wood wastes, and a number of sugar refineries are burning the cane wastes to supply all or most of their power. The burning of municipal refuse, which is largely wastepaper, for power production is another example (see Chapter 20). However, the burning of wastes must be weighed against the advantages of recycling or composting the materials.

Use of alcohol as fuel is being vigorously promoted in grain-growing regions of the United States. More than just energy is involved here, however. Alcohol is produced by fermenting grains or other starchy or sugary food products. Therefore it is being advocated largely to provide another market for grain, reduce surpluses, and thereby increase prices and generally improve the economy of the grain-growing region.

Where grain surpluses do not exist—and this includes most of the world—alcohol as a fuel is a dubious goal at best. For example, Brazil has promoted the large-scale production of alcohol from sugar cane. Consequently, Brazil's sugarcane production is at record levels, while food crops are down by 10 to 15 percent, despite widespread malnutrition and a rapidly growing population.

Another problem with alcohol is pollution. Alcohol is promoted as clean-burning, and it is. However, producing the alcohol generates much pollution because inexpensive, dirty-burning fuels such as soft coal are used for its distillation. The net energy yield of alcohol is modest because an amount of fuel equivalent to at least one-half gallon of alcohol is used for every gallon produced.

One biomass method that creates a valuable synergism between recycling and energy production is the anaerobic digestion of sewage sludge and manure. Recall from Chapter 13 that such digestion yields biogas, which is two-thirds methane, plus a nutrient-rich compost that is a good organic fertilizer. The utility of this concept is demonstrated by Richard Waybright, the operator of the Mason-Dixon Dairy farm located near Gettysburg, Pennsylvania. Manure from 2000 cows fed in a barn drops through a grated floor and slowly flows by gravity into anaerobic digesters (Fig. 23–21). The biogas produced (purification of the methane has proved unnecessary) fuels engines that drive generators supplying all the electrical power for the dairy and considerable excess, which is sold to the utility company. Waste heat from the engines is used to heat the digesters and also buildings. The nutrient-rich sludge remaining after digestion is complete is recycled back to the land in order to maintain the fertility of the fields growing forage for the cows. Between the savings on energy he does not have to buy and the income he receives from the energy he sells, Waybright estimates that he makes close to a hundred thousand dollars a year on

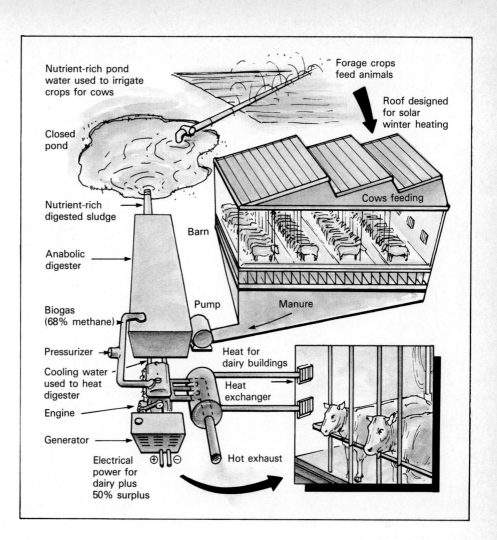

FIGURE 23–21
Dairy farm operated on cow manure. The total power needs for the Mason-Dixon Dairy located near Gettysburg, Pennsylvania, are obtained as a byproduct of cow manure, and nutrients are recycled in the process. Excess power, nearly half of what is produced, is sold to the local utility.

Labels in figure: Forage crops feed animals; Roof designed for solar winter heating; Nutrient-rich pond water used to irrigate crops for cows; Closed pond; Cows feeding; Barn; Nutrient-rich digested sludge; Anabolic digester; Pump; Manure; Biogas (68% methane); Pressurizer; Heat for dairy buildings; Cooling water used to heat digester; Heat exchanger; Engine; Generator; Hot exhaust; Electrical power for dairy plus 50% surplus

his energy system, plus there are additional savings on fertilizer. He estimates that the system cost about 80 cents per watt to install, which compares favorably with the $3 per watt for coal and $5 per watt for nuclear power.

If all dairy farms in the United States followed this example, we could be getting nearly as much electrical power from cows as we are currently getting from nuclear power at about one-fifth of the cost, plus the system would be sustainable and without adverse environmental consequences. Sewage sludges could be processed and recycled in a similar manner.

In China, millions of small farmers maintain a simple digester in the form of a sealed pit into which they put agricultural wastes. The biogas produced is used as a source of gas for cooking. This concept could be expanded throughout the Third World to provide an alternative to fuelwood.

Ocean Thermal Energy Conversion

Over much of the world's oceans, a thermal gradient of about 36 F° (20 C°) exists between surface water

heated by the sun and colder deep water. The concept of harnessing this temperature difference to produce power is known as **ocean thermal energy conversion** or **OTEC.** An OTEC power plant would be built on a barge that could go anywhere in the oceans. It would use the warm surface water to heat and vaporize a liquid having a low boiling point, such as ammonia. The increased pressure of the vaporized liquid would drive turbogenerators. The ammonia vapor leaving the turbines would then be condensed by cold water pumped up from as much as 300 feet (100 m) deep and returned to the start of the cycle. The electrical power from the generator could be used to produce hydrogen gas, which could be shipped to shore. Alternatively, an energy-intensive industry could be located on factory ships that would moor alongside the OTEC plant.

A few small demonstration OTEC plants have been built and tested. The problem is that the relatively small temperature difference between surface and deep water precludes a conversion efficiency of anything greater than 2–3 percent. By itself, this low efficiency is immaterial since the primary energy source—the temperature difference over most of the ocean—is free. However, coupled with the high capital costs of the plant and persistent maintenance problems—fouling of pipes and pumps with marine organisms—the low efficiency leads to net energy yields that are meager if even positive and certainly not economically competitive. However, this poor yield is a reflection of the current state of the OTEC technology. Some individuals and companies are maintaining an interest in the concept.

Additional Renewable Energy Options

Three additional renewable energy options have strong support from certain sectors and deserve our understanding and consideration: geothermal power, tidal power, and wave power.

GEOTHERMAL ENERGY

The molten core of the earth (see p. 111) offers, in theory, an unlimited, sustainable source of heat. The concept of **geothermal energy** is to use this heat for space heating or to boil water to drive conventional steam turbogenerators. The most convenient situations occur in volcanic regions, where hot rock is relatively close to the surface and where natural groundwater comes in contact with the hot rock. Such heated water may come to the surface through natural steam vents like those seen in Yellowstone National Park.

Where natural steam exists, essentially all that is required is to drill holes in the hot rock–groundwater structure and allow the rising steam to drive turbogenerators. The world's largest such facility is at a location known as The Geysers about 70 miles (110 km) north of San Francisco (Fig. 23–22). From the early 1970s electrical output from The Geysers was steadily expanded. In 1988 it reached about 2000 megawatts (equivalent to two large nuclear power plants), and it was projected that by the year 2000 the facility would be producing 3000 megawatts.

In fact, output from The Geysers has been steadily declining since 1988; in 1991 it was producing just 1500 megawatts, and it is now projected that by the end of the decade this number may be down to half of the peak 1988 level. The problem is that while the source of heat may be unlimited, the amount of groundwater isn't. What is occurring at The Geysers is that the "pot" is boiling dry. Attempts are being made to inject additional water, but so far results are mixed.

Similar geothermal facilities exist in the Philippines, Mexico, Italy, Iceland, and Japan, generating a total of about 3000 megawatts. With the experience at The Geysers, however, further expansion has been curtailed, and the outlook for the future of geothermal power is much less optimistic than it was a few years ago.

Additional potential for geothermal energy may lie in development of **hot dry rock** structures.

FIGURE 23–22
One of the 11 geothermal units operated by the Pacific Gas and Electric Company at The Geysers in Sonoma and Lake counties, California. The field was generating 2000 megawatts in 1988, but power output is now falling off because the field is running out of water. (Photograph courtesy of Pacific Gas and Electric Company.)

Whereas natural steam vents occur in only a few regions, hot dry rock can, theoretically, be found anywhere by drilling deeply enough. Two parallel holes can be drilled into the hot rock and fractures created between the two holes. Water forced down one hole will be heated as it seeps through the fractures and will come up the other hole as steam to drive turbines. Unfortunately, the technological problems of drilling deeply enough—3–4 miles (5–6 km) into hot rock and then causing fractures to occur as desired have proved difficult to overcome. While work is proceeding, the ultimate success of these endeavors is still speculative.

Geothermal power presents pollution problems as well. Hot steam and water brought to the surface are frequently heavily laced with salts and other contaminants, particularly sulfur compounds leached from minerals in the earth. These contaminants are highly corrosive to turbines and other equipment, and they cause air pollution if the steam is released into the atmosphere. Sulfur dioxide pollution from a geothermal plant may be equivalent to that from a plant burning high-sulfur coal. Hot brines released into steams or rivers may be ecologically disastrous.

TIDAL POWER

A great deal of energy is inherent in the twice-daily rise and fall of the tides, and many imaginative schemes have been proposed for capturing this eternal, pollution-free source of energy. The most straightforward is to build a dam across the mouth of a bay and mount turbines in the structure. The incoming tide flowing through the turbines generates power. As the tide shifts, the blades may be reversed so the outflowing water continues to generate power. Two large tidal plants are presently in operation— one in France and the other in Russia. A small plant is in operation on a tributary of the Bay of Fundy in Nova Scotia.

Tidal power sufficient to have any practical value requires a fluctuation between high and low tide of at least 20 feet (6 m). Otherwise, net energy yields may be zero or negative. There are only about 15 locations in the world that have tides of this magnitude, only one of which is in North America: the Bay of Fundy, where a major tidal power plant is under consideration.

Nor is tidal power without adverse environmental impacts. Far-reaching environmental effects would result because the dams would trap sediments, impede migration of marine organisms, change water circulation, and cause the mixing of fresh and salt water. With these restrictions in location plus the environmental impact, it is questionable whether additional tidal power plants will be built. Whether they are or not will depend on the availability of alternatives and on public acceptance of potential environmental impacts.

WAVE POWER

Waves are generated by wind and hence might be considered another form of indirect solar energy. However, we discuss wave power here because it has much in common with tidal power.

The energy inherent in waves holds the same general allure as the energy inherent in tides. However, building a machine that will harness that energy cost-effectively has proved difficult to say the least. One such device that seems workable is shown in Figure 23–23. Still, there are only a few locations

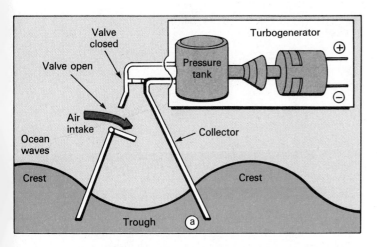

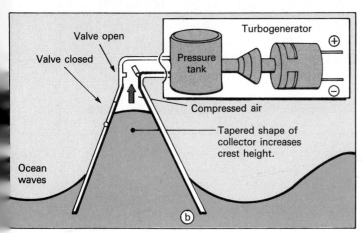

FIGURE 23–23
Wave power machine. (a) As the trough of the wave passes, air is drawn into the collector. (b) As the crest of the wave passes, air is compressed and forced into a pressure tank from where it drives a turbogenerator. The tapered shape of the collector causes the crest of the wave in the collector to reach a higher elevation and generate greater pressure than would otherwise be the case.

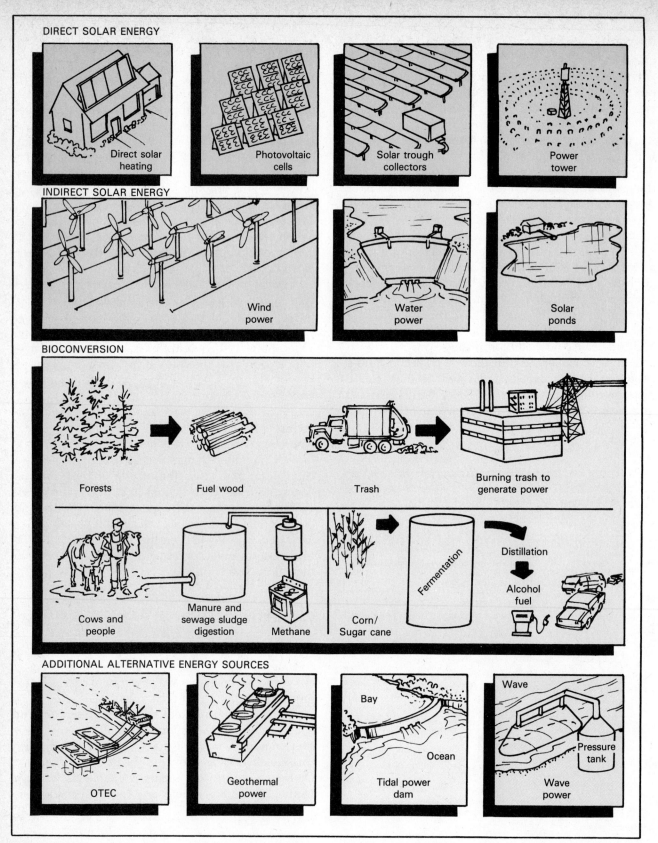

FIGURE 23–24
Renewable energy resources. At the present time, solar
heating, photovoltaic cells, solar trough collectors, wind
power, and production of methane from animal manure
and sewage sludges seem to offer the greatest potential
for supplying additional sustainable energy with a
minimum of environmental impact.

The potential availability of electrical power and alternative fuels from solar sources should not cause us to overlook the importance of conservation. If total energy use continues to grow with population growth as it has in the past, even massive installation of solar alternatives could barely keep pace with growing demand. It is difficult to see how solar facilities could be deployed rapidly enough to allow both an orderly phaseout of traditional fossil fuels and nuclear power and keep abreast of growing demand. Thus, regardless of which alternatives are chosen, conservation will remain a most important element in energy policy. It is also exceptionally cost-effective. Whatever the source of power or fuels, the costs will be less if fewer kilowatts or gallons need to be produced.

Most people still think of conservation in terms of turning the heat down, turning off lights, and so on. Thus, conservation still has the reputation among many as "freezing in the dark." Importantly, the greatest energy savings are not to be achieved by measures that cause inconvenience or discomfort; they are to be achieved by making heating, lighting, and transportation systems more efficient so that we can still have the same comfort, convenience, and transportation but use less energy in the process. By pursuing conservation through greater energy efficiency, the world can support a growing economy without demanding increasing amounts of energy.

Amory and Hunter Lovens, founders of the Rocky Mountain Institute (RMI) located near Aspen, Colorado, are recognized as world authorities on energy conservation. According to their analysis, universal use of today's best energy-saving technologies—high-mileage vehicles and more efficient appliances, motors, lights, windows, and insulation—could reduce U.S. oil and electricity consumption by 75 percent, saving $300 billion annually (about the size of the national debt) with no loss of services.

While many feel that this analysis is overly optimistic, numerous companies are taking advantage of RMI's consulting services and realizing tremendous savings. For example, The Boston Globe is achieving a savings of $350 000 per year through installing a high efficiency lighting, heating, and cooling system recommended by RMI. The World Bank headquarters in Washington, D.C. is now saving $500 000 per year on a one-time investment of $100 000 for more efficient lighting. The greatest success, however, is the following.

In 1988, Northeast Utilities proposed building a new power plant to meet Connecticut's growing demand. The Conservation Law Foundation (CLF), an environmental group, using information from RMI challenged the proposal in court claiming that power needs could be met through conservation. CLF won! The court ordered the utility to work with CLF to start a program of conservation. The program developed involves the utility spending the money that would have been spent on the new power plant for retrofitting homes, businesses, and factories with energy-efficient lighting, appliances, and insulation, and designing more energy-efficient new buildings. A key ingredient was the state changing utility regulations so that the utility could earn a higher rate of return on its money "invested" in conservation than on its money invested in traditional power production. This has created a win–win situation in which the utility is earning more on its investments, consumers are saving money on power, and the environment is not suffering the impacts of yet another power plant. Other utilities have now joined in the program and the concept is spreading.

With growth in power demand stemmed by implementing conservation, we can move toward replacing that demand with power from solar sources.

---

where waves are consistently high enough to make such machines cost-effective. To date, only England and Ireland, who have access to consistent high waves in the North Sea, are pursuing development of wave power, and its economic competitiveness is still to be determined.

The various solar and other renewable energy sources discussed are summarized in Figure 23–24.

Developing a Sustainable Energy Policy

Environmentalists frequently claim that the United States has no coherent energy policy. Others maintain that is not really true. It is more a matter of opinion as to what constitutes an energy policy. If we judge from their actions, the U.S. energy policy of the

Reagan and Bush administrations has been as follows:

○ First, promote drilling for and bringing into production every estimated reserve possible, including those potentially underlying national parks and refuges. Only consistent defense by environmentalists in and out of government has restrained, to some extent, the drilling in ecologically sensitive areas.

○ Second, increase imports of foreign oil and militarily defend those imports at all cost, and ignore the balance of trade deficit involved (Fig. 21–13).

○ Third, continue to support and promote nuclear power despite a lack of public acceptance. Over 1.5 billion dollars per year of government funds have been consistently allocated to nuclear research and development even though no new plants have been ordered since 1974. At the same time, budgets for research and development of renewable energy resources were cut by nearly 75 percent.

○ Finally, oppose measures aimed at conservation or otherwise cutting back on the use of fossil fuels despite a growing body of scientific information and world opinion that such cutbacks are necessary to avoid excessive global warming as well as to extend the useful life of existing oil reserves.

From what you have learned in these three chapters on energy, the shortcomings of this policy should be self-evident. The main features of a policy that could lead to a sustainable energy future should be equally clear. A number of solar alternatives have already reached the point of economic practicality. All that is needed is some added impetus to move in that direction.

One can conceive of any number of laws, from mandating increased mileage for cars to providing tax subsidies for installation of solar facilities. However, many ecologists feel that numerous separate laws and regulations create a bureaucratic nightmare, and one single law might do the trick: a **carbon tax** on all fossil fuels proportional to the amount of carbon dioxide that is released as they are burned. Rather than a tax, this money may be seen as paying for some of the hidden costs that we currently ignore. A number of

European countries have already adopted such a tax as a means of mitigating global warming.

The following effects are anticipated from a carbon tax. By increasing the costs of fossil fuels, it would provide economic incentive for all aspects of conservation, from consumers seeking more fuel-efficient cars all the way to improving home insulation and using solar energy for space and hot water heating. Since power companies would pay the tax on coal or other fossil fuels, it provides an incentive for them to build solar power plants. It would also provide incentive for power companies to promote conservation, as a number are already doing. Revenues from the carbon tax could be allocated to the development of renewable energy resources just as federal and state taxes on gasoline are currently allocated to highway construction.

An objection raised whenever a major technological shift is considered is: What about jobs? A study by Worldwatch Institute, "Jobs in a Sustainable Economy," points out that the building and installing of solar facilities—whether it is solar panels for individual homes or a commercial solar power plant—require the traditional skills of carpenters, plumbers, sheetmetal workers, and electricians. Furthermore, providing solar energy is basically more labor-intensive than providing fossil-fuel energy. For instance, coal mining and running a coal-fired power plant have become so automated that they provide very few jobs. A solar energy system of comparable energy output would involve many more jobs. Thus, moving toward a solar economy would provide additional jobs and generally improve the economy. Additional economic improvement would come from the reduction of what we currently pay both directly and indirectly for imported oil.

Many students ask: Why don't *they* develop solar? We should see that the obstacles are no longer technological or even economic. The main obstacle is people's reluctance to upset the status quo of our traditional energy-using habits. Each one of us, to a greater or lesser degree, contributes to that inertia. The question therefore needs to be rephrased as: Why don't *I* develop solar? Each one of us has numerous opportunities to help promote conservation, as well as solar and other renewable energy alternatives, as individuals, consumers, and voters. In Chapter 24, we shall examine, in more detail, the connections between lifestyle and sustainability.

 Review Questions

1. What are the three major virtues of solar energy?
2. How may solar energy be used for space heating and hot water heating?

3. What is the distinction between active and passive solar heating systems, and what are the advantages and disadvantages of each?

4. What has proved to be the most cost-effective method of providing for heat storage?

5. Does the occasional need for conventional backup heating mean that solar heating is impractical? Explain.

6. What are photovoltaic cells, and how do they work?

7. In what ways are photovoltaic cells currently being used, and what is their potential for providing energy in the next 10 to 15 years?

8. Describe what the solar trough system is, how it works, and its potential for providing power.

9. Why is it said that the storage of solar-produced power is not an important issue for at least the next 15 years?

10. How might hydrogen gas be produced using solar energy?

11. How might hydrogen gas be transported to cities and used as a vehicle fuel?

12. What is the potential for developing more wind power, and what would be the environmental impacts of such development?

13. What is the potential for developing more hydroelectric power in North America, and what would be the environmental impacts of such development?

14. What are four ways of converting biomass to useful energy?

15. What is the potential for and environmental impact of each of the four methods listed in question 14?

16. What is meant by geothermal energy, and how may it be harnessed?

17. What has been the recent experience with geothermal power in California, and how has this changed the outlook for geothermal energy?

18. What is the potential for developing tidal power and wave power in the United States?

19. What has been the de facto energy policy of the United States for the past 12 years, and how does it contrast with a policy for sustainable energy?

20. What actions (laws, taxes, subsidies) might move us in the direction of a sustainable energy future?

 Thinking Environmentally _____

1. Design what you feel should be the energy policy for the United States based on the total range of energy options, including fossil fuels and nuclear power as well as various solar alternatives. Which energy sources should be promoted and which should be discouraged? Give a rationale for each of your recommendations.

2. Suggest and give a rationale for laws, taxes, subsidies, and so on that might be used to bring the policy you suggested in question 1 to fruition.

3. Various solar energy alternatives seem to promise a sustainable energy future. Yet, the U.S. government seems locked into a policy of maintaining dependence on fossil fuels and nuclear power. Does this mean that solar advocates are wrong? Discuss what economic and political forces might be at work in maintaining the status quo.

4. Consider all the energy needs in your daily life. Describe how each might be provided by one or another solar option.

Lifestyle and Sustainability

LEARNING OBJECTIVES

When you have finished studying this chapter, you should be able to:

1. Define urban sprawl, exurban migration, urban decay, and gentrification.

2. Contrast the structure of urban areas prior to World War II with that which developed after World War II, describing specifically the role played by private ownership of cars.

3. Describe what the Highway Trust Fund is and how it has fostered urban sprawl.

4. Describe the environmental consequences of the car-dependent lifestyle.

5. List the public services that local governments provide and name the major source of revenue for those services.

6. Discuss three factors that cause the exurban migration and explain how exurban migration contributes to urban sprawl, gentrification, and urban decay.

7. Describe how exurban migration, urban sprawl, and urban decay become a vicious cycle.

8. Describe the key layout features of a hypothetical sustainable community.

9. Describe how existing suburbs might move toward becoming sustainable communities.

10. Discuss what laws and regulations would have to be enacted or modified to enable existing suburbs to evolve toward becoming sustainable communities.

11. Discuss the main features of a livable city using existing livable cities as examples.

528

12. Discuss four techniques and programs that are being used with some success to rehabilitate cities.

All the environmental problems we have discussed, from the loss of ecosystems and biodiversity through various forms of pollution to depletion of energy resources, have their origins in people and lifestyle. Recall the equation on page 127 showing that negative environmental impact is a function of population multiplied by the consumptiveness of lifestyle divided by environmental regard. Our first objective in this chapter is to study in more detail the connection between the typical, middle-class lifestyle found in the United States and, to a growing extent, in many other countries and the environmental problems exacerbated by this lifestyle. Second, we shall consider how lifestyle may be gradually modified over the years ahead to bring about a sustainable society.

Urban Sprawl and Car Dependency

The dominant, and environmentally most provocative, feature of the U.S. lifestyle is our near-total dependence on cars. We have developed an urban-suburban layout in which the locations where we live, work, go to school, and shop are widely separated. Thus, we have come to rely on private cars, which we drive an average of 200 miles (300 km) a week simply to meet our everyday needs, and multilane highways are more-or-less continuously congested with traffic. We shall refer to this lifestyle that re- volves around going everywhere by car as a *car-dependent lifestyle.*

This urban-suburban layout in which residential areas, shopping malls, industrial parks, and other facilities are spread out but laced together by multilane highways is referred to as **urban sprawl.** The word *sprawl* is used because, in large measure, the growth has occurred without any overall plan. Moreover, we can readily see that the process is continuing. Nearly everywhere we go around urban areas, we are confronted by farms and natural areas giving way to new developments, new highways being constructed, and old roadways being upgraded and expanded (Fig. 24–1).

FIGURE 24–1
A hallmark of urban sprawl is that new highways spawn more traffic, creating more congestion. Consequently, there is a constant need for new and expanded highways. Is this snowball effect sustainable? (David Lawrence/The Stock Market.)

THE ORIGINS OF URBAN SPRAWL

Most of you reading this text have grown up in the midst of urban sprawl and the car-dependent lifestyle. Therefore it may be difficult to imagine that things were ever different. However, urban sprawl really started only in the late 1940s when, following World War II, private ownership of cars became common. Even though cars had been developed around the turn of this century, they were unaffordable to most people until Henry Ford developed assembly line production in the 1920s. Just as that technological advance came about, however, ownership was sharply curtailed first by the Great Depression and then by World War II, at which time all production facilities were given over to the war effort. Therefore, before the end of World War II, a relatively small percentage of people owned cars.

Without cars, cities had developed in ways that allowed people to get to school, work, and shopping either by walking or by bicycling. (For most people, ownership of a horse in a city was as untenable then as now because of the expense and impracticality of maintaining one in the limited space of an urban property.) Walking distances were generally short; nearly every block had a small grocery, hardware, drug, and other stores and professional offices integrated with residences. Often a building had a store at the street level and residences above (Fig. 24–2). Public transportation—horse-drawn trolleys, cable cars, and, later, buses—provided for traveling longer distances within cities, but it did not change the compact structure because people still needed to walk to the transit line. At the ends of the transit lines, cities gave way abruptly to farms and open country. The small towns and villages surrounding cities were compact for the same reasons and mainly served farmers in the immediate area. This pattern held until the end of World War II; then it began to change dramatically.

Despite the ecological virtues of walking, bicycling, and public transit, many people found cities less-than-pleasant places to live. Especially in industrial cities, poor housing, inadequate sewage systems, inadequate refuse collection, pollution from home furnaces and industry, and generally congested, noisy conditions were much the norm. A decrease in services during World War II aggravated these problems. Hence, many people had the desire, the American Dream if you will, to live in their own house on their own piece of land away from the city.

Because of a lack of consumer goods during the war, both civilians and returning veterans had accumulated considerable savings. As mass production of cars commenced at the end of the war, people flocked to buy them. With private cars, people were no longer restricted to live within walking distance of where they worked or of a transit line. They could move out of their cramped city apartments and into homes of their own outside the city. Their cars would allow them to easily drive back and forth to their jobs, shopping, recreation, and so on.

Developers responded quickly to the demand for private homes. They bought farms wherever they could and put up houses, each with its own well and septic system. The government aided this trend by providing low-interest mortgages through the Veterans Administration and the Federal Housing Administration, and interest payments on mortgages were made tax deductible (rent continued to be non-

FIGURE 24–2
Before the widespread use of cars, cities had an integrated structure. A wide variety of small stores and offices on ground floors with residences above placed everyday needs within walking distance. This structure still exists in certain areas of some cities. (Geri Engberg/The Stock Market.)

FIGURE 24–3
Development and urban sprawl. The land being developed is prime agricultural land. The same is true for most development around all major U.S. cities. The scattered nature of development reflects farms that developers were able to buy. An overall plan for suburban development generally does not exist, a major factor resulting in sprawl. (USDA Soil Conservation Service Photo.)

tax-deductible). Property taxes in the suburbs were much lower than in the city. These financial factors meant that, for the first time in history, making monthly payments on one's own home in the suburbs was cheaper than paying rent for equivalent or less living space in the city.

Thus mushrooming development around cities did not proceed according to any plan; rather it proceeded wherever developers could acquire land (Fig. 24–3). In fact, planning was more than merely neglected; it was actively prevented by the fact that cities were surrounded by a maze of more-or-less autonomous local jurisdictions (towns, townships, counties, municipalities). No governing body existed to come up with an overall plan, much less enforce it. Local governments were simply thrown into a catch-up role of trying to provide schools, roads, sewers, water systems, and other public facilities to accommodate the uncontrolled growth. Urban sprawl was born.

In struggling to provide services for booming development, local, state, and federal governments contributed to urban sprawl. The influx of commuters into a previously rural area soon resulted in traffic congestion, creating a need for new and larger roads. To raise money to build and expand highways Congress passed the Highway Revenue Act of 1956, which created the **Highway Trust Fund.** This legislation placed a tax on gasoline and earmarked the revenues to be used exclusively for building new roads. (The fund still exists, but was modified in 1991 to allow a portion of the money to be used for other forms of transportation as will be described later.)

Perhaps you can already see how the Highway Trust Fund perpetuates urban sprawl. A new highway not only alleviates existing congestion, it also encourages further development and sprawl because time, not distance, is the limiting factor for commuters. (Listen to how people describe how far they live from work; it is almost always put in terms of

minutes, not distance.) The average person is willing to spend 20 to 40 minutes each way on a daily commute. Given this time limit, people who have to walk would have to live a maximum of 1 to 2 miles (1.5–3.0 km) from work. With cars and expressways, people can live 20 to 40 miles (30–60 km) from work and still get there in the same time. New highways do not get people to work faster; they simply allow people to live farther from where they work. Average commuting *distance* has doubled since 1960, while average commuting *time* has remained about the same. Thus, the Highway Trust Fund creates the vicious cycle illustrated in Figure 24–4.

Residential developments are followed (or sometimes led) by shopping malls and, more recently, industrial parks and office complexes. This has not lessened dependence on cars, however; it has only changed the direction of commuting. Whereas in the early days of suburban sprawl the major traffic flow was into and out of cities, it is now between suburban centers. Multilane highways connecting suburban centers are perpetually congested with traffic going in *both* directions. It is more than evident that new highways relieve congestion only temporarily. With the additional traffic they spawn, congestion is soon as bad as ever.

In broad perspective then, urban sprawl should be seen mainly as a process of **exurban migration,** that is, a movement of people and businesses in an outward direction from the city. Population, while it has been growing, has played a relatively minor part in the development of urban sprawl. The populations of many U.S. cities, excluding the suburbs, have actually declined in the last several decades as a result of exurban migration. For example, in the past 25 years, the population of the New York metropolitan region has grown by just 5 percent, while developed land area has increased by 61 percent. The exurban migration is continuing in a leap-frog fashion as peo-

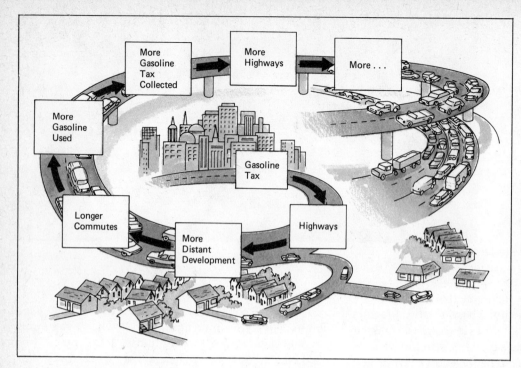

FIGURE 24-4
The vicious cycle spawned by the Highway Trust Fund.

ple from older suburbs move to **exurbs** (communities farther from cities than suburbs).

The "love affair" with cars is not just a U.S. phenomenon. Around the world, in both developed and less-developed countries, people aspire to own cars and adopt the car-dependent lifestyle. Thus, urban sprawl is occurring around many Third World cities as people become affluent enough to own cars. Love of the car-dependent lifestyle, however, is not sufficient to make it sustainable. In fact, this lifestyle is at the crux of a number of the unsustainable trends described in earlier chapters.

ENVIRONMENTAL IMPACT OF THE CAR-DEPENDENT LIFESTYLE

Depletion of Energy Resources

The car-dependent lifestyle imposes an ever-increasing demand for energy resources. The increased fuel efficiency of cars has been more than offset by the increasing number of cars. The oil shortages we face are in large measure a consequence of escalating demand for the crude oil needed to produce gasoline. In addition, individual suburban homes require 1.5 to 2 times more energy for heating and cooling than comparable attached city dwellings. Thus, a large part of the growth in demand for fuel oil, natural gas, and electrical power is a result of the shift to more spacious suburban living.

Air Pollution, Acid Rain, the Greenhouse Effect, and Stratospheric Ozone Depletion

Since vehicles are a major source of the pollutants that form photochemical smog, we can see how ground-level air pollution is largely a product of our car-dependent lifestyle. Energy demands for cars and spacious suburban homes also exacerbate the potential for global warming in that all the additional fossil fuels burned contribute carbon dioxide to the atmosphere, and acid-forming pollutants result in acid rain. Automobiles likewise are implicated in depletion of the stratospheric ozone shield since a major source of CFCs entering the atmosphere is that which escapes from vehicle air conditioners.

Degradation of Water Resources and Water Pollution

With all its roadways, parking lots, and driveways, urban sprawl results in increasing runoff and decreasing infiltration over large areas. Even suburban lawns increase runoff because they are more compacted than the original soils. Recall the consequences of increasing runoff—increased flooding, streambank erosion, and decreased water quality, to name a few—and the consequences of decreasing infiltration—depletion of groundwater, drying of springs and waterways, saltwater encroachment, and others. The large quantities of water used by suburbanites for washing cars, watering lawns and gardens, and filling pools hasten the depletion of water supplies. Chemicals used on lawns and gardens and for de-icing roadways further degrade water quality.

Negative Impact on Recreational Areas and Wildlife

New highways are frequently routed through parks or along stream valleys because such open areas provide the least expensive rights of way. The result is the sacrifice of aesthetic, recreational, and wildlife values in the very places where they are most important: metropolitan areas. Highway planners argue that a highway through a large park will take a relatively small portion of the total area. If a park is to provide humans with a measure of peace and tranquility, however, it is dubious that two halves split by a pollution-generating and noise-generating highway will still add up to the whole.

Wildlife is found to decline markedly as natural areas are fragmented by highways. The loss is even more conspicuous when wildlife is blocked from a source of drinking water, as frequently happens when highways run along rivers or stream valleys. Expanded and improved highways lead to increasing numbers of roadkills. Much more wildlife is killed by cars nowadays than by hunters.

Loss of Agricultural Land

Finally, and perhaps most serious, is the loss of prime agricultural land. You may have heard this problem minimized by arguments that the world has plenty of land. However, 65 percent of the world's land is deserts, tundra, rugged mountains, and hills, all of which preclude agriculture although about half of it supports forests. Another 24 percent is considered marginal land; it is excessively dry (savannas and grasslands) or wet (wetlands), but given suitable inputs, it might be used for agriculture. Only the remaining 11 percent is prime agricultural land.

Because food has always been a basic need, it was natural to locate cities in prime agricultural regions. Consequently, in the United States alone, sprawling development eats up 2.5 million acres (1 million hectares) per year of the best agricultural land, and production may be reduced on remaining farmland because of air pollution. Perhaps, for the time being, the United States has adequate agricultural land to tolerate this loss, but this is hardly the case in many Third World countries. For example, over the past decade, growing sprawl around Bangkok, Thailand, has consumed 8000 acres (3200 hectares) per year of prime agricultural land. Between this loss and a growing population, Thailand will soon be re-

quiring food imports. Many Third World countries already dependent on food imports are still losing land to urban sprawl. Limits to bringing more land into production or to increasing production on remaining acres (Chapter 8) should not go unnoticed.

These environmental impacts resulting from or exacerbated by the car-dependent lifestyle are summarized in Figure 24–5.

SOCIAL CONSEQUENCES OF URBAN SPRAWL

Urban sprawl and the car-dependent lifestyle have had more than just environmental impact. There has been sociological impact also, stemming from the fact that the driving force behind urban sprawl has been largely exurban migration. This migration has involved a segregation of society and deterioration of inner cities. We shall explore this further.

Exurban Migration and Economic and Ethnic Segregation

Historically, U.S. cities included a wide diversity of people with different economic and ethnic backgrounds. However, in general, moving to the suburbs requires the financial ability to afford a car and a home mortgage, and the ability to drive. Thus, exurban migration for the most part has excluded the poor, the elderly, and the handicapped. Generations of discrimination in education and jobs have made African-Americans and certain other minority ethnic groups disproportionately poor. Discriminatory lending practices by banks and discriminatory sales practices by real estate agents have kept African-Americans and other minorities out of the suburbs even when money was not a factor. Civil rights laws passed in the 1960s made such practices illegal, but there is much evidence that they still persist.

In short, exurban migration and urban sprawl have also led to **gentrification**—a segregation of the population into groups sharing common economic, social, and cultural backgrounds. Moving into the new suburban and exurban developments were the economically advantaged, while many areas of the cities and older suburbs were effectively abandoned to the economically depressed, who in large part were African-Americans, Hispanics, and other ethnic minorities, and the handicapped and elderly (Fig. 24–6).

New immigrants and people moving from small rural communities to large metropolitan areas have fallen into the same pattern of gentrification according to economic and ethnic status.

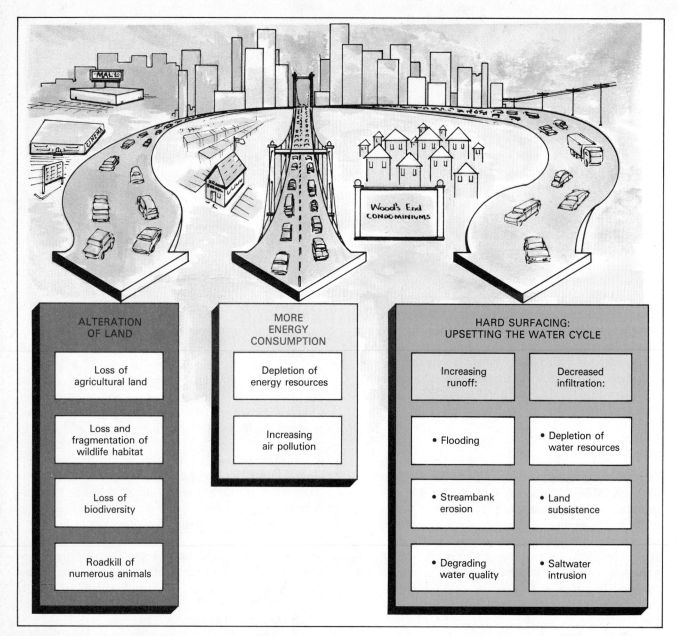

FIGURE 24–5
A summary of the environmental side effects of urban sprawl.

Declining Tax Base and Deterioration of Cities

Through economic segregation, exurban migration has created a downward spiral of urban decay that today is one of society's most vexing and pressing problems. To understand how this downward spiral results from exurban migration, we must note several points concerning local government.

First, local governments (usually city or county, though the particular entities vary from state to state) are responsible for providing public schools, maintenance of local roads, police and fire protection, ref-

use collection and disposal, public water and sewers, welfare services, libraries, and local parks. Second, a local government's major source of revenue to pay for these services is local property taxes, a yearly tax proportional to the market value of the land and home or other buildings on it. If property values increase or decrease, the property taxes are adjusted accordingly. Third, in most cases, the central city is a governmental jurisdiction separate from the surrounding suburbs.

Because of these three characteristics of local government, exurban migration and gentrification

(a)

(b)

FIGURE 24–6
Exurban migration, which has been the driving force behind urban sprawl, has also led to the segregation of the population along economic and racial lines. (a) An area of suburbia and (b) an area of inner-city Baltimore contrast the extremes of gentrification. (Photographs by BJN.)

have the following consequences. By the economic law of supply and demand, property values in the suburbs escalate with the influx of affluent newcomers. Thus, suburban jurisdictions enjoy increasing tax revenue with which they can improve and expand local services. At the same time, property values in the city decline because of decreasing demand. Also, property taxes create disincentive toward maintaining property. Often landlords allow property to deteriorate to reduce their taxes while keeping rents high to maximize their income. Often properties end up being abandoned by the owner for nonpayment of taxes. The city "inherits" such abandoned properties, but they are a liability for the city rather than a source of revenue. The declining tax revenue resulting from declining property values is referred to as **eroding tax base,** and it has been a serious hand-

icap for most U.S. cities since the exurban migration started in the late 1940s.

Vicious Cycle of Urban Sprawl and Inner City Decay

The result of an eroding tax base is **urban decay,** or **urban blight,** and it involves a vicious cycle. In the face of an eroding tax base, city governments are forced either to cut the quality and quantity of local services or to increase the tax rate. Generally they have done both. Hence, the property tax on a home in a city is often two to three times greater than on a comparably priced home in the suburbs, a difference that amounts to $2000 to $3000 per year in the tax bill on an averaged-priced home. At the same time schools, refuse collection, street repair, and other ser-

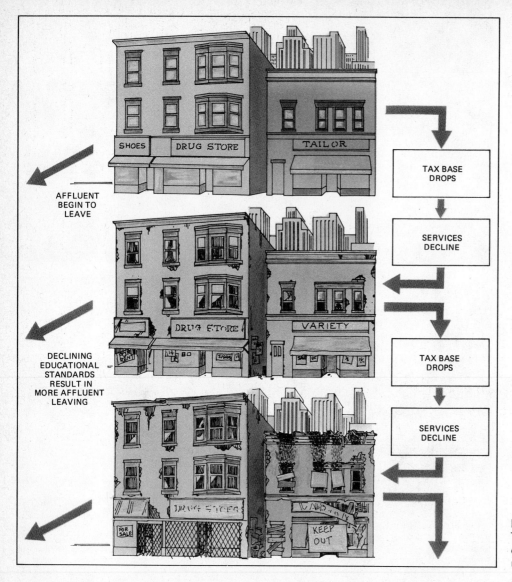

SHOES DRUG STORE TAILOR

AFFLUENT
BEGIN TO
LEAVE

DRUG STORE VARIETY

DECLINING
EDUCATIONAL
STANDARDS
RESULT IN
MORE AFFLUENT
LEAVING

DRUG STORE

FOR SALE!

KEEP
OUT

TAX BASE
DROPS

SERVICES
DECLINE

TAX BASE
DROPS

SERVICES
DECLINE

FIGURE 24–7
The vicious cycle of
exurban migration and
urban decay.

vices in the city are often far inferior to their suburban counterparts. This situation of increasing taxes and deteriorating services causes more people to leave the city and adopt the car-dependent lifestyle even if this lifestyle is not their primary choice. It is still only the relatively affluent who can exercise this choice, however, and thus the whole cycle is perpetuated. The tax base erodes further, the city is forced to cut services more and raise the tax rate, getting another group of taxpayers angry enough to leave, . . . (Fig. 24–7).

Stores, businesses, and professional offices follow their clientele. Thus, countless establishments have closed city stores and offices and moved to suburban shopping malls and commercial centers. The most recent trend is new industrial plants, which traditionally located in cities, locating in exurban **industrial parks,** spacious developments of business and light-industrial facilities. This exurban migration of businesses means further loss of tax revenue for

the city. Even more serious, it involves a loss of jobs for city people who are unable to commute—public transportation is often unavailable, and city people who can't afford reliable cars are stuck. Thus, urban unemployment rates are very high, while in the suburbs jobs frequently go begging.

To be sure, in the last two decades many cities have redeveloped core areas. Shops, restaurants, hotels, convention centers, entertainment centers, office buildings, residences, and places for walking and relaxing are all combining to bring a new spirit of life and hope, as well as the practical assets of taxes and employment, back into cities (Fig. 24–8). It is an encouraging new trend, so far as it goes.

However, we should not let such redevelopment overshadow urban blight, which is still all too real and in many cases worsening. While most of the country was enjoying economic prosperity during the 1980s, another 10 million people trapped between the suburbs and the revitalized urban centers sank

(a) (b)

FIGURE 24–8
High population densities make a city; they don't destroy it. Redeveloping key areas of cities with heterogeneous mixtures of office buildings, shopping, restaurants, nearby residences, and recreational facilities, all of which attract people, has helped to bring new life back into cities. (a) Inner Harbor, Baltimore. (b) Faneuil Hall, Boston. (c) Renaissance Center, Detroit. (d) Tom McCall Waterfront Park, Portland. [(a) BJN; (b)(c) Kunio Owaki/The Stock Market; (d) John M. Roberts/The Stock Market.]

(c)

(d)

Many environmentalists aspire to have their own piece of land where they live more self-sufficiently and "ecologically" by growing their own food, recycling wastes, utilizing solar energy, and so on. During the 1960s and 1970s, many people retreated to far reaches to do this. Living in the city and environmentalism seemed contradictory. However, there were also many people who realized that retreating to remote locations was no solution. There is simply not enough wilderness to accommodate more than a minute fraction of the population, and those who retreat there to carve out their own homestead are simply adding their impact on open space, which is already under extreme pressure. Then, there are many social and cultural ties to the city that we do not want to break. Thus, certain groups began considering: Is it possible to live more ecologically within the confines of the urban environment? Out of this was born the concept of the integral urban house, a typical urban residence that, by following all the ecological principles, is made environmentally sound and nearly self-sufficient in virtually all respects, including food production. The integral urban house is a concept that can be pursued by nearly all city and suburban residents and that would reduce our impact on the environment and lead toward a sustainable biosphere.

below the poverty line. Walk two blocks in any direction from the sparkling, revitalized core of most U.S. cities and you will find urban blight continuing unabated. In the words of renowned city planner and developer, James Rouse:

But right alongside [the redeveloped center] in almost every city is a second city of those who are not making it—the city of the poor. It is like a Third World city—the city of people who are struggling to survive in miserably unfit housing in wretched, disorderly neighborhoods— with too little food, too little health care, too little work and too little training for work, too little education, too little happiness, too little hope. It is like another nation where we are growing people who feel left out, abandoned, separated from the opportunities for the good life that abounds all around them. . . .

We have read the dismal figures, seen pictures of dilapidated housing in derelict neighborhoods, but most of us have not walked those streets, stepped inside those houses, climbed the stairs in those apartments; have not seen good people with clean, decent families huddled in that miserable housing, paying outrageous rents; have not looked into the saddened, sullen faces; felt the hopelessness, the distrust, suspicion, and separation that pervades their life and all around them [see Fig. 24–12].[1]

In the last few years sections of some cities have deteriorated so much that they are referred to as "war zones." Increasing numbers of the people who live in these areas, being economically trapped, see drug dealing and other illicit activities as a fast track to "the good life" and become part of a subculture where guns and killings are commonplace.

It is becoming increasingly apparent that the cardependent lifestyle and the urban sprawl it engenders are not sustainable on environmental grounds. Further, the urban blight that has been left in its wake is not sustainable on sociological grounds. The Los Angeles riot that occurred in May 1992 exemplifies its potential for explosion.

Moving Toward a Sustainable Society

Cars that get more miles per gallon and the other energy-conservation measures and alternative energy sources discussed in Chapters 21 and 23 may ease energy shortages, but they will not lessen most of the other environmental and sociological impacts of urban sprawl. To find sustainability, we must look beyond the car-dependent lifestyle and urban sprawl.

[1] J. Rouse, "Suffering in the Second City," *EPA Journal* 14(4) (May 1988), pp. 25–26.

THE FALSE OPTION OF HOMESTEADING

An appealing idea to a considerable number of young people is, individually or as small groups, to buy a piece of property in some remote area and "homestead" it, that is, build a simple cabin and take up subsistence agriculture much as U.S. pioneers did. However, modern-day homesteading is more an exercise in escapism than any practical avenue toward sustainability. Modern-day homesteaders still depend on cars to get to cities for many of their needs such as clothing, tools, and medical care. Therefore, all homesteading does is extend the perimeters of urban sprawl, doing nothing toward solving any of the problems. Any movement toward sustainability that is to succeed must involve existing communities and eventually entire cities, not just a few rugged individualists going off on their own.

SUSTAINSVILLE, A MODEL FOR A SUSTAINABLE COMMUNITY

To create sustainable communities, we must begin with a visualization of what one might look like. Figure 24–9 shows a model of a sustainable community of about 20 000 people that we have named Sustainsville. Sustainsville is not the only possible design; you may think of any number of variations. However, features that incorporate the principles of ecosystem sustainability are essential. By studying this figure and its key, you can see how our model incorporates the basic principles of ecosystem sustainability (Table 4–1).

The proximity of homes, schools, workplaces, and food-producing areas is such that virtually all transportation needs can be met by walking, bicycling, and very short hauls for most materials. With transportation needs thus minimized, the remaining energy needs can be met by a variety of solar energy options. Further, the proximity of things also facilitates the recycling of nutrients and other materials. Sustainsville produces virtually all its own food and wood products. Therefore the residents are aware of the need for a policy of stable population. Since they need land to grow food and wood, they can't turn their agricultural and woodland areas into more developments without destroying the essence of their sustainability. The stable population policy, careful forest management, and emphasis on biological pest control foster preservation of natural areas and biodiversity.

Sustainsville is not self-sufficient in every respect, nor is it necessary that real sustainable communities be entirely self-sufficient. It would be quite impractical, if not impossible, for each community to

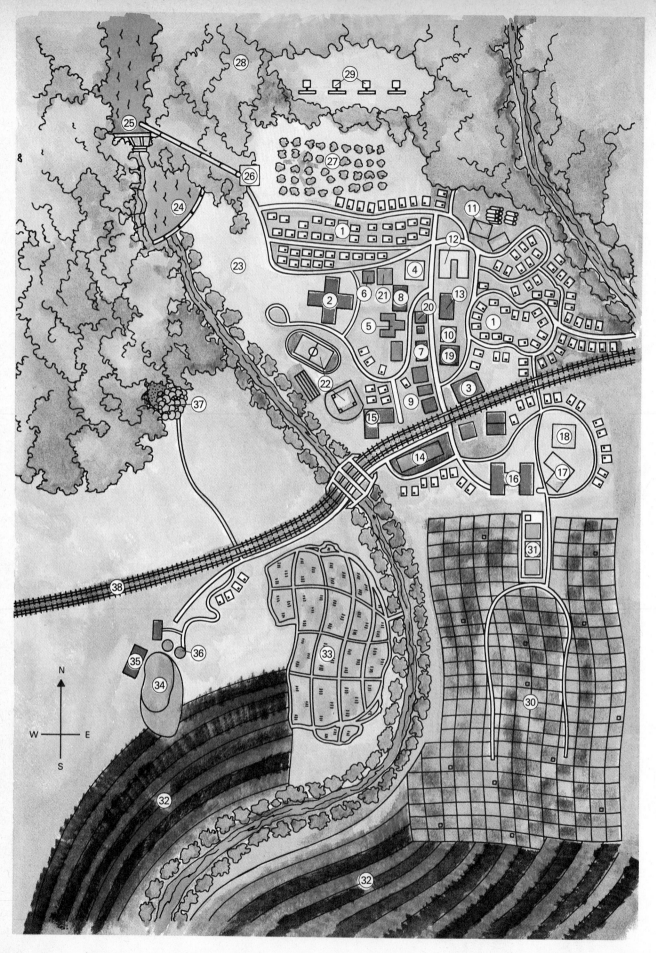

FIGURE 24–9

Sustainsville, a model of a sustainable community. A prime feature of Sustainsville is that essentially all transportation within the town is by walking and bicycling. Virtually the only vehicles are for emergency and deliveries of heavy items. Housing, shopping, offices, and workplaces are located so that walking/bicycling distances are short. Elimination of cars and the extensive roadways and parking lots necessary to accommodate them allows a much closer arrangement of facilities. Transportation between cities is provided by an electric rail system. Houses and other buildings are heated, cooled, and powered by a combination of passive solar energy and photovoltaic cells.

1. Housing
2. School
3. Shopping Mall, Train Station, and Post Office
4. Interfaith Center
5. Entertainment/Recreation Center
6. Daycare Center
7. Professional Offices
8. Library/Media Center
9. Stores
10. Healthcare Center
11. Sawmill and Wood Products
12. Paper Products and Paper Recycling
13. Newspaper, T.V. and Communications
14. Factory
15. Textile/Clothing Production

16. Food Processing, Canning; Cleaning and Refilling of Glass Containers
17. Leather Goods Production
18. Research and Development
19. Firehouse
20. Law Enforcement/Courthouse
21. Senior Center
22. Playing Fields
23. Park
24. Low Reservoir/Water Recreation
25. High Reservoir/Drinking Water and Power
26. Water Treatment Plant
27. Orchards with Bee Hives

28. Forests
29. Wind Power
30. Vegetable Crops and Small Animals
31. Fish Farm
32. Strip Cropping: Corn, Alfalfa, Cotton, Wheat
33. Rice Paddy
34. Waste Water Treatment Primary Settling, Artificial Wetland; Final Effluent Used for Irrigation
35. Dairy Barn and Pasture
36. Alcohol and Methane Production from Agricultural and Sewage Wastes
37. Stone Quarry
38. Inter-City Electric Rail System

produce such things as photovoltaic cells, computers, and telephones. We can visualize sustainable communities that will be linked together in an overall economy wherein each community specializes in the production of certain items, "exporting" the excess and "importing" things it doesn't make itself. And of course people would always be free to travel. The main distinction is that much less transportation and travel would be required in meeting everyday needs.

We are not proposing that communities such as Sustainsville be built from scratch; that approach would simply add to the urban sprawl. This model is presented to summarize the aspects of sustaina-bility and help you visualize how they might be integrated into a community.

The big question that needs to be addressed is, How do we move from our present urban sprawl and car-dependent lifestyle toward modern sustainable communities?

MAKING SUBURBS SUSTAINABLE

Given a chance to do so, innumerable existing suburbs and exurbs might evolve into sustainable communities. In many cases, though, archaic zoning laws

IN PERSPECTIVE
What Is Utopia?

You may observe that the relatively small-scale production of food, clothing, and other items in Sustainsville will require more hand labor. A trend of the past several decades has been toward more and more centralization to reduce labor and gain economies of size. However, are such economies of size and reduced labor costs any benefit to society if half of the population ends up out of work? It has been pointed out that we tend to think of Utopia as a place where no one works. If no one works, however, no one gets paid and everyone starves to death in Utopia. Are there certain economies and social benefits to returning to a situation where there is more hand labor?

FIGURE 24–10
A favorite location for new shopping malls and industrial parks is where there is good access to major traffic arteries. Such locations, far removed from any housing, assure that the only access by shoppers and workers is by car, forcing car-dependency, an anathema to the concept of sustainability. Shown here is White Marsh shopping mall located northeast of Baltimore. (Photograph by Richard Adelberg, Jr., Owings Mills, MD.)

and building restrictions force just the opposite. By requiring large lot sizes and detached housing, and the separation of residential and commercial or industrial areas, zoning laws ensure that the automobile will continue to rule in suburbia. A particularly poor design feature found in many developments is new industrial parks and shopping malls surrounded by low-density, expensive housing or even vacant land. Such a layout ensures that virtually all the workers and shoppers can get to the parks and malls only by car (Fig. 24–10).

Mass transit (bus or rail lines), a much-promoted idea, is often unworkable in suburbs because it makes no sense economically or environmentally to run a 40-passenger bus with only a few passengers. Spread-out suburban areas simply do not have sufficient population density to provide the ridership required to make mass transit viable, further forcing the reliance on private cars.

A study by the Worldwatch Institute, *Shaping Cities: The Environmental and Human Dimensions*, points out that a key to developing sustainable communities is changing zoning laws to allow stores, light-industrial plants, professional offices, and higher-density housing to be integrated so that more people can live within walking or bicycling distance of jobs. Most urban-suburban areas have existing open areas or areas to be redeveloped such that heterogeneity—even including agricultural production—could be built in if there were the will to do so.

A second point stressed in the Worldwatch report is the need to provide housing around industrial parks and shopping malls that is affordable to the people who work and shop in those places. Individual homes spread over the landscape are the most land- and resource-extravagant way of providing housing, and they ensure dependence on cars. Cluster development, because it is much more land- and resource-efficient, has the potential to provide more-affordable housing, and it need not be unattractive or lacking in privacy (Fig. 24–11). Again, however, zoning restrictions often preclude such development.

A third point stressed in the Worldwatch report is the need to provide bicycle and pedestrian access between residential areas and workplaces. The current situation in most cities and communities of the United States is that even when people live conveniently close to work, they are often loath to walk or bicycle because doing so means exposing themselves to the very real hazard of road traffic. Safe walk- and bicycle-ways, on or off the road, are seldom provided. Likewise few employers provide a convenient, safe place to keep a bicycle, although they spend a great deal of money providing parking spaces for cars.

In summary, urban sprawl has been promoted and supported by zoning regulations, tax incentives, and legislation such as the Highway Trust Fund. It follows that by simply changing these regulations and incentives, growth could be guided in the directions suggested above, which would lead to sustainability.

There is some good news in the area of providing better bicycle and pedestrian access. In late 1991 Congress passed the Intermodal Surface Transportation Efficiency Act, a bill environmentalists have been promoting for many years. Under this act almost half of the money levied by the Highway Trust Fund is now eligible for use on other modes of transportation, including cycling and walking as well as mass

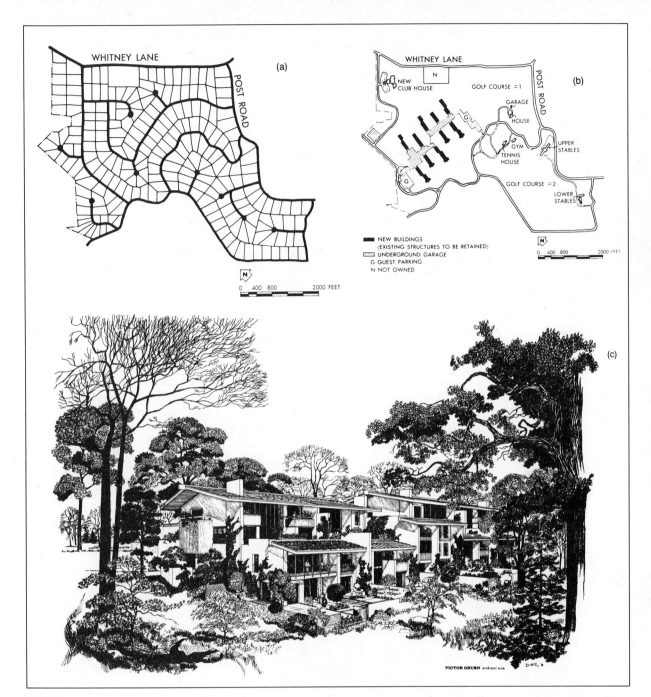

FIGURE 24–11
Alternative plans for developing the same tract of land. (a) Typical subdivision into half-acre lots. (b) The same number of units clustered into groups of attached dwellings. (c) The proposed design for the attached dwellings. The floor space of the homes and the final cost to the buyer were calculated to be the same for the two cases. You can see that clustering dwellings involves less total paving for road access. Providing water, sewer, trash collection, and other utilities and services is likewise more efficient. Most important, clustered development leaves most of the land available for other uses. Thus, clustered development has relatively less negative environmental impact than traditional development. (By permission from the Victor Gruen Center for Environmental Planning.)

transit facilities. The legislation also requires states to hire bicycle and pedestrian coordinators to oversee programs and to establish long-range pedestrian-bicycle plans. The act also requires that states spend a total of $3 billion over the next 6 years on "transportation enhancements," which include pedestrian and bicycle facilities. Whether this funding, which is now technically available, actually gets used to provide bicycle-pedestrian facilities in your area will still depend on citizen lobbying.

Additional legislation that would contribute to curbing urban sprawl is the carbon tax discussed in Chapter 23. In addition to providing an impetus toward greater fuel efficiency, a carbon tax would provide an incentive for keeping future developments more integrated and closer together to reduce commuting distances. The need to restrict the conversion of agricultural land, wetlands, and other natural areas to development speaks for itself.

Steps that curtail urban sprawl may also impede further exurban migration and serve to focus some attention and resources back on cities. However, the Los Angeles riot of 1992 served to point out that direct attention needs to be given to cities and the urban poor immediately. The only real debate is, What form should assistance take? To get at this question, we shall first consider some general factors concerning what makes cities livable.

WHAT MAKES CITIES LIVABLE?

Livability is a general index based on people's response to the question: Do you like living here, or would you rather live somewhere else? Crime; pollution; recreational, cultural, and professional opportunities; and many other social and environmental factors are summed up in the subjective answer.

As described above, many U.S. metropolitan centers have suffered increasing blight because of the exodus of affluent people followed by commerce and industry. City neighborhoods have been further dehumanized by being sliced by traffic arteries that bring nothing more than noise and pollution to city life. Additional city space has been leveled to provide parking. Sidewalk vendors, an important part of commerce for city dwellers, have been largely outlawed in most U.S. cities because they impede traffic flow. Many areas are blighted by abandoned residential, commercial, and industrial properties (Fig. 24–12). In short, planning and development of U.S. cities over the past several decades have been dominated by highway engineers intent on moving traffic at whatever cost to communities. In the words of city planner William Whyte, "It is difficult to design space that will not attract people. What is remarkable is how often this has been accomplished."

Many people assume that the social ills of the city are an outcome of high population densities, but in fact crime rates and other problems in many U.S. cities have climbed even as populations have dwindled as a result of the exurban migration. To quote Worldwatch's *Shaping Cities* report: "There is no scientific evidence of a link between population density per se and social ills. . . . Copenhagen and Vienna—two cities widely associated with urban charm and livability—each have relatively high 'activity density' (the number of residents and jobs in the city), 47 and 72 people per hectare (1 hectare = 2.5 acres) respectively. By contrast, low-density cities such as Phoenix (13 people per hectare) often are dominated by unwelcoming, car-oriented commercial strips and vast expanses of concrete and asphalt." William Whyte goes even further. His studies show that wherever urban population density drops below a certain point, there are no longer enough people to support restaurants, department stores, and so on, and the area "dies" as a result.

Looking at livable cities around the world, one

FIGURE 24–12
Many city's ills are because of too little population, not too much. When the exurban migration causes population to drop below a certain level, businesses are no longer supported and result is abandonment, the bane of urban blight. Show here is a section of Baltimore only a few blocks from the redeveloped Inner Harbor. (Photograph by BJN.)

finds that the common denominators are (1) maintaining a high population density; (2) preserving a heterogeneity of residences, businesses, stores, and shops; and (3) keeping layouts on a human dimension, where people can incidentally meet, visit, or conduct business over coffee in a sidewalk café.

The world's most livable cities are all ones that have taken measures to reduce outward sprawl, reduce automobile traffic, and improve access by foot and bicycle in conjunction with mass transit. For example, Geneva, Switzerland, prohibits car parking at workplaces in the city's center, forcing commuters to use the excellent public transportation system. Copenhagen bans all on-street parking in the downtown core. Paris has removed 200 000 parking places in the downtown area. Curitiba, Brazil, with a population of 1.5 million, is cited as the most livable city in all of Latin America. The achievement of Curitiba is due almost entirely to the efforts of Jaime Lerner who, serving as mayor since the early 1970s, has guided development in terms of mass transit rather than cars. The space saved by not building highways and parking lots has been put into parks and shady walkways, causing green area per inhabitant to increase from 4.5 ft^2 (0.5 m^2) in 1970 to 4500 ft^2 (50 m^2) today.

In Tokyo millions of people ride bicycles either all the way to work or to stations from which they catch fast, efficient subways or the bullet train to their destinations (Fig. 24–13). By sharply restricting development outside certain city limits, Japan has maintained population densities within cities and along metropolitan corridors that ensure the viability of commuter trains. Japan's cities have maintained a heterogeneous urban structure that mixes small shops, professional offices, and residences in such a way that a large portion of the population meets their needs without cars. In maintaining an economically active city it is probably no coincidence that street crime, vagrancy, and begging are virtually unknown in the vast expanse of Tokyo despite the seeming congestion.

Portland, Oregon, is one of the very few U.S. cities that have taken more than token steps to curtail car use. The first step was to encircle the city with an urban growth boundary, a line outside of which new development was prohibited. Thus compact growth rather than sprawl was ensured. Second, an efficient light rail and bus system was built, which now carries 43 percent of all commuters to downtown jobs. (In most U.S. cities, only 10 to 25 percent of commuters ride public transit systems.) By reducing traffic, Portland was able to convert a former expressway and huge parking lot into the now-renowned Tom McCall Waterfront Park. Portland is now ranked among the world's most livable cities.

The same factors that underlie livability also lead

FIGURE 24–13
Most people in Tokyo do not own cars but ride bicycles or walk to stations, from where they take fast, efficient, inexpensive subways to reach their destinations. Seen here is the bicycle parking lot at Nakano Station. (Kim Newton/Woodfin Camp & Associates.)

to sustainability. The reduction of auto traffic and greater reliance on foot and public transportation reduce energy consumption and pollution. Urban heterogeneity can facilitate recycling of materials. Improved insulation, passive solar heating of space and hot water, and use of landscaping for cooling, as described in Chapter 23, are all applicable to city residences and buildings. A number of cities are developing vacant or cleared areas into garden plots, and rooftop hydroponic gardens are becoming popular (Fig. 24–14). These gardens will never make cities agriculturally self-sufficient, of course, but they add to livability, provide an avenue for recycling compost, give a source of fresh vegetables, and have the potential of providing income for many unskilled workers (see In Perspective Box, p. 546). If urban sprawl is curbed, relatively close-in farms could provide most of the remaining food needs.

FIGURE 24–14
Urban garden plots on formerly vacant lots in Philadelphia. Urban gardening is being recognized as having a number of sociological, economic, environmental, and aesthetic benefits. (Blaine.)

REHABILITATING CITIES

The basic question is: How do we get from the blighted condition of cities today to the sustainable, livable cities of tomorrow? Again much may be done by changing tax structures to provide incentives. For example, Pittsburgh, Pennsylvania, and some other cities have brought urban decay to a halt simply by changing the tax code so that the bulk of the property tax is assigned to the land rather than to the building on it. Thus there is no financial penalty for improving dwellings and no reward in letting them deteriorate.

Another avenue is through the creation of what are called **enterprise zones.** This involves the government designating depressed areas with high unemployment as "enterprise zones," and then providing tax subsidies to businesses that locate or expand and hire one or more additional employees in the area. This program has been adopted by 36 states but incentive provided by rebates of only state and local taxes is limited. Currently there is much discussion regarding making this a massive program supported by the federal government.

A program known as **urban homesteading** originated in Wilmington, Delaware, and is now being used in cities nationwide. Under this program the city "sells" abandoned properties to individuals for the sum of $1. In signing the contract for the home, the individual agrees to fix it up and live in it for at least

---

IN PERSPECTIVE
Rooftop Gardens for Livability and Food

For years, workers at ECHO (Educational Concerns for Hunger Organization—a nonprofit organization helping to meet food needs of the poor) have been developing techniques for growing vegetables in model rooftops at ECHO's facilities in Fort Myers, Florida. Even though it would be possible to use hydroponic technology to accomplish this, ECHO staff decided to avoid traditional hydroponics because of the dependence on water or air pumps to keep roots aerated—technology that could be expensive and energy-dependent. Instead, they found that they could grow luxuriant vegetables in a "stagnant water" system only 1 inch deep. The shallow depth allowed oxygen to diffuse effectively to the roots.

ECHO workers also found that they could grow almost any vegetable in a shallow (3–4 inch) layer of compost or slightly decomposed ref-

use. No soil was used, and the beds were kept shallow to keep weight down (rooftops are often too weak to support heavy weights). In one Third World application of the technology, rooftops on a school and a hospital in Port-au-Prince, Haiti, were raising high-nutrient crops to feed students and patients. Water and nutrients were fed to the vegetables with a wick made from a sheet of polyester cloth. The cloth was spread out on the cement roof, vegetable plants were spaced out on the cloth, and a few inches of pine needles were used to fill in around the plants for lightweight support and protection against drying. A 5-gallon bucket was then filled with a complete, soluble fertilizer, a $\frac{3}{8}$-inch hole drilled in the lid, and the bucket was placed upside down on a clear section of the cloth. Water and nutrients simply wicked out to the vegetables through the cloth.

The workers found that if pine needles were unavailable, they could substitute crushed soda cans to occupy the space between plants and inhibit evaporation of water. They were even able to raise field corn with this system!

The lessons to be learned are straightforward: Plants really need light, water, and nutrients, but, surprisingly, little or no soil. These needs can be met in virtually any Third World city, by the literally acres of flat "land" in the form of rooftops in the city. This kind of urban "farming" also has the potential to rehabilitate cities in the industrialized countries, giving people the opportunity to raise some of their own food and enjoy living in the company of green plants. (For information write ECHO, 17430 Durrance Rd., Fort Myers, Florida 33917.)

This chapter has documented all too accurately the human costs of urban blight caused by the middle class and businesses moving to the suburbs. The car-dependent lifestyle that characterizes our country is so thoroughly embedded in our economy and social structure that it is almost bizarre to try to imagine how to do it differently, as we attempted to do with our model town, Sustainsville. Yet, as we have seen, there will be no environmental sustainability unless we make major moves away from our current urban-suburban divergence and rehabilitate our cities. There is another important reason for doing this, however, and that is *justice*.

Justice has to do with the distribution of things that are scarce or in relatively short supply: jobs, education, good housing, food, consumer goods, and the like. A sense of *distributive justice* is part of our human makeup; we (as individuals) believe that we should have a share of those things that are in short supply. History has proven that distributive justice is not accomplished when there is a free-for-all scramble for scarce things. It is only accomplished when law and order exist and people cooperate to maintain that order. Thus, an orderly society is one where people at least perceive that they are being dealt with justly. They have a stake in maintaining order. However, if people believe they are being treated unjustly, their stake in the orderly society is more tenuous. Sooner or later they become convinced that things are not going to change, and then they are prone to take action that is disruptive to that order—thus the riots we have seen in cities. Rioting and looting rarely take place in the suburbs for the perfectly good reason that middle-class people in the suburbs have much more to lose by such behavior than the inner city poor.

As we have seen, the exurban migration has left the inner city to minority groups, the elderly, and the handicapped, and society has become gentrified. The distribution of goods and services has become similarly divided, leading to the downward spiral of urban decay. The pathway to making the cities more livable is going to require a commitment on the part of all segments of society to distributive justice. It will take major structural changes to correct the injustices that exist. New programs and new sources of funding must be put in place.

It is important, however, that enhancing distributive justice also be put in a moral context. The case can be made that those who "have it all" are benefiting from the unjust distribution of goods and services that leaves others out. We will probably not make much progress until the outrage of the disadvantaged minorities is augmented by those who are well off and are also fair-minded enough to care about the plight of those who have been left out. The responsibility to contribute to distributive justice should be proportionate to the degree of well-being—in other words, those who have more should be more willing to work toward a just society. The moral imperative becomes even greater when we realize that we are also talking about environmental justice—as we make cities more livable we are also creating a more sustainable society for ourselves and the other creatures with whom we share the planet. Where do you fit in?

3 years. To aid the buyer in this effort, the city provides low-interest, long-term loans that effectively become a mortgage on the property.

This has proved to be a very successful program in which everyone wins. The "homesteaders" either end up with a nice home at reasonable cost or else make a considerable profit on selling after the 3 years. The city gets the property back on the tax rolls. Most important, as middle-income people come back into the city, so do stores and other businesses, which provide employment and additional taxes, helping to reverse the entire cycle of urban blight (Fig. 24–15). However, it has also been observed that urban redevelopment tends to drive up prices such that many of the poor are displaced into other decaying parts of the city or even into homelessness.

Thus, while each of these programs provides certain benefits, none provides a total solution. It is being increasingly recognized that addressing urban ills requires a broader perspective of both the problem and its solution.

A society—complex or simple—fundamentally comprises people exchanging goods and services with one another. Money is simply a medium that facilitates the exchange. A bizarre paradox is that cities suffer simultaneously from high rates of unemployment—over 50 percent in many cities—and from dilapidated housing, roadways and bridges in need of repair, a need for better schools and health care, and any number of other things begging for labor. In other words, what we see in many U.S. cities is a situation where society has ceased to function—a suitable exchange of goods and services is not occurring. If cities are to be made livable—and they must be if we are to achieve sustainability—it will be necessary to enable the unemployed of cities to provide the services and skills they can potentially offer.

Daunting as this task is, a project aimed at accomplishing it was launched in Baltimore in 1992. The site of the project, known as Sandtown, is a 72-block area of about 3000 decrepit housing units (some 600 abandoned). Of the 12 000 residents, over 50 percent are unemployed, and most of the rest have incomes below the poverty line. Crime and drugs are rife. The project involves the cooperative leadership of Habitat for Humanity, a worldwide charitable organization devoted to renovating housing for the poor; developer James Rouse, chairman of the Enterprise Foundation; Baltimoreans United in Leadership Development (BUILD); Mayor Kurt Schmoke; and others.

FIGURE 24–16
A major project to rehabilitate a blighted neighborhood in Atlanta involves teaching residents the skills necessary to renovate their own homes. Other residents will be trained as teachers and social workers, and in other trades. (Margaret Miller/Photo Researchers.)

FIGURE 24–15
A few years ago, this area was mostly decrepit abandoned homes. They have been refurbished through the Urban Homesteading program in Baltimore. (Photograph by BJN.)

The plan is to do much more than make Sandtown's housing stock fit and affordable. Residents will be trained in the process and will do most of the work (Fig. 24–16). At the same time, and also by training residents, the goal is to reform the neighborhood schools, health-care facilities, police protection, and other services. "Most important, the objective is to develop community leaders who can get the job done themselves. The goal is nothing less than a rehabilitation of the human spirit in Sandtown."[2]

It is important to observe in this project the cooperative efforts of government, charitable organizations, private enterprise, and most of all the people themselves. In the words of Gary Rodwell, leading organizer of BUILD, "People have to be a part of an effort like this fully. You can't 'service' people out of poverty. People have to be trained and developed and given opportunities to empower themselves to raise their standards of living." The final success of the Sandtown project is yet to be determined, but if it is successful it may well become a model for the rest of the nation.

[2] J. Block, "Reclamation Battle Line Drawn in W. Baltimore," *The Sun*, March 27, 1992.

Perhaps getting into the rehabilitation of poor urban neighborhoods seems like a far cry from nature and environmental science. However, we need only focus on our overall theme of sustainability. A sustainable relationship between the human species and the biosphere will necessarily involve our learning to live sustainably and cooperatively with one another, not only the different segments within our own society but throughout the world. Ultimately, we are one human society on Spaceship Earth. Ultimately, we will achieve sustainability together or suffer the consequences together.

 Review Questions _____

1. In terms of work, shopping, and residence locations, how did cities begin to change following World War II?
2. What was the prime factor that enabled the change described in question 1?
3. Why are the terms *car-dependent* and *urban sprawl* used to describe our current suburban lifestyle and urban layout?
4. How did the ownership of cars and the Highway Trust Fund foster urban sprawl?
5. What are five categories of environmental consequences stemming from the car-dependent lifestyle and urban sprawl?
6. How is suburban sprawl connected to exurban migration and gentrification?
7. What services do local governments provide, and what is their prime source of revenue for these services?
8. How is exurban migration connected to erosion of a city's tax base and growing urban blight?
9. How do exurban migration, urban sprawl, and urban decay become a vicious cycle?
10. What should be the key features of a hypothetical sustainable community?
11. What measures might existing suburbs adopt to move toward becoming sustainable communities?
12. What laws and regulations would have to be enacted or changed to foster the development of existing suburbs into sustainable communities?
13. What are three main features of a livable city? (Use existing livable cities as examples.)
14. How might the way in which property taxes are levied be changed to encourage improvements in inner-city housing?
15. What is meant by urban homesteading, and how does it work to improve cities?
16. What is meant by enterprise zones, and how are they expected to improve cities?
17. Is it necessary to involve the poor and unemployed in city improvements as opposed to simply providing more services and crime protection? Explain.
18. Is rehabilitating cities a job for just government? Who else needs to be involved? Explain.

 Thinking Environmentally _____

1. Describe ways in which tax structures might be changed to level the playing field between cities and sprawling suburbs.
2. What are the major benefits and problems of living in an urban area? In a suburban area? In a rural area? What do you consider to be the ideal living environment? Why?
3. Revitalization of inner cities has become a major concern in many large cities in the United States. Investigate an urban renewal project. Find out what has been accomplished and which renewal measures have been the most successful and why.
4. Imagine that, as an urban planner, you have been asked to draw up plans for the continuing growth and development of a city with a population of 50 000. This small city is located on the ocean and lies about 30 miles from a major city. How would you locate residential, commercial, and industrial facilities to encourage both livability and sustainability?
5. What changes in our economic and political systems would be necessary to make the transition to a sustainable society?
6. Is it feasible or desirable to rebuild or relocate our cities to encourage sustainability? What can be done in existing cities to increase sustainability?
7. What types of mass transit are available where you live? What systems were available 20 years ago? Have these changes been beneficial or harmful?

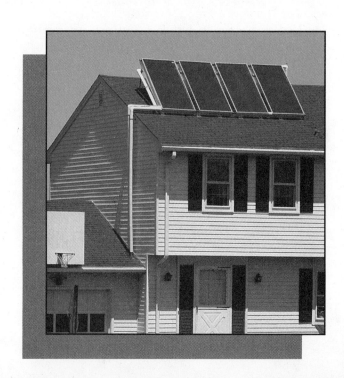

Making a Difference

1. Do not buy any item that might come from an endangered species (for example, turtle shell products, seashells and corals, animal skins, and many cactus species).

2. If you hunt or fish, observe all pertinent laws and kill only what you can use; alternatively, become a photographer of nature—your pictures can be a source of enjoyment to many.

3. When you visit parks and other natural areas, stay on the designated trails. Avoid making new trails that might lead to erosion, and never disturb nesting birds and wild animals raising their young.

4. Buy food products that are derived from extractive use of tropical forests rather than degradative use (for example, "Rainforest Crunch" ice cream, Brazil nuts, as opposed to tropical wood products).

5. Investigate how municipal solid waste is handled in your school or community and what the future plans are for waste management. If recycling is not at least in the planning stage, join or start a group to promote it. Recycle everything you can!

6. Organize a roadside litter cleanup, and solicit newspaper coverage to publicize the problem of roadside litter. If bottles and cans are a problem, join with others in promoting a bottle bill in your state.

7. Consider your consumption patterns: live as simply as possible, so as to minimize your impact on the environment; buy and use durable products, and minimize your use of disposables.

8. Examine your transportation uses: make a high MPG car a top priority; get behind car pools; use alternative transportation; keep your car tuned up and tires inflated and balanced properly.

9. Take advantage of services from your electric company on energy conservation, home energy audits, and subsidized compact fluorescents.

10. Call for strict adherence to NRC guidelines for all currently operating nuclear plants; if you do not support nuclear energy, join an organization that lobbies against it.

11. Study your own home (or plans for a new one) in terms of thermal efficiency (insulation, appliances, etc.) and the potential for using solar water or space heating. Improve your home's thermal efficiency and add solar in whatever ways you can.

12. Investigate land-use planning and policies in your region. Is urban sprawl or urban blight a problem? Become involved in zoning issues, and support zoning and policies that will support rather than hinder ecologically sound land-use.

13. Contact local conservation organizations or land trusts and offer your help in protecting and preserving open land in your region.

14. Find out if there is a Habitat for Humanity project in your area (see Appendix A), and if there is, volunteer some time in support of their work.

Epilogue

Throughout this text we have had to point out many unsustainable interactions between humankind and the environment that characterize this time in human history. These are the consequences of the Cornucopian world view, the predominant ideology behind the exploitation of the natural world that has led modern society for a long time. It should be evident to the reader that this world view has failed us. Things are not going well, not in the industrialized nations of the North nor in the less-developed and economically poor nations of the South.

We are living on borrowed time, and tragically, we are borrowing the time from coming generations—in effect, our children and grandchildren. The longer this current generation puts off coming to terms with the carrying capacity of the biosphere—living sustainably—the harder it will be for future generations to make it. We (your authors) are absolutely convinced that enormous changes are coming, that the environmental revolution will occur. The choice before us is whether the revolution will be an orderly transition to a sustainable future, or a turbulent, often violent period of economic decline and social instability as changes are imposed by environmental constraints.

In Chapter 1 we compared the opposing Cornucopian and environmental world views, and pointed out that these two world views map out radically different paths to the future. We are convinced that the overthrow of the Cornucopian world view is an essential element of the environmental revolution. In its place, we are recommending the environmental world view, one that is characterized by environmental stewardship and concern for economic justice. Stewardship implies care for the natural world, recognizing its great potential to meet human needs but also its limits, and treating nature not as resources for the taking but as a great gift to be conserved. Justice implies a concern for other humans, caring about those who are impoverished and hungry, and following up that concern and care with action and policies that go beyond mere economic self-interest and recognize a higher morality.

Throughout the text we have tracked environmental problems through the different levels of our social order: global actions leading to international accord, political and economic actions within nations leading to public policy, and individual actions leading to lifestyles. The environmental revolution will of necessity involve changes at all of these levels.

For a brief time in June 1992 the world's attention was focused on Rio de Janeiro, site of the United Nations Conference on Environment and Development (UNCED). The difficulties of forging international agreements on environmental concerns were evident in the outcome of the conference. Nations of the South have their sights set on enlisting the industrialized North to help them to develop more rapidly and in environmentally sustainable directions. Nations of the North, no longer motivated by cold-war rivalry, have already drastically cut aid programs and are clearly mostly concerned about their own economic growth and their own poor—Eastern Europe and the former Soviet republics. The environment has emerged as the bone of contention and at the same time the most promising base for cooperation. Global problems such as disappearing ozone and climate change require the concerted effort of all players, and nations of the South have given notice that they will hold out their cooperation in exchange for development aid. Although the long-lasting effects of the UNCED conference remain to be determined, the immediate impact was to dramatize environmental concerns as never before.

Within nations, public policy changes are already under way, although much remains to be accomplished. In the United States, the past 25 years have seen the establishment of regulatory agencies at all levels of government. Numerous nongovernmental organizations aimed at protecting the environment have been formed by citizens. Many environmental laws have been passed, and environmental concerns are now involved in political and economic decision making from local to national levels. However, all too often the outcome of those decisions continues to reflect Cornucopian attitudes. We are still reluctant to get at the root causes of environmental ills, and the decisions made at the national level persist in elevating economic concerns over the environment. It is encouraging to note that environmental groups are often quite effective in mobilizing public concern and organizing citizens to press for environmentally wise policies. They often provide a vital counter to the Cornucopian special interest groups.

What about individual actions? Your authors recognize that the problems seem so enormous and forces with vested interest in the status quo seem so entrenched that you may often feel that there is nothing you can do that will make any difference. Yet, everything you do already does make a difference. You cannot avoid the facts: The car you drive, the products you use, the wastes you throw away, and everything else you do contribute to the problems to a greater or lesser extent. Whatever you may do to lessen impacts on the environment is part of the solution. Consequently, it is not a matter of having an effect. It is simply a matter of asking yourself, *Will I be part of the problem or part of the solution?*

There are four levels on which you may participate to work toward a sustainable society (Fig. E–1):

○ Individual lifestyle change
○ Political involvement
○ Membership and participation in environmental organizations
○ Career choices

Lifestyle changes may involve such things as switching to a more fuel-efficient car; recycling paper, cans, and bottles; adapting your home to solar energy; composting and recycling food and garden wastes into your own lawn or garden; or living closer to your workplace. Political involvement means supporting and voting for candidates with strong envi-

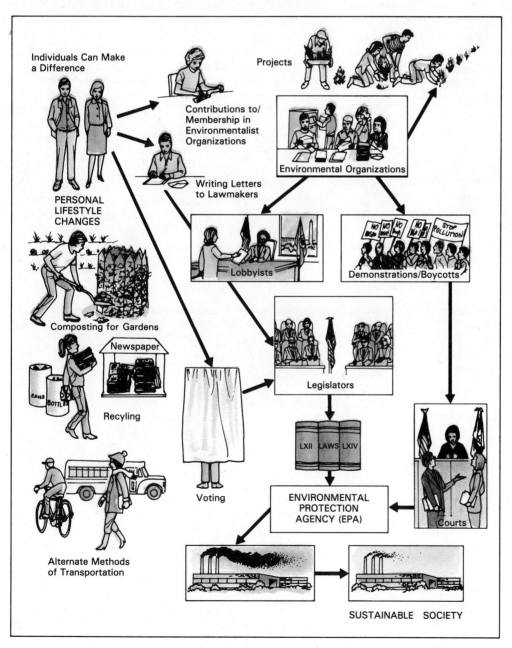

FIGURE E–1
Avenues of individual action.

ronmental positions and writing or making calls in support of environmental legislation. Environmental candidates don't get elected and environmental bills don't get passed unless constituents have *expressed* support for them.

Membership in nongovernmental environmental organizations can enhance both lifestyle change and political involvement. As a member of an environmental organization you will receive and may help disseminate information making you and others more aware of particular environmental problems and things you can do to help. Specifically, you will be informed regarding environmentally significant legislation so that you may focus your political efforts at the most effective time and place. Also, your membership and contributions serve to support lobbying efforts of the organization. A lobbyist representing only him or herself has relatively little impact on legislators. On the other hand, if the lobbyist represents a million-member organization that can follow up with that many phone calls and letters (and ultimately votes) the impact is considerable. Finally, where enforcement of existing law has been the weak link, some organizations such as Public Interest Research Groups, the Natural Resources Defense Council, and Environmental Defense Fund have been very influential in bringing polluters or the government to court to see that the law is upheld. Again, this can be done only with the support of members.

Finally you may choose to devote your career to promoting solutions to environmental problems. Environmental careers go far beyond the traditional occupations of wildlife or park management. There are any number of lawyers, journalists, teachers, scientists, medical personnel, entertainers, and others focusing their talents and training on environmental issues or hazards. There are innumerable business opportunities in pollution control, recycling, waste management, environmental monitoring and analysis, nonchemical pest control, production and marketing of organically grown produce, and so on. Some developers concentrate on rehabilitation and reversal of urban blight as opposed to contributing more to urban sprawl. Some engineers are working on plans for pollution-free vehicles to help solve the photochemical smog dilemma of our cities. Indeed, it is difficult to think of a vocation that cannot be focused on promoting solutions to environmental problems.

We would like to close with a statement made by Lester Brown, president of Worldwatch Institute, in *State of the World 1992*:

Until now the Environmental Revolution has been viewed by society much like a sporting event—one where thousands of people sit in the stands watching, while only a handful are on the playing field actively attempting to influence the outcome of the contest. Success in this case depends on erasing the imaginary sidelines that separate spectators from participants so we can all get involved. Saving the planet is not a spectator sport.

There are many more chapters of environmental science to be written, but they will be written by you, perhaps not in a formal way on paper, but by the things you do and accomplish toward bringing us all toward a sustainable society.

ABC Video Case Studies

Each of the following 18 ABC Video Case Studies will bring you face-to-face with a significant environmental issue somewhere in the world, if not in your own backyard. Through the cooperation and courtesy of Prentice Hall and ABC News, the authors of this text have selected video segments from such award-winning ABC news programs as **Nightline, 20/20** and **World News Tonight.** Each video segment has a written overview or abstract that appears in the following section of the text, along with several questions that have been designed to encourage you to focus on the poignant issues and controversies within each case study. Each written case study has a chapter and/or topic reference that serves to emphasize the relevance between the case study and specific sections of the text. We hope that each of these case studies will foment class discussion and debate, for environmental science is a dynamic field that should always accommodate diverse opinions and perspectives. People must talk about the environmental issues of our day; opinions and perspectives should be exchanged and examined.

Prentice Hall is making all of these ABC video segments available to your instructor for classroom use on two convenient long-playing video cassettes. We encourage you to explore these cases as you use this text.

Officially named the United Nations Conference on Environment and Development, the "Earth Summit" was held in Rio de Janeiro from June 3 to June 14, 1992. Three major items dominated the agenda: a global warming treaty, a treaty to protect biodiversity, and Agenda 21 (which sets forth a blueprint for sustainable development in the "South," or Third World). The first news clip highlights a confrontation between the Third World and the industrial nations by looking at India, the world's second most populous country. Basically, if India modernizes along the path of the industrial countries, the global impact will be enormous. The alternative? Development by way of nonpolluting technologies, which will require massive aid from the already developed nations.

The second clip features the involvement of the United States in the major treaties signed: President Bush did sign a watered-down global warming treaty, but he refused to sign the biodiversity treaty. By most accounts, the United States failed to give leadership at the Earth Summit and instead appeared to act primarily to protect the interests of sectors of the U.S. economy. (Not mentioned in the clip was the very important establishment of a U.N. Commission on Sustainable Development to oversee implementation of Agenda 21 projects).

Nebel/Wright reference: Epilogue, global warming in Chapter 16, biodiversity in Chapter 18, and sustainable development in many places, especially Chapters 7, 8, 9, 18, and 19.

QUESTIONS:

1. What are the complaints of the developing countries?
2. Why did President Bush refuse to sign the biodiversity treaty?
3. What was the significance of the Earth Summit meeting?

ABC CASE STUDY
Famine and Relief Efforts in Somalia

The world has watched while Somalia, a small East African nation, slid into political anarchy and famine. A two-year civil war and an extended drought combined to bring on conditions of hunger and starvation that were taking an estimated 5000 lives a day. ABC News "Nightline" featured the crisis in Somalia on January 29, 1992. A video clip from Carlos Mavroleon, a freelance journalist, documents the war and food shortages with graphic reality. Ted Koppel then interviews U.N. Undersecretary General James Jonah, Andrew Timpson of Save the Children UK, and Rakiya Omaar of Africa Watch, a human rights organization. The group discusses the basic issue of responsibility for coming to the aid of the Somalians.

Should the United Nations or the Organization of African Unity be responsible? In light of the dangers, how can aid be brought without a political resolution? The group has trouble agreeing on answers to these questions, and viewers are left with the prospect of a coming famine.

The second and shorter clip is from ABC News, August 13, 1992. The predicted famine has come, more than a million Somalians face starvation, and still there are difficulties getting food aid into Somalia. The news clip highlights the dangers of delivering food and the remorseless starvation bringing death to Somalian villagers. (Since that time, the United Nations has sent armed troops, the United States has airlifted food, and certainly many lives are being saved by the international effort—but the famine continues to bring untold human suffering and death.)

Nebel/Wright reference: Chapter 8, especially pp. 180–184.

QUESTIONS:

1. What was the political impasse discussed by Mr. Jonah and Ms. Omaar?
2. What was done to provide security for the relief program?
3. How could the international community have improved its response to the plight of the Somalians?

ABC CASE STUDY
Hazardous Wastes in the Wrong Place

Hazardous wastes have a way of turning up where they're least expected. The first news segment documents some of our past sins. Massachusetts fishermen have been catching barrels containing dangerous wastes in their nets. Thousands of barrels of radioactive wastes and toxic chemicals were dumped, legally, some 20 miles off the coast from the 1940s to the 1970s. The wastes were dumped in water only 300 feet deep, in areas only poorly identified. EPA is now involved in locating the wastes.

The second news clip presents a disturbing situation: Hazardous waste dumps and incinerators tend to be located in towns and neighborhoods where the residents are nonwhite. The issue is environmental racism, and the video describes some of the possible effects of the toxic wastes. In response to the issue, grass-roots minority groups are organizing and are beginning to be heard across the country.

Nebel/Wright reference: Chapter 14, pp. 330 and 331, especially Ethics Box: Racism and Hazardous Waste.

QUESTIONS:

1. How does George Perry (the atomic garbageman) figure in the problem of the drums in the ocean?
2. Recount the story told by Amos Favorite from his home in Louisiana. What action has he taken?
3. Why were toxic wastes dumped in the shallow ocean? Why were waste dumps sited near minority neighborhoods?

ABC CASE STUDY
Environmental Concerns under Communism

What would things be like if we simply ignored environmental concerns in the interest of promoting economic production? The answer is becoming clear now that the end of the Cold War has opened up the countries of eastern Europe. This video documents life in Bitterfeld, East Germany, as it was shortly after the Berlin wall came down: air choking with sulfur dioxide and chlorine gas, rivers fouled with industrial waste, land that no longer can grow trees, people dying young with cancer, children with skin diseases. Hans Zimmerman and Rainer Fromann describe how the East German government suppressed information on health and environmental conditions while promoting the "workers' and farmers' paradise" propaganda.

Journalist Bob Brown takes a walk on a Nature Trail that also bears a sign "Keep Out—Dangerous to Life." Zimmerman, now a member of the German parliament, speaks poignantly of shared guilt in response to a question of how people in charge could let things get so bad. The program ends with a sober estimate of what must be done and what it may cost to remedy the environmental problems in the industrial regions of former East Germany.

Nebel/Wright reference: Chapter 14 In Perspective Box, p. 323: Grass Roots in a Toxic Wasteland; Chapter 15, pp. 340–344.

QUESTIONS:

1. What kind of human health effects were described in the document? What chemicals might be responsible for such effects?
2. What attempt was made by the authorities to lessen the impact of life in Bitterfeld on the children?
3. Compare environmental regulation in the United States and East Germany in the light of grass roots action and government action.

ABC CASE STUDY
Global Warming and the Scientific Community

The Earth Summit meeting of June, 1992, had global warming at the top of its agenda. Citing the scientific uncertainty about global warming, the Bush administration effectively lobbied to water down the global warming treaty. This news segment investigates attempts to document the impacts of early global warming. Biologists examine the phenomena of coral bleaching as a response to warming ocean water and the shift northward of populations of some heat-sensitive plants in the United States.

The complexity of the issue is introduced: What effect will clouds have? Will the oceans absorb the excess heat? Will added smoke from volcanoes and burning forests keep us shaded? Scientists F. Fred Singer and Stephen Schneider give opposing views on the issue, and the continuing uncertainty suggests that the search for evidence of global warming is vital to the future of our civilization.

Nebel/Wright reference: Chapter 16, pp. 368–375, especially Ethics Box: Staking a Claim on the Moral High Ground, p. 374.

QUESTIONS:

1. What evidence for warming comes from the South Pacific?
2. What are the advantages and disadvantages of taking action to prevent global warming before we are absolutely certain it is happening?

ABC CASE STUDY
Break in the Ozone Shield

Sometimes, one individual can catalyze action in the environmental arena by careful work and subsequent advocacy. Rachel Carson comes to mind, as does chemist Sherwood Rowland. Rowland, featured in this news clip, has done more than any other single person to bring about a global response to the threat posed by chlorofluorocarbons (CFCs) to the ozone shield. The video briefly documents Rowland's progress from turning down professional basketball to go to graduate school, to initial publication of the atmospheric impacts of CFCs in 1974, to rejection of his claims by industry, to final acceptance by the global community.

The occasion of the broadcast was the directive by President Bush to speed up the phaseout of CFC production because of new evidence of ozone depletion in the Arctic. Rowland has now turned his attention to global warming and closes with a comment on the lesson we should have learned from our experience with the ozone shield and CFCs.

Nebel/Wright reference: Chapter 16, pp. 375–381, especially In Perspective Box: Lessons from the Ozone Treaty, p. 380.

QUESTIONS:

1. What is the lesson Rowland believes we should learn from this issue?
2. How does Rowland link the ozone depletion problem and global warming?

ABC CASE STUDY
The Place of Risk Analysis in Pollution Cleanup

Faced with a host of environmental problems, how does our society make decisions on which ones to address, given the budgetary constraints? The answer is: We have to rank the problems in order of their risk. This segment looks at environmental risks from two perspectives: scientific logic and public concern. The video points out that the experts and the public often do not agree in their ranking of environmental risks. EPA head William Reilly wants to rely more on scientific risk analysis than we have in the past, and he cites the work of his science advisory board in ranking risks.

Peter Sandman of Rutgers University takes the view that EPA's strategy will backfire if public concerns are not considered. Citizens in Louisiana are shown interacting with their state Department of Environmental Quality to help set the state's priorities in assessing environmental hazards, and the segment closes with the view that both risk analysis and public concern must be considered.

Nebel/Wright reference: Chapter 17, pp. 393–397.

QUESTIONS:

1. What are some risks ranked high by the public and low by the science advisers? Vice versa?
2. How can public concerns be integrated with science-based risk analysis in forming public policy?

ABC CASE STUDY
Council on Competitiveness: Bush's Shadow Government

Environmental policy is usually viewed as Congress making laws and administrative agencies setting up and enforcing regulations based on those laws. ABC News examines what appears to be a short-circuit in the system: the Council on Competitiveness, chaired by Vice President Quayle. The two segments of news explore the actions of the Council and different views of the basic politics involved.

The Council on Competitiveness was established "to relieve the private sector of regulatory burdens . . . create more opportunities and, thus . . . create more jobs," in the words of Quayle. The council reviews all federal agency regulations and has the power to change those regulations in response to complaints from the business sector. Critics point out that the council is unaccountable to Congress or the public, and it is in effect capable of undermining the work and authority of many federal agencies. A number of the council's rulings are highlighted in the video, especially its work on the new Clean Air Act.

Representative views from industry are also presented in support of the council's work, and its basic strategy—to cut the costs of regulations and keep jobs—is articulated by Quayle. Criticism of the council's work comes from EPA officials, state environmental officials, and members of Congress, all focusing on the severe impact of the council on environmental protection.

Nebel/Wright references: Chapter 17, Ethics Box: Conflict of Interest, p. 389; Chapter 19, p. 436 on wetlands policy.

QUESTIONS:

1. What is the basic strategy behind creation of the Council on Competitiveness?
2. What are three rulings by the council that have significantly turned back environmental regulations?
3. Discuss the legitimacy of the Council on Competitiveness in the context of establishing public policy.

ABC CASE STUDY
Biodiversity: What Can Be Done to Protect It?

The amazing diversity of species of plants, animals, and microbes spread over the earth is being affected by human activities; as species become extinct, we are losing forever some of the biological wealth of our planet. Two news clips address this concern. The first, on the eve of the twentieth Earth Day celebration, documents the impact of development on desert animals in Arizona and on many plants and birds on the Hawaiian Islands. Some positive effects of the Endangered Species Act are then shown, along with important limitations of the act. The end of the clip focuses on the worldwide disappearance of amphibian populations, which is compared with the canary in the coal mine—a warning of trouble ahead for us all.

The second clip presents a dispute between a group of botanists at the University of Wisconsin and the U.S. Forest Service. The immediate issue is how the Forest Service is managing two national forests in northern Wisconsin. The botanists have brought suit in a U.S. District Court against the Forest Service, claiming that the Service has not considered the maintenance of biodiversity in their management strategy, but has promoted a pattern of logging that would fragment the ecological community and lead to significant losses of biodiversity. The Forest Service does not want to maintain any areas in the forests that would be off-limits to logging and claims that they are capable of managing the entire forest with ecological sensitivity. By bringing suit,

the botanists are hoping to extend the requirements of environmental impact statements to include impacts on biodiversity for all federal projects.

Nebel/Wright reference: Chapter 18, pp. 409–416; see also Science, 257 (September 18, 1992), pp. 1618–1619.

QUESTIONS:

1. What two situations are responsible for major losses of biodiversity in the Hawaiian Islands?
2. Compare the management plans of the University of Wisconsin botanists with those of the U.S. Forest Service.
3. What are some possible reasons why the Forest Service doesn't include concerns for biodiversity in its management guidelines?

ABC CASE STUDY
Challenge on the Right—Environmental Movement As Subversive

"We intend to destroy the environmental movement once and for all" (Ron Arnold). "There's no ozone depletion, there's no crisis" (Rush Limbaugh). "We now face a global ecological crisis that is more serious than anything human civilization has ever faced" (Al Gore).

This newscast pits defenders of the environment like the Sierra Club and Senator Al Gore against critics of the environmental movement. The first segment hears complaints from Arnold and others that environmentalists are infringing on the rights of people to use nature, highlighting the philosophical conflict between those who want to make use of protected public lands and

those who believe that such lands should be kept in their natural state. Bruce Hamilton of the Sierra Club speaks out in favor of preservation.

The majority of the video is a lively debate, moderated by Ted Koppel, between Rush Limbaugh, radio talk show host, and Al Gore, Democratic vice-presidential candidate and author of *Earth in the Balance: Ecology and the Human Spirit.* Limbaugh believes that environmentalists are undermining the "American way of life" by pushing for radical measures that stifle capitalism and economic growth. Essentially, he says, the earth is not fragile and human actions are not really threatening the environment. Gore

counters with reasons why he is convinced that humans are now capable of doing serious damage to the global environment. He responds to Limbaugh's claims that ozone depletion is no problem and that environmentalists are simply antibusiness radicals who want to go back to the Stone Age.

Nebel/Wright reference: Chapter 16, 19, especially In Perspective Box: Environmental Backlash, Chapter 19, p. 440.

QUESTIONS:

1. How would you characterize the "wise-use" movement?

2. How does Gore deal with the disagreement in the scientific community over the link between CFCs and the ozone hole?

3. Limbaugh claims that the environmental crisis is a fiction being used by some people to expand the role of government and shut down business. How does Gore counter this claim?

ABC CASE STUDY
Mismanaging the National Forests

Do our national forests mean more to us than wood for the economy? The issue presented in this news clip is the pressure to increase logging on public lands, a pressure that has engendered serious protests from within the Forest Service. John Mumma, a regional forester in the service, refused to authorize higher timber quotas in his forest in Montana on grounds that he would be breaking environmental laws. Shortly after that, he was "reassigned" to another job. Mumma's contention is supported by a host of other employees and former employees. Their concern is twofold: Heavy pressure to log the forests is destroying crucial habitat for sensitive wildlife populations, such as the spotted owls, and the Forest Service actually loses money because of management practices. One study is cited that shows that the Forest Service actually loses money to the tune of $300 million a year. Forest Service spokesmen dispute that figure and maintain there there will always be controversy over the economic and environmental interests in the national forests.

Nebel/Wright reference: Chapter 19, pp. 438–440.

QUESTIONS:

1. What aspects of Forest Service management practice lead to lost revenues?
2. In your view, how should the Forest Service change its practices?

ABC CASE STUDY
Who Wants the Trash?

Some 15 million tons of trash cross state lines each year, and the states are helpless to do anything about it. The problem of where to put the growing heaps of trash is left in the hands of local towns. This case study explores two very different kinds of response to this traffic in trash. The first news clip starts in Hinsdale, in western Massachusetts, where a developer wants permission to put a huge landfill that will take in garbage from New York, New Jersey, and New England. Private landfills in Indiana and Virginia are also featured. It is clear that once town residents approve a private landfill, they basically lose control of how the land is filled; the film shows convoys of trash converging on Centerpoint, Indiana. (As a postscript, in the case of Hinsdale, a referendum was taken, the proposal was turned down in a vote of 830 to 59, and the area was designated as a state Area of Critical Environmental Concern, which precludes landfills.)

Taking an entirely different tack, Riverview, Michigan, has welcomed trash from surrounding municipalities and has built its landfill into a 125-foot-high mound ("Mount Trashmore"), which hosts a skilift in the winter. Profits from the landfill bring in $8 million a year, the town has built a new City Hall, fully funded its school programs, and cut property taxes of its residents. Expensive homes and a 27-hole golf course have been built next to the landfill. Clearly, the town has embraced the traffic in trash and reaped a continuing profit, an example not unnoticed by other towns and cities with their own trash problems.

Nebel/Wright reference: Chapter 20, especially In Perspective Box: Trash on the Move, p. 447.

QUESTIONS:

1. What happened in Centerpoint, Indiana, and Selma, Virginia, to turn local people against landfills?
2. List the benefits and costs of maintaining the landfill to the people of Riverview.
3. Why do you think the people of Hinsdale were so opposed to the landfill while Riverview people welcomed theirs?

ABC CASE STUDY
Oil and the Arctic National Wildlife Refuge

An ongoing battle in national energy policy pits defenders of the Alaskan wilderness against oil companies. The issue is whether to push ahead with exploration and development of oil suspected to lie under the Arctic National Wildlife Refuge (ANWR). Political and economic pressures to develop the oil are illustrated in this news clip: a new source of oil for the Alaska pipeline when Prudhoe Bay runs out, billions of dollars for the oil companies and the Inupiat Eskimos who live in northern Alaska, and some relief from our dependence on foreign oil. Against these interests is the concern for the fragile tundra wilderness and especially the large herd of caribou that use the land over the oil as crucial breeding grounds. Is more oil money better for the native peoples than continued use of the caribou for food and skins? Can hundreds of miles of new roads and pipelines and scores of drilling sites coexist with wildlife in a fragile wilderness?

The news clip makes it clear that behind this issue is the larger issue of a national energy policy that so far has been much more concerned with using oil than conserving it. The refuge is thought to hold 3.5 billion barrels of oil, about 6 months supply at the current rate of use in the United States. We are left with the question, Shouldn't there be a better way to address our energy needs than to sacrifice one of the remaining wilderness areas of our land (and a refuge at that)?

Nebel/Wright reference: Chapter 21, pp. 465–474, especially Ethics Box: Trading Wilderness for Energy in the Far North, p. 473.

QUESTIONS:

1. What are the reasons given for going ahead with development of the oil under ANWR?
2. What are the potential negative impacts of the exploration and development?
3. How does this conflict fit into the development of a long-term energy strategy for the United States?

ABC CASE STUDY
The Case for Energy Conservation

Environmentalists are accustomed to regarding the utility companies as part of the problem, but as these two news clips show, this does not have to be true. The first segment documents a Kraft General Foods ice cream plant near Boston that was in serious economic trouble 2 years prior to the report. Boston Edison came in with a complete energy renovation plan that cut energy costs by $400 000 a year and saved the plant from extinction. Utilities all over New England are aggressively pursuing conservation measures with their commercial and residential customers, providing energy audits, shower heads, and subsidizing use of energy-efficient fluorescent light bulbs. Instead of selling more power, the utilities are selling energy efficiency and making a genuine contribution toward solving our energy problems.

The second news clip crosses the country to California, where the focus is on energy efficiency in order to reduce the threat of global warming. Again, the utilities have realized that it is much less costly to invest in energy savings than to construct new power plants. In promoting energy conservation by helping people to cut their electricity bills, the utilities are reducing their use of fossil fuels and thus holding down global warming. Two power companies are also investing in solar and wind power as the best long-run answer to pollution. The main point: The protests coming from some businesses and the federal government about the economic impacts of cutting our use of fossil fuels appear to be largely unfounded, and the utility industry may well become major proponents of action to curb global warming.

Nebel/Wright reference: Chapter 21, p. 477.

QUESTIONS:

1. Why are utility companies, who sell electricity to their customers, trying to get those same customers to use less electricity?
2. What are the main steps being taken by the utility companies to help residents save energy?

ABC CASE STUDY
Fuel-Efficient Cars: Detroit versus Japan

In crafting the Clean Air Act of 1990, Congress and the administration refused to increase the fuel-efficiency standard for cars because of intense lobbying by the U.S. automobile industry, and so it remains locked at 27.5 miles per gallon (mpg). This news clip features the 1992 Honda Civic VX, which went from 33 mpg to 51 mpg in 1 year without a major cost increase or downsizing. Detroit carmakers have maintained that you can't have fuel efficiency in anything but the smallest cars; the Geo Metro, shown in the news clip, is their case in point. The Honda Civic has proven otherwise.

Detroit is also losing out in the race to develop useful electric cars. Los Angeles has had to turn to a Swedish firm in a joint venture to develop an electric car to meet the coming requirements of the law in California. The feature emphasizes the difference in basic attitudes of the Japanese and U.S. car manufacturers regarding better mileage and electric cars. The result? U.S. car companies will continue to play catch-up because of their reluctance to change.

Nebel/Wright reference: Chapter 15, pp. 353, 354; Chapters 21 and 24.

Nebel/Wright reference: Chapter 15, pp. 353, 354; Chapters 21 and 24.

QUESTIONS:

1. What did Honda do to get such high mileage in the Civic without sacrificing size and speed?
2. What are the attitudes of the two country's auto industries, according to Senator Bryan, and why do you suppose there is such a difference?

ABC CASE STUDY
Oil on the Sound: Alaskan Oil Spill

On the evening of March 23, 1989, the *Exxon Valdez* left the port of Valdez loaded with Prudhoe crude from the Alaska oil pipeline. The tanker had state-of-the-art navigational and safety equipment, the night was calm, and yet, at 12:04 A.M. the ship plowed into a well-known, shallow rock formation and ripped her hull, spilling 11 million gallons of crude oil. This "Nightline" newscast, almost 2 weeks after the spill, investigates the impacts of the oil, the attempts made to control the spill, and the connections between the oil spill and current energy policy. After documenting the spill and its consequences, Ted Koppel interviews Charles DiBona, president of the American Petroleum Institute, and George Frampton, president of The Wilderness Society.

Frampton presents the view that the spill highlights the almost total dependence of our economy on the oil industry. He emphasizes the need for a strategy based on conservation, use of alternative fuels, and investment in renewable energy resources. DiBona takes issue with Frampton's claim that we can get much more oil out of existing wells and argues that we need to develop our domestic sources of oil in spite of oil spills. Koppel stirs things up by proposing raising gasoline prices to European levels, and he probes DiBona for steps for the oil industry to ensure that such an oil spill doesn't happen again.

Nebel/Wright reference: Chapters 21 and 23.

Nebel/Wright reference: Chapters 21 and 23.

QUESTIONS:

1. Why did the *Exxon Valdez* go on the rocks?
2. What are George Frampton's views on how to reduce our dependency on oil?
3. According to Charles DiBona, what steps is the oil industry taking to prevent such spills from occurring? How does DiBona minimize the spill's impacts at the end of the program?

ABC CASE STUDY
The Nuclear Waste Dilemma

The focus of this newscast is on the growing stockpile of nuclear wastes from the nuclear power industry. Currently, the wastes are stored in protective pools and above-ground containers at the sites of power plants. Everyone agrees that this is only a temporary answer to the waste storage problem; the permanent answer is geologic disposal—deep burial in the earth. The site selected by the government and the industry is Yucca Mountain in the Nevada desert, shown in the newscast.

Of course, Nevadans, who don't have nuclear plants of their own, are unhappy with this site selection. A million dollars a day is being spent on studies of the Yucca Mountain site, while governmental action clears the way for the eventual disposal by the year 2010. An attempted state veto of the site appears to be doomed, and current work is focusing on the geological properties of the mountain, which may eventually hold 70 000 tons of high-level waste when it is finally developed.

It appears that Yucca Mountain is the only site being considered for waste disposal from nuclear power plants. If this site is blocked by political action or by scientific considerations, the waste will live on at the power plants.

Nebel/Wright reference: Chapter 22, pp. 491–494, especially Ethics Box: Showdown in the New West, p. 493.

Nebel/Wright reference: Chapter 22, pp. 491–494, especially Ethics Box: Showdown in the New West, p. 493.

QUESTIONS:

1. What are the arguments pro and con regarding the desire of the state of Nevada to block the disposal plan?
2. How is the nuclear power industry involved in this situation?
3. How important is this problem to the future of the nuclear power industry?

ABC CASE STUDY
Energy Alternatives in Our Future

One-third of all air pollution in the United States comes from burning coal, oil, and gas in electrical power generation. Two important alternatives to the fossil fuels are explored in this newscast: solar and wind power. Not only are they pollution-free, they are also limitless sources of energy. The clip visits California's Altamont Pass, where 4000 wind turbines now generate enough electricity to supply every home in San Francisco. Then we shift to the Mojave Desert, where solar thermal power is producing electricity with solar trough collectors. The costs of these two energy sources are compared with current costs of electricity from coal and gas, and they are getting very close.

The newscast mentions that federal subsidies of clean power are not forthcoming because the Energy Department is unwilling to interfere with the free enterprise system. In the meantime, Japanese and European firms are actively developing wind and solar technologies, as shown in a new California wind farm using Mitsubishi turbines. Sun and wind are seen as the best way to deal effectively with the problems of air pollution and global warming.

Nebel/Wright reference: Chapter 23, pp. 512–514, 517–519.

QUESTIONS:

1. What are costs of fossil fuel-generated electrical power that do not show up on your electric bill?
2. How could wind and solar power be made competitive today?
3. What benefits would come from all-out development of these energy sources?

Bibliography

General References

BROWN, LESTER R. "Launching the Environmental Revolution." *State of the World 1992*, Chapter 11. Washington, D.C.: Worldwatch Institute, 1992.

COUNCIL ON ENVIRONMENTAL QUALITY. *Environmental Quality: 22nd Annual Report*. Washington, D.C.: Council on Environmental Quality, 1992. Annual report on the state of the environment in the United States, with policies, issues, and data.

DALY, HERMAN E., AND JOHN B. COBB, JR. *For the Common Good: Redirecting the Economy toward Community, the Environment, and a Sustainable Future*. Boston: Beacon Press, 1989.

THE EARTHWORKS GROUP. *50 Simple Things You Can Do to Save the Earth*. Berkeley, CA: EarthWorks Press, 1989.

THE EARTH WORKS GROUP. *The Next Step: 50 More Things You Can Do to Save the Earth*. Kansas City, MO: Andrews and McMeel, 1991.

FRENCH, HILARY F. "After the Earth Summit: The Future of Environmental Governance." *Worldwatch Paper 107*. Washington, D.C.: Worldwatch Institute, 1992.

GORE, AL. *Earth in the Balance: Ecology and the Human Spirit*. Boston: Houghton Mifflin, 1992. A bold and visionary portrayal of the current environmental scene and an outline of what can be done to turn things around.

MATTHEWS, JESSICA TUCHMAN, ED. *Preserving the Global Environment: The Challenge of Shared Leadership*. New York: W. W. Norton, 1991.

PARKER, JONATHAN, AND CHRIS HOPE. "The State of the Environment: A Survey of Reports from Around the World." *Environment* 34 (Jan/Feb 1992), pp. 19–20, 39–45.

THE WORLD RESOURCES INSTITUTE. *World Resources 1992–93*. New York: Oxford University Press. An annual compilation of data and reports covering a wide range of environmental issues.

WORLDWATCH INSTITUTE. *State of the World 1992: A Worldwatch Institute Report on Progress toward a Sustainable Society*. Washington, D.C.: Worldwatch Institute. An annual collection of articles from the institute covering an array of environmental issues.

WRIGHT, RICHARD T. *Biology through the Eyes of Faith*. San Francisco: Harper/Collins, 1989. Chapter 9 ("Stewards of Creation") and Chapter 12 ("The Environmental Revolution") deal with environmental concerns from a Christian perspective.

Chapter 1. Introduction: The Aim of Environmental Science

APPENZELLER, TIM, ED. "Global Change." *Science* 256 (22 May 1992), pp. 1138–1147.

BABBITT, BRUCE. "Earth Summit." *World Monitor* (Jan 1992), p. 26.

BROWN, LESTER R. "Launching the Environmental Revolution." *State of the World 1992*, pp. 174–190. Washington, D.C.: Worldwatch Institute, 1992.

CAREY, JOHN, ET AL. "Fighting to Save a Fragile World." *International Wildlife* (March/April 1990), pp. 4–21.

CONNIFF, RICHARD. "Blackwater Country." *National Geographic* (April 1992), pp. 34–63.

DUNLAP, RILEY E. "Public Opinion in the 1980's: Clear Consensus, Ambiguous Commitment." *Environment* 33 (Oct 1991), pp 10–15.

DURNING, ALAN. "The Grim Payback of Greed." *International Wildlife* 21 (May/June 1991), pp. 36–39.

Earthwatch. "Biology and the Balance Sheet." (July/Aug 1992), pp. 6–10.

EHRLICH, ANNE, AND PAUL EHRLICH. "Thinking about Our Environmental Future." *EPA Journal* 16 (Jan/Feb 1990), pp. 40–45.

FEENEY, ANDY. "Gambling on Growth." *Environmental Action* 22 (July/Aug 1990), pp. 12–15.

FLATTAU, EDWARD. "Interview, Lester Brown—Measuring the Pulse of the Earth." *E Magazine* (Nov/Dec 1990), pp. 12–15.

GRANT, LINDSAY. *The Cornucopian Fallacies*. Washington, D.C.: The Environmental Fund, 1982.

International Wildlife. "Save Our Planet." (April/May 1990), p. 29.

MARTIN, AMY. "Environmental Careers." *Garbage* (Jan/Feb 1992), pp. 24–31.

MCCLOSKEY, J. MICHAEL, AND CAROL POPE. "Together in Time?" *Sierra* 77 (May/June 1992), pp. 96–99.

MCCOY-THOMPSON, MARI. "Three Plans for a Greener Tomorrow." *WorldWatch* (March/April 1990), pp. 7–8.

MCDONALD, NORRIS, ET AL. "Grass-Roots Groundswell." (a series of articles), *EPA Journal* (March/April 1992), pp. 45–53.

MEADOWS, DONELLA, ET AL. *Beyond the Limits*. Post Mills, VT: Chelsea Green Publishing, 1992.

MILIUS, SUSAN. "The Law with the Big Loophole." *National Wildlife* 30 (Aug/Sept 1992), pp. 40–43.

MILIUS, SUSAN, ET AL. "People Who Make a Difference." *National Wildlife* 29 (Oct/Nov 1991), pp. 40–47.

NASH, STEVE. "What Price Nature?" *BioScience* 41 (Nov 1991), pp. 677–681.

National Wildlife. "The 24th Environmental Quality Index." (Feb/March 1991), pp. 33–40.

National Wildlife. "The Environmental Quality Index." (Feb/March 1992), pp. 33–41.

National Wildlife. "Wildlife Digest." (Aug/Sept 1992), pp. 25–28.

ORIANS, GORDON H. "Ecological Concepts of Sustainability." *Environment* 32 (Nov 1990), pp. 21–24.

PONTING, CLIVE. "Historical Perspectives on Sustainable Development." *Environment* 32 (Nov 1990), pp. 31–33.

POST, JAMES E. "Managing As If the Earth Mattered." *Business Horizons* 34 (July/Aug 1991), pp. 32–38.

REES, WILLIAM E. "The Ecology of Sustainable Development." *The Ecologist* 20 (Jan/Feb 1990), pp. 18–23.

RENNER, MICHAEL. "Forging Environmental Alliances." *WorldWatch* (Nov/Dec 1989), pp. 8–15.

RUCKELSHAUS, WILLIAM D. "Toward a Sustainable World." *Scientific American* (Sept 1989), pp. 166–175.

RUSSEL, DICK. "Global Environmental Outlook—1990s: The Critical Decade." *E Magazine* Premier Issue (Jan/Feb 1990), pp. 30–35.

SCHNEIDER, STEPHEN H. "Can We Repair the Air?" *Discover* 13 (Sept 1992), pp. 28–32.

SMITH, ZACHARY A. *The Environmental Policy Paradox*. Englewood Cliffs, NJ: Prentice Hall, 1992.

STEINHART, PETER. "Essay: Waterway Watchdogs." *Audubon* (Nov 1990), pp. 26–31.

WANDERSEE, JAMES H. "Francis Bacon, Mastermind of Experimental Science." *JCST* (Nov 1987), pp. 120–123.

WEILBACHER, MIKE. "Education That Cannot Wait." *E Magazine* (March/April 1991), pp. 29–35.

WEISSKOPF, MICHAEL. "From Fringe to Political Mainstream, Environmentalists Set Policy Agenda." *The Washington Post* (19 April 1990).

WORLD COMMISSION ON ENVIRONMENT AND DEVELOPMENT. *Our Common Future*. New York: Oxford University Press, 1987.

WUERTHNER, GEORGE. "Rocky Mountain Refuge." *Nature Conservancy* 42 (Jan/Feb 1992), pp. 8–13.

YOUNG, JOHN E. "Mining the Earth." *Worldwatch Paper 109*. Washington, D.C.: Worldwatch Institute, 1992.

Chapter 2. Ecosystems: What They Are

ABATE, TOM. "Into the Northern Philippines Rainforest." *BioScience* 42 (April 1992), pp. 246–251.

ARMSTRONG, SUE. "When Paradise Is a Swamp." *International Wildlife* 21 (Nov/Dec 1991), pp. 44–50.

BARTLETT, DES, AND JEN BARTLETT. "Africa's Skeleton Coast." *National Geographic* (January 1992), pp. 54–86.

BREINING, GREG. "Rising from the Bogs." *Nature Conservancy* 42 (July/August 1992), pp. 24–29.

BRUEMMER, FRED. "Colors of My Arctic." *International Wildlife*, 22(2) (March/April 1992), pp. 4–11.

BURKE I. C., ET AL. "Regional Analysis of the Central Great Plains." *BioScience* 41 (Nov 1991), pp. 685–692.

CLARK, WILLIAM C. "Managing Planet Earth." *Scientific American* 261 (Sept 1989), pp. 46–57.

CLOUD, PRESTON. "The Biosphere." *Scientific American* 249 (Sept 1983), pp. 176–189.

CONOVER, ADELE. "He's Just One of the Bears." *National Wildlife* 30 (June/July 1992), pp. 30–37.

EHRLICH, PAUL, AND JONATHAN ROUGHGARDEN. *The Science of Ecology*. New York: Macmillan, 1987.

NABHAN, GARY PAUL. "Desert Rescuers." *World Monitor* 5(7) (July 1992), pp. 36–41.

RENNIE, JOHN. "Living Together." *Scientific American* 266 (Jan 1992), pp. 122–133.

SOLBRIG, OTTO T., AND MICHAEL D. YOUNG. "Toward a Sustainable and Equitable Future for Savannas." *Environment* 34 (April 1992), pp. 6–15.

TENNESEN, MICHAEL. "Kelp: Keeping a Forest Afloat." *National Wildlife* 30 (June/July 1992), pp. 4–11.

TERBORGH, JOHN. "Why American Songbirds Are Vanishing." *Scientific American* (May 1992), pp. 98–104.

WILLE, CHRIS. "Mystery of the Missing Migrants." *Audubon* (May 1990), pp. 80–87.

Chapter 3. Ecosystems: How They Work

BAZZAZ, FAKHRI, AND ERIC D. FAJER. "Plant Life in a CO_2-Rich World." *Scientific American* 266 (Jan 1992), pp. 68–77.

BALLAFANTE, GINIA. "New Wave Fish Farming." *Garbage* 3 (Jan/Feb 1991), pp. 16–17.

BERNER, ROBERT A., AND ANTONIO LASAGA. "Modeling the Geochemical Carbon Cycle." *Scientific American* (March 1989), pp. 74–81.

BRILL, W.J. "Biological Nitrogen Fixation." *Scientific American* 236 (March 1977), pp. 68–74, 79–81.

BROUGH, HOLLY. "Holy Cows, Unholy Trouble." *WorldWatch* 4 (Sept/Oct 1991), pp. 19–21.

CHERFAS, JEREMY. "Disappearing Mushrooms: Another Mass Extinction?" *Science* 254 (6 Dec 1991), p. 1458.

DURNING, ALAN B. "Fat of the Land." *WorldWatch* 4 (May/June 1991), pp. 10–17.

DURNING, ALAN B. "Junk Food, Food Junk." *WorldWatch* (Sept/Oct 1991), pp. 7–9.

FRIEDEN, EARL. "The Chemical Elements of Life." *Scientific American* (July 1972), pp. 52–60.

GALSTON, ARTHUR W. "Photosynthesis As a Basis for Life Support on Earth and in Space." *BioScience* 42 (Aug 1992), pp. 490–493.

GILLIS, ANNA MARIA. "Should Cows Chew Cheatgrass on Commonlands?" *BioScience* 41 (Nov 1991), pp. 668–676.

HOLMES, HANNAH. "Eating Low on the Food Chain." *Garbage* (Jan/Feb 1992), pp. 32–39.

RIFKIN, JEREMY. *"Beyond Beef, The Rise and Fall of the Cattle Culture."* New York: Dutton, Penguin Group, 1992.

ROUNICK, J.S., AND J.J. WINTERBOURN. "Stable Carbon Isotopes and Carbon Flow in Ecosystems." *BioScience* 36 (March 1986), pp. 171–177.

Scientific American 223 (Sept 1970). Issue devoted to articles describing nutrient cycles and energy flow in ecosystems.

SCRIMSHAW, NEVIN S. "Iron Deficiency." *Scientific American* 265 (Oct 1991), pp. 46–52.

SUNQUIST, FIONA. "Blessed Are the Fruit-Eaters." *International Wildlife* 22 (May/June 1992), pp. 4–11.

SUPLEE, CURT. "Planet in a Bottle, The Making of Biosphere 2." *The Washington Post* (21 Jan 1990), pp. 10–15, 23–27.

TUTTLE, MERLIN D. "Bats—The Cactus Connection." *National Geographic* (June 1991) p. 131.

Chapter 4. Ecosystems: What Keeps Them the Same? What Makes Them Change?

ANDERSON, BOB. "The Swamp Bear's Last Stand." *Nature Conservancy* (Sept/Oct 1991), pp. 16–21.

BARRETT, TODD. "Oh, Deer!" *National Wildlife* 29 (Oct/Nov 1991), pp. 16–21.

BILGER, RICHARD. "Battle for the Prairie." *Earthwatch* (July/Aug 1992), pp. 16–25.

BLOCH, NINI. "Invaders of the Last Ark." *Earthwatch* (Nov 1991), pp. 22–29.

BURKE, WILLIAM K. "Return of the Native: The Art and Science of Environmental Restoration." *E Magazine* (July/Aug 1992), pp. 38–44.

CONAWAY, JAMES. "Eastern Wildlife—Bittersweet Success." *National Geographic* (Feb 1992), pp. 66–89.

CONNIFF, RICHARD. "Fuzzy-Wuzzy Thinking about Animal Rights." *Audubon* (Nov 1990), pp. 120–135.

DAWIDOFF, NICHOLAS. "One for the Wolves." *Audubon* 94 (July/Aug 1992), pp. 38–45.

DURBIN, KATHIE. "From Owls to Eternity." *E Magazine* (March/April 1992), pp. 30–37, 64, 65.

ELEY, THOMAS, J., AND T.H. WATKINS. "In a Sea of Trouble." *Wilderness* 55 (Fall 1991), pp. 18–26.

GOODMAN, BILLY. "From Russia with Love." *Discover* 11 (May 1990), pp. 62–65.

HALPERN, SUE. "Losing Ground." *Audubon* 94 (July/Aug 1992), pp. 70–79.

HANSEN, A.J., ET AL. "Conserving Biodiversity in Managed Forests." *BioScience* 41 (June 1991), p. 382.

HOLDEN, CONSTANCE. "Animal Rightists Trash MSU Lab." *News and Comment* (13 March 1992), p. 1349.

HORNOCKER, MAURICE G. "Mountain Lions." *National Geographic* (July 1992), pp. 38–65.

KEENEY, JOHN. "Control of the Wild." *National Parks* 65 (Sept/Oct 1990), pp. 20–25.

KUNZIG, ROBERT. "These Woods Are Made for Burning." *Discover* 10 (March 1989), pp. 87–93.

LAYCOCK, GEORGE. "There Will Always Be Elms." *Audubon* (May 1990), pp. 58–65.

MAGNUSON, JOHN J., AND JENNIFER A. DRURY. "Global Change Ecology." *The World and I* (April 1991, pp. 304–311.

MILLS, JUDY. "Milking the Bear Trade." *International Wildlife* 22 (May/June 1992), pp. 38–45.

MLOT, CHRISTINE. "Restoring the Prairie." *BioScience* 40 (Dec 1990), pp. 804–809.

National Wildlife. "Damned If They Do, Damned If They Don't." (April/May 1991), pp. 34–37.

NEWHOUSE, JOSEPH R. "Chestnut Blight." *Scientific American* (July 1990), pp. 106–111.

NORRIS, RUTH. "Can Ecotourism Save Natural Areas?" *National Parks* 66 (Jan/Feb 1992), pp. 30–34.

POIANI, KAREN A., AND W. CARTER JOHNSON. "Global Warming and Prairie Wetlands." *BioScience* 41 (Oct 1991), pp. 611–618.

PRITCHARD, PAUL C. "The Best Idea America Ever Had." *National Geographic* (Aug 1991), pp. 36–59.

RALOFF, JANET. "From Tough Ruffe to Quagga." *Science News* 142 (25 July 1992), pp. 56–58.

ROBERTS, LESLIE. "Zebra Mussel Invasion Threatens U.S. Waters." *Science* 249 (21 Sept 1990), pp. 1370–1372.

SELLARS, RICHARD WEST. "Science or Scenery?" *Wilderness* 52 (Summer 1989), pp. 29–40.

SPENCER, CRAIG N., ET AL. "Shrimp Stocking, Salmon Collapse, and Eagle Displacement." *BioScience* 41 (Jan 1991), pp. 14–21.

STEINHART, PETER. "Essay: Humanity without Biology." *Audubon* (May 1990), pp. 24–29.

WILLE, CHRIS. "Race to Save a Green Giant." *National Wildlife* 29 (Oct/Nov 1991), pp. 24–28.

WUERTHNER, GEORGE. "The Flames of '88." *Wilderness* 52 (Summer 1989), pp. 41–54.

YOUNG, JAMES A. "Tumbleweed." *Scientific American* 264 (March 1991), pp. 82–87.

YOUNG, LOUISE B. "Eastern Island: Scary Parable," *World Monitor* 4 (Aug 1991) pp. 40–45.

Chapter 5. Ecosystems: Adapting to Change—or Not

ADLER, JERRY, AND MARY HAGER. "How Much Is a Species Worth?" *National Wildlife* 30 (April/May 1992), pp. 4–15.

BARINAGA, MARCIE. "Evolutionists Wing It with a New Fossil Bird." *Science* 255 (Feb 1992), p. 796.

BEGLEY, SHARON. "A Question of Breeding." *National Wildlife* 29 (Feb/March 1991), pp. 12–17.

BOLGIANO, CHRIS. "The Fall of the Wild." *Wilderness* 55 (Spring 1992), pp. 25–29.

BROECKER, WALLACE S., AND GEORGE DENTON. "What Drives Glacial Cycles?" *Scientific American* 262 (Jan 1990), p. 48.

COUTURIER, LISA. "The Endangered Endangered Species Act." *E Magazine* (March/April 1992), pp. 32–37.

DIAMOND, JARED. "The Great Leap Forward." *Discover* (May 1989), pp. 50–60.

DIAMOND, JARED. "Playing Dice with Megadeath." *Discover* 11 (April 1990), pp. 54–59.

DIAMOND, JARED. "A Question of Size." *Discover* 13 (May 1992), pp. 70–77.

Discover. "Animal Finds." (Jan 1991), pp. 56–59.

Discover. "Extinction Watch." (May 1990), p. 36.

Discover. "Human Origins." (Jan 1990), pp. 50–53.

EHRLICH, PAUL R., AND EDWARD O. WILSON. "Biodiversity Studies: Science and Policy." *Science* 253 (16 Aug 1991), pp. 758–761.

FINKBEINER, ANN. "Terra Infirma." *Discover* (Nov 1991), p. 18.

GASSER, CHARLES S., AND ROBERT T. FRALEY. "Transgenic Crops." *Scientific American* (June 1992), pp. 62–69.

GIBBONS, ANN. "Mission Impossible: Saving All Endangered Species." *Science* 256 (5 June 1992), p. 1386.

GRANT, PETER R. "Natural Selection and Darwin's Finches." *Scientific American* 265 (Oct 1991), pp. 82–87.

HARWOOD, MICHAEL. "What's Going On Here?" *Audubon* (March 1991), pp. 98–99.

HAUB, CARL. "New U.N. Projections Show Uncertainty of Future World." *Population Today* 20 (Feb 1992), pp. 6–7.

HOFFMAN, CAROL A. "Ecological Risks of Genetic Engineering of Crop Plants." *BioScience* (June 1990), pp. 434–437.

HOLDGATE, MARTIN W. "The Environment of Tomorrow." *Environment* 33 (July/Aug 1991), pp. 14–20, 40–42.

HORTON, TOM. "The Endangered Species Act: Too Tough, Too Weak, or Too Late?" *Audubon* (March/April 1992), pp. 68–75.

KNOLL, ANDREW H. "End of the Proterozoic Eon." *Scientific American* 265 (Oct 1991), pp. 64–73.

LIKENS, GENE. "Human Accelerated Environmental Change." *BioScience* 41 (March 1991), p. 130.

MAY, ROBERT M. "How Many Species Are There on Earth?" *Science* (16 Sept 1988), pp. 1441–1448.

MCNEELY, JEFFREY A., ET AL. "Strategies for Conserving Biodiversity." *Environment* 32 (April 1990), pp. 16–20, 36–40.

NEAL, DAN. "Free at Last." *National Wildlife* 30 (Feb/March 1992), pp. 4–7.

PACKER, CRAIG. "Captives in the Wild." *National Geographic* (April 1992), pp. 122–136.

PENNISI, ELIZABETH. "Planting the Seeds of a Nation." *National Wildlife* 28 (April/May 1990), p. 52.

PHILLIPS, KATHRYN. "Frogs in Trouble." *International Wildlife* (Nov/Dec 1990), pp. 4–11.

RANCOURT, LINDA M. "Saving the Endangered Species Act." *National Parks* (March/April 1992), pp. 28–33.

RHOADES, ROBERT E., AND L. JOHNSON. "The World's Food Supply at Risk." *National Geographic* (April 1991), pp. 74–105.

RYAN, JOHN C. "Conserving Biological Diversity." *State of the World*, pp. 9–26. New York: W.W. Norton, 1992.

RYAN, JOHN, C. "Life Support: Conserving Biological Diversity." *Worldwatch Paper 108*. Washington, D.C.: Worldwatch Institute, 1992.

SMITH, NIGEL J.H., ET AL. "Conserving the Tropical Cornucopia." *Environment* 33 (July/Aug 1991), pp. 6–9.

SOLBRIG, OTTO T. "The Origin and Function of Biodiversity." *Environment* 33 (June 1991), pp. 16–20, 34–38.

STOLZENBURG, WILLIAM. "Detectives of Diversity." *Nature Conservancy* 42 (Jan/Feb 1992), pp. 22–27.

THORNE, ALAN G., AND MILFORD H. WOLPOFF. "The Multiregional Evolution of Humans." *Scientific American* 266 (April 1992), pp. 76–83.

UDALL, JAMES R. "Launching the Natural Ark." *Sierra* 76 (Sept/Oct 1991), pp. 80–89.

VAUGHAN, DUNCAN, A., AND LESLEY A. SITCH. "Gene Flow from the Jungle to Farmers." *BioScience* 41 (Jan 1991), pp. 22–28.

VOLKMAN, JOHN M. "Making Room in the Ark: The Endangered Species Act and the Columbia River Basin." *Environment* 34 (May 1992), pp. 18–20, 37–43.

WATERS, TOM. "Almost Human." *Discover* 11 (May 1990), pp. 42–53.

WELLNHOFER, PETER. "Archaeopteryx." *Scientific American* 262 (May 1990), pp. 70–91.

WILLIAMS, TED. "Incite: Waiting for Wolves in Yellowstone." *Audubon* (Nov 1990), pp. 32–43.

WILLS, CHRISTOPHER. "Is Evolution Over for Us?" *Discover* 13 (Aug 1992), pp. 22–24.

WILSON, ALLAN C., AND REBECCA L. CANN. "The Recent African Genesis of Humans." *Scientific American* 266 (April 1992), pp. 66–73.

WILSON, EDWARD O. "The Diversity of Life." *Discover* 13 (Sept 1992), pp. 45–68.

WILSON, EDWARD O. "Threats to Biodiversity." *Scientific American* 261 (Sept 1989), pp. 108–117.

WILSON, EDWARD O., ET AL., EDS. *Biodiversity*. Washington, D.C.: National Academy Press, 1988.

ZIMMER, CARL. "Ruffled Feathers." *Discover* 13 (May 1992), pp. 44–54.

Chapter 6. The Population Explosion: Causes and Consequences

BAJPAI, SHAILAJA. "India's Lost Women." *Indian Express, World Press Review* (April 1991), p. 49.

BERREBY, DAVID. "The Numbers Game." *Discover* 11 (April 1990), pp. 42–49.

BROWN, LESTER, R., AND JODI L. JACOBSON. "Our Demographically Divided World." *Worldwatch Paper 74*. Washington, D.C.: Worldwatch Institute, 1986.

CRUICKSHANK, JOHN. "The Rise and Fall of the Third World." *Globe and Mail* of Toronto, *World Press Review* (Feb 1991), pp. 28–29.

DAVIS, KINGSLEY. "Population and Resources: Fact and Interpretation." *Resources, Environment and Population*, pp. 1–21. New York: Oxford University Press, 1991.

DURNING, ALAN B. "Life on the Brink." *WorldWatch* 3 (March/April 1990), pp. 22–30.

DURNING, ALAN B. "Poverty and the Environment: Reversing the Downward Spiral." *WorldWatch Paper 92*. Washington, D.C.: Worldwatch Institute, 1989.

DURNING, ALAN B. "Recycling out of Poverty." *WorldWatch* (March/April 1989), pp. 9–10.

EHRLICH, ANNE, AND PAUL EHRLICH. "The Population Explosion: Why Isn't Everyone As Scared As We Are?" *The Amicus Journal* 12 (Winter 1990), pp. 22–29.

FORNOS, WERNER. *Gaining People, Losing Ground*. Washington, D.C.: Population Institute, 1987.

HARDEN, BLAINE. "Africa's Great Black Hope." *World Monitor* (Aug 1990), pp. 30–43.

HARDIN, GARRETT. "Sheer Numbers—Can Environmentalists Grasp the Nettle of Population?" *E Magazine* 1 (Nov/Dec 1990), pp. 40–48.

HEISE, LORI. "Prophets of the Pavement." New York: U.N. World Food Council, *WorldWatch* (July/Aug 1989), pp. 36–38.

HENDERSON, KEITH. "Immigration As an Economic Engine." *The Christian Science Monitor* (27 March 1992), p. 9.

HENDRY, PETER. "Food and Population: Beyond Five Billion." *Population Bulletin* 43 (April 1988), pp. 1–39.

HIFT, FRED. "Urbanization Sweeps the Globe." *The Christian Science Monitor* (11 Dec 1990), p. 12.

HINRICHSEN, DON. "The Decisive Decade: What We Can Do about Population." *The Amicus Journal* 12 (Winter 1990), pp. 30–32.

JOHNSON, R.W. "Shantytown Survival." *World Press Review* (Nov 1991), p. 54.

KEYFITZ, NATHAN. "The Growing Human Population." *Scientific American* 261 (Sept 1989), pp. 118–127.

LEONARD, H. JEFFREY, ET AL. "Environment and the Poor: Development Strategies for a Common Agenda." *U.S.-Third World Policy Perspectives, No. 11*. The Overseas Development Council, New Brunswick, NJ: Transaction Books, 1989.

MCFALLS, JOSEPH, A., JR. "Population: A Lively Introduction." *Population Bulletin* (Oct 1991), Washington, D.C.: Population Reference Bureau, 1991.

MICHAELS, JULIA. "Mega-Cities, Points of the Compass." *The Christian Science Monitor* (11 Dec 1991), pp. 10–11.

MILLS, STEPHANIE. "Population." *Garbage* (May/June 1991), pp. 38–43.

MISCH, ANN. "Lost in the Shadow Economy." *WorldWatch* 5 (March/April 1992), pp. 18–25.

MISCH, ANN. "Purdah and Overpopulation in the Middle East." *WorldWatch* (Nov/Dec 1990), p. 10/34.

MURPHY, ELAINE, M. *World Population: Toward the Next Century.* Washington, D.C.: Population Reference Bureau, 1989.

The New York Times International. "Excerpts from the United Nations Declaration on Children." (1 Oct 1990), p. 12.

The New York Times International. "U.N. Summit: The Plight of the World" (2 Sept 1990).

Popline. A bimonthly publication of the Population Institute, 110 Maryland Avenue, N.E., Washington, D.C. 20002.

Population Reference Bureau, Inc. "1992 World Population Data Sheet," "Population Today," and other informational and educational materials. Washington D.C.: Population Reference Bureau.

SADIK, NAFIS. "1991 U.N. State of World Population Report." Condensed by *Popline* (May/June 1991), pp. 4–5.

SADIK, NAFIS. "World Population Continues to Rise." *The Futurist* (March/April 1991), pp. 9–14.

SANDERS, ALLAIN L. "Meanwhile, in Latin America." *Time* (3 Dec 1990), p. 50.

STONE, ROGER D., AND ARMAND G. ERPF FELLOW. "The View from Kilum Mountain." *World Wildlife Fund Letter, No. 4.* Washington, D.C.: World Wildlife Fund, 1989.

UNITED NATIONS POPULATION FUND. *Population and the Environment: The Challenges Ahead.* New York: United Nations Population Fund, 1991.

UNITED NATIONS WORLD FOOD COUNCIL. *Hunger and Malnutrition in the World: Situation and Outlook.* 5–8 June 1991.

YANAGISHITA, MACHIKO. "Japan's Declining Fertility: '1.53 Shock'." *Population Today* 20 (April 1992), pp. 3–4.

YOUNG, JOHN E. "Madagascar Teeters on the Brink." *WorldWatch* (March/April 1989), p. 10/38.

Chapter 7. Addressing the Population Problem

AMATO, MIA. "The Urban Garden." *Garbage* (Nov/Dec 1990), pp. 36–43.

AZIZ, QUTUBUDDIN. "Intensified Family Planning for Pakistan?" *Population Today* 19 (Sept 1991), Population Reference Bureau, p. 4.

BELLAFANTE, GINIA. "New Wave Fish Farming." *Garbage* 3 (Jan/Feb 1991), pp. 16–17.

BENJAMIN, MEDEA, AND ANDREA FREEDMAN. *Bridging the Global Gap: A Handbook to Linking Citizens of the First and Third Worlds."* Washington, D.C.: Seven Locks Press, 1989.

BILSBORROW, RICHARD, E., AND PAMELA F. DELARGY. "Land Use, Migration, and Natural Resource Deterioration: The Experience of Guatemala and the Sudan." *Resources, Environment, and Population,* pp. 125–147. New York: Oxford University Press, 1991.

BROUGH, HOLLY B. "A New Lay of the Land." *WorldWatch* 4 (Jan/Feb 1991), pp. 12–19.

CHEATER, MARK. "AIDS Zeroes in on Women and Children." *WorldWatch* (Nov/Dec 1990), pp. 34–35.

Discover. "Ethics: The Politics of the Abortion Pill." (Jan 1991), p. 86.

DONALDSON, PETER, J., AND AMY ONG TSUI. "The International Family Planning Movement." *Population Bulletin* 45 (Nov 1990).

DURNING, ALAN. "Groundswell at the Grass Roots." *WorldWatch* 2 (Nov/Dec 1989), pp. 16–23.

DURNING, ALAN B., AND HOLLY B. BROUGH. "Taking Stock: Animal Farming and the Environment." *Worldwatch Paper* 103. Washington, D.C.: Worldwatch Institute, 1991.

ECKHOLM, ERIK. "What Makes the Two Sexes So Vulnerable to Epidemic." *The New York Times International* (16 Sept 1990), p. 15.

ECKHOLM, ERIK, AND JOHN TIERNEY. "AIDS in Africa: A Killer Rages On." *The New York Times* (16 Sept 1990).

ENVIRONMENTAL DEFENSE FUND. "EDF Spurs World Bank Rehabilitation Loans to India." *Environmental Defense Fund Letter* (Oct 1989), pp. 1, 5.

FELDMAN, LINDA. "Two Voices on Abortion" and "Pro-Choice Chief Cites Self-Determination." *The Christian Science Monitor* (25 Nov 1991), pp. 12, 13.

FORNOS, WERNER. "Population Politics." *Technology Review* (Feb/March 1991), pp. 43–51.

GIBBONS, ANN. "Small Is Beautiful: Microlivestock for the Third World.' *Science* 253 (26 July 1991), p. 378.

GOLIBER, THOMAS J. "Africa's Expanding Population: Old Problems, New Policies." *Population Bulletin* 44 (Nov 1989).

HAGERMAN, ERIC. "As the Third World Turns." *WorldWatch* 4 (Sept/Oct 1991), pp. 5–7.

HARDIN, GARRETT. "Sheer Numbers—Can Environmentalists Grasp the Nettle of Population?" *E Magazine* (Nov/Dec 1990), pp. 40–47.

JACOBSON, JODI L. "Abortion in a New Light." *WorldWatch* 3 (March/April 1990), pp. 31–38.

JACOBSON, JODI L. "Baby Budget." *WorldWatch* (Sept/Oct 1989), pp. 21–31.

JACOBSON, JODI L. "The Global Politics of Abortion." *Worldwatch Paper* 97. Washington, D.C.: Worldwatch Institute, 1990.

JACOBSON, JODI L. "India's Misconceived Family Plan." *WorldWatch* 4 (Nov/Dec 1991), pp. 18–25.

JACOBSON, JODI L. "Paying Interest in Human Life." *WorldWatch* (July/Aug 1989), pp. 6–8.

JACOBSON, JODI L. "Reproductive Perestroika." *WorldWatch* (Sept/Oct 1990), pp. 29–37.

JACOBSON, JODI L. "Women's Reproductive Health: The Silent Emergency." *Worldwatch Paper* 102. Washington, D.C.: Worldwatch Institute, 1991.

KASLOW, AMY, AND GEORGE D. MOFFETT III. "World Bank Faces New Demands." *The Christian Science Monitor* (Aug 1991), p. 9.

LEIBMAN, DENA. "Default or Deliver?" *Sierra* (Sept/Oct 1990).

LENSSEN, NICHOLAS. "Cooked by the Sun." *WorldWatch* (March/April 1989), pp. 41–42.

LEVINE, ROBERT A., ET AL. "Women's Schooling and Child Care in the Demographic Transition: A Mexican Case Study." *Population and Development Review* 17 (Sept 1991), pp. 459–496.

LIPSCHUTZ, RONNIE D. "Redefining National Security." *E Magazine* (March/April 1991), pp. 54–55.

LIVERNASH, ROBERT. "The Growing Influence of NGOs in the Developing World." *Environment* 34 (June 1992), pp. 12–20.

MACNEILL, JIM. "Strategies for Sustainable Economic Development." *Scientific American* (Sept 1989), pp. 154–165.

MAURICE, JOHN. "Improvements Seen for RU-486 Abortions." *Science* 254 (11 Oct 1991), pp. 198, 200.

MCCLINTOCK, JOHN M. "Free-Trade Pact Raises Concerns of Mexico's Child Workers." *The Baltimore Sun* (15 Dec 1991), pp. A1, 10, 11.

MCNICOLL, GEOFFREY, AND MEAD CAIN, EDS. *Rural Development and Population, Institutions and Policy.* Food and Agriculture Organization of the United Nations report. New York: Oxford University Press, 1990.

MOSLEY, W. HENRY, AND PETER COWLEY. "The Challenge of World Health." *Population Bulletin* 46 (Dec 1991), pp. 1–39.

MUSOKE, MULINDE. "Reform Must Come from Within." *The New Vision* of Kampala, Uganda, *World Press Review* (Aug 1991), pp. 11–14.

NORSE, DAVID. "A New Strategy for Feeding a Crowded Planet." *Environment* 34 (June 1992), pp. 6–11.

PALCA, JOSEPH. "The Sobering Geography of AIDS." *Science* 252 (19 April 1991), pp. 372–373.

PATTERSON, ALAN. "Debt-for-Nature Swaps and the Need for Alternatives." *Environment* 32 (Dec 1990), pp. 4–13.

PRESS, ROBERT. "Big Payoff for Small Loans." *The Christian Science Monitor* (7 Jan 1992), p. 10.

REID, WALTER V.C. "Sustainable Development: Lessons from Success." *Environment* 31 (May 1989), pp. 7–9, 29–35.

RENNER, MICHAEL G. "Guns versus Grains." *WorldWatch* (March/April 1989), p. 38–40.

RENNER, MICHAEL G. "Forging Environmental Alliances." *WorldWatch* 2 (Nov/Dec 1989), pp. 8–15.

REPETTO, ROBERT. "Economic Aid and the Environment." *EPA Journal* 16 (July/August 1990), pp. 20–22.

REPETTO, ROBERT. "Population Resources, Environment: An Uncertain Future." *Population Bulletin* 42 (April 1989), Population Reference Bureau, 1989.

STETSON, MARNIE. "Giving Credit Where It's Due." *WorldWatch* 4 (March/April 1991), pp. 7–8.

TIEN, H. YUAN, ET AL. "China's Demographic Dilemmas." *Population Bulletin,* 47 (June 1992), Washington, D.C.: Population Reference Bureau, 1992.

UNITED NATIONS WORLD FOOD COUNCIL. "Focusing Development Assistance on Hunger and Poverty Alleviation." New York: U.N. World Food Council (5–8 June 1991).

UNITED NATIONS WORLD FOOD COUNCIL. "Food-Security Implications of the Changes in the Political and Economic Environment." New York: U.N. World Food Council, (5–8 June 1991).

UNITED NATIONS WORLD FOOD COUNCIL. "Meeting the Developing Countries' Food Production Challenges of the 1990s and Beyond." New York: U.N. World Food Council, (5–8 June 1991).

WORLD BANK. *World Development Report 1991: The Challenge of Development.* New York: Oxford University Press, 1988.

YOUTH, HOWARD. "Farming in a Fish Tank." *WorldWatch* 5 (May/June 1992), pp. 5–7.

Chapter 8. The Production and Distribution of Food

ALITERI, MIGUEL A., AND SUSANNA B. HECHT. *Agroecology and Small Farm Development.* New York: CRC Press, 1990.

BARBIER, EDWARD B. "Sustaining Agriculture on Marginal Land." *Environment* 31 (Nov 1989), pp. 12–40.

BREAD FOR THE WORLD. *Hunger 1990. A Report on the State of World Hunger.* Washington, D.C.: Bread for the World Institute on Hunger & Development, 1990.

BREAD FOR THE WORLD. *Hunger 1992. Second Annual Report on the State of World Hunger.* Washington, D.C.: Bread for the World Institute on Hunger & Development, 1991.

BROWN, LESTER, R. "The Changing World Food Prospect: The Nineties and Beyond." *Worldwatch Paper 85.* Washington, D.C.: Worldwatch Institute, 1988.

BROEN, LESTER, R., AND JOHN E. YOUNG. "Feeding the World in the Nineties." *State of the World 1990.* Washington, D.C.: Worldwatch Institute, 1990.

CROSSON, PIERRE R., AND NORMAN J. ROSENBERG. "Strategies for Agriculture." *Scientific American* 261 (Sept 1989), pp. 128–135.

DURNING, ALAN B., AND HOLLY W. BROUGH. "Taking Stock: Animal Farming and the Environment." *Worldwatch Paper 103.* Washington, D.C.: Worldwatch Institute, 1991.

The Ecologist 21 (1991). Issue devoted to world hunger.

EDWARDS, CLIVE, A., ET AL., EDS. *Sustainable Agricultural Systems.* Ankeny, IA: Soil and Water Conservation Society, 1990.

GASSER, CHARLES S., AND ROBERT T. FRALEY. "Transgenic Crops." *Scientific American* 266 (June 1992), pp. 62–69.

GRANT, JAMES P. *The State of the World's Children 1991.* New York: Oxford University Press, 1991.

JACKSON, WES. "Nature As the Measure for a Sustainable Agriculture." *Ecology, Economics, Ethics: The Broken Circle,* Chapter 4. New Haven, CT: Yale University Pres, 1991.

MELLOR, JOHN W. "The Intertwining of Environmental Problems and Poverty." *Environment* 30 (Nov 1988), pp. 8–13, 28–30.

PONTING, CLIVE. "Historical Perspectives on Sustainable Development." *Environment* 32 (Nov 1990), pp. 4–9, 31–33.

REGANOLD, JOHN P., ET AL. "Sustainable Agriculture." *Scientific American* 262 (June 1990), pp. 112–120.

RHOADES, ROBERT E. AND L. JOHNSON. "The World's Food Supply at Risk." *National Geographic* (April 1991), pp. 74–105.

VANDER ZEE, DELMAR, AND RON VOS. "Trends in Agriculture: Sustainability." *Pro Rege* 18 (March 1990), pp. 19–28.

WORLD RESOURCES INSTITUTE. "Part I: Sustainable Development" and "Food and Agriculture." *World Resources 1992–93,* Chapters 1–4 and 7. New York: Oxford University Press, 1992.

Chapter 9. Soil and the Soil Ecosystem

AMATO, MIA. "Backyard Restoration." *Garbage* (March/April 1991), pp. 50–55.

American Farmland (Spring 1991). Issue devoted to conservation for a sustainable agriculture.

AMIRAN, DORON, AND STEVEN E. SHERMAN. "Source Reduction

through Home Composting." *BioCycle* (April 1992), pp. 97–99.

BARASH, LEAH. "The Ethics of Good Farming." *National Wildlife* 30 (June/July 1992), p. 24.

CARTER, VERNON G., AND T. DALE. *Topsoil and Civilization.* Norman, OK: University of Oklahoma Press, 1974.

CLEMINGS, RUSSELL. "Mirage." *Earthwatch* (June 1991), pp. 14–21. Description of salinization.

CONACHER, ARTHUR J. "Salt of the Earth: Secondary Soil Salinization in the Australian Wheat Belt." *Environment* 32 (July/Aug 1990), pp. 4–9.

CROSSON, PIERRE R., AND NORMAN J. ROSENBERG. "Strategies for Agriculture." *Scientific American* (Sept 1989), pp. 128–135.

DURNING, ALAN B., AND HOLLY B. BROUGH. "Taking Stock: Animal Farming and the Environment." *Worldwatch Paper* 103. Washington, D.C.: Worldwatch Institute, 1991.

EIERA STAFF AND THE FOX GROUP ENVIRONMENTAL COMMUNITY. "Yard Waste As a Resource." *BioCycle* (April 1992), pp. 58–60.

EL-BAZ, FAROUK. "Desertification." *Geotimes* 29 (Feb 1991), pp. 52–54.

EL-BAZ, FAROUK. "Do People Make Deserts?" *New Scientist* 128 (Oct 1990), pp. 41–44.

ELLIS, WILLIAM S. "Africa's Stricken Sahel." *National Geographic* 172 (Aug 1987), pp. 40–179.

ELLIS, WILLIAM S. "Harvest of Change." *National Geographic* 179 (Feb 1991), pp. 49–73.

JONES, B.J. "Composting Food and Vegetative Waste." *BioCycle* 33 (March 1992), pp. 69–70.

KIEFER, MICHAEL. "Fall of the Garden of Eden." *International Wildlife* (July/Aug 1989), pp. 38–43.

KOHL, LARRY. "Heavy Hands on the Land." *National Geographic* 174 (Nov 1988), pp. 632–657.

LIPSKE, MICHAEL. "Natural Farming Harvests New Support." *National Wildlife* 28 (April/May 1990), pp. 18–23.

MADDOX, HARRY. *Your Garden Soil.* West Vancouver, BC: David and Charles, 1974.

MANN, R.D. "Time Running Out: The Urgent Need for Tree Planting in Africa." *The Ecologist* 20 (March/April 1990), pp. 48–53.

MARINELLI, JANET. "Composting." *Garbage* (July/Aug 1990), pp. 44–51.

McDERMOTT, JEANNE. "Alternative Agriculture Is Gaining Ground." *Smithsonian* 21 (April 1990), pp. 114–131.

MILLER, SUSAN KATZ. "Sermon on the Farm." *International Wildlife* 22 (March/April 1992), pp. 48–51.

MOLLISON, BILL. *Permaculture: A Practical Guide for a Sustainable Future.* Washington, D.C.: Island Press, 1990.

NORSE, DAVID. "A New Strategy for Feeding a Crowded Planet." *Environment* 34 (June 1992), pp. 6–11.

PONTING, CLIVE. "Historical Perspectives on Sustainable Development." *Environment* 32 (Nov 1990), pp. 4–9, 31–33.

POSTEL, SANDRA. "China Revives Lost Land." *WorldWatch* (March/April 1989), pp. 7–8.

POSTEL, SANDRA. "Land's End." *WorldWatch* 2 (May/June 1989), pp. 12–20.

REGANOLD, JOHN P., ET AL. "Sustainable Agriculture." *Scientific American* 262 (June 1990), pp. 112–120.

ROBBINS, GRACE JONES, ED. "*Alternative Agriculture.*" Washington, D.C.: National Academy Press, 1989.

SCHUELER, DONALD G. "Losing Louisiana." *Audubon* (July 1990), pp. 78–87.

SEITZ, CHRIS. "Breaking New Ground." *Environmental Action* 22 (Nov/Dec 1990), pp. 15–16.

SPAID, ELIZABETH LEVITAN. "All-Natural Agriculture Education." *The Christian Science Monitor* (16 Dec 1991), p. 12.

STANLEY, CHRISTOPHER. "New Solutions." *Organic Gardening* 37 (July/Aug 1990), pp. 63–66.

TSEO, GEORGE. "The Greening of China." *Earthwatch* (May/June 1992), pp. 20–26.

WEBER, PETER. "U.S. Farmers Cut Soil Erosion by One-Third." *WorldWatch* (July/Aug 1990), pp. 5–6.

WILLE, CHRIS. "Trees on Trial in Central America." *American Forests* (Sept/Oct 1990), pp. 21–24.

WORLD RESOURCES INSTITUTE. "Paying the Farm Bill: U.S. Agriculture Policy and the Transition to Sustainable Agriculture." Washington, D.C.: World Resources Institute, 1991.

WUERTHNER, GEORGE. "How the West Was Eaten." *Wilderness* 54 (Spring 1991), pp. 28–37.

ZORZI, G., ET AL. "Cocomposting Conditions for MSW Composting." *BioCycle* 33 (June 1992), pp. 72–75.

Chapter 10. Pests and Pest Control

BARINAGA, MARCIA. "Entomologists in the Medfly Maelstrom." *Science* 247 (9 March 1990), pp. 1168–1169.

BARRONS, KEITH C. "How Risky Are Pesticides?" *Science of Food and Agriculture* 5 (Jan 1988), pp. 21–25.

BEARD, JONATHAN D. "Bug Detectives Crack the Tough Cases." *Science* 254 (13 Dec 1991), pp. 1580–1581.

BULL, D. *Growing Problems: Pesticides and the Third World Poor.* Washington, D.C.: Sidney Kramer Books, 1984.

CARSON, RACHEL. *Silent Spring.* Boston: Houghton Mifflin, 1962.

CULOTTA, ELIZABETH. "Biological Immigrants under Fire." *Science* 254 (6 Dec 1991), pp. 1444–1447.

DEBACH, PAUL. *Biological Control by Natural Enemies.* London: Cambridge University Press, 1974.

DOVER, MICHAEL J., AND BRIAN A. CROFT. "Pesticide Resistance and Public Policy." *BioScience* 36 (Feb 1986), pp. 78–85.

FRIENDS OF THE EARTH. *How to Get Your Lawn and Garden off Drugs.* Ottawa, Ontario: Friends of the Earth, 1990.

HUSSEY, N.W., AND N. SCOPES. *Biological Pest Control.* Ithaca, NY: Cornell University Press, 1986.

HYNES, H. PATRICIA. *The Recurring Silent Spring.* New York: Pergamon Press, 1989.

KOURIK, ROBERT. "Combatting Household Pests without Chemical Warfare." *Garbage* 2 (March/April 1990), pp. 22–29.

LANDSBERG, HANS. "Reducing Pesticide Use." *Environment* 29 (July/Aug 1987), pp. 52–53.

MARCO, GINO J., ET AL., EDS. *Silent Spring Revisited.* Washington, D.C.: American Chemical Society, 1987.

NATIONAL VETERANS LEGAL SERVICES PROJECT. *Human Health Effects Associated with Exposure to Herbicides and/or Their Associated Contaminants—Chlorinated Dioxins.* Washington, D.C.: National Veterans Legal Services Project, 1990.

NATURAL RESOURCES DEFENSE COUNCIL. *A Report on Intolerable*

Risk: Pesticides in our Children's Food. Washington, D.C.: Natural Resources Defense Council, 1989.

OLKOWSKI, WILLIAM. "Update: Great Expectations for Non-Toxic Pheromones." *The IPM Practitioner* 10 (June/July 1988), p. 1.

PIMENTEL, DAVID. "The Dimensions of the Pesticide Question." *Ecology, Economics, Ethics: The Broken Circle*, Chapter 5. New Haven, CT: Yale University Press, 1991.

POSTEL, SANDRA. "Defusing the Toxics Threat: Controlling Pesticides and Industrial Waste." *Worldwatch Paper 79.* Worldwatch Institute, 1987.

REVKIN, ANDREW C. "March of the Fire Ants." *Discover* 10 (March 1989), pp. 71–76.

ROSENTHAL, GERALD A. "The Chemical Defenses of Higher Plants." *Scientific American* 254 (Jan 1986), pp. 94–99.

STONE, RICHARD. "Researchers Score Victory over Pesticides—and Pests—in Asia." *Science* 256 (29 May 1992), pp. 1272–1273.

STROBEL, GARY. "Biological Control of Weeds." *Scientific American* 265 (July 1991), pp. 72–78.

Chapter 11. Water, the Water Cycle, and Water Management

ABELSON, PHILIP H. "Desalination of Brackish Marine Water." *Science* (15 March 1991), p. 1289.

BROWN, LESTER R. "The Aral Sea: Going, Going . . ." *WorldWatch* 4 (Jan/Feb 1991), pp. 20–27.

CARRIER, JIM. "Water and the West: The Colorado River." *National Geographic* (June 1991), pp. 2–35.

COUSTEAU, JEAN. "War and Water." *Calypso Log* (Aug 1990), p. 3.

DURNING, ALAN B. "Apartheid's Other Injustice." *WorldWatch* (May/June 1990), pp. 11–17.

ELLIS, WILLIAM S. "California's Harvest of Change." *National Geographic* (Feb 1991), pp. 48–73.

FALKENMARK, MALIN. "Rapid Population Growth and Water Scarcity: The Predicament of Tomorrow's Africa." *Resources, Environment, and Population*, pp. 81–94. New York: Oxford University Press, 1991.

GORMAN, JAMES. "Wetlands? Wetlands? Whatever Happened to the Swamps?" *Audubon* 94 (May/June 1992), pp. 82–83.

GRAHAM, FRANK, JR. "Gambling on Water." *Audubon* 94 (July/Aug 1992), pp. 65–69.

GRAY, PAUL. "The Colorado." *Time* (22 July 1991), pp. 22–26.

HARGER, CINDY. "A Spoonful of Sugar Makes the Everglades Go Down." *Conservation* 90 (8 June 1990), pp. 7–10.

HOTTELET, RICHARD C. "The Drain on Middle East Water Reserves." *The Christian Science Monitor* (19 Aug 1991), p. 19.

HUTCHINSON, CHARLES F. "Development in Arid Lands: Lessons from Chad Lake." *Environment* 34 (July/Aug 1992), pp. 16–20.

KNICKERBOCKER, BRAD. "The Sun Is Setting on a Century of Concrete Waterways out West." *The Christian Science Monitor* (18 June 1991), pp. 1–2.

KOTLYAKOV, V.M. "The Aral Sea Basin: A Critical Environmental Zone." *Environment* 33 (Jan/Feb 1991), pp. 4–9.

KOURIK, ROBERT. "Cisterns Deliver Rainwater." *Garbage* (July/Aug 1992), pp. 42–47.

KOURIK, ROBERT. "Drip Irrigation." *Garbage* (May/June 1991), pp. 46–51.

KOURIK, ROBERT. "Greywater: Why Throw It Away?" *Garbage* (Jan/Feb 1990), pp. 41–45.

KOURIK, ROBERT. "Toilets: Low Flush/No Flush." *Garbage* (Jan/Feb 1990), pp. 16–23.

KOURIK, ROBERT. "Tracking the Big Drip." *Garbage* (March/April 1991), pp. 24–30.

LAYCOCK, GEORGE. "How to Save a Wetland." *Audubon* (July 1990), pp. 96–101.

MACLEISH, WILLIAM H. "Water, Water, Everywhere, How Many Drops to Drink?" *World Monitor* (Dec 1990), pp. 54–58.

MAURITS LA RIVIERE, J.W. "Threats to the World's Water." *Scientific American* 261 (Sept 1989), pp. 80–97.

MCDOWELL, JEANNE. "A Race to Rescue the Salmon." *Time* (2 March 1992), pp. 59–60.

MELODY, INGRID. "Solar Water Desalinization." *Solar Today* 4 (Nov/Dec 1990), pp. 14–17.

MISCH, ANN. "India's Wells Run Dry." *WorldWatch* (May/June 1991), pp. 9–10.

MITCHELL, JOHN G. "Our Disappearing Wetlands." *National Geographic* (Oct 1992), pp. 3–45.

MONKS, VICKI. "Engineering the Everglades." *National Parks* 65 (Sept/Oct 1990), pp. 32–36.

OKUM, DANIEL A. "A Water and Sanitation Strategy for the Developing World." *Environment* 33 (Oct 1991), pp. 16–20.

OSANN, EDWARD R., AND DAVID C. CAMPBELL. "Water Resources—The Central Utah Project Completion Act." *Environment* 34 (March 1992), pp. 2–3.

PATTERSON, ALAN. "Debt-for-Nature Swaps and the Need for Alternatives." *Environment* 32 (Dec 1990), pp. 4–13.

POSTEL, SANDRA. "Trouble on Tap." *WorldWatch* 2 (Sept/Oct 1989), pp. 12–20.

POWELL, BILL, AND HASSAN SHAHRIAR. "Bailing out Bangladesh." *Newsweek* (28 Aug 1989), p. 42.

REISNER, MARC. "Can Anyone Win the Water War?" *National Wildlife* 29 (June/July 1991), pp. 4–9.

RIDGEWAY, JAMES. "Watch on the Danube." *Audubon* 94 (July/Aug 1992), pp. 46–64.

SCHUELER, DONALD. "That Sinking Feeling." *Sierra* 75 (March/April 1990), pp. 42–51.

SERAYDARIAN, HARRY. "San Francisco Bay, Beset by Freshwater Diversion." *EPA Journal* 16 (Nov/Dec 1990), pp. 20–22.

SIMON, PAUL. "It's Time for a Breakthrough in Desalinization." *The Christian Science Monitor* (26 March 1991), p. 19.

SPENCER, CRAIG N., ET AL. "Shrimp Stocking, Salmon Collapse, and Eagle Displacement." *BioScience* 41 (Jan 1991), pp. 14–21.

STARR, JOYCE R. "Water Resources: A Foreign-Policy Flash Point." *EPA Journal* (July/Aug 1990), pp. 34–36.

STEINHART, PETER. "Essay: No Net Loss." *Audubon* (July 1990), pp. 18–21.

TONER, MIKE. "Fixing a Broken River." *National Wildlife* (April/May 1991), pp. 18–21.

TYSON, JAMES L. "Chinese Dam Threatens Massive Disruption." *The Christian Science Monitor* (18, 22, 23 July 1991).

VETTER, DON. "Teeming Oasis or Desert Mirage?" *Nature Conservancy* 41 (Sept/Oct 1991), pp. 22–29.

VIESSMAN, WARREN. "Water Management." *Environment* 32 (May 1990), pp. 11–15.

WAGGONER, PAUL E. "U.S. Water Resources versus an Announced but Uncertain Climate Change." *Science* 251 (1 March 1991), p. 1002.

WILEY, KEN. "Untying the Western Knot." *The Nature Conservancy* 40 (March/April 1990), pp. 4–13.

WOOD, DANIEL B. "California Launches Water Plan." *The Christian Science Monitor* (21 April 1992), p. 8.

WORLDWATCH IINSTITUTE. "Gandhian Greens Fight Dams." *WorldWatch* (July/Aug 1990), p. 5.

Chapter 12. Sediments, Nutrients, and Eutrophication

BAKER, WILLIAM, C., AND TOM HORTON. "Runoff and the Chesapeake Bay." *EPA Journal* 16 (Nov/Dec 1990), pp. 13–16.

CHRISTIAN, SHIRLEY. "Pollution in Paradise: Can Salmon Be the Cause?" *The New York Times* (15 Oct 1990), p. A4.

EPA JOURNAL (Nov/Dec 1991). Issue devoted to problems of nutrient, sediment, and chemical runoff from agricultural and urban areas.

EWEL, KATHERINE C. "Multiple Demands on Wetlands." *BioScience* 40 (Oct 1990), pp. 660–666.

GUP, TED. "Getting at the Roots of a National Obsession." "How to Maintain a Chemical-Free Lawn." *National Wildlife* 29 (June/July 1991), pp. 18–24.

HORTON, TOM, AND WILLIAM EICHBAUM. "Turning the Tide." Washington, D.C.: Island Press, 1991.

KUSLER, JON. "Wetlands Delineation: An Issue of Science or Politics?" *Environment* 34 (March 1992), pp. 6–11, 29–37.

LAYCOCK, GEORGE. "How to Save a Wetland." *Audubon* (July 1990), pp. 96–101.

LIPSKE, MICHAEL. "Floating in Controversy." *National Wildlife* 29 (Oct/Nov 1991), pp. 22–23.

LIPSKE, MICHAEL. "How Much Is Enough?" *National Wildlife* 28 (June/July 1990), pp. 18–24.

MAKAREWICZ, JOSEPH C., AND PAUL BERTRAM. "Evidence for the Restoration of the Lake Erie Ecosystem." *BioScience* (April 1991), pp. 216–224.

MARYLAND DEPARTMENT OF NATURAL RESOURCES, DEPARTMENT OF AGRICULTURE AND SSTATE DEPARTMENT OF EDUCATION. "Maryland, Restoring the Chesapeake, A Progress Report 1988." Annapolis, MD: Office of the Governor, 1988.

MILLER, SUSAN KATZ. "When Pollution Runs Wild." *National Wildlife* 30 (Dec/Jan 1992), pp. 26–28.

MITCHELL, JOHN G. "Our Disappearing Wetlands," *National Geographic* (Oct 1992), pp. 3–45.

SCHLOSSER, ISAAC J. "Stream Fish Ecology: A Landscape Perspective." *BioScience* 41 (Nov 1991), pp. 704–712.

SCHULTZ, WARREN. "Natural Lawn Care." *Garbage* (July/Aug 1990), pp. 28–34.

Chapter 13. Water Pollution Due to Sewage

BENCIVENGA, JIM. "Florida Wetlands Filter City Waste." *The Christian Science Monitor* (5 March 1992), pp. 16–17.

DOLIN, ERIC JAY. "Boston Harbor's Murky Political Waters." *Environment* 34 (July/Aug 1992), pp. 6–11.

GILLIS, ANNA MARIA. "Shrinking the Trash Heap." *BioScience* 42 (Feb 1992), pp. 90–93.

GOLDSTEIN, NORA. "Contractors Vary Approaches to Sludge Reuse." *BioCycle* 33 (July 1992), pp. 62–65.

GOLDSTEIN, NORA, AND DAVID RIGGIE. "Sludge Composting Sets Health Pace/Sludge Composting Projects in the United States." *BioCycle* (Dec 1991), pp. 30–37.

GORMAN, CHRISTINE. "Death in the Time of Cholera." *Time* (6 May 1991), pp. 58–61.

HAMMER, DONALD A., ED. *Constructed Wetlands for Wastewater Treatment.* Chelsea, MI: Lewis Publishers, 1989.

HOCHREIN, PETER, AND THOMAS OUTERBRIDGE. "Anaerobic Digestion for Soil Amendment and Energy." *BioCycle* 33 (June 1992), pp. 63–64.

LOGSDON, GENE. "Innovative Waste Treatment in a Midwest Cornfield." *BioCycle* (Jan 1992), pp. 46–48.

LOWE, MARCIA, D. "Down the Tubes." *WorldWatch* 2 (March/April 1989), pp. 22–29.

MARINELLI, JANET. "After the Flush the Next Generation." *Garbage* (Jan/Feb 1990), pp. 24–35.

MELENDEZ, JOSE. "Cholera Moves North." *Excelsior* of Mexico City, *World Press Review* (Sept 1991), p. 46.

OKUM, DANIEL A. "A Water and Sanitation Strategy for the Developing World." *Environment* 33 (Oct 1991), pp. 16–20.

ROSS, ELIZABETH. "Boston Recycles Sludge into Fertilizer." *The Christian Science Monitor* (22 Jan 1992), p. 15.

SCHERER, RON. "Sydney Pipes Sewage out to Sea." *The Christian Science Monitor* (22 Jan 1992), p. 14.

SMITH-VAAGO, LINDA. "New Possibilities for Wastewater Treatment." *Water Engineering and Management* 138 (March 1991), pp. 27–29.

SPENCER, ROBERT. Solar Aquatic Treatment of Septage." *BioCycle* (May 1990), pp. 66–71.

Water Engineering and Management. "UV Effect on Wastewater." (Dec 1990), pp. 15–23.

WATERMAN, MELISSA. "Pollution-Prevention Tactics in the Gulf of Maine." *EPA Journal* 16 (Nov/Dec 1990), pp. 17–19.

Chapter 14. Pollution from Hazardous Chemicals

AKULA, VIKRAM. "The New Indian Burial Ground." *WorldWatch* 5 (Aug 1992), p. 7.

ANDERSON, JOHN V., AND BIMLESHWAR P. GUPTA. "Solar Detoxification of Hazardous Waste." *Solar Today* 4 (Nov/Dec 1990), pp. 10–13.

BELL, LAUREN. "Once Aflame and Filthy, a River Shows Signs of Life." *National Wildlife* 28 (Feb/March 1990), p. 24.

BELLAFANTE, GINIA. "Bottled Water: Fads and Facts." *Garbage* (Jan/Feb 1990), 46–51.

BELLAFANTE, GINIA. "Minimizing Household Hazardous Waste." *Garbage* (March/April 1990), pp. 44–49.

BONNER, ROBIN. "The Untouchables." *National Wildlife EnviroAction* (Aug 1991), pp. 10–13.

BRIGHT, CHRIS. "The Poison Trade: America's Pesticide Exports." *E Magazine* (July/Aug 1990), pp. 30–36.

CHEPESIUK, RON. "From Ash to Cash: The International Trade in Toxic Waste." *E Magazine* 11 (July/Aug 1991), pp. 31–37, 63.

CHOLLAR, SUSAN. "The Poison Eaters." *Discover* 11 (April 1990), pp. 76–79.

CLEAN WATER ACTION NEWS. "Mercury Contaminates the Land of 10,000 Lakes." *Environment* 33 (May 1991), pp. 21–22.

COLBURN, THEO, AND RICHARD A. LIROFF. "Toxics in the Great Lakes." *EPA Journal* 16 (Nov/Dec 1990), pp. 5–8.

CUTLER, DANIEL S. "Chain Reaction, PCB Poisoning throughout the Great Lakes." *Buzzworm: The Environmental Journal* (Autumn 1989), pp. 32–39.

E Magazine. "Resettlement of Love Canal." (Jan/Feb 1991), p. 16.

EPA Journal (July/Aug 1991). Issue devoted to hazardous waste cleanup.

FRENCH HILARY F. "Green Revolutions: Environmental Reconstruction in Eastern Europe and the Soviet Union." *Worldwatch Paper 99.* Washington, D.C.: Worldwatch Institute, 1990.

FRENCH, HILARY F. "A Most Deadly Trade." *WorldWatch* (July/Aug 1990), pp. 11–17.

GEISER, KEN. "The Greening of Industry." *Technology Review* (Aug/Sept 1991), pp. 64–72.

GLASS, DAVID J. "Waste Management—Biological Treatment of Hazardous Wastes." *Environment* 33 (Nov 1991), pp. 5, 43–45.

GOLD, JACKIE. "The Pioneers." *Financial World* (23 Jan 1990).

GOLULKE, CLARENCE. "Solar Detoxification of Hazardous Wastes." *BioCycle* (Feb 1991), p. 6.

GRAUET, THANE. "Dishonorable Discharges—The Military's Peacetime War on Planet Earth." *E Magazine* (July/Aug 1990), pp. 43–45.

GROSSMAN, KARL. "Of Toxic Racism and Environmental Justice." *E Magazine* (May/June 1992), pp. 28–35.

HEDGES, STEPHEN J. "Enviro-Cops on the Prowl for Polluters." *U.S. News and World Report* (9 Oct 1989).

HINRICHSON, DON. "Those Danube Blues." *International Wildlife* 21 (Sept/Occt 1991), pp. 38–47.

KUNREUTHER, HOWARD, AND RUTH PATRICK. "Managing the Risks of Hazardous Waste." *Environment* 33 (April 1991), pp. 12–15.

LIROFF, RICHARD A. "Eastern Europe: Restoring a Damaged Environment." *EPA Journal* 16 (July/Aug 1990), pp. 50–55.

MACDONALD, JACQUELINE. "Report on Home Water Purifiers." *Garbage* (March/April 1991), pp. 31–37.

MICHAELS, JULIA. "South Americans Shut Door on Toxic Imports." *The Christian Science Monitor* (10 March 1992), p. 4.

MISCH, ANN. "The Amazon: River at Risk." *WorldWatch* 5 (Jan/Feb 1992), pp. 35–37.

NRIAGU, JEROME O. "Global Metal Pollution: Poisoning the Biosphere? *Environment* 32 (Sept 1990), pp. 6–11.

PEASE, WILLIAM S. "Chemical Hazards and the Public's Right to Know: How Effective Is California's Proposition 65?" *Environment* 33 (Dec 1991), pp. 12–20.

PENFIELD, WENDY. "Message from the Belugas." *International Wildlife* 20 (May/June 1990), pp. 40–44.

ROY, KIMBERLY. "Thermochemical Reduction Process Offers Absolute Detoxification." *Hazmat World* 3 (Oct 1990), pp. 50–51.

RUSSELL, DICK. "Lois Gibbs—On the Front Lines for Environmental Justice." *E Magazine* (July/Aug 1990), pp. 12–15.

RUSSELL, MILTON, ET AL. The U.S. Hazardous Waste Legacy." *Environment* 34 (July/Aug 1992), pp. 12–15.

SCHNEIDER, PAUL. "Other People's Trash." *Audubon* (July/Aug 1991), pp. 108–119.

SCHWARTZ, JOEL, AND RONNIE LEVIN. "Lead: Example of the Job Ahead." *EPA Journal* 18 (March/April 1992), pp. 42–44.

SIMONS, MARLISE. "Virus Linked to Pollution Is Killing Hundreds of Dolpins in Mediterranean." *The New York Times International* (28 Oct 1990), pp. 2–3.

STEINHART, PETER. "Waterway Watchdogs." *Audubon* 92 (Nov 1990), pp. 26–32.

STIGLIANI, WILLIAM M., ET AL. "Chemical Time Bombs: Predicting the Unpredictable." *Environment* 33 (May 1991), pp. 4–9.

STRANAHAN, SUSAN Q. "It's Enough to Make You Sick." *National Wildlife* (Feb/March 1990), pp. 8–15.

THOMPSON, JON. "East Europe's Dark Dawn." *National Geographic* (June 1991), p. 36–69.

TRUMBALL, MARK. "It's Easier Not to Pollute in the First Place." *The Christian Science Monitor* (6 March 1992), p. 7.

UNITED STATES ENVIRONMENTAL PROTECTION AGENCY. *Bioremediation in the Field.* Cincinnati, OH: Center for Environmental Research Information, 1991.

WATERS, TOM. "Ecoglasnost." *Discover* 11 (April 1990), pp. 50–53.

Chapter 15. Air Pollution and Its Control

AMERICAN MEDICAL ASSOCIATION. *Journal of the American Medical Association* 225 (28 Feb 1986). Issue devoted to effects of smoking tobacco.

ARTIS, THERESA. "Radon—Nature's Own Toxic Waste." *Discover* 9 (Oct 1988), pp. 85–91.

BRUNE, WILLIAM H. "Stalking the Elusive Atmospheric Hydroxyl Radical." *Science* 256 (22 May 1992), pp. 1154–1155.

CORCORAN, ELIZABETH. "Cleaning Up Coal." *Scientific American* 264 (May 1991), pp. 107–116.

FEDER, WILLIAM A. "Cumulative Effects of Chronic Exposure of Plants to Low Levels of Air Pollutants." *Air Pollution Damage to Vegetation*, pp. 21–30. Washington, D.C.: American Chemical Society of Washington, 1973.

FINDLAYSON-PITTS, BARBARA J., AND JAMES N. PITTS JR. *Atmospheric Chemistry: Fundamentals and Experimental Techniques.* New York: John Wiley & Sons, 1986.

FRENCH, HILARY F. "Clearing the Air: A Global Agenda." *Worldwatch Paper 94.* Washington, D.C.: Worldwatch Institute, 1990.

GLOBAL ENVIRONMENTAL MONITORING SYSTEM. "An Assessment of Urban Air Quality." *Environment* 31 (Oct 1989), pp. 6–13, 26–37.

GRAEDEL, THOMAS E., AND PAUL J. CRUTZEN. "The Changing Atmosphere." *Scientific American* 261 (Sept 1989), pp. 58–68.

HENSCHEL, BRUCE D., ET AL. *Radon Reduction Techniques for*

Detached Houses. Washington, D.C.: Environmental Protection Agency, 1988.

KERR, RICHARD. "Hydroxyl, the Cleanser That Thrives on Dirt." *Science* 253 (13 Sept 1991), pp. 1210–1211.

LEGGE, ALLAN H., AND SAGAR V. KRUPA, EDS. *Air Pollutants and Their Effects on the Terrestrial Ecosystem.* Corelo, CA: Island Press, 1986.

MACKENZIE, JAMES J., AND MOHAMED T. EL-ASHRY. *Ill Winds: Airborne Pollution's Toll on Trees and Crops.* World Resources Institute, Sept 1988.

MUMME, STEPHEN P. "Clearing the Air: Environmental Reform in Mexico." *Environment* 33 (Dec 1991), pp. 7–11, 26–30.

NATIONAL ACADEMY OF SCIENCES. *Air Pollution, the Automobile, and Human Health.* Washington, D.C.: National Academy Press, 1988.

NERO, ANTHONY V., JR. "Controlling Indoor Air Pollution." *Scientific American* 258 (May 1988), pp. 42–48.

NEWELL, REGINALD E., ET AL. "Carbon Monoxide and the Burning Earth." *Scientific American* 261 (Oct 1989), pp. 82–88.

SEINFELD, JOHN. *Atmospheric Chemistry and Physics of Air Pollution.* New York: John Wiley & Sons, 1986.

SEINFELD, JOHN. "Urban Air Pollution: State of the Science." *Science* 243 (10 Feb 1989), pp. 745–752.

SMITH, KIRK R. "Air Pollution: Assessing Total Exposure in the United States." *Environment* 30 (Oct 1988), pp. 10–15, 33–38.

SPETH, JAMES GUSTAVE. *Environmental Pollution: A Long-Term Perspective.* Washington, D.C.: World Resources Institute, 1988.

THOMPSON, ANNE M. "The Oxidizing Capacity of the Earth's Atmosphere: Probable Past and Future Changes." *Science* 256 (22 May 1992), pp. 1157–1165.

U.S. DEPARTMENT OF HEALTH AND HUMAN SERVICES. *The Health Consequences of Smoking.* Rockville, MD: U.S. Dept. of Health and Human Services, Annual.

U.S. ENVIRONMENTAL PROTECTION AGENCY. "Clean Air Act Amendments of 1990: Detailed Summary of Titles." Washington, D.C.: U.S. Environmental Protection Agency.

U.S. ENVIRONMENTAL PROTECTION AGENCY. "The New Clean Air Act." *EPA Journal* 17 (Jan/Feb 1991). Issue on the Clean Air Act of 1990.

Chapter 16. Major Atmospheric Changes

ABRAHAMSON, DEAN E., ED. *The Challenge of Global Warming.* Washington, D.C.: Island Press, 1989.

BAKER, LAWRENCE A., ET AL. "Acidic Lakes and Streams in the United States: The Role of Acidic Deposition." *Science* 252 (24 May 1991), pp. 1151–1154.

BAZZAZ, FAKHRI A., AND ERIC D. FAJER. "Plant Life in a CO_2-Rich World." *Scientific American* 266 (Jan 1992), pp. 68–74.

BENEDICK, RICHARD E. *Ozone Diplomacy: New Directions in Safeguarding the Planet.* Cambridge, MA: Harvard University Press, 1991.

EDGERTON, LYNNE T. *The Rising Tide: Global Warming and World Sea Levels.* Covelo, CA: Island Press, 1991.

ENVIRONMENTAL CANADA. *Downwind: The Acid Rain Story.* Ottawa: Ministry of Supply and Services, 1981.

FIROR, JOHN. *The Changing Atmosphere: A Global Challenge.* New Haven, CT: Yale University Press, 1990.

FISHMAN, ALBERT, AND ROBERT KALISH. *Global Alert: The Ozone Pollution Crisis.* New York: Plenum. 1990.

FLAVIN, CHRISTOPHER. "Slowing Global Warming: A Worldwide Strategy." *Worldwatch Paper 91.* Washington, D.C.: Worldwatch Institute, 1989.

GRAEDEL, THOMAS E., AND PAUL J. CRUTZEN. "The Changing Atmosphere." *Scientific American* 261 (Sept 1989), pp. 58–68.

HORDIJK, LEEN. "A Model Approach to Acid Rain." *Environment* 30 (March 1989), pp. 17–20, 40–41.

HOUGHTON, J.T., ET AL., EDS. *Climate Change: The IPCC Scientific Assessment.* New York: Cambridge University Press, 1990.

JONES, PHILIP D., AND TOM M.L. WIGLEY. "Global Warming Trends." *Scientific American* 263 (Aug 1990), pp. 84–91.

KERR, RICHARD A. "Greenhouse Science Survives Skeptics." *Science* 256 (22 May 1992), pp. 1138–1140.

KERR, RICHARD A. "Pollutant Haze Cools the Greenhouse." *Science* 255 (7 Feb 1992), pp. 682–683.

LIKENS, GENE E. "Toxic Winds: Whose Responsibility?" *Ecology, Economics, Ethics: The Broken Circle,* Chapter 9. New Haven, CT: Yale University Press, 1991.

MATHEWS, JESSICA T., ED. *Greenhouse Warming: Negotiating a Global Regime.* Washington, D.C.: World Resources Institute, 1991.

MOHNEN, VOLKER A. "The Challenge of Acid Rain." *Scientific American* 259 (Aug 1988), pp. 30–38.

NATIONAL ACADEMY OF SCIENCES. *Acid Deposition: Long-Term Trends.* Washington, D.C.: National Academy Press, 1986.

NATIONAL ACADEMY OF SCIENCES. *Policy Implications of Greenhouse Warming.* Washington, D.C.: National Academy Press, 1991.

ROWLAND, F. SHERWOOD. "Chlorofluorocarbons and the Depletion of Stratospheric Ozone." *American Scientist* 77, 1989, pp. 36–45.

RUBIN, EDWARD S., ET AL. "Realistic Mitigation Options for Global Warming." *Science* 257 (10 July 1992), pp. 148–149, 261–266.

SCHINDLER, D.W. "Effects of Acid Rain on Freshwater Ecosystems." *Science* 239 (8 Jan 1988), pp. 149–153.

SCHNEIDER, STEPHEN H. "Global Warming: Are We Entering the Greenhouse Century?" San Francisco: Sierra Club Books, 1989.

SCHWARTZ, S.E. "Acid Deposition: Unraveling a Regional Phenomenon." *Science* 243 (10 Feb 1989), pp. 753–761.

SHEA, CYNTHIA POLLACK. "Protecting Life on Earth: Steps to Save the Ozone Layer." *Worldwatch Paper 87.* Washington, D.C.: Worldwatch Institute, 1988.

STOLARSKI, RICHARD, ET AL. "Measured Trends in Stratospheric Ozone." *Science* 256 (17 April 1992), pp. 342–349.

TOON, OWEN B., AND RICHARD P. TURCO. "Polar Stratospheric Clouds and Ozone Depletion." *Scientific American* 264 (June 1991), pp. 68–74.

WHITE, ROBERT M. "The Great Climate Debate." *Scientific American* 263 (July 1990), pp. 36–43.

WHITE, ROBERT M. "Our Climatic Future: Science, Technology, and World Climate Negotiations." *Environment* 33 (March 1991), pp. 18–20, 38–41.

YOUNG, LOUISE B. *Sowing the Wind: Reflections on the Earth's Atmosphere*. New York: Prentice Hall, 1990.

Chapter 17. Risks and Economics of Pollution

BELL, LAUREN. "The High Cost of Neglecting Wildlife." *National Wildlife* 27 (March 1989), pp. 4–8.

BORELLI, PETER, ET AL. *Crossroads: Environmental Priorities for the Future*. Washington, D.C.: Island Press, 1989.

CARPENTER, RICHARD A., AND JOHN A. DIXON. "Ecology Meets Economics: A Guide to Sustainable Development." *Environment* 27 (June 1985), pp. 6–11, 27–32.

CLARKE, LEE. *Acceptable Risk? Making Decisions in a Toxic Environment*. Berkeley: University of California Press, 1989.

COHEN, BERNARD L. "How to Assess the Risks You Face." *Consumer's Research* (June 1992), pp. 11–16.

COUNCIL ON ENVIRONMENTAL QUALITY. "Making the Environment Count: Costs, Benefits, Goals and Tools." *Environmental Quality: 21st Annual Report*, Chapter 2. Washington, D.C.: Council on Environmental Quality, 1990.

GRAHAM, J.D., ET AL., EDS. *In Search of Safety: Chemicals and Cancer Risk*. Cambridge, MA: Harvard University Press, 1988.

KNEESE, ALLAN V. *Measuring the Benefits of Clean Air and Water*. Resources for the Future, 1984.

KRUPNICK, ALAN J., AND PAUL R. PORTNEY. "Controlling Urban Air Pollution: A Benefit-Cost Assessment." *Science* 252 (26 April 1991), pp. 522–528.

KUNREUTHER, HOWARD, AND RUTH PATRICK. "Managing the Risks of Hazardous Waste." *Environment* 33 (April 1991), pp. 12–15, 31–36.

LAVE, LESTER B. *Risk Assessment and Management*. New York: Plenum, 1987.

MARSHALL, ELIOT. "A Is for Apple, Alar and . . . Alarmist?" *Science* 254 (4 Oct 1991), pp. 20–22.

MORGENSTERNER, RICHARD, AND STUART SESSIONS. "Weighing Environmental Risks: EPA's Unfinished Business." *Environment* 30 (July/Aug 1988), pp. 15–17, 34–39.

NATIONAL ACADEMY OF SCIENCES. *Improving Risk Communication*. Washington, D.C.: National Academy Press, 1989.

PERROW, CHARLES. *Normal Accidents: Living with High-Risk Technologies*. New York: Basic Books, 1985.

ROBERTS, LESLIE. "Counting on Science at EPA." *Science* 249 (10 Aug 1990), pp. 616–618.

RUSSELL, MILTON, AND MICHAEL GRUBER. "Risk Assessment in Environmental Policy-Making." *Science* 236 (17 April 1987), pp. 286–290.

SANDMAN, PETER M. "Risky Business." *Rutgers Magazine* (April/June, 1989), pp. 36, 37.

STAVINS, ROBERT. "Harnessing Market Forces to Protect the Environment." *Environment* 31 (Jan/Feb 1989), pp. 4–7, 28–38.

SUMMERS, EMMANUEL. "Environmental Hazards Show No Respect for National Boundaries." *Environment* 29 (June 1987), pp. 7–9, 31–33.

U.S. ENVIRONMENTAL PROTECTION AGENCY. *Reducing Risk: Setting Priorities and Strategies for Environmental Protection.* Washington, D.C.: Science Advisory Board, U.S. EPA, 1990.

U.S. ENVIRONMENTAL PROTECTION AGENCY. "Setting Environmental Priorities: The Debate about Risk." *EPA Journal* 17 (March/April 1991). Issue about risk analysis.

WENZ, PETER S. *Environmental Justice*. Albany, NY: State University of New York Press, 1988.

WILSON, RICHARD, AND E.A.C. CROUCH. "Risk Assessment and Comparisons: An Introduction." *Science* 236 (17 April 1987), pp. 267–270.

ZECKHAUSER, RICHARD J., AND W. KIP VISCUSI. "Risk within Reason." *Science* 248 (4 May 1990), pp. 559–564.

Chapter 18. Wild Species: Diversity and Protection

BUTLER, WILLIAM A. "Incentives for Conservation." *Ecology, Economics, Ethics: The Broken Circle*, Chapter 11. New Haven, CT: Yale University Press, 1991.

COUNCIL ON ENVIRONMENTAL QUALITY. "Linking Ecosystems and Biodiversity." *Environmental Quality: 21st Annual Report*, Chapter 4. Washington, D.C.: Council on Environmental Quality, 1990.

DiSILVESTRO, ROGER L. "Saga of AC-9, The Last Free Condor." *Audubon* 89 (July 1987), pp. 12–14.

EHRLICH, PAUL R., AND EDWARD O. WILSON. "Biodiversity Studies: Science and Policy." *Science* 253 (16 Aug 1991), pp. 758–761.

ERWIN, TERRY L. "An Evolutionary Basis for Conservation Strategies." *Science* 253 (1991), pp. 750–752.

HARGROVE, EUGENE C. *Foundations of Environmental Ethics*. Englewood Cliffs, NJ: Prentice Hall, 1989.

KARR, PAUL. "The Scourge of Suburbia." *Massachusetts Audubon* 31 (Fall 1991), pp. 5–7.

LAZELL, JAMES. "The Evils of Exotics." *Massachusetts Audubon* 31 (Fall 1991), pp. 12–13.

MANN, CHARLES C. "Are Ecologists Crying Wolf?" *Science* 253 (16 Aug 1991), pp. 736–738.

McNEELY, JEFFREY A., ET AL. "Strategies for Conserving Biodiversity." *Environment* 32 (April 1990), pp. 16–20, 36–41.

MOROWITZ, HAROLD J. "Balancing Species Preservation and Economic Considerations." *Science* 253, (1991), pp. 752–754.

MYERS, NORMAN. "Biological Diversity and Global Security." *Ecology, Economics, Ethics: The Broken Circle*, Chapter 2. New Haven, CT: Yale University Press, 1991.

MYERS, STEVEN LEE. "Wild Turkeys Roar Back from Near-Extinction." *The New York Times* 24 (Nov, 1991), p. 5.

REID, WALTER V., AND KENTON R. MILLER. *Keeping Options Alive: The Scientific Basis for Conserving Biodiversity*. Washington, D.C.: World Resources Institute, 1989.

ROBERTS, LESLIE. "Zebra Mussel Invasion Threatens U.S. Waters." *Science* 249 (21 Sept 1990), pp. 1370–1372.

ROLSTON, HOLMES, III. *Environmental Ethics: Duties to and Values in the Natural World*. Philadelphia: Temple University Press, 1988.

RYAN, JOHN C. "Life Support: Conserving Biological Diversity." *Worldwatch Paper 108*. Washington, D.C.: Worldwatch Institute, 1992.

SOLBRIG, OTTO T. "The Origin and Function of Biodiversity." *Environment* 33 (June 1991), pp. 16–20, 34–39.

SOULÉ, MICHAEL E. "Conservation: Tactics for a Constant Crisis." *Science* 253 (1991), pp. 744–750.

TERBORGH, JOHN. "Why American Songbirds Are Vanishing." *Scientific American* 266 (May 1992), pp. 98–104.

U.S. ENVIRONMENTAL PROTECTION AGENCY. *Ecology and Management of the Zebra Mussel and Other Introduced Aquatic Nuisance Species.* EPA/600/3-91/003. Washington, D.C.: USEPA, 1991.

U.S. FISH AND WILDLIFE SERVICE. *Endangered and Threatened Species Recovery Program: Report to Congress.* Washington, D.C.: U.S. Dept. of the Interior, 1990.

WILSON, EDWARD O., ED. *Biodiversity.* Washington, D.C.: National Academy Press, 1988.

WILSON, EDWARD O. "Biodiversity, Prosperity, Value." *Ecology, Economics, Ethics: The Broken Circle,* Chapter 1. New Haven, CT: Yale University Press, 1991.

Chapter 19. Ecosystems as Resources

ALLARD, WILLIAM, A., AND LOREN MCINTYRE. "Rondonia: Brazil's Imperiled Rain Forest." *National Geographic* 174 (Dec 1988), pp. 772–799.

BERGER, JOHN J., ED. *Environmental Restoration: Science and Strategies for Restoring the Earth.* Washington, D.C.: Island Press, 1990.

BOTKIN, DANIEL B. *Discordant Harmonies: A New Ecology for the Twentyfirst Century.* New York: Oxford University Press, 1990.

BUSCHBACHER, ROBERT J. "Tropical Deforestation and Pasture Development." *Bioscience* 30 (Jan 1989), pp. 22–29.

CETACEAN RESEARCH UNIT. *The Whale Watcher's Companion.* Gloucester, MA: Cetacean Research Unit, 1990.

FISHER, RON, ET AL. *The Emerald Realm: Earth's Precious Rain Forests.* Washington, D.C.: National Geographic Society, 1990.

GOLDSMITH, EDWARD, ET AL. *Imperiled Planet: Restoring Our Endangered Ecosystems.* Cambridge, MA: The MIT Press, 1990.

GOODLAND, ROBERT, AND GEORGE LEDEC. "Wildlands: Balancing Conversion with Conservation in World Bank Projects." *Environment* 31 (Nov 1989), pp. 6–11, 27–35.

GRADWOHL, JUDITH, AND RUSSELL GREENBERG. *Saving the Tropical Forests.* Washington, D.C.: Smithsonian Institution, 1988.

HARDIN, GARRETT, AND JOHN BADIN, EDS. *Managing the Commons.* San Francisco: W.H. Freeman and Co., 1977. Contains Hardin's original paper on the Tragedy of the Commons and many other papers relating to the concept of the commons.

HINCK, JON. "The Tangled Web: Driftnets and the Decline of the North Pacific." *Greenpeace* 12 (April/June 1987), pp. 6–9.

JORDAN, WILLIAM R., III, ET AL. *Restoration Ecology: A Synthetic Approach to Ecological Research.* Cambridge, U.K.: University Press, Cambridge, 1987.

KUSLER, JON A. "Wetlands Delineation: An Issue of Science or Politics?" *Environment* 34 (March 1992), pp. 7–11, 29–37.

KUSLER, JON A., AND MARY E. KENTULA. *Wetland Creation and Restoration: The Status of the Science.* Washington, D.C.: Island Press, 1990.

LIPSKE, MICHAEL. "Who Runs America's Forests?" *National Wildlife* (Oct/Nov 1990), pp. 24–48.

LUGO, ARIEL E. "The Future of the Forest: Ecosystem Rehabilitation in the Tropics." *Environment* 30 (Sept 1988), pp. 16–20, 41–45.

MCNEELY, JEFFREY A., ET AL. *Conserving the World's Biological Resources: A Primer on Principles and Practice for Development Action.* Washington, D.C.: World Resources Institute, 1989.

MOORE, PETER D. "The Exploitation of Forests." *Science and Christian Belief,* 2, pp. 131–140, 1990.

MYERS, NORMAN. "The World's Forests and Human Populations: The Environmental Interconnections." *Resources, Environment and Population: Present Knowledge, Future Options,* pp. 237–251. New York: Oxford University Press, 1991.

NORSE, ELLIOTT A. *Ancient Forests of the Pacific Northwest.* Washington, D.C.: Island Press, 1990.

OELSCHLAEGER, MAX. *The Idea of Wilderness: From Prehistory to the Age of Ecology.* New Haven, CT: Yale University Press, 1991.

POSTEL, SANDRA, AND LORI HEISE. "Reforesting the Earth." *Worldwatch Paper 83.* Washington, D.C.: Worldwatch Institute, 1988.

REPETTO, ROBERT. *The Forest for the Trees? Government Policies and the Misuse of Forest Resources.* Washington, D.C.: World Resources Institute, 1988.

REPETTO, ROBERT, ET AL. *Wasting Assets: Natural Resources in the National Income Accounts.* Washington, D.C.: World Resources Institute, 1989.

REVKIN, ANDREW. *The Burning Season: The Murder of Chico Mendes and the Fight for the Amazon Rain Forest.* Boston: Houghton Mifflin, 1990.

ROSS, MICHAEL R. *Recreational Fisheries of Coastal New England.* Amherst, MA: University of Massachusetts Press, 1991.

U.S. FISH AND WILDLIFE SERVICE. *Restoring America's Wildlife, 1937–1987.* Washington, D.C.: U.S. Dept. of the Interior, 1987.

VINCENT, JEFFREY R. "The Tropical Timber Trade and Sustainable Development." *Science* 256 (19 June 1992), pp. 1651–1655.

WILLIAMS, MARLA. "Save the People: New, Militant Antienvironmentalists Fight to Return Nature to a Back Seat." *The Boston Globe* (13 Jan 1992), p. 3.

WORLD RESOURCES INSTITUTE. *World Resources 1992–93.* New York: Oxford University Press, 1992. Excellent resource containing separate chapters on different habitats and extensive data tables.

Chapter 20. Converting Trash to Resources

CONNETT, PAUL H. "The Disposable Society." *Ecology, Economics, Ethics: The Broken Circle,* Chapter 7. New Haven, CT: Yale University Press, 1991.

DENISON, RICHARD A., AND JOHN RUSTON. *Recycling and Incineration: Evaluating the Choices.* Covelo, CA: Island Press, 1990.

FROSCH, ROBERT A., AND NICHOLAS E. GALLOPOULOS. "Strategies for Manufacturing." *Scientific American* 261 (Sept 1989), pp. 144–152.

GRASSY, JOHN. "Bottle Bills." *Garbage* 4 (Jan/Feb 1992), pp. 44–48.

KHARBANDA, O.P., AND E.A. STALLWORTHY. *Waste Management: Toward a Sustainable Society.* New York: Auburn House, 1990.

LEVENSON, HOWARD. "Wasting Away: Policies to Reduce Trash, Toxicity and Quantity." *Environment* 32 (March 1990), pp. 10–15, 31–36.

LYHUS, RANDY. "Composting at the Landfill." *Environment* 30 (Oct 1988), pp. 21–22.

MARTIN, AMY. "Recycling, Inc." *Garbage* 4 (March/April 1992), pp. 26–31.

NATURAL RESOURCES DEFENSE COUNCIL. *A Solid Waste Blueprint for New York State.* New York: Natural Resources Defense Council, 1988.

NEAL, HOMER A., AND J.R. SCHUBEL. *Solid Waste Management and the Environment—The Mounting Garbage and Trash Crisis.* Englewood Cliffs, NJ: Prentice Hall, 1987.

OFFICE OF TECHNOLOGY ASSESSMENT, U.S. CONGRESS. *Facing America's Trash: What Next for Municipal Solid Waste?* OTA-0-424. Washington, D.C.: U.S. Govt. Printing Office, 1989.

O'LEARY, PHILIP R., ET AL. "Managing Solid Waste." *Scientific American* 259 (Dec 1988), pp. 36–42.

PLATT, BRENDA, ET AL. *Beyond 40 Percent: Record-Setting Recycling and Composting Programs.* Washington, D.C.: Island Press, 1991.

POLLACK, CYNTHIA. "Mining Urban Wastes: The Potential for Recycling." *Worldwatch Paper 76.* Washington, D.C.: Worldwatch Institute, 1987.

SCHWARTZ, ANNE. "Poisons in Your Home: A Disposal Dilemma." *Audubon* (May 1987), pp. 12–17.

U.S. ENVIRONMENTAL PROTECTION AGENCY. "The Garbage Crisis: Understanding It, Finding Answers." *EPA Journal* 15 (March/April 1989). Issue about current solid waste problems.

U.S. ENVIRONMENTAL PROTECTION AGENCY. *The Solid Waste Dilemma, An Agenda for Action.* EPA/530-SW-89-019. Washington, D.C.: U.S. EPA, Office of Solid Waste, 1989.

WOLF, NANCY, AND ELLEN FELDMAN. *America's Packaging Dilemma.* Covelo, CA: Island Press, 1990.

YOUNG, JOHN E. "Discarding the Throwaway Society." *Worldwatch Paper 101.* Washington, D.C.: Worldwatch Institute, 1991.

YOUNG, JOHN E. "Mining the Earth." *Worldwatch Paper 109.* Washington, D.C.: Worldwatch Institute, 1992.

Chapter 21. Energy Resources and the Energy Problem

BABBIT, BRUCE. "Will America Join the Waste Watchers?" *World Monitor* 496 (June 1991), pp. 34–35.

BAUMGARTEN, FRED. "America's Arctic Refuge—At the Crossroads." *Audubon Activist* (April 1991), pp. 1, 4.

BEGLEY, SHARON. "The Road Not Taken." *National Wildlife* 29 (Aug/Sept. 1991), pp. 10–15.

CANINE, CRAIG. "Home Energy." *Garbage* (Nov/Dec 1989), pp. 20–27.

CHEMICAL MANUFACTURERS ASSOCIATION. "Earthships—Homes for the Future?" *Chemecology* 19 (Dec/Jan 1991).

COMMONER, BARRY. "Toward a Sustainable Energy Future." *E Magazine* (May/June 1991), pp. 54–55.

CORCORAN, ELIZABETH. "Cleaning Up Coal." *Scientific American* 264 (May 1991), pp. 106–117.

DILLIN, JOHN. "Natural Gas Seeps into Spotlight." *The Christian Science Monitor* (8 March 1991), p. 3.

DREW, LISA. "Truth and Consequence along Oiled Shores." *National Wildlife* 28 (June/July 1990), pp. 34–43.

DREYFUS, DANIEL, AND ANNE ASHBY. "Fueling Our Global Future." *Environment* 32 (May 1990), p. 16.

FICKETT, ARNOLD. "Efficient Use of Electricity." *Scientific American* 263 (Sept 1990), pp. 65–74.

FLAVIN, CHRISTOPHER. "The Bridge to Clean Energy." *WorldWatch* 5 (Aug 1992), pp. 10–18.

FLAVIN, CHRISTOPHER. "Building a Bridge to Sustainable Energy." *State of the World*, pp. 3–45. New York: W.W. Norton, 1992.

FLAVIN, CHRISTOPHER. "Conquering U.S. Oil Dependence." *WorldWatch* (Jan/Feb 1991), p. 28.

FLAVIN, CHRISTOPHER. "Yankee Utilities Learn to Love Efficiency." *WorldWatch* 3 (March/April 1990), pp. 5–6.

FREEDMAN, DAVID H. "Batteries Included." *Discover* 13 (March 1992), pp. 90–99.

GEVER, JOHN, ET AL. *Beyond Oil: The Threat to Food and Fuel in the Coming Decades.* Cambridge, MA: Ballinger Publishing, 1986.

GWIN, HOLLY L., ET AL. "Strategies for Energy Use." *Scientific American* (Sept 1989), pp. 136–143.

HARRIS, MARK. "Energy-Efficient Appliances: Lowering the Cost of Plugging In." *E Magazine* (Nov/Dec 1990), pp. 58–61.

HEED, RICHARD, AND ROBERT BISHOP. "Corporate Wealth through Waste." *Sierra* (July/Aug 1991), pp. 16–18.

HIRST, ERIC. "Boosting U.S. Energy Efficiency through Federal Action." *Environment* 33 (March 1991), pp. 6–11.

HIRST, ERIC. "Demand-Side Management: An Underused Tool for Conserving Electricity." *Environment* 32 (Jan/Feb 1990), pp. 4–9.

HUBBARD, HAROLD M. "The Real Cost of Energy." *Scientific American* 264 (April 1991), pp. 36–43.

JACKSON, J.B.C., ET AL. "Ecological Effects of a Major Oil Spill on Panamanian Coastal Marine Communities." *Science* 243 (6 Jan 1989), pp. 37–43.

KNICKERBOCKER, BRAD. "Smolts, Volts Charge Dam Renewals." *The Christian Science Monitor* (16 Jan 1992), p. 8.

LEE, WILLIAM S. "Energy for Our Globe's People." *Environment* 32 (Sept 1990), pp. 12–15.

LEWIS, THOMAS A. "The Heat Is On." *National Wildlife* 28 (April/May 1990), pp. 38–41.

LOVINS, AMORY B. "Energy, People, and Industrialization." *Resources, Environment, and Population*, pp. 95–124. New York: Oxford University Press, 1991.

LOVINS, AMORY B. "The Negawatt Revolution." *Across the Board* (Sept 1990), pp. 18–23.

LUOMA, JON R. "Shouldn't You Switch?" *Audubon* (May 1991), pp. 80–85.

LYMAN, FRANCESCA. "The Future of Transportation in America." *E Magazine* (Sept/Oct 1990), pp. 34–41.

MARINELLI, JANET. "Cars." *Garbage* (Nov/Dec 1989), pp. 28–37.

MASTERS, C.D., ET AL. "Resource Constraints in Petroleum Production Potential." *Science* 253 (12 July 1991), pp. 146–152.

MILLER, WILLIAM H. "Balance Sought: Energy, Environment, Economy." *Industry Week* (1 April 1991), pp. 62–65, 68–70.

MUCKLESTON, KEITH W. "Salmon vs. Hydropower: Striking a Balance in the Pacific Northwest." *Environment* (Jan/Feb 1990), pp. 10–15.

NIXON, WILL. "Energy for the Next Century." *E Magazine* (May/June 1991), pp. 31–39.

NORBERG-BOHM, VICKI. "From Inside Out: Reducing CO_2 Emissions in the Buildings Sector." *Environment* (April 1991), pp. 16–20.

ROSEN, YERETH. "Alaskans Reluctantly Face Drop in Oil Boom." *The Christian Science Monitor* (22 April 1992), p. 9.

SCHNEIDER, STEPHEN H. "Our Waste-Free House." *World Monitor* 4 (June 1991), pp. 36–41.

Scientific American (Sept 1990). "Energy for Planet Earth." Special issue.

STARR, C., ET AL. "Energy Sources: A Realistic Outlook." *Science* (15 May 1992), pp. 981–986.

STEVENS, MARY OTIS. "Design for Living 1991." *World Monitor* 4 (June 1991), pp. 42–45.

UDALL, JAMES R. "Prophets of an Energy Revolution." *National Wildlife* 30 (Dec/Jan 1992), pp. 10–13.

WOODBURY, RICHARD. "The Great Energy Bust." *Time* (16 March 1992), pp. 50–51.

WORLDWATCH INSTITUTE. "California Cuts Its Electric Bill." *WorldWatch* (Jan/Feb 1991), p. 5.

YERGIN, DANIEL. "The Age of Hydrocarbon Man." *World Monitor* 4 (Feb 1991), pp. 54–59.

Chapter 22. Nuclear Power: Promise and Problems

AMATO, IVAN. "Lab of the Rising Microsuns." *Science* 254 (25 October 1991). pp. 515–517. Article on cluster impact fusion.

ANSPAUGH, LYNN R., ET AL. "The Global Impact of the Chernobyl Reactor Accident." *Science* 242 (16 Dec 1988), pp. 1513–1519.

CONN, ROBERT W., ET AL. "The International Thermonuclear Experimental Reactor." *Scientific American* 266 (April 1992), pp. 103–110.

ENERGY IINFORMATION ADMINISTRATION. *Annual Energy Review 1990*. DOE/EIA-0384(90). Washington, D.C.: U.S. Dept. of Energy, 1991.

ENERGY IINFORMATION ADMINISTRATION. *International Energy Outlook 1991: A Post-War Review of Energy Markets*. DOE/EIA-0484(91). Washington, D.C.: U.S. Dept. of Energy, 1991.

FLAVIN, CHRISTOPHER. "Reassessing Nuclear Power: The Fallout from Chernobyl." *Worldwatch Paper 75*. Washington, D.C.: Worldwatch Institute, 1987.

FORSBERG, C.W., AND A.M. WEINBERG. "Advanced Reactors, Passive Safety, and Acceptance of Nuclear Energy." *An-

nual Review of Energy* 15, pp. 133–152. Palo Alto, CA: Annual Reviews, Inc., 1990.

FRIEDLANDER, GERHART, ET AL. *Nuclear and Radiochemistry*, 3d ed. New York: John Wiley & Sons, 1981.

GLASSTONE, SAMUEL. *Nuclear Power and Its Environmental Effects*. Chicago: Wesley Publishing, 1987.

GOLAY, MICHAEL W., AND NEIL E. TODREAS. "Advanced Light-Water Reactors." *Scientific American* 262 (April 1990), pp. 82–89.

HÄFELE, WOLF. "Energy from Nuclear Power." *Scientific American* 263 (Sept 1990), pp. 137–144.

HOHENEMSER, CHRISTOPH, ET AL. "Institutional Aspects of the Future Development of Nuclear Power." *Annual Review of Energy* 15, pp. 173–200. Palo Alto, CA: Annual Reviews, Inc., 1990.

HOLDREN, J.P. "Safety and Environmental Aspects of Fusion Energy." *Annual Review of Energy and the Environment* 16, pp. 235–258. Palo Alto, CA: Annual Reviews, Inc., 1991.

LENSSEN, NICHOLAS. "Nuclear Waste: The Problem That Won't Go Away." *Worldwatch Paper 106*. Washington, D.C.: Worldwatch Institute, 1991.

MEDVEDEV, ZHORES A. *The Legacy of Chernobyl*. New York: W.W. Norton, 1990.

MORONE, JOSEPH G., AND EDWARD J. WOODHOUSE. *The Demise of Nuclear Energy? Lessons for Democratic Control of Technology*. New Haven, CT: Yale University Press, 1989.

RAZAVI, HOSSEIN, AND FEREIDUN FESHARAKI. "Electricity Generation in Asia and the Pacific: Historical and Projected Patterns of Demand and Supply." *Annual Review of Energy and the Environment* 16, pp. 275–294. Palo Alto, CA: Annual Reviews, Inc., 1991.

ROSSIN, A.D. "Experience of the U.S. Nuclear Industry and Requirements for a Viable Nuclear Industry in the Future." *Annual Review of Energy* 15, pp. 153–172. Palo Alto, CA: Annual Reviews, Inc., 1990.

SLOVIC, PAUL, ET AL. "Lessons from Yucca Mountain." *Environment* 33 (April 1991), pp. 7–11, 28–31.

SLOVIC, PAUL, ET AL. "Perceived Risk, Trust, and the Politics of Nuclear Waste." *Science* 254 (13 Dec 1991), pp. 1603–1608.

VOGELSANG, W.F., AND H.H. BARSCHALL. "Nuclear Power." *The Energy Sourcebook: A Guide to Technology, Resources, and Policy*, pp. 127–152. New York: American Institute of Physics, 1991.

Chapter 23. Solar and Other "Renewable" Energy Resources

THE AIA RESEARCH CORPORATION. *Solar Dwelling Design Concepts*. Washington, D.C.: U.S. Department of Housing and Urban Development, Office of Policy Development and Research, May 1976.

AITKEN, DONALD W. "Sustained Orderly Development of the Solar-Electric Technologies." *Solar Today* 6 (May/June 1992), pp. 20–22.

ANDERSON, BRUCE, WITH MICHAEL RIORDAN. *"The New Solar Home Book."* Andover, MA: Brick House Publishing, 1987.

CHILES, JAMES R. "Tomorrow's Energy Today." *Audubon* (Jan 1990), pp. 59–72.

CONKLING, MARK. "A Solar Community: Eldorado at Santa Fe." *Solar Today* 5 (Nov/Dec 1991), pp. 18–19.

CUSHMAN, JANET H., ET AL. "Energy Crops for Biofuels." *The World and I* (Aug 1991), pp. 334–341.

DILLIN, JOHN. "Congress Weighs Policy Choices Aimed at More Energy Security." *The Christian Science Monitor* (23, 26, 29, 30 Aug 1991).

DOSTROVSKY, ISRAEL. "Chemical Fuels from the Sun." *Scientific American* 265 (Dec 1991), pp. 102–107.

ECKHOLM, ERIK. "The Other Energy Crisis: Firewood." *Worldwatch Paper 1* (Sept 1975), Washington, D.C.: Worldwatch Institute, 1975.

EVANS, LORI C. "Wind Energy in Europe." *Solar Today* 6 (May/June 1992), pp. 32–34.

FLAVIN, CHRISTOPHER. "Yankee Utilities Learn to Love Efficiency." *WorldWatch* 3 (March/April 1990), pp. 5–6.

FLAVIN, CHRISTOPHER, AND NICHOLAS LENSSEN. "Beyond the Petroleum Age: Designing a Solar Economy." *Worldwatch Paper 100*. Washington, D.C.: Worldwatch Institute, 1990.

FLAVIN, CHRISTOPHER, AND NICHOLAS LENSSEN. "Here Comes the Sun." *WorldWatch* (Sept/Oct 1991), pp. 10–18.

FLAVIN, CHRISTOPHER, ET AL. "Sustainable Energy." Washington, D.C.: Renew America, 1989.

GOLD, ALLAN R. "Quebec Indians Ponder True Cost of Electricity." *The New York Times International* (12 Oct 1990), p. A10.

GOTTFRIED, KURT, AND JONATHAN DEAN. "Which Energy Future Will America Choose?" *Nucleus* 13 (Winter 1991–92), pp. 1, 4, 5.

GREENBERG, DAVID A. "Modeling Tidal Power." *Scientific American* 257 (Nov 1987), pp. 128–133.

HOLING, DWIGHT. "America's Energy Plan: Missing in Action." *The Amicus Journal* 13 (Winter 1991), pp. 12–20.

KERR, RICHARD A. "Geothermal Tragedy in the Commons." *Science* 253 (12 July 1991), pp. 134–135.

KNICKERBOCKER, BRAD. "Renewables and Conservation Are Finding a Place in the Sun." *The Christian Science Monitor* (Aug 1991).

LENSSEN, NICHOLAS. "California's Wind of Industry Takes Off," *WorldWatch* 3 (July/Aug 1990), pp. 38–39.

LENSSEN, NICHOLAS. "Cooked by the Sun." *WorldWatch* (March/April 1989), pp. 41–42.

LENSSEN, NICHOLAS. "Third World PVs Hit the Roof." *WorldWatch* 5 (May/June 1992), pp. 7–8.

LENSSEN, NICHOLAS, AND JOHN E. YOUNG. "Filling Up in the Future." *WorldWatch* (May/June 1990), pp. 18–26.

LIPSKE, MICHAEL. "Playing for Power in Quebec's North." *International Wildlife* 21 (May/June 1991), pp. 10–17.

LOTKER, MICHAEL, AND DAVID KEARNEY. "Solar Thermal Electric Performance and Prospects: The View From Luz." *Solar Today* (May/June 1991), pp. 10–13.

LYND, LEE R., ET AL. "Fuel Ethanol from Cellulosic Biomass." *Science* 251 (15 March 1991), pp. 1318–1323.

MACLEOD, ALEXANDER. "Europe Harnesses the Wind." *The Christian Science Monitor* (22 April 1992), p. 12.

MUCKLESTON, KEITH W. "Salmon vs. Hydropower: Striking a Balance in the Pacific Northwest." *Environment* 32 (Jan/Feb 1990), pp. 10–15.

NIXON, WILL. "Energy for the Next Century." *E Magazine* (May/June 1991), pp. 30–39.

NORDMANN, THOMAS, ET AL. "Innovative 100 KW Grid-Connected PV Installation." *Solar Today* 5 (Aug 1991), pp. 11–14.

OGDEN, JOAN M., AND ROBERT H. WILLIAMS. *Solar Hydrogen, Moving beyond Fossil Fuels.* Washington, D.C.: World Resources Institute, 1989.

OSBORN, DONALD E. "Using Solar Energy at the Sacramento Municipal Utility District." *Solar Today* 6 (July/Aug 1992), pp. 11–14.

PECK, LOUIS. "What Ever Happened to Solar?" *Garbage* (Jan/Feb 1991), pp. 24–31.

PIMENTEL, DAVID. *Grains for Food or Fuel.* Energy Policy Research and Information Program, Publication Series No. 81-3. West Lafayette, IN: Purdue University, 1981.

REECE, NANCY S. "On the Road to an Alternative Fueled Future." *Solar Today* 5 (March/April 1991), pp. 10–12.

ROSENBAUM, MARC. "Solar Hot Water for the 90s." *Solar Today* 5 (Sept/Oct 1991), pp. 20–22.

RUSSELL, MILES C., AND EDWARD C. KERN, JR. "Utility Interactive Residential PV: Lessons Learned." *Solar Today* 6 (May/June 1992), pp. 23–25.

SHAPIRO, ANDREW M. *The Homeowner's Complete Handbook for Add-On Solar Greenhouses and Sunspaces.* Emmaus, PA: Rodale Press, 1985.

SIMONS, MARLISE. "Ocean Waves Power a Generator." *New York Times* (25 Sept 1990), pp. C1, C6.

THURSTON, HARRY. "Power in a Land of Remembrance." *Audubon* (Nov/Dec 1991), pp. 52–59.

WILLIAMS, NEVILLE. "Solar Serendipity: Photovoltaic Rural Electrification in Sri Lanka." *Solar Today* (Nov/Dec 1991), pp. 22–24.

Chapter 24. Lifestyle and Sustainability

BOCK, JAMES. "Reclamation Battle Line Drawn in W. Baltimore." *The Baltimore Sun* (27 March 1992), pp. 1, 6.

BROWN, LESTER R., ET AL. "Earth Day 2030." *WorldWatch* 3 (March/April 1990), pp. 12–21.

THE CONSERVATION FOUNDATION. "Will We Live in Accidental Cities or Successful Communities?" *Conservation Foundation Letter.* Washington, D.C.: The Conservation Foundation, 1987.

DURNING, ALAN THEIN. "Native Americans Stand Their Ground." *WorldWatch* 4 (Nov/Dec 1991), pp. 10–17.

ERICSON, JODY. "Island of Hope." *Nature Conservancy* 42 (Jan/Feb 1992), pp. 14–21.

FULLER, MILLARD, AND DIANE SCOTT. *No More Shacks.* Waco, TX: Word Books, 1986.

GRAVITZ, ALISA, ET AL. "Creating a Sustainable Society" (and following articles). *Co-op America Quarterly* (Summer 1992), pp. 11–24.

GUTTMAN, ASTRID. "Urban Releaf." *WorldWatch* (Nov/Dec 1989), pp. 5–6.

HISS, TONY, ET AL. "Special Section: Cities." *The Amicus Journal* 14 (Summer 1992), pp. 11–36.

HOLDGATE, MARTIN W. "The Environment of Tomorrow." *Environment* 33 (July/Aug 1991), pp. 14–20.

HOLMSTROM, DAVID. "Facing Squarely the Need to Rebuild Communities." *The Christian Science Monitor* (8 June 1992), pp. 1, 4.

KAY, JANE HOLTZ. "Waterfront Renaissance." *World Monitor* (Aug 1989), pp. 38–43.

LAWREN, BILL. "Singing the Blues for Songbirds." *National Wildlife* 30 (Aug/Sept 1992), pp. 4–11.

LOWE, MARCIA D. "Alternatives to the Automobile: Transport for Livable Cities." *Worldwatch Paper 98*. Washington, D.C.: Worldwatch Institute, 1990.

LOWE, MARCIA D. "The Bicycle: Vehicle for a Small Planet." *Worldwatch Paper 90*. Washington, D.C.: Worldwatch Institute, 1989.

LOWE, MARCIA D. "China's Shrinking Cropland." *World-Watch* 2 (July/Aug 1989), p. 9/36.

LOWE, MARCIA D. "City Limits." *WorldWatch* 5 (Jan/Feb 1992), pp. 18–25.

LOWE, MARCIA D. "Pedal-Powered Development." *WorldWatch* 3 (Jan/Feb 1990), pp. 28–37.

LOWE, MARCIA D. "Reclaiming Cities for People." *WorldWatch* 5 (Aug 1992), pp. 19–25.

LOWE, MARCIA D. "Shaping Cities: The Environmental and Human Dimensions." *Worldwatch Paper 105*. Washington, D.C.: Worldwatch Institute, 1991.

LYMAN, FRANCESCA. "The Future of Transportation in America." *E Magazine* (Sept/Oct 1990), pp. 34–41.

MARGOLIS, MAC. "A Third-World City That Works." *World Monitor* 5 (March 1992), pp. 42–51.

MILLS, JUDY. "Rails-to-Trails: An Exercise in Linear Logic." *Smithsonian* 21 (April 1990), pp. 132–145.

MORRIS, DAVID. *Getting from Here to There: Building a Rational Transportation System*. Washington, D.C. Institute for Local Self-Reliance, 1992.

National Wildlife EnviroAction (March 1991), pp. 9–14. "Transportation Policy Reform: The Road Less Traveled," Washington, D.C.: National Wildlife Federation Office of Legislative Affairs.

NIXON, WILL, AND FRANCESCA LYMAN. "The Greening of Our Cities." *E Magazine* (Sept/Oct 1991), pp. 30–39.

PENDLETON, SCOTT. "Financing for Low-Income Housing at Risk." *The Christian Science Monitor* (30 April 1991), p. 6.

POOLE, WILLIAM. "In Land We Trust." *Sierra* 77 (March/April 1992), pp. 52–58.

POSTEL, SANDRA, ET AL. "Earth Day 2030." *WorldWatch* (March/April 1990), pp, 12–21.

RENNER, MICHAEL. "Creating Sustainable Jobs in Industrial Countries," *State of the World 1992*. New York: Worldwatch Institute, W.W. Norton & Company, Inc., 1992, pp. 138–154.

RENNER, MICHAEL. "Saving Earth, Creating Jobs." *World-Watch* 5 (Jan/Feb 1992), pp. 10–17.

RIGGLE, JAMES D. "A Good Investment." *American Farmland* (Winter 1991/1992), p. 16.

ROSS, ELIZABETH. "'Fed Up' with Crime and Grime, Inner-City Dwellers Take Charge." *The Christian Science Monitor* (12 March 1991), p. 9.

SCHWARTZ, DAVID M. "Pond Crazy in Britain." *International Wildlife* 21 (May/June 1991), pp. 20–24.

STEGNER, PAGE. "Let It Be Woods." *Sierra* 76 (Sept/Oct 1991), pp. 54–61.

VON TSCHARNER, RENATA, AND RONALD LEE FLEMING. "Making Cities Memorable." *World Monitor* (Feb 1990), pp. 34–45.

WHYTE, WILLIAM H. *"City. Rediscovering the Center."* New York: Doubleday, 1988.

WOOD, DANIEL B. "California's Central Valley." *The Christian Science Monitor* (19 Feb 1992).

Epilogue

FRENCH, HILARY F. "After the Earth Summit: The Future of Environmental Governance." *Woldwatch Paper 107*. Washington, D.C.: Worldwatch Institute, 1992.

RUDIG, WOLFGANG. "Green Party Politics around the World." *Environment* 33 (Oct 1991), pp. 6–9.

APPENDIX A

Environmental Organizations

This is a list of nongovernmental organizations active in environmental matters. Included here are national organizations as well as some small, specialized ones. These organizations offer a variety of fact sheets, brochures, newsletters, publications, educational materials, and annual reports. Requests for information should be specific and are best made by phone to determine what information is available. Some organizations have internship positions available for those wishing to do work for an environmental group. A more complete listing of environmental organizations can be found in the *Conservation Directory* put out by National Wildlife Federation (address below). This directory includes local, regional, national and international organizations. The cost is $18.00 plus $4.50 for shipping and handling.

AMERICAN LUNG ASSOCIATION. Research, education, legislation, lobbying, advocacy: indoor and outdoor air pollution effects and means of control. 1740 Broadway, New York, NY 10019. (212)315-8700.

AMERICAN RIVERS, INC. Mission is "to preserve and restore America's rivers' systems and to foster a river stewardship ethic." Education, litigation, lobbying: wild and scenic rivers, hydropower relicensing. Information on specific rivers available. 801 Pennsylvania Avenue, S.E., Suite 400, Washington, D.C. 20003. (202)547-6900.

BREAD FOR THE WORLD. National advocacy group that lobbies for hunger-related legislation. 802 Rhode Island Avenue, N.E., Washington, D.C. 20018. (202)269-0200.

CENTER FOR ENVIRONMENTAL INFORMATION. Provides training for corporations and individuals on environmental risks, offers environmental law courses, generates publications, and has library open to the public. 46 Prince Street, Rochester, NY 14607. (716)271-3550.

CENTER FOR SCIENCE IN THE PUBLIC INTEREST. Research and education: alcohol policies, food safety, health, nutrition, organic agriculture. 1875 Connecticut Avenue, N.W., Suite 300, Washington, D.C. 20009. (202)332-9110.

CHESAPEAKE BAY FOUNDATION. Research, education, and litigation: environmental defense and management of Chesapeake Bay and surrounding area. 162 Prince Georges Street, Annapolis, MD 21401. (301)268-8816.

CLEAN WATER ACTION PROJECT. Lobbying, education, research: water quality. 1320 18th Street, N.W., Suite 300, Washington, D.C. 20036. (202)457-1286.

COMMON CAUSE. Lobbying: government reform, energy reorganization, clean air. 2030 M Street, N.W., Washington, D.C. 20036. (202)833-1200.

COMMUNITY TRANSPORTATION ASSOCIATION OF AMERICA. Provides technical assistance for rural and specialized transportation systems. 725 15th Street, N.W., Suite 900, Washington, D.C. 20005. (800)527-8279.

CONCERN, INC. Environmental education. 1794 Columbia Road, N.W., Washington, D.C. 20009. (202)328-8160.

CONGRESS WATCH. Lobbying: consumer health and safety, pesticides. 215 Pennsylvania Avenue, S.E., Washington, D.C. 20003. (202) 546-4996.

CONSERVATION INTERNATIONAL. Education and research: rain forests. 1015 18th Street, N.W., Suite 1000, Washington, D.C. 20036. (202)429-5660.

CRITICAL MASS ENERGY PROJECT. Research and education: alternative energy and nuclear power. 215 Pennsylvania Avenue, S.E., Washington, D.C. 20003. (202)546-4996.

DEFENDERS OF WILDLIFE. Research, education and lobbying: endangered species. 1244 19th Street, N.W., Washington, D.C. 20036. (202)659-9510.

EARTHWATCH EXPEDITIONS, INC. Environmental research is encouraged by organizing teams to make expeditions to various locations all over the world. Team members contribute to the expenses and spend several weeks to months on site. 680 Mount Auburn Street, P.O. Box 403, Watertown, MA 02272. (617)926-8200.

ENVIRONMENTAL ACTION, INC. Lobbying, education, grass roots organizing: energy efficiency and conservation, toxics reduction, right-to-know laws, transportation, solid waste, solar energy, deposit legislation. Offer internships. 6930 Carroll Avenue, Suite 600, Takoma Park, MD 20912. (301)891-1100.

ENVIRONMENTAL DEFENSE FUND. Research, litigation, and lobbying: cosmetics safety, drinking water, energy, transportation, pesticides, wildlife, air pollution, cancer prevention, radiation. 257 Park Avenue South, New York, NY 10010. (212)505-2100.

ENVIRONMENTAL LAW INSTITUTE. Training, educational workshops, and seminars for environmental professionals, lawyers, and judges on institutional and legal issues affecting the environment. 1616 P Street, N.W., Suite 200, Washington, D.C. 20036. (202)328-5150.

FREEDOM FROM HUNGER. Develop programs for the elimination of hunger worldwide. 1644 DaVinci Court, Davis, CA 95617. (916)758-6200.

FRIENDS OF THE EARTH. Research, lobbying: all aspects of energy development, preservation, restoration, and rational use of the earth. 218 D Street, S.E., Washington, D.C. 20036. (202)544-2600.

GREENPEACE, USA, INC. "An international environmental organization dedicated to protecting the planet through nonviolent, direct action providing public education, scientific research, and legislative lobbying. Greenpeace campaigns to free the Earth of nuclear and toxic pollution, to protect marine ecology, and to end atmospheric destruction." 1436 U Street, N.W., Washington, D.C. 20009. (202)462-1177.

HABITAT FOR HUMANITY INTERNATIONAL. Fosters home building and ownership for the poor. Education on housing issues. 121 Habitat Street, Americus, GA 31709-3498. (912)924-6935.

INSTITUTE FOR LOCAL SELF-RELIANCE. Research and education: appropriate technology for community development. 2425 18th Street, N.W., Washington, D.C. 20009. (202)232-4108.

IZAAK WALTON LEAGUE OF AMERICA, INC. Research, education, endowment grants: conservation, air and water quality, streams. 1401 Wilson Boulevard, Level-B, Arlington, VA 22209. (703)528-1818.

LAND TRUST ALLIANCE. Works with local and regional land trusts to enhance their work. 900 17th Street, N.W., Suite 410, Washington, D.C. 20006. (202)785-1410.

LEAGUE OF CONSERVATION VOTERS. Political arm of the environmental community. Works to elect candidates to the U.S. House and Senate who will vote to protect the nation's environment, and holds them accountable by publishing the *National Environmental Scorecard* each year, which can be orderd for $6 and is free to students. 1707 L Street, N.W., Suite 550, Washington, D.C. 20036. (202)785-8683.

LEAGUE OF WOMEN VOTERS OF THE U.S. Education and lobbying, general environmental issues. Publications on groundwater, agriculture and farm policy, and other topics available. 1730 M Street, N.W., Suite 1000, Washington, D.C. 20036. (202)429-1965.

MONITOR CONSORTIUM OF CONSERVATION AND ANIMAL WELFARE ORGANIZATIONS. Lobbying and networking: conservation issues, animal issues, endangered species and their habitats. 1506 19th Street, N.W., Washington, D.C. 20036. (202)234-6576.

NATIONAL AUDUBON SOCIETY. Research, lobbying, education, litigation, and citizen action: broad-based environmental issues. 700 Broadway, New York, NY 10003. (212)546-9100.

NATIONAL PARK FOUNDATION. Education, land acquisition, management of endowments, grant making: National Parks, 1101 17th Street, N.W., Suite 1102, Washington, D.C. 20036. (202)785-4500.

NATIONAL PARKS AND CONSERVATION ASSOCIATION. Research and education: parks, wildlife, forestry, general environmental quality. 1776 Massachusetts Avenue, N.W., Suite 200, Washington, D.C. 20036. (202)223-6722.

NATIONAL RESOURCES DEFENSE COUNCIL. Research and litigation: water and air quality, land use, energy, pesticides, toxic waste, 40 West 20th Street, New York, NY 10011. (212)727-2700.

NATIONAL WILDLIFE FEDERATION. Research, education, lobbying: general environmental quality, wilderness, and wildlife. 1400 16th Street, N.W., Washington, D.C. 20036. (202)797-6800.

NATURE CONSERVANCY. "Preserve plants, animals, and natural communities that represent the diversity of life on Earth by protecting the land and the water they need to survive." 1815 North Lynn Street, Arlington, VA 22209. (703)841-5300.

OXFAM AMERICA. Funding agency for projects to benefit the "poorest of the poor" in South America, Africa, India, Central America, the Caribbean, and the Philippines.

Provide whatever resources are needed. 115 Broadway, Boston, MA 02116. (617)482-1211.

PLANNED PARENTHOOD FEDERATION OF AMERICA. Education, services, and research: fertility control, family planning. 810 7th Avenue, New York, NY 10019. (212)541-7800.

THE POPULATION INSTITUTE. Education, research, and speaking engagements: population control. 107 2nd Street, N.E., Washington, D.C. 20002. (202)544-3300.

POPULATION REFERENCE BUREAU, INC. Organization engaged in collection and dissemination of objective population information. Excellent publications. 1875 Connecticut Avenue, N.W., Suite 520, Washington, D.C. 20009. (202)483-1100.

RACHEL CARSON COUNCIL, INC. Publication and distribution of information on pesticides and toxic substances, educational conferences, and seminars. 8940 Jones Mill Road, Chevy Chase, MD 20815. (301)652-1877.

RAIN FOREST ACTION NETWORK. Information and educational resources: world's rain forests. 450 Sansome Street, Suite 700, San Francisco, CA 94111. (415)398-4404.

RAINFOREST ALLIANCE. Education, medicinal plants project, timber project to certify "smart wood," and news bureau in Costa Rica. 279 Lafayette Street, Suite 512, New York, NY 10012. (212)941-1900.

RENEW AMERICA. "A nationwide clearinghouse for environmental solutions by seeking out and promoting successful programs. We offer positive, constructive models to help communities meet environmental challenges." Twenty different environmental categories addressed. 1400 16th Street, N.W., Suite 710, Washington, D.C. 20036. (202)232-2252.

RESOURCES FOR THE FUTURE. Research and education: conservation of natural resources, environmental quality. 1616 P Street, N.W., Washington, D.C. 20036. (202)328-5000.

SCIENTISTS INSTITUTE FOR PUBLIC INFORMATION. Education: scientists of full range of disciplines provide information for the public. 355 Lexington Avenue, New York, NY 10017. (212)661-9110.

SIERRA CLUB. Education and lobbying: broad-based environmental issues. 730 Polk Street, San Francisco, CA 94109. (415)776-2211.

TRUST FOR PUBLIC LAND. Works with citizen groups and government agencies to acquire and preserve open space. 116 New Montgomery, 4th Floor, San Francisco, CA 94105. (415)495-4014.

U.S. ENVIRONMENTAL PROTECTION AGENCY, PUBLIC INFORMATION CENTER. Provides general information about environmental topics. 401 M Street, S.W., Washington, D.C. 20460. (202)260-7751.

U.S. PUBLIC INTEREST RESEARCH GROUP. Research, education, and lobbying: alternative energy, consumer protection, utilities regulation, public interest. 215 Pennsylvania Avenue, S.E., Washington, D.C. 20003. (202)546-9707.

WATER ENVIRONMENT FEDERATION. Research, education, and lobbying. 601 Wythe Street, Alexandria, VA 22314. (703)684-2400.

WILDERNESS SOCIETY. Research, education and lobbying: wilderness, public lands. 900 17th Street, N.W., Washington, D.C. 20006. (202)833-2300.

WORLD RESOURCES INSTITUTE. Research and publish reports on environmental issues that are sent to educators, policy makers, and organizations. 1709 New York Avenue, N.W., Washington, D.C. 20006. (800)822-0504.

WORLD WILDLIFE FUND. Preservation of wildlife habitats and protection of endangered species. 1250 24th Street, N.W., Washington, D.C. 20036. (202)293-4800.

WORLDWATCH INSTITUTE. Research and education: energy, food, population, health, women's issues, technology, the environment. 1776 Massachusetts Avenue, N.W., Washington, D.C. 20036. (202)452-1999.

ZERO POPULATION GROWTH, INC. Public education, lobbying, and research: population. 1400 16th Street, N.W., Suite 320, Washington, D.C. 20036. (202)332-2200.

Units of Measure

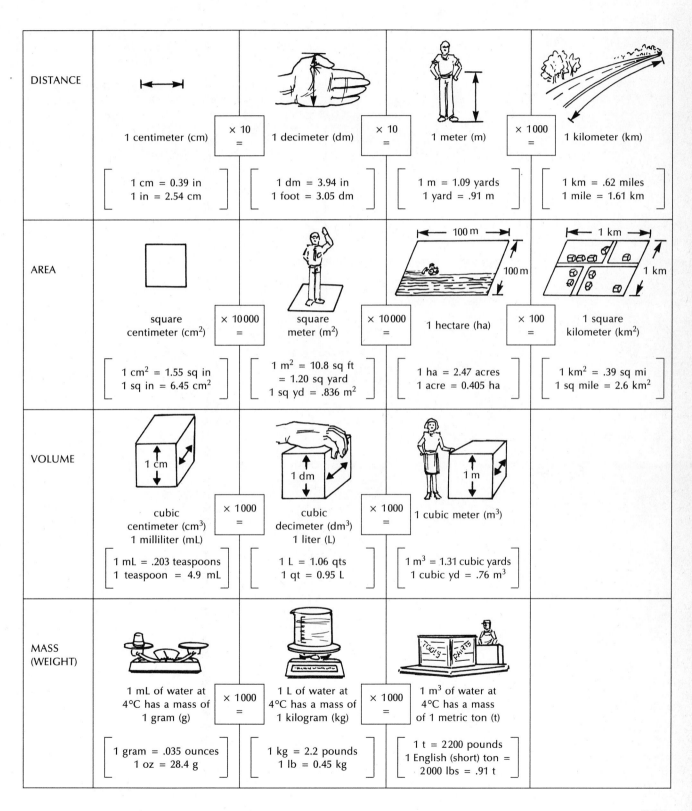

DISTANCE	1 centimeter (cm)	× 10 =	1 decimeter (dm)	× 10 =	1 meter (m)	× 1000 =	1 kilometer (km)
	1 cm = 0.39 in 1 in = 2.54 cm		1 dm = 3.94 in 1 foot = 3.05 dm		1 m = 1.09 yards 1 yard = .91 m		1 km = .62 miles 1 mile = 1.61 km

AREA	square centimeter (cm²)	× 10 000 =	square meter (m²)	× 10 000 =	1 hectare (ha)	× 100 =	1 square kilometer (km²)
	1 cm² = 1.55 sq in 1 sq in = 6.45 cm²		1 m² = 10.8 sq ft = 1.20 sq yard 1 sq yd = .836 m²		1 ha = 2.47 acres 1 acre = 0.405 ha		1 km² = .39 sq mi 1 sq mile = 2.6 km²

VOLUME	cubic centimeter (cm³) 1 milliliter (mL)	× 1000 =	cubic decimeter (dm³) 1 liter (L)	× 1000 =	1 cubic meter (m³)		
	1 mL = .203 teaspoons 1 teaspoon = 4.9 mL		1 L = 1.06 qts 1 qt = 0.95 L		1 m³ = 1.31 cubic yards 1 cubic yd = .76 m³		

MASS (WEIGHT)	1 mL of water at 4°C has a mass of 1 gram (g)	× 1000 =	1 L of water at 4°C has a mass of 1 kilogram (kg)	× 1000 =	1 m³ of water at 4°C has a mass of 1 metric ton (t)		
	1 gram = .035 ounces 1 oz = 28.4 g		1 kg = 2.2 pounds 1 lb = 0.45 kg		1 t = 2200 pounds 1 English (short) ton = 2000 lbs = .91 t		

Energy Units and Equivalents

1 Calorie, food calorie, or kilocalorie—The amount of heat required to raise the temperature of one kilogram of water one degree Celsius (1.8°F).

1 BTU (British Thermal Unit)—The amount of heat required to raise the temperature of one pound of water one degree Fahrenheit.

1 Calorie = 3.968 BTU's
1 BTU = 0.252 calories

1 therm = 100,000 BTU's
1 quad = 1 quadrillion BTU's

1 watt standard unit of electrical power

   1 watt-hour (wh) = 1 watt for 1 hr. = 3.413 BTU's

1 kilowatt (kw) = 1000 watts

   1 kilowatt-hour (kwh) = 1 kilowatt for 1 hr. = 3413 BTU's

1 megawatt (Mw) = 1,000,000 watts

   1 megawatt-hour (Mwh) = 1 Mw for 1 hr. = 34.13 therms

1 gigawatt (Gw) = 1,000,000,000 watts or 1,000 megawatts

   1 gigawatt-hour (Gwh) = 1 Gw for 1 hr. = 34,130 therms

1 horsepower = .7457 kilowatts; 1 horsepower-hour = 2545 BTU's

1 cubic foot of natural gass (methane) at atmospheric pressure = 1031 BTU's

1 gallon gasoline = 125,000 BTU's

1 gallon No. 2 fuel oil = 140,000 BTU's

1 short ton coal = 25,000,000 BTU's

1 barrel (oil) = 42 gallons

APPENDIX C

Some Basic Chemical Concepts

Atoms, Elements, and Compounds

All matter, whether gas, liquid, or solid, living or non-living, organic or inorganic, is comprised of fundamental units called **atoms.** Atoms are extremely tiny. If all the world's people, about 5,000,000,000 of us, were reduced to the size of atoms, there would be room for all of us to dance on the head of a pin. In fact, we would only occupy a tiny fraction (about $\frac{1}{10,000}$) of the pin's head. Given the incredibly tiny size of atoms, even the smallest particle which can be seen with the naked eye consists of billions of atoms.

The atoms comprising a substance may be all of one kind; or they may be of two or more different kinds. If the atoms are all of one kind, the substance is called an **element.** If the atoms are of two or more different kinds bonded together, the substance is called a **compound.**

Through countless experiments, chemists have ascertained that there are only 96 distinct kinds of atoms which occur in nature. They are listed in Table C–1 with their chemical symbols. By scanning Table C–1, you can see that a number of familiar substances such as aluminum, calcium, carbon, oxygen, and iron are elements; that is, they are a single distinct kind of atom. However, most of the substances with which we interact in every-day life, such as water, stone, wood, protein, and sugar, are not on the list. Their absence from the list is indicative that they are not elements; rather they are compounds, which means they are actually comprised of two or more different kinds of atoms bonded together.

Atoms, Bonds, and Chemical Reactions

In chemical reactions, atoms are neither created, nor destroyed, nor is one kind of atom changed into another. What occurs in chemical reactions, whether mild or explosive, is simply a rearrangement of the ways in which the atoms involved are bonded together. An oxygen atom, for example, may be combined and recombined with different atoms to form any number of different compounds, but a given oxygen atom always has been, and always will be, an oxygen atom. The same can be said for all the other kinds of atoms. In order to understand how atoms may bond and rearrange to form different compounds, it is necessary to first have some concepts concerning the structure of atoms.

STRUCTURE OF ATOMS

In every case, an atom consists of a central core called the nucleus (not to be confused with the cell nucleus). The nucleus of the atom contains one or more **protons** and, except for hydrogen, one or more **neutrons** as well. Surrounding the nucleus are particles called **electrons.** Each proton has a positive (+) electric charge and each electron has an equal but opposite negative (−) electric charge. Thus, the charge of the protons may be balanced by an equal number of electrons making the whole atom neutral. Neutrons have no charge.

Atoms of all elements have this same basic structure consisting of protons, electrons, and neutrons. The distinction among atoms of different elements is in the number of protons. The atoms of each element have a characteristic number of protons which is known as the **atomic number** of the element (see Table C–1). The number of electrons characteristic of the atoms of each element also differs corresponding to the number of protons. The general structure of the atoms of several elements is shown in Figure C–1.

The number of protons and electrons, i.e., the atomic number of the element, determines the chemical properties of the element. However, the number of neutrons may also vary. For example, most carbon atoms have six neutrons in addition to the six protons as indicated in Figure C–1. But some carbon atoms have eight neutrons. Atoms of the same element which have different numbers of neutrons are known as **isotopes** of the element. The total number of protons plus neutrons is used to define different isotopes. For example, the usual isotope of carbon is referred to as carbon-12 while the isotope noted above is referred to as carbon-14. The chemical reactivity of different isotopes of the same element is identical. However, certain other properties may differ. Many isotopes of various elements prove to be radioactive as is carbon-14.

589

TABLE C-1

The Elements

Element	Symbol	Atomic Number	Element	Symbol	Atomic Number
Actinium	Ac	89	Neodymium	Nd	60
Aluminum	Al	13	Neon	Ne	10
Americium	Am	95	Neptunium	Np	93
Antimony	Sb	51	Nickel	Ni	28
Argon	Ar	18	Niobium	Nb	41
Arsenic	As	33	Nitrogen	N	7
Astatine	At	85	Nobelium	No	102
Barium	Ba	56	Osmium	Os	76
Berkelium	Bk	97	Oxygen	O	8
Beryllium	Be	4	Palladium	Pd	46
Bismuth	Bi	83	Phosphorus	P	15
Boron	B	5	Platinum	Pt	78
Bromine	Br	35	Plutonium	Pu	94
Cadmium	Cd	48	Polonium	Po	84
Calcium	Ca	20	Potassium	K	19
Californium	Cf	98	Praseodymium	Pr	59
Carbon	C	6	Promethium	Pm	61
Cerium	Ce	58	Protoactinium	Pa	91
Cesium	Cs	55	Radium	Ra	88
Chlorine	Cl	17	Radon	Rn	86
Chromium	Cr	24	Rhenium	Re	75
Cobalt	Co	27	Rhodium	Rh	45
Copper	Cu	29	Rubidium	Rb	37
Curium	Cm	96	Ruthenium	Ru	44
Dysprosium	Dy	66	Samarium	Sm	62
Einsteinium	Es	99	Scandium	Sc	21
Erbium	Er	68	Selenium	Se	34
Europium	Eu	63	Silicon	Si	14
Fermium	Fm	100	Silver	Ag	47
Fluorine	F	9	Sodium	Na	11
Francium	Fr	87	Strontium	Sr	38
Gadolinium	Gd	64	Sulfur	S	16
Gallium	Ga	31	Tantalum	Ta	73
Germanium	Ge	32	Technetium	Tc	43
Gold	Au	79	Tellurium	Te	52
Hafnium	Hf	72	Terbium	Tb	65
Helium	He	2	Thallium	Tl	81
Holmium	Ho	67	Thorium	Th	90
Hydrogen	H	1	Thulium	Tm	69
Indium	In	49	Tin	Sn	50
Iodine	I	53	Titanium	Ti	22
Iridium	Ir	77	Tungsten	W	74
Iron	Fe	26	Unnilennium	Une	109
Krypton	Kr	36	Unnilhexium	Unh	106
Lanthanum	La	57	Unniloctium	Uno	108
Lawrencium	Lr	103	Unnilseptium	Uns	107
Lead	Pb	82	Uranium	U	92
Lithium	Li	3	Vanadium	V	23
Lutetium	Lu	71	Xenon	Xe	54
Magnesium	Mg	12	Ytterbium	Yb	70
Manganese	Mn	25	Yttrium	Y	39
Mendelevium	Md	101	Zinc	Zn	30
Mercury	Hg	80	Zirconium	Zr	40
Molybdenum	Mo	42			

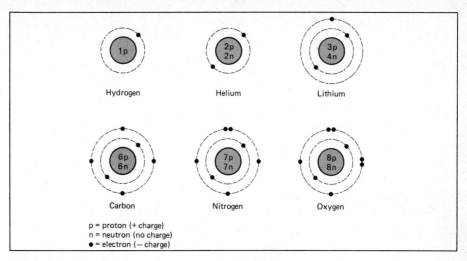

FIGURE C-1
Structure of atoms. All atoms consist of fundamental particles: protons (P), which have a positive electric charge, neutrons (n), which have no charge, and electrons, which have a negative charge. Protons and neutrons are located in a central core, the nucleus. The positive charge of the protons is balanced by an equal number of electrons, which occupy various levels or orbitals around the nucleus. The uniqueness of each element is given by its atoms having a distinct number of protons, its atomic number.

BONDING OF ATOMS

The chemical properties of an element are defined by the ways in which its atoms will react and form bonds with other atoms. By examining how atoms form bonds, we shall see how the number of electrons and protons determines these properties. There are two basic kinds of bonding: (1) **covalent bonding,** and (2) **ionic bonding.**

In both kinds of bonding, it is first important to recognize that electrons are not randomly distributed around the atom's nucleus. Rather, there are, in effect, specific spaces in a series of layers, or **orbitals,** around the nucleus. If an orbital is occupied by one or more electrons but not filled, the atom is unstable; it will tend to react and form bonds with other atoms to achieve greater stability. A stable state is achieved by having all the spaces in the orbital filled with electrons. But, it is also important to keep the charge neutral, i.e., the total number of electrons equal to that of the protons.

Covalent Bonding

These two requirements, filling all the spaces and keeping the charge neutral, may be satisfied by adjacent atoms sharing one or more pairs of electrons as shown in Figure C-2. The sharing of a pair of elec-

trons holds the atoms together in what is called a **covalent bond.**

Covalent bonding, by satisfying the charge-orbital requirements, leads to discrete units of two or more atoms bonded together. Such units of two or more covalently bonded atoms are called **molecules.** A few simple but important examples are shown in Figure C-2.

A chemical formula is simply a shorthand description of the number of each kind of atom in a given molecule. The element is given by the chemical symbol and a subscript following the symbol gives the number present, no subscript being understood as one. A molecule with two or more different kinds of atoms may also be called a compound, but a molecule comprised of a single kind of atom, oxygen (O_2) for example, is still defined as an element.

Only a few elements, namely carbon, hydrogen, oxygen, nitrogen, phosphorus and sulfur, have configurations of electrons which lend readily to the formation of covalent bonds. But, carbon specifically, with its ability to form four covalent bonds, can produce long, straight or branched chains, or rings (Fig. C-3). Thus, an infinite array of molecules can be formed by using covalently bonded carbon atoms as a "backbone" and filling in the sides with atoms of hydrogen or other elements. Thus, it is covalent bonding among atoms of carbon and these few other

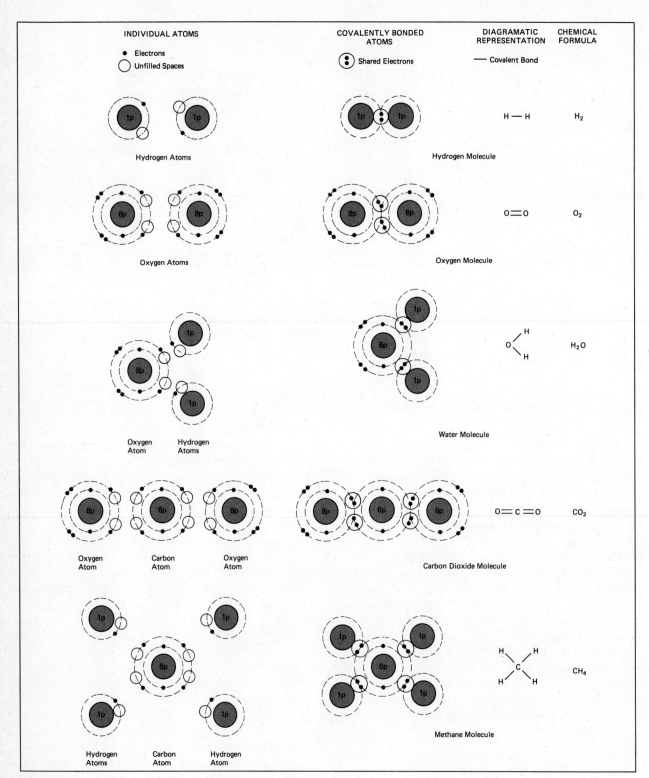

INDIVIDUAL ATOMS

• Electrons
○ Unfilled Spaces

COVALENTLY BONDED ATOMS

⦿ Shared Electrons

DIAGRAMATIC REPRESENTATION

— Covalent Bond

CHEMICAL FORMULA

Hydrogen Atoms

Hydrogen Molecule

H — H

H_2

Oxygen Atoms

Oxygen Molecule

O═O

O_2

Oxygen Atom   Hydrogen Atoms

Water Molecule

H_2O

Oxygen Atom   Carbon Atom   Oxygen Atom

Carbon Dioxide Molecule

O═C═O

CO_2

Hydrogen Atoms   Carbon Atom   Hydrogen Atom

Methane Molecule

CH_4

FIGURE C–2

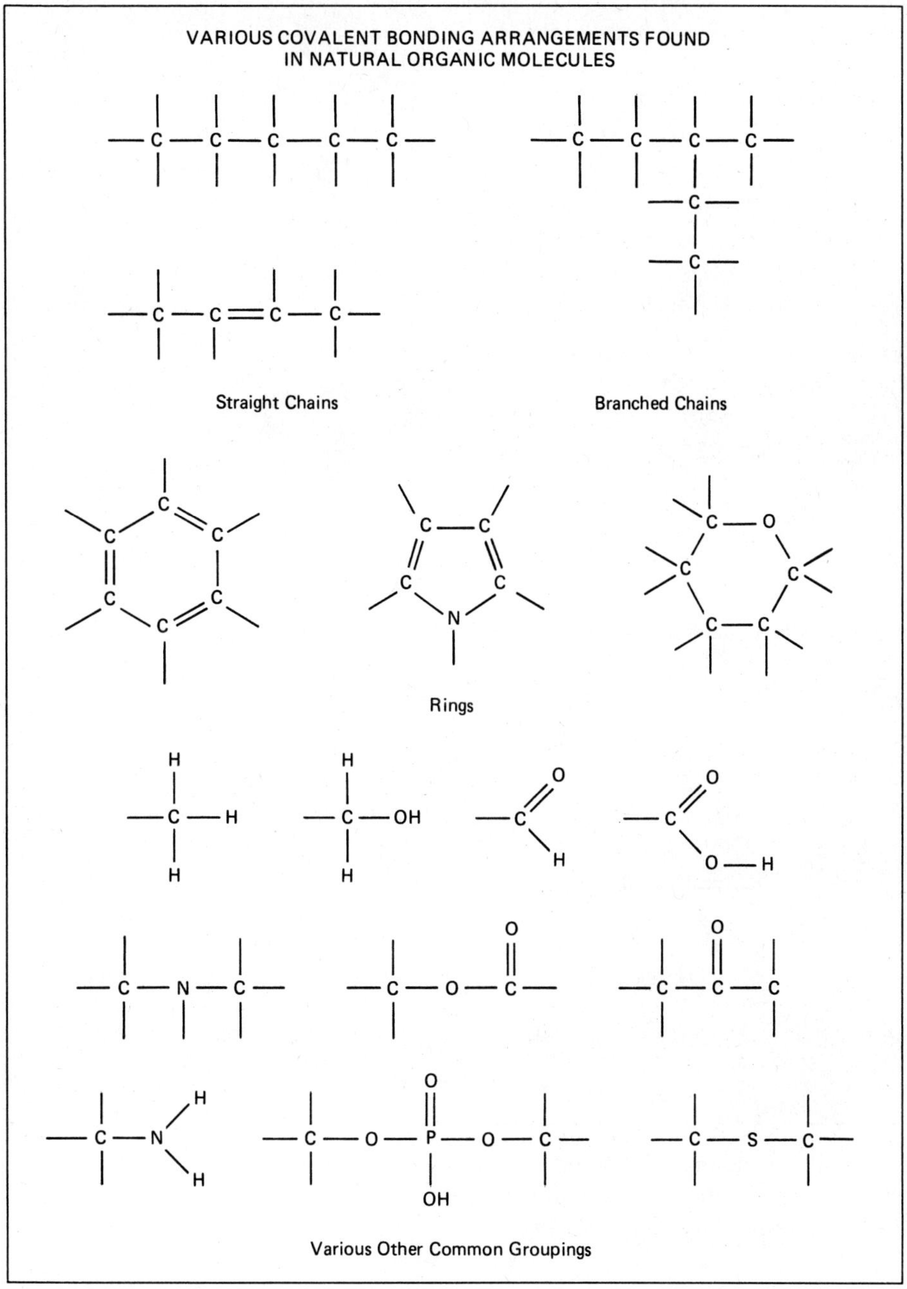

VARIOUS COVALENT BONDING ARRANGEMENTS FOUND IN NATURAL ORGANIC MOLECULES

Straight Chains

Branched Chains

Rings

Various Other Common Groupings

FIGURE C–3
Covalent bonding and organic molecules. The ability of carbon and a few other elements to readily form covalent bonds leads to an infinite array of complex molecules, organic molecules, which constitute all living things. A few major kinds of groupings are shown here. Note that each element forms a characteristic number of bonds: carbon, 4; nitrogen, 3; oxygen, 2; hydrogen, 1; sulfur, 2; phosphorus, 5. Bonds (dashed lines) left "hanging" indicate attachments to other atoms or groups of atoms.

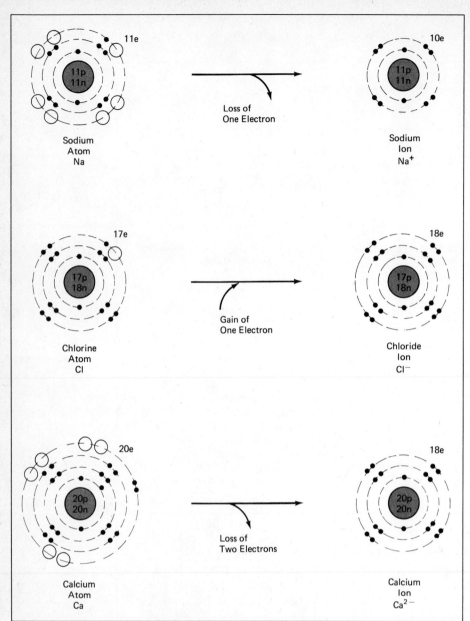

FIGURE C–4
Formation of ions. Many atoms will tend to gain or lose one or more electrons in order to achieve a state of complete (electron-filled) orbitals. In doing so they become positively or negatively charged ions as indicated.

elements that produces all natural organic molecules, those molecules that comprise all the tissues of living things, and also synthetic organic compounds such as plastics.

TABLE C–2

Ions of Particular Importance to Biological Systems

Negative (−) Ions		Positive (+) Ions	
Phosphate	PO_4^{3-}	Potassium	K^+
Sulfate	SO_4^{2-}	Calcium	Ca^{2+}
Nitrate	NO_3^-	Magnesium	Mg^{2+}
Hydroxyl	OH^-	Iron	Fe^{2+}, Fe^{3+}
Chloride	Cl^-	Hydrogen	H^+
Bicarbonate	HCO_3^-	Ammonium	NH_4^+
Carbonate	CO_3^{2-}	Sodium	Na^+

Ionic Bonding

Another way in which atoms may achieve a stable electron configuration is to gain additional electrons to complete the filling of an orbital, or lose electrons which are over a completed orbital. In general, the maximum number of electrons that can be gained or lost by an atom is three. Therefore, an element's atomic number determines whether one or more electrons will be lost or gained. If an atom's outer orbital is one to three electrons short of being filled, it will always tend to gain additional electrons. Conversely, if an atom has one to three electrons over its last complete orbital it will always tend to give them away.

Of course gaining or losing electrons results in the number of electrons being greater or less than the number of protons, and the atom consequently hav-

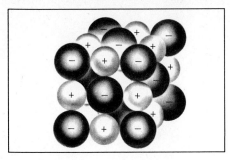

FIGURE C–5
Positive and negative ions bond together by their mutual attraction.

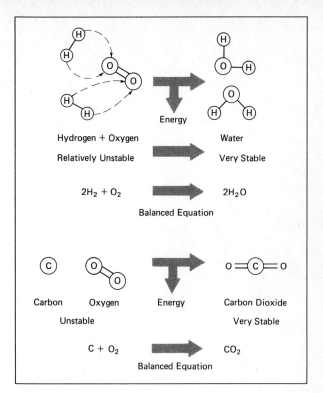

FIGURE C–6
Some bonding arrangements are more stable than others. Chemical reactions will go spontaneously toward more stable arrangements, releasing energy in the process. But, reactions may be driven in the opposite direction with suitable energy inputs.

ing an electric charge. The charge will be one negative for each electron gained or one positive for each electron lost (Fig. C–4). A covalently bonded group of atoms may acquire an electric charge in the same way. An atom or group of atoms which has acquired an electric charge in this way is called an **ion,** positive or negative. Ions are designated by a superscript following the chemical symbol giving the number of positive or negative charges. Absence of superscripts indicates that the atom or molecule is neutral. Some important ions are listed in Table C–2.

Since unlike charges attract, positive and negative ions tend to join and pack together in dense clusters in such a way as to neutralize the overall electric charge. This joining together of ions through the attraction of their opposite charges is called **ionic bonding.** The result is the formation of hard, brittle, more or less crystalline substances of which all rocks and minerals are examples (Fig. C–5).

It is significant to note that whereas covalent bonding leads to discrete molecules, ionic bonding does not. Any number and combination of positive and negative ions may enter into an ionically bonded cluster to produce crystals of almost any size. The only restriction is that the overall charge of positive ions is balanced by that of negative ions. Thus, ionically bonded substances are properly called compounds but not molecules. When chemical formulas are used to describe such compounds, they define the ratio of various elements involved, not specific molecules.

CHEMICAL REACTIONS AND ENERGY

While atoms themselves do not change, the bonds between atoms may be broken and reformed with different atoms producing different compounds and/ or molecules. This is essentially what occurs in all chemical reactions. What determines whether a given chemical reaction will occur or not? We noted above that atoms form bonds because they achieve a greater stability by doing so. But some bonding arrangements

may provide greater overall stability than others. Consequently substances with relatively unstable bonding arrangements will tend to react to form one or more different compounds which have more stable bonding arrangements. Common examples are the reaction between hydrogen and oxygen to produce water, and the reaction between carbon and oxygen to produce carbon dioxide (Fig. C–6).

Additionally, energy is always released in the process of gaining greater overall stability as indicated in Figure C–6. Thus, energy being released from a chemical reaction is synonymous with the atoms achieving more stable bonding arrangements. Thus, it may be said that chemical reactions always tend to go in a direction that releases energy as well as one which gives greater stability.

However, chemical reactions can be made to go in a reverse direction. With suitable energy inputs and under suitable conditions, stable bonding arrangements may be broken and less stable arrangements formed. As described in Chapter 2, this is the basis of photosynthesis occurring in green plants. Light energy is brought to bear on splitting the highly stable hydrogen-oxygen bonds of water and forming less stable carbon-hydrogen bonds thus creating high-energy organic compounds.

Glossary

A

Abiotic. Pertaining to factors or things that are separate and independent from living things; nonliving.

Absolute poverty. The lack of sufficient income in cash or exchange items for meeting the most basic human needs for food, clothing, and shelter.

Acid. Any compound that releases hydrogen ions when dissolved in water. Also, a water solution that contains a surplus of hydrogen ions.

Acid deposition. Any form of acid precipitation and also fallout of dry acid particles. (See **acid precipitation**)

Acid dew. Acidic dew; results from water vapor condensing on dry acid fallout.

Acid fallout. Molecules of acid formed from reactions involving nitrogen and sulfur oxides and water vapor settling out of the atmosphere without additional water.

Acid precipitation. Includes acid rain, acid fog, acid snow, and any other form of precipitation that is more acidic than normal, i.e., less than pH 5.6. Excess acidity is derived from certain air pollutants, namely sulfur dioxide and oxides of nitrogen.

Active safety. Referring to nuclear reactors, active safety features are those that rely on operator-controlled reactions, external power sources, and other features that are capable of failing. (See **passive safety**)

Activated charcoal. A form of carbon that readily adsorbs organic material. Therefore it is frequently used in air and/or water filters to remove organic contaminants. It does not remove ions such as those of the heavy metals.

Activated sludge. Sludge made up of clumps of living organisms feeding on detritus that settles out and is recycled in the process of secondary wastewater treatment.

Activated sludge system. A system for removing organic wastes from water. The system uses microorganisms and active aeration to decompose such wastes. The system is used most as a means of secondary sewage treatment following the primary settling of materials.

Active solar heating system. Solar heating system using pumps and/or blowers to transfer the heat from the collector to the place of use.

Adaption (ecological or evolutionary). A change in structure or function that produces better adjustment of an organism to its environment and hence enhances its ability to survive and reproduce.

Adsorption. The process of chemicals (ions or molecules) sticking to the surface of other materials.

Advance disposal fee. A fee added to the purchase price of all products in glass, metal, and plastic cans or bottles to cover the cost of disposing of the container.

Advanced treatment (sewage treatment). Any of a variety of systems that follow secondary treatment and that are designed to remove one or more nutrients, such as phosphate, from solution.

Aeration. *Soil:* The property of a soil relating to its ability to allow the exchange of oxygen and carbon dioxide, which is necessary for the respiration of roots. *Water:* The bubbling of air or oxygen through water to increase the dissolved oxygen.

Age structure. Within a population proportions of people who are old, middle-aged, young adults, and children.

Agricultural district. A region in which the state or county limits subdivision and development in order to preserve the agricultural viability of the region.

Agroforestry. Production of tree crops in a manner similar to agriculture. Also, production of trees along with regular crops.

Air. The mixture of gases, namely 78 percent nitrogen, 21 percent oxygen, and 0.035 percent carbon dioxide, making up the atmosphere. Water vapor and various pollutants may also be present.

Air pollution disaster. Short-term situation in industrial cities where intense industrial smog brings about significant increase in human mortality.

Air toxics. A category of air pollutants including radioactive materials and other toxic chemicals that are present at low concentrations but are of concern because they are often carcinogenic.

Alga, pl. **algae.** Any of numerous kinds of photosynthetic plants that live and reproduce entirely immersed in water. Many species, the planktonic forms, exist as single or small groups of cells that float freely in the water. Other species, the "seaweeds," may be large and attached.

Algal bloom. A relatively sudden development of a heavy growth of algae, especially planktonic forms. Algal blooms generally result from additions of nutrients, whose scarcity is normally limiting.

Alleles. The two or more variations of a gene for any particular characteristic, e.g., blue and brown are alleles of the gene for eye color.

Alternative farming. Farming methods designed to minimize the use of agricultural chemicals.

Ambient standards. Levels of air pollutants set by law as the maximum tolerated in order to maintain environmental and human health.

Anaerobic. Oxygen-free.

Anaerobic digestion. The breakdown of organic material by microorganisms in the absence of oxygen. The process results in the release of methane gas as a waste product.

Anaerobic respiration. Respiration carried on by certain bacteria in the absence of oxygen. Methane, which can be used as fuel gas (it is the same as natural gas), may be a byproduct of the process.

Annuals. Plants that grow from seed, flower, set seed, and die, thus completing their life cycle in a single year.

Anthropogenic. Referring to pollutants and other forms of impacts on natural environments that can be traced to human activities.

Appropriate technology. Technology that seeks to increase the efficiency and productivity of hand labor without displacing workers. That is, it seeks to enable people to improve their well-being without disrupting the existing social and economic system.

Aquaculture. A propagation and/or rearing of any aquatic (water) organism in a more-or-less artificial system.

Aquifer. An underground layer of porous rock, sand, or other material that allows the movement of water between layers of nonporous rock or clay. Aquifers are frequently tapped for wells.

Artesian aquifer. An aquifer in which the groundwater is under such pressure that it comes to the surface when a well is drilled into it.

Artificial selection. Plant and animal breeders' practice of selecting individuals with the greatest expression of desired traits to be the parents of the next generation.

Asbestos fibers. Crystals of asbestos, a natural mineral, which have the form of minute fibers.

Assimilate. To incorporate into the natural working or functioning of the system as, for example, natural organic wastes are assimilated (broken down and incorporated) into the nutrient cycles of the ecosystem.

Atom. The fundamental unit of all elements.

Autotroph. Any organism that can synthesize all its organic substances from inorganic nutrients, using light or certain inorganic chemicals as a source of energy. Green plants are the principal autotrophs.

B

Background radiation. Radioactive radiation that comes from natural sources apart from any human activity. We are all exposed to such radiation.

Bacterium, pl. **bacteria.** Any of numerous kinds of microscopic organisms that exist as simple, single cells that multiply by simple division. Along with fungi, they constitute the decomposer component of ecosystems. A few species cause disease.

Balanced herbivory. A diversified plant community held in balance by various herbivores specific to each plant species.

Bar screen. A set of iron bars about an inch apart used to screen debris out of wastewater.

Base. Any compound that releases hydroxyl (OH^-) ions when dissolved in water. A solution that contains a surplus of OH^- ions.

Basic science. Science conducted purely for the purpose of gaining understanding as opposed to applied science where a particular application is the purpose.

Bedload. The load of coarse sediment, mostly coarse silt and sand, that is gradually moved along the bottom of a river bed by flowing water rather than being carried in suspension.

Benefit-cost. Used to describe an analysis and/or comparison of the value benefits in contrast to the costs of any particular action or project.

Benthic plants. Plants that grow under water attached to or rooted in the bottom. For photosynthesis, they depend on light penetrating the water.

Best management practice. Farm management practices that serve best to reduce soil and nutrient runoff and subsequent pollution.

Bioaccumulation. The accumulation of higher and higher concentrations of potentially toxic chemicals in organisms. It occurs in the case of chemicals that are ingested but cannot be excreted or broken down.

Biochemical oxygen demand (BOD). The amount of oxygen that will be absorbed or "demanded" as wastes are being digested or oxidized. This includes both biological and chemical processes. Potential impacts of wastes are commonly measured in terms of the BOD.

Biocide. Applies to any pesticide or other chemical that is toxic to many, if not all, kinds of living organisms.

Bioconversion (energy). The use of biomass as fuel. Burning materials such as wood, paper, and plant wastes directly to produce energy, or converting such materials into fuels such as alcohol and methane.

Biodegradable. Able to be consumed and broken down to natural substances such as carbon dioxide and water by biological organisms, particularly decomposers. Opposite: **nonbiodegradable.**

Biodiversity. The diversity of living things to be found in the natural world. The concept usually refers to the different species, but also includes ecosystems and the genetic diversity within a given species.

Biogas. The mixture of gases, about two-thirds methane, one-third carbon dioxide and small portions of foul-smelling compounds, resulting from the anaerobic (without air) digestion of organic matter. The methane content enables biogas to be used as a fuel gas.

Biological control (pest control). Control of a pest population by introduction of predatory, parasitic, or disease-causing organisms.

Biological treatment (sewage treatment). (See **secondary treatment**)

Biomagnification. Bioaccumulation occurring through several levels of a food chain.

Biomass. Mass of biological material. Usually the total mass of a particular group or category; for example, biomass of producers.

Biomass energy or **biomass fuels.** Energy, or fuels such as alcohol and methane, produced from current photosynthetic production of biological materials. (See **bioconversion**)

Biomass pyramid. Refers to the structure that is obtained when the respective biomasses of producers, herbivores, and carnivores in an ecosystem are compared. Producers have the largest biomass, followed by herbivores and then carnivores.

Biome. A group of ecosystems that are related by having a similar type of vegetation governed by similar climatic conditions. Examples include prairies, deciduous forests, arctic tundra, deserts, and tropical rainforests.

Bioremediation. Refers to the use of microorganisms for the decontamination of soil or groundwater. Usually involves injecting organisms and/or oxygen into contaminated zones.

Biosphere. The overall ecosystem of the earth. It is the sum total of all the biomes and smaller ecosystems, which are ultimately all interconnected and interdependent through global processes such as water and atmospheric cycles.

Biota. Refers to any and all living organisms and the ecosystems in which they exist.

Biotic. Living or derived from living things.

Biotic potential. The potential of a species for increasing its population and/or distribution. The biotic potential of every species is such that, given optimum conditions, its population will increase. Contrast **environmental resistance.**

Biotic structure. The organization of living organisms in an ecosystem into groups such as producers, consumers, detritus feeders and decomposers.

Birth control. Any means, natural or artificial, that may be used to reduce the number of live births.

BOD. (See **Biochemical oxygen demand**)

Borrowed time. Time preceding a predictable and inevitable collapse or failure of a system during which nothing is done to avert the end result despite awareness of it.

Bottle bill. A law that provides for the recycling or reuse of beverage containers, usually by requiring a returnable deposit at the purchase of the item.

Breeder reactor. A nuclear reactor that in the course of producing energy also converts nonfissionable uranium-238 into fissionable plutonium-239, which can be used as fuel. Hence, a reactor that produces as much nuclear fuel as it consumes, or more.

Broad-spectrum pesticides. Chemical pesticides that kill a wide range of pests. They also kill a wide range of nonpest and beneficial species; therefore, they may lead to environmental upsets and resurgences. The opposite of narrow-spectrum pesticides and biorational pesticides.

BTU (British Thermal Unit). A fundamental unit of energy in the English system. The amount of heat required to raise the temperature of 1 pound of water 1 degree Fahrenheit.

Buffer. A substance that will maintain the pH of a solution by reacting with the excess acid. Limestone is a natural buffer that helps to maintain water and soil at a pH near neutral.

Buffering capacity. Refers to the amount of acid that may be neutralized by a given amount of buffer.

C

Calorie. A fundamental unit of energy. The amount of heat required to raise the temperature of 1 gram of water 1 degree Celsius. All forms of energy can be converted to heat and measured in calories. Calories used in connection with food are kilocalories, or ''big'' calories, the amount of heat required to raise the temperature of 1 liter of water 1 degree Celsius.

Capillary water. Water that clings in small pores, cracks, and spaces against the pull of gravity, like water held in a sponge.

Carbon dioxide effect. (See **greenhouse effect**)

Carbon monoxide. A highly poisonous gas, the molecules of which consist of a carbon atom with one oxygen attached. Not to be confused with nonpoisonous carbon dioxide, a natural gas in the atmosphere.

Carbon tax. A tax levied on all fossil fuels in proportion to the amount of carbon dioxide that is released as they burn.

Carcinogenic. Having the property of causing cancer, at least in animals and by implication in humans.

Carnivore. An animal that feeds more-or-less exclusively on other animals.

Carrying capacity. The maximum population of a given animal or humans that an ecosystem can support without being degraded or destroyed in the long run. The carrying capacity may be exceeded, but not without lessening the system's ability to support life in the long run.

Catalyst. A substance that promotes a given chemical reaction without itself being consumed or changed by the reaction. Enzymes are catalysts for biological reactions. Also catalysts are used in some pollution control devices, e.g., the **catalytic converter.**

Catalytic converter. The device used by U.S. automobile manufacturers to reduce the amount of carbon monoxide and hydrocarbons in the exhaust. The converter contains a catalyst that oxidizes these compounds to carbon dioxide and water as the exhaust passes through.

Cell (biological). The basic unit of life, the smallest unit that still maintains all the attributes of life. Many microscopic organisms consist of a single cell. Large organisms consist of trillions of specialized cells functioning together.

Cell respiration. The chemical process that occurs in all living cells wherein organic compounds are broken down to release energy required for life processes. Higher plants and animals require oxygen for the process as well and release carbon dioxide and water as waste products, but certain microorganisms do not require oxygen. (See **anaerobic respiration**)

Cellulose. The organic macromolecule that is the prime constituent of plant cell walls and hence the major molecule in wood, wood products, and cotton. It is composed of glucose molecules, but since it cannot be digested by humans its dietary value is only as fiber, bulk, or roughage.

Cell wall. A more-or-less rigid wall, composed mainly of cellulose, which surrounds plant cells and provides the supporting structure of plant tissues.

Center pivot irrigation. An irrigation system consisting of a spray arm several hundred meters long supported by wheels pivoting around a central well from which water is pumped.

CFCs. (See **chlorofluorocarbons**)

Chain reaction (nuclear). Reaction wherein each atom that fissions (splits) causes one or more additional atoms to fission.

Channelization/Channelized. The straightening and deepening of stream or river channels to speed water flow and reduce flooding. A waterway so treated is said to be channelized.

Chemical barrier. In reference to genetic pest control, a chemical aspect of the plant that makes it resist pest attack.

Chemical energy. The potential energy that is contained in certain chemicals; most importantly, the energy contained in organic compounds such as food and fuels, which may be released through respiration or burning.

Chemical technology. Applied to the control of agricultural pests, this refers to the use of pesticides and herbicides to control or eradicate the pests.

Chemosynthesis. The ability of some microorganisms to utilize the chemical energy contained in certain inorganic chemicals such as hydrogen sulfide for the production of organic material. Such organisms are producers.

Chlorinated hydrocarbons. Synthetic organic molecules in which one or more hydrogen atoms have been replaced by chlorine atoms. They are extremely hazardous compounds, because they tend to be nonbiodegradable, they tend to bioaccumulate, and many have been shown to be carcinogenic. Also called organochlorides.

Chlorination. The process of adding chlorine to drinking water or sewage water in order to kill microorganisms that may cause disease.

Chlorofluorocarbons (CFCs). Synthetic organic molecules that contain one or more of both chlorine and fluorine atoms.

Chlorophyll. The green pigment in plants responsible for absorbing the light energy required for photosynthesis.

Clean Air Act of 1970. Amended in 1977 and 1990, this is the foundation of U.S. air pollution control efforts.

Clean Water Act of 1972. The cornerstone federal legislation addressing water pollution.

Clearcutting. Cutting every tree, leaving the area completely clear.

Climate. A general description of the average temperature and rainfall conditions of a region over the course of a year.

Climax ecosystem. The last stage in ecological succession. An ecosystem in which populations of all organisms are in balance with each other and existing abiotic factors.

Clustered development. The development pattern in which homes and other facilities are arranged in dense clusters on a relatively small portion of the land considered for development, allowing the rest of the land to remain open.

Co-composting. A technique of composting sewage sludge and shredded paper together.

Cogeneration. The joint production of useful heat and electricity. For example, furnaces may be replaced with gas turbogenerators that produce electricity while the hot exhaust still serves as a heat source. An important avenue of conservation, it effectively avoids the waste of heat that normally occurs at centralized power plants.

Command-and-control strategy. The basic strategy behind most air and water pollution public policy. It involves setting limits on pollutant levels and specifying control technologies that must be used to accomplish those limits.

Compaction. Packing down. *Soil:* Packing and pressing out air spaces present in the soil. Reduces soil aeration and infiltration and thus reduces the capacity of the soil to support plants. *Trash:* Packing down trash to reduce the space that it requires.

Commons, common pool resources. Resources (usually natural ones) owned by many people in common, or, as in the case of the air or the open oceans, owned by no one but open to exploitation.

Compliance schedule. A timetable for reducing pollutants by certain amounts by certain dates. Such schedules are arrived at through negotiations between companies and regulatory agencies.

Composting/Compost. The process of letting organic wastes decompose in the presence of air. A nutrient-rich humus or compost is the resulting product.

Composting toilet. A toilet that does not flush wastes away with water but deposits them in a chamber where they will compost. (See **composting**)

Compound. Any substance (gas, liquid, or solid) that is made up of two or more different kinds of atoms bonded together. Contrast **element.**

Comprehensive Environmental Response, Compensation, and Liability Act of 1980 (Superfund). Cornerstone legislation aimed at protecting groundwater from abandoned chemical waste sites' leakage.

Condensation. The collecting together of molecules from the vapor state to form the liquid state, as for example, water vapor condenses on a cold surface and forms droplets. Opposite: **evaporation.**

Confusion technique (pest control). Applying a quantity of sex attractant to an area so that males become confused and are unable to locate females. The actual quantities of pheromones applied are still very small because of their extreme potency.

Conservation. The management of a resource in such a way as to assure that it will continue to provide maximum benefit to humans over the long run. Conservation may include various degrees of use or protection, depending on what is necessary to maintain the resource over the long run. *Energy:* Saving energy. It not only entails cutting back on use of heating, air conditioning, lighting, transportation, and so on, but also entails increasing the efficiency of energy use. That is, developing and instigating means of doing the same jobs, e.g., transporting people, with less energy.

Conservation district. A region in which the state or county limits subdivision and development in order to preserve/conserve the natural environment and its values.

Consumers. In an ecosystem, those organisms that derive their energy from feeding on other organisms or their products.

Consumptive (water use). Use of water for such things as irrigation, where it does not remain available for potential purification and reuse.

Containment building (of nuclear power plant). Reinforced concrete building housing the nuclear reactor. Designed to contain an explosion should one occur.

Contour farming. The practice of cultivating land along the contours across rather than up and down slopes. In combination with strip cropping it reduces water erosion.

Contraceptive. Any device or drug that is designed to allow sexual intercourse but prevent pregnancies.

Control group. The group in an experiment that is the same as and is treated like the experimental group in every way except for the particular factor being tested. Only by comparison with a control group can one gain specific information concerning the effect of any test factor.

Controlled experiment. An experiment with adequate control groups. (See **control group**)

Control rods (nuclear power). Part of the core of the reactor, the rods of neutron-absorbing material that are inserted or removed as necessary to control the rate of nuclear fissioning.

Convection currents. Wind or water currents promoted by the fact that warming causes expansion, decreases density, and thus causes the warmer air or water to rise. Conversely, the cooler air or water sinks.

Cooling tower. A massive tower designed to dissipate waste heat from a power plant (or other industrial process) into the atmosphere.

Cooperative energy use. Utilizing the waste heat from one process as the source of heat for another process that requires a lesser temperature.

Cornucopianism. Dominant worldview that embodies the assumption that all parts of the environment are natural resources to be exploited for the advantage of humans.

Cosmetic damage (of fruits and vegetables). Damage to the surface that affects appearance but does not otherwise affect taste, nutritional quality, or storability.

Cosmetic spraying. Spraying of pesticides that is done to control pests that damage only the surface appearance.

Cost-benefit ratio/Benefit-cost ratio. The value of the benefits to be gained from a project divided by the costs of the project. If the ratio is greater than 1, the project is economically justified; if less than 1, it is not economically justified.

Cost-effective. Pertaining to a project or procedure that produces economic returns or benefits that are significantly greater than the costs.

Council on Competitiveness. An administrative council headed by the vice president that reviews all new regulatory rules to determine their impact on the business community. The council has turned back a number of proposed environmental regulations at the request of business interests.

Covalent bond. A chemical bond between two atoms, formed by sharing a pair of electrons between the two atoms. Atoms of all organic compounds are joined by covalent bonds.

Criteria pollutants. Certain pollutants the level of which is used as a gauge for the determination of air (or water) quality.

Critical level. The level of one or more pollutants above which severe damage begins to occur and below which few if any ill effects are noted.

Critical number. The minimum number of individuals of a given species that is required to maintain a healthy, viable population of the species. If a population falls below its critical number its extinction will almost certainly occur.

Crop rotation. The practice of alternating the crops grown on a piece of land. For example, corn one year, hay for two years, then back to corn.

Crude birth rate. Number of births per 1000 individuals per year.

Crude death rate. Number of deaths per 1000 individuals per year.

Crystallization. The joining together of molecules or ions from a liquid (or sometimes gaseous) state to form a solid state.

Cultivar. A cultivated variety of a plant species. All individuals of the cultivar are genetically highly uniform.

Cultural control (pest control). A change in the practice of growing, harvesting, storing, handling, or disposing of wastes that reduces the susceptibility or exposure to pests. For example, spraying the house with insecticides to kill flies is a chemical control; putting screens on the windows to keep flies out is a cultural control.

Cultural eutrophication. The accelerated process of natural eutrophication caused by humans. (See **eutrophication**)

D

DDT (dichlorodiphenyltrichloroethane). The first and most widely used of the synthetic organic pesticides belonging to the chlorinated hydrocarbon class.

Debt crisis. Refers to the fact that many less-developed nations are so heavily in debt that they may not be able to meet their financial obligations, e.g., interest payments. Failing to meet such obligations could have severe economic impacts on the entire world.

Debt-for-nature swap. Trading portions of the foreign debt that Third World countries owe us for their saving portions of their natural environment, e.g., tropical rain forests.

Declining tax base. The loss of tax revenues that occurs as a result of affluent taxpayers and businesses leaving an area and a subsequent decline of property values. It has been especially severe in inner cities as a result of migration to suburbs and exurbs.

Decommissioning (of nuclear power plants). Refers to the inevitable need to take nuclear power plants out of service after 25–35 years because the effects of radiation will gradually make them inoperable.

Decomposers. Organisms the feeding action of which results in decay or rotting of organic material. The primary decomposers are fungi and bacteria.

Degrade. To lower the quality or usefulness of.

Demographer. A person who studies population trends (growth, movement, development, and so on) and makes projections accordingly.

Demographic transition. The transition from a condition of high birth rate and high death rate through a period of declining death rate but continuing high birth rate finally to low birth rate and low death rate. This transition may result from economic development.

Density-dependent (factors). In reference to population balance, factors such as parasitism that increase and decrease in intensity corresponding to population density.

Deoxyribonucleic acid. (See **DNA**)

Derelict lands. Land areas so degraded that they have lost the entire ecosystem that once occupied them.

Desertification. Declining productivity of land due to mismanagement. Overgrazing and overcultivation allowing erosion and salinization are the major causes.

Desertified. Land for which productivity has been significantly reduced (25 percent or more) due to human mismanagement. Erosion is the most common cause.

Desalinization plants. Plants that purify seawater into high-quality drinking water via distillation or other techniques.

Department of Transportation Regulations (DOT Regs). Regulations intended to reduce the risk of spills, fires, and poisonous fumes by specifying the kinds of containers and methods of packing to be used in transporting hazardous materials.

Detritus. The dead organic matter, such as fallen leaves, twigs, and other plant and animal wastes, that exists in any ecosystem.

Detritus feeders. Organisms such as termites, fungi, and bacteria that obtain their nutrients and energy mainly by feeding on dead organic matter.

Deuterium (^{2}H). A stable, naturally occurring isotope of hydrogen. It contains one neutron in addition to the single proton normally in the nucleus.

Developed countries. Industrialized countries, United States, Canada, Western Europe, Japan, Australia, and New Zealand, in which the gross national product exceeds $7000 per capita.

Developing countries. All free-market countries in which the gross national product is less than $7000 per capita.

Development rights. Legal documents that grant permission to develop a given piece of property. They must be owned by the developer before development can occur. They can be bought and sold apart from the property itself.

Differential reproduction. Refers to the fact that within a population certain individuals reproduce much more than others.

Digest. The biological breakdown of organic material into simpler molecules.

Dioxin. A synthetic organic chemical of the chlorinated hydrocarbon class. It is one of the most toxic compounds known to humans, having many harmful effects, including induction of cancer and birth defects, even in extremely minute concentrations. It has become a widespread environmental pollutant because of the use of certain herbicides that contain dioxin as a contaminant.

Direct solar energy. (See **solar energy**)

Disinfection. The killing (as opposed to removal) of microorganisms in water or other media where they might otherwise pose a health threat. For example, chlorine is commonly used to disinfect water supplies.

Dissolved oxygen (DO). Oxygen gas molecules (O_2) dissolved in water. Fish and other aquatic organisms are dependent on dissolved oxygen for respiration. Therefore concentration of dissolved oxygen is a measure of water quality.

Distillation. A process of purifying water or other liquids by boiling the liquid and recondensing the vapor. Contaminants remain behind in the boiler.

District heating. The heating of an entire community or city area through circulating heat (e.g., steam) from a central source; particularly, utilizing waste heat from a power plant or from incineration of refuse.

Diversion (of water). Taking some or all of the flow of a natural waterway and carrying it to other places for uses such as municipal water supplies or irrigation.

DNA (deoxyribonucleic acid). The natural organic macro-molecule that carries the genetic or hereditary information for virtually all organisms.

DO. (See **dissolved oxygen**)

Domestic solid wastes. Wastes that come from homes, offices, schools, and stores, as opposed to wastes that are generated from agricultural or industrial processes.

Dose (of exposure to a hazardous material). A consideration of the concentration times the length of exposure. For any given material or radiation, effects correspond to the product of these two factors.

Doubling time. The time it will take a population to double in size assuming the continuation of current rate of growth.

Drift-netting. The practice of harvesting marine fish and squid by laying down miles of gill nets across the open seas. The nets collapse around larger organisms and kill many whales, dolphins, seals, marine birds, and turtles.

Drip irrigation. Supplying irrigation water through tubes that literally drip water onto the soil at the base of each plant.

E

Easement. In reference to land protection, an easement is an arrangement where a landowner gives up development rights into the future but retains ownership.

Ecological pest management. Control of pest populations through understanding the various ecological factors that provide natural control and so far as possible utilizing these factors as opposed to using synthetic chemicals.

Ecological collapse. When virtually everything in an ecosystem dies as a result of sudden or relatively drastic changes.

Ecological regard. Taking into consideration the environmental impact, direct and indirect, of one's actions and lifestyle. Adjusting actions and lifestyle to minimize the impact as much as possible.

Ecological succession. Process of gradual and orderly progression from one ecological community to another.

Ecological upset. A drastic, relatively sudden change in an ecosystem, some species benefiting and/or becoming much more abundant while others are eliminated. Such upsets are most often caused by humans altering some biotic or abiotic factors.

Ecologists. Scientists who study ecology, i.e., the ways in which organisms interact with each other and their environment.

Ecology. The study of any and all aspects of how organisms interact with each other and/or their environment.

Economic threshold (in reference to pest management). A certain level of pest damage that, to be reduced further, would require an application of pesticides that is more costly than the economic damage caused by the pests.

Ecosystem. A grouping of plants, animals, and other organisms interacting with each other and their environment in such a way as to perpetuate the grouping more-or-less indefinitely. Ecosystems have characteristic forms such as deserts, grasslands, tundra, deciduous forests, and tropical rain forests.

Ectoparasite. (See **parasites**)

Ecotone. A transitional region between two ecosystems

that contains some of the species and characteristics of the two adjacent ecosystems and also certain species characteristic of the transitional region.

Ecotourism. The enterprises involved in promoting tourism of unusual or interesting ecological sites.

Electrolysis (of water). The use of electrical energy to split water molecules into their constituent hydrogen and oxygen atoms. Hydrogen gas and oxygen gas result.

Electrons. Fundamental atomic particles that have a negative electrical charge but virtually no mass. They surround the nuclei of atoms and thus balance the positive charge of protons in the nucleus. A flow of electrons in a wire is synonymous with an electrical current.

Element. A substance that is made up of one and only one distinct kind of atom. Contrast **compound.**

Embrittlement. Becoming brittle. Pertains especially to the reactor vessel of nuclear power plants gradually becoming prone to breakage or snapping as a result of continuous bombardment by radiation. It is the prime factor forcing the decommissioning of nuclear power plants.

Emergency response teams. Teams of people, generally associated with police or fire departments, specially trained to handle accidents involving hazardous materials.

Emergent vegetation. Aquatic plants whose lower parts are under water but upper parts emerge from the water.

Endangered species. A species the total population of which is declining to relatively low levels, a trend that if continued will result in extinction.

Endangered Species Act. The federal legislation that mandates protection of species and their habitats that are determined to be in danger of extinction.

Endoparasite. (See **parasites**)

Energy. The ability to do work. Common forms of energy are light, heat, electricity, motion, and chemical bond energy inherent in compounds such as sugar, gasoline, and other fuels.

Enrichment. With reference to nuclear power, signifies the separation and concentration of uranium-235 so that, in suitable quantities, it will sustain a chain reaction.

Entomologist. A scientist who studies insects, their life cycles, physiology, behavior, and so on.

Entropy. Refers to the degree of disorder: increasing entropy means increasing disorder.

Environment. The combination of all things and factors external to the individual or population of organisms in question.

Environmental impact. Effects on the natural environment caused by human actions. Includes indirect effects through pollution, for example, as well as direct effects such as cutting down trees.

Environmental impact statement. A study of the probable environmental impacts of a development project. The National Environmental Policy Act of 1968 (NEPA) requires such studies prior to proceeding with any project receiving federal funding.

Environmental movement. Refers to the upwelling of public awareness and citizen action regarding environmental issues that occurred during the 1960s.

Environmental regard. A factor that may moderate negative environmental impacts, such as suitable attention to conservation or recycling.

Environmental resistance. The totality of factors such as adverse weather conditions, shortage of food or water, predators, and diseases that tend to cut back populations and keep them from growing or spreading. Contrast **biotic potential.**

Environmental science. The branch of science concerned with environmental issues.

Environmentalism. Embodies the assumptions that what we generally view as natural resources are products of the natural environment and can be maintained only insofar as the natural environment is maintained.

EPA. U.S. Environmental Protection Agency. The federal agency responsible for control of all forms of pollution and other kinds of environmental degradation.

Epidemiological study. Determination of causes of disease conditions (e.g., lung cancer) through the study and comparison of large populations of people living in different locations or following different lifestyles and/or habits (e.g., smoking versus nonsmoking).

Epiphytes. Air plants that are not parasitic but "perch" on tree branches, where they can get adequate light.

Erosion. The process of soil particles being carried away by wind or water. Erosion moves the smaller soil particles first and hence degrades the soil to a coarser, sandier, stonier texture.

Estimated reserves. (See **reserves**)

Estuary. A bay open to the ocean at one end and receiving freshwater from a river at the other. Hence, mixing of fresh and salt water occurs (brackish).

Euphotic zone. The layer or depth of water through which there is adequate light penetration to support photosynthesis.

Eutrophic. Refers to a body of water characterized by nutrient-rich water supporting abundant growth of algae and/or other aquatic plants at the surface. Deep water has little or no dissolved oxygen.

Eutrophication. The process of becoming eutrophic.

Evaporation. Molecules leaving the liquid state and entering the vapor or gaseous state as, for example, water evaporates to form water vapor.

Evapotranspiration. The combination of evaporation and transpiration.

Evolution. The theory that all species now on earth are descended from ancestral species through a process of gradual change brought about by natural selection.

Evolutionary succession. The succession of different species that have inhabited the earth at different geological periods, as revealed through the fossil record. The process of new species coming in through the process of speciation while other species pass into extinction.

Experimental group. The group in an experiment that receives the experimental treatment in contrast to the control group, used for comparison, which does not receive the treatment. Synonym: test group.

Exotic species. A species introduced to a geographical area where it is not native.

Exponential increase. The growth produced when the base population increases by a given *percentage* (as opposed to a given amount) each year. It is characterized by doubling again and again, each doubling occurring in the same period of time. It produces a J-shaped curve.

Externality. Any effect of a business process not included in the usual calculations of profit and loss. Pollution of air or water is an example of a *negative* externality—one that imposes a cost on society that is not paid for by the business itself.

Extinction. The death of all individuals of a particular species. When this occurs, all the genes of that particular line are lost forever.

Extractive reserves. As now established in Brazil, forest lands that are protected for native peoples and others who harvest natural products of the forests such as latex and Brazil nuts.

Exurban migration. Refers to the pronounced trend since World War II of people relocating homes and businesses from the central city and older suburbs to more-outlying suburbs.

F

Famine. A severe shortage of food accompanied by a significant increase in the local or regional death rate.

FAO. Food and Agriculture Organization of the United Nations.

Fecal. Refers to the solid excretory wastes of animals. Consists of undigested material passing through the gut and bacteria that have begun to feed on it.

Fecal coliform test. A test for the presence of *Escherichia coli*, the bacterium that normally inhabits the gut of humans and other mammals. A positive test indicates sewage contamination and the potential presence of disease-causing microorganisms carried by sewage.

Federal Manual for Identifying and Delineating Wetlands. The interagency manual that specifies the criteria to be used for determining whether a given area may qualify as a wetlands and thus come under wetlands protection legislation.

Fermentation. A form of respiration carried on by yeast cells in the absence of oxygen. It involves a partial breakdown of glucose (sugar) that yields energy for the yeast and the release of alcohol as a byproduct.

Fertility rate. (See **total fertility rate**)

Fertilization. The application of fertilizer. (See **fertilizer**)

Fertilizer. Material applied to plants or soil to supply plant nutrients, most commonly nitrogen, phosphorus, and potassium but may include others. Organic fertilizer is natural organic material such as manure, which releases nutrients as it breaks down. Inorganic fertilizer, also called chemical fertilizer, is a mixture of one or more necessary nutrients in inorganic chemical form.

Field capacity. A measure of the maximum volume of water that a soil can hold by capillary action, i.e., against the pull of gravity.

Field scouts (regarding pest management). Persons trained to survey crop fields and determine whether applications of pesticides or other procedures are actually necessary to avert significant economic loss.

FIFRA. Federal Insecticide, Fungicide, and Rodenticide Act.

Filtration. The passing of water (or other fluid) through a filter to remove certain impurities.

Fire climax ecosystems. Ecosystems that depend on the recurrence of fire to maintain the existing balance.

First basic principle of ecosystem sustainability. Resources are supplied and wastes are disposed of by recycling all elements.

First-generation pesticides. Toxic inorganic chemicals that were first used to control insects, plant diseases, and other pests. Included mostly compounds of arsenic and cyanide and various heavy metals such as mercury and copper.

First Law of Thermodynamics. The fact based on irrefutable observations that energy is never created or destroyed but may be converted from one form to another, e.g. electricity to light. (See also **Second Law of Thermodynamics**)

Fishery. Fish species being exploited, or a limited marine area containing commercially valuable fish.

Fission. The splitting of a large atom into two atoms of lighter elements. When large atoms such as uranium or plutonium fission, tremendous amounts of energy are released.

Fission products. Any and all atoms and subatomic particles resulting from splitting atoms in nuclear reactors. All or most such products are highly radioactive.

Flare. The burning of natural gas and/or other gaseous, flammable byproducts when there is no economical means available for utilizing it.

Flat-plate collector (solar energy). A solar collector that consists of a stationary, flat, black surface oriented perpendicular to the average sun angle. Heat absorbed by the surface is removed and transported by air or water (or other liquid) flowing over or through the surface.

Flood irrigation. Technique of irrigation where water is diverted from rivers through canals and flooded through furrows in fields.

Food aid. Food of various forms that is donated or sold below cost to needy people for humanitarian reasons.

Food chain. The transfer of energy and material through a series of organisms as each one is fed upon by the next.

Food security. For families, the ability to meet the food needs of everyone in the family, providing freedom from hunger and malnutrition.

Food web. The combination of all the feeding relationships that exist in an ecosystem.

Fossil fuels. Mainly crude oil, coal, and natural gas, that are derived from prehistoric photosynthetic production of organic matter on earth.

Fourth principle of ecosystem sustainability. Biodiversity must be maintained.

Freshwater. Water that has a salt content of less than 0.01 percent (100 parts per million).

Fuel assembly (nuclear power). The assembly of many rods containing the nuclear fuel, usually uranium, positioned close together. The chain reaction generated in the fuel assembly is controlled by rods of neutron-absorbing material between the fuel rods.

Fuel elements (nuclear power). The pellets of uranium or other fissionable material that are placed in tubes, which, with the control rods, form the core of the reactor.

Fuel rods. (See **Fuel elements**)

Fungus, pl. **fungi**. Any of numerous species of molds, mushrooms, brackets, and other forms of nonphotosynthetic plants. They derive energy and nutrients by consuming other organic material. Along with bacteria they form the decomposer component of ecosystems.

Fusion. The joining together of two atoms to form a single atom of a heavier element. When light atoms such as hydrogen are fused, tremendous amounts of energy are released.

G

Gasohol. A blend of 90 percent gasoline and 10 percent alcohol, which can be substituted for straight gasoline. It serves to stretch gasoline supplies.

Gene pool. The sum total of all the genes that exist among all the individuals of a species.

Genes. Segment of DNA that codes for one protein, which in turn determines a particular physical, physiological, or behavioral trait.

Genetic bank. The concept that natural ecosystems with all their species serve as a tremendous repository of genes that is frequently drawn upon to improve domestic plants and animals and to develop new medicines, among other uses.

Genetic control (pest control). Selective breeding of the desired plant or animal to make it resistant to attack by pests. Also, attempting to introduce harmful genes—for example, those that cause sterility—into the pest populations.

Genetic engineering. The artificial transfer of specific genes from one organism to another.

Genetic makeup (of an individual). Refers to all the genes that an individual possesses and that determine all his/her/its inherited characteristics.

Genetics. The study of heredity and the processes by which inherited characteristics are passed from one generation to the next.

Gentrification. The trend seen in modern society of people moving into more-or-less isolated communities with others of similar economic, ethnic, and social backgrounds.

Geothermal. Refers to the naturally hot interior of the earth. The heat is maintained by naturally occurring nuclear reactions in the earth's interior.

Geothermal energy. Useful energy derived from the naturally hot interior of the earth.

Global warming. The term given to the possibility that the earth's atmosphere is gradually warming because of the greenhouse effect of carbon dioxide and other gases. Global warming is thought by many to be the most serious global environmental issue facing our society. (See also **greenhouse effect** and **greenhouse gases**)

Glucose. A simple sugar, the major product of photosynthesis. Serves as the basic building block for cellulose and starches and as the major "fuel" for the release of energy through cell respiration in both plants and animals.

Gravitational water. Water that is not held by capillary action in soil but percolates downward by the force of gravity.

Gray water. Wastewater, as from sinks and tubs, that does not contain human excrements. Such water can be reused without purification for some purposes.

Green manure. A legume crop such as clover that is specifically grown to enrich the nitrogen and organic content of soil.

Greenhouse effect. An increase in the atmospheric temperature because of increasing amounts of carbon dioxide and certain other gases absorbing and trapping heat radiation that normally escapes from the earth.

Greenhouse gases. Gases in the atmosphere that absorb infrared energy and contribute to the air temperature. These gases are like a heat blanket and are important in insulating the earth's surface. They include carbon dioxide, water vapor, methane, nitrous oxide, and chlorofluorocarbons and other halocarbons.

Green Revolution. Refers to the development and introduction of new varieties of wheat and rice (mainly) that increased yields per acre dramatically in some countries.

Grit chamber (of wastewater treatment plants). Part of preliminary treatment, a swimming-pool-like tank where the velocity of the water is slowed enough to let sand and other gritty material settle.

Grit settling tank. Same as **grit chamber**.

Gross national product per capita. The total value of all goods and services exchanged in a year in a country, divided by its population. A common indicator for the average level of development and standard of living for a country.

Groundwater. Water that has accumulated in the ground, completely filling and saturating all pores and spaces in rock and/or soil. Groundwater is free to move more-or-less readily. It is the reservoir for springs and wells, and is replenished by infiltration of surface water.

Groundwater remediation. The repurification of contaminated groundwater by any of a number of techniques.

Growth momentum (of the human population). Refers to the fact that the human population will continue to grow for some time even after the fertility rate is reduced to 2.0 because there is currently such an excessive number of children moving into the reproductive age brackets.

Gully erosion. Gullies, large or small, resulting from water erosion.

H

Habitat. The specific environment (woods, desert, swamp) in which an organism lives.

Half-life. The length of time it takes for half of an unstable isotope to decay. The length of time is the same regardless of the starting amount. Also refers to the amount of time it takes compounds to break down in the environment.

Halogenated hydrocarbon. Synthetic organic compound containing one or more atoms of the halogen group, which includes chlorine, fluorine, and bromine.

Hard water. Water that contains relatively large amounts of calcium and/or certain other minerals that cause soap to precipitate.

Hazard. Anything that can cause (1) injury, disease, or death to humans; (2) damage to property; or (3) degradation of the environment.

Hazard assessment. The process of examining evidence linking a particular hazard to its harmful effects.

Hazardous materials (HAZMAT for short). Any material having one or more of the following attributes: ignitability, corrosivity, reactivity, toxicity.

Heavy metals. Any of the high atomic weight metals such as lead, mercury, cadmium, and zinc. All may be serious pollutants in water or soil because they are toxic in relatively low concentrations and they tend to bioaccumulate.

Herbicide. A chemical used to kill or inhibit the growth of undesired plants.

Herbivore, adj., **herbivorous.** An organism such as rabbit or deer that feeds primarily on green plants, or plant products such as seeds or nuts. Synonym: **primary consumer.**

Herbivory. The feeding on plants that occurs in an ecosystem. The total feeding of all plant-eating organisms.

Heterotroph, adj., **heterotrophic.** Any organism that consumes organic matter as a source of energy.

Highway Trust Fund. The monies collected from the gasoline tax designated for construction of new highways.

Hormones and **pheromones.** Natural chemical substances that control development, physiology, and/or behavior of an organism. Hormones are produced internally and affect only that individual. Pheromones are secreted externally and affect the behavior of other individuals of the same species, as, for example, attracting mates. Both hormones and pheromones are coming into use in pest control. (See **natural chemical control**)

Host. In feeding relationships, particularly parasitism, refers to the organism that is being fed upon, i.e., supporting the feeder.

Host-parasite relationship. The combination of a parasite and the organism it is feeding on.

Host-specific. When insects, fungal diseases, and other parasites are unable to attack species other than their specific host.

Humidity. The amount of water vapor in the air. (See also **relative humidity**)

Humus. A dark brown or black, soft, spongy residue of organic matter that remains after the bulk of dead leaves, wood, or other organic matter has decomposed. Humus does oxidize, but relatively slowly. It is extremely valuable in enhancing physical and chemical properties of soil.

Hunger. A general term referring to the lack of basic food required for meeting nutritional and energy needs, such that the individual is unable to lead a normal, healthy life.

Hunter-gatherers (referring to early humans). Humans surviving by hunting and gathering seeds, nuts, berries, and other edible things from the natural environment.

Hybrid. A plant or animal resulting from a cross between two closely related species that do not normally cross.

Hybridization. Cross-mating between two more-or-less closely related species.

Hydrocarbon emissions. Exhaust of various hydrogen-carbon compounds due to incomplete combustion of fuel. They are a major contribution to photochemical smog.

Hydrocarbons. *Chemistry:* Natural or synthetic organic substances that are composed mainly of carbon and hydrogen. Crude oil, fuels from crude oil, coal, animal fats, and vegetable oils are examples. *Pollution:* A wide variety of relatively small carbon-hydrogen molecules resulting from incomplete burning of fuel and emitted into the atmosphere. (See **volatile organic compounds**)

Hydroelectric dam. A dam and associated reservoir used to produce electrical power by letting the high-pressure water behind the dam flow through and drive a turbo-generator.

Hydroelectric power. Electrical power that is produced from hydroelectric dams or, in some cases, natural waterfalls.

Hydrogen bonding. A weak attractive force that occurs between a hydrogen atom of one molecule and, usually, an oxygen atom of another molecule. It is responsible for holding water molecules together to produce the liquid and solid states.

Hydrogen ions. Hydrogen atoms that have lost their electrons. Chemical symbol, H^+.

Hydrological cycle. (See **water cycle**)

Hydroponics. The culture of plants without soil. The method uses water with the required nutrients in solution.

Hydroxyl radical. The hydroxyl group (OH^-) missing the electron. It is a natural cleansing agent of the atmosphere. It is highly reactive and readily oxidizes many pollutants on contact and thus contributes to their removal.

Hypothesis. An educated guess concerning the cause behind an observed phenomenon that is then subjected to experimental tests to prove its accuracy or inaccuracy.

I

Indicator organism. An organism, the presence or absence of which indicates certain conditions. Example: The presence of *E. coli* indicates water is contaminated with fecal wastes, and pathogens may be present; the absence indicates the water is free of pathogens.

Indirect products (of air pollution). Air pollutants that are not contained in emissions but are formed as a result of such compounds undergoing various reactions in the atmosphere.

Indirect solar energy. (See **solar energy**)

Industrial smog. The grayish mixture of moisture, soot, and sulfurous compounds that occurs in local areas where industries are concentrated and coal is the primary energy source.

Industrialized agriculture. Using fertilizer, irrigation, pesticides, and energy from fossil fuels to produce large quantities of crops and livestock with minimal labor for domestic and foreign sale.

Infant mortality. The number of babies that die before age 1.

Infiltration/Infiltrate. The process of water soaking into soil as opposed to its running off the surface.

Infiltration-runoff ratio. The ratio between the amount of water soaking into the soil and that running off the surface. (The ratio is given by dividing the first amount by the second.)

Infrared radiation. Radiation of somewhat longer wavelengths than red light, the longest wavelengths of the visible spectrum. Such radiation manifests itself as heat.

Infrastructure. The sewer and water systems, roadways, bridges, and other facilities that underlie the functioning of a city and that are owned, operated, and maintained by the city.

Inherently safe reactor. In theory, a nuclear reactor that is designed in such a way that any accident would be automatically corrected and no radioactivity released.

Inorganic compounds or **molecules.** *Classical definition:* All things such as air, water, minerals, and metals that are neither living organisms nor products uniquely produced by living things. *Chemical definition:* All chemical compounds or molecules that do not contain carbon atoms as an integral part of their molecular structure. Contrast **organic.**

Insecticide. Any chemical used to kill insects.

Instrumental value. Living organisms or species are considered to be worthwhile if their existence or use benefits people.

Insurance spraying (in reference to use of pesticides). Spraying that is done when it is not really needed in the belief that it will insure against loss due to pests.

Integral urban house. A house in an urban setting that utilizes ecological principles, including water and materials conservation and recycling, solar energy, and intensive cultivation of food plants in so far as possible.

Integrated pest management (IPM). Two or more methods of pest control carefully integrated into an overall program designed to avoid economic loss from pests. The objective is to minimize the use of environmentally hazardous, synthetic chemicals. Such chemicals may be used in IPM, but only as a last resort to prevent significant economic losses.

Integrated waste management. The approach to municipal solid waste that provides for several options for dealing with wastes, including recycling, composting, waste reduction, as well as landfilling and incineration where unavoidable.

Interim permits (referring to discharge of waste materials into air and water). Permission to discharge undesirable amounts of certain wastes into air and/or water until pollution control facilities can be installed. Such permits are granted to prevent forcing the closure of certain industrial plants.

Intrinsic value. Living organisms or species are considered to be worthwhile in their own right; they do not have to be useful to have value.

Inversion. (See **temperature inversion**)

Ion. An atom or group of atoms that has lost or gained one or more electrons and consequently acquired a positive or negative charge. Ions are designated by + or − superscripts following the chemical symbol.

Ion-exchange capacity (soil). (See **Nutrient-holding capacity**)

Ionic bond. The bond formed by the attraction between a positive and a negative ion.

IPM. (See **integrated pest management**)

Irrigation. Any method of artificially adding water to crops.

Isotope. A form of an element in which the atoms have more (or less) than the usual number of neutrons. Isotopes of a given element have identical chemical properties, but they differ in mass (weight) as a result of the additional (or lesser) neutrons. Many isotopes are unstable and give off radioactive radiation. (See **radioactive decay, radioactive emissions,** and **radioactive materials**)

J

Juvenile hormone. The insect hormone sufficient levels of which preserve the larval state. Pupation requires diminished levels; hence artificial applications of the hormone may block development.

K

Kerogen. A hydrocarbon material contained in oil shale that vaporizes when heated and can be recondensed into a material similar to crude oil.

Keystone species. A species whose role is essential for the survival of many other species in an ecosystem.

Kinetic energy. The energy inherent in motion or movement including molecular movement (heat) and movement of waves, hence radiation including light.

L

Landfill. A site where wastes (municipal, industrial, or chemical) are disposed of by burying them in the ground or placing them on the ground and covering them with earth. Also used as a verb meaning to dispose of a material in such a way.

Land subsidence. The phenomenon of land gradually sinking in elevation. It may result from removing groundwater or oil, which is frequently instrumental in supporting the overlying rock and soil.

Land trusts. Land that is purchased and held by various organizations specifically for the purpose of protecting its natural environment and biota that inhabit it.

Larva, pl. **larvae,** adj. **larval.** A free-living immature form that occurs in the life cycle of many organisms and that is structurally distinct from the adult. For example, caterpillars are the larval stage of moths and butterflies.

Law of Conservation of Energy. (See **First Law of Thermodynamics**)

Law of Conservation of Matter. In chemical reactions, atoms are neither created, changed, nor destroyed. They are only rearranged.

Law of Limiting Factors. Also known as Liebig's Law of Minimums. A system may be limited by the absence or minimum amount (in terms of that needed) of any required factor. (See **limiting factor**)

Leachate. The mixture of water and materials that are leaching.

Leaching. The process of materials in or on the soil gradually dissolving and being carried by water seeping through the soil. It may result in valuable nutrients being removed from the soil, or it may result in wastes buried in the soil being carried into and contaminating groundwater.

Legumes. The group of land plants that is virtually alone in its ability to fix nitrogen (see **nitrogen fixation**). The legume group includes such common plants as peas, beans, clovers, alfalfa, and locust trees but no major cereal grains.

Lethal mutation. A mutation that results in such severe abnormalities that the organism cannot survive.

Liebig's Law of Minimums. (See **Law of Limiting Factors**)

Life cycle. The various stages of life, progressing from the adult of one generation to the adult of the next.

Limiting factor. A factor primarily responsible for limiting the growth and/or reproduction of an organism or a population. The limiting factor may be a physical factor such as temperature or light, a chemical factor such as a particular nutrient, or a biological factor such as a competing species. The limiting factor may differ at different times and places.

Limitists. Those who believe that our present economic-social system may be sharply curtailed in the future by resource shortages. Consequently, those who believe that changes in our system must be instigated now in order to accommodate future resource shortages.

Limits of tolerance. The extremes of any factor, e.g., temperature, that an organism or a population can tolerate and still survive and reproduce.

Lipids. A class of natural organic molecules that includes animal fats, vegetable oils, and phospholipids, the last being an integral part of cellular membranes.

Litter. In an ecosystem, the natural cover of dead leaves, twigs, and other dead plant material. This natural litter is subject to rapid decomposition and recycling in the ecosystem, whereas human litter, such as bottles, cans, and plastics, is not.

Loam. A solid consisting of a mixture of about 40 percent sand, 40 percent silt, and 20 percent clay.

Longevity. The average life span of individuals of a given population.

M

Macromolecules. Very large, organic molecules such as proteins and nucleic acids that constitute the structural and functional parts of cells.

MACT (maximum achievable control technology). The best technologies available for reducing the output of especially toxic industrial pollutants.

Malnutrition. The lack of essential nutrients such as vitamins, minerals, and amino acids. Malnutrition ranges from mild to severe and life-threatening.

Mariculture. The propagation and/or rearing of any marine (saltwater) organism in more-or-less artificial systems.

Marine environment. An ocean environment that supports a more-or-less distinctive array of seaweeds, plankton, fish, shellfish, and other marine organisms depending on temperature, water depth, nature of the bottom, and concentrations of nutrients and sediments.

Mass number. The number that accompanies the chemical name or symbol of an element or isotope. It represents the number of neutrons and protons in the nucleus of the atom.

Materials recycling facility (MRF). A processing plant where regionalized recycling is facilitated. Recyclable municipal solid waste, usually presorted, is prepared in bulk for the recycling market.

Matter. Anything that occupies space and has mass. Refers to any gas, liquid, or solid. Contrast **energy.**

Maximum sustainable yield (renewable resources). The maximum amount that can be taken year after year without depleting the resource. It is the maximum rate of use or harvest that will be balanced by the regenerative capacity of the system—for example, the maximum rate of tree cutting that can be balanced by tree regrowth.

Meltdown (nuclear power). The event of a nuclear reactor getting out of control or losing its cooling water so that it melts from its own production of heat. The melted reactor would continue to produce heat and could melt its way out of the reactor vessel and eventually down into groundwater, where it would cause a violent eruption of steam that could spread radioactive materials over a wide area.

Metabolism. The sum of all the chemical reactions that occur in an organism.

Methane. A gas, CH_4. It is the primary constituent of natural gas. It is also produced as a product of fermentation by microbes. Methane from ruminant animals is thought to be responsible for the rise in atmospheric methane, of concern because methane is one of the greenhouse gases.

Microbe. A term used to refer to any microscopic organism, primarily bacteria, viruses, and protozoans.

Microclimate. The actual conditions experienced by an organism in its particular location. Due to numerous factors such as shading, drainage, and sheltering, the microclimate may be quite distinct from the overall climate.

Microfiltration or **reverse osmosis.** A process for purifying water in which water is forced under very high pressure through a membrane that filters out ions and molecules in solution.

Microorganism. Any microscopic organism, particularly bacteria, viruses, and protozoans.

Midnight dumping. The wanton illicit dumping of materials, particularly hazardous wastes, frequently under the cover of darkness.

Minamata disease. A "disease" named for a fishing village in Japan where an "epidemic" was first observed. Symptoms, which included spastic movements, mental retardation, coma, death, and crippling birth defects in the next generation, were found to be the result of mercury poisoning.

Mineral. Any hard, brittle, stonelike material that occurs naturally in the earth's crust. All consist of various combinations of positive and negative ions held together by ionic bonds. Pure minerals, or crystals, are one specific combination of elements. Common rocks are composed of mixtures of two or more minerals. (See also **ore**)

Mineralization (soil science). The process of gradual oxidation of the organic matter (humus) present in soil that leaves just the gritty mineral component of the soil.

Mixture (in elements). Means there is no chemical bonding between the molecules involved. Example: Air contains (is a mixture of) oxygen, nitrogen, and carbon dioxide.

Mobilization. In soil science, the bringing into solution of normally insoluble minerals. Presents a particular problem when the elements of such minerals have toxic effects.

Moderator. In a nuclear reactor, the moderator is any material that slows down neutrons from fission reactions so that they are traveling at the right speed to trigger another fission. Water and graphite represent two types of moderators.

Molecule. A specific union of two or more atoms held together by covalent bonds. The smallest unit of a compound that still has the characteristics of that compound.

Monocropping. The practice of growing the same crop year after year on the same land. Contrast **crop rotation.**

Monoculture. The practice of growing a single crop over very wide areas, for example, thousands of square kilometers of wheat, and only wheat, grown in the Midwest.

Montreal Protocol. An agreement made in 1987 by a large group of nations to cut back the production of chlorofluorocarbons by 50 percent by 2000 in order to protect the ozone shield. A 1990 amendment calls for the complete phaseout of these chemicals by 2000 in developed nations and by 2010 in less-developed nations.

Municipal solid waste. The entirety of refuse or trash generated by a residential and business community. The refuse that a municipality is responsible for collecting and disposing of, distinct from agricultural and industrial wastes.

Mutagenic. Causing mutations.

Mutation. A random change in one or more genes of an organism. Mutations may occur spontaneously in nature, but their number and degree are vastly increased by exposure to radiation and/or certain chemicals. Mutations generally result in a physical deformity and/or metabolic malfunction.

Mutualism. Refers to a close relationship between two organisms in which both organisms benefit from the relationship.

Mycelia. The threadlike feeding filaments of fungi.

Mycorrhizae. The mycelia of certain fungi that grow symbiotically with the roots of some plants and provide for additional nutrient uptake.

N

NASA. National Aeronautics and Space Administration.

National Forests. Administered by the National Forest Service, these are public forest and woodlands that are managed for multiple uses, such as logging, mineral exploitation, livestock grazing, and recreation.

National Parks. Administered by the National Park Service, National Parks are lands and coastal areas of great scenic, ecological or historical importance. They are managed with the dual goals of protection and providing public access.

National Priorities List (NPL). A list of the chemical waste sites presenting the most immediate and severe threats. Such sites are scheduled for cleanup ahead of other sites.

Natural (to describe a substance or factor). Occurring or produced as a normal part of nature apart from any activity or intervention of humans. Opposite of artificial, synthetic, human-made, or caused by humans.

Natural chemical control (pest control). The use of one or more natural chemicals such as hormones or pheromones to control a pest.

Natural control methods (pest control). Any of many techniques of controlling a pest population without resorting to the use of synthetic organic or inorganic chemicals. (See **biological control, cultural control, genetic control, hormones** and **pheromones**)

Natural enemies. All of the predators and/or parasites that may feed on a given organism. Organisms used to control a specific pest through predation or parasitism.

Natural increase (in populations). The number of births minus the number of deaths. It does not consider immigration and emigration.

Natural laws. Derivations from our observations that matter, energy, and certain other phenomena apparently always act (or react) according to certain "rules."

Natural organic compounds. (See **organic compounds** or **molecules**)

Natural rate of change (for a population). The percent of growth (or decline) of a population during a year. It is found by subtracting the crude death rate from the crude birth rate and changing the result to a percent. It does not include immigration or emigration.

Natural resources. As applied to natural ecosystems and species, this term indicates that they are expected to be of economic value and may be exploited. Likewise the term applies to particular segments of ecosystems such as air, water, soil, and minerals.

Natural selection. The process whereby the natural factors of environmental resistance tend to eliminate those members of a population that are least well adapted to cope and thus, in effect, select those best adapted for survival and reproduction.

Natural services. Functions performed by natural ecosystems such as control of runoff and erosion, absorption of nutrients, and assimilation of air pollutants.

Natural succession. (See **ecological succession**)

NEPA. National Environmental Policy Act.

Net energy yield. The amount of energy produced minus the amount that is expended in production and transmission to consumers.

Neutron. A fundamental atomic particle found in the nuclei of atoms (except hydrogen) and having one unit of atomic mass but no electrical charge.

Niche (ecological). The total of all the relationships that bear on how an organism copes with both biotic and abiotic factors it faces.

NIMBY. Acronym for Not In My Back Yard. NIMBY refers to a common attitude regarding undesirable facilities such as incinerators, nuclear facilities, and hazardous waste treatment plants, whereby people do everything possible to prevent the location of such facilities nearby.

Nitric acid (HNO_3). One of the acids in acid rain. Formed by reactions between nitrogen oxides and the water vapor in the atmosphere.

Nitric oxide. (See **nitrogen oxides**)

Nitrogen dioxide. (See **nitrogen oxides**)

Nitrogen fixation/Nitrogen fixing. The process of chemically converting nitrogen gas (N_2) from the air into compounds such as nitrates (NO_3^-) or ammonia (NH_3) that can be used by plants in building amino acids and other nitrogen-containing organic molecules.

Nitrogen oxides (NOx). A group of nitrogen-oxygen compounds formed as a result of some of the nitrogen gas in air combining with oxygen during high-temperature combustion, they are a major category of air pollutants. Along with hydrocarbons, they are a primary factor in the production of ozone and other photochemical oxidants that are the most harmful components of photochemical smog. They also contribute to acid precipitation (see **nitric acid**). Major nitrogen oxides are: nitric oxide, NO; nitrogen dioxide, NO_2; nitrogen tetroxide, N_2O_2.

Nitrogen tetroxide. (See **nitrogen oxides**)

Nitrous oxide. A gas, N_2O. Nitrous oxide comes from biomass burning, fossil fuel burning, and the use of chemical fertilizers and is of concern because in the troposphere it is a greenhouse gas and in the stratosphere it contributes to ozone destruction.

NOAA. National Oceanic and Atmospheric Administration.

Nonbiodegradable. Not able to be consumed and/or broken down by biological organisms. Nonbiodegradable substances include plastics, aluminum, and many chemicals used in industry and agriculture. Particularly dangerous are nonbiodegradable chemicals that are also toxic and tend to accumulate in organisms. (See **biodegradable, bioaccumulation**)

Nonconsumptive (water use). Refers to the use of water for such purposes as washing and rinsing where the water, albeit polluted, remains available for further uses. With suitable purification, such water may be recycled indefinitely.

Nonpersistent (with respect to pesticides and other chemicals). Chemicals that break down readily to harmless compounds, as, for example, natural organic compounds break down to carbon dioxide and water.

Nonpoint sources (of pollution). Pollution from general runoff of sediments, fertilizer, pesticides, and other materials from farms and urban areas as opposed to pollution from specific discharges (contrast **point sources**).

Nonrenewable resources. Resources such as ores of various metals, oil, and coal that exist as finite deposits in the earth's crust and that are not replenished by natural processes as they are mined.

Nontidal wetlands. Inland wetlands not affected by tides.

No-till agriculture. The farming practice in which weeds are killed with chemicals (or other means) and seeds are planted and grown without resorting to plowing or cultivation. The practice is very effective in reducing soil erosion.

Nuclear power. Electrical power that is produced by using a nuclear reactor to boil water and produce steam which, in turn, drives a turbogenerator.

Nuclear Regulatory Commission. The agency within the Department of Energy that sets and enforces safety standards for the operation and maintenance of nuclear power plants.

Nuclear winter. A pronounced global cooling that would occur as a result of a large-scale nuclear conflict. It is based on theoretical projections concerning the amount of dust and smoke that would be ejected into the atmosphere and the resulting decrease in solar radiation.

Nucleic acids. The class of natural organic macromolecules that function in the storage and transfer of genetic information.

Nucleus. *Physics:* The central core of atoms, which is made up of neutrons and protons. Electrons surround the nucleus. *Biology:* The large body contained in most living cells that contains the genes or hereditary material, DNA.

Nutrient. *Plant:* An essential element in a particular ion or molecule that can be absorbed and used by the plant. For example, carbon, hydrogen, nitrogen, and phosphorus are essential elements; carbon dioxide, water, nitrate (NO_3^-), and phosphate (PO_4^{3-}) are respective nutrients. *Animal:* Materials such as protein, vitamins, and minerals that are required for growth, maintenance, and repair of the body and also materials such as carbohydrates that are required for energy.

Nutrient cycles. The repeated pathway of particular nutrients or elements from the environment through one or more organisms back to the environment. Nutrient cycles include the carbon cycle, the nitrogen cycle, the phosphorus cycle, and so on.

Nutrient-holding capacity. The capacity of a soil to bind and hold nutrients (fertilizer) against their tendency to be leached from the soil.

O

Observations. Things or phenomena that are perceived through one or more of the basic five senses in their normal state. In addition, to be accepted as factual, the observations must be verifiable by others.

Ocean thermal energy conversion (OTEC). The concept of harnessing the temperature difference between surface water heated by the sun and colder deep water to produce power.

Oil field. The area in which exploitable oil is found.

Oil shale. A natural sedimentary rock that contains a material, kerogen, that can be extracted and refined into oil and oil products.

Oligotrophic. Refers to a lake the water of which is nutrient-poor. Therefore, it will not support phytoplankton, but it will support submerged aquatic vegetation, which get nutrients from the bottom.

Omnivore. An animal that feeds more or less equally on both plant material and other animals.

OPEC. Organization of Petroleum Exporting Countries.

Optimal range. With respect to any particular factor or combination of factors, the maximum variation that still supports optimal or near-optimal growth of the species in question.

Optimum. The condition or amount of any factor or combination of factors that will produce the best result. For example, the amount of heat, light, moisture, nutrients, and so on that will produce the best growth. Either more or less than the optimum is not as good.

Optimum population (resources). The population that will provide the maximum sustainable yield. The yield is reduced at higher or lower populations.

Ore. A mineral rich in a particular element such as iron, aluminum, or copper that can be economically mined and refined to produce the desired metal. High-grade ore contains a relatively high percentage and low-grade ores contain a relatively low percentage of the desired element.

Organic. *Classical definition:* All living things and products that are uniquely produced by living things, such as wood, leather, and sugar. *Chemical definition:* All chemical compounds or molecules, natural or synthetic, that contain carbon atoms as an integral part of their structure. Contrast **inorganic.**

Organic compounds or **molecules.** Chemical compounds or molecules, the structure of which is based on bonded carbon atoms with hydrogen atoms attached.

Organic fertilizer. (See **fertilizer**)

Organic gardening or **farming.** Gardening or farming without the use of inorganic fertilizers, synthetic pesticides, or other human-made materials.

Organic molecules. (See **organic compounds** or **molecules**)

Organic phosphate. Phosphate (PO_4^{-3}) bonded to an organic molecule.

Organism. Any living thing—plant, animal, or microbe.

Osmosis. The phenomenon of water diffusing through a semipermeable membrane toward an area where there is more material in solution (where there is a relatively lower concentration of water). Has particular application regarding salinization of soils where plants are unable to grow because of osmotic water loss.

Overland flow system. An alternative method of wastewater treatment that involves allowing water to percolate through a field of grass or other vegetation.

Outbreak (of pests). A population explosion of a particular pest. Often caused by an application of pesticides that destroys the pest's natural enemies.

Overgrazing. The phenomenon of animals grazing in greater numbers than the land can support in the long run. There may be a temporary economic gain in the short run, but the grassland (or other ecosystem) is destroyed, and its ability to support life in the long run is vastly diminished.

Overreproduction. Refers to the fact that all species produce far more offspring, eggs, or seeds than would seem necessary to sustain their population.

Oxidation. A chemical reaction process that generally involves breakdown through combining with oxygen. Both burning and cellular respiration are examples of oxidation. In both cases, organic matter is combined with oxygen and broken down to carbon dioxide and water.

Ozone. A gas, O_3, that is a pollutant in the lower atmosphere, but necessary to screen out ultraviolet radiation in the upper atmosphere. May also be used for disinfecting water.

Ozone hole. First discovered over the Antarctic, this is a region of stratospheric air that is severely depleted of its normal levels of ozone during the Antarctic spring because of CFCs from anthropogenic sources.

Ozone shield. The layer of ozone gas (O_3) in the upper atmosphere that screens out harmful ultraviolet radiation from the sun.

P

PANs (peroxyacetylnitrates). A group of compounds present in photochemical smog that are extremely toxic to plants and irritating to eyes, nose, and throat membranes of humans.

Parasites. Organisms (plant, animal, or microbial) that attach themselves to another organism, the host, and feed on it over a period of time without killing it immediately but usually doing harm to it. Commonly divided into *ectoparasites,* those that attach to the outside, and *endoparasites,* those that live inside their hosts.

Parent material. The rock material, the weathering and gradual breakdown of which is the source of the mineral portion of soil.

Particulates. (See **suspended particulate matter** and **PM-10**)

Parts per million (ppm). A frequently used expression of concentration. It is the number of units of one substance present in a million units of another. For example, 1 g of phosphate dissolved in 1 million grams ($= 1$ ton) of water would be a concentration of 1 ppm.

Passive safety. In reference to nuclear facilities, passive safety features are those that involve processes that are not vulnerable to operator intrusion or electrical power failures. They contribute to a higher degree of safety for nuclear reactors. (See **active safety**).

Passive solar heating system. A solar heating system that does not use pumps or blowers to transfer heated air or water. Instead natural convection currents are used, or the interior of the building itself acts as the solar collector.

Pasteurization. The process of applying heat to kill pathogens.

Pastoralist. One involved in animal husbandry, usually in subsistence agriculture.

Pathogen, adj., **pathogenic.** An organism, usually a microbe, that is capable of causing disease.

PCBs (polychlorinated biphenyls). A group of very widely used industrial chemicals of the chlorinated hydrocarbon class. They have become very serious and widespread pollutants, contaminating most food chains on earth, because they are extremely resistant to breakdown and are subject to bioaccumulation. They are known to be carcinogenic.

Percolation. The process of water seeping through cracks and pores in soil or rock.

Perennials. Plants that survive and grow year after year as opposed to annuals, which die and must be restarted from seed each year.

Permafrost. The ground of arctic regions that remains permanently frozen. Defines tundra since only small herbaceous plants can be sustained on the thin layer of soil that thaws each summer.

Persistent (with respect to pesticides or other chemicals). Nonbiodegradable and very resistant to breakdown by other means. Such chemicals therefore remain present in the environment more-or-less indefinitely.

Pesticide. A chemical used to kill pests. Pesticides are further categorized according to the pests they are designed to kill—for example, herbicides to kill plants, insecticides to kill insects, fungicides to kill fungi, and so on.

Pesticide treadmill. Refers to the fact that use of chemical pesticides simply creates a vicious cycle of "needing more pesticides" to overcome developing resistance and secondary outbreaks caused by the pesticide applications.

Pest-loss insurance. Insurance that a grower can buy that will pay in the event of loss of crop due to pests.

Petrochemical. A chemical made from petroleum (crude oil) as a basic raw material. Petrochemicals include plastics, synthetic fibers, synthetic rubber, and most other synthetic organic chemicals.

pH. Scale used to designate the acidity or basicity (alkalinity) of solutions or soil. pH 7 is neutral; values decreasing from 7 indicate increasing acidity; values increasing from 7 indicate increasing basicity. Each unit from 7 indicates a tenfold increase over the preceding unit.

Pheromone. A chemical substance secreted by certain members that affects the behavior of other members of the population. The most common examples are sex attractants, which female insects secrete to attract males.

Phosphate. An ion composed of a phosphorus atom with 4 oxygen atoms attached. PO_4^{3-}. It is an important plant nutrient. In natural waters it is frequently the limiting factor. Therefore, additions of phosphate to natural water are frequently responsible for algal blooms.

Photochemical oxidants. A major category of air pollutants, including ozone, that are highly toxic and damaging especially to plants and forests. Formed as a result of interreactions between nitrogen oxides and hydrocarbons driven by sunlight.

Photochemical smog. The brownish haze that frequently forms on otherwise clear sunny days over large cities with significant amounts of automobile traffic. It results largely from sunlight-driven chemical reactions among nitrogen oxides and hydrocarbons, both of which come primarily from auto exhausts.

Photosynthesis. The chemical process carried on by green plants through which light energy is used to produce glucose from carbon dioxide and water. Oxygen is released as a byproduct.

Photosynthetic organism. An organism capable of carrying on photosynthesis. Opposite: nonphotosynthetic.

Photovoltaic cells. Devices that convert light energy into an electrical current.

Physical barrier (regarding genetic control of pests). The presence of a genetic feature on a plant, such as sticky hairs, that physically blocks attack by pests.

Phytoplankton. Any of the many species of algae that consist of single cells or small groups of cells that live and grow freely suspended in the water near the surface. Given abundant nutrients, they may become so numerous as to give the water a green "pea soup" appearance and/or form a thick green scum over the surface.

Plankton, adj., **planktonic.** Any and all living things that are found freely suspended in the water and that are carried by currents as opposed to being able to swim against currents. It includes both plant (phytoplankton) and animal (zooplankton) forms.

Plant community. The array of plant species, including numbers, ages, distribution, that occupies a given area.

PM-10. The new standard criteria pollutant for suspended particulate matter. PM-10 refers to particles smaller than 10 micrometers in diameter. Such particles are readily inhaled directly into the lungs.

Point sources (of pollution). Pollutants coming from specific points such as discharges from factory drains or outlets from sewage treatment plants (contrast **nonpoint sources**).

Pollutant. A substance the presence of which is causing pollution.

Pollution. Contamination of air, water, or soil with undesirable amounts of material or heat. The material may be a natural substance, such as phosphate, in excessive quantities, or it may be very small quantities of a synthetic compound such as dioxin that is exceedingly toxic.

Pollution avoidance/Pollution prevention. A strategy of encouraging development of techniques that would not generate pollutants.

Polyculture. The growing of two or more species together. Contrast **monoculture**.

Poor. Economically unable to afford adequate food and/or housing.

Population. A group within a single species, the individuals of which can and do freely interbreed. Breeding between populations of the same species is less common because of differences in location, culture, nationality, and so on.

Population density. The numbers of individuals per unit of area.

Population explosion. The exponential increase observed to occur in a population when or if conditions are such that a large percentage of the offspring are able to survive and reproduce in turn. Frequently leads, in turn, to overexploitation, upset, and eventual collapse of the ecosystem.

Population momentum. Refers to the fact that a rapidly growing human population may be expected to grow for 50–60 years after replacement fertility (2.1) is reached because of increasing numbers entering reproductive age.

Population profile. A bar graph that shows the number of individuals at each age or in each five-year age group.

Population structure. Refers to the proportion of individuals in each age group. For example, is the population predominantly made up of young people, old people, or a more-or-less even distribution of young and old?

Potential energy. The ability to do work that is stored in some chemical or physical state. For example, gasoline is a form of potential energy; the ability to do work is stored in the chemical state and is released as the fuel is burned in an engine.

ppm. (See **parts per million**)

pOH. The negative logarithm of the concentration of hydroxyl ions. Like pH, the scale ranges from 0 to 14, each unit representing a tenfold increase over the preceding unit. The lower the pOH, the higher the concentration of hydroxyl ions.

Practical availability (nonrenewable resources). The fraction of a material such as copper that can be obtained from the earth's crust given restraints of unacceptable pollution, expenditure of energy, and other costs associated with production from low-grade ores or remote sources.

Precipitation. Any form of moisture condensing in the air and depositing on the ground.

Predator. An animal that feeds on another.

Predator-prey relationship. A feeding relationship existing between two kinds of animals. The predator is the animal feeding on the prey. Such relationships are frequently instrumental in controlling populations of herbivores.

Preliminary treatment (of sewage). The removal of debris and grit from wastewater by passing the water through a coarse screen and grit settling chamber.

Prey. In a feeding relationship, the animal that is killed and eaten by another.

Primary air pollutants. The major air pollutants that are the direct products of combustion and evaporation: suspended particulate matter, volatile organic compounds, carbon monoxide, nitrogen oxides, sulfur dioxide, and lead and other heavy metals.

Primary clarifiers. Large tanks used in primary treatment of water to remove particulate organic material.

Primary consumer. An organism such as a rabbit or deer that feeds more-or-less exclusively on green plants or their products such as seeds and nuts. Synonym: **herbivore.**

Primary energy sources. Fossil fuels, radioactive material, and solar, wind, and water and other energy sources that exist as natural resources.

Primary standard. The maximum tolerable level of a pollutant that is believed to protect human health with some margin of safety.

Primary succession. (See **succession**)

Primary treatment (of sewage). The process that follows preliminary treatment. It consists of passing the water very slowly through a large tank, which permits the particulate organic material in the water to settle out. The settled material is raw sludge.

Producers. In an ecosystem, those organisms, mostly green plants, that use light energy to construct their organic constituents from inorganic compounds.

Production (of oil). The withdrawing of oil reserves.

Profligate growth. Growth characterized by extravagant and wasteful use of resources.

Property taxes. Taxes that the local government levies on privately owned properties, generally a few dollars per hundred of property value. This is the major source of revenue for local governments.

Protein. The class of organic macromolecules that is the major structural component of all animal tissues and that functions as enzymes in both plants and animals.

Proton. Fundamental atomic particle with a positive charge, found in the nuclei of atoms. The number of protons present equals the atomic number and is distinct for each element.

Protozoan, pl. **protozoa.** Any of a large group of microscopic organisms that consist of a single, relatively large complex cell or in some cases small groups of cells. All have some means of movement. Amoebae and paramecia are examples.

Proven reserves. Amounts of mineral resources (including oil, coal, and natural gas) that have been discovered and surveyed and are available for exploitation.

Punctuated evolution. "Step" model of evolution in which there is little change while an ecosystem is in a balanced state but a shift alters selective pressures and sets into motion fairly rapid charges in almost all, if not all, species in the ecosystem until a new balance is reached.

Pure science. (See **basic science**)

Q

Qualitative. Refers to issues involving purity (is quality sufficient?)

Quantitative. Refers to issues involving numbers (are quantities sufficient?)

R

RACT (reasonably available control technology). Applied to the goals of the Clean Air Act, EPA-approved forms of technology that will reduce the output of industrial air pollutants. (See also **MACT**)

Radioactive decay. The reduction of radioactivity that occurs as an unstable isotope (radioactive substance) gives off radiation and becomes stable.

Radioactive emissions. Any of various forms of radiation and/or particles that may be given off by unstable isotopes. Many such emissions have very high energy and may destroy biological tissues or cause mutations leading to cancer or birth defects.

Radioactive materials. Substances that are or that contain unstable isotopes and that consequently give off radioactive emissions. (See **isotope, radioactive emissions**)

Radioactive wastes. Waste materials that are or contain or are contaminated with radioactive substances. Many materials used in the nuclear industry become wastes because of their contamination with radioactive substances.

Radioisotope. An isotope of an element that is unstable and may tend to gain stability by giving off radioactive emissions. (See **isotope** and **radioactive decay**)

Radon. A radioactive gas produced by natural processes in the earth that is known to seep into buildings. It can be a major hazard within homes and is a known carcinogen.

Rain shadow. The low-rainfall region that exists on the leeward (downwind) side of mountain ranges. It is the result of the mountain range causing the precipitation of moisture on the windward side.

Range of tolerance. The range of conditions within which an organism or population can survive and reproduce, for example, the range from the highest to lowest temperature that can be tolerated. Within the range of tolerance is the optimum, or best, condition.

Raw sewage. The sum total of all the wastewater collected from homes and buildings before any treatment has occurred. Also called raw wastewater.

Raw sludge. The untreated organic matter that is removed from sewage water by letting it settle. It consists of organic particles from feces, garbage, paper, and bacteria.

Raw wastewater. (See **raw sewage**)

Reactor vessel. Steel-walled vessel that contains the nuclear reactor.

Recharge area. With reference to groundwater, the area over which infiltration and resupply of a given aquifer occurs.

Recruitment. With reference to populations, the maturation and entry of young into the adult breeding population.

Recycling. The practice of processing wastes and using them as raw material for new products as, for example, scrap iron is remelted and made into new iron products. Compare **reuse.**

Relative humidity. The percentage of moisture in the air compared with how much the air can hold at the given temperature.

Remediate. To return groundwater to its original uncontaminated state.

Renewable energy. Energy sources, namely solar, wind, and geothermal, that will not be depleted by use.

Renewable resources. Biological resources such as trees that may be renewed by reproduction and regrowth. Conservation to prevent overcutting and protection of the environment are still required, however.

Replacement capacity (biological resources). The capacity of a system to recover to its original state after a harvest or other form of use.

Replacement fertility or **level.** The fertility rate that will just sustain a stable population.

Reproductive rate. The rate at which offspring, eggs, or seed are produced.

Reserves (of a mineral resource). The amount remaining in the earth that can be exploited using current technologies and at current prices. Usually given as *proven reserves,* those that have been positively identified, and *estimated reserves,* those that have not yet been discovered but that are presumed to exist.

Resources Conservation and Recovery Act of 1976 (RCRA). The cornerstone legislation to control indiscriminate land disposal of hazardous wastes.

Respiration. (See **Cell respiration**)

Restoration ecology. The branch of ecology devoted to restoring degraded and altered ecosystems to their natural state.

Resurgence (of populations, especially of pests). The rapid comeback of a population after a severe dieoff, especially populations of pest insects, after being largely killed off by pesticides, returning to even higher levels than before the treatment.

Reuse. The practice of reusing items as opposed to throwing them away and producing new items, as, for example, bottles can be collected and refilled. Compare **recycling.**

Reverse osmosis. A process used by small desalination plants by which seawater is forced under great pressure through a membrane that is fine enough to filter out the salt.

Risk. The probability of suffering injury, disease, death, or other loss as a result of exposure to a hazard.

Risk analysis. The process of evaluating the risks associated with a particular hazard before taking some action (Often called risk *assessment*).

Risk characterization. The process of determining a risk and its accompanying uncertainties after the process of hazard assessment, dose-response assessment, and exposure assessment have been accomplished.

Risk management. The task of regulators, involving reviewing the risk data and making regulatory decisions based on the evidence. Often, the process is influenced by considerations of costs and benefits, as well as by public perception.

Risk perception. People's intuitive judgments about risks, which are often not in agreement with the level of risk as judged by experts.

Rivulet erosion. A treelike pattern of numerous tiny (less than 15 cm) gullies caused by water erosion.

Runoff. That portion of precipitation which runs off the surface as opposed to soaking in.

S

Safe Drinking Water Act of 1974. Legislation to protect the public from the risk of toxic chemicals contaminating drinking water supplies. Mandates regular testing of municipal water supplies.

Salinization. The process of soil becoming more and more salty until finally the salt prevents the growth of plants. It is caused by irrigation because salts brought in with the water remain in the soil as the water evaporates.

Saltwater intrusion/Saltwater encroachment. The phenomenon of seawater moving back into aquifers or estuaries. It occurs when the normal outflow of freshwater is diverted or removed for use.

Sand. Mineral particles 0.2–2.0 mm in diameter.

Sanitary sewers. Separate drainage system used to receive all the wastewater from sinks, tubs, and toilets.

Scientific fact. An observation regarding some object or phenomenon that has been repeated and confirmed by the scientific community.

Scientific method. The methodology by which scientific information is generated. Involves observing, formulating specific questions and hypotheses regarding the question's answer, then testing the hypotheses through experimentation.

SCS. U.S. Department of Agriculture—Soil Conservation Service.

Secondary air pollutants. Air pollutants resulting from reactions of primary air pollutants while resident in the atmosphere. These include ozone, other reactive organic compounds, and sulfuric and nitric acids. (See **Ozone, PANs,** and **photochemical oxidants**)

Secondary consumer. An organism such as a fox or coyote that feeds more-or-less exclusively on other animals that feed on plants.

Secondary energy source. A form of energy such as electricity that must be produced from primary energy source such as coal or radioactive material.

Secondary pest outbreak. The phenomenon of a small, and therefore harmless, population of a plant-eating insect suddenly exploding to become a serious pest problem. Often caused by the elimination of competitors through pesticide use.

Secondary succession. (See **succession**)

Secondary treatment (of sewage). Also called biological treatment. A process that follows primary treatment. Any of a variety of systems that remove most of the remaining organic matter by enabling organisms to feed on it and oxidize it through their respiration. Trickling filters and activated-sludge systems are the most commonly used methods.

Second basic principle of ecosystem sustainability. Ecosystems run on solar energy, which is exceedingly abundant, nonpolluting, constant, and everlasting.

Second-generation pesticides. Synthetic organic compounds used to kill insects and other pests. Started with the use of DDT in the 1940s.

Second Law of Thermodynamics. The fact based on irrefutable observations that in every energy conversion, e.g., electricity to light, some of the energy is converted to heat and some heat always escapes from the system because it always moves toward a cooler place. Therefore, in every energy conversion, a portion of energy is lost. Therefore, since energy cannot be created (First Law) the functioning of any system requires an energy input.

Secured landfill. A landfill with suitable barriers, leachate drainage, and monitoring systems such that there is deemed adequate security against hazardous wastes in the landfill contaminating groundwater.

Sediment. Soil particles, namely sand, silt, and clay, being carried by flowing water. The same material after it has been deposited. Because of different rates of settling, deposits are generally pure sand, silt, or clay.

Sediment trap. A device for trapping sediment and holding it on a development or mining site.

Sedimentation. The filling in of lakes, reservoirs, stream channels, and so on with soil particles, mainly sand and silt. The soil particles come from erosion, which generally results from poor or inadequate soil conservation practices in connection with agriculture, mining, and/or development. Also called siltation.

Seep. Where groundwater seeps from the ground over some area as opposed to a spring, which is the exit as a single point.

Selection pressure (with reference to evolution). An environmental factor that results in individuals with certain traits, which are not the norm for the population, surviving and reproducing more than the rest of the population. This results in a shift in the genetic makeup of the population. For example, the presence of insecticides provides a selection pressure toward increasing pesticide resistance in the pest population.

Selective breeding. The breeding of certain individuals because they bear certain traits and the exclusion from breeding of others.

Selective cutting. The cutting only of particular trees in a forest. Contrast **clearcutting.**

Sex attractant. A natural chemical substance (pheromone) secreted by the female of many insect species that serves to attract males for the function of mating. Sex attractants may be used in traps or for the **confusion technique** to aid in the control of insect pests.

Sexual reproduction. Reproduction involving segregation and recombination of chromosomes such that the offspring bear some combination of genetic traits from the parents. Contrast with asexual reproduction, where all the offspring are exact genetic copies of the parent.

Shadow pricing. In cost-benefit analysis, a technique used to estimate benefits where normal economic analysis is ineffective. For example, people could be asked how much they might be willing to pay monthly to achieve some improvement in their environment.

Sheet erosion. The loss of a more-or-less even layer of soil from the surface due to the impact and runoff from a rainstorm.

Shelterbelts. Rows of trees around cultivated fields for the purpose of reducing wind erosion.

Sidestream smoke. Smoke inhaled by nonsmokers that is produced by smokers in enclosed spaces.

Silt. Soil particles between the size of sand particles and clay particles; namely, particles 0.002 to 0.2 mm in diameter.

Siltation. (See sedimentation)

Sinkhole. A large hole resulting from the collapse of an underground cavern.

Slash-and-burn agriculture. The practice, commonly exercised throughout tropical regions, of cutting and burning jungle vegetation to make room for agriculture. The process is highly destructive of soil humus and may lead to rapid degradation of soil.

Sludge cake. Treated sewage sludge that has been dewatered to make it a moist solid.

Sludge digesters. Large tanks in which raw sludge (removed from sewage) is treated through anaerobic digestion by bacteria.

Smog. (See **industrial smog** and **photochemical smog**)

Snow White syndrome. The human attribute that expects or demands cosmetic perfection in produce even though this is often obtained only by using huge quantities of environmentally hazardous pesticides.

Soft water. Water with little or no calcium, magnesium, or other ions in solution that will cause soap to precipitate (form a curd that makes a "ring around the bathtub").

Soil. A dynamic system involving three components: mineral particles, detritus, and soil organisms feeding on the detritus.

Soil aeration. (See **aeration**)

Soil erosion. The loss of soil caused by particles being carried away by wind and/or water.

Soil fertility. Soil's ability to support plant growth, often refers specifically to the presence of proper amounts of nutrients. The soil's ability to fulfill all the other needs of plants is also involved.

Soil profile. A description of the different, naturally formed layers within a soil.

Soil structure. The phenomenon of soil particles (sand, silt, and clay) being loosely stuck together to form larger clumps and aggregates, generally with considerable air spaces in between. Structure enhances infiltration and aeration. It develops as a result of organisms feeding on organic matter in and on the soil.

Soil texture. The relative size of the mineral particles that make up the soil. Generally defined in terms of the sand, silt, and clay content.

Solar cells. (See **photovoltaic cells**)

Solar energy. Energy derived from the sun. Includes direct solar energy (the use of sunlight directly for heating and/or production of electricity) and indirect solar energy (the use of wind, which results from the solar heating of the atmosphere, and biological materials such as wood, which result from photosynthesis).

Solid waste. The total of materials that are discarded as "trash" and handled as solids, as opposed to those that are flushed down sewers and handled as liquids.

Solubility. The degree to which a substance will dissolve and enter into solution.

Solution. A mixture of molecules (or ions) of one material in another. Most commonly, molecules of air and/or ions of various minerals in water. For example, seawater contains salt in solution.

Specialization. With reference to evolution, the phenomenon of species becoming increasingly adapted to exploit one particular niche but, thereby, becoming less able to exploit other niches.

Speciation. The evolutionary process whereby populations of a single species separate and, through being exposed to differing forces of natural selection, gradually develop into distinct species.

Species. All the organisms (plant, animal, or microbe) of a single kind. The "single kind" is determined by similarity of appearance and/or by the fact that members do or potentially can mate and produce fertile offspring. Physical, chemical, or behavioral differences block breeding between species.

Splash erosion. The destruction and compaction of soil structure that results from rainfall impacting bare soil.

Spores. Reproductive cells produced by fungi, some bacteria, and lower plants.

Springs. Natural exits of groundwater.

Standards (air or water quality). Set by the federal or state governments, the maximum levels of various pollutants that are to be legally tolerated. If levels go above the standards, various actions may be taken.

Standing biomass. That portion of the biomass (population) that is not available for consumption but that must be conserved to maintain the productive potential of the population.

Starvation. The failure to get enough calories to meet energy needs over a prolonged period of time. It results in a wasting away of body tissues until death occurs.

Sterile male technique (pest control). Saturating the area of infestation with sterile males of the pest species that have been artificially reared and sterilized by radiation. Matings between normal females and sterile males render the eggs infertile.

Steward/Stewardship. A steward is one to whom a trust has been given. In reference to natural lands, stewardship is an attitude of active care and concern for nature.

Stomas. Microscopic pores in leaves, mostly in the undersurface, that allow the passage of carbon dioxide and oxygen into and out of the leaf and that also permit the loss of water vapor from the leaf.

Storm drains. Separate drainage systems used for collecting and draining runoff from precipitation.

Stormwater. In cities, the water that results directly from rainfall, as opposed to municipal water and sewage water piped to and from homes, offices, and so on. The exten-

sive hard surfacing in cities creates a vast amount of stormwater runoff, which presents a significant management problem.

Stormwater management. Policies and procedures for handling stormwater in acceptable ways to reduce the problems of flooding and erosion of stream banks.

Stormwater retention reservoirs. Reservoirs designed to hold stormwater temporarily and let it drain away slowly in order to reduce problems of flooding and stream bank erosion.

Stratosphere. The layer of the earth's atmosphere between 10 and 30 miles above the surface that contains the ozone shield. This layer mixes only slowly; pollutants that enter may remain for long periods of time. (See also **troposphere**)

Stream bank erosion. The eating away of the stream bank by flowing water. It is greatly aggravated by flooding.

Strip cropping. The practice of growing crops in strips alternating with grass (hay) at right angles to prevailing winds or slopes in order to reduce erosion.

Strip mining. The mining procedure in which all the earth covering a desired material such as coal is stripped away with huge power shovels in order to facilitate removal of the desired material.

Submerged aquatic vegetation (SAV). Aquatic plants rooted in bottom sediments growing under water. Depend on light penetrating through the water for photosynthesis.

Subsistence farming. Farming that meets the food needs of the farmers and their families but little more. It involves hand labor and is practiced extensively in the Third World.

Subsoil. In a natural situation, the soil beneath topsoil. In contrast to topsoil, subsoil is compacted and has little or no humus or other organic material, living or dead. In many cases, topsoil has been lost or destroyed as a result of erosion or development, and subsoil is at the surface.

Succession (ecological or natural). The gradual, or sometimes rapid, change in species that occupy a given area, with some species invading and becoming more numerous while others decline in population and disappear. Succession is caused by a change in one or more abiotic or biotic factors benefiting some species but at the expense of others. *Primary succession:* The gradual establishment, through a series of stages, of a climax ecosystem in an area that has not been occupied before, e.g., a rock face. *Secondary succession:* The reestablishment, through a series of stages, of a climax ecosystem in an area from which it was previously cleared.

Sulfur dioxide (SO_2). A major air pollutant, this toxic gas is formed as a result of burning sulfur. The major sources are burning coal (coal-burning power plants) that contains some sulfur and refining metal ores (smelters) that contain sulfur.

Sulfuric acid (H_2SO_4). The major constituent of acid precipitation. Formed as a result of sulfur dioxide emissions reacting with water vapor in the atmosphere. (See also **sulfur dioxide**)

Superfund. The popular name for the Comprehensive Environmental Response, Compensation, and Liability Act of 1980. This act provides the mechanism and funding for the cleanup of potentially dangerous hazardous waste sites.

Surface water. Includes all bodies of water, lakes, rivers, ponds, and so on that are on the surface of the earth in contrast to groundwater, which lies below the surface.

Survival of the fittest. The concept that individuals best adapted to cope with both biotic and abiotic factors in their environment are the "fittest" and most likely to survive and reproduce.

Survival rate. (See **survivorship**)

Survivorship. The proportion of individuals in a specified group alive at the beginning of an interval, e.g., a five-year period, who survive to the end of the period.

Survivorship graph. A graph that shows the probability of an individual's surviving from birth to any particular age. Survivorship graphs are constructed (and may differ) for specific categories, such as sex, race, nationality, economic status, and so on.

Suspended particulate matter (SPM). A category of major air pollutants consisting of solid and liquid particles suspended in the air. (See also **PM-10**)

Suspension. With reference to materials contained in or being carried by water, materials kept "afloat" only by the water's agitation that settle as the water becomes quiet.

Sustainability. Refers to whether a process can be continued indefinitely.

Sustainable agriculture. Agriculture that maintains the integrity of soil and water resources such that it can be continued indefinitely. Much of modern agriculture is depleting these resources and is, hence, not sustainable.

Sustainable development. Development that provides people with a better life without sacrificing or depleting resources or causing environmental impacts that will undercut future generations.

Symbiosis. The intimate living together or association of two kinds of organisms, especially in a way that provides a mutual benefit to both organisms.

Synergism/Synergistic effect/Synergistic interactions. The phenomenon in which two factors acting together have a very much greater effect than would be indicated by the sum of their effects separately—as, for example, modest doses of certain drugs in combination with modest doses of alcohol may be fatal.

Synfuels/Synthetic fuels. Fuels similar or identical to those that come from crude oil and/or natural gas, produced from coal, oil shale, or tar sands.

Synthetic. Human-made as opposed to being derived from a natural source. For example, synthetic organic compounds are those produced in chemical laboratories, whereas natural organic molecules are those produced by organisms.

Synthetic organic compounds. (See **organic compounds**)

T

Tar sands. Sedimentary material containing bitumine that can be "melted out" using heat and then refined in the same way as crude oil.

Taxonomy. The science of identification and classification of organisms according to evolutionary relationships.

Technological assessment. A study aimed at projecting the environmental and social impacts of introducing a new technology into a less-developed nation. Choices between alternative development projects may be made accordingly.

Technology. The application of scientific information to solve practical problems or achieve desired goals.

Tectonic plates. Huge slabs of rock that make up the earth's crust.

Temperature inversion. The weather phenomenon in which a layer of warm air overlies cooler air near the ground and prevents the rising and dispersion of air pollutants.

Teratogenic. Causing birth defects.

Terracing. The practice of grading sloping farmland into a series of steps and cultivating only the level portions in order to reduce erosion.

Territoriality. The behavioral characteristic seen in many animals, especially birds and mammalian carnivores, to mark and defend a given territory against other members of the same species.

Tertiary treatment. Third stage of sewage treatment, following primary and secondary, designed to remove one or more of the nutrients, usually nitrogen and/or phosphate, from the wastewater. Necessary to reduce the problem of eutrophication.

Test group. (See **experimental group**)

Texture. With reference to solids, the sizes of the particles, sand, silt, and/or clay, that make up the mineral portion.

Theory. A conceptual formulation that provides a rational explanation or framework for numerous related observations.

Thermal pollution. The addition of abnormal and undesirable amounts of heat to air or water. It is most significant with respect to discharging waste heat from electric generating plants, especially nuclear power plants, into bodies of water.

Third basic principle of ecosystem sustainability. Large biomasses cannot be supported at the end of long food chains. The size of consumer populations is maintained such that overgrazing does not occur.

Third World. Nations in Latin America, Africa, and Asia that are in various stages of economic development but not fully industrialized. Because these nations are mostly in the Southern Hemisphere, they are often referred to as the South, in contrast to the industrialized nations of the Northern Hemisphere.

Threatened species. A species the population of which is declining precipitously because of direct or indirect human impacts.

Threshold level. The maximum degree of exposure to a pollutant, drug, or other factors that can be tolerated with no ill effect. The threshold level will vary depending on the species, the sensitivity of the individual, the length of exposure, and the presence of other factors that may produce synergistic effects.

Tidal wetlands. Areas of marsh grasses and reeds along coasts and estuaries where the ground is covered by high tides but drained at low tide.

Tie-in strategy. In connection with global warming, the idea that society should take actions that not only deal with global warming but also have other beneficial effects. For example, energy conservation not only reduces carbon dioxide emissions, but also saves money, reduces acid deposition, and lowers our dependency on foreign oil.

Topsoil. The surface layer of soil, which is rich in humus and other organic material, both living and dead. As a result of the activity of organisms living in the topsoil, it generally has a loose, crumbly structure as opposed to being a compact mass. In many cases, because of erosion, development, or mining activity, the topsoil layer may be absent.

Total fertility rate. The average number of children that would be born alive to each woman during her total reproductive years, assuming she follows the average fertility at each age.

Total watershed planning. A consideration of the entire watershed and planning development and other activities so as to maintain the overall water flow characteristics of the area.

Trace elements. Those essential elements that are needed in only very small or trace amounts.

Tradeoffs. The things that are given up to get or achieve something else that is valued.

Traditional farming. Current farming methods involving intensive use of fertilizers, pesticides, and other chemicals.

Tragedy of the commons. The overuse or overharvesting and consequent depletion and/or destruction of a renewable resource that tends to occur when the resource is treated as a commons, that is, when it is open to be used or harvested by any and all with the means to do so.

Trait (genetic). Any physical or behavioral characteristic or talent that an individual is born with.

Transpiration. The loss of water vapor from plants. Evaporation of water from cells within the leaves and exiting through stomas.

Trapping technique (of pest control). The use of sex attractants to lure male insects into traps.

Treated (sewage) sludge. Solid organic material that has been removed from sewage and treated so that it is non-hazardous.

Trickling filter system. System where wastewater trickles over rocks or a framework coated with actively feeding microorganisms. The feeding action of the organisms in a well-aerated environment results in the decomposition of organic matter. Used in secondary or biological treatment of sewage.

Tritium (^{3}H). An unstable isotope of hydrogen that contains two neutrons in addition to the usual single proton in the nucleus. It does not occur in significant amounts naturally but is human-made.

Trophic level. Feeding level with respect to the primary source of energy. Green plants are at the first trophic level, primary consumers at the second, secondary consumers at the third, and so on.

Troposphere. The layer of the earth's atmosphere from the surface to about 10 miles in altitude. The *tropopause* is the boundary between the troposphere and the stratosphere

above. This layer is well mixed and is the site and source of our weather, as well as the primary recipient of air pollutants. (See also **stratosphere**)

Turbid. Refers to water purity; means cloudy.

Turbine. A sophisticated "paddle wheel" driven at a very high speed by steam, water, or exhaust gases from combustion.

Turbogenerator. A turbine coupled to and driving an electric generator. Virtually all commercial electricity is produced by such devices. The turbine is driven by gas, steam, or water.

Turnover rate. The rate at which a population is replaced by the next generation.

U

Ultraviolet radiation. Radiation similar to light but with wavelengths slightly shorter than violet light and with more energy. The greater energy causes it to severely burn and otherwise damage biological tissues.

Undernutrition. A form of hunger where there is a lack of adequate food energy as measured in calories. Starvation is the most severe form of undernutrition.

Upset (ecological). A vast shift in the relative size of one or more populations within an ecosystem. It may result in tremendous damage or even total destruction of the original ecosystem. It may be caused by any factor that changes the normal balances between species, such as introduction of a foreign species or eliminating a predator that was instrumental in controlling the population of an herbivore. Especially the phenomenon of an economically insignificant insect becoming a serious threat when its predators are killed off by pesticide treatment.

Urban decay. General deterioration of structures and facilities such as buildings and roadways, and also the decline in quality of services such as education, that has occurred in inner city areas as growth has been focused on suburbs and exurbs.

Urban sprawl. The rapid expansion of metropolitan areas through building housing developments and shopping centers farther and farther from urban centers and lacing them together with more and more major highways. Widespread development that has occurred without any overall land-use plan.

USDA. U.S. Department of Agriculture.

UST legislation. Amendments to the Resources Conservation and Recovery Act of 1976, passed in 1984 to address the mounting problem of leaking underground storage tanks (USTs).

Utterly Dismal Theorem. The view that supplying free food to food-poor nations undercuts the recipients' agricultural systems and worsens the problem it is intended to solve.

V

Verification. The checking of observations and/or experiments by others to determine their accuracy.

Vitamin. A specific organic molecule that is required by the body in small amounts but that cannot be made by the body and therefore must be present in the diet.

Volatile organic compounds. A category of major air pollutants present in the air in vapor state, including fragments of hydrocarbon fuels from incomplete combustion and evaporated organic compounds such as paint solvents, gasoline, and cleaning solutions. They are major factors in the formation of photochemical smog.

W

Waste inventories. Rosters of kinds and quantities of various waste products produced by various industries in a region. The information is compiled and distributed to facilitate recycling.

Water. Chemically, H_2O. All naturally occurring water contains additional materials in solution.

Water balance. Refers to the capacity and necessity of all organisms to control the relative volume of water inside versus outside their cells.

Water cycle. The movement of water from points of evaporation through the atmosphere, through precipitation, and through or over the ground, returning to points of evaporation.

Water-holding capacity (soil). The property of a soil relating to its ability to hold water so that it will be available to plants.

Waterlogging (of soil). The total saturation of soil with water. Results in plant roots not being able to get air and dying as a result.

Watershed. The total land area that drains directly or indirectly into a particular stream or river. The watershed is generally named from the stream or river into which it drains.

Watershed management. Controlling activities in a watershed to maintain water quality in the reservoir.

Water table. The upper surface of groundwater. It rises and falls with the amount of groundwater.

Water vapor. Water molecules in the gaseous state.

Weathering. The gradual breakdown of rock into smaller and smaller particles, caused by natural chemical, physical, and biological factors.

Wetlands. Areas that are constantly wet and are flooded at more-or-less regular intervals. Especially, marshy areas along coasts that are regularly flooded by tide.

Wetland systems (for wastewater). An aquatic system to remove nutrients biologically when irrigation is not feasible.

Wilderness Act of 1964. Federal legislation that provides for the permanent protection of undeveloped and unexploited areas so that natural ecological processes can operate freely. Most uses are excluded from such areas, which now total 90 million acres in the United States.

Wildlife management. Humans attempting to provide balance for wildlife in whatever ways possible after the natural balance has been upset.

Wind farms. Arrays of numerous, modestly sized wind turbines for the purpose of producing electrical power.

Windrows. Piles of organic material extended into long rows to facilitate turning and aeration to enhance composting.

Wind turbines. "Windmills" designed for the purpose of producing electrical power.

Work (physics). Any change in motion or state of matter. Any such change requires the expenditure of energy.

Workability. With reference to soils, the relative ease with which a soil can be cultivated.

World View. A set of assumptions that a person holds regarding the world and how it works.

X

Xeroscaping. Landscaping with drought-resistant plants that need no watering.

Y

Yard wastes. Grass clippings and other organic wastes from lawn and garden maintenance.

Z

Zones of stress. Regions where a species finds conditions tolerable but suboptimal. Where a species survives but under stress.

Index

621

Migration:
    exurban migration, 533
    meaning of, 109
Minamata disease, 316–17
Mineralization, soil, 197, 203
Minerals:
    nature of, 48
    in soil, 189–90
Mining, environmental impact, 279
Mixtures, air as, 48
Moderator, nuclear reactor, 487
Molecules:
    inorganic molecules, 49
    organic molecules, 48, 49
Molina, Mario, 378
Monoculture, 210, 225
    meaning of, 80
Mono Lake, 250–51
Montreal Protocol, 379–80
Moss, 86
Mount St. Helens, 112
Muller, Paul, 217
Municipal water use, 246, 247–58
Mutations, 100
    causes of, 100
    lethal mutation, 100
    negative aspects of, 100, 102
Mutualism, 29–30
    example of, 29–30
Myocorrhizae, 196

National Acid Precipitation Assessment
        Program, 364
National Audubon Society, 207
National Emission Standards for Hazardous Air
        Pollutants, 349
National forests, 438–40
    logging in, 438–39
    mismanagement of, 439–40
National Forest Service, 438, 439
National Parks, 438
National Park Service, 438
National Priorities List, 325–26, 445
National Resources Defense Council, 5
National Surface Water Survey, 364
National Wetlands Policy Forum, 436
Natural chemical pest control, 229–31
    confusion technique, 230
    goal of, 229
    natural enemies, use of, 76, 226–28, 423
    trapping technique, 230
Natural gas, 460, 475
    advantages of, 460, 475
    solar energy as substitute for, 516
Natural laws:
    predictive power of, 8–9
    types of, 8
Natural organic compounds, 48
Natural resources, 2
    common pool resources, 426–27
    conservation and preservation of, 426
    ecosystems as, 424–25
    maximum sustainable yield, 426–28
    renewable resources, 426
    tragedy of the commons, 425–26
Natural Resources Defense Council, 235, 396
Natural selection, 98–108
    and adaptation, 100, 103
    and alleles, 98
    and changes to species, 98–100
    and selective pressure, 105
Natural services, of ecosystems, 422–24
Natural species:
    and agriculture, 406
    benefits of, 405
    commercial value of, 407
    instrumental versus intrinsic value, 404–5
    intrinsic value of, 408–9
    medical applications, 407
    recreational and aesthetic value, 407
    See also Biodiversity
Natural species management:
    Endangered Species Act, 417–20

and game animals, 417
    Lacey Act, 417–18
Nature Conservancy, 440
Nematodes, 196
Nevada, nuclear waste plant, 493, 494
New Mexico, nuclear waste plant, 493
Niche:
    and competition, 32
    meaning of, 32
Nitrogen cycle, 62–64
    maintenance of, 423
Nitrogen fixation, 62
    atmospheric, 63–64
Nitrogen oxides:
    as air pollutant, 340, 344, 348
    and air traffic, 376
    and global warming, 371
Nonbiodegradable compounds, 69
Nonbiodegradable materials, 269
    halogenated hydrocarbons, 314
    nonbiodegradable synthetic organics, 314
Nonconsumptive water, 245
Nonfeeding relationships:
    competition, 31–32
    mutualism, 29–30
    symbiosis, 30–31
Non-Indigenous Aquatic Nuisance Prevention
        and Control Act, 414
Nonpersistent pesticides, 223
    negative aspects of, 223
Nonpoint sources, pollution, 206
Nontidal wetlands, 283
Norplant, 158, 166
Nuclear power:
    compared to coal power, 488–89
    economic problems related to, 496–98
    future view, 500–501
    hazards from, 489–91
    process in release of, 483–87
    public opinion about, 490, 500
Nuclear power plants, 463
    decline in building of, 482–83
    double loop design, 487
    fuel for, 484–85
    lifespan of, 497–98
    loss-of-coolant accident, 487–88
    meltdown, 488
    nuclear reactor, 486–87
    other countries use of, 482–83
    potential for accidents, 494–96
    safety standards for, 496
Nuclear reactors:
    advanced light water reactors, 496
    breeder reactors, 498
    fusion reactors, 498–500
    operation of, 486–87
Nuclear Regulatory Commission, 482
Nuclear waste, transportation issue, 495
Nutrient-holding capacity, soil, 190–91, 200
Nutrients, eutrophication-causing, 273–75,
        281–83

Oakland, California, fires, 91
Observation, in scientific method, 6–7
Ocean fisheries, 433–36
    drift-netting, 434–35
    extent of catch from, 433–34
    overfishing and decline of species, 434
    whale harvesting, 435–36
Ocean thermal energy conversion, 521–22
    power plant in, 521
    problems related to, 521–22
Ogallala aquifer, 252
Oil:
    declining U.S. reserves, 467–68
    early uses, 459, 460
    energy independence efforts by U.S., 470–71
    exploration for, 465
    future of reserves, 472
    OPEC oil embargo, 468–69
    and Persian Gulf War, 472
    prices, 471
    production process, 466–67

Oil shales, 476
Oligotrophic condition, aquatic systems, 273
Omnivores, 24
Optimal population, 80, 145
Optimum, 32
Optimum range, 32
Oral rehydration therapy, 154
Orchid bees, 415
Organically grown food, 231–32
    growth of market, 232
    methods in, 232
Organic farming:
    Amish farms, 232–33
    nature of, 184
    productivity and economic return in, 232–33
Organic fertilizer, 190, 206
    and sustainable agriculture, 208
    treated sludge as, 304
Organic molecules, 23, 49
Organic phosphate, 60
Organisms:
    consumers, 24–25
    decomposers, 25, 27
    detritus feeders, 25
    energy, use of, 54–60
    feeding relationships, 27–29
    nonfeeding relationships, 29–32
    producers, 23–24
Organization of Petroleum Exporting Countries
        (OPEC)
    oil embargo, 468–69
    U.S. dependence on, 471, 472
Origin of Species by Natural Selection, The
        (Darwin), 98, 103
Outrage factor, in public perception, 393–95
Overcultivation, and soil erosion, 200–201
Overgrazing, 173
    meaning of, 65
    and soil erosion, 202–3
    versus sustainable livestock management, 185
Overland flow system, sewage treatment, 305,
        306
Oxidation, process of, 58
Oxidation pond system, sewage treatment, 305
Oxygen:
    and cell respiration, 58
    and photosynthesis, 56
Oxygen cycle, maintenance of, 423
Ozone:
    as air pollutant, 340, 343, 347–48
    atmospheric, 359–60
    control strategies, 352
    in disinfection of sewage, 299
Ozone depletion, 375–80
    and chloroflurocarbons, 377–78, 380
    international agreements related to, 379–80
    ozone hole, 378–79
    United States remedial action, 380–81
Ozone shield:
    formation of, 377
    importance of, 375

Pacific Gas and Electric Company, 511
Pacific yew, 407
Packaging materials, recycling of, 454
Pangaea, 111, 112
Paper:
    rate of biodegradation, 445
    reprocessing of, 452
Parasites, 24–25
    host-parasite relationship, 24
    host-specific, 80–81
    and population balance, 76–77
    types of, 25
Parathion, 223
Parent material, 189
Paris, France, 545
Particulates, reduction of, 349
Passive safety, nuclear power plants, 496
Passive solar heating, 504–5
Pasteurization, sewage, 305
Pastoralists, 180
Pathogens:

Replacement fertility rate, 136, 137
Replacement level, 74
Reprocessing, recycled materials, 451–52
Reproduction:
  genetic transmission in, 96, 97
  selective pressure, 97, 102
Reproductive rate, 73
Reptiles, and evolution, 115–16
Reservoirs, 254
  as water source, 247–48
Resistance, of pests to pesticides, 218–19
Resources Conservation and Recovery Act, 328
  limitations of, 329
  provisions, 328
Resurgence, and pesticides, 219, 221
Retention ponds, 261–62
Reverse osmosis, 255
Rhizobium, 62
Rhythm method, 156
Risk:
  definition of, 391
  public perception of, 393–95
Risk analysis, 390–95
  categories of risks, 393
  definition of, 391
  dose-response assessment, 392
  Environmental Protection Agency, 391, 392, 393, 394, 395
  exposure assessment, 392
  hazard assessment, 391–92
  risk characterization, 392–93
Risk characterization, 392–93
Risk communication, 395
Risk management, 395–96
  comparative risk analysis, 395–96
  setting environmental priorities, 395–96
  steps in, 395
Rivers:
  and dams, 247, 250–51, 254–55, 516–17
  watershed of, 258–59
  as water source, 245, 247
Roberts, Walter Orr, 373
Rock, weathering, 189, 190
Rodale Press, 207
Rodwell, Gary, 548
Rosy periwinkle, 407
Rouse, James, 539, 548
Rowland, Sherwood, 378
RU486, 166
Runoff, 244
Rural areas, population problems of, 128–29

Safe Drinking Water Act, 324
  limitations, 324
  provisions, 324
Sahel region famine, 180, 182
Salinization:
  avoidance of, 205–6
  and irrigation, 203, 205, 206
Saltwater intrusion, and dropping water table, 253
San Andreas fault, 112
Sand County Almanac, A (Leopold), 403
Sandhill cranes, 419
Sandman, Peter, 393
Sandtown, 545
Sanitary sewers, 294
Sanitation, as pest control, 224–25
Sao Paulo, Brazil, 129
Scale insects, control of, 216–17, 220, 226
Schmoke, Kurt, 548
Science:
  controversy in, 11–12
  instruments used in, 9–10
  nature of, 5–6
  and value judgments, 12
Scientific method, 6–9
  controlled experiments, 7–8
  hypothesis in, 7
  and natural laws, 8–9
  observation in, 6–7
  theory formulation, 8
Screwworm flies, 229

Scrubbers, 366, 367
S curve, 85
Seawater, desalination, 255
Secondary consumers, 24
Secondary detritus feeders, 25
Secondary pest outbreak, and pesticides, 219, 221
Secondary pollutants, air pollutants, 347–48
Secondary succession, 86, 89
Second-generation pesticides, 217
Secure landfill, 320
Sediments:
  bedload, 281
  control of, 285–87
  and eutrophication, 271, 278–81
  and soil erosion, 203, 279, 281
  sources of, 279
Sediment trap, 285
Seeps, of groundwater, 244
Selective breeding, and changes to species, 97–98
Selective pressure, 97, 102
  and ecosystem development, 105
  and natural selection, 105
Septic systems, sewage treatment, 307
Sewage, 291–309
  control of, historical view, 292, 294
  monitoring for contamination, 293
  pathogens carried by, 291–92
  raw sewage, 294
Sewage treatment:
  activated-sludge system, 297
  advanced treatment, 299
  composting, 304–5
  disinfection in, 299
  drying of sludge, 305
  greenhouse system, 305–7
  impediments to progress for alternative systems, 308–9
  individual septic systems, 307
  irrigating with effluents, 299–301
  overland flow system, 305, 306
  oxidation pond system, 305
  pasteurization, 305
  preliminary treatment, 295
  primary treatment, 297
  secondary treatment, 297–99
  sludge digesters, 303–4
  trickling-filter system, 297
  and water conservation, 303
Sexually transmitted disease:
  AIDS, 155
  protection from, 157, 159
Shadow government, 389
Shadow pricing, 387–88
Shaping Cities report, 542, 544
Shoreham Nuclear Power plant, 482
Sierra Club, 207, 440
Silent Spring (Carson), 217–18
Sinkholes, 252, 253
Sixth of November Cooperative, 153
Sludge:
  raw sludge, 297, 303
  sludge digesters, 303–4
  toxic metal contamination, 308–9
  treated sludge, 303–4
Smog:
  industrial smog, 335–36
  photochemical smog, 336
Soil:
  aeration, 192
  clay in, 192–93, 195, 200
  components of, 188
  composting, 195–97
  fertilization, 206
  humus, 194–97
  irrigation, 203–6
  loam, 192–93, 194
  mineralization of, 197, 203
  mineral nutrients, 189–90
  nutrient-holding capacity, 190–91, 200
  and pesticides, 206
  pH, 192

soil fertility, 189
  soil structure, 195
  texture, 192–94
  water-holding capacity, 192
  water in, 191–92
  workability of, 193
Soil Conservation Act, 207
Soil Conservation Service, 207
Soil erosion, 198–203
  and deforestation, 203
  desertification, 200
  and overcultivation, 200–201
  and overgrazing, 202–3
  protection from, 198
  and sediments, 203, 279, 281
Solar energy, 503–16
  active solar heating, 504
  for automobiles, 512, 515–16
  collection equipment, 504–6
  cost savings in, 507
  and ecosystem sustainability, 65, 69–70
  for electricity, 509–14
  and landscaping, 507
  as natural gas substitute, 516
  passive solar heating, 504–5
  photovoltaic cells, 509–12
  problems related to, 503, 507, 509
  scope of use, 509, 516
  solar ponds, 514
  solar tower, 514
  solar trough collectors, 512, 514–15
  storage of, 505–6
Solar trough collectors, 512, 514–15
Solid waste:
  extent of waste generated, 443–44
  nature of, 443
Solid waste management:
  composting, 454–55
  costs of, 447–48
  incineration, 444, 446–47
  integrated waste management, 456
  landfill, 444–46
  recycling, 450–54
  refuse-to-energy conversion, 455
  waste reduction, 448–50
Solution, water as, 48
Southern Appalachian Man and the Biosphere Cooperative, 438
Spaceship Earth, 17
Speciation, 103–5
  examples of, 103–5
  process of, 103
Species:
  definition of, 17, 105
  gene pool for, 96–101
  variation among, 95–96
Spermicides, 158
Speth, Gustave, 332
Spotted knapweed, 83–84
Spotted owl, 419, 439
Springs, 244
Spruce budworm, 223
Standing biomass, 65
Stellwagen Bank, 435–36
Stewardship attitude, 425
Storm drains, 294
Stormwater management, 261–62
  methods of, 261–62
Stratosphere, 359–60
Streams:
  channelized, 261
  effects in change of flow, 259–60
  streambank erosion, 260, 261
Submerged aquatic vegetation, 272, 275
Subsistence agriculture, 172–73
  nature of, 172–73
  negative aspects of, 173
Subsoil, 195
Suburbs:
  sustainable suburbs, 541–44
  See also Urban sprawl
Sulfur dioxide:
  as air pollutant, 343, 344, 523
  control strategies, 351–52, 367, 368

The Authors of this text and Prentice Hall are proud to contribute, at your direction, a portion of the purchase price of this book to one of these environmental organizations:

Bread for the World

Habitat for Humanity International

Nature Conservancy

Natural Resources Defense Council

World Wildlife Fund

We invite you to turn to page xxi (following the Preface) for details about how you can specify to which of the above organizations you would like us make a contribution.